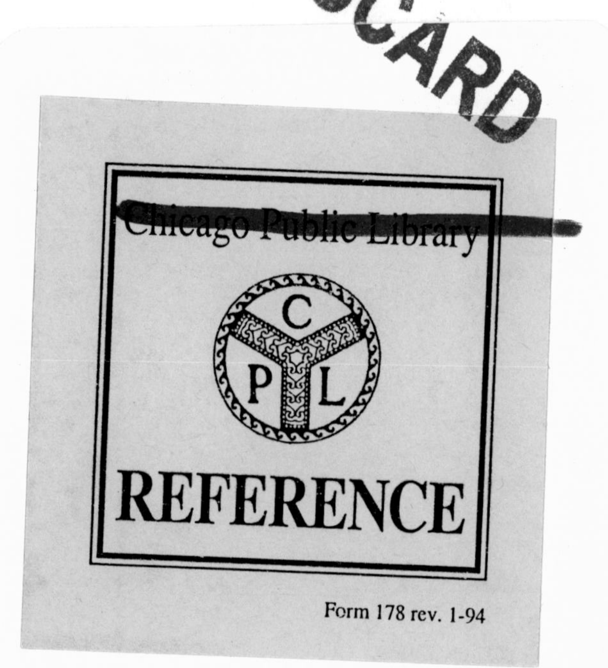
Chicago Public Library
C
P
L
REFERENCE
Form 178 rev. 1-94

HANDBOOK OF STATISTICAL METHODS FOR ENGINEERS AND SCIENTISTS

Other McGraw-Hill Books of Interest

Bothe • MEASURING PROCESS CAPABILITY

Chopey • HANDBOOK OF CHEMICAL ENGINEERING CALCULATIONS

Hicks • STANDARD HANDBOOK OF ENGINEERING CALCULATIONS

Hicks • HANDBOOK OF MECHANICAL ENGINEERING CALCULATIONS

Kazmierski• STATISTICAL PROBLEM SOLVING IN QUALITY ENGINEERING

Kurtz • HANDBOOK OF APPLIED MATHEMATICS FOR ENGINEERS AND SCIENTISTS

Ross • TAGUCHI TECHNIQUES FOR QUALITY ENGINEERING

Seidman, et al • HANDBOOK OF ELECTRIC POWER CALCULATIONS

Tuma & Walsh • ENGINEERING MATHEMATICS HANDBOOK

HANDBOOK OF STATISTICAL METHODS FOR ENGINEERS AND SCIENTISTS

Harrison M. Wadsworth, Jr. Editor

Second Edition

MCGRAW-HILL
New York San Francisco Washington, D. C. Auckland Bogotá
Caracas Lisbon London Madrid Mexico City Milan
Montreal New Delhi San Juan Singapore
Sydney Tokyo Toronto

Library of Congress Cataloging-in-Publication Data

Handbook of statistical methods for engineers and scientists/
Harrison M. Wadsworth, Jr., editor. — 2nd ed.
p. cm.
Includes index.
ISBN 0-07-067678-X (alk. paper)
1. Engineering—Statistical methods. 2. Science—Statistical methods. I. Wadsworth, Harrison M.
TA340. H34 1997
519.5—dc21 97-38029
CIP

McGraw-Hill

*A Division of The **McGraw-Hill** Companies*

1 2 3 4 5 6 7 8 9 0 DOC/DOC 9 0 2 1 0 9 8

ISBN 0-07-067678-X

The sponsoring editor for this book was Harold B. Crawford, the editing supervisor was Paul R. Sobel, and the production supervisor was Sherri Souffrance. It was set in Times Roman by Compuvision.

Printed and bound by R. R. Donnelley & Sons Company

McGraw-Hill books are available at special quantity discounts to use as premiums and sales promotions, or for use in corporate training programs. For more information, please write to the Director of Special Sales, McGraw-Hill, 11 West 19th Street, New York, NY 10011. Or contact your local bookstore.

This book is printed on recycled, acid-free paper containing a minimum of 50% recycled, de-inked fiber.

CONTENTS

CONTRIBUTORS

Benjamin M. Adams
Department of Management Science and Statistics
University of Alabama
Tuscaloosa, AL 35487-0226
CHAPTER 7

Frank B. Alt
College of Business and Management
University of Maryland
College Park, MD 20742
CHAPTERS 19 AND 21

Sigmund J. Amster
20 Pilgrim Way
Colts Neck NJ 07722-1227
CHAPTER 9

Jerry Banks
School of Industrial and Systems Engineering
Georgia Institute of Technology
Atlanta, GA 30332-0205
CHAPTER 13

Thomas W. Calvin
13 Malstrome Road
Wappingers Falls, NY 12590-3013
CHAPTER 11

John S. Carson, II
Auto Simulations, Inc.
1355 Terrell Mill Road
Bldg. 1470, Suite 20
Marietta, GA 30067
CHAPTER 13

Keith R. Eberhardt
Statistical Engineering Division
National Institute of Standards and Technology
Gaithersburg, MD 20899-0001
CHAPTER 10

Subir Ghosh
Department of Statistics
University of California
Riverside, CA 92521-0138
CHAPTER 20

Jean Dickinson Gibbons
4400 North A1A-No. 301
North Hutchinson Island, FL 34949
CHAPTER 12

A. Blanton Godfrey
Juran Institute, Inc.
II River Road
Wilton, CT 06897
CHAPTER 3

David Goldsman
School of Industrial and Systems Engineering
Georgia Institute of Technology
Atlanta, GA 30332-0205
CHAPTER 13

Gerald J. Hahn
Corporate Research and Development
General Electric Company
Schenectady, NY 12301-0008
CHAPTER 5

Jeffrey H. Hooper
283 King George Road
Lucent Technologies
Warren, NJ 07059
CHAPTER 9

Ken Hung
Institute of Business Management
National Dong Hwa University
Hualieh, Taiwan, R.O.C.
CHAPTER 19

Kamlesh Jain
School of Business
Golden Gate University
536 Mission Street
San Francisco, CA 94105-2968
CHAPTER 21

Raghu N. Kacker
Statistical Engineering Division
National Institute of Standards and Technology
Gaithersburg, MD 20899-0001
CHAPTER 20

Douglas C. Montgomery
Department of Industrial Engineering
College of Engineering and Applied Sciences
Arizona State University
Tempe, AZ 85287-5906
CHAPTER 16

Peter R. Nelson
Department of Mathematical Sciences
Clemson University
Clemson, SC 29634-1907
CHAPTER 15

Elizabeth A. Peck
The Coca-Cola Company
Atlanta, GA 30301
CHAPTER 16

Peter J. Rousseeuw
University of Antwerp (UIA)
Universite itsplein I, B-2610, Antwerp, Belgium
CHAPTER 17

Thomas P. Ryan
Department of Statistics
Case Western Reserve University
Cleveland, OH 44106-7054
CHAPTER 14

Samuel S. Shapiro
Department of Statistics
Florida International University
Miami, FL 33199-0001
CHAPTER 6

James R. Simpson
United States Air Force Academy
Department of Mathematics
Colorado Springs, CO 80840
CHAPTER 16

Nancy D. Smith
Food and Drug Administration
Division of Biometrics
Rockville, MD 820857
CHAPTER 21

James J. Swain
Department of Industrial and Systems Engineering
The University of Alabama in Huntsville,
Huntsville, AL 35899
CHAPTER 18

Harrison M. Wadsworth, Jr.
Professor Emeritus
School of Industrial and Systems Engineering
Georgia Institute of Technology
Atlanta, GA 30332-0205
CHAPTERS 1, 2, 4 AND 8

William H. Woodall
Department of Management Science and Statistics
University of Alabama
Tuscaloosa, AL 35487-0226
CHAPTER 7

Lap-Ming Wun
National Institute of Health
9000 Rockville Pike
Bethesda, MD 20892
CHAPTER 19

PREFACE

The first edition of this handbook was published in 1990. It was first proposed in 1983 and the first contract was signed in that year. It was recognized at that time that there was a need for a better understanding of statistical methods by practicing engineers and scientists. In the intervening years this need has become greater with the advent of higher technology in industry. For example, the automobile industry now requires that all of their suppliers demonstrate a knowledge of statistical methods and use them to control the quality of their product.

In addition, the Accreditation Board for Engineers and Technology (ABET) recommends that all engineering curricula contain course work in statistics. While this training is certainly appropriate, it is unlikely to result in engineers that are able to use statistical methods without some help. Since its publication this handbook has provided this help to many engineers and scientists.

Robert Hauserman, the final editor at McGraw-Hill of the first edition of the handbook first suggested a second edition in early 1994. All of the contributors to the first edition were contacted and responded favorably. This, in itself was quite a task in that many of them had changed positions or retired since the first edition was completed. The main purpose of the revision is to update the material in the first edition. That is, to introduce any new techniques or concepts that might have been introduced in the literature in the intervening years. The scope of the handbook remains to produce a book that is reasonable in size and still with enough detail to be useful to the reader. A new chapter on acceptance sampling has been added to this edition. Even though, we have never suggested that acceptance sampling should be used to control product quality, there is still a place for this topic in a book of this type.

The book is in four parts. The first part contains elementary material that can be understood by persons with no statistical background. The second part contains basic statistical methods that are most useful to practicing engineers and scientists. The third part contains material of intermediate difficulty. The chapters in this part are intended for persons that wish to delve a little more deeply into the material covered in the second part. The fourth part contains material of a more specialized nature that should be particularly useful to persons with some knowledge of statistics but that lack an understanding of these particular topics

Each chapter covers a specific topic in varying depth, depending on its place in the book. However each chapter is written so that it can be read without referring to other, earlier chapters. Each is written by one or more expert in that particular field of statistics.

The authors of each chapter are listed in the roster of contributors. We are extremely grateful to all of them for their diligent and high quality work while engaged in their own full-time occupations. We are particularly grateful to those who have retired because many of them no longer have secretarial support. There are many other persons who helped in the preparation of the book. Those who helped with the initial planning of the first edition include A. Blanton Godfrey, Russell G. Heikes, John S. Ramberg, Fred C. Leone, and others who reviewed the initial proposal.

Several reviewers were called upon to review individual chapters after they were written. This review stage was most prevalent during the writing of the first edition and particularly when substantial changes were made to the content of a chapter, continued during the writing of this edition. These reviewers were Mark E. Johnson, Russell G. Heikes, Robert G. Easterling, A. Blanton Godfrey, and several unknown reviewers who reviewed the first edition in preparation for the second. Editorial review of the first edition was done by Ingeborg M. Stochmal. Editorial review of the second edition was done by other persons within McGraw-Hill. Their invaluable assistance is greatly appreciated.

Even though I have retired from the Georgia Institute of Technology, they have allowed me to maintain my office there and have provided some secretarial support as well as use of their copy machines. For this assistance I am extremely grateful. Each of the contributors also obtained secretarial support from their employers. This support is also gratefully acknowledged.

We also wish to express our appreciation to those publishers who gave their permission for the inclusion of various materials needed by the authors for their presentations. In particular, Tables A.7 through A.17 have been reprinted with permission from *Nonparametric Methods for Quantitative Analysis* (2d edition) by Jean Dickinson Gibbons, © 1976, 1985 by American Sciences Press, Inc., 20 Cross Road, Syracuse, N.Y. 13224-2144.

Thanks also go to the editors at McGraw-Hill for their patience with me and with the individual chapter authors. Both editions took much longer than originally planned. Three editors were involved in the first edition. These were, successively, Patricia Allen-Browne, Betty Sun, and Robert W. Hauserman. Two editors were involved in the second edition. These were Robert W. Hauserman and Harold B. Crawford. In addition Paul Sobel has been involved in the final editing and publication of the book.

Finally, and most importantly, I wish to express my appreciation to my wife, Irene. She and my two children and seven grandchildren keep asking when I am really going to retire. My grandchildren have had to stop playing games on my computer so Grandaddy can work on "the book." Irene has provided patience, encouragement, and help, including assistance with the editing of the manuscript and reading page proofs. Without this assistance and encouragement the book would never have been published.

Harrison M. Wadsworth, Jr.

HANDBOOK OF STATISTICAL METHODS FOR ENGINEERS AND SCIENTISTS

P • A • R • T • 1

THE FUNDAMENTALS

CHAPTER 1
INTRODUCTION AND USE OF THE HANDBOOK

Harrison M. Wadsworth, Jr.
Georgia Institute of Technology, Atlanta, GA

1.1 OVERVIEW

This handbook consists of a selection of topics which the editor and many reviewers agreed are most useful for engineers and scientists in their work. As with most handbooks, various authors have contributed chapters to the handbook. Each chapter is intended to be virtually self-contained so that a reader does not need to read earlier chapters in order to understand those appearing later on. However, suitable cross-references are included if additional information is needed.

The language used is such that an engineer or scientist will be able to understand what is said. He or she should not need to be educated in statistical theory beyond, perhaps, a basic course in statistical methods. Someone without such a course could study the early chapters and gain enough understanding to proceed with later ones. Such a reader would have to accept many things on faith since it would be inappropriate to include proofs of mathematical concepts in a handbook such as this. However, references are included where proofs may be found.

Users of this handbook include anyone who might wish a quick overview of statistical techniques that are available. The authors of the various chapters have included, along with their discussions of the topics, many cautions that must be observed in the use of the techniques about which they write. I hope this handbook will prove useful to practitioners in any field of engineering or of the sciences. It should prove useful to persons in other fields, such as the biological sciences or the social and behavioral sciences. Such persons may have to think of different applications and examples than those discussed here. However, this should not present a serious problem.

The authors of each chapter have included extensive bibliographies so that the interested reader may be able to obtain further information regarding topics discussed in that chapter as well as related topics. These bibliographies may be one of the most important contributions of the handbook.

1.2 DESCRIPTION OF THE HANDBOOK

The handbook is divided into four parts. Part 1 contains introductory material, which describes the book, how it may be used, and how data may be summarized and interpreted for further statistical analysis.

A description of the book and suggestions for its use are included in the present chapter. Chapter 2 contains some elementary probability concepts along with suggestions for the summarization and interpretation of data from an analytical point of view. Chapter 3 presents procedures and suggestions for summarizing data using graphic procedures.

Part 2 contains some basic statistical methods and concepts useful throughout the remainder of the book. This part also contains three chapters. Chapter 4 is a basic discussion of some of the principles of statistical inference, preceded by an introduction to many of the probability distributions to which other chapters will refer. Chapter 5 discusses the concepts of estimation and predictions to which engineers and scientists are forever referring. The chapter points out how estimates and predictions may be properly made along with the concept of tolerances. Chapter 6 discusses methods of determining, from a sample, the probability distribution from which it might have come.

Part 3 contains intermediate methods and constitutes a major part of the handbook. Nine chapters are included in this part. These chapters contain the major applications of statistical techniques which engineers and scientists are most likely to need. The topics found in this part of the handbook include statistical process control, acceptance sampling, reliability, survey sampling, bayesian methods, nonparametric methods, statistical simulation, linear regression, and design of experiments. The chapters are somewhat interdependent in that, for example, statistical tests for parameter differences are found in different chapters. The tests, however, are approached differently. While Chap. 4 discusses such tests based on an underlying normal distribution, Chap. 12 discusses the same tests with no such assumption. Chapter 11 looks at such tests from a bayesian point of view.

Part 4 contains a review of some more advanced statistical concepts. There are six chapters in this part. The first three extend some of the concepts initially presented in Chap. 14 on regression analysis. These include multicollinearity, robustness, and nonlinear regression. The last three chapters extend other concepts initially presented in earlier chapters. These chapters deal with time series models, robust design methods, and multivariate methods.

Applications of all of these concepts are presented immediately following the introduction of each technique. These applications include numerical examples.

1.3 STATISTICAL NOTATION AND CONVENTIONS

Standard statistical notation will be used throughout the handbook. There are many glossaries of statistical terms available. Some are in the form of standards, such as ISO 3534, Parts 1, 2, and 3, ANSI/ISO/ASQC A3534, Parts 1 and 2, and ANSI/ISO/ASQC A8402. Others are in the form of glossaries, such as those published by the American Society for Quality Control (ASQC) and the European Organization for Quality Control (EOQC). A partial list of some of these standards and glossaries is included at the end of this chapter. The standards and glossaries are all in substantial agreement. The fact that many of them are available should present no serious problem to statisticians. Unfortunately some statisticians have chosen to invent their own terminology. The authors of the chapters in this handbook have been asked to refrain from this practice.

1.4 HOW TO USE THE HANDBOOK

There are two ways to find things in the book, the table of contents and the index. The table of contents at the beginning of the book lists the chapters in the order of their appearance. Pages are double numbered within each chapter. Thus, for example, page 3.4 means the fourth page in Chap. 3. Tables, figures, and equations are also double numbered, that is, Table 5.2, Fig. 6.1, and Eq. (8.4) refer to the second table in Chap. 5, the first figure in Chap. 6, and the fourth equation in Chap. 8, respectively.

In order to use the table of contents the user would need to have some idea of the kind of statistical procedure that is needed. An alternative method of using the book is by means of the index. The index contains, in addition to page references for specific techniques, suggested pages to which the reader may go to find alternative methods of solutions for a specific problem. It is believed that this will provide additional utility for the handbook. As mentioned previously, sometimes several techniques are available to solve a particular problem. Each, however, has its own set of assumptions that must be met. The reader can thus look into each technique and decide which set of assumptions most nearly fits his or her situation.

REFERENCES

ANSI/ISO/ASQC 3534-1, "Statistics — Vocabulary and Symbols — Probability and General Statistical Terms," American Society for Quality Control, Milwaukee, Wis., 1993.

ANSI/ISO/ASQC A3534-2, "Statistics — Vocabulary and Symbols — Statistical Quality Control," American Society for Quality Control, Milwaukee, Wis., 1993.

ANSI/ISO/ASQC A8402, "Quality Management and Quality Assurance — Vocabulary," American Society for Quality Control, Milwaukee, Wis., 1994.

ANSI Z94, "Industrial Engineering Terminology," Institute of Industrial Engineers, Atlanta, Ga., 1988. (Section 8 of this standard contains definitions of many statistical terms.)

ASQC, "Glossary and Tables for Statistical Quality Control," 3d ed., American Society for Quality Control, Milwaukee, Wis., 1996.

ASTM E456-87b, "Standard Terminology Relating to Statistics," American Society for Testing and Materials, Philadelphia, Pa., 1987.

EOQC, "Glossary of Terms Used in the Management of Quality," 5th ed., European Organization for Quality Control, Berne, Switzerland, 1981.

IEV-191, "Terminology, Reliability, Maintainability and Quality of Service," International Electrotechnical Commission, Geneva, Switzerland, 1985.

ISO 8402, "Quality Assurance—Vocabulary," International Organization for Standardization, Geneva, Switzerland, 1994.

ISO 3534-1, "Statistics — Vocabulary and Symbols — Probability and General Statistical Term" International Organization for Standardization, Geneva, Switzerland, 1993.

ISO 3534-2, "Statistics — Vocabulary and Symbols — Statistical Quality Control," International Organization for Standardization, Geneva, Switzerland, 1993.

ISO 3534-3, "Statistics — Vocabulary and Symbols — Design of Experiments," International Organization for Standardization, Geneva, Switzerland, 1997.

CHAPTER 2
SUMMARIZATION AND INTERPRETATION OF DATA

Harrison M. Wadsworth, Jr.
Georgia Institute of Technology, Atlanta, GA

2.1 INTRODUCTION

The subject of this handbook is statistics. Therefore, these first few chapters have been written for the reader who is not proficient in the basic concepts of this field. Statistics includes the collection, analysis, and interpretation of data. These chapters present methods of summarizing and interpreting data along with some fundamental analysis procedures. Later chapters will describe how these fundamental procedures are used to solve specific problems dealing with the analysis of data.

Statistical analysis procedures involve the use of probability concepts. Therefore, we will start with a brief survey of some of these concepts. This will be followed by a discussion of the application of these concepts to the analysis of data. Chapter 4 then provides a review of some statistical concepts in as few pages as possible. It is hoped that this survey will provide sufficient knowledge for the reader to proceed through the remainder of the handbook. The reader is referred to any good text in statistics to learn more about the general area, and is reminded that the present book is really not such a text. A sample of such texts may be found in the references to this chapter. These, however, are not to be construed as the only ones that are useful for this purpose.

2.2 PROBABILITY

Probability is a measure that describes the chance, or likelihood, that an event will occur. It is a dimensionless number that ranges from 0 to 1, with 0 meaning the event cannot happen and 1 meaning the event is certain to occur. A probability of 0.5 means that it is just as likely to occur as not to occur, and a probability of 0.9 means it has 9 chances in 10 of occurring. In other words, the measure may be assumed to be linear.

We may measure the probability of an event occurring in several ways. The most obvious is the concept of a priori probability. This is the ratio of the number of ways an event may occur to the total number of possible outcomes of the experiment under consideration. The assumption here is that all outcomes are equally likely.

The set of all possible outcomes is called the *sample space*. The denominator in this ratio is, therefore, the total number of points in the sample space. For an event A we can write

$$P(A) = \frac{n_A}{N} \tag{2.1}$$

where $P(A)$ = probability that event A will occur
n_A = number of ways A can occur
N = total number of possible outcomes

This equation could, for example, be used to determine the likelihood of a part, selected at random, not conforming to specifications when 3 parts in a lot of 1000 such parts are known to be nonconforming. The probability, from Eq. (2.1), will be

$$P(d) = \frac{3}{1000} = 0.003$$

Conversely, if we do not know the number of nonconforming pieces in our lot, we might look at a sample selected from the lot and use the sample results to estimate the fraction nonconforming in the lot. This assumes that the sample is representative of the lot. We cannot ever be absolutely sure that the sample results are indicative of the lot. However, if the sample is a random sample, that is, all possible samples have an equal chance of being selected, we can estimate the parameters of the lot and also measure the amount of error likely to be in our estimate. These concepts are discussed later in this chapter and again in more detail in Chaps. 4, 5, and 10.

Now to return to our example, suppose we decide to take a sample of 10 items from the lot of 1000. If we find one nonconforming piece in our sample, we would infer that the fraction nonconforming is 0.10. This is a far cry from the a priori probability that we calculated. Before we can make a statement, based on this sample, that there are 100 nonconforming items in the lot, we must understand the probabilistic nature of our experiment. If we were to take a second sample of 10, we might get no bad parts. Our estimate of the fraction nonconforming would then be $^1/_{20}$, or 0.05.

This discussion raises many questions, which we shall attempt to answer in this and subsequent chapters. Some of these are

1. What constitutes an adequate sample size?
2. What is the best way to select the items that make up the sample?
3. How close is our estimate to the true parameter value?
4. What is the best way to estimate the parameter?

None of these questions have simple answers. In this discussion we have used the word "lot" to refer to the entire population, or universe, from which the sample is selected. The word "parameter" refers to some measure of the population, or lot, such as the population mean or, in this case, the fraction nonconforming.

The above is centered on some fairly simple concepts, predicting the outcome of a future sample from a priori knowledge of a lot, or population, or taking a sample from a lot and estimating the lot fraction nonconforming from the sample. Before we pursue these concepts further, let us look at some basic rules of probability which enable us to compute the probability of occurrence of multiple events.

Rule 1. Addition rule for mutually exclusive events

$$P(A \text{ or } B) = P(A) + P(B) \tag{2.2}$$

Rule 2. Multiplication rule for independent events

$$P(A \text{ and } B) = P(A)P(B) \tag{2.3}$$

Rule 3. General addition rule

$$P(A \text{ or } B) = P(A) + P(B) - P(A \text{ and } B) \tag{2.4}$$

Rule 4. General multiplication rule

$$P(A \text{ and } B) = P(A)P(B|A) \tag{2.5}$$

The first rule applies to the case where either of two events A or B can occur, but not both simultaneously. For example, the diameter of a steel rod may be too small or too large, but it cannot be both. Thus if we are producing 1% too small and 2% too large, the percentage with nonconforming diameters is 3%, and we could say, the probability of a rod selected at random being nonconforming is 0.03.

Rule 2 applies to the situation where the two events could both occur, but the occurrence of one does not influence the occurrence of the other. Suppose, for example, we are concerned with the probability of the diameter of a rod being too large or the rod being out of round. If the occurrence of one type of defect does not influence the probability of the other occurring, the probability that a rod selected at random will be both too large and out of round is, if 2% are too large and 3% are out of round,

$$P(AB) = 0.02(0.03) = 0.0006$$

Thus only six parts in 10,000 would be expected to have both types of nonconformities.

Rule 3 applies to the either/or situation, but in this case the alternative outcomes are not necessarily mutually exclusive. Thus, to continue the above example, the probability that a part will be either too large or out of round will be

$$P(A \text{ or } B) = 0.02 + 0.03 - 0.0006 = 0.0494$$

The amount subtracted is necessary to assure us that we count the joint portion of the probabilities of the two events only once.

Figure 2.1 is an example of a Venn diagram to illustrate rule 3. The entire rectangle in Fig. 2.1 represents the population. Circle A represents the fraction of the population corresponding to event A, and circle B that portion corresponding to event B. The overlap (shaded in the figure) represents that portion corresponding to both A and B. From the figure we see that $P(A)$ is represented by the area of circle A divided by the area of the rectangle. $P(B)$ is similarly represented. $P(AB)$ is the shaded area divided by the area of the rectangle. If the two circles did not

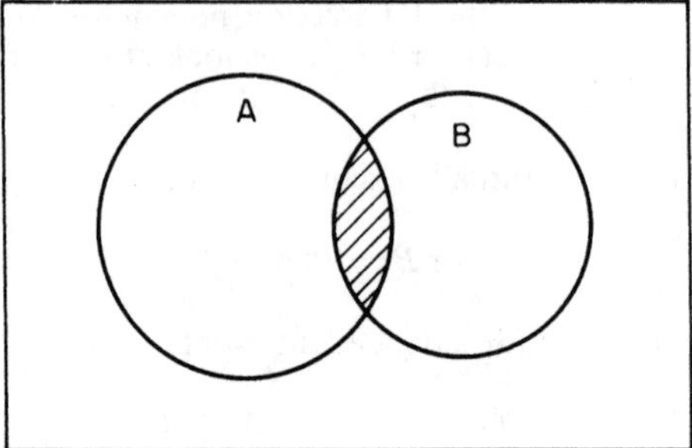

FIGURE 2.1 Venn diagram of two events.

overlap, rule 1 would apply and $P(A$ or $B)$ would be the sum of the areas of both circles divided by the area of the rectangle. When the circles overlap, as in Fig. 2.1, after adding the areas of both circles we have added the overlap, that is, the shaded area, twice. Thus we must subtract the overlapping area to get the area corresponding to event A or B.

Rule 4 is used to compute the joint probability of two events occurring when they are not independent, that is, the probability of one event occurring is different depending on whether the other has occurred. Suppose, for example, event A refers to the supplier of a certain part, with A_1 being one supplier and A_2 being another, and the percentage of parts nonconforming from A_1 is 2% while from A_2 it is 1%. Further, suppose that 40% of our parts come from A_1 while 60% are from A_2. If event B is the event that a part is nonconforming, the probability that a part will be from A_1 and be nonconforming is

$$P(A_1)P(B|A_1) = 0.40(0.02) = 0.008$$

The total percentage nonconforming in this mixture would be

$$\begin{aligned} P(d) &= P(A_1)P(B|A_1) + P(A_2)P(B|A_2) \\ &= 0.40(0.02) + 0.60(0.01) = 0.014, \text{ or } 1.4\% \end{aligned}$$

2.2.1 Counting

There are often a great number of ways an event, such as a certain type of nonconformance to specifications, can occur. Thus, we need some rules to assist us in counting. For example, we might be interested in the number of ways we can obtain a sample of 10 items with one nonconforming item from a manufacturing process. In other words, there must be nine conforming and one nonconforming items in our sample. Since the nonconforming item may occur as any of the 10 sample selections, the answer is of course 10. Formally, we can calculate this by saying there are 10! (10 factorial) ways to arrange the 10 parts in the sample, that is, any of the 10 parts could have been selected first, leaving nine for the second selection, eight for the third, and so on. The number of ways, then, is $10 \cdot 9 \cdot 8 \cdots 2 \cdot 1$, or 10!

The 10 parts may be divided into two classes, good and bad. There are 9! ways to arrange the good parts for every position of the nonconforming one. Thus, the number of ways of obtaining one bad and nine good items in a sample of 10 items is given by

$$\frac{10!}{1!9!} = 10$$

This is usually called the number of *combinations* of 10 things taken one at a time or, in general, n things taken x at a time. The word "combination" means a selection of objects without regard to order. Symbolically we write this as the binomial coefficient

$$\binom{n}{x} = \frac{n!}{x!(n-x)!} \tag{2.6}$$

This example was so obvious that we could have computed it without the use of Eq. (2.6). Another example, the solution of which is not so obvious, is the number of ways of getting two nonconforming items in a sample of 50 items. Using Eq. (2.6) this would be

$$\binom{50}{2} = \frac{50!}{2!48!} = 1225$$

The use of this binomial coefficient assumes that there are only two possible outcomes. If there are k possible outcomes, or categories, into which the result of an experiment may be divided, we can extend Eq. (2.6) to the multinomial coefficient

$$\binom{n}{x_1 x_2 \cdots x_k} = \frac{n!}{x_1! x_2! \cdots x_k!} \tag{2.7}$$

where $x_1 + x_2 + \cdots + x_k = n$.

For example, there may be four ways in which a part may fail to conform to specifications. If there are 50 items in our sample and two have the first type of nonconformance, five the second, three the third, six the fourth, and the remaining 34 have no nonconformance, the number of possible arrangements of this result is

$$\frac{50!}{2!5!3!6!34!} = 9.936 \times 10^{19}$$

2.3 SUMMARIZATION OF DATA

Most of the discussion in Sec. 2.2 related to computing probabilities when the form of the distribution is known. This is not usually the case in engineering and scientific investigations. We therefore need a procedure to enable us to compute probabilities when we are dealing with unknown populations. We might take a large sample or, if the population is not too large, observe the entire population to determine the nature of the variable under consideration. The purpose of this section is to review some procedures which might be used to describe the distribution using the data available.

This description may be analytical in form, or it may be graphic. It may even be a combination of both. In this section we will concentrate on analytical procedures, but will also introduce some graphic procedures. The next chapter will be devoted entirely to graphic means to describe data. Other, more advanced, ways to describe data are presented in later chapters.

2.3.1 Frequency Distributions

The first thing we might do with a large amount of data which come from a study of some sort is to tabulate them in the form of a frequency distribution. As its name implies, a frequency distribution is used to tabulate the number of occurrences of each class of data. If the data are qualitative, the identity of the classes may be obvious. If the data are continuous, we must define classes as intervals.

Suppose, for example, we have 100 measurements on parts selected from a process. We might just list the numbers in the order in which they were selected, or we might order them from smallest to largest. In either case the resulting table is very difficult to read. We could determine the range of the data and the middle of it if they are ordered. However, two different sets of data could have the same minimum and maximum values and even the same middle value and still be quite different.

In order to construct a frequency distribution we must first select the class intervals. It is most convenient for further evaluation if the width of the intervals is constant. We must decide on the width of the intervals, or the number of intervals, and the location of them. Some considerations that should be used to select the number of intervals are as follows:

1. There should be enough intervals to show the variability in the data. This may require anywhere from 6 to 18 intervals. A rule of thumb is to use approximately as many intervals as the square root of the number of observations. Ishikawa (1982) suggests the numbers in Table 2.1 for such a rule. This table, however, agrees pretty well with the square root of N rule.

TABLE 2.1 Number of Class Intervals

Number of observations	Number of classes
Under 50	5-7
50-100	6-10
100-250	7-12
Over 250	10-20

2. There should not be so many intervals that many of them are empty or that adjacent intervals have widely different frequencies.
3. The intervals should all be of the same width. The only exceptions to this are open-ended intervals, which may be necessary at the ends of the distribution when there are a few extreme values present.

A graph of a frequency distribution is called a *histogram*. These rules are most readily observed if a histogram is drawn of the frequency distribution. The location of the intervals should be such that the data cluster roughly around the center of as many intervals as possible.

As an example, consider the data of Table 2.2, which contains 100 measurements in millimeters of the diameter of a manufactured part.

TABLE 2.2 100 Measurements in Millimeters

22	25	15	13	27	30	18	10	16	12
19	27	24	22	27	27	30	18	19	23
15	20	20	27	25	29	17	15	26	24
32	14	20	20	27	21	15	22	16	19
25	27	18	13	23	25	25	27	24	32
27	30	22	24	16	19	23	25	30	30
27	22	21	22	24	29	17	19	22	26
23	21	24	26	30	32	15	19	20	20
17	22	20	27	29	19	26	30	16	20
17	23	16	22	24	23	24	23	22	22

Table 2.2 may be examined, but not much meaningful information may be easily obtained from it in its present form. The 100 observations might be ordered, thus allowing us to observe more easily the smallest and largest numbers as well as the middle of the data set. If we did that, we would see that the numbers range from 10 to 32, with most of the measurements in the 20 to 27 range. The middle number, or median, is 22. The 25th percentile, that is, the number below which 25% of the data lie, is 19, while the 75th percentile is 26, that is, the middle 50% of the data is in the interval between 19 and 26.

Table 2.3 shows the same data grouped in the form of a frequency distribution. The distribution has eight intervals of 3 mm in width. In the table f refers to the frequency, or number of observations, in each interval.

TABLE 2.3 Frequency Table

Class interval	Midpoint	f
$9 < x \leq 12$	11	2
$12 < x \leq 15$	14	8
$15 < x \leq 18$	17	12
$18 < x \leq 21$	20	18
$21 < x \leq 24$	23	26
$24 < x \leq 27$	26	21
$27 < x \leq 30$	29	10
$30 < x \leq 33$	32	3
Total		100

Figure 2.2 illustrates the histogram of the data. Thus, it is a picture of the frequency table, Table 2.3. The figure illustrates the nature of the distribution of measurements. It indicates that the measurements are centered in the 21 to 24 interval with a slightly skewed pattern of variation. The skewness results because of the occurrence of a few more small measurements than large ones. Such a pattern, if extreme, might indicate the occurrence of extreme values, or outliers, that should be investigated for the presence of measurement errors or some other cause. Outliers, and their effect on data analyses, are discussed in later chapters, such as Chap. 17.

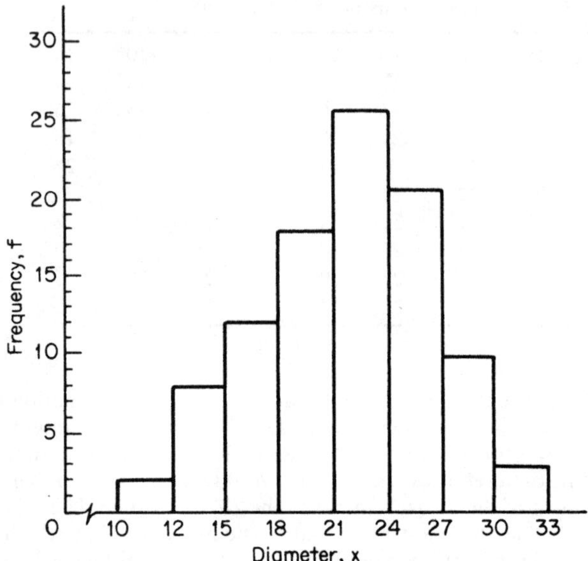

FIGURE 2.2 Histogram of 100 measurements.

2.4 MEASURES OF CENTRAL TENDENCY

In the example of the previous section we may observe that the measurements in Table 2.2 seem to be clustered around 22 or 23. In order to make a more definite statement we might actually compute the value about which the data are centered. Such a value is called a *measure of central tendency,* or *average.* There are several types of such measures available to us, all of which have their uses. The most common one is the *arithmetic mean.* In fact, this one is so common that it is often called simply the *sample average.* The symbol for this average is $\overline{x}$. The mean of a population from which the sample is taken is μ. The sample average $\overline{x}$ is an estimate of μ, which is usually unknown. In fact, all measures of central tendency may be considered to be estimates of μ. The population mean μ is a parameter that describes the population, whereas $\overline{x}$, the sample average, is a statistic that describes the sample.

The arithmetic mean of a sample of size n may be computed as

$$\overline{x} = \frac{1}{n} \sum_{i=1}^{n} x_i \tag{2.8}$$

that is, it is the sum of the observations divided by the sample size. For the 100 measurements in Table 2.2, the arithmetic mean is

$$\overline{x} = \frac{2234}{100} = 22.34$$

One disadvantage of the arithmetic mean is that extreme values have an effect on them. Suppose, for example, that one of the measurements was inadvertently read as 220 instead of 22. The value for $\overline{x}$ would then become

$$\bar{x} = \frac{2432}{100} = 24.32$$

Another measure of central tendency is the *median*. This is the middle number in an ordered set of numbers. If there is an odd number of observations, it is the middle number, while in the case of an even value for n it is halfway between the two middle numbers. A simple rule is that it is the $(n + 1)/2$th number. Thus, if n is 5 (an odd number), the median is the 3rd value. If n is 6 (an even number), it is the 3.5th value, or halfway between the 3rd and the 4th numbers.

For our example n is 100, so the median is halfway between the 50th and the 51st values. Since both of these values are 22, the median is 22. If, as above, one of the 22s was recorded as 220, the 50th smallest observation would be 22, while the 51st would be 23. Thus the median would become 22.5. Compare this change with the effect on $\bar{x}$. If our distribution is symmetric, the values of these two measures would be the same. If skewness is present, they will differ, with $\bar{x}$ being more affected by the skewness.

Another measure of central tendency is the *mode*. This is the most frequently occurring number. For the data in Table 2.2 there are 11 22s and 11 27s. Thus there are two modes for these data.

In the case of grouped data, such as those in Table 2.3, the arithmetic mean is computed by considering the average for all the observations in each class to be the midpoint of that interval. This leads us to Eq. (2.9),

$$\bar{x} = \frac{1}{n} \sum_{i=1}^{k} f_i x_i \tag{2.9}$$

where x_i = midpoint of ith interval

k = number of intervals

f_i = number of observations in interval i

n = sample size,

$$n = \sum_{i=1}^{k} f_i$$

For our example we find from Table 2.4 that

$$\bar{x} = \frac{2228}{100} = 22.28 \text{ mm}$$

Note that this result, 22.28, is somewhat different from the earlier result using the exact 100 numbers. However, the difference is not appreciable.

When a random sample of n observations is taken from a finite population whose size is $N > n$, the sample mean approximates the population mean with N substituted for n in Eq. (2.8) and the summation carried over all values in the population. The population mean μ is rarely known, since the entire population is often inaccessible. Equation (2.8) simply finds the average of all the observations that were taken and gives an estimate, $\bar{x}$, of what the population mean would have been had we been able to measure the entire population. As mentioned earlier, unless we have a badly skewed distribution, the estimate $\bar{x}$ is the estimate of μ that makes the most sense. It is the maximum-likelihood estimate, that is, from statistical theory, $\bar{x}$ is the value of the unknown population mean that would have given the greatest likelihood of having a sample from it turn out exactly as observed. It

TABLE 2.4 Calculation of $\bar{x}$

Class interval	x_i	f_i	$f_i x_i$
$9 < x \le 12$	11	2	22
$12 < x \le 15$	14	8	112
$15 < x \le 18$	17	12	204
$18 < x \le 21$	20	18	360
$21 < x \le 24$	23	26	598
$24 < x \le 27$	26	21	546
$27 < x \le 30$	29	10	290
$30 < x \le 33$	32	3	96
Totals		100	2228

is also the value such that, if repeated samples of size n were observed, μ is the value the average of the $\bar{x}$'s would be expected to approach. This last property, unbiasedness, is shared by other estimates of μ such as the median. As is discussed in Sec. 2.6.1, the expected value, or long-run average, of $\bar{x}$ is μ. If this is true for any estimator of a parameter, we say the estimator is *unbiased.*

Another result is that the accuracy of $\bar{x}$ as an estimate of μ tends to increase as the square root of n. This means that doubling the sample size decreases the error by a factor of $\sqrt{2}$, or 41%.

2.5 MEASURES OF DISPERSION

The dispersion of a set of observations (a sample) is most often measured by the deviation of the observations from their average. The usual average measure used is $\bar{x}$. Thus the statistic $x_i - \bar{x}$ is the deviation to which we are referring. The sum of these deviations will be zero, but the sum of squares of the deviations is positive (or zero if all the x_i's are equal). The average of these squared deviations is called the *variance*, and the square root of the variance is called the *standard deviation.*

The population variance is denoted by σ^2. This, along with μ, is another parameter, or measure, of the population. The positive square root of σ^2, denoted by σ, is called the *population standard deviation.* The corresponding measures of a sample, that is, statistics, are the sample variance s^2 and the sample standard deviation s. Equation (2.10) expresses the procedure used to calculate the sample variance,

$$s^2 = \frac{\Sigma(x_i - \bar{x})^2}{n-1} = \frac{\Sigma x^2 - (\Sigma x)^2/n}{n-1} \tag{2.10}$$

In Eq. (2.10), since we do not know μ, we substitute $\bar{x}$ for it. Statistical theory tells us that dividing the sum of the squared deviations by n (as would be the normal averaging procedure) results in an estimate of σ^2 which is biased, that is, too small on the average. If many samples from the same population were taken, the average of their squared deviations (dividing by n) would be too small by a factor of $(n-1)/n$. Therefore, so that sample variances will not be biased, we define the sample variance as shown in Eq. (2.10).

Other measures of dispersion have applications in engineering statistics. One of these is the range. This is merely the difference between the largest and the smallest observations. Applications of this statistic will be found in quality control work and are discussed in Chap. 7.

Extreme values, such as those caused by a bad reading or a misplaced decimal, will have a large effect on the sample variance because the deviations are squared. They would also affect the sample range since the extreme value will be one of the numbers used to calculate the range. If the median of the squared deviations rather than the average squared deviation were used, we would have a measure that is less affected by the presence of outliers. Such approaches are discussed in Chaps. 12 and 17.

Other measures of dispersion may be found that have specialized uses. Where appropriate they will be presented later in this handbook.

If our sample data are in the form of a frequency distribution such as Table 2.3, the sample standard deviation may be calculated by the use of Eq. (2.11),

$$s = \sqrt{\frac{\Sigma f_i(x_i - \bar{x})^2}{n-1}} = \sqrt{\frac{\Sigma f_i x_i^2 - (\Sigma f_i x_i)^2/n}{n-1}} \tag{2.11}$$

As an illustration of the use of Eq. (2.11), let us again refer to the previous example (Table 2.3). Table 2.5 illustrates the calculation of s from the 100 observations. Using the information in Table 2.5 and Eq. (2.11), we calculate s,

$$s = \sqrt{\frac{51{,}910 - (2228)^2/100}{99}} = 4.79$$

2.5.1 Properties of These Estimators

The sample standard deviation s is a biased estimator of the population standard deviation σ. However, the sample variance is an unbiased estimate of the population variance σ^2. Equations (2.12) and (2.13) express these facts,

$$E(s^2) = \sigma^2 \tag{2.12}$$

$$E(s) = c_4\sigma \tag{2.13}$$

In Eq. (2.13) c_4 is a measure of the bias in s and is a function of the sample size, assuming the samples are all the same size and are drawn from the same normal

TABLE 2.5 Calculation of s

Class interval	x_i	f_i	$f_i x_i$	fx_i^2
$9 < x \le 12$	11	2	22	242
$12 < x \le 15$	14	8	112	1,568
$15 < x \le 18$	17	12	204	3,468
$18 < x \le 21$	20	18	360	7,200
$21 < x \le 24$	23	26	598	13,754
$24 < x \le 27$	26	21	546	14,196
$27 < x \le 30$	29	10	290	8,410
$30 < x \le 33$	32	3	96	3,072
Totals		100	2,228	51,910

population. E is the expected-value operator and is discussed in Sec. 2.6.1. The expression for c_4 is given by

$$c_4 = \sqrt{\frac{2}{n-1}}\,\frac{\left(\frac{n-2}{2}\right)!}{\left(\frac{n-3}{2}\right)!} \tag{2.14}$$

Thus we see that as an estimate of σ we can use the value s/c_4. To improve on this estimate, we often take a series of k samples and average the sample standard deviations to get the statistic

$$\bar{s} = \frac{1}{k}\sum_{i=1}^{k} s_i$$

We then use, as an estimate of σ,

$$\hat{\sigma} = \frac{\bar{s}}{c_4} \tag{2.15}$$

As Eq. (2.14) indicates, c_4 can be tabulated by the sample size n (see Chap. 7, Table 7.1).

This estimate of σ is often used in control chart work (Chap. 7), where the s chart is presented for the control of process variability. Another estimate of σ used more frequently in other statistical work takes advantage of the unbiased property of s^2 as an estimate of the population variance σ^2. Thus if we have k samples from a population with unknown standard deviation σ, another useful estimate of σ is

$$\hat{\sigma} = \sqrt{\frac{1}{k}\sum_{i=1}^{k} s_i^2} \tag{2.16}$$

Equation (2.16) may be used if all the k samples are of the same size. If the sample size varies, we must weight the sample variances according to their sample sizes, as shown in Eq. (2.17),

$$\hat{\sigma} = \sqrt{\frac{(n_1-1)s_1^2 + (n_2-1)s_2^2 + \cdots + (n_k-1)s_k^2}{n_1 + n_2 + \cdots + n_k - k}} \tag{2.17}$$

These last two estimates do not use the factor c_4. They are called *pooled estimates* of σ.

2.6 PROBABILITY DISTRIBUTIONS

We can often describe the distribution of the values of a variable in terms of a theoretical distribution instead of a frequency distribution, as we did in Sec. 2.3. This is particularly helpful when it is not practical to obtain large samples as in Table 2.2. Thus we find that it is useful to be able to write mathematical equations, graphs, or tables to describe the population. Such equations, graphs, or tables show the possible values of a variable and the probabilities that each value will occur. If a variable has such a set of probabilities, we refer to it as a *random variable*. The distribution of the probabilities is called a *probability distribution*, and the equation, graph, or table describes the probability distribution. Since the random variable may be either discrete or continuous, the probability distribution may also be discrete or continuous. Continuous probability distributions are often called *density functions*.

There are several properties that probability distributions must have in order to qualify. If we have a discrete distribution, for example, the number of nonconforming pieces in a sample taken from a manufacturing process, we let p_i be the probability of outcome x_i. For continuous distributions we let $f(x)$ be the density of the probabilities of outcomes of a continuous variable x, such that the integral of $f(x)$ over an interval is the probability that x lies in that interval. The properties then are

1. For x discrete,

$$p_i = \text{probability that } x = x_i$$
$$p_i \geq 0, \qquad \text{all } i$$
$$\Sigma p_i = 1$$

2. For x continuous,

$$\int_a^b f(x)\,dx = \text{probability of } a < x < b$$
$$f(x) \geq 0, \qquad -\infty < x < \infty$$
$$\int_{-\infty}^{\infty} f(x)\,dx = 1$$

For example, suppose we are interested in the probability distribution associated with the number of heads obtained in tossing three coins. For the random variable x = number of heads, the probability distribution is discrete and may be described in tabular form as follows:

x_i	p_i
0	1/8
1	3/8
2	3/8
3	1/8
Total	1

The distribution in this table may also be described by the equation

$$p_i = \binom{3}{i}\left(\frac{1}{2}\right)^3, \qquad i = 0, 1, 2, 3$$

This is a binomial distribution, which is discussed in Chap. 4. This probability distribution is shown as a graph in Fig. 2.3*a*.

An example of a continuous distribution might be described by the equation

$$f(x) = 6x(1-x), \qquad 0 \leq x \leq 1$$

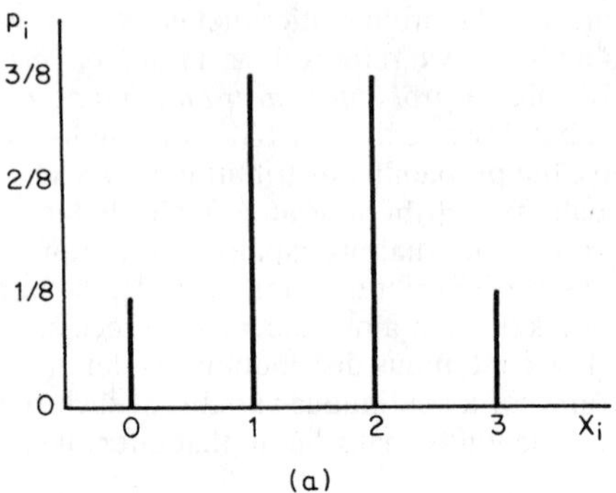

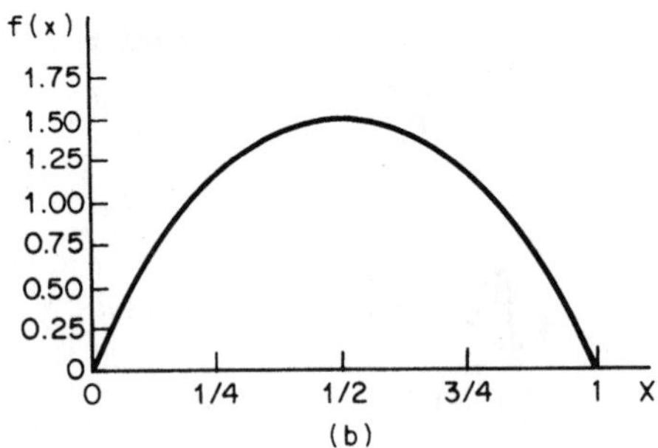

FIGURE 2.3 Graphs of probability distributions. (*a*) Discrete. (*b*) Continuous.

The probability that x lies between 0 and $^1/_4$, for example, is

$$\int_0^{1/4} 6x\,(1-x)dx = \frac{5}{32}$$

For this density function we see that

$$\int_0^1 6x(1-x)\,dx = 1$$

and for all x, $0 \le x \le 1$, $f(x)$ is nonnegative.

This distribution is shown as a graph in Fig. 2.3*b*. In the figure the entire area under the curve is 1 and the area under the curve between 0 and $^1/_4$ is $^5/_{32}$. Similar areas, or probabilities, may be calculated.

Continuous distributions are useful when we are dealing with measurement data. Even though our measuring device may make the data seem like discrete data, they are theoretically continuous and we treat them as such. Measurement data would include dimensional data, time data such as the time to failure of a device, and many other phenomena. On the other hand, count-type data, such as the number of failures in a time interval, are clearly discrete. Many useful theoretical probability distributions are described in Chap. 4. Some applications are suggested there, but these and other applications are described more fully in subsequent chapters.

2.6.1 Expected Values and Variances

The mean of a distribution is often called the *expected value*. The symbol $E(x)$ is used to mean the expected value of x, where $E(x) = \mu$. This may be more formally defined as

$$E(x) = \begin{cases} \sum_{\text{all } i} x_i p_i, & x \text{ discrete} \\ \int_{-\infty}^{\infty} x f(x) dx, & x \text{ continuous} \end{cases} \tag{2.18}$$

Similarly, the variance σ^2 of a random variable described by a probability distribution may be defined as $E(x - \mu)^2$ or, equivalently, $E(x^2) - \mu^2$, where $E(x^2)$ is defined as

$$E(x^2) = \begin{cases} \sum_{\text{all } i} x_i^2 p_i, & x \text{ discrete} \\ \int_{-\infty}^{\infty} x^2 f(x) dx, & x \text{ continuous} \end{cases} \tag{2.19}$$

and μ^2 is $[E(x)]^2$. The standard deviation is the positive square root of the variance.

The expected value and the variance of the number of heads obtained in three tosses of a coin may be computed by the following table:

x_i	p_i	$x_i p_i$	$x_i^2 p_i$
0	1/8	0	0
1	3/8	3/8	3/8
2	3/8	6/8	12/8
3	1/8	3/8	9/8
		12/8	24/8

Thus the expected number of heads is $E(x) = 12/8 = 1.5$, and the variance is

$$\sigma^2 = \sum_{i=0}^{3} x_i^2 p_i - \left(\sum_{i=0}^{3} x_i p_i\right)^2 = \frac{24}{8} - \left(\frac{12}{8}\right)^2 = 0.75$$

The standard deviation is $\sigma = \sqrt{0.75} = 0.87$.

Similarly we may compute the mean, the variance, and the standard deviation for the continuous distribution used as an example earlier,

$$E(x) = \mu = \int_0^1 x f(x) dx = \int_0^1 6x^2 (1 - x)\, dx = 0.50$$

$$E(x^2) = \int_0^1 6x^3 (1 - x) dx = 0.30$$

$$\sigma^2 = 0.30 - (0.5)^2 = 0.05$$

$$\sigma = \sqrt{0.05} = 0.22$$

This section has discussed probability distributions in general. There are many specific distributions that all meet the conditions discussed here and which are particularly useful in statistical applications in science and engineering. Perhaps the most important continuous distribution is presented next. Other discrete and continuous distributions are discussed in Chap. 4.

2.7 THE NORMAL DISTRIBUTION

The distribution most frequently used by scientists and engineers is the normal distribution. It was developed by Gauss to describe the distribution of measurement errors and has since been used in many other applications. As a result of Gauss' work it is sometimes called the *gaussian distribution.* It is quite often a good model for the distribution of variables representing many natural phenomena which may be expected to be reasonably symmetric. An obvious exception is in the case of life-test data, where there is a lower bound of zero. Appropriate distributions for life-test data are presented in Chap. 9.

The normal distribution is a symmetric, continuous distribution, theoretically having a range including all real numbers. The density function can be written as

$$f(x) = \frac{1}{\sqrt{2\pi}\sigma} e^{-(x-\mu)^2/2\sigma^2}, \qquad -\infty < x < \infty \tag{2.20}$$

where the parameters μ and σ^2 are the mean and the variance, respectively.

As can be seen from Eq. (2.20), the normal distribution is really a family of distributions, each member with unique values of μ and σ^2. To facilitate tabulation of this distribution, we may transform any of these normal distributions to a normal distribution with a mean of 0 and a variance of 1. If we set $\mu = 0$ and $\sigma^2 = 1$ in Eq. (2.20), we get

$$f(x) = \frac{1}{\sqrt{2\pi}} e^{-x^2/2} \tag{2.21}$$

This distribution is called the *standard normal* and is plotted in Fig. 2.4. As indicated in Fig. 2.4, virtually the entire distribution is within three standard deviations of the mean.

Any normal distribution may be transformed into the standard normal by means of the transformation indicated in Eq. (2.22),

$$z = \frac{x - \mu}{\sigma} \tag{2.22}$$

where z = transformed value of x

μ = mean of x values

σ = standard deviation of x

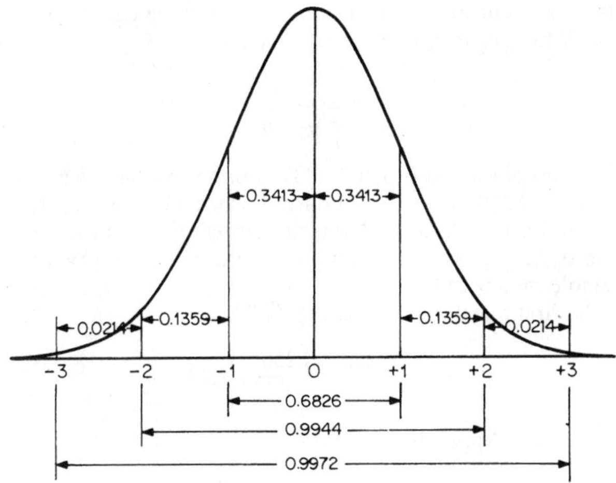

FIGURE 2.4 Standard normal distribution.

Thus, for example, if we have a normal distribution with mean 60 and standard deviation 2, an x value of 63 would give a z of 1.50.

The standard normal distribution is tabulated in the Appendix, Table A.1. The values in the body of the table are the areas from $-\infty$ to z, which we shall call $\Phi(z)$. Since the total area under the curve is 1, these areas represent the probability that a random observation from the standard normal distribution will be less than the corresponding value of z. For the above example, the probability of randomly selecting a value of x corresponding to a z of 1.50 is 0.9332. The probability of selecting one with a z greater than 1.50 is 1 – 0.9332, or 0.0668, that is, $P(z > 1.50) = 1 - \Phi(1.50) = 0.0668$. Because of the symmetrical nature of the normal distribution, $\Phi(z) = 1 - \Phi(-z)$.

2.7.1 Distribution of the Sample Mean

If a random variable X has a normal distribution with mean μ and variance σ^2, the distribution of arithmetic means $\bar{x}$ of random samples of size n taken from the population is also a normal distribution with mean μ, but with variance σ^2/n. Furthermore, a well-known theorem, known as the *central limit theorem*, tells us that the distribution of sample means converges to this normal distribution as n increases, even if the underlying distribution is not normal. The convergence is faster for distributions that are similar to the normal, such as being symmetric, or at least having but one mode.

For most distributions found in practice reasonable convergence is attained for very small sample sizes, such as 4 or 5. About the only time such convergence may not occur this quickly is for distributions that contain mixtures of two different populations. In such a case we should analyze the two populations separately. Even here, though, if the two populations are reasonably equal in size, the distributions of sample means will tend to converge to a normal distribution with a mean value at a point in between the two distributions.

As a result of the central limit theorem, the following equation may be used to compute probabilities regarding the sample mean:

$$z = \frac{\bar{x} - \mu}{\sigma/\sqrt{n}} \tag{2.23}$$

As an illustration of the use of Eq. (2.23), suppose we have a measurement that has a mean value of 0.250 in with a standard deviation of 0.004 in. If we draw samples of size 5 from this population, what is the probability that the sample mean will be greater than 0.252 in? We might also ask, within what interval may we expect 95% of the sample means to fall?

To answer the first question, we use Eq. (2.23):

$$z = \frac{0.252 - 0.250}{0.004/\sqrt{5}} = 1.12$$

From Table A.1 in the Appendix we find that

$$P(\bar{x} > 0.252) = 1 - \Phi(1.12) = 1 - 0.8686 = 0.1314$$

Approximately 13% of the sample means drawn from this population may be expected to be larger than 0.252 in. We may expect $100(1 - \alpha)\%$ of the sample means to fall in the interval $\mu \pm z_{\alpha/2}(\sigma/\sqrt{n})$. For this example, for $\alpha = 0.05$, this interval will be $0.250 \pm 1.96(0.004/\sqrt{5}) = 0.250 \pm 0.0035$, that is, we may expect 95% of the means of samples of size 5 to be in the interval of 0.2465 to 0.2535. Recall that, under the assumption of normality, 95% of the individual measurements will be expected to be in the interval

$$\mu \pm z_{\alpha/2}\sigma = 0.250 \pm 1.96(0.004) = 0.250 \pm 0.0078, \text{ or } 0.2422 \text{ to } 0.2578$$

This is illustrated in Fig. 2.5, in which the larger distribution is the distribution of individuals and the smaller, more compact one is the distribution of sample averages. The use of the normal distribution to describe the distribution of sample averages is the basis of the control chart for averages described in Chap. 7. A further detailed discussion of the normal distribution and its applications is given in succeeding chapters.

2.8 PARETO DIAGRAMS

The Pareto principle was first used to deal with economic data. However, Juran (1964), in his book *Managerial Breakthrough*, urges the use of the principle, by means of Pareto diagrams, to enable managers to focus on the "critical few rather than the trivial many." Juran gives many examples where, for example, relatively few defect types account for most of the defects or a few sales accounts account for 80% of the orders. In fact the Pareto principle is found to occur with striking regularity in many industrial and commercial processes. Industrial engineers have long used it in inventory analysis to show the distribution of the value, by turnover, by mass, or by size, to isolate the relatively few items deserving attention.

The principle states that, for example, a relatively large number of quality problems (say, 80-90%) will be caused by a few (10-20%) of the related factors or defect types. In fact, some authors refer to this as the "80-20 rule." The principle may be

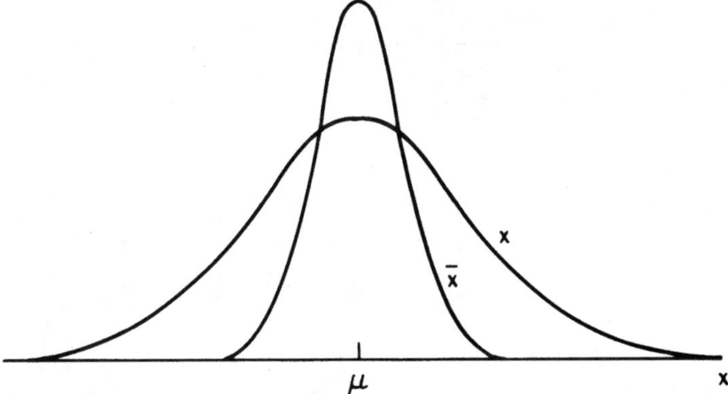

FIGURE 2.5 Distribution of averages and individuals.

illustrated by means of a diagram called the *Pareto diagram*, where the factors, such as defect types or sales accounts, are first ordered by the number, or the percentage, of occurrences. The highest number is plotted on a bar chart, such as Fig. 2.6, as the first bar, and so on. An example taken from Wadsworth et al. (1986) lists the causes of breakdown on rejection of truck cylinder blocks at a foundry ordered by percentage of occurrence:

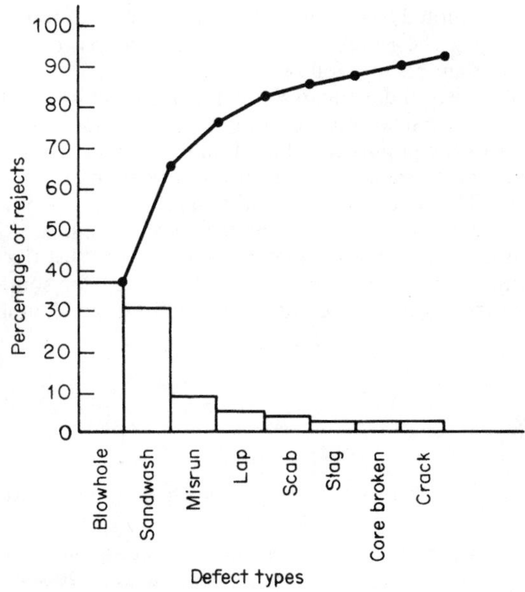

FIGURE 2.6 Pareto diagram of foundry rejections.

Defect cause	Percentage
Blowhole	37.8
Sandwash	30.2
Misrun	9.4
Lap	5.7
Scab	3.7
Slag	1.9
Core broken	1.9
Crack	1.9
Miscellaneous	7.5

Except for the miscellaneous category, the rejection causes are plotted as a Pareto diagram in Fig. 2.6. Note that, in addition to the bars, the cumulative distribution is shown by a line. This shows that blowholes and sandwash account for 68% of the defects, while the remainder account for 32%.

Other examples illustrating the Pareto principle and the use of Pareto diagrams are given in Chaps. 3 and 7. It has been presented here to emphasize its importance in engineering work. Since most problems in engineering are caused by a few sources, the engineer needs a rapid way to separate these sources from the remainder.

2.9 SUMMARY

This chapter has attempted to give the reader an overview of some of the basic concepts of probability as they apply to the statistical analysis of data. This material will be used in the chapters that follow.

The nature of statistical data analysis deals principally with statistical inference which, in turn, means that we are inferring something about a population (often about parameters of the population) based on the results of a sample taken from that population. There are two forms of statistical inference, estimation and hypothesis testing. The first deals with estimating a parameter's value based on a sample, and the second deals with testing the reasonableness of a hypothesis, or statement, about the value of a parameter. In order to discuss the underlying concepts dealing with statistical inference we must first develop some other theoretical probability distributions. These are developed in Chap. 4, along with the concepts of inference.

REFERENCES

Bowker, A. H., and G. J. Lieberman: *Engineering Statistics*, 2d ed., Prentice-Hall, Englewood Cliffs, NJ, 1972.

Guttman, I., S. S. Wilks, and J. S. Hunter: *Introductory Engineering Statistics*, 3d ed., Wiley, New York, 1982.

Hines, W. W., and D. C. Montgomery: *Probability and Statistics in Engineering and Management Science*, 3d ed., Wiley, New York, 1990.

Ishikawa, K.: *Guide to Quality Control*, Asian Productivity Organization, UNPUB, New York, 1982.

Juran, J. M.: *Managerial Breakthrough*, McGraw-Hill, New York,1964.

Milton, J. S., and J. C. Arnold: *Probability and Statistics in the Engineering and Computing Sciences*, McGraw-Hill, New York, 1986.

Scheaffer, R. L., and J. T. McClave: *Probability and Statistics for Engineers*, 2d ed., Duxbury Press, Boston, MA, 1986.

Vardemam, S. B.: *Statistics for Engineering Problem Solving*, PWS Publishing Company, Boston, 1994.

Wadsworth, H. M., K. S. Stephens, and A. B. Godfrey: *Modern Methods for Quality Control and Improvement*, Wiley, New York, 1986.

CHAPTER 3

GRAPHIC METHODS FOR QUALITY IMPROVEMENT

A Case Study in Quality Improvement

A. Blanton Godfrey
Juran Institute, Inc., Wilton, CT

3.1 INTRODUCTION

The following is a case study in quality improvement. We have changed the names of the company and even the name of the product to protect the guilty and the innocent.

The company, which we call here the XOFF Company, produces a wide range of products. In this particular manufacturing plant many different raw materials are turned into parts, the parts into subassemblies, and the subassemblies into final products, which are shipped to customers. The X7 is a subassembly. The cost for each X7 was $1.28. This is the transfer cost, the cost the shop charges the final assembly shop for its subassemblies. At the time of this study the plant was producing 9.7 million items per year at a total cost of $12,400,000. A competitor had gotten management's attention by offering to make a very similar subassembly for $1.00 each. Even with additional handling (approximately $700,000 a year for packaging, unpackaging, and storage), the total cost would only be $10,400,000, so management was faced with an opportunity for a $2 million cost reduction. Of course, the 100 workers who would be laid off if the company stopped making the subassembly did not have exactly the same views as management. The XOFF Company situation is summarized in Table 3.1.

Management and engineering also felt somewhat reluctant to stop making the subassembly. If they did not make the subassembly themselves, a key part of the final product that they sold, they could lose a competitive edge and later might even lose the market of the final product to the competitor.

TABLE 3.1 XOFF Company Situation

Cost per X7 subassembly	$1.28
Annual production	9,700,000
Total cost	$12,400,000
Competitor's bid per X7	$1.00
Additional handling	$700,000
Total cost	$10,400,000

3.2 THE X7 PROCESS

The first step in the breakthrough process (Juran, 1964; Juran and Gryna, 1988) for quality improvement is proof of the need. For this assembly line, the need was obvious. If they could not reduce the cost of the X7, they had no jobs. The next step is to analyze the symptoms. During this step we often first make sure we understand the process. The section chief and several engineers drew a simple flowchart of the X7 process (Fig. 3.1).

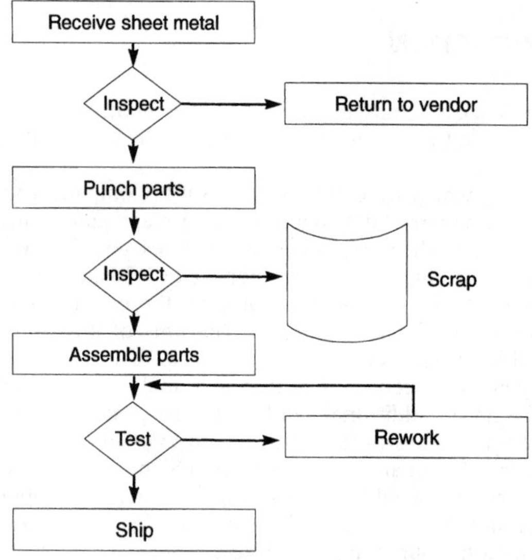

FIGURE 3.1 The X7 process.

The basic steps were as follows. The punch press operation received rolls of sheet metal which were inspected. Some rolls were found to be defective and were returned to the vendor, but most were the right material with the right specifications, and these rolls were fed continuously into large punch presses, which stamped out small parts. These parts were inspected and some were found defective and scrapped, but most were sent to the assembly shop. They were sent in large batches of thousands.

The assembly operation was fairly simple. Three parts that came from the other shop were assembled with small screws, some wiring, and much testing. Of these subassemblies, 25% failed the test and were sent to another production line which reworked them, made some adjustments, and sometimes took them apart and reassembled them. Most of these subassemblies were then able to pass tests and were shipped. A few were scrapped.

3.3 PROJECT IDENTIFICATION

After having studied the flowchart, the section chief and his engineers discovered the following facts. The workers who assembled the subassemblies and other parts into the final product thought that the final quality of the subassemblies was good. There were 100 employees involved in the subassembly process. There were four assembly lines, three making the product the first time, the fourth assembly line making the product the second time. Out of these 100 employees, eight were engaged in full-time inspection, eight were engineers and managers, and there were 84 assembly line workers, 21 on each production line.

Sometimes we call the assembly line that is doing nothing but repair work the "hidden factory." If all the assemblies worked right the first time, we could shut down this hidden factory and save almost 25% of our costs. This is illustrated in Fig. 3.2.

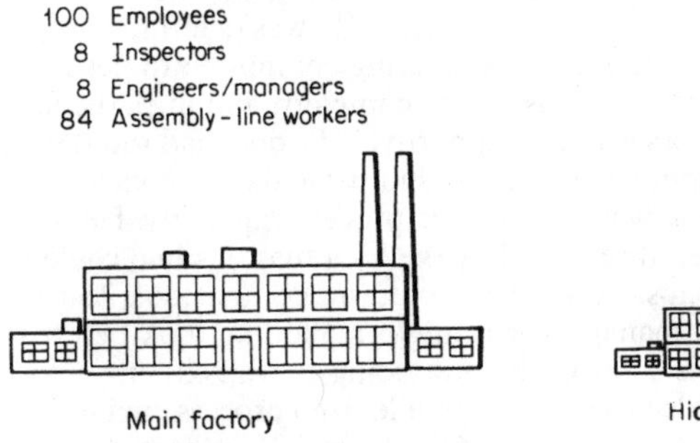

FIGURE 3.2 The hidden factory.

In Fig. 3.3 the section chief and his engineers have used a simple graphic method for identifying the most important problem. This method is the Pareto diagram, a simple set of ordered bar graphs showing which problems are the "critical few" versus the "useful many" (see Sec. 2.8).

If we take the simplest view, as first cut at the poor-quality cost, we count the entire fourth assembly line, 21 people out of 100, as rework. This is the 21% poor-quality cost for rework. Inspection is simply the eight inspectors out of 100 people working on this line. Scrap is roughly a little bit less than 5% of the total cost of the product. There were a few other costs. It is very clear that the number one thing to focus the quality improvement effort on is rework. If they can get the yield of the product up from 75% to 90 or 95%, they can save enormous costs on this product.

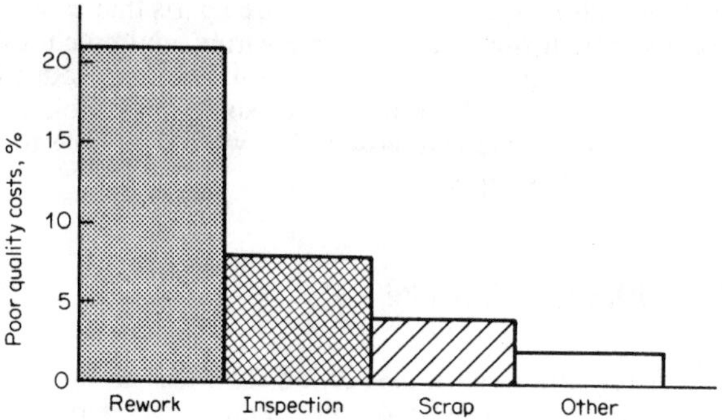

FIGURE 3.3 Pareto diagram for ranking importance.

3.4 ORGANIZING FOR SUCCESS

The next step in making a breakthrough in quality is organizing for success. It is important to get members from the "ailing departments," the departments feeling the pain. In this case the assembly line had the pain of reworking 25% of the product. It is also important to have members from the "suspect departments," those departments who might be causing the pain.

The hero in this process was a young section chief with a worn and tattered copy of a quality control handbook lying on his desk. He was determined to improve the yield of this product line. He formed a steering committee to get the support he needed from the different functions of the company, and in particular to get the support of his boss. His boss was not supportive. His boss had run the same assembly line when he was a young engineer, and in those days the yield was only 50%. His boss thought that 75% was pushing the process capability. He seemed a little defensive and maybe even threatened by the idea that this line could run at 90 or 95% —threatened, perhaps, because it would make him look bad in retrospect when he was the section manager, or it might make him look bad that the other production lines under his control were not doing so well either.

The section manager also tried to get people from process engineering involved, but they were very reluctant members of the team. The process had been in operation for many years, and it had had many minor improvements. The process engineers were not interested in revamping or improving an old process. They had many more important things to do, such as designing new processes for new products. The people from the product design departments were such reluctant members of the team that they never came to the meetings. They would occasionally, when forced, read the project's reports and sometimes comment on them. When we talked to the product designers, they explained to us that they were very busy with new products, that this old product, although made literally in the millions, was not really of that much interest to them, and that they did not think it could be improved much. Besides, they knew that their specifications were correct. After all, the product designers had set them, and this constant carping and complaining about the specifications by the production workers was really based on no fact, just on opinions of people who really did not understand engineering.

The section manager had much more luck with his own team. They were all employees in his section. He made himself team leader, selected two of his section

TABLE 3.2 Organizing for Success

Steering committee	
Section manager	
His boss	
Process engineering	
Product design	
Quality improvement team	
Team leader:	Section manager
Team members:	2 section engineers
	2 checkers
	4 line managers

engineers as team members, took two people off the assembly line to make them checkers (or sorters, as we will see later), and also invited the four assembly line managers. The steering committee and the quality improvement team are listed in Table 3.2.

3.5 THE DIAGNOSTIC JOURNEY

The first step for the quality improvement team was to review all relevant data (analyze the symptoms). The next step on the diagnostic journey was to theorize on causes. As with most teams, it was very natural and very easy to list all the reasons why other people were causing them to make bad products. The first things they thought about were the three parts that they were given from the punch press shop, and why it must be problems with these parts that caused the assemblies not to be good. On the first part they looked at the dimensions. They said the height of the part could be wrong, the diameter could be wrong (it was a circular, doughnut-shaped part), or maybe the thickness of the material was wrong. Or maybe just the material was wrong; the steel bought by the punch press operation to make the parts may be the wrong steel or the wrong consistency, or it may have the wrong electrical properties.

They also suspected that part 2, which was very similar to part 1 and formed the other half of the doughnut when assembled, might have the same problems. They suspected that part 3, a magnet that went in between the two pieces, could also have dimensional problems: height, diameter, or thickness. They thought it could also have the wrong magnetic flux or something could be wrong in the heat treatment process for this part.

Someone else suggested that maybe everything was done right by the punch press people, but the handling of the parts between the punch press shop and the assembly line caused parts to be bent or to get dirt on them, or maybe they were even bent during assembly and things were just beaten up so badly that they did not work.

Someone suggested that maybe they were not making 25% bad parts at all, they were actually fixing good parts. This could be because the test specifications were wrong; after all, the product designers had not looked at this product in a long time or perhaps the inspectors were really rejecting many good subassemblies.

At this time somebody said, "Well, maybe some problems were caused because the operators really had very little experience and no formal training, so maybe

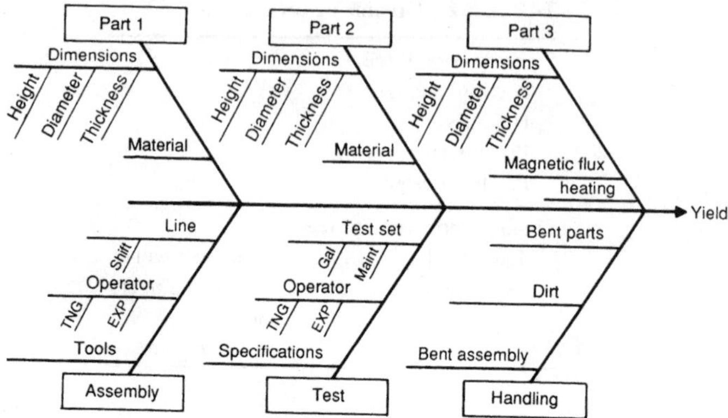

FIGURE 3.4 Theories on causes.

they did not know how to operate the test sets." Then somebody suggested that maybe the test set was wrong, that the calibration had not been done regularly at all, and that the test sets were very poorly maintained.

The quality improvement team finally faced the possibility that they might have something to do with the failure rate, too. Their tools were very outdated, poorly maintained, and not always used properly. Their own operators had very little formal training, and many had little experience on a production line. They also were running three shifts and they wondered if there were variations in the quality produced by the different shifts.

After the brainstorming the quality improvement team used the *Ishikawa diagram* (Wadsworth, Stephens, and Godfrey, 1986) to organize their thoughts. The Ishikawa diagram is sometimes called the *fishbone diagram* because of how it looks, or the *cause-and-effect diagram* because of its use. The results of their study are shown in Fig. 3.4.

3.5.1 Test and Narrow List of Theories

The next step in the quality improvement process is to test and narrow down the list of theories. They did this very much by a brute force method. They created a large wall next to the assembly line and posted 22 scatter plots on the wall. On each scatter plot they plotted one quality characteristic on the *Y* axis and plotted the day of manufacture on the *X* axis. The key variable that they wanted to track was yield, the percentage of good product made by the assembly operations. A sample of some of these 22 scatter plots is shown in Fig. 3.5.

The yield varied from around 75% on the average to sometimes as high as 90% and sometimes as low as 50%. They did a very unsophisticated, but effective eyeball test of their theories. They kept looking at all the patterns on the charts, trying to find patterns that in some way correlated with the yield. They finally threw out most of the plots and concentrated on three—the flux of the magnets and the distance away from nominal, the center point in the specs, for the heights of parts 1 and 2.

In the final examination over these 10 days they failed to see any correlation between the flux, measured in maxwells, and the percentage of good product made.

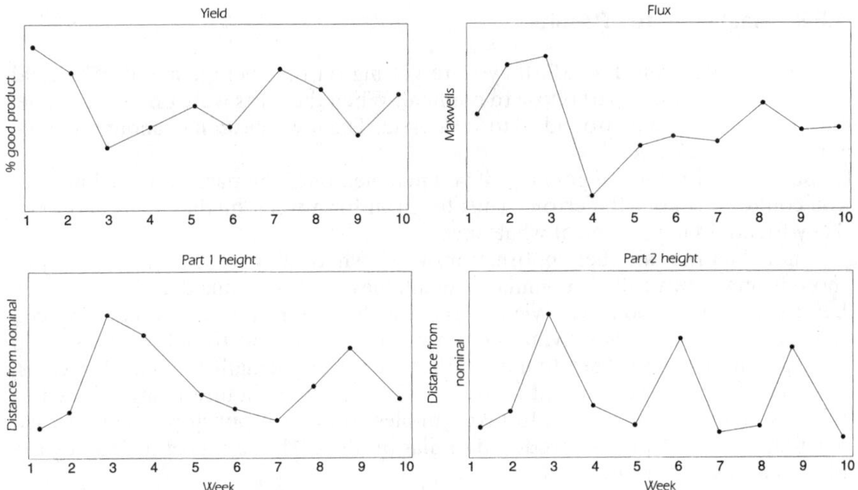

FIGURE 3.5 Scatter plots on the walls.

But when they looked at the distance away from nominal of parts 1 and 2, they started to see some interesting patterns. They saw that part 1 in weeks 3 and 9 was quite a bit away from the center of the spec, and in those weeks they saw a low yield. Weeks 3 and 9 have a distinct dip. Part 2 also had problems in weeks 3 and 9, as well as in week 6, which showed a slight dip in the yield for the final assembly. Now they were ready to conduct an experiment to see whether the cause of the problem really was the heights of the parts they were using.

3.5.2 The Experiment

The experiment was very simple. They used three large boxes, actually carts on wheels, to segregate the parts (Fig. 3.6). They set their own specifications for the punch parts. Those parts that were very close to nominal, the centerline of the specification, they called nominal and put in the large cart in the middle. Those parts that were still within spec but outside the tight range they called nominal and on the low side; they put them into the box called "low." The parts on the high side they put into the box called "high." They then ran for a week only the parts out of the cart called "nominal," then for a week the parts on the high side, and finally for a week the parts on the low side. The results are shown on Fig. 3.7.

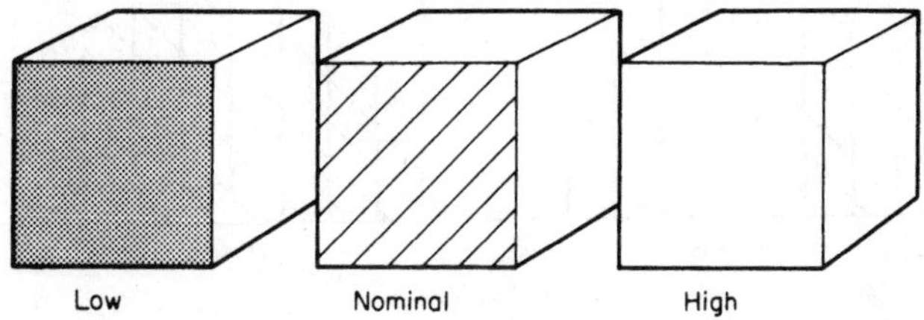

FIGURE 3.6 The boxes.

3.5.3 Analyzing the Results

As can be seen from Fig. 3.7, they were getting on the average about 95% yield when they were using parts close to nominal. When the parts were on the high side, they dropped down to around 50 to 70% yield. The low ones ran in about the same range as the high ones.

Someone else had a theory that if you matched one high part with one low part, you could cancel out the errors. Thus they ran for a week mixing highs and lows. They found no improvement whatsoever.

They then ran another confirmatory week where they went back and ran just those from the box called "nominal." Again, they got the results that they had seen before, averaging about 98% yield. Armed with these results, and very convinced that the punch press was giving them bad parts, they went to talk to the section manager in that shop. They found out that there were actually two punch presses, and that the parts were mixed and stored in large bins before being shipped to their assembly. They decided to take samples from each punch press and to see whether the punch presses produced similar product. The results of their study are plotted as histograms in Fig. 3.8.

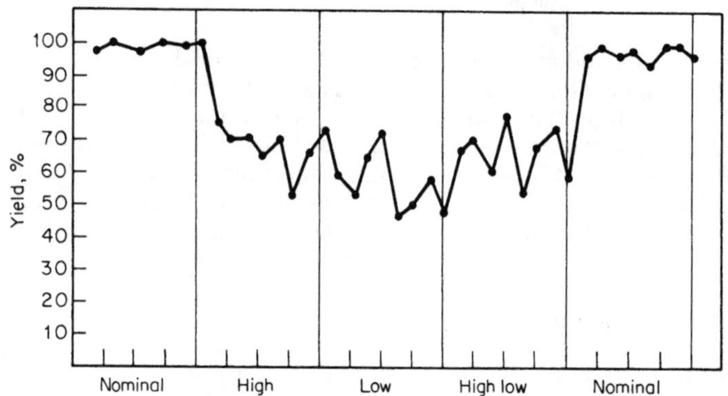

FIGURE 3.7 Results of experiment.

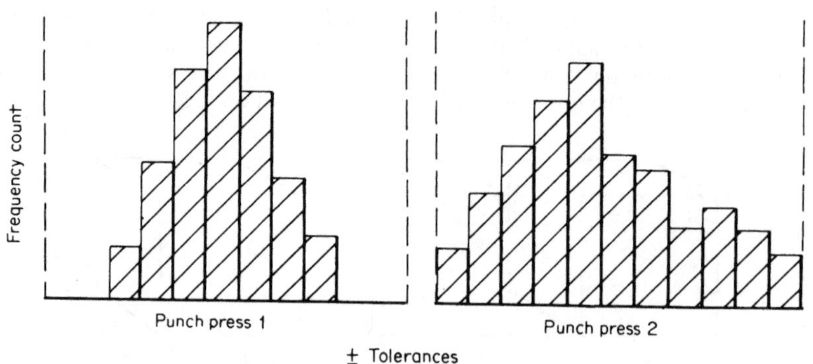

FIGURE 3.8 Results of punch press study.

3.6 JOURNEY TO THE REMEDY

Punch press 1 had everything within specification and the heights were very tightly grouped around the nominal. Punch press 2 had a very wide spread, with things all the way out to the specification lines. Actually the punch press operators were doing quite a bit of inspection, and anything outside of specification was scrapped. Only parts within specification (but sometimes right up to the line of the specification) were shipped to the assembly line. All parts shipped conformed to specification.

When they showed these results to the punch press section manager, he just shrugged and said, "So what." He explained that last year they had asked for capital budget to rebuild both of the punch presses since they were both worn out. The budget request had been for $120,000, but it was cut in half. So they only rebuilt punch press 1. They knew that punch press 2 was worn out, and he was not surprised at all at the results.

The assembly line section manager then sat down and did some quick calculations. He estimated that if they could run the yield at 95% or better, they could save the cost of an entire assembly line, or over $3 million a year. Armed with this information, the assembly line section manager and the punch press section manager went to see their only common boss—the plant manager. When presented with the possible cost reduction of $3 million for a capital investment of $60,000, the plant manager naturally approved.

3.7 RESULTS

The results are shown in Table 3.3. The yield was improved to greater than 95%, the cost per part was reduced to $0.95 (you remember this is $0.05 below the competitor's bid), and the savings were $3.2 million per year.

TABLE 3.3 Results

Yield improved to > 95%
Cost per X7 reduced to $0.95
Savings per year $3,200,000

There was also an investment made (Table 3.4). The following estimates are fairly accurate and definitely accurate enough for this study. The rebuilding of the punch press was $60,000. Their estimate of engineering time, the time of the section chief and the other engineers involved in this quality improvement project, was about $60,000 using their loaded salaries. The estimates of loaded salaries for people who were pulled off the assembly line and did parts sorting and checking were about $40,000 for the duration of this project period. They had also generated quite a bit of scrap during this project since they had sent back one whole load of punch parts that did not meet their specifications and thrown away other things beyond the usual amount for about $30,000. There were other costs, say, $10,000 a year. So the total investment was $200,000. Given a return for the first year of $3.2 million, the return on this investment was 16:1.

TABLE 3.4 Investment

One time	
Rebuilding of punch press	$ 60,000
Engineering time	60,000
Parts sorting and checking	40,000
Scrap	30,000
Miscellaneous	10,000
Total	$200,000
Annual	
Training	
Parts scrap	
Process checking	
Machine maintenance	

Not only is this an outstanding return for the first year of this project, but if this gain is held year after year, they could expect an even better return on investment because the ongoing costs would be much less and the savings, assuming that they still produced 9.7 million items per year, would continue to be in the over $3 million range per year. The costs would include training new people who came on the assembly line, some parts scrap as they maintain their tight control, some cost for process checking to make sure that they were still running at the 95% level and that the piece parts that they were receiving were still of the quality that they needed, and some annual costs associated with machine maintenance so that as the punch presses wore out in the future they could continue to be rebuilt when needed.

3.8 SUMMARY

To continue to hold these gains, another graphic method would be needed, the control chart. Control charts should be used on all critical product and process variables to provide early warning of any undue variation in the process. Then problems can be corrected quickly before scrap and rework are generated. Chapter 7 contains a discussion of control charts.

In this chapter we have given a quick introduction to the basic graphic methods used frequently in quality improvement. We have also shown how these methods can be used during problem solving to quickly identify the reasons for the problem and help focus attention on the remedy.

REFERENCES

Juran, J. M.: *Managenial Breakthrough,* McGraw-Hill, New York, 1964.

Juran, J. M., Ed., and F. M. Gryna, Assoc. Ed.: *Juran's Quality Control Handbook,* 4th ed., McGraw-Hill, New York, 1988.

Wadsworth, H. M., K. S. Stephens, and A. B. Godfrey: *Modern Methods for Quality Control and Improvement,* Wiley, New York, 1986.

P • A • R • T • 2

BASIC STATISTICAL METHODS

CHAPTER 4
PROBABILITY DISTRIBUTIONS AND STATISTICAL INFERENCE

Harrison M. Wadsworth, Jr.
Georgia Institute of Technology, Atlanta, GA

4.1 INTRODUCTION

This topic was introduced in Chap. 2 along with the most important distribution, the normal. Several more distributions will be introduced at this time. These distributions, along with the normal, will be referred to repeatedly throughout the handbook. After a short discussion of each distribution, we discuss the basic concepts of statistical inference on a very elementary and quite general level. In later chapters these elementary concepts are used for specific situations.

The distributions described here may be divided into two general classes, discrete and continuous. Discrete distributions are most often used in situations where the data refer to some sort of count. For example, we may be dealing with the number of items which have a certain characteristic. Continuous distributions, on the other hand, often deal with measurement-type data. These may be measurements of distance, weight, time, and so on, which would be considered to be on a continuous scale.

We look first at four discrete distributions: hypergeometric, binomial, Poisson, and geometric. Then we present four continuous distributions. The t, chi square, and F distributions are related to the normal distribution. The remaining one, the exponential, is used most often in reliability or life-testing situations. It is presented here only briefly. More detailed discussions of this and other distributions with similar applications are found in Chap. 9.

Along with the equation for each distribution we present a brief discussion of its properties and some of its applications. These applications make use of the concepts of statistical inference, which are presented later in this chapter. Statistical inference may be thought of in two different situations, estimation and hypothesis testing. Both involve inferring something about a population from a sample taken from that population. Estimation usually involves estimating a parameter of the population based on the value of one or more statistic calculated from the sample. Hypothesis testing, on the other hand, starts with a statement, or hypothesis, about the value of a parameter. A statistic, calculated from a sample taken from the population, is then compared to the hypothesized parameter value. The hypothesis is then either accepted or rejected, depending on how close the statistic is to the hypothesized parameter. Obviously the two procedures are related. We could, for example, set up a reasonable interval about a statistic (estimation) and see whether the hypothesized parameter value lies in the interval (hypothesis test). Detailed discussion as to the procedure will follow in this chapter after we have discussed the pertinent distributions.

4.2 DISCRETE DISTRIBUTIONS

The first few distributions presented are the hypergeometric, binomial, Poisson, and geometric distributions. They are used when we are dealing with the number of items in a population which have some characteristic (such as not conforming to specifications) versus the number that do not. Thus each observation has two possible outcomes, and the variable we are considering is the number of observations that have the characteristic in question. The hypergeometric distribution deals with the situation where our population is of finite size relative to the sample size. Thus the probability an observation has the characteristic in question depends on the results of previous observations. If, on the other hand, the population is of infinite size, or at least much larger than the sample, the observations can be considered to be independent of each other. This calls for the binomial distribution.

4.2.1 The Hypergeometric Distribution

When sampling from a population of finite size N containing a specified number of nonconforming items D, the probability of, say, the fourth item selected being nonconforming depends on the number of nonconforming pieces among the first three items selected. This statement assumes that the first three items selected were not replaced after they were observed (sampling without replacement). If the items were replaced, they could theoretically be selected again, and our observations would be independent. The question of sampling with or without replacement becomes less important as the ratio of sample size to population size becomes smaller. As a rule of thumb, we usually say it makes no important difference when this ratio is less than 10%, that is, $N \geq 10n$, where n is the sample size.

However, if we consider the situation where the lot, or population, size is less than 10 times the sample size, the hypergeometric distribution is the appropriate distribution to evaluate probabilities of particular outcomes. For example, suppose we receive cartons of 24 light bulbs each and we wish to test five bulbs from each carton to determine whether they operate. We propose to accept each carton for which we find no defective bulbs among the five tested. The probability of acceptance of a particular carton will be given by $P(x = 0)$, where x is the number of defective bulbs in the sample.

For a particular carton, D is the number of nonoperating (and thus defective) bulbs, N is the carton size (24), and n is the sample size (5). We can then compute the probability of x defective bulbs in the carton as

$$P(x) = \frac{\binom{D}{x}\binom{N-D}{n-x}}{\binom{N}{n}} \tag{4.1}$$

The numerator of Eq. (4.1) is the number of ways we can get x bulbs from among the D defective bulbs times the number of ways we can get $n - x$ of the $N - D$ good bulbs. The denominator is the total number of possible samples of size n we can get from a population of N.

For our example, if we have a carton with two bad bulbs among the 24, the probability of not finding any of them in a random sample of five would be

$$P(x = 0) = \frac{\binom{2}{0}\binom{22}{5}}{\binom{24}{5}} = 0.62$$

Similarly, the probability of one bad bulb in our sample is

$$P(x = 1) = \frac{\binom{2}{1}\binom{22}{4}}{\binom{24}{5}} = 0.34$$

The probability that our sample contains both bad bulbs would be

$$P(x = 2) = \frac{\binom{2}{2}\binom{22}{3}}{\binom{24}{5}} = 0.04$$

The mean, or expected, value of x is

$$E(x) = n\left(\frac{D}{N}\right) \tag{4.2}$$

and the variance of x is

$$\sigma_x^2 = n\left[\frac{D}{N}\left(1 - \frac{D}{N}\right)\right]\left(\frac{N-n}{N-1}\right) \tag{4.3}$$

If D/N is set equal to θ, the preceding equations may be written as

$$E(x) = n\theta \tag{4.4}$$

and

$$\sigma_x^2 = n[\theta(1 - \theta)]\left(\frac{N-n}{N-1}\right) \tag{4.5}$$

This shows the relationship between this distribution and the binomial distribution, which is discussed next.

As another example, suppose a manager has 10 chemical engineers on her staff, four of whom are women. The engineers are equally qualified for a particular project requiring three of them. Assuming random selection, we are interested in the probability that none of the three selected engineers are women. The solution may be determined using the hypergeometric distribution with the parameter θ representing the number of women on the staff, N the number of engineers on the staff, and n the number selected for the project. The number of women selected for the project would be denoted by x,

$$P(x = 0) = \frac{\binom{4}{0}\binom{6}{3}}{\binom{10}{3}} = 0.1667, \text{ or } \frac{1}{6}$$

The expected number of women to be selected is

$$E(x) = 3\left(\frac{4}{10}\right) = 1.2$$

4.2.2 The Binomial Distribution

Continuing with the case where there are two possible outcomes for each trial, that is, any item inspected may conform or not conform to specifications, we now consider the case where the results of each trial do not depend on previous results. If the population is very large relative to the sample size, this may be said to occur. The appropriate distribution in this case is the binomial. Many times we consider the population size to be infinite. This occurs if the sample is selected randomly from a continuous process that is assumed to continue at the same level for the foreseeable future. Stated another way, the limit of the hypergeometric distribution as $N \to \infty$ and $D/N = \theta$ remains fixed is the binomial distribution.

Suppose, for example, that the above conditions are met and we are interested in the probability of getting x nonconforming parts in a sample of size n for which the probability of each part being nonconforming is θ. This parameter θ might be the number of nonconforming parts randomly distributed in a large lot divided by the lot size, that is, D/N. It might also be the constant probability of a process producing a nonconforming part. We may use the multiplication rule from Chap. 2 [Eq. (2.3)] to say the probability of obtaining x nonconforming items and $n - x$ conforming items is

$$P(x) = \binom{n}{x} \theta^x (1 - \theta)^{n-x}, \quad x = 0, 1, 2, \ldots, n \tag{4.6}$$

Let us return to the light bulb example in Sec. 4.2.1, in which we had a batch of light bulbs with two out of 24 nonoperating. We said that we were only concerned with one particular box of 24 bulbs which had two defectives. A more realistic approach would be to say that this ratio ($^2/_{24}$) is a long run process average for all such cartons of bulbs. Some cartons have no bad bulbs, some one, some two, some three, and so on. We could then say that the average proportion of the process that is defective is $\theta = {^2/_{24}} = 0.0833$. We can now reason that our previous sample of five bulbs was taken from a process operating at a quality level of 8.33% nonoperating bulbs.

We would then compute the probability of getting all good bulbs in our sample of five as

$$P(x = 0) = \binom{5}{0} (0.0833)^0 (0.9167)^5 = 0.65$$

Thus we see that we now reason that 65% of all such samples would have no nonoperating bulbs, whereas if we knew there were exactly two bad bulbs in the carton, the previous section would indicate that 62% of such samples would contain five operating bulbs. Similarly we could calculate from Eq. (4.6) the probability of any number of bad bulbs in our sample. In this case we can get as many as five defective bulbs in a sample of five. Thus, for example,

$$P(x = 1) = \binom{5}{1} (0.0833)^1(0.9167)^4 = 0.29$$

The probabilities of the remaining possible results to our sampling problem are

$$P(x = 2) = \binom{5}{2} (0.0833)^2(0.9167)^3 = 0.05$$

$$P(x = 3) = \binom{5}{3} (0.0833)^3(0.9167)^2 = 0.005$$

$$P(x = 4) = \binom{5}{4} (0.0833)^4(0.9167)^1 = 0.0002$$

$$P(x = 5) = \binom{5}{5} (0.0833)^5(0.9167)^0 = 0.000004$$

As with the hypergeometric distribution, the mean value of x is

$$E(x) = n\theta \tag{4.7}$$

but the variance is

$$\sigma^2 = n\theta(1 - \theta) \tag{4.8}$$

As a second example, suppose that 80% of all printers used on home computers operate correctly when they are first installed. The others require some adjustment. A particular dealer installs 15 printers during a given month. He has decided to price them on the assumption that each month one printer will require adjustment. The probability that this goal is met is

$$P(x = 1) \binom{15}{1} (0.20)(0.80)^{14} = 0.13$$

Similarly, the probability that none of the printers will require adjustment is $(0.80)^{15} = 0.04$. In other words, the probability that two or more printers will require adjustment is $1 - 0.17 = 0.83$.

A random variable which has a binomial distribution might also be presented as an estimate of θ. Let us call this estimate p, where

$$p = \frac{x}{n} \tag{4.9}$$

Since x has the binomial distribution and n is a constant, the mean and the variance of the random variable p are

$$E(p) = \theta \tag{4.10}$$

$$\sigma_p^2 = \frac{\theta(1 - \theta)}{n} \tag{4.11}$$

The relationship between Eqs. (4.7) and (4.8) and Eqs. (4.4) and (4.5), which deal with the hypergeometric distribution, should be noted.

4.2.3 The Poisson Distribution

Another distribution that will prove useful in this handbook is the Poisson distribution. One way to think of this distribution is the "rare-event" situation, that is, we have a large sample for which the probability of each piece not conforming to specifications is very small, but the overall number of nonconforming pieces per sample is of moderate size. In fact, it can be shown mathematically that the limit of the binomial distribution as $n \to \infty$ and $\theta \to 0$, while $n\theta$ remains constant, is the Poisson distribution. This distribution is given by Eq. (4.12) for the probability of x nonconforming pieces in a sample when there are an average of $\lambda = n\theta$ nonconforming items in the population from which the sample was drawn,

$$P(x) = \frac{\lambda^x}{x!} e^{-\lambda}, \qquad x = 0, 1, 2, \ldots \tag{4.12}$$

Now returning to our light bulb example, we may consider the probability of a carton containing exactly two nonoperating bulbs. Recall that the average percentage of the process that is defective (from Sec. 4.2.2) was 8.33%. The sample size would now be 24 (assuming the cartons are randomly filled). The product λ will thus be

$$\lambda = 24(0.0833) = 2$$

The probability of a case containing exactly two defective bulbs would then be

$$P(x = 2) = \frac{2^2}{2!} e^{-2} = 0.27$$

The cumulative Poisson distribution is tabulated in the Appendix, Table A.2. The table shows for various values of λ the probability that $x \le c$. In other words, the values in the table are

$$\sum_{x=0}^{c} \frac{\lambda^x}{x!} e^{-\lambda} \tag{4.13}$$

The mean value of x when x has the Poisson distribution is λ, and the variance is also λ. The Poisson distribution has many other applications in scientific and engineering work. For example, if calls come into a switchboard in a random manner with λ calls per time unit on the average, the probability that x calls occur in any time unit often follows the Poisson distribution. We call this a Poisson process.

The distribution is often used as an approximation to the binomial distribution. It is a close approximation when n is large and θ small. Another useful approximation to the binomial is the normal distribution, which is discussed next.

4.2.4 The Normal Approximation to the Binomial

When the use of the Poisson distribution as an approximation to the binomial is not valid, the normal distribution discussed in Chap. 2 may be used. In particular, if the

TABLE 4.1 Probability $x \leq c$ for Samples Drawn from Finite Lots

θ	N	n	c	Hyper-geometric	Binomial	Poisson	Normal
0.02	50	10	1	1.000	0.984	0.982	0.998
	100	10	1	0.991	0.984	0.982	0.998
	1000	10	1	0.985	0.984	0.982	0.998
0.05	100	10	1	0.923	0.914	0.910	0.927
	1000	10	1	0.915	0.914	0.910	0.927
	1000	100	2	0.106	0.118	0.125	0.125
0.10	100	10	1	0.739	0.736	0.736	0.701
	1000	10	1	0.737	0.736	0.736	0.701
	1000	100	2	0.000	0.002	0.003	0.006
0.15	100	10	1	0.538	0.544	0.558	0.500
	1000	10	1	0.544	0.544	0.558	0.500
	1000	100	10	0.088	0.099	0.118	0.104

value of θ exceeds 0.10 or if the product $n\theta > 5$, the normal distribution is generally better than the Poisson. This is illustrated in Table 4.1.

The binomial may be approximated by a normal distribution with $\mu = n\theta$ and $\sigma^2 = n\theta(1 - \theta)$, the mean and the variance of the binomial given in Eqs. (4.7) and (4.8). This is the distribution of x, the number of items in the sample with the attribute in question. If, instead, we are interested in the random variable $p = x/n$, the fraction of the sample with the attribute, we use the normal distribution with mean θ and variance $\theta(1- \theta)/n$ from Eqs. (4.10) and (4.11). To illustrate this, suppose we have a process operating at a 5% nonconforming rate. We wish to determine the probability that a sample of 50 pieces will contain more than five nonconforming pieces. This may be computed by first using Eq. (4.14) or Eq. (4.15) to calculate z and then using Table A.1 in the Appendix,

$$z = \frac{(x \pm 0.5) - n\theta}{\sqrt{n\theta(1 - \theta)}} \tag{4.14}$$

$$z = \frac{\left(p \pm \dfrac{1}{2n}\right) - \theta}{\sqrt{\dfrac{\theta(1 - \theta)}{n}}} \tag{4.15}$$

The 0.5 added or subtracted in Eqs. (4.14) and (4.15) is to correct for the fact that a continuous distribution is being used to approximate a discrete distribution. For example, five nonconforming pieces would be represented in the normal distribution as the area between 4.5 and 5.5.

Using Eq. (4.14) to answer the question posed in the previous paragraph, we get

$$z = \frac{5.5 - 50(0.05)}{\sqrt{50(0.05)\,(0.95)}} = 1.95$$

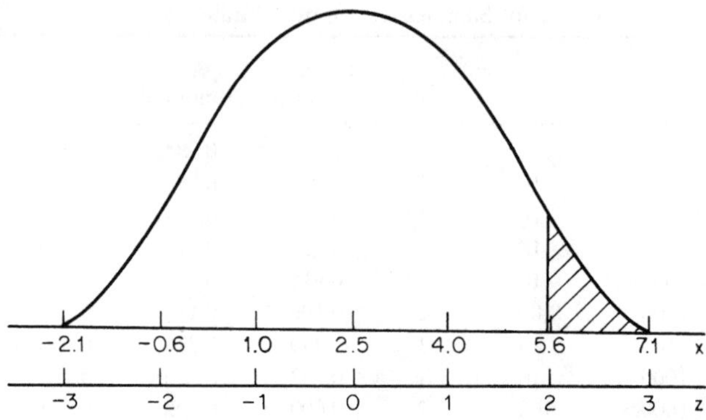

FIGURE 4.1 Normal approximation for binomial probability.

This is illustrated in Fig. 4.1, where the shaded area is the desired probability. From Table A.1 we find the probability that there will be more than five nonconforming pieces in our sample to be 0.0256.

We may also use Eq. (4.15) to answer this question. Here we consider $p = x/n = 5/50 = 0.10$,

$$z = \frac{(0.10 + 1/100) - 0.05}{\sqrt{0.05\ (0.95)/50}} = 1.95$$

This, of course, yields the same answer we obtained using Eq. (4.14). The correct answer, using the binomial distribution, is

$$P(x > 5) = \sum_{x=6}^{50} \binom{50}{x} (0.05)^x (0.95)^{50-x}$$

$$= 1 - \sum_{x=0}^{5} \binom{50}{x} (0.05)^x (0.95)^{50-x}$$

$$= 1 - 0.962 = 0.038$$

Of course we cannot get less than zero nonconforming pieces in our sample. As Fig. 4.1 indicates, the normal distribution has a positive probability for negative values of x. If θ or n were larger, this would not occur and our approximation would improve. Table 4.1 illustrates this point.

4.2.5 The Geometric Distribution

The geometric distribution is of practical value when we have a phenomenon that follows the binomial distribution, that is, we have independent samples with a constant probability of one of two outcomes occurring. The geometric distribution describes the number of trials taken before the first trial for which the outcome with probability θ occurs. For example, if we are inspecting items following an assembly operation, we might be interested in the number of items to be inspected before we encounter a nonconforming one.

If x is the trial on which the first nonconforming part occurs and θ is the probability of a part being nonconforming, the probability that the first nonconforming part will occur on trial x is

$$P(x) = (1 - \theta)^{x-1}\theta, \qquad x = 1, 2, \ldots \tag{4.16}$$

Equation (4.16) expresses the fact that we have $x - 1$ conforming parts followed by a nonconforming one. The cumulative distribution function is

$$P(x \leq n) = \sum_{x=1}^{n} (1 - \theta)^{x-1}\theta = 1 - (1 - \theta)^n \tag{4.17}$$

This is the probability that the first nonconforming part will be found somewhere among the first n items observed.

The geometric distribution has the property of being memoryless, that is, if the outcome under consideration has not occurred in the first n trials, the probability it will not occur during the next n trials is the same as the probability it will not occur in the first n trials. This property is shared by several other distributions, such as the exponential distribution to be discussed later.

The mean of the geometric distribution is $1/\theta$ and the variance is $(1 - \theta)/\theta^2$. Returning to our light bulb example, recall that θ was given as 0.0833. If we inspect the bulbs one at a time, the probability that the first nonconforming bulb would be encountered on, say, trial number 10 would be

$$P(x = 10) = (1 - 0.0833)^9(0.0833) = 0.0381$$

The probability that the first bad bulb will occur somewhere among the first 10 bulbs inspected will be, from Eq. (4.17),

$$P(x \leq 10) = 1 - (1 - 0.0833)^{10} = 0.5809$$

Note that this result is the same as we would obtain using the binomial distribution and finding the probability of one or more defective bulbs among the first 10 tested.

The mean trial on which the first nonoperating bulb would occur is

$$\mu = \frac{1}{\theta} = \frac{1}{0.0833} = 12$$

that is, the average time between the occurrence of a defective bulb is 12 trials. The memoryless property tells us that even if we have already tested several bulbs, we would still expect to find the next defective bulb to be the 12th bulb tested.

To further illustrate this distribution, recall the second example of Sec. 4.2.2, which dealt with printers that needed adjustment. The dealer might wish to know the probability that the first printer needing adjustment is, for example, the fifth one installed. Thus the probability that $x = 5$ is

$$P(x = 5) = (0.8)^4(0.2) = 0.0819$$

The expected number of printers to be installed before one will need adjustment is $1/0.2 = 5$, that is, on the average our dealer would be expected to adjust one out of every five printers installed.

4.2.6 Comparison of Discrete Distributions

To summarize the discussion of discrete distributions, we may review a few points. In all of these distributions, namely, hypergeometric, binomial, and Poisson, x is an integer. If we are sampling without replacement from a finite-sized population, the correct distribution is the hypergeometric. If the ratio $n/N \leq 0.10$, the binomial is a good approximation. If the sample size is large (but still less than 10% of N) and the probability associated with each observation θ is small but the product $n\theta$ is of moderate size, the Poisson distribution is a good approximation of the binomial. If $\theta \geq 0.10$ or $n\theta > 5$, the normal is usually a better approximation to the binomial than is the Poisson.

Table 4.1 illustrates these ideas for several values of θ, N, and n. It gives the probability that x is less than or equal to c using the hypergeometric, binomial, Poisson, and normal distributions. While these approximations have been suggested here, modern computational aids such as computers and hand-held calculators have lessened the need for them.

4.3 CONTINUOUS DISTRIBUTIONS

The most important continuous distribution, the normal, was presented in Chap. 2. Here we first present three distributions which may be thought of as being closely related to the normal distribution, the t, chi square, and F distributions. This will be followed by another frequently used continuous distribution, the exponential. While there are many other distributions that are useful in statistics, their applications are more specialized, and their introduction to this handbook will be saved until they are needed by that application.

4.3.1 The Chi Square Distribution

We observed in Chap. 2 that the sample mean $\bar{x}$ of observations from a normal population follows a normal distribution. The sample variance, however, does not. The sample mean is a linear function of the observations, and any linear function of normal random variables is also a normal random variable. The sample variance, however, is a quadratic function, so we need to determine the distribution of squared normal random variables.

To be more precise, the chi square distribution may be generated by considering the sum of squares of independent standard normal ($\mu = 0, \sigma^2 = 1$) random variables, that is, if we add the squares of n independent standard normal random variables, we get a chi square distribution with n degrees of freedom,

$$\sum_{i=1}^{n} z_i^2 = \chi_{(n)}^{\ 2} \tag{4.18}$$

Proof of the relationship expressed in Eq. (4.18) may be found in many texts on mathematical statistics.

Degrees of freedom refers here to the number of free choices we have in calculating our statistic. If the observations are independent, the knowledge of some of them, or even all but one of them, will not give us any knowledge of the remainder. However, in the next example, our chi square statistic contains the statistic $\bar{x}$. Since

$\bar{x}$ includes the sum of the observations, if we know $\bar{x}$ and all but one of the observations, we can determine the last one. Thus the statistic has $n - 1$ degrees of freedom. Degrees of freedom is the single parameter of the chi square distribution as well as of several others that will follow.

The distribution of the statistic $(n - 1)s^2/\sigma^2$ follows a chi square distribution with $n - 1$ degrees of freedom, where σ^2 is the population variance and s^2 is the sample variance. This assumes that the sample is a random sample of size n drawn from a normal population. Recall that s^2 is calculated according to Eq. (2.10) as

$$s^2 = \sum_{i=1}^{n} \frac{(x_i - \bar{x})^2}{n - 1} \tag{4.19}$$

The mean of the chi square is its degrees of freedom and the variance is twice the degrees of freedom. From this we see that

$$E\left[\frac{(n-1)s^2}{\sigma^2}\right] = n - 1$$

or

$$E(s^2) = \sigma^2 \tag{4.20}$$

and

$$\text{Var}\left[\frac{(n-1)s^2}{\sigma^2}\right] = 2(n - 1)$$

or

$$\text{Var}(s^2) = \frac{2\sigma^4}{n - 1} \tag{4.21}$$

Thus we can say that the sample variance is an unbiased estimate of the population variance and its sampling error, given by $\text{Var}(s^2)$, gets smaller as the sample size increases.

The sample standard deviation s is the positive square root of an unbiased estimator of the population variance, but it is not an unbiased estimator of the population standard deviation σ. This fact was discussed in Chap. 2, where the constant c_4 measured the bias in the random variables. Equation (2.14) is used to compute c_4. Thus, the mean and the variance of s are

$$E(s) = c_4\sigma \tag{4.22}$$

$$\text{Var}(s) = \sigma^2(1 - c_4^2) \tag{4.23}$$

If we substitute the value of c_4 into Eq. (4.23), we find that the variance of s is approximately

$$\text{Var}(s) \doteq \frac{\sigma^2}{2n} \tag{4.24}$$

Compare this result with the variance of σ^2 given in Eq. (4.21). The chi square distribution may be used to make probability statements about the sample variance or the standard deviation. To illustrate this, suppose we take samples of size 10 from a normal population with a known standard deviation of 0.020 in. The probability that the sample standard deviation will be larger than 0.030 in would be determined by first evaluating the chi square statistic as

$$\chi_{(9)}^2 = \frac{9(0.030)^2}{(0.020)^2} = 20.25$$

Table A.3 in the Appendix gives the chi square distribution. The rows indicate the degrees of freedom and the columns the probability that the value of chi square given in the body of the table will be exceeded. Thus, if we observe the row for 9 degrees of freedom, we find that the area to the right of 20.25 is approximately 0.02. In other words, there is approximately a 2% chance that the sample standard deviation of 10 observations would be as large as 0.030 if the true standard deviation is 0.020.

4.3.2 The Student's *t* Distribution

In Chap. 2 we indicated that the statistic $(\bar{x} - \mu)/(\sigma/\sqrt{n})$ has a standard normal distribution when x has a normal distribution with mean μ and variance σ^2. We have also seen in the previous section that $(n - 1)s^2/\sigma^2$ has a chi square distribution with $n - 1$ degrees of freedom. Using these two relationships, we find that the expression $(x - \mu)/(s/n)$ may be said to be the ratio of a standard normal of the square root of a chi square divided by its degrees of freedom, that is,

$$t = \frac{z}{\sqrt{\chi^2/df}} = \frac{(\bar{x} - \mu)/(\sigma/\sqrt{n})}{\sqrt{(n-1)s^2/\sigma^2(n-1)}} = \frac{\bar{x} - \mu}{s/\sqrt{n}} \tag{4.25}$$

Thus we see that by substituting the estimate s for σ in the expression for the standard normal, we get a t, or Student's t, distribution.

The t distribution, like the standard normal, has expected value zero, but the variance is greater than 1. In other words, it is more spread out than the standard normal, but this difference disappears as the sample size increases. There is a single parameter, again called the degrees of freedom, associated with this distribution. The distribution is tabulated in Table A.4 in the Appendix. The number of degrees of freedom for the t statistic in Eq. (4.25) are $n - 1$, the degrees of freedom associated with the chi square.

As an illustration of the use of this distribution, consider a normal population. Now suppose we take samples of size 10 from this population. The values of $t = (\bar{x} - \mu)/(s/\sqrt{10})$ would form a t distribution with 9 degrees of freedom. The probability that $t_{(9)}$ is less than –2.262 or greater than +2.262 is 0.05, that is, from Table A.4 we see that there is 2.5% of the $t_{(9)}$ distribution to the right of +2.262 and another 2.5% to the left of –2.262. In other words, 95% of the distribution is between –2.262 and +2.262. If we now calculate the mean and the standard deviation of our sample of 10 observations to be $\bar{x} = 60$ and $s = 2$, we can construct an interval around $\bar{x} = 60$,

$$\bar{x} \pm t_{(0.025,9)} \frac{s}{\sqrt{n}} = 60 \pm \frac{2.262(2)}{\sqrt{10}}$$

$$= 60 \pm 1.43, \text{ or } 58.57 \text{ to } 61.43$$

This is commonly called a *confidence interval estimate* of the mean μ. We could thus say that we have a confidence of 95% that the true population mean is between 58.57 and 61.43. Confidence interval estimates are discussed further in Sec. 4.4 and Chap. 5.

4.3.3 The *F* Distribution

The ratio of two chi square distributions divided by their respective degrees of freedom generates a distribution commonly called an *F distribution*. For example, consider the ratio

$$F_{v_1,v_2} = \frac{\chi_1^2/v_1}{\chi_2^2/v_2} = \frac{(n_1-1)s_1^2/\sigma_1^2(n_1-1)}{(n_2-1)s_2^2/\sigma_2^2(n_2-1)} = \frac{s_1^2/\sigma_1^2}{s_2^2/\sigma_2^2} \tag{4.26}$$

where v_1 and v_2 are the degrees of freedom and the subscripts 1 and 2 refer to the respective sample and population variances associated with two random samples from normal populations. If both samples were chosen from populations with the same variances ($\sigma_1^2 = \sigma_2^2$), Eq. (4.26) could be written as

$$F = \frac{s_1^2}{s_2^2} \tag{4.27}$$

and this ratio would generate an F distribution with $n_1 - 1$ and $n_2 - 1$ degrees of freedom. Note that the F statistic has a pair of degrees of freedom, one from each chi square variable.

The F distribution is tabulated in the Appendix, Table A.5. The probability that the tabulated statistic F will be exceeded is shown in the heading of each page of the table. The degrees of freedom associated with the numerator are shown across the top and those associated with the denominator in the left margin of each table. As an example of the use of this table, we may conclude that if two samples of 5 and 10 observations, respectively, are drawn from the same population, the ratio of their sample variances will be no larger than $F_{(0.95;4,9)} = 3.63$ with probability 0.95. Other applications are presented later in the handbook.

4.3.4 The Exponential Distribution

There are many situations where the number of occurrences of an event in a specified time follows a Poisson distribution, that is, if x is the number of occurrences in time t, the probability of x occurrences is

$$P(x) = \frac{(\lambda t)^x}{x!} e^{-\lambda t} \tag{4.28}$$

In Eq. (4.28), λ is the rate of occurrences per time unit and t is the number of time units, such as hours. Equation (4.28) is of course the Poisson distribution with parameter λt. From Eq. (4.28) we can say that the probability of no occurrences, that is, $x = 0$, is

$$P(0) = e^{-\lambda t} \tag{4.29}$$

From this expression we could write the cumulative distribution function as the probability that the first occurrence will be by time t,

$$F(t) = 1 - e^{-\lambda t} \tag{4.30}$$

By differentiating Eq. (4.30) we get the density function,

$$f(t) = \lambda e^{-\lambda t} \tag{4.31}$$

Equation (4.31) is commonly called the *exponential distribution*. As indicated, if the probability of x occurrences in time t follows the Poisson distribution, the probability distribution of the time to the first occurrence follows the exponential distribution. If the occurrence to which we refer is a failure of some device, this deals with the subject of reliability, which is further discussed in Chap. 9. Here λ would be called the failure rate. The mean and the variance of the exponential distribution are

$$E(t) = \frac{1}{\lambda} = \mu \tag{4.32}$$

$$\text{Var}(t) = \frac{1}{\lambda^2} = \mu^2 \tag{4.33}$$

The mean can be thought of as the mean time between occurrences of an event (such as a failure).

4.4 STATISTICAL INFERENCE

Statistical inference deals with inferring something about a population based on the results of a sample. In most cases this inference is about the value of a parameter. Statistical inference may be based on an assumption about the form of the underlying distribution from which the sample was taken, that is, normal. In other cases no such assumption is made. In the latter case we say we are dealing with nonparametric statistical inference. This subject is covered in Chap. 12. The present section is restricted to the case where we are sampling from a normal population.

There are two forms of statistical inference: estimation and hypothesis testing. These two related, but different concepts were discussed briefly in Sec. 4.1. As indicated there, the first deals with estimating a parameter, while the second deals with a test of the reasonableness of a hypothetical value of a parameter. These two forms of inference are now discussed in turn.

4.4.1 Estimation

As we said in the last section, the purpose of estimation is to make an inference about the value of an unknown parameter. We might, for example, be interested in the mean thickness μ of a lot of steel washers or the true fraction θ of a lot of material which is nonconforming. If we take a sample in each case and compute $\bar{x}$ in the first case and p in the second, we will probably use these statistics to estimate the respective parameters. This certainly makes sense because we know them to be unbiased and reasonably efficient estimators of the parameters. However, we do not know whether they are correct. Although we cannot make a statement expressing such correctness, we can determine how close they are likely to be to the unknown parameter. This determination depends on the dispersion of the statistics, measured by the variance or the standard deviation. We use what is commonly called a *confidence interval* to express this measure of closeness to the parameter. The standard deviation of the statistic, such as $\bar{x}$ or p, is often called the *standard error* to distinguish it from the population standard deviation.

In order to express our parameter estimate with a measure of its closeness to the parameter, we often state it in the form of an interval within which we believe, with a stated confidence, that the unknown parameter lies. From our discussion of the normal distribution in Chap. 2 we know that $100(1 - \alpha)\%$ of the sample averages from a normal population would fall within an interval of $\mu \pm z_{\alpha/2}\sigma/\sqrt{n}$, where α is some small fraction and $z_{\alpha/2}$ is the value of the standard normal distribution exceeded by an area of $\alpha/2$. When we do not know μ but do know the parameter σ, we use $\bar{x}$ as an estimate of μ. We then might construct a similar interval about $\bar{x}$ and infer that this interval contains the parameter μ with a confidence of $100(1 - \alpha)\%$ of being correct, that is, if we took many samples of the same size n and constructed such an interval about each sample average, we would expect $100(1 - \alpha)\%$ of the intervals to contain μ. The resulting interval would be

$$\bar{x} \pm z_{\alpha/2} \frac{\sigma}{\sqrt{n}} \tag{4.34}$$

If the population standard deviation were unknown, we might calculate the sample standard deviation s and use it as an estimate of σ. We would also need to substitute appropriate values from the t distribution for the normal deviate z in Eq. (4.34). For this situation, Eq. (4.34) would become

$$\bar{x} \pm t_{(\alpha/2;n-1)} \frac{s}{\sqrt{n}} \tag{4.35}$$

where $t_{(\alpha/2;n-1)}$ is the value of the t distribution with $n - 1$ degrees of freedom exceeded by an area of $\alpha/2$. Examples illustrating the use of these intervals and others will follow. Other types of intervals, such as tolerance intervals and prediction intervals, along with their relationships to confidence intervals, are discussed in Chap. 5.

If we wish to estimate the true standard deviation, or variance, of a population, we would use the fact that the statistic $(n - 1)s^2/\sigma^2$ has a chi square distribution (see Sec. 4.3.1). Thus we can make the following probability statement about σ^2:

$$P\left[\frac{(n-1)s^2}{\chi^2_{(1-\alpha/2;n-1)}} < \sigma^2 < \frac{(n-1)s^2}{\chi^2_{(\alpha/2;n-1)}}\right] = 1 - \alpha \tag{4.36}$$

We can therefore use Eq. (4.36) to determine a confidence interval estimate of the population variance or the standard deviation. We could say that, for the population standard deviation, a $100(1 - \alpha)\%$ confidence interval is

$$L = \frac{s\sqrt{n-1}}{\sqrt{\chi^2_{(1-\alpha/2;n-1)}}} \tag{4.37}$$

$$U = \frac{s\sqrt{n-1}}{\sqrt{\chi^2_{(\alpha/2;\,n-1)}}}$$

where L and U are the lower and upper bounds of our confidence interval. Note that this interval, unlike those for the mean, is not symmetrical about our estimate s.

For confidence interval estimates of θ, for example, the process or lot fraction nonconforming (a binomial parameter), we would use $p = x/n$ as our initial point estimate, where x is the number of nonconforming pieces in a sample of size n. If binomial tables are available, appropriate confidence intervals may be computed as follows:

$$\sum_{i=0}^{x} \binom{n}{i} \theta_L^i (1 - \theta_L)^{n-i} = \frac{\alpha}{2}$$

$$\sum_{i=x}^{n} \binom{n}{i} \theta_U^i (1 - \theta_U)^{n-i} = 1 - \frac{\alpha}{2} \tag{4.38}$$

where θ_L and θ_U are the lower and upper bounds, respectively, to our $100(1 - \alpha)\%$ confidence interval for θ. Since x, the number of nonconforming pieces in the sample, must be an integer, the desired values of $\alpha/2$ and $1 - \alpha/2$ are probably not available. However, we could select probabilities as close to them as possible. Curves to facilitate the use of Eq. (4.38) have been developed (Clopper and Pearson, 1934). A sample of such curves, for $\alpha = 0.01$ and $\alpha = 0.05$, adapted from Pearson and Hartley (1970), is shown in the Appendix, Table A.6.

For large sample sizes we may make use of the normal approximation to the binomial discussed in Sec. 4.2.4. Thus a $100(1 - \alpha)\%$ confidence interval estimate of θ would be

$$p \pm z_{\alpha/2} \sqrt{\frac{p(1-p)}{n}} \tag{4.39}$$

where p is the estimate x/n of θ and $z_{\alpha/2}$ is the standard normal value exceeded by $\alpha/2$. Care must be exercised in the use of Eq. (4.39) since the binomial distribution is not symmetric except when $\theta = 0.50$. Equation (4.39) might, for small p and n, result in a negative value for the lower confidence bound. Since a negative value of θ is impossible, we would set the lower bound at zero. As a rule of thumb, Eq. (4.39) may be used when $p \geq 0.10$ and $n \geq 50$.

4.4.2 Tests of Hypotheses

To test a hypothesis is to make a decision regarding the reasonableness of a statement, or hypothesis, about the value of a parameter. For example, we may say that a process manufactures pins with a desired diameter of 0.750 in. We might take a sample from the process and measure the diameter of each of the pins. If the average diameter is close to 0.750 in, we would conclude that the process is satisfactory. In carrying out this decision process, we incur two possible types of errors:

1. We can say the hypothesis is false when it is really correct.
2. We can say the hypothesis is correct when it is really false.

The first is commonly called a type 1 error. The probability of making this type of error is given the symbol α. The second is called a type 2 error. The probability of making this type of error is given the symbol β. In quality control work α is often called the "producer's risk" and β is called the "consumer's risk." In an ideal situation both α and β should be small. Of course, they can only be zero when we examine the entire population.

The value of β depends on the true parameter value. It is therefore a conditional probability. Its value is often shown, therefore, as a function of the parameter value. As such it is called an *operating characteristic function*, or OC curve, when it is plotted against the parameter values. Thus β, or the OC function, is the probability of accepting the hypothesis, called the *null hypothesis*, versus possible values of the parameter about which the hypothesis is made.

The appropriate test is based on the underlying distribution of the statistic used to estimate the parameter. Thus, if the parameter is the mean μ of a normal population, the sample average $\bar{x}$ would be used to estimate it, and the test statistic shown in Eq. (4.40) or (4.41) would be used, depending on whether σ is known or unknown,

$$z = \frac{\bar{x} - \mu_0}{\sigma/\sqrt{n}} \tag{4.40}$$

$$t = \frac{\bar{x} - \mu_0}{s/\sqrt{n}} \tag{4.41}$$

The test statistic in Eq. (4.40) would be compared to values in the standard normal distribution, while the test statistic in Eq. (4.41) would be compared to values in the t distribution with $n - 1$ degrees of freedom. In both cases μ_0 is the value of the hypothesized mean.

If we wish to test a hypothesis about the variance of a population, or standard deviation, we would use the chi square distribution, and our test statistic would be

$$\chi^2 = \frac{(n-1)s^2}{\sigma_0^2} \tag{4.42}$$

where σ_0^2 is the hypothesized value of the population variance and s^2 is the sample variance. The test statistic in Eq. (4.42) is compared to the chi square distribution with $n - 1$ degrees of freedom.

If we wish to compare two processes, we might wish to test the hypothesis that the two have equal means. For this test, the test statistic shown in Eq. (4.43) or (4.44) would be used, depending on whether the population variances are known or unknown,

$$z = \frac{\bar{x}_1 - \bar{x}_2}{\sqrt{\sigma_1^2/n_1 + \sigma_2^2/n_2}} \tag{4.43}$$

$$t = \frac{\bar{x}_1 - \bar{x}_2}{s_p\sqrt{1/n_1 + 1/n_2}} \tag{4.44}$$

where $\bar{x}_i$ = sample mean of sample i, $i = 1, 2$

σ_i^2 = population variance for population i, $i = 1, 2$

n_i = sample size, $i = 1, 2$

$$s_p^2 = \frac{(n_1 - 1)s_1^2 + (n_2 - 1)s_2^2}{n_1 + n_2 - 2}$$

with s_i^2 being the sample variance of sample i, $i = 1, 2$.

The statistic s_p^2 is the weighted average of the population variances, using the degrees of freedom of each variance as its weight. The square root of s_p^2 is s_p, and this is used in Eq. (4.44). This averaging, or pooling, should only be done when the two populations can be assumed to have equal variances. Thus the sample variances are two estimates of the same thing, and averaging them makes sense.

To determine whether the two populations have equal variances, we should test the hypothesis that this is true, that is, $\sigma_1^2 = \sigma_2^2$. The test statistic in Eq. (4.45) is used for this test,

$$F = \frac{s_1^2}{s_2^2} \tag{4.45}$$

This test statistic should be compared to appropriate values from an F distribution with $n_1 - 1$ and $n_2 - 1$ degrees of freedom.

Finally, if we wish a test concerning the process fraction nonconforming (θ in a binomial distribution), we might make use of the normal approximation to the binomial and use the normal test statistic shown in either Eq. (4.46) or Eq. (4.47), depending on whether we are using x or $p = x/n$ as our estimate,

$$z = \frac{(x \pm 0.5) - n\theta_0}{\sqrt{n\theta_0(1 - \theta_0)}} \tag{4.46}$$

$$z = \frac{\left(p \pm \dfrac{1}{2n}\right) - \theta_0}{\sqrt{\dfrac{\theta_0(1 - \theta_0)}{n}}} \tag{4.47}$$

In Eqs. (4.46) and (4.47) θ_0 is the hypothesized value of θ and x is the number of nonconforming items in a sample of n items. The test statistic z is compared to the standard normal distribution.

We could also test whether two processes have the same fraction nonconforming by taking samples from each. The test statistic then would be

$$z = \frac{p_1 - p_2}{p\sqrt{1/n_1 + 1/n_2}} \tag{4.48}$$

where $p_i = x_i/n_i$, $i = 1, 2$, and

$$p = \frac{x_1 + x_2}{n_1 + n_2}$$

Equations (4.46) through (4.48) make use of the normal approximation to the binomial. Tests could also be devised using the binomial or Poisson distributions. As needed, these will be found in later chapters in this handbook or in standard texts on statistics.

4.4.3 Examples

Example 4.1. A process manufacturing steel pins is supposed to be operating with a mean pin diameter of 0.7500 in and a standard deviation not greater than 0.0020 in. A random sample of 15 pins was taken from the process with the following results:

$$\bar{x} = 0.7560 \text{ in}$$

$$s = 0.0030 \text{ in}$$

The first hypothesis to be tested was that $\sigma = 0.0020$ in. If this hypothesis is not rejected, the value of $\sigma = 0.0020$ will be used to test the mean diameter using Eq. (4.40). Thus we are testing the hypothesis that $\sigma = 0.0020$ versus the alternative that $\sigma > 0.0020$, that is,

$$H_0: \quad \sigma = 0.0020 \text{ in}$$

$$H_1: \quad \sigma > 0.0020 \text{ in}$$

where H_0 is the null hypothesis and H_1 is the alternative. If H_0 is rejected, we conclude H_1 is true. For our test it was decided that α should be 0.05. Thus, we should reject H_0 if the test statistic in Eq. (4.42) exceeds $\chi^2_{(0.05;14)} = 23.68$ (from Table A.3). For this example,

$$\chi^2 = \frac{14(0.0030)^2}{(0.0020)^2} = 31.50$$

Since 31.50 is greater than 23.68, the critical value from the chi square table, we would conclude that $\sigma > 0.0020$, based on this sample.

Next we wish to test the mean. However, we cannot now assume that the standard deviation σ is 0.0020, or even 0.0030, since that estimate is based on only 15 observations. We then need to use the test statistic in Eq. (4.41), which uses the t distribution to test the hypothesis that $\mu = 0.7500$ in. In this case we are interested in whether the mean diameter is either larger or smaller than 0.7500 in. Thus we would use a two-sided alternative, that is, we would reject the null hypothesis that $\mu = 0.7500$ if the test statistic indicates it to be either larger or smaller. If we again choose α to be 0.05, our test would be

$$H_0: \quad \mu = 0.7500 \text{ in}$$

$$H_1: \quad \mu \neq 0.7500 \text{ in}$$

and we would reject H_0 in favor of H_1 if $t < -t_{(0.025;14)}$ or $t > t_{(0.025;14)}$. From Table A.4 we find this value of t to be 2.145. For our example we compute the test statistic to be

$$t = \frac{\bar{x} - \mu_0}{s/\sqrt{n}} = \frac{0.7560 - 0.7500}{0.0030/\sqrt{15}} = 7.75$$

Since 7.75 is greater than 2.145, we would conclude that the process mean μ is not equal to 0.7500 in.

We might now wish to estimate the values of μ and σ by means of confidence intervals. A 95% confidence interval for σ is obtained using Eq. (4.37) as

$$L = \frac{s\sqrt{n-1}}{\sqrt{\chi^2_{(0.025;14)}}} = \frac{0.0030\sqrt{14}}{\sqrt{26.12}} = 0.0022 \text{ in}$$

$$U = \frac{s\sqrt{n-1}}{\sqrt{\chi^2_{(0.975;14)}}} = \frac{0.0030\sqrt{14}}{\sqrt{5.63}} = 0.0047 \text{ in}$$

Similarly, using Eq. (4.35), we get a 95% confidence interval estimate of μ as

$$\bar{x} \pm t_{(0.025;14)} \frac{s}{\sqrt{n}} = 0.7560 \pm \frac{2.145(0.0030)}{\sqrt{15}}$$

$$= 0.7560 \pm 0.0017, \text{ or } 0.7543 \text{ to } 0.7577 \text{ in}$$

Note that neither confidence interval contains the hypothesized values of μ = 0.7500 and σ = 0.0020.

Example 4.2. A similar process at another location is also manufacturing these pins. A sample of 20 pins was selected from this process with the following results:

$$\bar{x} = 0.7470 \text{ in}$$

$$s = 0.0025 \text{ in}$$

We now wish to test the hypothesis that these two processes have the same mean and variance. In order to determine whether we would be justified in pooling the variances, we first test the hypothesis that the two processes have equal variances. Our hypotheses are

$$H_0: \quad \sigma_1^2 = \sigma_2^2$$

$$H_1: \quad \sigma_1^2 > \sigma_2^2$$

Here we wish to assume that if we reject H_0, the first process has more variability than does the second. Recall that the specified value of the standard deviation is 0.0020 in, and both sample standard deviations exceed this value. We have shown in Example 4.1 that it is very unlikely that the first process standard deviation meets the requirement. In order to test the above hypothesis we use the test statistic in Eq. (4.45),

$$F = \frac{s_1^2}{s_2^2} = \frac{(0.0030)^2}{(0.0025)^2} = 1.44$$

This ratio of 1.44 is compared to the value of $F_{(0.05;14,19)}$ found in Table A.5 of the Appendix. This value is 2.25. Since our test statistic does not exceed 2.25, we conclude that the two processes may have the same variability, and therefore the two sample standard deviations may be pooled. We next pool the two sample standard deviations to get

$$s_p = \left[\frac{(n_1-1)s_1^2 + (n_2-1)s_2^2}{n_1+n_2-2}\right]^{1/2} = \left[\frac{14(0.0030)^2 + 19(0.0025)^2}{15+20-2}\right]^{1/2} = 0.0027$$

The test statistic t is computed using Eq. (4.44),

$$t = \frac{\bar{x}_1 - \bar{x}_2}{s_p \sqrt{\frac{1}{n_1} + \frac{1}{n_2}}} = \frac{0.7560 - 0.7470}{0.0027 \sqrt{\frac{1}{15} + \frac{1}{20}}} = 9.68$$

If we are again using a two-sided test with significance level α of 0.05, the hypothesis to be tested and its alternative are

$$H_0\colon \quad \mu_1 = \mu_2$$
$$H_1\colon \quad \mu_1 \neq \mu_2$$

The critical value from Table A.4 is $t_{(0.025;33)} = 2.04$. Since 9.68 exceeds 2.04, we conclude that the two locations have different mean diameters.

We might next obtain a confidence interval estimate of the true difference in mean diameters. For such a $100(1 - \alpha)\%$ confidence interval we have

$$(\bar{x}_1 - \bar{x}_2) \pm t_{(\alpha/2;n_1+n_2-2)} s_p \sqrt{\frac{1}{n_1} + \frac{1}{n_2}} \tag{4.49}$$

For our example a 95% confidence interval for the difference in the two process means is

$$(0.7560 - 0.7470) \pm 2.04(0.0027) \sqrt{\frac{1}{15} + \frac{1}{20}} = 0.0090 \pm 0.0019, \text{ or } 0.0071 \text{ to } 0.0109$$

Note that zero is not in our interval. This is consistent with the results of our significance test.

Example 4.3. If the requirements for the pin diameters in Examples 4.1 and 4.2 are 0.7500 ± 0.0100 in, the sample results may be used to get an estimate of the fraction nonconforming for each process location. For the first process, assuming the measurements follow a normal distribution and using the pooled estimate of the standard deviation (0.0027), we find

$$z_U = \frac{0.7600 - 0.7560}{0.0027} = 1.48$$

$$z_L = \frac{0.7400 - 0.7560}{0.0027} = -5.93$$

From Table A.1 in the Appendix we see that the fraction above the upper specification limit is 0.0694. The fraction below the lower limit is zero, so the total fraction nonconforming is 0.0694, or 6.94%.

For the second process location we get

$$z_U = \frac{0.7600 - 0.7470}{0.0027} = 4.81$$

$$z_L = \frac{0.7400 - 0.7470}{0.0027} = 2.59$$

From Table A.1 we determine that there are no pins above the upper limit and the fraction below the lower limit is 0.0048, or 0.48%.

Example 4.4. It was required that both processes be operating at a 3% nonconforming level or better. We might wish to test the hypothesis that the first process location is operating at this level. Using the test statistic in Eq. (4.47), we test the following null hypothesis and its alternative:

$$H_0: \quad \theta \le 0.03$$

$$H_1: \quad \theta > 0.03$$

If we again use a significance level α of 0.05, we would reject the null hypothesis if the value of the test statistic exceeded $z_{0.05} = 1.645$. Using Eq. (4.47) without the correction since our estimate of p is not based on x/n, we find

$$z = \frac{p - \theta_0}{\sqrt{\dfrac{\theta_0(1 - \theta_0)}{n}}} = \frac{0.0649 - 0.03}{\sqrt{\dfrac{0.03(0.97)}{15}}} = 0.7924$$

Since this value is less than 1.645, we cannot assume that the percentage of the process nonconforming is greater than 3% based on this very small sample.

4.5 CONCLUSION

This chapter has discussed several of the most important probability distributions. Other distributions are also quite important, but their applications are such that it seems more appropriate to introduce them in later chapters along with their applications.

We have also introduced the basic concepts of statistical inference, which is really at the heart of statistical applications. Both estimation and tests of hypotheses have been presented in a quite general way. Again, more specific applications will be presented in later chapters. In those chapters, different hypothesis tests will be used, but they are all based on the elementary principles presented here.

REFERENCES

Clopper, C. J., and E. S. Pearson: "The Use of Confidence or Fiducial Limits," *Biometrika*, vol. 26, p. 410, 1934.

Pearson, E. S., and H. O. Hartley: *Biometrika Tables for Statisticians*, vol. 1, 3ed, Cambridge University Press, Cambridge, England, 1970.

CHAPTER 5
STATISTICAL INTERVALS*

Gerald J. Hahn
Corporate Research and Development,
General Electric Company, Schenectady, NY

5.1 INTRODUCTION

Engineers and scientists have come to appreciate that few things in life are known exactly. The most they can do is to use the results of a random sample from a given population to obtain an estimate of some quantity of interest and construct an interval (i.e., range of uncertainty) to contain the unknown quantity with a specified degree of confidence. This chapter describes three types of statistical intervals which arise in many practical applications and indicates when each should be used. The three intervals are:

1. A *confidence interval* to contain the mean of the sampled population (or some other population value, such as its standard deviation)
2. A *tolerance interval* to contain a specified proportion of the sampled population
3. A *prediction interval* to contain the values of all of a specified number of future observations from the sampled population

For example, an airline might have data available on the delay in arrival time for a "random sample" of 20 past flights between two locations. From the data one might wish to construct:

- A confidence interval to contain the average delay time for the population of flights from which the 20 flights can be regarded to be a random sample
- A tolerance interval to contain at least 95% of the delay times of the entire population of such flights
- A prediction interval to contain the delay time for a future flight between these two destinations

*This chapter has been adapted from H. E. Hill and J. W. Prane, Eds., *Applied Techniques in Statistics for Selected Industries: Coatings, Paints and Pigments*, Wiley, New York, 1984, Chap. 7.

Many nonstatistical users of statistics are well acquainted with confidence intervals for the population mean and, perhaps, also for the population standard deviation. Some are also aware of tolerance intervals, but most know very little about prediction intervals, despite their practical importance. A frequent mistake is to calculate a confidence interval to contain the population mean when the problem actually calls for a tolerance interval or a prediction interval. At other times, a tolerance interval is used when a prediction interval is needed. Such confusion is understandable, since most texts on statistics discuss confidence intervals in detail, make limited reference to tolerance intervals, and almost never cover prediction intervals (except in regression analysis applications). This is unfortunate because tolerance intervals and prediction intervals are needed almost as frequently in industrial applications as confidence intervals, and the procedure for constructing such intervals, given the required tabulations, is no more difficult than that for constructing confidence intervals.

In this chapter, procedures for constructing the various types of intervals are described and their uses illustrated. The types of situations in which each interval applies are emphasized.

5.2 ASSUMPTIONS

5.2.1 Overview

It is assumed that one is dealing with a random sample from a normally distributed population (normal population, for short) or process. The sample mean, or arithmetic average, $\bar{x}$ and the sample standard deviation s are calculated from n given observations $x_1, x_2, \ldots, x_n$, using the well-known expressions

$$\bar{x} = \sum_{i=1}^{n} \frac{x_i}{n} \tag{5.1}$$

$$s = \left[\frac{\sum_{i=1}^{n} (x_i - \bar{x})^2}{n-1} \right]^{1/2}. \tag{5.2}$$

From this information it is desired to make probability statements about the population from which the sample was selected or about the results that one might expect in some future random sample from the same population, or both. In constructing prediction intervals, it is further assumed that the future sample is selected randomly from the given normal population and independently of the first sample. These assumptions are now considered further.

5.2.2 Assumption of a Random Sample from a Specified Population

The assumption that the given data are a random sample from some population or process is of paramount importance, since statistical inferences or predictions are only valid for the population from which the sample was randomly selected. This first requires a precise definition of the population, or process, of interest. Thus in the flight delay example, the population of interest might be the delay time for all runs of International Airlines Flight 100, operating daily from New York to London during a specified 3-year period.

Once the population has been defined, random sampling implies, among other things, that the sample is selected from the same population, or process, as that about which inferences are to be drawn. A simple random sample (the only type to be considered here) implies that every possible sample (of the specified size) has an equal probability of being selected in the sample. This might well not be the case in the flight delay example since the available data may be on past flight delays, and the real interest is in the delays of current or future flights. Another situation would be where the available data are from the past production of a prototype unit, and one wants to draw inferences about future production units. In these cases, and in many others, the sampled population clearly differs from the population with regard to which inferences are desired. Nevertheless, the statistical inferences that we shall consider will require that these two populations be identical. This, in turn, requires some nonstatistical assumptions that must be clearly stated and understood. Thus, in the example it implies that the airline has not changed its practices, that there have been no changes in traffic or the number of runways at the airport, and so on. [All of this is closely related to the differentiation that has been consistently emphasized by W. Edwards Deming (1975), the eminent statistician and quality-control guru, between enumerative and analytical studies.] See also Hahn and Meeker (1993).

The assumption of a random sample also implies independence of the observations. This assumption would be highly questionable in the flight delay example if the available data were from flights on 20 consecutive days because of the likelihood of similar weather conditions and air traffic patterns, and for various other reasons.

It is clear from the preceding discussion that the requirement of a random sample from the specified population, or process, of interest is often not met strictly in practice. It needs, therefore, to be emphasized that the statistical interval reflects only sampling uncertainty and not the variability introduced by the failure to select a random sample. Even in those cases in which the sample is not strictly random, the intervals described in this chapter might still be useful. They generally provide a lower bound on the uncertainty associated with an estimate (i.e., the uncertainty that would apply if one could assume a random sample). Thus the uncertainty suggested by the statistical interval can, as a minimum, provide a lower bound on the true uncertainty.

5.2.3 Assumption of Normality

As previously indicated, the procedures given in this chapter are based on the assumption of normality of the measured characteristic. (This may be a poor assumption for the flight delay example. However, normality might well be achieved by working with the logarithms of the delay times after, adding a constant to all of the observations to take into account early and on-time arrivals.)

The normality assumption is generally not critical in the construction of a confidence interval to contain the population mean (due to the central limit theorem). It is important in constructing a tolerance interval to include a specified proportion of the population and a prediction interval to include all members of a future sample. It is also important in various other situations not discussed in detail in this chapter, such as obtaining a confidence interval to include the population standard deviation. When the assumption of normality cannot be met, one might wish to use a "distribution-free" method. These methods are generally not as informative as those based on the assumption of an underlying model, such as the normal distribution. However, they have the distinct advantage of not requiring such an assumption. Specific distribution-free methods, such as the construction of distribution-free tolerance intervals, are discussed in Dixon and Massey (1969), Natrella (1966), and, in detail, by Hahn and Meeker (1991).

5.3 CONFIDENCE INTERVALS

5.3.1 Confidence Interval to Contain the Population Mean

The most commonly employed value to describe a population is its mean μ. It is frequently used to characterize product performance, or as a standard to compare competing processes. This second use is especially appropriate when it can be assumed that each of the processes has the same statistical variability (as measured by the standard deviation). In that case, differences between normal populations can be described completely by the differences between their means.

The sample mean $\bar{x}$ is an estimate of the unknown population mean μ but, because of sampling fluctuations, differs from μ. However, it is possible to construct a statistical interval, known as a confidence interval, to contain μ. A confidence interval provides a range which one can claim, with a specified degree of confidence, to contain an unknown value. More specifically, if one constructs 95% confidence intervals to contain the value of μ in many independent applications, one will, in the long run, correctly include the true value of μ in 95 cases out of 100. However, in 5 cases out of 100 the calculated confidence interval will fail to contain μ. The degree of confidence, 95 percent in the preceding statement, is known as the associated confidence level. In addition to the 95 percent confidence level, 90 percent and 99 percent confidence levels are also often used (see below).

A 95 percent confidence interval to contain the mean μ for a normal population is calculated as

$$\bar{x} \pm c_M(n)s \tag{5.3}$$

where $c_M(n)$ is given in the second column of Table 5.1 as a function of the sample size n. [Those with previous training in statistics will recognize that the factor $c_M(n)$ equals $t_{(n-1;0.975)}/\sqrt{n}$, where $t_{(n-1;0.975)}$ is the 97.5th percentile of the t distribution with $n - 1$ degrees of freedom.]

Example 5.1. The following readings of product performance have been obtained from a random sample of five observations from a normal population: 51.4, 49.5, 48.7, 49.3, and 51.6. The sample mean and the sample standard deviation are calculated from these values as

$$\bar{x} = \frac{51.4 + \cdots + 51.6}{5} = 50.10$$

$$s = \left[\frac{(51.4 - 50.10) + \cdots + (51.6 - 50.10)}{5-1}\right]^{1/2} = 1.31.$$

TABLE 5.1 Factors for Calculating Two-Sided 95% Statistical Intervals for a Normal Distribution*

Number of given observations	Factors for confidence interval to contain population mean μ	Factors for tolerance interval to contain at least 90%, 95%, and 99% of population			Factors for prediction interval to contain values of all of 1, 2, 5, 10, and 20 future observations				
n	$c_M(n)$	$c_{T,90}(n)$	$c_{T,95}(n)$	$c_{T,99}(n)$	$c_{P,1}(n)$	$c_{P,2}(n)$	$c_{P,5}(n)$	$c_{P,10}(n)$	$c_{P,20}(n)$
4	1.59	5.37	6.37	8.30	3.56	4.41	5.56	6.41	7.21
5	1.24	4.28	5.08	6.63	3.04	3.70	4.58	5.23	5.85
6	1.05	3.71	4.41	5.78	2.78	3.33	4.08	4.63	5.16
7	0.92	3.37	4.01	5.25	2.62	3.11	3.77	4.26	4.74
8	0.84	3.14	3.73	4.89	2.51	2.97	3.57	4.02	4.46
9	0.77	2.97	3.53	4.63	2.43	2.86	3.43	3.85	4.26
10	0.72	2.84	3.38	4.43	2.37	2.79	3.32	3.72	4.10
11	0.67	2.74	3.26	4.28	2.33	2.72	3.24	3.62	3.98
12	0.64	2.66	3.16	4.15	2.29	2.68	3.17	3.53	3.89
15	0.55	2.48	2.95	3.88	2.22	2.57	3.03	3.36	3.69
20	0.47	2.31	2.75	3.62	2.14	2.48	2.90	3.21	3.50
25	0.41	2.21	2.63	3.46	2.10	2.43	2.83	3.12	3.40
30	0.37	2.14	2.55	3.35	2.08	2.39	2.78	3.06	3.33
40	0.32	2.05	2.45	3.21	2.05	2.35	2.73	2.99	3.25
60	0.26	1.96	2.33	3.07	2.02	2.31	2.67	2.93	3.17
∞	0	1.64	1.96	2.58	1.96	2.24	2.57	2.80	3.03

*A two-sided 95% interval is $\bar{x} \pm c(n)s$, where $c(n)$ is the appropriate tabulated value and $\bar{x}$ and s are the mean and the standard deviation of the given sample of size n.

For this example, $c_M(5) = 1.24$ from Table 5.1 and the 95 percent confidence interval for μ is

$$50.10 \pm 1.24(1.31).$$

Consequently, one can state with 95 percent confidence that the interval of 48.48 to 51.72 contains the unknown value of μ.

5.3.2 Use of Other Confidence Levels

The second column of Table 5.1 provides factors for constructing a 95 percent confidence interval to contain μ. Standard texts in statistics, such as Dixon and Massey (1969), or a statistical handbook, such as Natrella (1966), and also Hahn and Meeker (1991) provide procedures for obtaining similar confidence intervals based on confidence levels other than 95 percent.

The selection of the appropriate confidence level depends on the specific application and the importance of any decision that might be made based on the calculated confidence interval. For example, at the outset of a research project, one might be willing to take a reasonable chance of drawing incorrect conclusions and, thus, use a relatively low confidence level, such as 90 percent, or even 80 percent, since one's initial conclusions will be verified by subsequent analyses. On the other hand, if one is about to report the final results of the project, and perhaps recommend building an expensive new plant or releasing a new product, one would generally wish to have a high degree of confidence. Large samples tend to lead to small confidence intervals and small samples result in large confidence intervals. Thus, in practice, one may be inclined to use relatively high confidence levels with large samples and lower confidence levels with small samples. This implies that the more data one has, the surer one would like to be of the conclusions.

In selecting a confidence level, one must, keep in mind that, as previously indicated, it expresses only the *statistical uncertainty*, that is, the uncertainty resulting from sampling variability and not that due to possible failures in the assumptions upon which the method is based. Thus when the validity of these assumptions may be questionable, it might not seem very reasonable to employ a high statistical confidence level, such as 99 percent, since this might lead to a feeling of reverence in the results that is not warranted by nonstatistical considerations.

5.3.3 One-Sided Confidence Bounds

In the preceding discussion, both a lower and an upper bound were of interest. Thus we obtained a two-sided confidence interval. In some situations only one of these bounds matters; thus only a lower confidence bound *or* only an upper confidence bound is appropriate. For example, a manufacturer who needs to establish a warranty on the average weight of a packaged product, or on the mean time to failure of a system, usually requires only a lower confidence bound. The concept of confidence level remains the same, regardless of whether one is dealing with a two-sided confidence interval or a one-sided confidence bound, although details of the calculation differ slightly.

In particular, the lower endpoint of a 95 percent confidence interval provides a one-sided lower 97.5 percent confidence bound and the upper endpoint of a 95 percent confidence interval provides an upper 97.5 percent confidence bound. Thus, in

Example 5.1 one can state with 97.5 percent confidence that the true value of μ exceeds 48.48; similarly, one can state with 97.5 percent confidence that the true value of μ is less than 51.72.

One-sided confidence bounds to contain μ for confidence levels other than 97.5 percent can be obtained using procedures described in standard texts in statistics, such as Dixon and Massey (1969), in a statistical handbook, such as Natrella (1966), or in Hahn and Meeker (1991).

5.3.4 Confidence Intervals on Other Quantities

The preceding discussion has dealt with confidence intervals to contain the mean μ of a normal population. The same concepts apply for constructing confidence intervals for other population parameters, such as a confidence interval to include the standard deviation σ of a normal population or a percentile of the sampled population. Additional cases are summarized in Sec. 5.8.

5.4 TOLERANCE INTERVALS

5.4.1 Tolerance Intervals to Contain a Specific Percentage of a Normal Population

Many applications require an interval to contain a specific percentage of the population, instead of, or in addition to, a confidence interval to contain the population mean μ. Such an interval is required, for example, by a manufacturer of a mass-production item who needs to establish limits to contain at least a specified (usually large) percentage of the product with a high degree of confidence. Let us assume readings have been obtained on some performance characteristic on a random sample of 25 integrated circuits from a particular production process. Based on the resulting data, a tolerance interval provides limits to contain, with a known degree of confidence, the readings of at least a specified percentage (e.g., 90 percent) of the population of units from the process.

For a normally distributed population, if μ and σ are known, it is also known that 90 percent of the population values are located in the interval

$$\mu \pm 1.64\sigma$$

(and 95 percent of the population values are in the interval $\mu \pm 1.96\sigma$). However, if only the sample estimates $\bar{x}$ and s of the population values μ and σ are available, added uncertainty is introduced. The best that one can do is construct an interval which, one can claim with a specified degree of confidence, say 95 percent, contains at least a specified percentage, such as 90%, 95%, or 99% of the population values. Such an interval is called a 95 percent tolerance interval and can be calculated for a normal population by using the factors $c_{T,90}(n)$, $c_{T,95}(n)$, and $c_{T,99}(n)$, shown in columns 3, 4, and 5 of Table 5.1 for population percentages of 90 percent, 95 percent, and 99 percent, respectively. For example, it can be stated with 95 percent confidence that the interval

$$\bar{x} \pm c_{T,90}(n)s \tag{5.4}$$

contains at least 90 percent of a normally distributed population.

The fact that two percentages are associated with a tolerance interval should not confuse the user. One of these refers to the percentage of the population that the interval is to contain. The second specifies the confidence level associated with the calculated interval. If μ and σ were known exactly, there is no longer any uncertainty associated with the population percentage contained in the interval, and the "confidence level" associated with the interval is 100 percent.

Example 5.2. A 95 percent tolerance interval to contain 90 percent of the sampled normal population may be calculated for the data in Example 5.1 as

$$50.10 \pm 4.28(1.31) \quad \text{or} \quad 44.49 \text{ to } 55.71, \text{ using } c_{T,90}(5) = 4.28.$$

Thus one may state with 95 percent confidence that at least 90 percent of the sampled population is contained in the interval 44.49 to 55.71. Similarly,

1. A 95 percent tolerance interval to contain 95 percent of the population is

$$50.10 \pm 5.08(1.31)$$

or

$$43.45 \text{ to } 56.75, \quad \text{using } c_{T,95}(5) = 5.08.$$

2. A 95 percent tolerance interval to contain 99 percent of the population is

$$50.10 \pm 6.63(1.31)$$

or

$$41.41 \text{ to } 58.79, \quad \text{using } c_{T,99}(5) = 6.63.$$

5.4.2 Use of Other Confidence Levels in Constructing Tolerance Intervals and One-Sided Tolerance Bounds

Although a 95 percent tolerance interval is used perhaps more frequently than any other, one sometimes wishes to construct tolerance intervals involving other confidence levels. Tabulations of factors for 90 percent and 99 percent tolerance intervals for a normal population are given in Dixon and Massey (1969), Hahn and Meeker (1991), and Natrella (1966). Also, frequently, one requires a lower tolerance bound only, that is, a lower limit on the population values, or an upper tolerance bound only, rather than a two-sided tolerance interval. Such one-sided tolerance bounds cannot be constructed exactly from the tabulations for two-sided tolerance intervals. Instead, specially constructed factors for obtaining one-sided tolerance bounds for a normal population should be used. Such factors are tabulated in Hahn (1970) and Hahn and Meeker (1991).

5.5 PREDICTION INTERVALS FOR FUTURE OBSERVATIONS

5.5.1 Prediction Intervals to Contain All of *k* Future Observations

Sometimes instead of requiring an interval to include a specified proportion of the population (tolerance interval), one desires an interval that will contain the next observation, or each of a small number of future observations, from the given population with a high probability. This requires a third type of interval, a prediction interval.

Such an interval may be thought of as an astronaut's interval. A typical astronaut has been assigned to undertake a specific number of space flights, and needs to decide how much fuel to take along each time. He or she is not interested in what will happen on the average in the population of all such flights, of which his or hers happens to be a random sample (confidence interval on the population mean), or even what will happen on at least 95 percent or 99 percent of such flights (tolerance interval). This person's main concern is in the one flight, or the small number of flights, in which he or she will be involved personally. Similarly, a systems manufacturer who needs to warrant the performance of each unit in an order of three units, based on past experience with units of the same type, would wish a prediction interval to contain the values of a specified performance parameter for all three units with a high probability. Thus in contrast to tolerance limits, which are frequently required by a manufacturer of a mass-production item, prediction intervals are of most interest to a manufacturer of a relatively small number of units per order. Such intervals are also desired by the typical customer who purchases one or a small number of units of a given product. He or she is concerned with predicting the performance of the particular units that were purchased (in contrast to the long-run performance of the process from which the sample has been selected).

Prediction intervals to contain all of k future observations from a normal population, based on the results of n past observations from the same population, can be calculated using the factors in the last five columns of Table 5.1. These provide the values $c_{P,k}(n)$ such that one can claim with 95 percent confidence that all of k future observations from the previously sampled normal population will be located in the interval

$$\bar{x} \pm c_{P,k}(n)s \tag{5.5}$$

for $k = 1, 2, 5, 10,$ and 20.

Example 5.3. Consider again the random sample of five observations from the normal population of Example 5.1. Now assume that it is desired to obtain a 95 percent prediction interval to contain the values of both of two further randomly selected observations from the same population. For this case $k = 2$ and $n = 5$; from Table 5.1 the factor $c_{P,2}(5) = 3.70$. Thus one can claim with 95 percent confidence that both of two future units from the sampled population will be located in the interval

$$50.10 \pm 3.70(1.31)$$

or

$$45.25 \text{ to } 54.95.$$

Similarly, a 95 percent prediction interval is found to have the following values if it contains:

1. A single future observation,

$$50.10 \pm 3.04(1.31)$$

or

$$46.12 \text{ to } 54.08, \quad \text{using } c_{P,1}(5) = 3.04.$$

2. All of 10 future observations,

$$50.10 \pm 5.23(1.31)$$

or

$$43.25 \text{ to } 56.95, \quad \text{using } c_{P,10}(5) = 5.23.$$

From these results, and also from inspection of the factors in Table 5.1, it is noted that, as one might expect, the shortest prediction interval is the one to contain a single future observation. The size of the interval increases as the number of future observations to be contained increases.

5.5.2 Use of Other Confidence Levels in Constructing Prediction Intervals and One-Sided Prediction Bounds

One sometimes wishes to construct prediction intervals involving confidence levels other than 95 percent. Tabulations of factors for 99 percent prediction intervals for a normal population are given in Hahn (1970) and Hahn and Meeker (1991).

Also, frequently one requires only a lower prediction bound, or an upper prediction bound, to contain all of k future observations. One-sided prediction bounds cannot be constructed directly from the tabulations for two-sided prediction intervals except for the special case of $k = 1$ (i.e., a single future observation). Instead, specially constructed factors for obtaining one-sided prediction bounds should be used. Such factors are tabulated in Hahn (1970) and Hahn and Meeker (1991).

5.5.3 Some Further Prediction Intervals

Prediction intervals that contain the values of all of k future observations are only one kind of prediction interval. In some applications, other kinds of prediction intervals are required. Some examples are:

1. A prediction interval to contain the mean of k future observations with a specified probability; see Hahn (1970) and Hahn and Meeker (1991)
2. A prediction interval to contain the standard deviation of k future observations with a specified probability; see Hahn (1970) and Hahn and Meeker (1991)
3. A prediction interval to contain at least k' out of k future observations with a specified probability, where $k' < k$; see Hall and Prairie (1973)

5.6 COMPARISON OF RESULTS FROM EXAMPLES

The relative lengths of the interval in Example 5.1 and the first interval in Examples 5.2 and 5.3 are compared in Fig. 5.1. It is seen that for a sample of 5, the

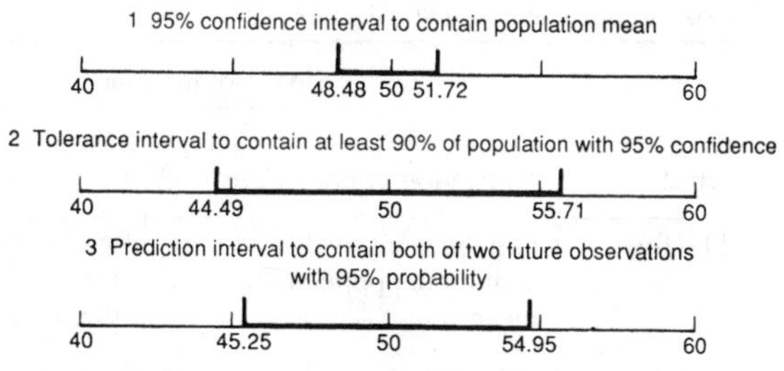

FIGURE 5.1 Comparison of lengths of statistical intervals for examples.

95 percent confidence interval to contain the population mean is appreciably smaller than both the tolerance interval to contain at least 90 percent of the population and the 95 percent prediction interval to include both of two future observations. Also, the 95 percent tolerance interval to include at least 90 percent of the population is somewhat larger than the 95 percent prediction interval to contain both of two future observations.

Inspection of the tabulations of the factors in Table 5.1 indicates that a 95 percent confidence interval to contain the population mean is always smaller than a tolerance interval and a prediction interval. The relative sizes of tolerance and prediction intervals depend on the proportion of the population to be covered by the tolerance interval and the number of future observations to be contained in the prediction interval. Also, unlike the other two intervals, the length of a confidence interval approaches zero as the sample size increases; in fact, the interval converges to μ.

5.7 WHICH INTERVAL DO I USE?

As illustrated, statisticians have developed a variety of intervals for helping to answer relevant questions from given data. However, the user must decide what the relevant questions are in a given application. Once the questions to be answered have been clearly stated, it should be relatively easy to decide on the correct interval. Thus the decision of which interval or intervals to use must be made by the analyst based on an understanding of the problem.

For example, say one is concerned with the number of miles per gallon of gasoline obtained by a particular type of automobile under specified conditions. If all that is needed is an estimate of average gasoline consumption, perhaps to use as a standard, then a confidence interval to include the population mean would seem most appropriate. On the other hand, if the manufacturer must pay a penalty for each automobile with gasoline consumption below a value to be specified, then a lower tolerance bound would be desired. Finally, the purchaser of one of the automobiles would be most interested in a prediction interval to contain the gasoline consumption of a single future automobile.

Table 5.2 provides a categorization to help in selecting the proper interval for a particular application. The intervals are classified by area of inference and characteristic of interest. Thus one must decide whether the main concern is in drawing

TABLE 5.2 Categorization of Some Statistical Intervals

	Area of inference	
Characteristic of interest	Description of population	Prediction of results of future sample
Location (measured by the mean)	Confidence interval to contain the population mean	Prediction interval to contain the mean of k future observations
Spread (measured by the standard deviation)	Confidence interval to contain the population standard deviation	Prediction interval to contain the standard deviation of k future observations
Enclosure region	Tolerance interval to contain a population proportion	Prediction interval to contain all of k future observations

inferences about the population from which the sample was selected or in predicting the results of a future sample from the same population. Also, one must decide whether the major interest is in characterizing the location or spread of the population or of a future sample, or in developing some form of enclosure region.

5.8 FURTHER INTERVALS

Standard books on elementary engineering statistics such as Dixon and Massey (1969) and handbooks such as Natrella (1966) provide detailed discussions of confidence intervals, and some also consider tolerance intervals. A survey of prediction intervals is given in Hahn and Nelson (1973). In addition, Hahn (1970) provides a comprehensive comparison of statistical intervals for a normal population and a discussion of methods for constructing the various intervals. More detailed tabulations than the ones given here and further references to more extensive tabulations are also provided. These topics are considered in depth in Hahn and Meeker (1991).

As we have suggested, there are numerous statistical intervals beyond those discussed in this chapter. These include:

- Distribution-free confidence, tolerance, and prediction intervals, discussed briefly in Sec. 5.2.3
- Confidence, tolerance, and prediction intervals for distributions other than the normal, such as exponential and Weibull distributions, or situations involving proportions (binomial distribution) or counts (Poisson distribution)
- Confidence, tolerance, and prediction intervals for situations other than those involving a single population, such as regression models and comparisons between two or more populations
- Confidence intervals on some population characteristic other than the population mean, such as its standard deviation or a population percentile

- Confidence intervals on the percentage of the population below, or above, some specified threshold value

These cases and others are discussed further in Hahn and Meeker (1991).

The last of these intervals, a confidence interval on the population percentage below or above a specified threshold value, based on a random sample from the population, is frequently required in practical applications. Thus in the example presented at the beginning of the chapter, we might wish to construct an interval that we can claim with 90 percent confidence includes the percentage of flights in the sampled population that arrive within 30 min of the scheduled arrival time. This question can be thought of as the reverse of a one-sided tolerance interval. In constructing a one-sided tolerance interval, one wishes to determine the threshold value that one can claim exceeds, or is exceeded by, a prespecified proportion of the sampled population. Therefore, in the one-sided tolerance interval problem, the percentage to be included in the interval is given, and the threshold value is sought. In the problem that we are now suggesting, the threshold value is given and the included percentage is sought. Like other problems in this chapter, the desired interval can be obtained exactly by simple probability methods if the population of interest is completely defined. For example, if we knew that the flight arrival times were lognormally distributed, with known values of μ and σ, the percentage of the population above or below the threshold could be determined by elementary methods. In practice, all that is generally available is a random sample from the population of interest, and this calls for a statistical interval.

Hahn and Meeker (1991) present two different methods for handling the preceding problem. The first does not require any assumptions about the distribution of the sampled population and is based on a count of the number of observations in the sample that exceed or are exceeded by the specified threshold value. This calls for a confidence interval involving go–no go data (binomial distribution). Such intervals are described in numerous standard books, including Dixon and Massey (1969) and Natrella (1966). The second method assumes a normal population and uses Odeh and Owen (1980, table 7).

In addition, there are a variety of other concepts or methods for constructing statistical intervals. These include bayesian intervals [see Gelman et al.(1995)] and intervals constructed via bootstrap and related methods [see Efron and Tibshirani (1993)]. The latter have become especially popular in recent years and provide simple, intuitively attractive alternatives for the standard intervals, and for other, possibly interactive, approaches. Hahn and Meeker (1991) again provide some brief discussion and references.

REFERENCES

Deming, W. E.: "Probability as a Basis for Action," *Am. Stat.*, vol. 29, pp. 146–152, 1975.

Dixon, W. J., and F. J. Massey, Jr.: *Introduction to Statistical Analysis*, 3d ed., McGraw-Hill, New York, 1969.

Efron, B., and R. J. Tibshirani: *An Introduction to the New Bootstrap*, Chapman and Hall, New York, 1993.

Gelman, A., J. B. Carlin, H. S. Stern, and D. B. Rubin: *Bayesian Data Analysis*, Chapman and Hall, New York, 1995.

Hahn, G. J.: "Statistical Intervals for a Normal Population," *J. Qual Technol.*, vol. 2, pp. 115–125, 195–206, 1970.

_____ and W. Q. Meeker: *Statistical Intervals: A Guide for Practitioners*, Wiley, New York, 1991.

_____ and _____: "Assumptions for Statistical Inference," *Am. Stat.*, vol. 47, pp. 1—11, 1993.

_____ and W. B. Nelson: "A Survey of Prediction Intervals and Their Applications," *J. Qual. Technol.*, vol. 5, pp. 178–188, 1973.

Hall, I. J., and R. R. Prairie: "One-Sided Prediction Intervals to Contain at Least *m* of *k* Future Observations," *Technometrics*, vol. 15, pp. 897–914, 1973.

Natrella, M. G.: *Experimental Statistics*, Handbook 91, National Bureau of Standards, U.S. Govt. Printing Office, Washington, D C, 1966.

Odeh, R. E., and D. B. Owen: *Tables for Normal Tolerance Limits, Sampling Plans, and Screening*, Marcel Dekker, New York, 1980.

CHAPTER 6

SELECTION, FITTING, AND TESTING STATISTICAL MODELS

Samuel S. Shapiro
Florida International University, Miami, FL

6.1 INTRODUCTION

In order to analyze a set of data an engineer will often be required to choose a *statistical model* (probability distribution) that represents the population from which the data arose. For example, in lifetime studies an exponential distribution is assumed, and based on this assumption, the probability that the lifetime of a piece of equipment will exceed the warranty period is calculated. The accuracy of such calculations depends on whether the chosen model is appropriate; choice of a wrong model can lead to serious errors. The purpose of this chapter is to give engineers some guidelines for choosing a statistical model and suggest some techniques that can be used to assess the adequacy of the selection. There are several comprehensive books available which deal with the mathematical aspects of statistical models in greater detail and at a more theoretical level than this chapter. Those readers who desire a more rigorous treatment of this subject are referred to Johnson and Kotz (1970).

Prior to discussing the details of selecting and testing statistical models, it is first necessary to define and introduce the concept of a model.

Definition. A *model* is a mathematical expression relating the output (a measurement) to one or more factors that determine the value observed.

Consider a simple experiment where the length of a fixed-size bar is measured many times. The output, or measurement, Y_i on the ith trial can be represented by the model

$$Y_i = \mu + e_i$$

where μ is the true size of the bar and e_i is the measurement error for the ith trial. This model tells us that a specific measurement is made up of two components, a constant μ representing the quantity we are interested in estimating and an error which varies with each measurement. The constant μ is called a *parameter* and e_i is called a *random* variable. It is not an equation in the common sense since μ is unknown and e_i is a random variable, not a number; hence Y_i cannot be calculated. This is a simple mathematical model, and only a limited amount of information can be obtained from it. If we wish to know something about how the values are spread out over the range of the random variable, it is necessary to describe the behavior of e_i by means of a *statistical model.* We say that e_i is a random variable whose behavior and properties can be described by a statistical model.

Definition. A *statistical model* is a mathematical expression that describes in terms of probabilities how the values of a random variable are spread out over its range.

Thus a statistical model can be used to answer such questions as what is the probability that the strength of a connector will be greater than some specified minimum value, or the temperature of a chemical reaction will exceed some critical value, or the life of a component will exceed its mission time.

Statistical models are also useful for summarizing data. If an adequate statistical model can be found that describes the data obtained from a random sample, then the model equation can be used to summarize the description of the population. This eliminates the need for displaying the individual sample values, and a plot of the model equation yields a smooth graphic representation of the population. This equation also allows one to interpolate within and, with caution, to extrapolate beyond the range of the data. For example, by means of a statistical model it is possible to use the failure times from an accelerated life test to extrapolate and predict the lifetime characteristics under normal operating conditions.

There are several ways of selecting a model. A common method is by *past experience.* The phrase, "we have always used the normal distribution for this measurement" is common in many applications. Perhaps this method is the most frequently used since it is always easier to rely on someone's prior work than to seek new grounds and "gamble" on a new model. Another method of choosing a model which is useful when the basic mechanism giving rise to the measurement is understood is to derive it from *first principles*. Thus if it can be shown that at any time the growth of a flaw is a random proportion of its size at the previous instant of time, that these random variables are independent, and that failure occurs when the flaw reaches a given size, then the lognormal can be shown to be the appropriate time-to-failure model. Another method for making a model selection is to use a flexible *empirical distribution* that has enough parameters so that it fits most data sets.

Once the selection process is completed (especially if an arbitrary model was selected), it is important to verify that the selection was reasonable. The subject, *test of distributional assumptions*, deals with the techniques that permit making this evaluation based on a random sample of observations. Sometimes a graphic, subjective procedure is used, while at other times an objective, analytic test of hypothesis is used. It is often of value to do both. The result of such an analysis is the rejection or nonrejection of the model. If the model is rejected, the process of selection is repeated. These procedures of selection and testing are discussed in the following sections of this chapter.

Once the model is selected, the fitting procedure is one of estimating the parameters of the model. Most statistical texts discuss and derive the appropriate techniques, and the reader is referred to these for the details. There are usually several methods of doing the estimation; however, this chapter will present only the one for complete samples that is commonly used by engineers and does not require extensive tables. Procedures for estimating parameters from censored samples or for the cases where statistical precision is important can be found in the cited references. There are also several methods for testing the adequacy of a selected model. However, only a graphic procedure and one analytic test are given (for those models where appropriate), and this latter is selected based on the biases of the author.

6.2 TESTS OF GOODNESS OF FIT

There are many procedures for assessing whether a statistical model is an appropriate one for the data. Two general classes of test procedures are presented. The first is a graphic procedure known as *probability plotting*. This technique can be used with models that do not have shape parameters, or with models having a shape parameter that can be converted by a transformation to one without a shape parameter. In these latter cases the graph paper is scaled to perform the transformation so that the data can be plotted directly. Graphic techniques are extremely useful in statistical analyses and are especially useful for testing model assumptions since they give the analyst a clear picture of the data and, if the model is rejected, clues as to the possible alternatives. Thus if a test for a symmetrical distribution is rejected, the plot can indicate whether a left skewed or a right skewed alternative or another symmetric model should be considered. Also it can indicate whether the rejection was caused by one or two outliers. Extensive descriptions of these procedures can be found in Hahn and Shapiro (1967), Nelson (1979), and Shapiro (1980). The technique known as probability plotting yields a picture that allows the user to assess whether the selected model reasonably "fits the data" and, if so, yields estimates of the parameters. The procedure consists basically in obtaining a specially scaled paper corresponding to the model being fitted, ordering and plotting the data, and determining whether a reasonably straight line is obtained. If the assumed model is correct, the plot will be approximately linear. There will be deviations from linearity due to random sampling fluctuations, but the bigger the sample size, the greater the tendency toward a linear plot. However, if the assumed model is not correct, the plotted points will fluctuate around a nonlinear curve, and this will result in a systematic departure from linearity. Thus the hypothesis of a specific model is rejected or not rejected based on the subjective assessment of the linearity of the plot. Examples of plots where the model is and is not appropriate are given in Hahn and Shapiro (1967).

The plotting procedure starts by obtaining a sheet of probability paper for the model being tested. There are several commercially available sources for the paper, and it can be purchased in most technical supply stores. The next step is to order the observations from the smallest, Y_1, to the largest, Y_n. Let i denote the rank or order number; the smallest sample value corresponds to $i = 1$ and the largest to $i = n$. Next compute the quantities

$$P_i = \frac{100\left(i - \frac{1}{2}\right)}{n} \tag{6.1}$$

There are several different recommendations for calculating the P_i values, some of which are model-dependent. However, the values specified above are widely used and are a good compromise for most distributions. A table of P_i values for sample sizes between 5 and 50 can be found in Nelson (1979). Next the ordered observations are plotted versus P_i. Thus the fourth smallest value, corresponding to $i = 4$, is plotted versus P_4. The axes used for the plotting depend on the model being tested and the brand of paper chosen. One axis will have a prelabeled scale, with values between 0 and 100.0, such as from 0.01 to 99.99, and this is used for locating the P_i values. The other axis has no labeled scale and is used for locating the ordered observed values. It is necessary for the user to put a scale on this axis. The scale chosen should permit easy plotting of the data and should cover most of the sheet of paper. The next step is to plot the observations. It is not necessary to have all the observations to construct the plot. Only the order number (i values) must be known for each of the data values recorded. Therefore this technique can be used with censored samples where the highest values have not been recorded (this would occur with a life test which is terminated before all of the units have failed), where the lowest values have not been recorded, or where both situations occur.

The next step is to examine the plot. It is helpful to use a clear ruler in evaluating whether the plot is linear. Lay the ruler on top of the plotted points and attempt to fit "by eye" a "best" line through the points. More formal techniques for finding the best line are usually not worth the effort. Simple linear regression is inappropriate since the ordered observations are not independent. Since this is a subjective procedure, the use of generalized least squares is not recommended. Draw the line on the paper. Such a line will highlight systematic departures from linearity. In evaluating the plot, the following points should be noted.

- The plotted values are the realization of random variables and hence will never lie perfectly on the line.
- The ordered observations are not independent because they have been ranked. Therefore, if one observation is above the line, there is a tendency for the next to be above the line too. This results in runs of points above and below the line.

The variances of the extreme points in the unbounded tails of the distribution (the largest and smallest) are much higher than the variances of the points in the middle, and hence a greater discrepancy from a straight line should be allowed at the extremes. The variance of the points near a lower or upper bound will be smaller than that of points in the other regions. When drawing the straight line through the points, greater weight should be given to the points where the variance is smaller. The model parameters can be estimated from the plot for those cases where the hypothesis is not rejected. The slope of the line provides an estimate of the scaling parameter, and the intercept provides an estimate of the location parameter. Since the estimation technique varies for each distribution, the details will be discussed with each model.

Probability plots can be used to obtain estimates of cumulative probabilities directly from the graph. To find the probability that a random variable will be less than some value b, we locate b on the observational scale, find its intersection with the plotted line, and then determine the corresponding value on the P scale. This is the desired probability. Distribution percentiles can also be obtained in the inverse manner. The percentile of the distribution is obtained by locating the value of p on the P scale, finding its intersection with the plotted line, and determining the corresponding value on the observation scale.

The one major disadvantage of probability plotting is that a subjective decision must be made whether to reject or not to reject a model. Therefore the plotting technique is sometimes augmented by an "objective" test of hypothesis procedure.

There are many such objective tests. Since the format for each test depends on the hypothesized model, these will be described when the model is subsequently introduced.

6.3 THE HAZARD RATE

One of the most frequent applications for a statistical model is to represent the time to failure of a device. In such applications a positive-value random variable model is used. One of the criteria used to select a specific model in lifetime studies is the conditional failure rate, or *hazard rate*. This quantity tells us what is the probability of failure in the next instant of time given that the equipment is presently operating. An equipment can have an increasing hazard rate corresponding to a situation where the probability of failure in the next instant of time increases with time, a decreasing hazard rate corresponding to the probability of failure in the next instant of time decreasing as the equipment gets older, or a constant hazard rate where the rate does not change with time. The hazard rate is defined by

$$h(t) = \frac{f(t)}{1 - F(t)} \tag{6.2}$$

where $f(t)$ is the density function and $F(t)$ is the distribution function. There is a direct relationship between the hazard rate and the density function,

$$h(t) = \frac{d \ln [1 - F(t)]}{dt}$$

$$\int_0^t h(s)\, ds = \ln [1 - F(t)] \tag{6.3}$$

$$F(t) = 1 - \exp \left[- \int_0^t h(s)\, ds \right]$$

Thus given $h(t)$, the density function can be derived, and hence, knowing the shape of the hazard rate function can assist in the selection of a model.

6.4 NORMAL DISTRIBUTION

The most commonly used statistical model is the normal distribution. This model is used in many engineering situations to represent the measurement of a physical characteristic. The usefulness of the normal distribution can be explained by the *central limit theorem*, which states that it is the appropriate model when the measurement is produced by the sum of many random variables. Thus if an observation can be thought of as arising from the cumulative effect of a large number of factors, then the normal model may be appropriate. For example, in a machining operation the width of a groove represents the accumulated effect of a large number of small effects, such as the sharpness of the cutter, the hardness of the material, or the temperature of the coolant. In such cases the use of the normal model yields a good approximation to the distribution of the groove width. Another example where the

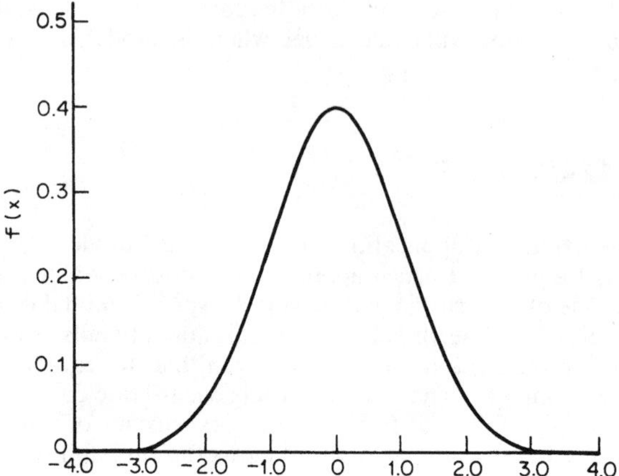

FIGURE 6.1 Standard normal density function; $\mu = 0$, $\sigma = 1$.

normal model is useful is for describing the output of a complex system, where the system output is determined by a combination of the outputs from a large number of subsystems. The central limit theorem is derived in most mathematical statistics books, and the reader is referred to these texts for its derivation.

The normal distribution has a single shape (bell shape) and is defined on the interval $(-\infty, \infty)$. A plot of the normal density function is shown in Fig. 6.1. The equation for the normal distribution is

$$f(x) = \frac{1}{\sqrt{2\pi\sigma^2}} \exp\left[-\frac{1}{2}\left(\frac{x-\mu}{\sigma}\right)^2\right], \qquad -\infty < x < \infty,\ -\infty < \mu < \infty,\ \sigma > 0 \tag{6.4}$$

where μ and σ are the parameters of the distribution. Fitting the normal model to a set of data simply means estimating these two quantities, which represent the mean and the standard deviation of the distribution. The estimates of μ and σ are given by

$$\bar{x} = \frac{1}{n}\Sigma x_i$$

$$s = \sqrt{\frac{n\Sigma x_i^2 - (\Sigma x_i)^2}{n(n-1)}} \tag{6.5}$$

where n is the sample size. The following example illustrates the technique.

Consider the data in Table 6.1, which represent 33 measurements of the weight of talc obtained from containers labeled 2 lb. The calculations and the resulting estimates of the parameters obtained using Eq. (6.5) are also given. The population of talc weight can be approximated by a normal distribution with a mean of 2.007 and a standard deviation of 0.093. Using this approximation, we can answer questions about this set of data or, more specifically, about the population from which the measurements were drawn. For example, if we wish to estimate the proportion

TABLE 6.1 Weights of Material in 2-lb Containers of Talc

2.052	1.971	2.055	2.061	1.944
2.164	1.951	2.024	2.002	2.114
1.852	2.125	2.001	2.052	2.010
1.819	2.180	1.927	2.015	1.883
2.103	1.993	2.041	1.972	1.972
2.023	2.047	2.106	1.984	
1.833	2.148	1.934	1.872	

$$n = 33$$

$$\overline{x} = \frac{1}{n}\Sigma x_i = 2.007$$

$$s^2 = \frac{n\Sigma x^2 - (\Sigma x)^2}{n(n-1)} = 0.008649$$

$$s = 0.093$$

of the measurements that will fall between the specifications 1.80 and 2.20, we consult Table A.1 in the Appendix for the standard normal distribution ($\mu = 0, \sigma = 1$) and use it to find the probability that a normal variable lies in the specification interval. In order to use this table, it is necessary to convert our values to z scores (values from a standard normal). This is done by subtracting the mean from each value of interest (the upper and lower specification limits) and dividing by the standard deviation,

$$z = \frac{x - \mu}{\sigma}$$

Since we have only estimates of the parameters, use of the normal table is approximate and sample sizes should be 30 or more. In this case we use

$$z = \frac{x - \overline{x}}{s}$$

In this example the two values representing the z scores corresponding to the upper and lower limits are

$$z_1 = \frac{1.800 - 2.007}{0.093} = -2.23$$

$$z_2 = \frac{2.200 - 2.077}{0.093} = 2.08$$

The corresponding probability values which are obtained from Table A.1 are $P(z < -2.23) = 0.013$ and $P(z < 2.08) = 0.981$, and the proportion of measurements within the specifications is $0.981 - 0.013 = 0.968$.

Let us prepare a probability plot for the normal distribution using the data in Table 6.1 to determine whether our choice of this model was reasonable. Figure 6.2 shows the specific paper chosen. On this paper the P scale is on the X axis and the

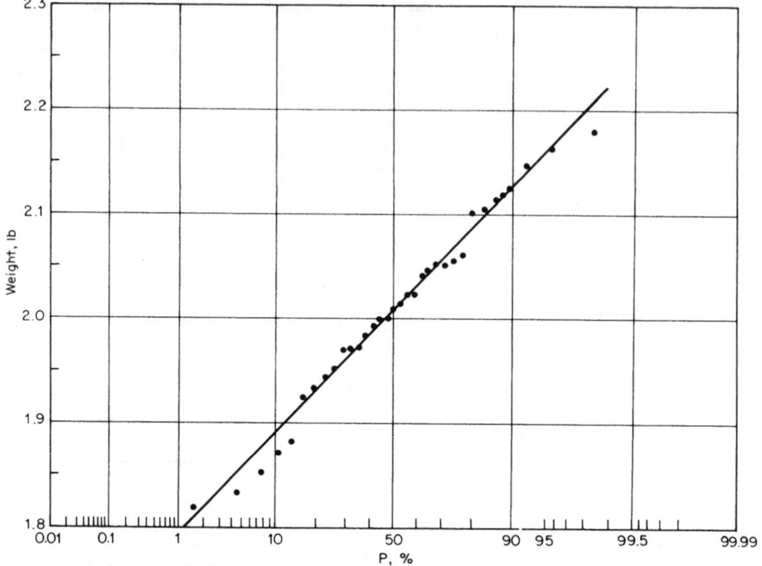

FIGURE 6.2 Normal probability plot of container weights.

observational scale is on the Y axis. The P_i values are computed for $n = 33$, using Eq. (6.1). The resulting plot is shown along with the best line. The plotted points clearly fall in a linear pattern, and there is no evidence that the normal distribution is not appropriate. Since the model is not rejected, it is possible to obtain estimates of the parameters from the plot. An estimate of the location parameter μ is obtained from Y_{50}, the Y value corresponding to a P value of 50. (This is the intercept, corresponding to a z value of zero.) This is read from the plot as follows. Find the line corresponding to the value of 50 on the P scale, determine the point where this line intersects the plotted line, and then determine the point on the Y scale corresponding to this intersection. Using this procedure, we obtain 2.008 as the estimate of μ. The estimate of σ is obtained from the slope of the line. However, the values shown on the P scale cannot be used to compute the slope since this is a transformed scale. The estimate is obtained from the fact that for the normal distribution, σ is equal to $^2/_5$ the difference between Y_{90} and Y_{10}. Thus

$$\sigma^* = \frac{2}{5}(Y_{90} - Y_{10})$$

The 90th and the 10th percentiles are obtained in the same manner as the 50th in the procedure for estimating μ. For this example $\sigma^* = {}^2/_5\,(2.127 - 1.891) = 0.0944$. Compare these values with those obtained previously. These graphic estimates do not have the good statistical properties of the previous estimates and should be used only where high accuracy is not important. Since many statistical computer packages have routines for preparing these plots and also give efficient estimates of μ and σ, the graphic values have lost their attractiveness. Cumulative probabilities and percentiles can now be obtained without the use of a normal table. For example, if we desire to determine the probability that the weight of talc is less than 1.88 lb, we locate this value on the Y axis, follow this value line until it reaches the plotted line, and then read the corresponding value from the *P* scale. In this case the probability is 0.096.

The *analytic test for normality* is known as the *W* test. Details of this procedure can be found in Shapiro and Wilk (1965). The first step in calculating the test statistic is the same as in the probability plotting procedure; the data must be ordered from the smallest to the largest (x_1 to x_n). Next the total sum of squares about the mean is obtained (from either the ordered or the unordered values). Denoting this quantity as S^2, we compute

$$S^2 = \Sigma x_i^2 - \frac{1}{n} (\Sigma x_i)^2 \tag{6.6}$$

Next calculate the quantity

$$b = \sum_{i=1}^{k} a_{n-i+1} (x_{n-i+1} - x_i) \tag{6.7}$$

where $k = n/2$ if n is even, while $k = (n - 1)/2$ if n is odd, and the a_{n-i+1} are found in Table A.20. Then the test statistic

$$W = \frac{b^2}{S^2} \tag{6.8}$$

is computed. In order to evaluate the hypothesis of normality, it is first necessary to select α (the probability of a type 1 error). The corresponding critical values for tests using an α of 1, 2, 5, and 10% are given in Table A.18. The *W* test is a lower-tailed test; thus for a chosen level of the type 1 error, if the computed value is smaller than the critical value given in the corresponding column of Table A.18 for the given sample size *n*, then the hypothesis of normality is rejected. Thus if the calculated value of *W* is less than the tabled value in the column labeled 5%, then the hypothesis of normality is rejected at the 5% level. Using the talc data of Table 6.1 with Eqs. (6.6) to (6.8) we find that

$$S^2 = 0.278$$
$$b = 0.519$$
$$W = 0.975$$

Choosing a type 1 error of 0.05, we find that $W_{0.05} = 0.931$. Since the computed value of *W* is larger than this, we conclude that there is no evidence to reject the null hypothesis of normality.

The constants and percentiles for the *W* test are limited to sample sizes of less than 50. The following version can be used for samples up to size 100. Shapiro and Francia (1972) extended the *W* test for the normal distribution by simplifying the calculation of the constants needed for the test. These constants, denoted by b_i (see Table A.19), are used in place of a_i in Eq. (6.7). The test again is lower-tailed and the percentiles can be found in Table A.21. Details are given in Shapiro and Francia (1972) and Shapiro (1980). The modified test was denoted by *W*'.

6.5 EXPONENTIAL DISTRIBUTION

The simplest of the lifetime models corresponds to the case where the hazard rate is constant. This model is used for equipment that "will not wear out" over its operating lifetime. In other words, the probability of failure in a fixed period of time, given the equipment is operating, is constant regardless of the equipment's age. This model is called the *exponential distribution* and is commonly used in the study of lifetime data. It can be derived directly from the fact that its hazard rate is constant. Let β be this constant. Then from Eq. (6.3),

$$F(t) = 1 - \exp\left[\int_0^t \beta\, ds\right] = 1 - \exp(-\beta t)$$

$$f(t) = \frac{d[F(t)]}{dt} = \beta \exp(-\beta t), \quad t \geq 0, \quad \beta \geq 0 \tag{6.9}$$

The exponential distribution has been used in many applications because of its simplicity and the ease of determining percentiles, estimators, hypothesis tests, and confidence intervals. It can be shown (Shapiro and Gross, 1981) that if the number of arrivals in a period of time has a Poisson distribution, then the time to the first arrival or the time between arrivals has an exponential distribution. Some examples of applications are as follows: the waiting time to the arrival of the next customer or the time between arrivals in queuing problems, the time to failure for a complex system of nonredundant components, and the time to complete a task. The exponential distribution has a single shape and is defined on the interval $(0, \infty)$. The distribution is illustrated in Fig. 6.3. The exponential has a single parameter, β, which is the reciprocal of the mean. The mean is often referred to as the *mean time to failure* or the *mean time between occurrences*, depending on the application. The parameter β is often called the *failure rate.*

The two-parameter exponential distribution can be defined on the interval (a, ∞), where a is the guaranteed, or minimum, life. The density function is defined as

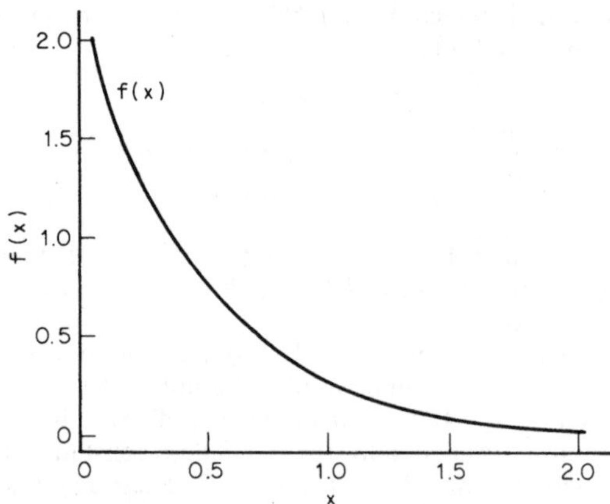

FIGURE 6.3 Exponential density function; β = 2.

$$f(t) = \beta \exp[-\beta(t-a)], \quad t \geq a, \beta \geq 0$$

In this formulation a represents the minimum possible value of the random variable.

The exponential is fit to a set of data by estimating its parameter β. The estimator for the single-parameter model with complete samples is

$$\beta^* = \frac{n}{\Sigma t_i} \tag{6.10}$$

In the two-parameter exponential the estimator of parameter a is the smallest sample value, t_1. The estimator of β is

$$\beta^* = \frac{n}{\Sigma t_i - nt_1} \tag{6.11}$$

In order to illustrate the technique we use the 25 times between arrivals shown in Table 6.2. The parameter is estimated, using Eq. (6.10), to be 0.252 (see also Table 6.2). Using this value we now can estimate, for example, the probability that

TABLE 6.2 Times between Arrivals of Customers (Minutes)

4.16	1.80	4.06	4.11	4.23
3.85	3.15	3.43	5.17	4.99
4.65	4.62	3.48	3.92	2.93
4.53	4.84	4.37	2.89	4.34
3.57	3.03	4.91	3.98	4.13

$n = 25$

$$\beta^* = \frac{n}{\Sigma t} = \frac{25}{99.15} = 0.252$$

after receiving a job, the next arrival occurs in less than 2.0 min. Since the equation for the exponential density can be integrated directly, no table of percentiles is needed. Thus the probability that T will be less than t minutes is given by

$$F(t) = 1 - \exp(-\beta t) \tag{6.12}$$

In the example illustrated, the probability that T is less than 2.0 min is $1 - 0.604 = 0.396$. Estimating techniques for censored samples can be found in many textbooks; see, for example, Mann et al. (1974).

Testing for the assumption of an exponential distribution using the probability plotting technique requires obtaining a sheet of exponential or chi square, 2 degrees of freedom (df) paper. Three-cycle semilogarithmic paper can also be used (see Shapiro, 1980). In this example the data in Table 6.2 are plotted on chi square paper. Again the location of the P scale and the observational scale will depend on the specific paper selected. However, the comments made here in connection with normal probability plots apply. A sheet of chi square (2 df) probability paper is shown in Fig. 6.4. For this brand of paper the prelabeled P scale is on the X axis

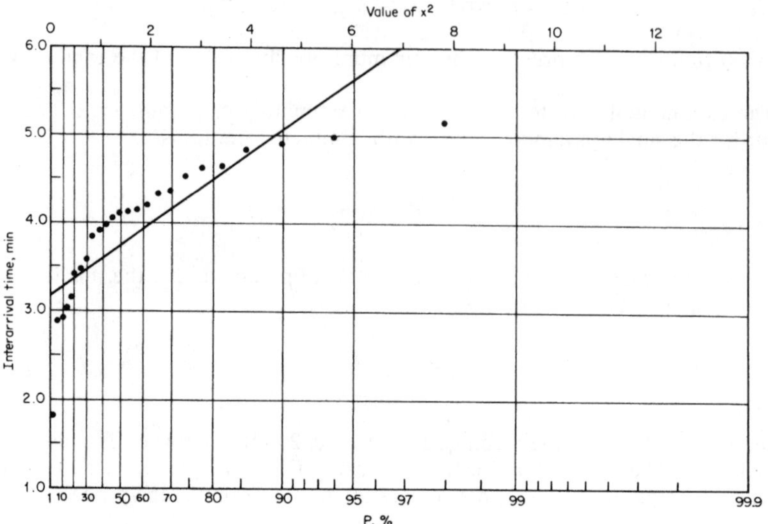

FIGURE 6.4 Exponential probability plot of customer arrival times.

and the observational scale on the Y axis. Using the same procedure as described previously, the data are ordered and plotted in Fig. 6.4 versus the P_i values of Eq. (6.1). The next step is to assess whether the plotted points fall in a linear pattern. Again using a clear ruler, we draw a best line through the points (Fig. 6.4). There is clear evidence that there are systematic departures from linearity, and hence the exponential model is inappropriate.

The parameters of the exponential can be estimated from the plot if the model is not rejected. The estimate of β is, for the one-parameter case,

$$\beta^* = \frac{0.693}{T_{50}}$$

where T_{50} is the 50th percentile and is obtained from the plot as described previously. The estimator of β for the two-parameter case is

$$\beta^* = \frac{1.504}{T_{80} - T_{10}}$$

The estimator for a is the value of T corresponding to the point where the plotted line crosses the axis scaled for the observations. On the paper shown in Fig. 6.4 this corresponds to the Y axis. Since we have rejected the model, it does not make sense to estimate any of the parameters from our data.

The W test for the exponential distribution is similar to that of the normal and is fully described in Shapiro and Wilk (1972). The first step is to order the observations from smallest to largest. Next S^2 is computed from Eq. (6.6) as before. The value of b is obtained from

$$b = \sqrt{\frac{n}{n-1}}\,(\overline{T} - T_1) \tag{6.13}$$

where $\overline{T}$ is the sample mean and T_1 is the smallest sample value.

Next W is obtained as the ratio

$$W = \frac{b^2}{S^2} \tag{6.14}$$

The exponential version of the W test is *two-tailed*, that is, the reject region has two parts corresponding to high and low values of W. The hypothesis of exponentiality is rejected if the computed value of W is either *higher than the upper-tail* critical value or *less than the lower-tail* critical value for a chosen value of α, the type 1 error. Percentiles of W are given in Table A.22 for tests of size 1, 2, 5, and 10%. When the value of the location parameter a (minimum value) is known, a special version of the test due to Stephens (1978) should be used. The W statistic is calculated as before, except that a is included as an observation and the sample size is $n + 1$.

Let us assume that the minimum value a is not known for the data in Table 6.2. The calculation of the W statistic is as follows. The value of S^2 is obtained from Eq. (6.6) as 14.9784. The smallest value is 1.80 and the sample mean is 3.966. Thus

$$W = \frac{25}{24}\,\frac{(3.966 - 1.80)^2}{14.9784} = 0.3263$$

from Eqs. (6.13) and (6.14). Assume we wish to run a 0.05 level test. Referring to Table A.22 for a sample of size 25, we see that the 0.025 and 0.975 critical values of W are 0.0218 and 0.0836. Since the computed value of W is larger than the upper critical value, the exponential hypothesis is rejected and we now have to find an alternative model.

In order to illustrate the test procedure when the *origin is known*, let us assume that the smallest possible value a is zero. We now recompute the values of S^2 and the mean using the value of zero as an observation. The value of S^2 is 30.10 and the mean is 3.813, which results in $b = 3.889$ and $W = 0.502$. Referring to Table A.22 for a sample of size 26, we see that the upper and lower critical values corresponding to a 0.05 level test are 0.0213 and 0.0791, respectively. Thus we reject the exponential hypothesis with an origin of zero.

6.6 WEIBULL DISTRIBUTION

The exponential distribution has limited application as a time-to-failure model because the assumption of no "wearout" or aging is not realistic for many systems. The *Weibull distribution* has the property that it can represent any of the three possible types of hazard rates: increasing, decreasing, and constant. If we define the hazard rate as the simple polynomial

$$h(t) = \alpha\beta t^{\alpha - 1}$$

then it follows directly from Eq. (6.3) that

$$f(t) = \alpha\beta t^{\alpha-1} \exp(-\beta t^{\alpha}), \quad t \geq 0, \quad \alpha \geq 0, \quad \beta \geq 0 \tag{6.15}$$

The major use for the Weibull distribution is as a time-to-failure model since by proper choice of its parameters it can represent the lifetime characteristics of a wide diversity of equipment. It can also be shown that it can be used as a model of the distribution of a minimum of a large number of random variables whose density functions are bounded from the left. (See Sec. 6.7 for a discussion of extreme-value distributions.) The Weibull model has been used as a lifetime model for relays and ball bearings and the years to failure for some businesses.

The Weibull distribution has two parameters, α and β. The parameter α determines the shape of the distribution; when α is greater than 1, the hazard rate *increases* with time (unbounded), when it is less than 1, it *decreases* (to zero), and when it equals 1, the exponential distribution is obtained with its constant hazard rate. The density function is shown in Fig. 6.5 for three values of the parameter α.

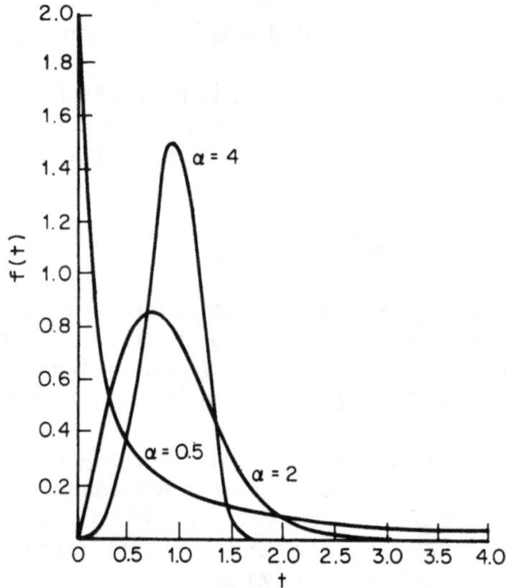

FIGURE 6.5 Weibull density functions; $\beta = 1$.

The parameter β is the scaling parameter and determines the spread of the values. One major drawback to the use of the Weibull model is the difficulty in estimating the parameters. Use of the maximum-likelihood technique requires the solution of two nonlinear equations. The reader is referred to Mann et al. (1974), Gross and Clark (1975), and Shapiro and Gross (1981) for the details. The estimates are obtained by the solution of the equations

$$\beta^* = \frac{n}{\Sigma T_i^{\alpha^*}} \tag{6.16}$$

$$\left(\frac{n}{\Sigma T_i^{\alpha^*}}\right) (\Sigma T_i^{\alpha^*} \ln T_i) = n(\alpha^*)^{-1} + \Sigma \ln T_i \tag{6.17}$$

TABLE 6.3 Times to Failure for Light Bulbs (Months)

1.25	1.17	0.42	0.96	1.03
1.37	0.65	1.39	0.45	0.67
0.28	1.00	0.82	1.61	0.48
0.53	0.66	0.57	0.31	0.29
0.98	1.76	1.71	0.95	0.25

Estimates can be obtained more easily from a probability plot, as indicated below. The data in Table 6.3 are the times to failure of 25 lamp bulbs placed on an accelerated test. The computer program UNIFIT was used to solve Eqs. (6.16) and (6.17), and the resulting estimates were $\alpha^* = 2.03$ and $\beta^* = 0.977$. Cumulative probabilities can be obtained without the use of a table since the density function can be integrated easily. Thus, the cumulative distribution function is

$$F(t) = 1 - \exp(-\beta t^{\alpha}) \qquad (6.18)$$

Therefore to estimate the probability that a light bulb fails the accelerated life test before 0.20 month we use the parameter estimates obtained above and Eq. (6.18) to calculate $F(0.2) = 0.04$. (As noted in the normal case, this is an approximate procedure.) Thus the probability is 0.04 that a bulb will fail the test prior to 0.20 month.

Although the Weibull distribution has a shape parameter, it is still possible to test the appropriateness of the model by means of a probability plot. The reason for this is that the logarithmic transformation converts a Weibull random variable to an extreme-value random variable which has no shape parameter. Thus when the natural logarithms of the data are plotted on extreme-value paper, the resulting plot will be linear if the underlying model is Weibull, regardless of the value of the shape parameter. Weibull probability paper is scaled so that it is not necessary to perform the transformation; the observations are plotted on a log scale. A sheet of Weibull probability paper is shown in Fig. 6.6. On this brand of paper the observational scale is on the horizontal axis and the P scale is on the vertical axis. The data in Table 6.3 are used to illustrate the methodology. The ordered data are plotted in Fig. 6.6 versus the P_i values given by Eq. (6.1). The line drawn by eye indicates that the model chosen is reasonable. The parameter α is estimated from

$$\alpha^* = \frac{3.084}{\ln T_{90} - \ln T_{10}} \qquad (6.19)$$

and the parameter β from

$$\beta^* = (T_{63})^{-\alpha^*} \qquad (6.20)$$

The values of the percentiles can be read from the plot in Fig. 6.6 and substituted into Eqs. (6.19) and (6.20), as described previously. These values are

$$T_{90} = 1.60, \qquad T_{10} = 0.26, \qquad T_{63} = 0.98$$

The estimate of α is

$$\alpha^* = \frac{3.084}{\ln(1.60) - \ln(0.26)} = 1.7$$

and the estimate of β is

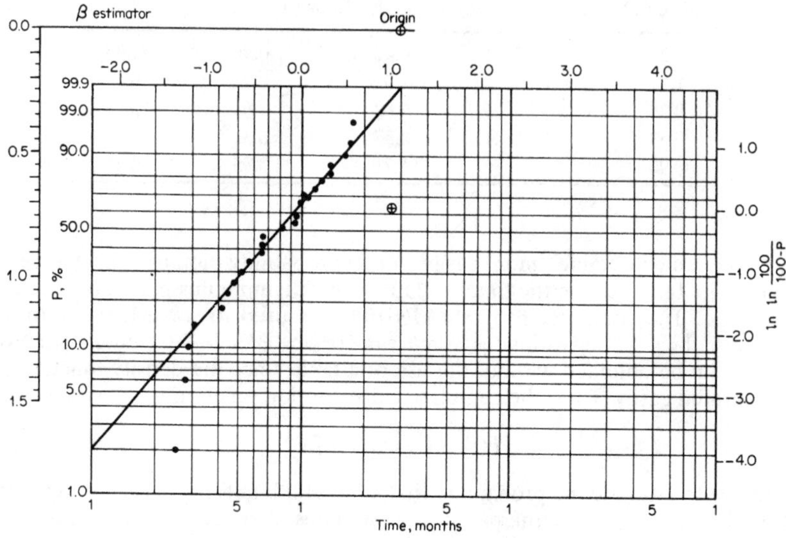

FIGURE 6.6 Weibull probability plot of bulb failure times.

$$\beta^* = (0.98)^{-1.7} = 1.03$$

Compare these estimates with those obtained previously. The probability that a bulb will not survive 2.0 hours on the accelerated life test can be read from the plot by locating the number 2 on the observational scale, finding where the value line intersects the plotted line, and reading off the corresponding value on the P scale. This probability is 0.965. The probability that it will fail prior to 0.2 hour is estimated as 0.065, which is close to the result of 0.040 obtained using the maximum-likelihood estimators.

The *analytic test* for the Weibull distribution was formulated by Shapiro and Brain (1987) and has a format similar to the previous W tests. The quantities S^2, b, and W are computed; however, the data must be transformed prior to the calculations. The first step in the procedure is to order the observations and obtain their natural logarithms. Next S^2 is calculated from Eq. (6.6) using the logarithms of the data. The value of b is obtained from the following set of equations:

$$b = \frac{0.6079L_2 - 0.2570L_1}{n} \tag{6.21}$$

where
$$L_1 = \sum_{i=1}^{n} w_i T_i$$

$$L_2 = \sum_{i=1}^{n} w_{n+i} T_i$$

$$w_i = \ln\left(\frac{n+1}{n+1-i}\right), \quad i = 1, 2, \ldots, n-1$$

$$w_n = n - \sum_{i=1}^{n-1} w_i$$

$$w_{n+i} = w_i\,(1 + \ln w_i) - 1, \quad i = 1, 2, \ldots, n - 1$$

$$w_{2n} = 0.4228n - \sum_{i=1}^{n-1} w_{n+i}$$

where the subscript identifies the order number. The test statistic is obtained from

$$W = \frac{nb^2}{S^2} \tag{6.22}$$

This W test is a two-tailed procedure (see discussion on the exponential distribution). The critical values corresponding to 1, 5, and 10% level tests are obtained from the equation given in Table 6.4. This equation is used, with the constants

TABLE 6.4 Percentiles for Weibull Distribution Test W

Percentage point p	Regression coefficients		
	β_0	β_1	β_2
0.005	0.10102	0.04249	0.005882
0.025	0.11787	0.08550	– 0.002048
0.050	0.13200	0.10792	– 0.006487
0.950	1.46218	– 0.21111	0.012914
0.975	1.64869	– 0.26411	0.016840
0.995	1.91146	– 0.31361	0.017669

$$W_p = \beta_0 + \beta_1 \ln (n) + \beta_2 (\ln n)^2$$

appropriate for the chosen level of the test and the sample size n, to compute the upper and lower critical values. Applying this procedure to the data on bulb failure times, using a 0.10 level test, yields the following:

$$L_1 = 4.3085 \qquad b = 0.4895 \qquad W = 0.707$$

$$L_2 = 21.9539 \qquad S^2 = 8.4690$$

The value of W obtained is compared to the 0.05 and 0.95 critical values computed with the equation in Table 6.4. These numbers are obtained by inserting the β_0, β_1, and β_2 values corresponding to $p = 0.05$ and 0.95 into the equation in Table 6.4. This yields

$$W_{0.05} = 0.412$$

$$W_{0.95} = 0.973$$

Since the value of W lies between the two numbers, there is no evidence to reject the hypothesis of the Weibull distribution.

6.7 EXTREME-VALUE DISTRIBUTION

In some modeling problems the random variable of interest represents the occurrence of an extreme. For example, the time to failure of a parallel connected circuit made up of many redundant components will occur when the last of the items fails. Thus the random variable representing the *maximum time to failure* of the components will describe the lifetime of the circuit. On the other hand the time to failure of a large series-connected circuit will correspond to the time at which the first component fails; therefore the random variable of interest is the *minimum time to failure* of the components. Similar modeling problems arise in many contexts, such as in the design of a safety system to handle the maximum daily load, or a dam to handle the maximum water flow.

In these circumstances the analyst wishes to use a random variable which can represent the minimum or maximum of a set of random variables. This distribution will depend on the form of the individual random variables and their number. In the following we assume that the number is large and the random variables are independent. For a more explicit discussion of this topic see Gumbel (1958) or Hahn and Shapiro (1967). There are several types of extreme-value distributions which are discussed in the literature. Those covered here are called type I for maxima, type I for minima, and type III for minima. The latter is the Weibull distribution discussed previously. The former are sometimes referred to as *extreme-value distributions*, a term used below to describe both of these distributions. The type I maximum model arises as the limiting distribution of the maximum of n independent random variables that are unbounded in the right-hand tail, are of the exponential type, and where n is approaching infinity. Almost all of the distributions discussed in this chapter have these characteristics. Similarly the type I minimum model is derived from the same limiting process, except that it pertains to the minimum of a set of n independent variables whose left-hand tail is unbounded. The normal and the extreme-value distributions of this chapter satisfy this criterion. Exponential-type distributions are those whose cumulative distribution function approaches unity at least as rapidly as that of the exponential distribution.

Each of the type I extreme-value distributions has a *single shape*, and one is the mirror image of the other. The largest-element extreme-value distribution is skewed to the right, and the other is skewed to the left. The density functions are shown in Fig. 6.7, and the equations for the maximum and the minimum are

$$
\begin{aligned}
f(t) &= \frac{1}{\sigma}\exp\left\{-\frac{1}{\sigma}(t-\mu) - \exp\left[-\frac{1}{\sigma}(t-\mu)\right]\right\} \\
f(t) &= \frac{1}{\sigma}\exp\left\{\frac{1}{\sigma}(t-\mu) - \exp\left[\frac{1}{\sigma}(t-\mu)\right]\right\}
\end{aligned}
\tag{6.23}
$$

where $-\infty < t < \infty$, $-\infty < \mu < \infty$, and $\sigma > 0$.

The two constants μ and σ are the location and scale parameters and $\exp(Y) = e^Y$. The parameter μ is the mode of the distribution. There is a relationship between the Weibull and the type I extreme-value distribution. If Y has a Weibull distribution, then $T = \ln(Y)$ has an extreme-value minimum distribution. This prop-

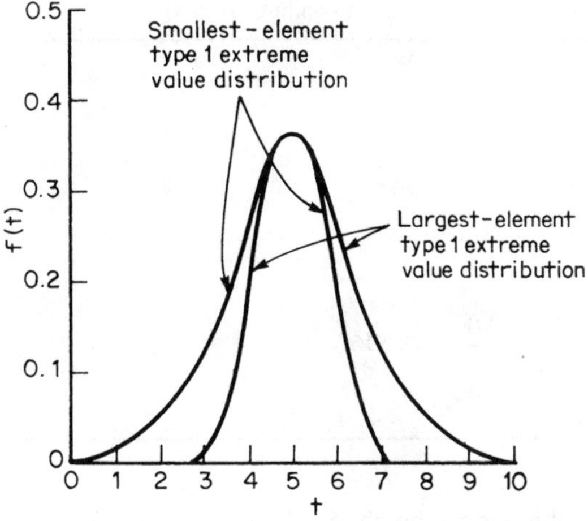

FIGURE 6.7 Type 1 extreme-value densities; $\mu = 5$, $\sigma = 1$.

erty was used in the construction of the probability plotting paper for the Weibull distribution. Techniques for estimating the two parameters of the distribution can be found in Mann et al. (1974). If high precision is not required and the sample is not censored, the technique of moment matching can be used. The variance of both distributions is $1.645\sigma^2$. The mean of the maximum-value distribution is $\mu + 0.577\sigma$; the mean of the minimum is $\mu - 0.577\sigma$. In the matching of moment technique the sample mean and the variance are set equal to the corresponding expressions for the population, and the resulting equations are solved for the unknown parameters. Thus

$$\sigma^{*2} = \frac{s^2}{1.645} \tag{6.24a}$$

and

$$\mu^* = \overline{T} - 0.577\sigma^* = \overline{T} - 0.450s \tag{6.24b}$$

for the maximum value and

$$\mu^* = \overline{T} + 0.450s \tag{6.24c}$$

for the minimum value, where $\overline{T}$ is the sample mean and s^2 is the sample variance from Eq. (6.5). A probability plot can also be used to estimate these parameters, as explained below. The moment matching technique is illustrated using the data in Table 6.5, which are the times to failure for 20 parallel-connected networks, each made up of five redundant circuits. The calculation of the sample mean and the variance is shown at the bottom of the table. The estimates of μ and σ are, using Eqs. (6.24),

$$\sigma^{*2} = \frac{(9.46)^2}{1.646} = 54.4, \qquad \sigma^* = 7.38$$

$$\mu^* = 54.3 - 0.450(9.46) = 50.04$$

TABLE 6.5 Network Times to Failure (Hours)

77.8	58.8	47.6	56.8
62.0	54.9	42.9	53.6
57.7	48.4	65.1	64.2
42.4	60.0	42.1	59.0
44.6	40.9	54.8	52.0

$$\overline{T} = \frac{1}{n}\Sigma t = 54.3$$

$$s^2 = \frac{n\Sigma t^2 - (\Sigma t)^2}{n(n-1)} = 89.4916$$

$$s = 9.46$$

Cumulative probabilities can be obtained directly by integrating the density functions. The functions for maximum and minimum are

$$F(t) = \exp\left\{-\exp\left[-\frac{1}{\sigma}(t-\mu)\right]\right\}$$
$$F(t) = 1 - \exp\left\{-\exp\left[\frac{1}{\sigma}(t-\mu)\right]\right\} \tag{6.25}$$

Suppose that, based on the data in Table 6.5, we wish to estimate the probability that the time to failure for the parallel-connected circuit is less than 40 hours. Substituting the estimated parameters in the first expression of Eqs. (6.25) yields a probability of 0.020.

A test of *goodness of fit* of the extreme-value model can be made with the use of extreme-value probability paper. A sheet of this paper is shown in Fig. 6.8. The horizontal axis has the P scale and the observations are plotted on the vertical axis. The same paper is used for both types of extreme-value distributions; however the plotting procedures are slightly different. The plot for the maximum value is done as previously described for the other models. The data are ordered from smallest to largest and plotted versus the P_i values obtained from Eq. (6.1). The plot for the minimum value requires that the observations be ranked from largest to smallest. Thus $i = 1$ corresponds to the largest value and $i = n$ to the smallest value. The ordered values are then plotted versus the P_i values. Thus in this latter case the largest observation is plotted against $(1 - 1/2)/n$. See Hahn and Shapiro (1967) for further details. The parameters can be estimated from the plot. The estimate of the parameter σ is obtained from the slope of the line and is estimated by

$$\sigma^* = \frac{Y_{90} - Y_{10}}{3.08} \quad \text{for the maximum}$$
$$\sigma^* = -\frac{Y_{90} - Y_{10}}{3.08} \quad \text{for the minimum} \tag{6.26}$$

The estimate of the parameter μ is obtained from the intercept, the point where $-\ln[-\ln F(t)]$ equals zero. This occurs when $-\ln F(t)$ equals 1, which corresponds to

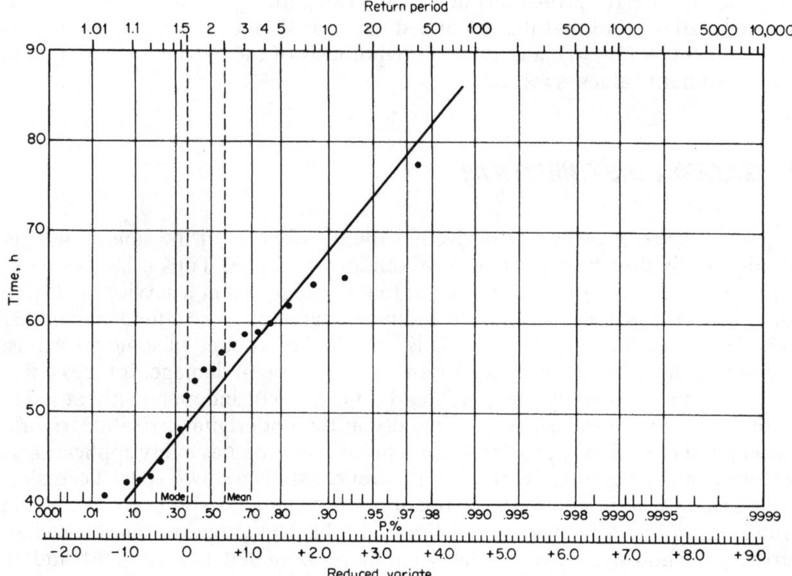

FIGURE 6.8 Type 1 maximum extreme-value probability plot of network failure times.

the 0.368 percentile, $Y_{36.8}$. On some probability papers this percentage point is drawn on the paper as a dashed line (Fig. 6.8). The estimate of μ is

$$\mu^* = Y_{36.8} \tag{6.27}$$

for both types of extreme-value distributions. To illustrate the procedure let us prepare a probability plot for the data in Table 6.5. Since these data represent the maximum lifetime of the components, a plot for the maximum extreme-value distribution is prepared. The data are ordered from lowest to highest and plotted against P_i. The plotted data and the best line are shown in Fig. 6.8. The estimated percentiles are $Y_{90} = 67.6$, $Y_{10} = 41.5$, and $Y_{36.8} = 48.7$. Using Eqs. (6.26) and (6.27) the estimates of the parameters are

$$\sigma^* = \frac{67.6 - 41.5}{3.08} = 8.47$$

$$\mu^* = 48.7$$

Compare these estimates with those obtained from the matching of moments.

The analytic distributional test is identical to the W test for the Weibull distribution except that no logarithmic transformation is made. The test statistic is calculated directly from the observations. Computing the W statistic for the data in Table 6.5 using Eqs. (6.6), (6.21), and (6.22) yields the following results:

$$L_1 = 1259.22 \qquad b = 9.948 \qquad W = 1.165$$

$$L_2 = 859.66 \qquad S^2 = 1698.97$$

Setting the type 1 error probability at 0.05 and using the equations of Table 6.4 with a sample of 20 we find that the calculated value of W does not fall between 0.355 ($W_{0.025}$) and 1.009 ($W_{0.975}$), and thus the hypothesis of the extreme-value distribution of maximum values is rejected.

6.8 GAMMA DISTRIBUTION

The gamma distribution can be derived as the distribution of the sum of independent, identically distributed exponential random variables. Thus it has been used to represent the *waiting-time* distribution to the kth arrival at a service facility, the failure time of a system made up of redundant units where redundant units are switched on when the operating unit fails, and the failure time of some mechanism or organism where failure is caused by an accumulation of damage. It takes a fixed number of attacks to reach this point, and attacks occur independently at a fixed rate. In all of these instances it is assumed that the underlying arrival distribution is exponential (to a first approximation). This distribution has many applications in inventory control, queuing theory, and theoretical statistics; two special cases of it are the Erlang k and the chi square distributions. The hazard rate for the gamma distribution is *bounded* (Shapiro and Gross, 1981). This property can be used as a criterion for choosing between the Weibull (unbounded hazard rate) and the gamma distribution. A plot of the empirical hazard rate can sometimes be used to distinguish between these two models; see Nelson (1979) and Gross and Clark (1975).

The density function of the gamma distribution is

$$f(t) = \frac{\beta^{\delta}}{\Gamma(\delta)}\, t^{\delta-1} \exp(-\beta t), \qquad t \geq 0, \quad \beta > 0, \quad \delta > 0 \tag{6.28}$$

The distribution has two parameters, β and δ. The parameter δ determines the shape of the distribution. When δ is less than 1, the density has an asymptote at the point $t = 0$ and its hazard rate decreases to a lower bound. When δ equals 1, the exponential model is obtained with a constant failure rate. When δ is greater than 1, the density function is zero at the origin, rises to a single mode, and decreases to zero as t goes to infinity. Its hazard rate increases to an upper bound. The constant β is a scaling parameter. Examples of the density function are shown in Fig. 6.9. Maximum-likelihood estimation procedures are not straightforward and require numerical solutions on a computer. Details of these procedures can be found in Gross and Clark (1975) and Mann et al. (1974). Where precision is not important, the method of moments can be used. Using this method the estimates of the parameters are

$$\begin{aligned} \beta^* &= \frac{\overline{T}}{s^2} \\ \delta^* &= \beta^* \overline{T} \end{aligned} \tag{6.29}$$

where $\overline{T}$ and s^2 are obtained from Eq. (6.5). Unless δ is an integer, it is not possible to integrate the gamma density function analytically. Therefore in these cases it is necessary to refer to tables of the incomplete gamma function. Two sources of these tables are Harter (1964) and Pearson and Hartley (1966). When δ is an inte-

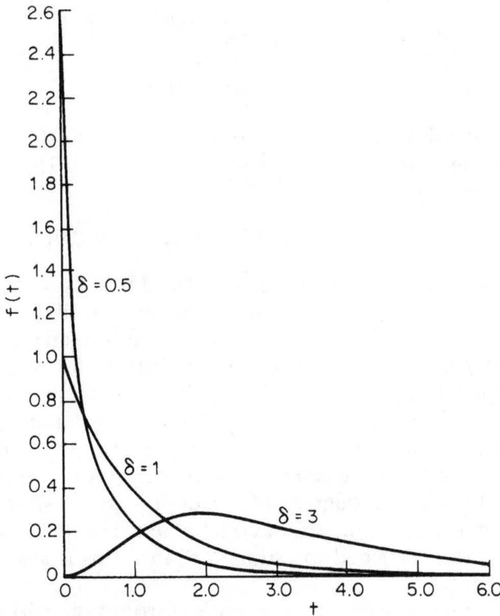

FIGURE 6.9 Gamma density functions; $\beta = 1$.

ger, cumulative probabilities can be obtained from the well-known relationship between the Poisson and gamma distributions. Thus

$$F(t) = 1 - P(X \leq \delta - 1) \tag{6.30}$$

where X is a Poisson variable with parameter βt. The Poisson probability on the right-hand side of Eq. (6.30) can be obtained from a cumulative Poisson table, Table A.2. Let us illustrate these concepts using the data in Table 6.6, which are the

TABLE 6.6 Times to Fourth Arrival (Minutes)

5.32	11.71	4.72	3.25
6.83	4.83	6.38	5.10
3.21	7.08	2.31	4.87
6.18	2.63	7.21	14.81
7.76	10.03	5.04	10.07

$$\overline{T} = \frac{1}{n}\Sigma t = 6.467$$

$$s = 3.179$$

times to the fourth arrival after start-up at a service station on 20 consecutive days. The values of $\overline{T}$ and s are also shown in the table. In this case we know that the appropriate value of δ is 4, but for illustrative purposes suppose that this is unknown. The estimates from Eq. (6.29) are

$$\beta^* = \frac{6.467}{10.106} = 0.640$$

$$\delta^* = 0.640(6.467) = 4.14$$

If we assume that $\delta = 4$, then the estimate of β is $\delta/\overline{T} = 4/6.467 = 0.62$. The probability that the waiting time to the fourth arrival will exceed 15 min, assuming that $\delta = 4$, is obtained from Eq. (6.30) as

$$P(T > 15) = 1 - F(15) = P(X \leq 3) = 0.017$$

where X is a Poisson variable with parameter $\beta t = 15(0.62) = 9.3$.

Since the gamma distribution has a shape parameter, there is no general probability paper for testing for distributional fit. There is available chi square paper for specific degrees of freedom. A *chi square* distribution is a special case of a gamma distribution where $\beta = {}^1/_2$ and $\delta = \theta/2$. The parameter θ is called *degrees of freedom*. Thus if we know the shape parameter and it is an integer or half-integer between 0.5 and 2.5 (this corresponds to integer degrees of freedom for a chi square variable from 1 to 5), we can use chi square paper to prepare a probability plot. A sheet of chi square 2 degrees of freedom paper is shown in Fig. 6.4. The technique for estimating β in the exponential case is used for the gamma distribution. Of course there is no need to estimate δ since this is assumed known when you selected the paper.

There are few analytic tests for the gamma distribution, and the ones available are probably marginal at best. The procedure presented here was proposed by Locke (1976). Locke's procedure makes use of a unique property of the gamma distribution. If X_1 and X_2 are independent, nondegenerate random variables, then $X_1 + X_2$ and X_1/X_2 are independent if and only if they are gamma distributed. In the suggested procedure the sample is broken up into $n/2$ bivariate pairs $[U, V]$, where

$$U_i = X_{2i-1} + X_{2i}$$

$$V_i = \max\left(\frac{X_{2i-1}}{X_{2i}}, \frac{X_{2i}}{X_{2i-1}}\right)$$

Then Kendall's (1970) rank correlation test is used to check for independence. The following example was taken from Shapiro and Brain (1982) and refers to the data in Table 6.7, the time to mill 20 slots. Since the procedure requires that we have paired data, there is some arbitrariness in the test. In this example the observations will be paired in the order of occurrence, which is equivalent to the order listed. Thus,

$$u_1 = 1.53 + 2.47 = 4.00 \qquad v_1 = \frac{2.47}{1.53} = 1.61$$

$$u_2 = 1.82 + 1.76 = 3.58 \qquad v_2 = \frac{1.82}{1.76} = 1.03$$

$$\vdots \qquad\qquad \vdots$$

$$u_{10} = 1.79 + 1.83 = 3.62 \qquad v_{10} = \frac{1.83}{1.79} = 1.02$$

TABLE 6.7 Times to Mill a Slot

Slot	Time	Slot	Time
1	1.53	11	1.83
2	2.47	12	1.94
3	1.82	13	1.97
4	1.76	14	1.62
5	1.94	15	1.91
6	1.98	16	1.85
7	1.86	17	1.89
8	2.20	18	1.73
9	1.68	19	1.79
10	1.79	20	1.83

Next the rank correlation test is used to test the hypothesis of independence. Details of this procedure can be found in Kendall (1970). The first step is to rank the u's and v's. In the case of ties, the tied scores are assigned the average of the ranks they would have had if there were no ties. The next step is to rearrange the pairs so that the ranks of the u's are in ascending order (i.e., 1, 2, ..., 10). If there are ties within the ranks of the u's, then the order of the ranks of the v's determines the order of the tied scores. Next a comparison is made in the ranks of the v's. The rank of each v is compared to the ranks of each of the v's below it. If the subsequent ranks are higher, the pair is said to be in natural order; if they are lower, they are in reverse natural order. The number of pairs in each order are recorded. If the two ranks being compared are the same, or if the corresponding ranks of the u's are the same, the pair is omitted from the count. The results for the milling data are given in Table 6.8.

TABLE 6.8 Results for Milling Data

Pair number	Rank of u	Rank of v	Pairs of v ranks in natural order	Pairs of v ranks in reverse natural order
5	1	6	4	5
2	2	3.5	5	2
7	3	9	1	6
10	4.5	1.5	4	0
9	4.5	7	2	3
8	6	3.5	3	1
6	7	5	2	1
3	8	1.5	2	0
1	9	10	0	1
4	10	8	–	–
			23	19

Kendall's $\hat{\tau}$ test statistic is

$$\hat{\tau} = \frac{2S}{\sqrt{[n(n-1) - T_u]\,[n(n-1) - T_v]}}$$

where S is the difference of the total number of pairs in natural order minus the total number of pairs in reverse natural order, n is the number of pairs of scores, and T_u and T_v are defined as

$$T_u = \Sigma t_u(t_u - 1)$$

$$T_v = \Sigma t_v(t_v - 1)$$

where t_u and t_v are the number of u and v scores, respectively, tied at a given rank. If there are no tied scores within the u's or the v's, the test statistic reduces to

$$\hat{\tau} = \frac{2S}{n(n-1)}$$

In this example we have $S = 4$, $T_u = 2$, $T_v = 4$, and $\hat{\tau} = 0.092$.

To test the significance of $\hat{\tau}$, the statistic S and its standard deviation σ_s are needed. For the case of tied scores the variance of S is given by

$$\sigma_S^2 = \frac{1}{18}[n(n-1)(2n+5) - \Sigma t_u(t_u - 1)(2t_u + 5) - \Sigma t_v(t_v - 1)(2t_v + 5)]$$
$$+ \frac{1}{9n(n-1)(n-2)}[\Sigma t_u(t_u - 1)(t_u - 2)][\Sigma t_v(t_v - 1)(t_v - 2)]$$
$$+ \frac{1}{2n(n-1)}[\Sigma t_u(t_u - 1)][\Sigma t_v(t_v - 1)]$$

For this example, $\sigma_S^2 = 122.044$, so $\sigma_s = 11.047$.

The standard normal approximation for the statistic S with the correction for continuity, as given in Kendall (1970), is

$$z = \frac{S - 1}{\sigma_s} = 0.272$$

Using a two-tailed test, $P(|Z| > 0.272) = 0.79$. Clearly the hypothesis of an underlying gamma distribution is not rejected.

6.9 LOGNORMAL DISTRIBUTION

The lognormal model has many uses in engineering. It can be derived as the appropriate lifetime model where failure of a unit occurs only when damage to it has reached a specific level. The times at which damage occurs are the random variables $T_1, T_2, \ldots, T_n$, where T_n is the failure time. Further, t_i is the time between the

ith and the $(i-1)$th occurrences of damage, which is a random proportion of the $(i-1)$th occurrence time, and this random proportion is independent of the time of occurrence. Then it follows that the time to failure has a lognormal distribution. This model has been used to model stress failure mechanisms, such as when a failure is caused by rupture. In such models it is assumed that the rupture was caused when a crack in the structure reached a given size and the growth of the crack at any instant of time is a random proportion of its size at that time. This model has also been used to represent the sizes of fragments from a breakage process, distribution of income, distribution of a variety of biological phenomena, and as a lifetime model for electronic and electromechanical components. If T has a lognormal distribution, then $X = \ln T$ has a normal distribution. The density function can be derived from this relationship and is given by

$$f(t) = \frac{1}{\sqrt{2\pi}\sigma t} \exp\left[-\frac{1}{2}\left(\frac{\ln t - \mu}{\sigma}\right)^2\right], \quad t \geq 0, \quad \sigma > 0, \quad -\infty < \mu < \infty \tag{6.31}$$

The lognormal densities are defined on the range $(0, \infty)$, have a value of zero at $T = 0$, rise to a single mode, and decrease to zero as T goes to infinity. Figure 6.10 shows a variety of shapes for the model. The shape is determined by the parameter σ and the scaling by the parameter μ. These parameters can be estimated by using the relationship with the normal distribution. The procedure used is to obtain the natural logarithm of the data and to use Eqs. (6.5) to estimate the parameters. Likewise cumulative probabilities can be obtained from a normal table since $P(T < t)$ is equivalent to $P(X < \ln t)$, where $X = \ln T$. The procedure for fitting a lognormal distribution is illustrated using the data in Table 6.2. The first step is to take the natural logarithms of the 25 observations. These are shown in Table 6.9. Use of Eqs. (6.5) yields the estimates $\overline{T} = 1.355$ and $s = 0.2302$. If we wish to obtain

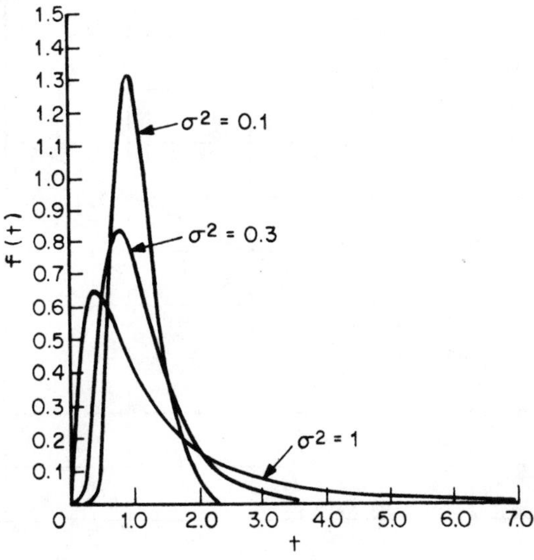

FIGURE 6.10 Lognormal density function; $\mu = 0$.

TABLE 6.9 Logarithms of Data in Table 6.2

1.43	0.59	1.40	1.41	1.44
1.35	1.15	1.23	1.64	1.61
1.54	1.53	1.25	1.37	1.08
1.51	1.58	1.47	1.06	1.46
1.27	1.11	1.59	1.38	1.42

$\overline{T} = 1.355$

$s = 0.2302$

an estimate that the waiting time exceeds 5.1 min, we use the estimated parameters and the normal table (Table A.1) to find the probability that a normal variate will exceed ln (5.1) = 1.629. The corresponding z score is

$$z = \frac{1.629 - 1.355}{0.2302} = 1.19$$

The probability that z exceeds 1.19 is 0.117 (Table A.1). Probability plots can be made using the relationship between lognormal and normal distributions. Thus a plot of the natural logarithms of the data on normal probability paper can be used to test the lognormal distributional hypothesis. There are special papers which are scaled so that it is not necessary to make the transformation. The parameters are estimated in the same manner as for the normal distribution. In order to illustrate the technique we shall use the data in Table 6.2. The resulting plot on normal probability paper using the logarithms given in Table 6.9 is shown in Fig. 6.11. The

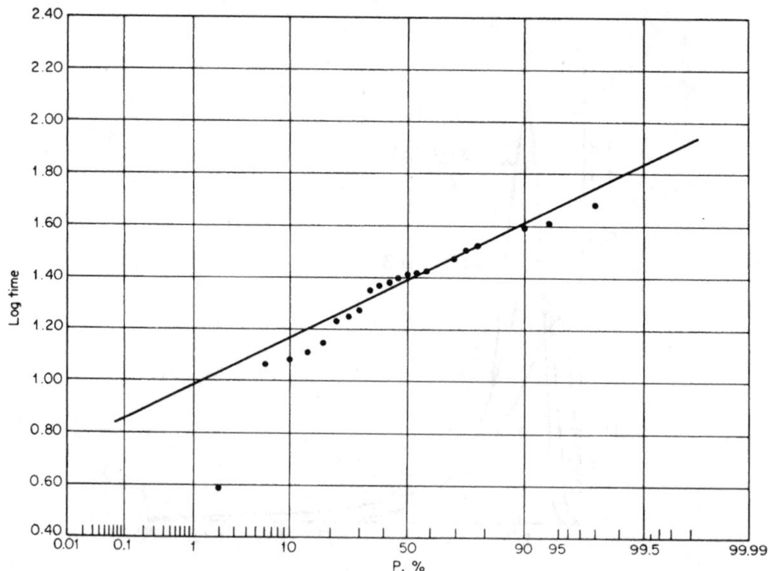

FIGURE 6.11 Lognormal probability plot of customer arrival times.

resulting plot indicates that the model is reasonable. Therefore we can use the techniques of Sec. 6.4 to estimate the parameters. Using $Y_{90} = 1.61$, $Y_{10} = 1.16$, and $Y_{50} = 1.38$ yields

$$\sigma^* = \frac{2}{5}(1.61 - 1.16) = 0.18$$

$$\mu^* = 1.38$$

These values are close to those obtained using the maximum-likelihood estimation technique. Note that μ and σ are not the mean and the standard deviation of the lognormal distribution; for details see Hahn and Shapiro (1967).

The W test for normality can be used for this model by using the natural logarithm of the observations with Eqs. (6.6) to (6.8). These equations yield, for the data in Table 6.9, $S^2 = 1.272$, $b = 1.054$, and $W = 0.8732$. Setting the type 1 error to 0.10, we find that the value computed is smaller than $W_{0.10} = 0.931$, and hence the hypothesis of lognormality is rejected. This contradiction with the result of the probability plot illustrates why both distributional assessment techniques should be used. Examination of the plot indicates that the smallest value is quite far from the plotted line. This possible outlier is possibly the reason why the W test rejected the null hypothesis of lognormality. Thus before proceeding with an analysis of the data the possible errant point should be investigated to see whether there is an explanation that would indicate whether or not to include this value.

6.10 EMPIRICAL MODELS

In some applications the engineer is more interested in summarizing the data and perhaps interpolating or extrapolating from the data than in developing a statistical model. In such instances one is not interested in inferences about the parameters of the model (confidence intervals or tests of hypotheses), but one is only concerned with probability calculations concerning the random variable. In such cases instead of using one of the prior models, a flexible *empirical model* is used. There are three commonly used families of empirical models. The oldest is the *Pearson* family; see Johnson, Nixon, and Amos (1963) for details. This system is harder to use than the other two and is not covered in this book. The newest is the *generalized lambda* system, which is described in Ramberg et al. (1979). This system is still under development and has problems with estimating its parameters and with the existence of density functions for various combinations of the parameters. The third, the *Johnson* system, was proposed by Johnson (1949) and is described here.

This system contains three groups of distributions, which are generated by transforming a standard normal variable. Each group has a different transformation. Details can be found in Shapiro and Gross (1981). The groups are called S_U, S_B, and S_L. The latter is a *three-parameter lognormal*, which differs from the lognormal in the previous section in that the random variable is defined on the interval (a, ∞) rather than starting at zero. The first two distributions each have four parameters. The S_B group is defined on a *bounded range* and two of the parameters define the boundary values. The S_U group is defined on an *unbounded range.* The steps in fitting the Johnson model require that the user first select the appropriate group and then use the estimating equations for the group chosen. A procedure for selecting the group and estimating its parameters was developed by Slifker and Shapiro (1980). The groups and the selection procedure are described in the following.

The systems proposed by Johnson are generated from the transformation

$$z = \tau + \delta K_i(x; \beta, \epsilon) \tag{6.32}$$

where z is a standard normal variable and the functions K_i are chosen to cover a wide range of possible shapes. The interesting point in the three functions chosen by Johnson is that all the S_B distributions lie in the upper half of the (β_1, β_2) plane, all the S_U distributions lie in the lower half, and the S_L distributions fall on a line separating the two. [$\sqrt{\beta_1}$ is the third standardized moment, and β_2 is the fourth standardized moment, which are often used to describe distributions; see Shapiro and Gross (1981) for a discussion of this subject.] Johnson's functions are

$$K_1 = \sinh^{-1}\left(\frac{x - \epsilon}{\beta}\right) \tag{6.33}$$

which defines the S_U distribution,

$$K_2 = \ln\left(\frac{x - \epsilon}{\beta + \epsilon - x}\right) \tag{6.34}$$

which defines the S_B distribution, and

$$K_3 = \ln\left(\frac{x - \epsilon}{\beta}\right) \tag{6.35}$$

which defines the S_L distribution.

The selection procedure involves estimating percentiles from the data. Let X_{3z}, X_z, X_{-z}, and X_{-3z} be the percentiles corresponding to $3z$, z, $-z$, and $-3z$ under the transformation defined in Eq. (6.32). Let

$$\begin{aligned} m &= X_{3z} - X_z \\ n &= X_{-z} - X_{-3z} \\ p &= X_z - X_{-z} \end{aligned} \tag{6.36}$$

Then if

$\dfrac{mn}{p^2} > 1$ select the S_U distribution

$\dfrac{mn}{p^2} < 1$ select the S_B distribution

$\dfrac{mn}{p^2} = 1$ select the S_L distribution

The following steps are necessary to implement the procedure. First choose a value of $z > 0$. This choice is motivated by the number of data points being fit. In general, for moderate size samples a value of z less than 1 should be used. A choice of z greater than 1 would make it difficult to estimate the percentiles corresponding to $\pm 3z$. A value of z close to 0.5 is often used. For example, the choice of $z = 0.524$ would mean that $3z = 1.572$ and the procedure requires estimating the

70th and 94.2th percentiles from the upper half of the data and the 30th and 5.8th percentiles from the lower half. It is possible to use higher values of z with larger data sets since it is then possible to estimate more extreme percentiles. Another criterion that is used in selecting z is to choose that value which will give a good fit in the region for which percentile estimates are desired subject to the ability to estimate these from the data. Therefore if we wish to estimate the 95th percentile, we should choose z such that $3z = 1.645$ since 1.645 is the 95th percentile of the normal distribution. The reason for this is that when fitting the distribution, the Johnson approximating distribution will fit the data exactly at the chosen percentiles. After choosing z the next step is to find the cumulative probabilities of the normal distribution (from Table A.1) corresponding to each of the selected z values. For example, if $3z = 1.645$, then $F(3z) = 0.95$. These values are obtained for each of the four points corresponding to $3z$, z, $-z$, and $-3z$. Next, for each of these probabilities, the corresponding percentile of the data must be estimated. This is usually done by solving the equation $(i - 1/2)/n = F(.)$ for the value of i, where $F(.)$ corresponds to one of the four selected points. The i value is the order number of the data percentile. The smallest ordered value corresponds to $i = 1$, while the largest ordered value corresponds to $i = n$, the sample size. Usually the value of i obtained from the above equation will not be an integer, and therefore it is necessary to interpolate. The four data values obtained correspond to X_{3z}, X_z, X_{-z}, and X_{-3z} in Eqs. (6.36).

The final step in the selection process is to use Eqs. (6.36) to select the appropriate group. It should be noted that since the Xs are continuous random variables, the probability that $mn/p^2 = 1$ is virtually zero. If one wishes to have the S_L distribution as a possible alternative, it will be necessary to allow a tolerance about 1. Once the proper group is selected, then the values of m, n, and p can be used to estimate the parameters. The formulas for the S_U group are as follows:

$$\delta^* = \frac{2z}{\cosh^{-1}\left[\dfrac{1}{2}\left(\dfrac{m}{p} + \dfrac{n}{p}\right)\right]}, \qquad \delta^* > 0 \tag{6.37a}$$

$$\tau^* = \delta^* \sinh^{-1}\left[\frac{\dfrac{n}{p} - \dfrac{m}{p}}{2\left(\dfrac{m}{p}\dfrac{n}{p} - 1\right)^{1/2}}\right] \tag{6.37b}$$

$$\beta^* = \frac{2p\left(\dfrac{m}{p}\dfrac{n}{p} - 1\right)^{1/2}}{\left(\dfrac{m}{p} + \dfrac{n}{p} - 2\right)\left(\dfrac{m}{p} + \dfrac{n}{p} + 2\right)^{1/2}}, \qquad \beta^* > 0 \tag{6.37c}$$

$$\epsilon^* = \frac{x_z + x_{-z}}{2} + \frac{p\left(\dfrac{n}{p} - \dfrac{m}{p}\right)}{2\left(\dfrac{m}{p} + \dfrac{n}{p} - 2\right)} \tag{6.37d}$$

The estimators for the S_B group are as follows:

$$\delta^* = \frac{z}{\cosh^{-1}\left\{\frac{1}{2}\left[\left(1+\frac{p}{m}\right)\left(1+\frac{p}{n}\right)\right]^{1/2}\right\}}, \qquad \delta^* > 0 \tag{6.38a}$$

$$\tau^* = \delta^* \sinh^{-1}\left[\frac{\left(\frac{p}{n}-\frac{p}{m}\right)\left[\left(1+\frac{p}{m}\right)\left(1+\frac{p}{n}\right)-4\right]^{1/2}}{2\left(\frac{p}{m}\frac{p}{n}-1\right)}\right] \tag{6.38b}$$

$$\beta^* = \frac{p\left\{\left[\left(1+\frac{p}{m}\right)\left(1+\frac{p}{n}\right)-2\right]^2-4\right\}^{1/2}}{\frac{p}{m}\frac{p}{n}-1}, \qquad \beta^* > 0 \tag{6.38c}$$

$$\epsilon^* = \frac{x_z + x_{-z}}{2} - \frac{\beta^*}{2} + \frac{p\left(\frac{p}{n}-\frac{p}{m}\right)}{2\left(\frac{p}{m}\frac{p}{n}-1\right)} \tag{6.38d}$$

The estimators for the S_L group are as follows:

$$\delta^* = \frac{2z}{ln\left(\frac{m}{p}\right)} \tag{6.39a}$$

$$\hat{\tau}^* = \delta^* \ln\left[\frac{\frac{m}{p}-1}{p\left(\frac{m}{p}\right)^{1/2}}\right] \tag{6.39b}$$

$$\epsilon^* = \frac{x_z + x_{-z}}{2} - \frac{m+p}{2\left(\frac{m}{p}-1\right)} \tag{6.39c}$$

where $\hat{\tau} = \tau - \delta \ln \beta$ and $z = \hat{\tau} + \delta \ln (x - \epsilon)$.

The following example, taken from Hahn and Shapiro (1967), will illustrate the procedure. The data are given in Table 6.10 and represent the resistances of 500 1/2-ohm resistors. A value of $z = 0.5483$ was selected so that the empirical distribution would fit the data perfectly at the 0.05 and 0.95 percentiles since $3z = 1.645$. This

TABLE 6.10 Resistance of 500 $\frac{1}{2}$-Ohm Resistors

Cell midpoint	Number of observations	S_B fit
< 0.4	4	6.5
0.425	33	36.1
0.475	78	74.1
0.525	99	93.8
0.575	87	90.4
0.625	76	73.0
0.675	51	52.0
0.725	32	33.7
0.775	21	19.9
0.825	7	10.9
0.875	5	5.5
> 0.9	7	4.1
Totals	500	500.0

choice required that the 5th, 29.2th, 70.8th, and 95th percentiles be estimated from the data. In order to obtain $X_{1.645}$ we must solve for i in $(i - \frac{1}{2})/n = F(X_{1.645})$. Since $n = 500$, this yields

$$i = 500(0.95) + \frac{1}{2} = 475.5$$

If we had the actual numbers, we would choose the value midway between the 475th and the 476th highest observations. However, in this example the actual data are not given, and hence it is necessary to interpolate within the cell. Interpolating in the table we find the following:

$$X_{1.645} = 0.786$$
$$X_{0.5483} = 0.635$$
$$X_{-0.5483} = 0.516$$
$$X_{-1.645} = 0.432$$

Next the group is chosen by computing

$$\frac{mn}{p^2} = 0.896$$

Since the value is less than 1.0, the S_B distribution is selected. Equations (6.38) yield the following estimates:

$$\delta^* = \frac{0.5483}{\cosh^{-1}\left[\frac{1}{2}\sqrt{1.788(2.417)}\right]} = 1.959$$

$$\tau^* = 1.959 \sinh^{-1}\left[\frac{(1.417 - 0.788)\sqrt{1.788(2.417) - 4}}{2[0.788(1.417) - 1]}\right] = 2.373$$

$$\beta^* = \frac{0.119\sqrt{[1.788(2.417) - 2]^2 - 4}}{0.788(1.417) - 1} = 1.203$$

$$\epsilon^* = \frac{0.635 + 0.516}{2} - \frac{1.203}{2} + \frac{0.119(1.417 - 0.788)}{2[0.788(1.417) - 1]} = 0.295$$

The S_B distribution fits the data reasonably well, as is shown in Table 6.10.

The probability plotting procedures and the W test for the normal distribution can be used to assess the goodness of fit of the Johnson distribution simply by converting the observations to z scores (standard normal) by means of Eq. (6.32) with the appropriate function for K_i.

REFERENCES

Gross, A. J., and V. A. Clark: *Survival Distributions: Reliability Applications in the Biomedical Sciences*, Wiley, New York, 1975.

Gumbel, E. J.: *Statistics of Extremes*, Columbia University Press, New York, 1958.

Hahn, G., and S. S. Shapiro: *Statistical Models in Engineering*, Wiley, New York, 1967.

Harter, H. L.: "New Tables of the Incomplete Gamma-Function Ratio and of Percentage Points of the Chi Square and Beta Distributions," Aerospace Research Labs., U.S. Air Force, Dayton, OH, 1964.

Johnson, N. L.: "Systems of Frequency Curves Generated by Methods of Translation," *Biometrika*, vol. 36, pp. 274–282, 1949.

_____ and S. Kotz: *Distributions in Statistics: Continuous Univariate Distributions*, vols. 1 and 2, Houghton Mifflin, Boston, 1970.

_____ E. Nixon, and D. E. Amos: "Table of Percentage Points of Pearson Curves for Given $\sqrt{\beta_1}$, β_2, Expressed in Standard Measure," *Biometrika*, vol. 50, p. 459, 1963.

Kendall, M. G.: *Rank Correlation Methods*, Charles Griffin, London, 1970.

Locke, C.: "A Test for the Composite Hypothesis that a Population Has a Gamma Distribution, " *Commun. Stat.*, vol. 4, pp. 351–364, 1976.

Mann, N. R., R. E. Schafer, and N. D. Singpurwalla: *Methods for Statistical Analysis of Reliability and Life Data*, Wiley, New York, 1974.

Nelson, W.: *How to Analyze Data with Simple Plots*, ASQC Basic References in Quality Control: Statistical Techniques, Am. Soc. Quality Control, Milwaukee, WI., 1979.

Pearson, E. S., and H. O. Hartley: *Biometrika Tables for Statisticians*, vol. 1, Cambridge Univ. Press, Cambridge, England, 1966.

Ramberg, J. S., P. R. Tadikamalla, E. J. Dudewicz, and E. F. Mykytka: "A Probability Distribution and Its Uses in Fitting Data," *Technometrics*, vol. 21, pp. 201–214, 1979.

Shapiro, S. S.: *How to Test Normality and Other Distributional Assumptions*, ASQC Basic References in Quality Control: Statistical Techniques, Am. Soc. Quality Control, Milwaukee, WI., 1980.

_____ and C. W. Brain: "A Review of Distributional Testing Procedures," *Am. J. Math. Manage. Sci.*, vol. 2, pp. 175–221, 1982.

_____ and _____ : "*W* Test for the Weibull Distribution," *Commun. Stat.*, vol. 16, pp. 209–219, 1987.

_____ and R. S. Francia: "Approximate Analysis of Variance Test for Normality," *J. Am Stat. Assoc.*, vol. 67, pp. 215–216, 1972.

_____ and A. J. Gross: *Statistical Modeling Techniques*, Marcel Dekker, New York, 1981.

_____ and M. B. Wilk: "Analysis of Variance Test for Normality (Complete Samples)," *Biometrika*, vol. 52, pp. 591–611, 1965.

_____ and _____: "An Analysis of Variance Test for the Exponential Distribution," *Technometrics*, vol. 14, pp. 355–370, 1972.

Slifker, J. F., and S. S. Shapiro: "The Johnson System: Selection and Parameter Estimation," *Technometrics*, vol. 22, pp. 239–246, 1980.

Stephens, M. A.: "On the *W* Test for Exponentiality with Origin Known," *Technometrics*, vol. 20, pp. 33–35, 1978.

P • A • R • T • 3

INTERMEDIATE METHODS

CHAPTER 7

STATISTICAL PROCESS CONTROL

William H. Woodall and Benjamin M. Adams
University of Alabama, Tuscaloosa, AL

7.1 INTRODUCTION

This chapter describes some of the basic principles and techniques of statistical process control which can be applied either to the manufacture of goods or to services. Manufacturing applications will be emphasized in this chapter. Readers interested in examples of applications to services are referred to Rosander (1985) or Deming (1986).

Dr. Walter A. Shewhart of Bell Telephone Laboratories is considered to be the founder of statistical quality control with the publication in 1931 of his now classic book, *Economic Control of Quality of Manufactured Product.* In recent years companies have shown strong renewed interest in the concept of quality and in the subject of statistical quality control. This interest has been due, in large part, to increased international competition and to the recognition of the importance of quality in achieving success in the marketplace.

Much of the competition facing U.S. industries today comes from Japanese and other companies with reputations for the high quality of their products. The Japanese were influenced by the ideas of several American statisticians, including W. Edwards Deming, who first lectured there in the 1950s. In recognition of Deming's contributions, the Union of Japanese Scientists and Engineers established the Deming Prize in 1951. This award has been presented annually to the Japanese companies with the most outstanding achievements in quality control and improvement.

Japanese companies have traditionally made greater use of statistical process control than U.S. companies. For example, Garvin (1983) found that all Japanese manufacturers of room air-conditioners relied heavily on statistical process control for monitoring their production processes. In the United States only the plant with the lowest defect rate among U.S. plants used these techniques. Garvin's study also showed that the quality of air-conditioners was uniformly much higher for the Japanese companies.

It is impossible to inspect or test quality into a product. The product and the production process must first be well designed, and then the production process must be controlled carefully. "Quality by design" refers to the decisions related to the type of product to be made, its design, and the design of the production process. This topic, which often includes the use of experimental design, is covered in Chaps. 15 and 20. "Quality of conformance" refers to the uniformity of the manufactured items and to the assurance that they meet the requirements and specifications of the design. Increased uniformity of the product leads to a wide variety of benefits, including higher customer satisfaction, increased yields, and reductions in scrap, rework, and inspection. Attention to quality can thus lead to increased productivity and to a substantial reduction of costs.

The subject of this chapter is the use of statistical process control techniques, primarily control charts, to achieve higher levels of quality of conformance. Computer software is often used in practical applications. See Klaus (1997) for a review of available software. Since the subject of statistical process control is too broad to be covered completely in a single chapter, an extensive list of references is provided for those who want more information. As Deming (1986) emphasizes, the use of statistics is not sufficient to achieve high quality, but it is a necessary part of a company-wide quality program. For information on the type of philosophy and management system necessary to achieve continuing quality improvement, the reader is referred to Juran (1964), Feigenbaum (1983), Ishikawa (1985), or Deming (1986).

7.2 GENERAL PRINCIPLES OF CONTROL CHARTS

In order to control a production process, it is necessary to identify and quantify characteristics which reflect the quality of the product. Such quality characteristics can be either discrete or continuous variables. For example, the number of air bubbles in a sheet of glass is a discrete variable. The weight of a container of cooking oil is considered to be a continuous variable, although measurements are always rounded to a specified number of decimal places. The type of control chart used to monitor a process varies, depending on whether the variable of interest is discrete or continuous. Control charts for continuous quality characteristics are considered in Sec. 7.3, those for discrete quality characteristics in Sec. 7.4.

Whether quality characteristics are discrete or continuous, some variability in the observed values is often unavoidable. In the theory of control charts, this variation is considered to be due to either random causes or assignable causes. Deming (1986) and others refer to random and assignable causes as common and special causes, respectively. An assignable cause is an influence considered to be significant, unusual, and capable of being removed from the process. Such causes may be due to human error, variation in raw materials, or the need for machine adjustment. Random causes are smaller, uncontrollable influences which cannot be removed from the process without fundamental changes in the process itself. Such changes usually require management action. This classification of influences can change over time. For example, changes in humidity can be considered to be a random cause, but later, after switching to a humidity-controlled environment, reclassified as an assignable cause.

An important tool used to understand and reduce process variation is the control chart. A control chart is a plot of values of a statistic, such as the sample mean, as a function of time. The statistic summarizes measurements of the quality characteristic from randomly chosen items during production. Control limits or other rules are used to determine whether the variability of the statistic over time appears to be due to random variation only, or if an assignable cause is present.

Regardless of the rules used, a good understanding of the process is indispensable in interpreting the chart and in finding and removing any assignable causes.

One purpose of control charts is to establish "statistical control" of a process. Statistical control means that the assignable causes of variation have been removed, at least temporarily, from the process and only the random causes are present. Once such stability has been achieved, the random variability can be modeled by a probability distribution, and the control charts can be used to monitor future production to detect the occurrence of assignable causes. The effect of a change in the process is usually represented by a change in a parameter of the probability distribution used to represent the random background variation of the process. Control charts are designed so that a signal indicating an assignable cause in the presence of only random variation, that is a false alarm, is unlikely. Responding frequently to random variation leads to overadjustment of the process, increasing the variability of the quality characteristic and destroying operator confidence in the chart itself. As Deming (1986) states, the purpose of a control chart is to prevent underadjustment and overadjustment of a process.

Reaching a state of statistical control may require great effort, but it is only the initial step. A control chart should then be used as a baseline to study the effect of planned changes in the production process in an effort to further improve the quality of the product. Evolutionary operation (EVOP), covered in Chap. 15, is the type of on-line experimental design approach often used at this stage to find more optimal operating conditions for the process. Any change to the process, whether planned or not, should be indicated on the chart when it occurs so that the time and effect of the change can be clearly seen.

For the use of control charts to be most effective, all items within a given sample should be produced under essentially the same conditions. The effect of an assignable cause should result in a change between samples, not within samples. For example, it may be preferable to apply a separate chart for each of several operators rather than combining their outputs and sampling to form only one chart. Such judicious choice of when and where to apply control charts is referred to as the use of *rational subgroups*. A good discussion of rational subgroups can be found in Wheeler (1995).

Unless the measurement process is automated, control charts can be used only on a relatively small percentage of quality characteristics which may be of interest. New processes or characteristics which exhibit excessive variability and thus cause quality problems should be considered first. Pareto analysis can help to establish the priority of the variables to be considered. Once processes have been improved, charts must be updated. As improvement continues, charts can eventually be removed in many cases and the emphasis switched to other variables.

Statistical process control differs from adaptive process control, which uses feed-forward and feedback schemes to predict future values of the process variables. Adaptive process control specifies whether action is required and specifies the amount of correction to keep the process variables on target. Temperature, flow rate, pressure, and pH concentration are examples of variables which are often controlled this way. Variables that are controlled automatically are frequently cyclical and will not appear to be in "statistical control." Box, Jenkins, and Reinsel (1994) provide the theory of adaptive control while Johnson (1982) gives a more practical guide to automatic control. Adaptive process control can be applied when the process variables are quantifiable, controllable, and affect the quality characteristic in a known way. Despite fundamental differences, the purpose of adaptive process control is the same as statistical process control, the reduction of variability associated with the manufacturing process. MacGregor (1987) discusses in more detail the relationship between engineering process control and statistical process control.

7.3 CONTROL CHARTS FOR CONTINUOUS VARIABLES

7.3.1 The Use of Specification Limits

Many quality characteristics can be measured on a continuous scale, such as weight, volume, viscosity, thickness, tensile strength, and voltage. In our notation, the particular variable of interest is represented by the symbol X. It is assumed that the measurement process is adequate. For information on how to establish the validity of the measurement process, the reader is referred to ASQC Automotive Division (1986).

In many applications a target value, denoted by μ_0, is given for the quality characteristic. Often specification limits S_L and S_U ($S_U > S_L$) are also given. These specification limits are usually, but not always, equidistant from μ_0. Meeting such specification limits has frequently been considered to be sufficient in Western industry, with an item considered to be conforming if $S_L < X < S_U$, and nonconforming otherwise. Although meeting specification limits is an important first step, Sullivan (1984), Deming (1986), Wheeler and Chambers (1986), and others emphasize the importance of concentrating the density of the quality characteristic as tightly as possible about the target value. As Deming (1982, p. 78) expresses it, "The aim in production should be to continually improve the process to the point where the distribution of the chief quality-characteristic of parts and materials is so narrow that the specifications are lost beyond the horizon." Kacker (1985) discusses the approach of Dr. Genichi Taguchi, which assumes that the cost of a deviation from target is proportional to $(X - \mu_0)^2$, not a step function with increases at the specification limits. This more realistic cost function provides a justification for continuous process improvement, whereas conformance to specifications does not. Some of Taguchi's methods are explained in Chap. 20.

The capability of a process can be measured by the capability index, defined as

$$C_p = \frac{S_U - S_L}{6\sigma} \tag{7.1}$$

where σ is the standard deviation of the quality characteristic. A method of estimating σ, and thus C_p, is given in the next section. Under the assumption of normality for X, the value $C_p = 1$ implies that roughly 997 out of 1000 items can be produced within the specification limits if the process is properly centered at the target value. To meet the goal of continuous process improvement, the value of C_p should be increased steadily by decreasing the variability of the quality characteristic. For example, the ASQC Automotive Division's *Statistical Quality Control Manual* states that General Motors requires an initial target value of $C_p = 1.33$, but stresses continuous improvement once this goal has been met. For more information on capability indices, the reader is referred to Rodriguez (1992).

7.3.2 Establishing Statistical Control Using $\overline{X}$ and R Charts

Efforts to establish a state of statistical control must occur before control charts can be used effectively to either monitor current production or study the effect of changes to the process. Data must be collected and any assignable causes must be found and eliminated so that only random causes remain.

For the required data, m samples of n measurements each are collected on the quality characteristic. It is usually recommended that m be at least 20 and that n be small, often 4 or 5. We will let X_{ij} represent the jth measurement in the ith sample, $i = 1, 2, \ldots, m$ and $j = 1, 2, \ldots, n$. For the ith sample we calculate the sample mean

$$\bar{X}_i = \frac{1}{n}\sum_{j=1}^{n} X_{ij} \tag{7.2}$$

and the sample range, $n \geq 2$,

$$R_i = \max_j X_{ij} - \min_j X_{ij} \tag{7.3}$$

for $i = 1, 2, \ldots, m$. The case $n = 1$ is covered in Sec. 7.3.6.

The sample means are used in an $\bar{X}$ chart to detect changes in location of the distribution of X while an R chart based on the sample ranges is often used to detect changes in variability. Each control chart has control limits which contain most, if not all, of the points if only random variation is present. These Shewhart-type charts are based on the fact that it is unlikely that the value of a statistic will be more than 3 standard errors from its mean.

Under the assumptions that the process is in statistical control and that the observations are independent and normally distributed with mean μ and variance σ^2, denoted by $N(\mu, \sigma^2)$, the sample means are independent and identically distributed with distribution $N(\mu, \sigma^2/n)$. In addition, the mean and the standard deviation of the sample range are $d_2\sigma$ and $d_3\sigma$, respectively, where the constants d_2 and d_3, depend only on n and are shown in Table 7.1. Thus an unbiased estimate of μ is the overall mean

$$\bar{\bar{X}} = \frac{1}{m}\sum_{i=1}^{m} \bar{X}_{i\cdot} \tag{7.4}$$

An unbiased estimate of the process standard deviation is

$$\hat{\sigma} = \frac{\bar{R}}{d_2} \tag{7.5}$$

where $\bar{R}$ is the average range,

$$\bar{R} = \frac{1}{m}\sum_{i=1}^{m} R_{i\cdot} \tag{7.6}$$

The control limits of the $\bar{X}$ chart are

$$\bar{\bar{X}} \pm \frac{3\bar{R}}{d_2\sqrt{n}} \tag{7.7}$$

or

$$\bar{\bar{X}} \pm A_2\bar{R}$$

where values of $A_2 = 3/(d_2\sqrt{n})$ are given in Table 7.1. The control limits for the R chart are at

TABLE 7.1 Table of Control Chart Constants

Observations in sample n	Averages			Standard deviations						Ranges						
	Factors of control limits			Factors for centerline		Factors for control limits				Factors for centerline		Factors for control limits				
	A	A_2	A_3	c_4	$1/c_4$	B_3	B_4	B_5	B_6	d_2	$1/d_2$	d_3	D_1	D_2	D_3	D_4
2	2.121	1.880	2.659	0.7979	1.2533	0	3.267	0	2.606	1.128	0.8865	0.853	0	3.686	0	3.267
3	1.732	1.023	1.954	0.8862	1.1284	0	2.568	0	2.276	1.693	0.5907	0.888	0	4.358	0	2.574
4	1.500	0.729	1.628	0.9213	1.0854	0	2.266	0	2.088	2.059	0.4857	0.880	0	4.698	0	2.282
5	1.342	0.577	1.427	0.9400	1.0638	0	2.089	0	1.964	2.326	0.4299	0.864	0	4.918	0	2.114
6	1.225	0.483	1.287	0.9515	1.0510	0.030	1.970	0.029	1.874	2.534	0.3946	0.848	0	5.078	0	2.004
7	1.134	0.419	1.182	0.9594	1.0423	0.118	1.882	0.113	1.806	2.704	0.3698	0.833	0.204	5.204	0.076	1.924
8	1.061	0.373	1.099	0.9650	1.0363	0.185	1.815	0.179	1.751	2.847	0.3512	0.820	0.388	5.306	0.136	1.864
9	1.000	0.337	1.032	0.9693	1.0317	0.239	1.761	0.232	1.707	2.970	0.3367	0.808	0.547	5.393	0.184	1.816
10	0.949	0.308	0.975	0.9727	1.0281	0.284	1.716	0.276	1.669	3.078	0.3249	0.797	0.687	5.469	0.223	1.777
11	0.905	0.285	0.927	0.9754	1.0252	0.321	1.679	0.313	1.637	3.173	0.3152	0.787	0.811	5.535	0.256	1.744
12	0.866	0.266	0.886	0.9776	1.0229	0.354	1.646	0.346	1.610	3.258	0.3069	0.778	0.922	5.594	0.283	1.717
13	0.832	0.249	0.850	0.9794	1.0210	0.382	1.618	0.374	1.585	3.336	0.2998	0.770	1.025	5.647	0.307	1.693
14	0.802	0.235	0.817	0.9810	1.0194	0.406	1.594	0.399	1.563	3.407	0.2935	0.763	1.118	5.696	0.328	1.672
15	0.775	0.223	0.789	0.9823	1.0180	0.428	1.572	0.421	1.544	3.472	0.2880	0.756	1.203	5.741	0.347	1.653
16	0.750	0.212	0.763	0.9835	1.0168	0.448	1.552	0.440	1.526	3.532	0.2831	0.750	1.282	5.782	0.363	1.637
17	0.728	0.203	0.739	0.9845	1.0157	0.466	1.534	0.458	1.511	3.588	0.2787	0.744	1.356	5.820	0.378	1.622
18	0.707	0.194	0.718	0.9854	1.0148	0.482	1.518	0.475	1.496	3.640	0.2747	0.739	1.424	5.856	0.391	1.608
19	0.688	0.187	0.698	0.9862	1.0140	0.497	1.503	0.490	1.483	3.689	0.2711	0.734	1.487	5.891	0.403	1.597
20	0.671	0.180	0.680	0.9869	1.0133	0.510	1.490	0.504	1.470	3.735	0.2677	0.729	1.549	5.921	0.415	1.585
21	0.655	0.173	0.663	0.9876	1.0126	0.523	1.477	0.516	1.459	3.778	0.2647	0.724	1.605	5.951	0.425	1.575
22	0.640	0.167	0.647	0.9882	1.0119	0.534	1.466	0.528	1.448	3.819	0.2618	0.720	1.659	5.979	0.434	1.566
23	0.626	0.162	0.633	0.9887	1.0114	0.545	1.455	0.539	1.438	3.858	0.2592	0.716	1.710	6.006	0.443	1.557
24	0.612	0.157	0.619	0.9892	1.0109	0.555	1.445	0.549	1.429	3.895	0.2567	0.712	1.759	6.031	0.451	1.548
25	0.600	0.135	0.606	0.9896	1.0105	0.565	1.435	0.559	1.420	3.931	0.2544	0.708	1.806	6.056	0.459	1.541

Source: Reproduced from ASTM-STP 15D by permission of the American Society for Testing and Materials.

$$\overline{R} \pm 3d_3 \frac{\overline{R}}{d_2} \tag{7.8}$$

or

$$D_3\overline{R},\ D_4\overline{R}$$

where the constants D_3 and D_4 also appear in Table 7.1. The lower control limit for the R chart will be zero for small values of n. Note that the control limits of both the $\overline{X}$ and the R charts depend on the sample size. Thus if the sample size varies over time, so will the values of the control limits.

All points falling outside the control limits of the $\overline{X}$ or R charts should be investigated. Points outside the R chart control limits should be considered first because a change in variability can affect values in the $\overline{X}$ chart. If the assignable causes resulting in points outside the limits can be found and removed, then the points should be deleted from the data and the control limits recalculated. Any points outside the revised limits should then be investigated. This iterative process continues until only a few or no points are outside the control limits. If a process is in statistical control, then it is unlikely that more than a few points will fall outside the control limits or that the points will show unusual or repetitive patterns. The interpretation of patterns is discussed in the next section.

It is very important to note that the process can be in a state of statistical control, but at an unacceptable level. For example, $\overline{\overline{X}}$ may not be close to μ_0. In this case it is necessary to adjust the process to bring the mean closer to the target value. Alternatively, the process mean could be on target, but the variability may be too great. If the process is in statistical control, with the mean at the target value, the capability index can be estimated by

$$\hat{C}_p = \frac{S_U - S_L}{6\hat{\sigma}} \tag{7.9}$$

where

$$\hat{\sigma} = \frac{\overline{R}}{d_2}.$$

If the value of $\hat{C}_p$ is too low, then the process is "not capable," and changes in the process are required. An estimate of the proportion of items outside specification limits is

$$1 - \Phi\left(\frac{S_U - \overline{\overline{X}}}{\hat{\sigma}}\right) + \Phi\left(\frac{S_L - \overline{\overline{X}}}{\hat{\sigma}}\right) \tag{7.10}$$

where $\Phi(\cdot)$ is the standard normal cumulative distribution function (cdf).

7.3.3 Analysis of Patterns on $\overline{X}$ and R Charts

Analysis of unusual patterns in the points on a control chart is often useful in identifying assignable causes. Unusual patterns such as cycles or several consecutive values near a control limit can indicate the presence of nonrandom variation because such patterns are unlikely to occur due to chance. If the quality characteristic is normally distributed and the process is in statistical control, then the points on the control chart should exhibit only random variation.

Many rules have been suggested to be used as indicators of assignable causes. The rules most commonly used in practice are discussed in this section along with some likely assignable causes associated with each one. As more and more rules are used on a chart, the rate of false alarms will increase steadily. It is important to look for patterns based on one's knowledge of the process, but it is also important to realize that it is very easy to "see" patterns in points resulting from purely random variation. To discuss the supplementary rules, it is helpful to break the control chart into the regions shown in Fig. 7.1. The dashed lines separating the regions on each side of the control chart are equally spaced between the centerline and the control limits.

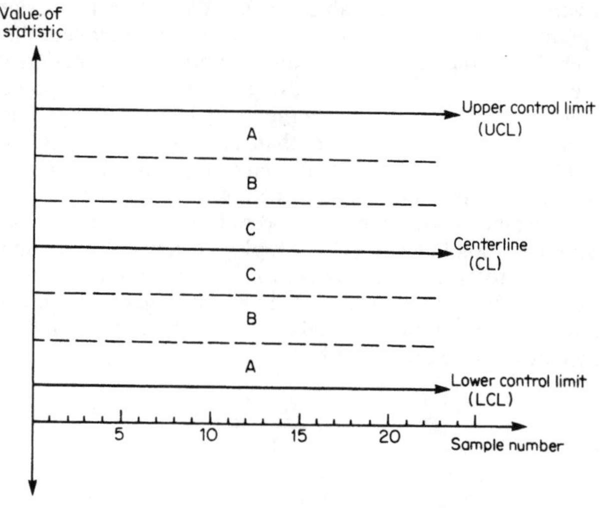

FIGURE 7.1 Control chart regions for interpretation of patterns.

Commonly applied runs rules, which are useful with an $\overline{X}$ chart in detecting a small sustained shift in the mean, are the following:

Rule 1. Two out of three consecutive points on the same side of the centerline and in region *A*.

Rule 2. Four out of five consecutive points on the same side of the centerline and in region *B* or beyond.

Rule 3. Eight consecutive points on the same side of the centerline.

Champ and Woodall (1987) show that these runs rules increase the sensitivity of the $\overline{X}$ chart to small shifts in the mean, even if the control limits of the $\overline{X}$ chart are narrowed to give an equivalent false alarm rate. The charts remain less sensitive, however, to small sustained shifts in the mean than the cumulative sum chart which is discussed in Sec. 7.3.7.

Other rules that are often recommended are the following:

Rule 4. Fifteen consecutive points on either side of the centerline, but all in region C.

This pattern would be desirable if coupled with an R chart showing a decrease in variability; otherwise the control limits may have been miscalculated or there may be a sampling problem. For example, this pattern would result if observations from two sources were combined for each sample and one source (gauge or inspector) gave consistently higher results than the other.

Rule 5. Eight consecutive points beyond region C on either side of the centerline.

This pattern, particularly if cyclical, could be due to regular differences between shifts or operators or due to regular changes in the environment.

Rule 6. Fourteen consecutive points alternating up and down.

This sawtooth pattern often results from overadjustment of the process by an operator who is responding to purely random variation.

Rule 7. Six consecutive points, either all increasing or decreasing.

This pattern could represent a trend due to tool wear or operator fatigue. Davis and Woodall (1988) show, however, that many trends in the underlying mean will be camouflaged by the random variation until a signal results from Rules 1-3 or from a point falling outside a control limit.

Rules 1-3 can also be used with an R chart, although somewhat more evidence is usually required to indicate a decrease in variability because the distribution of the sample range is skewed to the right. Some type of runs rule must be used to detect decreases in variability if the lower limit of the R chart is zero.

If larger values of the sample mean on the $\overline{X}$ chart tend to be paired with the larger values on the R chart, then this may indicate that the distribution of X is skewed to the right. Minor violations of the assumption of normality often do not cause problems with the use of the $\overline{X}$ and R charts, but the use of Eq. (7.10) to estimate the percentage of items outside specification limits becomes very dangerous.

Ideally, if the $\overline{X}$ chart exhibits cycles or other systematic fluctuation, then the source of the pattern should be found and removed. In some cases this may not be possible. For example, cycles may be due to temperature changes and these may not be controllable. As pointed out by Alwan and Roberts (1988) Montgomery and Mastrangelo (1991), Woodall and Faltin (1993), and Wardell, Moskowitz, and Plante (1994), the interpretation of standard control charts may not be meaningful if the data are autocorrelated. In such cases, the pattern should be modeled by a time series model (see Chap. 19), and a control chart should be applied to the residuals of the fitted model to detect assignable causes.

For more information on the interpretation of control chart patterns, refer to Western Electric's *Statistical Quality Control Handbook* (1956, pp. 149–180) or to Nelson (1985). Additional information coneerning autocorrelated data can be found in Lin and Adams (1996), Stinson and Mastrangelo (1996), and Vander Wiel (1996).

7.3.4 Using $\overline{X}$ and R Charts for Process Monitoring

After the process appears to be in a state of statistical control $\overline{X}$ and R charts can be used to monitor current production. Points outside the control limits are investigated, as well as any unusual patterns, such as those listed in the previous section.

Assignable causes are to be identified and eliminated as continued efforts are made to reduce the variability of the process. Reductions in variability will result in lower values for the control limits of the R chart and narrower limits on the $\overline{X}$ chart.

The performance of a basic $\overline{X}$ chart, without any supplementary rules, is often measured by its operating-characteristic function. The operating-characteristic value is the probability of not signaling out of control with a single sample when the underlying mean is $\mu = \mu_0 + \delta\sigma$. With control limits assumed to be established at $\mu_0 \pm 3\sigma/\sqrt{n}$, the operating characteristic value of the $\overline{X}$ chart is

$$\beta = \Phi(3 - \delta\sqrt{n}) - \Phi(-3 - \delta\sqrt{n}) \tag{7.11}$$

where again $\Phi(\cdot)$ represents the standard normal cumulative distribution function. As with all control charts, small shifts in the process are detected more quickly with larger sample sizes.

If supplementary rules or other types of charts are used, it is necessary to measure performance in terms of the average run length (ARL), where the run length is the number of samples required to obtain a signal. Under the assumptions of independence and a constant mean, the run length of the basic $\overline{X}$ chart is a geometric random variable with expected value

$$\text{ARL} = \frac{1}{1 - \beta}. \tag{7.12}$$

With control limits at $\mu_0 \pm 3\sigma/\sqrt{n}$, the probability of a false alarm is 0.0027 at each sampling stage, giving an average run length of 370.4. If the control limits common in British practice are used, $\mu_0 \pm 3.09\sigma/\sqrt{n}$, then the probability of a false alarm is 0.0020, giving an in-control average run length of 500. These run-length properties are calculated under the idealized conditions of normality and independence, but comparisons of the relative performance of control charts are not possible without these or similar assumptions.

To illustrate the use of $\overline{X}$ and R charts, consider the sample means and ranges shown in Table 7.2. These statistics were computer-generated using samples of size

TABLE 7.2 Sample Means and Ranges ($n = 4$)

Sample i	$\overline{X}_i$	R_i	Sample i	$\overline{X}_i$	R_i
1	0.03	3.05	16	0.65	3.16
2	0.09	1.41	17	–0.01	0.51
3	0.74	1.75	18	0.07	3.13
4	– 1.12	2.93	19	0.08	2.62
5	1.16	1.46	20	0.43	4.81
6	– 0.01	1.25	21	0.46	1.56
7	– 0.19	2.05	22	1.20	2.41
8	0.10	1.69	23	1.54	2.81
9	0.36	3.61	24	0.13	2.45
10	0.45	3.73	25	0.62	1.66
11	0.27	2.11	26	0.73	1.81
12	– 0.60	3.30	27	1.20	2.05
13	1.10	1.91	28	0.46	0.69
14	– 0.62	1.13	29	1.04	2.65
15	– 0.28	1.88	30	0.93	2.26

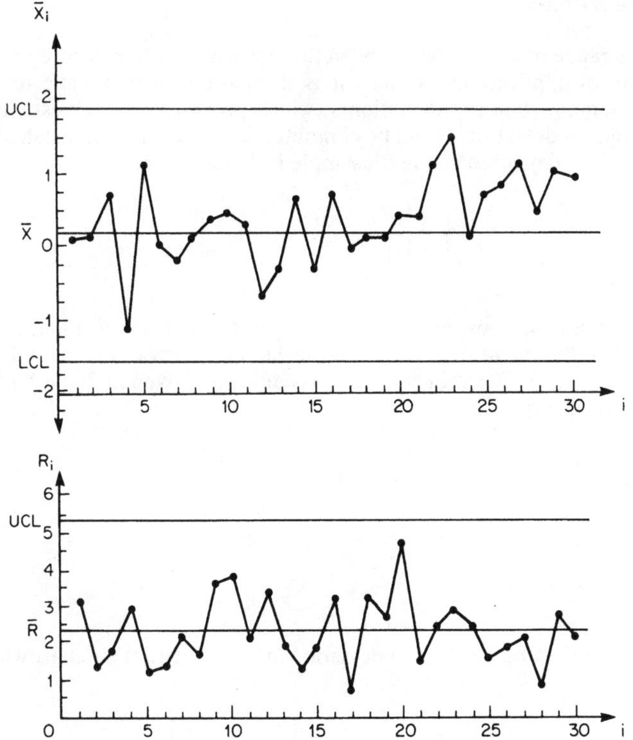

FIGURE 7.2 $\overline{X}$ and R charts for data in Table 7.2.

$n = 4$. The first 20 samples were from the $N(0, 1)$ distribution, while the last 10 were from $N(0.5, 1)$.

In this example the control limits for the $\overline{X}$ and R charts are based on the first 20 samples (Fig. 7.2). The control limits of the $\overline{X}$ chart are

$$\overline{\overline{X}} \pm A_2\overline{R}$$

$$0.135 \pm 0.729(2.3745)$$

$$(-1.596, 1.866).$$

For $n = 4$, the R chart has no lower control limit and the upper control limit is $D_4\overline{R} = 2.282(2.3745) = 5.419$. All of the first 20 points are inside the control limits for both the $\overline{X}$ and the R charts and no unusual patterns are present. Since the process appears to be in a state of statistical control, these control limits can be applied to monitor future production.

A shift in the process mean has occurred for samples 21 to 30. All 10 points on the $\overline{X}$ chart and on the R chart fall inside the control limits, but Rule 7 signals on sample 23 because six consecutive points on the $\overline{X}$ chart are all increasing.

7.3.5 The *S* Chart

The sample range has traditionally been the statistic used to measure variability in quality control applications because it is the easiest such statistic to calculate. Today the sample standard deviation s, which provides a better estimate of the process standard deviation σ, can be computed just as easily with a calculator. The sample standard deviation of the ith sample is defined to be

$$s_i = \left[\frac{1}{n-1}\sum_{j=1}^{n}(x_{ij}-\bar{x}_i)^2\right]^{1/2} \tag{7.13}$$

for $i = 1, 2, \ldots, m$.

Under the assumptions that the process is in statistical control and X is distributed $N(\mu, \sigma^2)$, the mean of s_i is $c_4\sigma$ and its standard error is $c_5\sigma$, $i = 1, 2, \ldots, m$, where $c_5 = (1 - c_4^2)^{1/2}$. The values of the constant c_4 are given in Table 7.1. An unbiased estimate of σ is

$$\hat{\sigma} = \frac{\bar{s}}{c_4} \tag{7.14}$$

where

$$\bar{s} = \frac{1}{m}\sum_{i=1}^{m} s_i. \tag{7.15}$$

The values of the sample standard deviation are plotted on the s chart with control limits set at

$$\bar{s} \pm \frac{3\bar{s}c_5}{c_4} \tag{7.16}$$

or

$$B_3\bar{s},\ B_4\bar{s} \tag{7.17}$$

where the values of the constants B_3 and B_4 are given in Table 7.1.

When an s chart is used, the control limits for the $\bar{X}$ chart become

$$\bar{\bar{X}} \pm \frac{3\bar{s}}{c_4\sqrt{n}} \tag{7.18}$$

or

$$\bar{\bar{X}} \pm A_3\bar{s} \tag{7.19}$$

where the values of A_3 are also given in Table 7.1.

As with the $\bar{X}$ and R charts, the control limits should be considered as trial limits until the process is in a state of statistical control. The control limits must be revised if assignable causes can be found for points falling outside the limits or for unusual patterns of points on the charts.

It is usually recommended that the s chart be used in place of the R chart if $n \geq 10$. The s chart is more efficient than the R chart for $n > 2$, but the increase in efficiency is small if n is small. For a detailed comparison of the efficiency of the s chart relative to the R chart, the reader is referred to Lowry, Champ, and Woodall (1995).

7.3.6 Control Charts for Individuals (X Chart)

In some cases each sample consists of a single observation, that is, $n = 1$. This situation can occur for a number of reasons. Sampling may be costly, the division of items into samples may be arbitrary, or each measurement may represent a batch, of relatively homogeneous material. If the manufacturing process is automated, then observations may be made on all items. In this case, however, one needs to be especially wary of autocorrelation among the observations.

The problem with having single observations, $X_1, X_2, X_3, \ldots, X_m$, is in estimating the process variability. The most common method of estimating σ is to use moving ranges of successive observations. Defining the ith moving range to be

$$MR_i = |X_i - X_{i-1}|, \qquad i = 2, 3, \ldots, m \tag{7.20}$$

then

$$\hat{\sigma} = \frac{\overline{MR}}{d_2} = 0.8865\overline{MR} \tag{7.21}$$

using d_2 for $n = 2$, where

$$\overline{MR} = \frac{1}{m-i}\sum_{i=2}^{m} MR_i. \tag{7.22}$$

The individual observations are plotted on the X chart with control limits

$$\overline{X} \pm 2.6595\overline{MR} \tag{7.23}$$

where

$$\overline{X} = \frac{1}{m}\sum_{i=1}^{m} X_i. \tag{7.24}$$

The moving ranges are plotted on the MR chart with upper limit $3.267\overline{MR}$ and no lower control limit. While use of the X and MR charts parallels that of the $\overline{X}$ and R charts, the moving ranges are correlated, so the interpretation of patterns on the MR chart is difficult. The reader is referred to Rigdon, Cruthis, and Champ (1994) for more details.

7.3.7 The Cumulative Sum Chart

The cumulative sum (CUSUM) chart provides an alternative to the use of the $\overline{X}$ chart with supplementary runs rules for detecting a shift in the process mean. First, $\overline{X}$ and R charts should be used to put the process in a state of statistical control on target at $\mu = \mu_0$ and to obtain an accurate estimate of σ. The CUSUM chart can then be represented either by a V mask or by the simultaneous use of two decision interval charts.

To use the V mask representation, the cumulative sums

$$S_i' = \sum_{j=1}^{i} (\bar{x}_j - \mu_0), \qquad i = 1, 2, 3, \ldots \tag{7.25}$$

are plotted against the sample number. The point A of the V mask as shown in Fig. 7.3 is placed on the most recent value of the cumulative sum. A shift in the process mean is indicated by any point falling outside an arm of the V mask. It is not the point that falls outside the mask that is suspect, but the pattern of increases or decreases in the subsequent points. The slope of the points over a specified time period measures the average deviation from the target value. The plot in Fig. 7.3 is based on the data in Table 7.2 using $\mu_0 = 0$ and $\sigma = 1$. The V mask is plotted for the 27th sample because this V mask is the first to signal.

Referring to Fig. 7.3, the length d is the lead distance of the mask and θ is the angle of each arm of the mask from the horizontal. To be used easily, the V mask can be drawn on a transparency. It is usually recommended that the chart be scaled such that the unit distance on the horizontal axis corresponds to the length used to represent a distance of $w = 2\sigma/\sqrt{n}$ on the vertical axis.

The V-mask representation has two major disadvantages. First, the cumulative sums in Eq. (7.25) will tend to wander considerably, perhaps off a sheet of graph paper, even if the process mean is on target. Second, it is not clear how many of the previous values of the cumulative sum must be remembered.

To overcome these and other disadvantages, we recommend use of the following two cumulative sums to represent the CUSUM procedure:

$$S_i = \max\,(0, S_{i-1} + \bar{X}_i - \mu_0 - k) \tag{7.26}$$

$$T_i = \min\,(0, T_{i-1} + \bar{X}_i - \mu_0 + k) \tag{7.27}$$

i = 1, 2, 3, ..., where the initial values are often $S_0 = T_0 = 0$. The reference value k is usually $\Delta/2$, where Δ is the smallest change in the mean considered important enough to be detected quickly. The CUSUM chart signals that the process is out of control as soon as $S_i > h$ or $T_i < -h$, where $h \geq 0$ is referred to as the decision

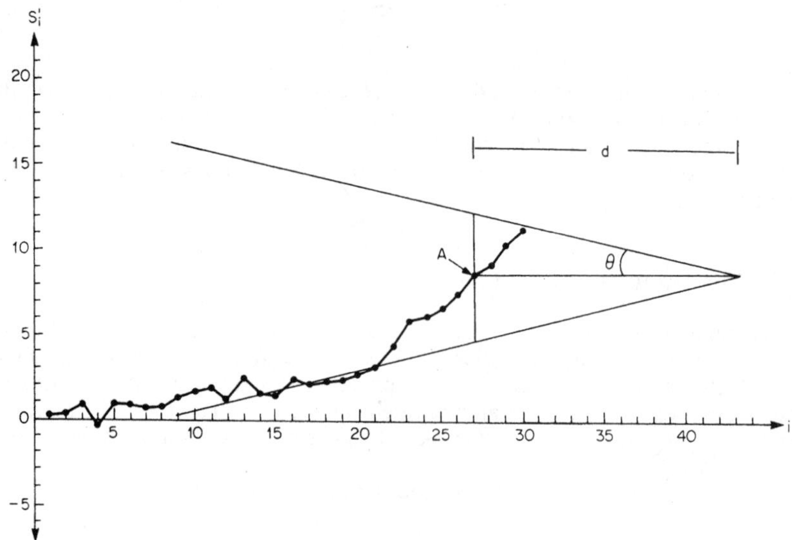

FIGURE 7.3 *V*-mask representation of CUSUM chart.

interval. It is convenient to define k^* and h^* such that $k = k^*\sigma/\sqrt{n}$ and $h = h^*\sigma/\sqrt{n}$. The basic Shewhart $\overline{X}$ chart is a special case of the CUSUM chart with $k^* = 3$ and $h^* = 0$.

The sums in Eq. (7.26) detect increases in the process mean, whereas those in Eq. (7.27) detect decreases. The use of the two decision interval rules is equivalent to the use of a V mask if

$$k = w \tan \theta \tag{7.28}$$

and

$$h = wd \tan \theta. \tag{7.29}$$

As an example, consider the case for which $h^* = 8$, $k^* = 0.5$, and $\sigma/\sqrt{n} = 0.5$. It can be assumed, without loss of generality, that $\mu_0 = 0$. Thus we have $h = 4$ and $k = 0.25$ for the decision interval charts. With the recommended scaling, the corresponding V mask has $\theta \approx 14°$ and $d = 16$. The graph of the decision interval charts for the data in Table 7.2 is shown in Fig. 7.4, and the corresponding V mask is the one illustrated in Fig. 7.3.

The easiest way to design a CUSUM chart is to determine first $k = \Delta/2$. Then k^* can be used with ARL curves such as those given by Gan (1991) to determine h^* to give the desired in-control average run length. The decision interval is then $h = h^* \sigma/\sqrt{n}$, and the values of d and θ for the corresponding V mask, if desired, can be determined from Eqs. (7.28) and (7.29).

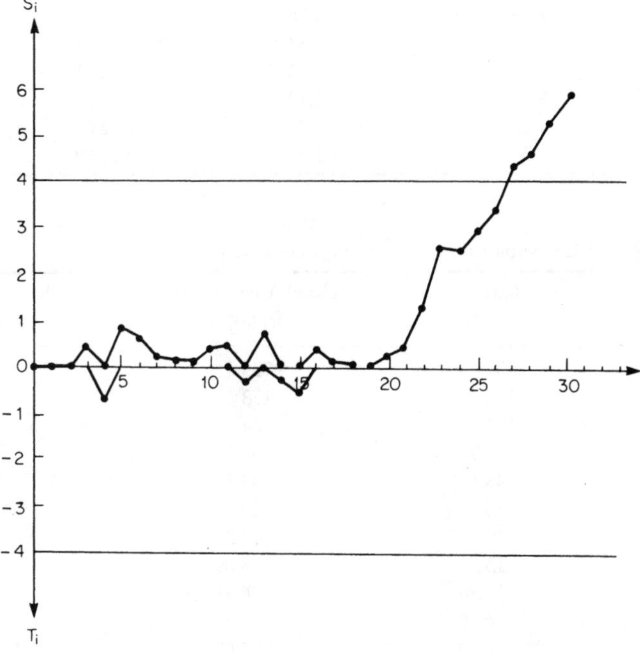

FIGURE 7.4 Decision interval representation of CUSUM chart.

The most commonly used CUSUM charts have $k^* = 0.5$ and $h^* = 4$ or 5. The average run lengths of these charts as given by Crosier (1986) are shown in Table 7.3, where $d = \sqrt{n}\,(\mu - \mu_0)/\sigma$ is the shift in the mean measured in units of the standard error.

A comparison of the ARL performance of the three types of charts is given in Table 7.4. The charts are designed so that their false alarm rates are almost the same. The Shewhart $\overline{X}$ chart is most effective for detecting large shifts in the mean whereas the CUSUM chart is the most effective for smaller shifts.

Various modifications have been suggested for improving the performance of the CUSUM procedure. Lucas and Crosier (1982) recommended using head start values $S_0 = h/2$ and $T_0 = -h/2$ to detect initial out-of-control conditions more quickly. Lucas (1982) adds a Shewhart control limit to the CUSUM procedure, which signals for large deviations from the target even if the CUSUM chart itself does not produce a signal. This modification improves the performance of the CUSUM chart for large shifts in the process mean.

TABLE 7.3 ARL Values for Two CUSUM Charts ($k^* = 0.5$)

Shift d	$h^* = 4$	$h^* = 5$
0.00	168	465
0.25	74.2	139
0.50	26.6	38.0
0.75	13.3	17.0
1.00	8.38	10.4
1.50	4.75	5.75
2.00	3.34	4.01
2.50	2.62	3.11
3.00	2.19	2.57
4.00	1.71	2.01
5.00	1.31	1.69

TABLE 7.4 ARL Comparison for Three Types of Charts

Shift d	$\overline{X}$ chart, $\mu_0 \pm 2.84\sigma/\sqrt{n}$	Usual $\overline{X}$ chart with Rule 1	CUSUM, $k^* = 0.5$, $h^* = 4.29$
0.0	222	225	226
0.2	188	178	116
0.4	126	104	44
0.6	77.9	57.9	21.3
0.8	48.1	33.1	12.9
1.0	30.4	20.0	8.96
1.2	19.8	12.8	6.83
1.4	13.3	8.69	5.52
1.6	9.30	6.21	4.64
1.8	6.70	4.66	4.01
2.0	4.99	3.65	3.54
3.0	1.77	1.68	2.37

Although most often used with sample means, CUSUM procedures can also be used to monitor variability or they can be used with attribute data. For more information on these applications, the reader is referred to Bourke (1991), Chang and Gan (1995), and Gan (1993).

7.3.8 Exponentially Weighted Moving-Average Chart

The exponentially weighted moving-average (EWMA) chart is another chart that can be used effectively to monitor the process mean. This chart is sometimes referred to as the geometric moving-average chart.

It is assumed, as with the CUSUM chart, that the process is in statistical control with $\mu = \mu_0$. The EWMA chart is based on the moving averages

$$M_i = \lambda \overline{X}_i + (1 - \lambda) M_{i-1} \tag{7.30}$$

$i = 1, 2, 3, \ldots$, where $M_0 = \mu_0$ and λ $(0 < \lambda \leq 1)$ is a specified constant. The moving averages can also be expressed as a weighted average of past sample means,

$$M_i = \sum_{j=0}^{i-1} \lambda(1 - \lambda)^j \overline{X}_{i-j} + (1 - \lambda)^i \mu_0 \tag{7.31}$$

$i = 1, 2, 3, \ldots$. It can be seen from Eq. (7.31) that the weights assigned to the sample means decrease geometrically with age.

The EWMA chart signals that a shift in the process mean has occurred if M_i falls outside control limits set at

$$\mu_0 \pm 3[\mathrm{Var}\,(M_i)]^{1/2} \tag{7.32}$$

where

$$\mathrm{Var}\,(M_i) = \frac{\sigma^2}{n} \frac{\lambda}{2 - \lambda} [1 - (1 - \lambda)^{2i}] \tag{7.33}$$

$i = 1, 2, 3, \ldots$. These control limits very quickly approach the limiting values

$$\mu_0 \pm 3\sigma \left[\frac{\lambda}{(2 - \lambda)n} \right]^{1/2}. \tag{7.34}$$

More weight is given to the present sample mean for larger values of λ. In fact, the $\overline{X}$ chart is a special case of the EWMA chart with $\lambda = 1$. Montgomery, Gardiner, and Pizzano (1987) recommend values of λ in the range of $0.05 \leq \lambda \leq 0.5$, with smaller values of λ being more effective in detecting smaller shifts in the mean. Lucas and Saccucci (1990) show that the statistical performance of the EWMA chart is very close to that of the CUSUM chart. Yashchin (1987) points out that the EWMA statistic can be very near one control limit when a shift in the other direction occurs, resulting in the possibility of very slow detection of the shift. This problem can be partially alleviated by using a Shewhart rule in combination with the EWMA chart, an enhancement discussed in Lucas and Saccucci (1990).

As an example, consider again the data in Table 7.2. The plot of the EWMA chart with $\lambda = 0.2$ is shown in Fig. 7.5. This EWMA chart signals on the 23rd sample that a shift in the process mean has occurred.

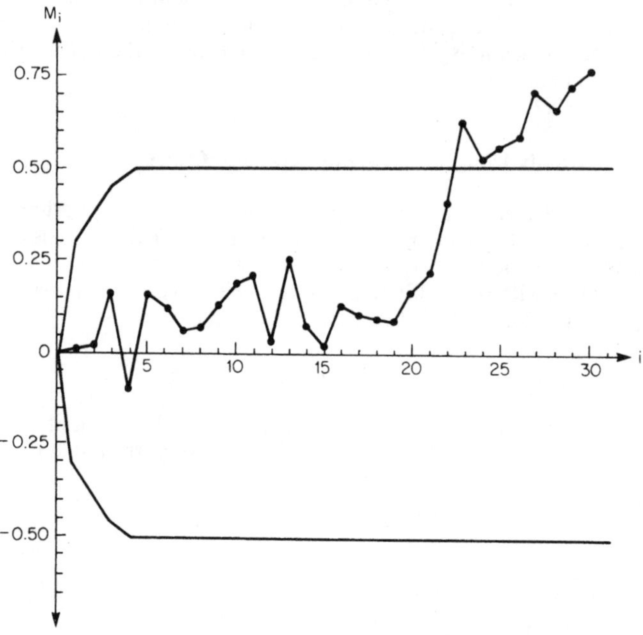

FIGURE 7.5 Exponentially weighted moving-average chart.

7.4 CONTROL CHARTS FOR ATTRIBUTES

7.4.1 Introduction

This section describes control charts for attribute data. For a more complete review of this topic, see Woodall (1977). We will be interested in either the number of nonconforming items in a sample or the number of nonconformities per item. A nonconforming item is defined as one that does not meet certain well-defined standards. If these standards correspond to specification limits, it is better to make a continuous measurement, if possible, and not use attribute data. The use of continuous observations leads to tighter control of the process, smaller sample sizes, and information on the degree and direction of any nonconformance. In some cases, however, it may be difficult or impossible to make a measurement on a continuous variable.

The use of p charts to monitor the percentage of nonconforming items is described in Sec. 7.4.2. In Sec. 7.4.3 the use of c charts and u charts is discussed for monitoring the number of nonconformities per item and the average number of nonconformities per unit, respectively. For each type of chart, the initial control limits are trial limits based on m preliminary samples. As with the $\overline{X}$ and R charts, points outside the control limits or unusual patterns lead to searches for assignable causes. Control limits are updated and used for future production after the process appears to be in a state of statistical control.

7.4.2 Control of Proportion Nonconforming (*p* Chart)

It is assumed that each item can be classified as either conforming or nonconforming. In some cases an item may be functional or nonfunctional, such as a semiconductor chip. In other cases an item may be classified as nonconforming if it does not meet specification limits. This practice might be dangerous, however, since a problem may not be detected until after it has become serious. In such a case, it is better to classify items as nonconforming on the basis of limits tighter than the specification limits. Often go–no–go gauges are used for this classification of items.

A p chart is used to monitor the proportion of nonconforming items. If items can be nonconforming in several different ways, these can be grouped together for one p chart. However, then the chart is harder to interpret since it does not detect changes in the proportion of each type of nonconformity. It is better to use separate charts, unless the individual proportions are very small, for example, less than 0.5%.

Suppose we have m preliminary samples of possibly varying sample sizes $n_1, n_2, \ldots, n_m$. The number of nonconforming items in the ith sample is D_i, $i = 1, 2, \ldots, m$, so the proportion of nonconforming items in the ith sample is

$$\hat{p}_i = \frac{D_i}{n_i}, \quad i = 1, 2, \ldots, m. \tag{7.35}$$

Each sample may represent the total production over a time period or during a production run.

The process is in statistical control if the underlying probability of a nonconforming item p is constant over time and nonconforming items occur independently. Under these conditions, D_i is a binomial random variable with parameters n_i and p. It follows that the mean of $\hat{p}_i$ is p and its variance is $p(1-p)/n_i$, $i = 1, 2, \ldots, m$. Since the value of p is usually unknown, it can be estimated by the overall proportion of nonconforming items,

$$\overline{p} = \frac{\sum_{i=1}^{m} D_i}{\sum_{i=1}^{m} n_i}. \tag{7.36}$$

For the p chart, the values of $\hat{p}_i$, $i = 1, 2, \ldots, m$, are plotted against the sample number with control limits set at

$$\overline{p} \pm 3\left[\frac{\overline{p}(1-\overline{p}}{n_i}\right]^{1/2}. \tag{7.37}$$

These limits will vary with the sample size, since they are drawn closer together for the larger sample sizes. The average sample size

$$\overline{n} = \frac{1}{m}\sum_{i=1}^{m} n_i$$

can be used in place of n_i in Eq. (7.37) to determine control limits if the sample sizes are relatively close, for example within 15% of the average sample size.

The sample sizes for p charts are much larger than those required for the $\overline{X}$ and R charts. Montgomery (1996) gives several rules which have been suggested for determining the sample size. One rule is to use a value of n so that the p chart will have a positive lower boundary. With this criterion we must have

$$n > \frac{9(1-\hat{p})}{\hat{p}}$$

where $\hat{p}$ is some estimate of the true in-control value of p.

The goal of continuous improvement requires a continuing effort to reduce the value of p. As progress is made, both control limits of the p chart will move closer to zero. A disadvantage of the p chart is that larger and larger samples are needed as the value of p decreases. Downward shifts in the p chart may indicate improved quality, but one must also be careful of inspection errors and weakening standards.

If the sample size is n for each sample, then a chart equivalent to the p chart can be based on the number of nonconforming items per sample, $D_1, D_2, \ldots, D_m$. This chart is referred to as an np chart and it has control limits at

$$n\overline{p} \pm 3[n\overline{p}(1-\overline{p})]^{1/2}. \tag{7.38}$$

7.4.3 Control of the Number of Nonconformities (*c* Chart and *u* Chart)

If a nonconforming item is defined to be one that has at least one nonconformity, then in some cases the proportion of nonconforming items will be close to 100%. In such cases it is more reasonable to count the number of nonconformities per item, such as the number of blemishes in a roll of cloth or the number of minor automobile defects.

Two cases are considered. In the first case the number of items or the amount of material inspected remains constant for each sample and the chart is based on the number of nonconformities in this inspection unit. This leads to a c chart. In the second case the amount of material inspected varies, or the average number of nonconformities is to be expressed in terms of some unit different from the inspection unit. In this second case a u chart is required.

In the first case suppose that the number of nonconformities observed in the ith inspection unit is c_i, $i = 1, 2, \ldots, m$. It is assumed that the number of nonconformities can be represented by a Poisson random variable. If the process is in statistical control, then the parameter of the Poisson distribution remains constant at a value c. This implies that the mean and the variance of c_i are both c, $i = 1, 2, \ldots, m$. For the c chart, the number of nonconformities is plotted against the sample number with control limits set at

$$\overline{c} \pm 3\sqrt{\overline{c}} \tag{7.39}$$

where

$$\overline{c} = \frac{1}{m}\sum_{i=1}^{m} c_i. \tag{7.40}$$

The u chart can be considered to be a generalization of the c chart. The observations are the average number of nonconformities per unit,

$$u_i = \frac{c_i}{n_i} \tag{7.41}$$

where c_i is the number of nonconformities observed in an examination of n_i items or units of material, $i = 1, 2, \ldots, m$. In this case the area of opportunity for nonconformities can vary over time, such as different areas of sheet metal or volumes of chemical compounds. Note that n_i does not have to be integer-valued. For the u chart, the values of $u_1, u_2, \ldots, u_m$ are plotted against the sample number with varying control limits set at

$$\overline{u} \pm 3\sqrt{\frac{\overline{u}}{n_i}}, \qquad i = 1, 2, \ldots, m \tag{7.42}$$

where

$$\overline{u} = \frac{\sum_{i=1}^{m} c_i}{\sum_{i=1}^{m} n_i}.$$

As with all the other control charts discussed in this chapter, the c chart and the u chart are used with future production once the process appears to be in a state of statistical control. Continuing efforts should be made to improve the process by reducing the number of nonconformities.

7.5 OTHER TOPICS

7.5.1 Allowing for Variation in the Process Mean

Thus far the use of the $\overline{X}$ chart and other control charts for monitoring the process mean assumes a single target value for the mean. Within this framework, any shift in the mean from the target value is to be detected quickly. In some cases this definition of control can be unrealistically tight and a number of alternatives have been suggested. The modified control chart proposed by Hill (1956) and the acceptance control chart introduced by Freund (1957) are both used to widen the control limits of the $\overline{X}$ chart when the value of the capability index Cp is quite large. In this case some variation in the process mean can occur without a significant increase in the percentage of items falling outside specification limits. The acceptance control chart is the more general of the two charts since it considers not only the false alarm rate, but also the power of the chart to detect shifts for which the corresponding proportion of nonconforming items exceeds a specified limit. Wadsworth, Stephens, and Godfrey (1998) offer a complete description of these techniques. With the goal of continuous reduction in the variability of the process, however, even small shifts in the mean will not be tolerated if the corresponding assignable cause can be identified and removed. This is true even if the small shift does not cause the percentage of nonconforming items to increase measurably. Deming (1986, p. 369) warns strongly against the use of modified control limits, evidently for this reason.

The approach of Freund (1957) is more general than commonly stated since the widened control limits do not have to be tied to the specification limits. Control limits should be widened when the specification of a single target value is too restrictive and some of the between-sample variability is considered to be due to uncontrollable random causes, not assignable causes. For example, in chemical batch processes with long delays between batches, there will be unavoidable differences between batches. In order to prevent control charts from signaling the presence of this unavoidable variation, it is necessary to allow some variation in the mean. Hahn and Cockrum (1987), Wheeler and Chambers (1986, pp. 224–229), and Woodall and Thomas (1995) solve this problem, in part by using the sample means representing batches in an X chart.

Tool wear is another example in which it is necessary to allow some changes in the mean to occur. If tool wear is expected, then the control limits of the $\overline{X}$ chart can be modified to change over time to reflect the trend in the underlying mean. Grant and Leavenworth (1996, pp. 386–391) discuss this application in detail.

7.5.2 Multivariate Control Charts

It is common to take multiple measurements on items, especially if the measurement process is automated. As the number of quality characteristics monitored by separate standard control charts increases, the rate of false alarms increases steadily. The use of a multivariate control chart is recommended when several correlated variables are being measured. The use of multivariate control charts is beyond the scope of this chapter and the reader is referred to Chap. 21, Jackson (1985), Alt (1984), Wierda (1994), and Adams (1994).

7.5.3 Economic Design of Control Charts

The control charts discussed in this chapter have been designed and compared on the basis of their statistical performance. The Shewhart-type charts have been constructed with standard 3σ limits. An alternative is to use the exact distribution of the chart statistic to determine the sample size and the control limits to obtain a specified probability of a false alarm and a specified power. This use of the distribution of the statistic provides greater flexibility in chart design. A different approach to determining the time interval between samples, the sample size, and the decision rule for each type of chart is based on an economic model. In the economic approach, a dozen or so cost and time parameters associated with the process must be specified. These include the cost of sampling, the cost of defective items, the average time before the process shift, the expected change in the quality level, and so forth. Based on the assumptions of the economic model, the control chart design is determined which maximizes the expected profit per time period. Ho and Case (1994) and Lorenzen and Vance (1986) provide detailed information on the economic approach to the design of control charts. Woodall (1987) points out the importance of checking the statistical performance of the economically designed charts. Saniga (1989) presents a reasonable compromise between economic and statistical design methods.

7.5.4 Control Charts Based on Standards Given

Many references suggest that in some cases one may want to detect shifts in the process from specified parameter values without the analysis of any preliminary data.

For data on continuous variables the mean and the standard deviation of the quality characteristic may be given to be μ_0 and σ_0. Then the control limits of the $\bar{X}$,R, and S charts are

$$\mu_0 \pm A\sigma_0$$
$$D_1\sigma_0,\ D_2\sigma_0$$
$$B_5\sigma_0,\ B_6\sigma_0$$

where the control chart constants $A = 3/\sqrt{n}$, $D_1 = d_2 - 3d_3$, $D_2 = d_2 + 3d_3$, $B_5 = c_4 - 3c_5$, and $B_6 = c_4 + 3c_5$ are given in Table 7.1.

The control limits of the p, c, and u charts become

$$p_0 \pm 3\left[\frac{p_0(1-p_0)}{n_i}\right]^{1/2}, \quad i = 1, 2, 3,...$$

$$c_0 \pm 3\sqrt{c_0}$$

$$u_0 \pm 3\sqrt{\frac{u_0}{n_i}}, \quad i = 1, 2, 3,...$$

where p_0, c_0, and u_0 are the specified parameter values.

If the process is not in a state of statistical control at the specified parameter values, then a signal from the control chart may not be meaningful. This is why the description of these charts is usually accompanied by strong warnings. It is not recommended that control charts be used with standards given in hopes that responding to signals will eventually result in the standards being achieved. Such an approach is ineffective because the charts cannot distinguish between assignable causes and random causes.

7.6 CONCLUSION

This chapter contains an introduction to the principles of statistical process control. For more information, the reader is referred to the publications listed in the References. The book by Wheeler and Chambers (1986) contains many examples, good practical advice, and detailed case studies. The textbooks by Montgomery (1996), Duncan (1986), Ryan (1989), and Wadsworth, Stephens, and Godfrey (1999) are also highly recommended. The American Society for Quality (ASQ) through its Quality Press publishes many excellent books on the subject of quality. For details, the reader may contact the ASQ at 611 E. Wisconsin Avenue, P.O. Box 3005, Milwaukee, WI 53201-3005.

REFERENCES

Adams, B. M.: "The Multivariate Control Web," *Qual. Eng.*, vol. 6, no. 4, pp. 533–546, 1994.

Alt, F. B.: "Multivariate Control Charts," in *Encyclopedia of Statistical Sciences*, S. Kotz and N. L. Johnson, Eds., Wiley, New York, 1984.

Alwan, L. C., and H. V. Roberts: "Time Series Modeling for Statistical Process Control," *J. Bus. Econ. Stat.*, vol. 6, pp. 87–95, 1988.

ASQC Automotive Division: *Statistical Process Control Manual*, Am. Soc. Quality Control, Milwaukee, WI, 1986.

Bourke, P. D.: "Detecting a Shift in Fraction Nonconforming Using Run-Length Control Charts with 100% Inspection," *J. Qual. Technol.*, vol. 23, pp. 225–238, 1991.

Box, G. E. P., G. M. Jenkins, and G. C. Reinsel: *Time Series Analysis — Forecasting and Control*, 3d ed., Prentice-Hall, Englewood Cliffs, NJ, 1994.

Champ, C. W., and W. H. Woodall: "Exact Results for Shewhart Control Charts with Suplementary Runs Rules," *Technometrics*, vol. 29, pp. 393–399, 1987.

Chang, T. C., and F. F. Gan: "A Cumulative Sum Control Chart for Monitoring Process Variance," *J. Qual. Technol.*, vol. 27, pp. 109–119, 1995.

Crosier, R. B.: "A New Two-Sided Cumulative Sum Quality Control Scheme," *Technometrics*, vol. 28, pp. 187–194, 1986.

Davis, R. B., and W. H. Woodall: "Performance of the Control Chart Trend Rule under Linear Shift," *J. Qual. Technol.*, vol. 20, pp. 260–262, 1988.

Deming, W. E.: *Quality, Productivity and Competitive Position*, Center for Advanced Engineering Study, M.I.T., Cambridge, MA, 1982.

_____: *Out of the Crisis*, Center for Advanced Engineering Study, M.I.T., Cambridge, MA, 1986.

Duncan, A. J.: *Quality Control and Industrial Statistics,* 5th ed., Richard D. Irwin, Homewood, IL, 1986.

Feigenbaum, A. V.: *Total Quality Control*, McGraw-Hill, New York, 1983.

Freund, R. A.: "Acceptance Control Charts," *Ind. Qual. Contr.*, vol. 14, pp.13–23, 1957.

Gan, F. F.: "An Optimal Design of CUSUM Control Charts," *J. Qual. Technol.*, vol. 23, pp. 279–286, 1991.

_____: "An Optimal Design of CUSUM Control Charts for Binomial Counts," *J. Appl. Stat.*, vol. 20, pp. 445–460, 1993.

Garvin, D. A.: "Quality on the Line," *Harvard Bus. Rev.*, pp. 65–75, 1983.

Grant, E. L., and R. S. Leavenworth: *Statistical Quality Control*, McGraw-Hill, New York, 1996.

Hahn, G. J., and M. B. Cockrum: "Adapting Control Charts to Meet Practical Needs: A Chemical Processing Application," *J. Appl. Stat.*, vol. 14, pp. 33–50, 1987.

Hill, D.: "Modified Control Limits," *Appl. Stat.,* vol. 5, pp. 12–19, 1956.

Ho, C., and K. E. Case: "Economic Design of Control Charts: A Literature Review for 1981–1991," *J. Qual. Technol.*, vol. 26, pp. 39–53, 1994.

Hu, J. S. and C. Roan, "Change Patterns of Time Series-Based Control Charts," *J. Qual. Technol.*, vol. 28, pp. 302–311, 1996.

Ishikawa, K: *What Is Total Quality Control?— The Japanese Way*, Prentice-Hall, Englewood Cliffs, NJ, 1985.

Jackson, J. E.: "Multivariate Quality Control," *Commun. Stat. Theory Methods*, vol. 14, pp. 2657–2688, 1985

Johnson, C. D.: *Process Control Instrumentation Technology*, Wiley, New York, 1982.

Juran, J. M.: *Managerial Breakthrough,* McGraw-Hill, New York 1964.

Kacker, R. N.: "Off-Line Quality Control, Parameter Design, and Taguchi Method," *J. Qual. Technol.*, vol. 17, pp. 176–188, 1985.

Klaus, L. A.: "Quality Progress' 14th Annual QA/QC Software Directory," *Quality Progress*, vol. 30, no. 4, pp. 31–66, 1997.

Lin, S. W. and B. M. Adams: "Combined Control Charts for Forecast-Based Monitoring Schemes," *J. Qual. Technol.*, vol. 28, pp 289–301, 1996.

Lorenzen, T. J., and L. C. Vance: "The Economic Design of Control Charts: A Unified Approach," *Technometrics*, vol. 28, pp. 3–10, 1986.

Lowery, C. A., C. W. Champ, and W. H. Woodall: "Performance of Control Charts for Monitoring Process Variation," *Communications in Statistics —Simulation and Computation*, vol. 24, pp. 409–437, 1995.

Lucas, J. M. "Combined Shewhart–CUSUM Quality Control Schemes," *J. Qual. Technol.*, vol. 14, pp. 51–59, 1982.

_____ and R. B. Crosier: "Fast Initial Response for CUSUM Quality Control Schemes," *Technometrics*, vol. 24, pp. 199–205, 1982.

_____ and M. S. Saccucci: "Exponentially Weighted Moving Average Control Schemes: Properties and Enhancements," (with discussion), *Technometrics*, vol. 32, 1990.

MacGregor, J. F.: "Interfaces between Process Control and On-Line Statistical Process Control," *AIChE Comput. Sys. Technol. Div. Commun.*, vol. 10, pp. 9–20, 1987.

Montgomery, D. C.: Introduction to Statistical Quality Control, 2d ed., Wiley, New York, 1991.

_____: "The Use of Statistical Process Control and Design of Experiments in Product and Process Improvement," *IIE Transactions*, vol. 24, pp. 4–17, 1992.

_____ and Mastrangelo, C. M.: "Some Statistical Process Control Methods for Autocorrelated Data," (with discussion), *J. Qual. Technol.*, vol. 23, pp. 179–204, 1991.

_____, J. S. Gardiner, and B. A. Pizzano: "Statistical Process Control Methods for Detecting Small Process Shifts," in *Frontiers in Statistical Process Control*, vol. 3, H.-J. Lenz, G. B. Wetherill, and P.-Th. Wilrich, Eds., Physica-Verlag, Heidelberg, West Germany, 1987, pp. 163–178.

_____ and W. H. Woodall (co-editors): "A Panel Discussion on Statistically-Based Process Monitoring and Control, "*J. Qual. Technol.*, vol. 29, pp. 121–162, 1997.

Nelson, L. S.: "Interpreting Shewhart $\overline{X}$ Control Charts," *J. Qual. Technol.*, vol. 17, pp. 114–116, 1985.

Rigdon, S. E., E. N. Cruthis, and C. W. Champ: "Design Strategies for Individuals and Moving Range Control Charts," *J. Qual. Technol.*, vol. 26, pp. 274–287, 1994.

Rodriguez, R. N.: "Recent Developments in Process Capability Analysis," *J. Qual. Technol.*, vol. 24, pp. 176–187, 1992 .

Rosander, A. C.: *Applications of Quality Control in the Service Industries*, Dekker, New York, 1985.

Ryan, T. P.: "*Statistical Methods for Quality Improvement*, Wiley, New York, 1989.

Saniga, E.: "Economic Statistical Control Chart Designs with an Application to $\overline{X}$ and R Charts," *Technometrics*, vol. 31, pp. 313–320, 1989.

Shewhart, W. A.: *Economic Control of Quality of Manufactured Product*, Van Nostrand, New York, 1931; available from Am. Soc. *Quality Control*, Milwaukee, WI.

Stinson, W. A. and C. M. Mastrangelo: "Monitoring Serially Dependent Processes with Attribute Data," *J. Qual. Technol.*, vol. 28, pp. 279–288, 1996.

Sullivan, L. P.: "Reducing Variability: A New Approach to Quality," *Qual Prog.*, vol. 17, pp. 15–21, 1984.

Vander Wiel, S.: "Monitoring Processes that Wander Using Integrated Moving Average Models," *Technometrics*, vol. 38, pp. 139–151, 1996.

Wadsworth, H. M., K. S. Stephens, and A. B. Godfrey: *Modern Methods for Quality Control and Improvement*, Wiley, New York, 1986.

Wardell, D. G., H. Moskowitz, and R. D. Plante; "Run Length Distributions of Special-Cause Charts for Correlated Data," (with discussion), *Technometrics*, vol. 36, no. 1, pp. 3–27, 1994.

Western Electric Corp.: *Statistical Quality Control Handbook*, Select Code 700–444, 1956; available from AT&T, P.O. Box 19901, Indianapolis, IN 46219.

Wheeler, D. J.: *Advanced Topics in Statistical Process-The Power of Shewhart's Charts*, SPC Press, Knoxville, TN, 1995.

Wheeler, D. J., and D. S. Chambers: *Understanding Statistical Process Control*, Statistical Process Controls, Inc., Knoxville, TN., 1986.

Wierda, S. J.: "Multivariate Statistical Process Control–Recent Results and Directions for Future Research," *Statistica Neerlandica*, vol. 48, pp. 147–168, 1994.

Woodall, W. H.: "Conflicts between Deming's Philosophy and the Economic Design of Control Charts," in *Frontiers in Statistical Process Quality Control*, vol. 3, H.-J. Lenz, G. B. Wetherill and P.-Th. Wilrich, Eds., Physica-Verlag, Heidelberg, West Germany, 1987, pp. 242–248.

_____: "Control Charting Based on Attribute Data: Bibliography and Review," *J. Qual. Technol.*, vol. 29, pp. 172–183, 1997.

_____ and B. M. Adams: "The Statistical Design of CUSUM Charts," *Qual. Eng.*, vol. 5, no. 4, pp. 559–570, 1993.

_____ and F. W. Faltin: "Autocorrelated Data and SPC," ASQC *Statistics Division Newsletter*, vol. 13, no. 4, pp. 18–21, 1993.

_____ and E. V. Thomas: "Statistical Process Control with Several Components of Common Cause Variability," *IIE Trans.*, vol. 27, pp. 328–332, 1995.

Yaschin, E.: "Some Aspects of the Theory of Statistical Control Schemes," *IBM J. Res. Dev.*, vol. 31, pp. 199–205, 1987.

_____: "Statistical Control Schemes: Methods, Applications and Generalizations," *Intern. Stat. Rev.*, vol. 61, pp. 41–66, 1993.

CHAPTER 8
ACCEPTANCE SAMPLING

Harrison M. Wadsworth, Jr.
Georgia Institute of Technology, Atlanta, GA

8.1 INTRODUCTION

This chapter contains procedures that may be used to select and apply sampling inspection plans to determine if a lot or batch meets specified quality requirements. Its purpose is to do the following:

- Provide guidance to the user when selecting a sampling procedure that will meet specified risk requirements
- Enable the user to develop a sampling plan for individual items in a scientific manner

The sampling plans in this chapter apply to inspection management programs for receiving, in-process, or final inspection of production items or for any other purpose for which acceptance sampling is required. The chapter emphasizes attribute sampling procedures for single sampling. Single sampling plans are the easiest to use and will probably be used most often. If average sample size is a critical consideration, item-by-item sequential sampling will result in the most savings and is thus recommended. In terms of average sample size, double and multiple sampling plans are superior to single plans but not as good as sequential plans. They will therefore not be discussed in this chapter. For a detailed discussion of other acceptance sampling procedures see Schilling (1982), Duncan (1986), or Wadsworth, Stephens, and Godfrey (1998).

Sampling procedures for attributes may be used when the extent of nonconformity is expressed in terms of either proportion (or percent) of nonconforming items of the number of nonconformities per item (or per 100 items) when each item may have more than one nonconformity.

The sampling plans are based on the assumption that nonconformities occur randomly and with statistical independence. If it is known that one nonconformity in an item could be caused by a condition also likely to cause other nonconformities, the items should be considered just as conforming or not and multiple nonconformities should be ignored.

Plans are included that may be used for single, isolated lots, or for a small number of lots. Other procedures are included for a continuing series of lots that are sufficient to allow switching rules to be applied. The latter procedures provide the following:

- An automatic protection to the consumer if a deterioration in quality is detected (by switching to tightened inspection)
- An incentive to reduce inspection costs by switching to reduced inspection if consistently good quality is achieved

The first set of procedures is for developing sampling plans that may be used for an isolated lot or a small number of lots. For a continuing series of lots, the second set of procedures contains a discussion of:

- Single sampling procedures found in ISO 2859-1 (ANSI/ASQC Z1.4)
- Item-by-item sequential plans that match the plans in ISO 2859-1 found in ISO 8422

8.2 DEFINITIONS OF TERMS USED FOR SINGLE SAMPLING

The following definitions are taken from ISO 3524-2 (ANSI/ISO/ASQC A3534-2) and ISO 8402 (ANSI/ISO/ASQC A8402). Some of the definitions are slightly modified from those in the standards but their meaning is unchanged.

8.2.1 Acceptable Quality Level (*AQL*)

When a continuing series of lots is considered, the *AQL* is the quality level that, for the purpose of sampling inspection, is the limit of the satisfactory process average. See Sec. 8.5 for a further discussion of this concept.

8.2.2 Defect

Nonfulfillment of an intended usage requirement. NOTE: The term defect is appropriate for use when a quality characteristic is evaluated in terms of its use as contrasted to conformance to specifications.

8.2.3 Lot

Definite quantity of items, each having uniform characteristics, collected together and submitted for examination.

8.2.4 Nonconforming Item

Item with one or more nonconformities.
NOTE: Nonconforming items generally are classified by their degree of seriousness, such as:

- Class A. An item that contains one or more nonconformities of class A and may also contain nonconformities of class B and/or class C.

- Class B. An item that contains one or more nonconformities of class B and may also contain nonconformities of class C, but contains no nonconformities of class A.

8.2.5 Nonconformities per 100 Items

One hundred times the number of nonconformities in a sample (one or more nonconformities being possible in any item) divided by the total number of items in the sample.

8.2.6 Nonconformity

Nonfulfillment of a specified requirement.
NOTES: In some instances, specified requirements may coincide with customer usage requirements (see defect). In other situations, they might not coincide, being either more or less stringent.

Nonconformities generally are classified according to their degree of seriousness as:

- Class A. The most serious nonconformities. Such nonconformities will be assigned a very small *AQL* value.
- Class B. Less serious nonconformities. These can be assigned a higher *AQL* value than those in class A and a smaller *AQL* value than those in class C, if a third class exists.

The user is cautioned that adding characteristics and classes of nonconformities generally will affect the overall probability of acceptance of a lot.

8.2.7 Percent Nonconforming

One hundred times the number of nonconforming items divided by the total number of items inspected.

8.2.8 Sample

One or more items taken from a lot intended to provide information about the lot.

8.2.9 Sample Size

Number of items in the sample.

8.2.10 Sampling Plan

Specific plan that states the sample size to be used and the associated criteria for accepting the lot. NOTE: A distinction should be made between the terms sampling plan, sampling scheme, and sampling system.

8.2.11 Sampling Scheme

Combination of sampling plans with rules for changing from one plan to another. Sampling schemes are used when there is a continuing series of lots to be submitted for acceptance inspection purposes.

8.2.12 Sampling System

Collection of sampling schemes, each with its own rules for changing plans, together with criteria by which appropriate schemes may be chosen. ISO 2859-1 (ANSI/ASQC Z1.4) is such a system.

8.3 PARAMETERS AND VARIABLES USED IN SEQUENTIAL SAMPLING

8.3.1 Acceptance Number for a Corresponding Single Sampling Plan (*Ac*)

The acceptance number for a single sampling plan that has essentially the same operating characteristic as the sampling plan under consideration.

8.3.2 Acceptance Number for Sequential Sampling (*A*)

Value calculated from the specified parameters of the sampling plan and the cumulative sample size. The cumulative count is compared to the acceptance number after each item is inspected to determine if the lot may be accepted.

8.3.3 A_t

Acceptance number corresponding to the curtailed value of the cumulative sample size.

8.3.4 Cumulative Count (*c*)

When sampling inspection from a lot is performed sequentially, c is the total number of nonconforming items found during inspection, counting from the start of inspection and including the last item inspected.

8.3.5 Cumulative Sample Size (n_{cum})

When sampling inspection of items is performed sequentially, n_{cum} is the total number of items inspected, starting from the start of inspection and including the last item inspected.

8.3.6 h_A

Constant that is used to determine the acceptance number; h_A is the intercept of the acceptance line.

8.3.7 h_R

Constant that is used to determine the rejection number; h_R is the intercept of the rejection line.

8.3.8 Multiplier of the Cumulative Sample Size (g)

Multiplier used to determine the acceptance and rejection numbers; g is the slope of the acceptance and rejection lines.

8.3.9 n_{av}

Average sample size.

8.3.10 n_s

Sample size for a single sampling plan that has essentially the same operating characteristic as the sequential sampling plan under consideration.

8.3.11 n_t

Curtailment value of the cumulative sample size.

8.3.12 p

Lot quality level in proportion nonconforming.

8.3.13 Rejection Number for Sequential Sampling (R)

Value calculated from the specified parameters of the sampling plan and the cumulative sample size. The cumulative count is compared to the rejection number after each item is inspected to determine if the lot should be rejected.

8.3.14 R_t

Rejection number corresponding to the curtailed value of the cumulative sample size.

8.4 EXPRESSION OF NONCONFORMITY

The extent of nonconformity may be expressed in terms of either percent nonconforming or nonconformities per 100 items. The procedures assume that nonconformities occur randomly and with statistical independence. An item may be nonconforming for a number of reasons depending on the system and the needs of the user. These nonconformities may be classified as class A, class B, class C, etc.

8.5 USE AND APPLICATION OF ACCEPTABLE QUALITY LEVEL (AQL)

The AQL, together with the sample size code letter, is used for indexing the sampling plans and schemes provided by ISO 2859-1 (ANSI/ASQC Z1.4) and ISO 8422 which are discussed later in this chapter. The AQL has no meaning for isolated lot inspection.

8.5.1 *AQL* Designation

To use sampling schemes, an AQL must be designated for each class of nonconformities. The effect of this designation is that the majority of lots submitted with quality levels (percent nonconforming or nonconformities per 100 items) equal to or less than this value will be accepted.

The sampling plans provided by ISO 2859-1 and ISO 8422 are arranged in such a way that the probability of acceptance at the designated AQL value depends upon the sample size (and thus the lot size) being higher for large samples than for small ones.

8.5.2 *AQL* and Process Average

The AQL is a parameter of the sampling scheme and should not be confused with the process average, which describes the operating level of a process. The process average is expected to be less than the AQL to avoid excessive rejections.

The designation of an AQL does not imply that the supplier has the right to knowingly supply any nonconforming items. The term AQL does not apply to isolated lot sampling plans. For such plans, the producer's risk and the producer's risk quality along with the consumer's risk and the consumer's risk quality are used. These concepts are discussed in Sec. 8.8.

8.6 TYPES OF SAMPLING PLANS

8.6.1 Attributes and Variables Sampling Plans

Sampling plans may be classified as attributes and variables plans. This chapter contains a detailed discussion of attributes plans only. For a discussion of variables plans see Wadsworth, Stephens, and Godfrey (1998) or the variables standards, ISO 3951 and ANSI/ASQC Z1.9. For attributes plans, each item inspected is classified as conforming or not conforming to specifications. The decision as to acceptability is then based on the number of items found to be nonconforming or the number of nonconformities found in the sample.

Variables plans require the calculation of one or more statistics, such as the average and standard deviation. They also require an assumption regarding the underlying distribution of measurements. They will, in general, require a much smaller sample size.

Attributes inspections are usually easier to make because they may be of the "go-no-go" variety, or they may be visual inspections.

8.6.2 Single Attribute Sampling Plans

The simplest type of sampling plans are those involving a single sample. That is, in the attributes case, a sample of size n is selected from the lot, and the lot is accepted if no more that Ac nonconforming items are found in the sample. Otherwise the lot is rejected.

Procedures for selecting single sampling plans from isolated lots are found in Sec. 8.10. Procedures for selecting single plans for a continuing series of lots are found in Sec. 8.11.

8.6.3 Sequential Attributes Sampling Plans

Item-by-item sequential sampling requires that one of the following three decisions be made after each item is inspected.

- Accept the lot.
- Reject the lot.
- Inspect another item.

This process is continued until the lot is either accepted or rejected.

8.7 CONSIDERATIONS WHEN CHOOSING AN ATTRIBUTE SAMPLING PROCEDURE

The use of sequential sampling results in an average sample size savings over single plans. However, sequential sampling is harder to administer, so that administrative costs must be balanced against cost savings due to decreased sample size. The difficulty with sequential sampling is that the sample size is unknown in advance of inspection. However, this can be overcome by selecting the curtailed sample size in advance. We discuss this in more detail later.

8.8 SAMPLING RISKS AND COSTS

Whenever sampling inspection is used (as opposed to 100 percent inspection), certain risks of making errors are always incurred. In situations where 100 percent inspection is not used, we cannot know everything about the lot. We only know about the items in the sample. Thus we incur the risks of making two kinds of errors in our decision to accept or to not accept the lot.

We can reject a lot that should have been accepted. The risk of doing this is called the producer's risk. The second type of error is that we can accept a lot that should have been rejected. The risk of doing this is called the consumer's risk. Although sampling always incurs both types of risks, they can be measured and controlled if proper statistical procedures are used.

Often the producer's risk is stated in terms of a probability or risk level denoted by the Greek letter alpha (α). The producer's risk is the probability that a lot with a quality level (e.g., the fraction nonconforming) at a certain acceptable level or better is not accepted by the sampling procedure. The symbol *PR* will be used for this risk. The producer's risk quality is the quality level for which the probability of not accepting a lot is just equal to the producer's risk. The producer's risk quality is denoted by the symbol *PRQ*.

The consumer's risk is sometimes denoted by the Greek letter beta (β). However, we will denote it by the symbol *CR*. The consumer's risk is the probability that a nonconforming lot of a certain quality is accepted. The quality level associated with the consumer's risk is called the consumer's risk quality and is denoted by the symbol *CRQ*.

8.9 ALLOWABLE NONCONFORMITIES AND RISKS ASSOCIATED WITH COMPLETE INSPECTION

Both the consumer's and producer's risks should be small, but they can only be zero when 100 percent inspection is used. Both the *PRQ* and the *CRQ* should also be small. The *PRQ* cannot be zero without the use of 100 percent inspection, and the *CRQ* must be larger than the *PRQ*. In general, the smaller the values of these risks and quality levels and the closer the two quality levels are to each other, the larger the sample must be. That is, we must pay for additional discrimination in our sampling procedure.

The fact that the *PRQ* (our definition of acceptable quality) must be greater than zero does not mean that a certain fraction of the lot may be nonconforming to specifications. It means that we are recognizing that, since we have not completely inspected a lot, it is possible for a certain number of nonconforming items to be present. If a series of lots are submitted with quality level equal to or better than the *PRQ*, the sampling procedure will reject only a small fraction (*PR*) or less of them.

If the items being inspected are of a critical nature and the inspection is not destructive, complete inspection may be necessary. Even complete inspection may not be enough, however, to assure us that only conforming items are present. This is because of errors in inspection. If a large number of items must be inspected, the chance of such inspection errors increases rapidly. For this reason, complete inspection may actually result in more nonconforming items being passed than under sampling, when more care may be taken during the inspection of each item and the risks are known.

8.10 SAMPLING PLANS FOR SINGLE LOTS

The sampling procedures described in Secs. 8.10.1 and 8.10.3 are for use in situations in which all of the items are received at one time. In cases where there is a continuing series of technically equivalent lots from the same supplier, sampling systems such as those described in Sec. 8.11 are preferred. The procedures in Sec. 8.11 provide a means to

- Reduce the sampling intensity when a supplier regularly submits superior quality items.
- Increase the sampling intensity when a supplier has a history of submitting a product that does not meet specifications.

When the number of lots to be submitted is not sufficient to permit the use of switching rules to change the sampling intensity, the procedures described in Secs. 8.10.1 and 8.10.3 should be used in preference to merely selecting a plan from one of the tables in Sec. 8.11.

Single sampling plans are described in Sec. 8.10.1 Item-by-item sequential plans are described in Sec. 8.10.3. In both cases it is necessary for the user to have chosen a definition of good and bad quality (i.e., *PRQ* and *CRQ*), along with producer's and consumer's risks.

8.10.1 Single Sampling Plans

The sampling procedures described here are the simplest plans that can be used when an isolated lot is submitted for inspection. A single sampling plan is described by a sample size n and an acceptance number Ac. A sample of n items is selected from the lot submitted for inspection. If no more than Ac of the items inspected are nonconforming, the lot is accepted.

To choose a sampling plan, the user must decide on the risks he or she is willing to take of rejecting a lot that should be accepted or accepting a lot that should be rejected. To simplify the process of selecting sampling plans and for the purpose of this chapter, a producer's risk (*PR*) of 5 percent and a consumer's risk (*CR*) of 10 percent will be used. The user must choose the producer's risk quality *PRQ* and the consumer's risk quality *CRQ*. In other words the user must decide what percent nonconforming levels should be accepted 95 percent of the time, and what percent nonconforming levels should be accepted only 10 percent of the time. These two quality levels will then define the sampling plan to be used.

A single sampling plan may be derived using Table 8.1. The following steps illustrate the use of Table 8.1.

1. First, we choose a *PRQ* and a *CRQ*. Suppose we choose a *PRQ* of 4 percent and a *CRQ* of 12.5 percent. That is, we wish to find a sampling plan that will accept lots that are 4 percent or less nonconforming 95 percent of the time and accept lots that are 12.5 percent or more nonconforming 10 percent of the time.
2. We determine the ratio of *CRQ* to *PRQ*.

   $$CRQ/PRQ = 0.125/0.04 = 3.125$$

3. Next, we look in the *CRQ*/*PRQ* column of Table 8.1 and locate the value nearest to our calculated value. From the table we find that 3.21 is the nearest to 3.125.
4. We locate the acceptance number in Table 8.1 by looking at the Ac value that corresponds to 3.21. Our acceptance number is 6.

TABLE 8.1 Values for Single Sampling Plans

Ac	$n \times CRQ$	$n \times PRQ$	CRQ/PRQ
0	2.303	0.0513	44.84
1	3.890	0.355	10.96
2	5.322	0.818	6.51
3	6.681	1.366	4.89
4	7.994	1.970	4.06
5	9.274	2.613	3.55
6	10.532	3.286	3.21
7	11.771	3.981	2.96
8	12.995	4.695	2.77
9	14.207	5.426	2.62
10	15.407	6.169	2.50
11	16.598	6.924	2.40
12	17.782	7.690	2.31
13	18.958	8.464	2.24
14	20.128	9.246	2.18
15	21.292	10.035	2.12

Source: Used with permission from F. E. Grubbs, "On Designing Single Sampling Inspection Plans." *Annals of Mathematical Statistics*, vol. XX, p. 256, 1949.

5. Next, we determine the sample size by dividing $n \times CRQ$ by the CRQ or $n \times PRQ$ by PRQ (rounding up).

$$n = (n \times CRQ)/CRQ = 10.532/0.125 = 85$$

$$n = (n \times PRQ)/PRQ = 3.286/0.04 = 83$$

6. Finally we choose the sampling plan. We use the highest result of the above calculations for our sampling plan:

$$n = 85$$

$$Ac = 6$$

We sample 85 items. If no more than 6 of the items are nonconforming, the lot is accepted.

8.10.2 Operating Characteristic (*OC*)

The operating characteristic is the probability of acceptance as a function of the fraction of the lot that is nonconforming. This function can be used to determine the likelihood that a lot of any given quality will be accepted. The chart in Fig. 8.1 can be used to determine the probability of acceptance of any lot quality for any single attribute sampling plan. The abscissa of the chart is the product np of the sample size and the fraction nonconforming. This product is the average number of nonconforming items in submitted lots.

Each curve on the chart is an acceptance number Ac. The ordinate of the chart is the probability of acceptance expressed in percent. For the example used earlier,

$$n \times PRQ = 85(0.04) = 3.4$$

$$n \times CRQ = 85(0.125) = 10.625$$

The curve for $Ac = 6$ goes through the two points 3.4, 95 and 10.625, 10. We also can determine from Fig. 8.1 that, with the plan $n = 85$, $Ac = 6$, an np of 6.7 gives a p of 6.7/85 = 7.9 percent nonconforming and has a 50 percent chance of being accepted by the plan.

8.10.3 Item-by-Item Sequential Sampling Plans

The procedures and tables in this section are based on ISO 8422. To obtain complete sequential sampling procedures, refer to that standard.

When using a sequential sampling plan for attributes, items are selected at random and subjected to inspection one at a time. A cumulative count is kept of the number of nonconforming items (or number of nonconformities) and the number of items inspected. After each item is inspected, this count is used to assess whether sufficient information exists to either accept or reject the lot.

As with single sampling, to evaluate the consumer's risk and the producer's risk, the producer's and consumer's risk qualities must be chosen. We again use a producer's risk of 5 percent and a consumer's risk of 10 percent. This leaves only the PRQ and CRQ to be selected by the user.

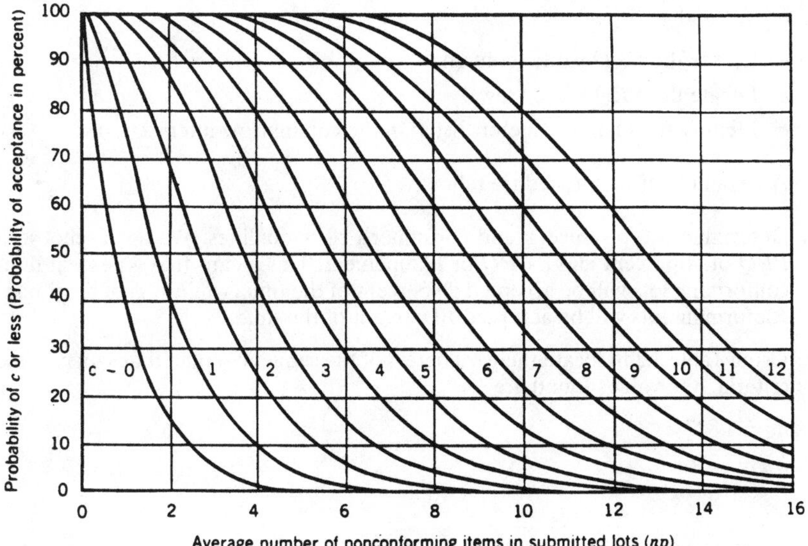

FIGURE 8.1 Curves for determining probability of acceptance.

The average sample size is the average number of items for a series of lots, all of the same quality, that will be inspected before a decision can be made to accept or reject the lot. Using a sequential plan leads to a smaller average sample size than a single sampling plan with the same operating characteristic. For very good lots this saving may reach or exceed 50 percent.

On the other hand, it is possible for the actual number of items inspected for a particular lot to exceed that required by a single sampling plan with the same operating characteristic. In fact, although very unlikely, a sequential sampling plan may not reach a decision until the entire lot has been inspected. To preclude such events, a curtailment rule has been established. This is discussed later in this section.

Because the ultimate sample size when using sequential sampling is not known in advance, selecting items for inspection may present operational difficulties. The user must decide if a smaller average sample size outweighs a sample size that is not known in advance. This becomes a matter of evaluating inspection costs versus the cost of selecting items for inspection.

It should be pointed out that the type of sampling plan must be decided before sampling begins. If a switch is made during the inspection of a lot, the operating characteristic of the resulting plan may be changed drastically because the inspection results would influence the choice of acceptance criteria.

Selecting Sequential Sampling Plan Parameters. The general procedure described here is to be used when both the producer's risk quality and the consumer's risk quality have been designated. Thus the sampling plan will have approximately the same operating characteristic as the single sampling plan developed in Sec. 8.10.1 using the same two points ($PR = 5$ percent and $CR = 10$ percent).

1. The first step is to choose the values of *PRQ* and *CRQ*.
2. The next step is to use Table 8.2 to determine the values of the plan parameters h_A, h_R, and g.
   a. Locate the *CRQ* value (column).
   b. Locate the *PRQ* value (row).
   c. Identify the plan parameters listed in the column/row intersection.

An example of the procedure follows:

1. Determine the producer's and consumer's risks qualities. We again choose a *PRQ* of 4 percent and a *CRQ* of 12.5 percent. This means that 4 percent nonconforming lots will be accepted 95 percent of the time, while 12.5 percent nonconforming lots will be accepted 10 percent of the time.
2. Using Table 8.2 to locate the parameters of the sequential plan that satisfy these criteria, the values found are

$h_A = 1.827$

$h_R = 2.346$

$g = 0.0752$

Determining the Curtailment Value. A curtailment rule ensures that a decision to reject or accept a lot while using sequential sampling is made before the entire

TABLE 8.2 Sequential Sampling Plans for Inspection for Percent Nonconforming for Producer's Risk = 0.05 and Consumer's Risk = 0.10

PRQ	Parameters	Consumer's risk quality level CRQ																
		0.80	1.00	1.25	1.60	2.00	2.50	3.15	4.00	5.00	6.30	8.00	10.00	12.50	16.00	20.00	25.00	31.50
0.100	h_A	1.079	0.974	0.887	0.808	0.747	0.694	0.647	0.604	0.568	0.535	0.504	0.478	0.454	0.429	0. 408	0.388	0.367
	h_R	1.385	1.250	1.139	1.037	0.959	0.891	0.830	0.775	0.729	0.687	0.647	0.614	0.583	0.551	0.524	0.498	0.472
	g	0.00337	0.00391	0.00456	0.00543	0.00637	0.00750	0.00891	0.0107	0.0127	0.0152	0.0185	0.0222	0.0267	0.0330	0.0402	0.0494	0.0616
0.125	h_A	1.208	1.078	0.973	0.878	0.806	0.746	0.691	0.642	0.602	0.565	0.531	0.502	0.475	0.448	0.425	0.403	0.381
	h_R	1.551	1.384	1.249	1.127	1.035	0.957	0.887	0.825	0.773	0.726	0.682	0.644	0.610	0.575	0.546	0.518	0.489
	g	0.00364	0.00421	0.00490	0.00580	0.00679	0.00797	0.0094	0.0113	0.0134	0.0160	0.0194	0.0232	0.0279	0.0344	0.0419	0.0513	0.0638
0.160	h_A	1.393	1.223	1.089	0.972	0.885	0.812	0.748	0.691	0.645	0.602	0.564	0.531	0.501	0.471	0.446	0.422	0.398
	h_R	1.789	1.570	1.399	1.247	1.136	1.042	0.960	0.887	0.828	0.774	0.724	0.682	0.644	0.605	0.572	0.542	0.511
	g	0.00398	0.00459	0.00531	0.00627	0.00731	0.00855	0.0101	0.0120	0.0142	0.0170	0.0205	0.0245	0.0294	0.0362	0.0439	0.0536	0.0666
0.200	h_A	1.617	1.392	1.221	1.075	0.970	0.883	0.808	0.742	0.689	0.641	0.597	0.561	0.528	0.494	0.466	0.440	0.414
	h_R	2.076	1.787	1.568	1.381	1.245	1.134	1.037	0.952	0.884	0.823	0.767	0.720	0.677	0.635	0.599	0.565	0.532
	g	0.00433	0.00498	0.00574	0.00675	0.00784	0.00915	0.0108	0.0128	0.0151	0.0180	0.0216	0.0257	0.0308	0.0378	0.0458	0.0559	0.0692
0.250	h_A	1.926	1.615	1.390	1.204	1.074	0.968	0.878	0.801	0.739	0.684	0.635	0.594	0.557	0.520	0.489	0.460	0.432
	h_R	2.473	2.074	1.785	1.546	1.378	1.243	1.128	1.028	0.949	0.879	0.815	0.762	0.715	0.667	0.626	0.591	0.555
	g	0.00473	0.00541	0.00622	0.00729	0.00844	0.00981	0.0115	0.0136	0.0160	0.0190	0.0228	0.0271	0.0324	0.0397	0.0479	0.0583	0.0721
0.315	h_A	2.403	1.937	1.622	1.374	1.207	1.075	0.966	0.873	0.800	0.736	0.679	0.632	0.591	0.549	0.515	0.483	0.452
	h_R	3.085	2.487	2.083	1.764	1.549	1.381	1.240	1.121	1.028	0.945	0.872	0.812	0.758	0.705	0.661	0.620	0.580
	g	0.00521	0.00593	0.00679	0.00792	0.00914	0.0106	0.0124	0.0146	0.0171	0.0202	0.0242	0.0287	0.0342	0.0418	0.0503	0.0611	0.0753
0.40	h_A	3.229	2.441	1.961	1.610	1.385	1.214	1.076	0.962	0.875	0.799	0.732	0.678	0.630	0.583	0.545	0.509	0.475
	h_R	4.146	3.134	2.518	2.067	1.778	1.559	1.382	1.236	1.123	1.026	0.940	0.871	0.809	0.749	0.700	0.654	0.610
	g	0.00577	0.00655	0.00747	0.00867	0.00996	0.0115	0.0134	0.0157	0.0184	0.0217	0.0258	0.0305	0.0363	0.0441	0.0530	0.0642	0.0790
0.50	h_A	4.759	3.224	2.437	1.917	1.606	1.381	1.205	1.064	0.958	0.868	0.790	0.727	0.673	0.619	0.576	0.537	0.498
	h_R	6.110	4.140	3.129	2.461	2.062	1.774	1.548	1.366	1.231	1.114	1.014	0.934	0.863	0.795	0.740	0.689	0.640
	g	0.00638	0.00722	0.00819	0.00947	0.0108	0.0125	0.0145	0.0169	0.0197	0.0232	0.0275	0.0324	0.0384	0.0466	0.0558	0.0674	0.0827
0.63	h_A	9.357	4.834	3.256	2.390	1.926	1.611	1.377	1.196	1.064	0.953	0.860	0.786	0.723	0.662	0.613	0.568	0.526
	h_R	12.013	6.206	4.180	3.069	2.472	2.069	1.768	1.535	1.366	1.224	1.104	1.009	0.928	0.849	0.787	0.729	0.675
	g	0.00712	0.00801	0.00905	0.0104	0.0119	0.0136	0.0157	0.0183	0.0212	0.0249	0.0294	0.0346	0.0408	0.0494	0.0590	0.0710	0.0868
0.80	h_A		9.999	4.994	3.210	2.425	1.946	1.614	1.371	1.200	1.062	0.947	0.858	0.783	0.712	0.656	0.605	0.557
	h_R		12.837	6.411	4.122	3.113	2.499	2.073	1.760	1.541	1.363	1.215	1.102	1.006	0.914	0.842	0.777	0.715
	g		0.00896	0.0101	0.0115	0.0131	0.0149	0.0172	0.0200	0.0231	0.0269	0.0317	0.0371	0.0437	0.0526	0.0626	0.0751	0.0916
1.00	h_A			9.976	4.729	3.201	2.417	1.925	1.589	1.364	1.188	1.046	0.939	0.850	0.767	0.702	0.644	0.590
	h_R			12.808	6.071	4.110	3.103	2.472	2.040	1.751	1.525	1.343	1.205	1.091	0.984	0.901	0.827	0.757
	g			0.0112	0.0128	0.0144	0.0164	0.0188	0.0217	0.0250	0.0290	0.0341	0.0397	0.0466	0.0559	0.0684	0.0794	0.0965

TABLE 8.2 *(Continued)*

PRQ	Parameters	Consumer's risk quality level CRQ																
		0.80	1.00	1.25	1.60	2.00	2.50	3.15	4.00	5.00	6.30	8.00	10.00	12.50	16.00	20.00	25.00	31.50
1.25	h_A				8.990	4.713	3.189	2.386	1.890	1.580	1.348	1.168	1.036	0.929	0.830	0.755	0.688	0.627
	h_R				11.543	6.052	4.095	3.063	2.426	2.028	1.731	1.500	1.331	1.193	1.066	0.969	0.884	0.805
	g				0.0142	0.0160	0.0180	0.0206	0.0237	0.0272	0.0314	0.0367	0.0427	0.0499	0.0597	0.0706	0.0841	0.1018
1.60	h_A					9.908	4.943	3.247	2.392	1.917	1.586	1.343	1.171	1.036	0.915	0.824	0.745	0.674
	h_R					12.721	6.346	4.169	3.072	2.461	2.036	1.724	1.504	1.330	1.175	1.058	0.957	0.865
	g					0.0179	0.0202	0.0229	0.0262	0.0299	0.0345	0.0401	0.0464	0.0540	0.0643	0.0758	0.0899	0.1084
2.00	h_A						9.863	4.830	3.154	2.376	1.888	1.553	1.329	1.157	1.008	0.899	0.806	0.723
	h_R						12.663	6.202	4.049	3.051	2.424	1.994	1.706	1.485	1.294	1.154	1.035	0.928
	g						0.0224	0.0253	0.0289	0.0328	0.0376	0.0436	0.0503	0.0582	0.0690	0.0810	0.0958	0.1150
2.50	h_A							9.467	4.637	3.131	2.335	1.843	1.535	1.311	1.123	0.989	0.878	0.780
	h_R							12.155	5.953	4.019	2.998	2.367	1.971	1.683	1.441	1.269	1.127	1.001
	g							0.0281	0.0319	0.0361	0.0412	0.0475	0.0546	0.0630	0.0743	0.0869	0.1023	0.1223
3.15	h_A								9.089	4.677	3.100	2.289	1.832	1.521	1.274	1.104	0.967	0.850
	h_R								11.669	6.005	3.980	2.939	2.353	1.953	1.635	1.417	1.242	1.091
	g								0.0356	0.0401	0.0455	0.0522	0.0597	0.0686	0.0805	0.0937	0.1099	0.1307
4.00	h_A									9.637	4.705	3.060	2.295	1.827	1.481	1.256	1.083	0.938
	h_R									12.372	6.040	3.929	2.947	2.346	1.902	1.613	1.390	1.204
	g									0.0448	0.0507	0.0578	0.0658	0.0752	0.0879	0.1018	0.1187	0.1406
5.00	h_A										9.193	4.484	3.013	2.255	1.750	1.445	1.220	1.039
	h_R										11.803	5.757	3.868	2.895	2.247	1.855	1.566	1.333
	g										0.0563	0.0639	0.0724	0.0824	0.0957	0.1103	0.1281	0.1509
6.30	h_A											8.753	4.482	2.987	2.162	1.714	1.406	1.171
	h_R											11.238	5.754	3.835	2.776	2.201	1.805	1.503
	g											0.0712	0.0802	0.0908	0.1049	0.1204	0.1390	0.1629
8.0	h_A												9.184	4.535	2.871	2.132	1.675	1.352
	h_R												11.792	5.822	3.686	2.737	2.151	1.735
	g												0.0897	0.1010	0.1160	0.1323	0.1520	0.1771
10.0	h_A													8.958	4.177	2.776	2.049	1.585
	h_R													11.501	5.363	3.564	2.631	2.035
	g													0.1121	0.1280	0.1452	0.1660	0.1922

lot is inspected. If the sample size of the single sampling plan that matches the sequential plan is known, the curtailment number is 1.5 times the single sample size rounded up to the nearest integer. (It was determined in Sec. 8.10.1 that a single sample size of 85 items and an acceptance number of 6 would meet the criteria described in Sec. 8.10.3.)

For this example, the curtailment values (n_t) is

$$n_t = (1.5)n$$
$$= (1.5)(85) = 128.$$

If the sample size of the equivalent single sampling plan is unknown, the curtailment value may be calculated as

$$n_t = (2h_A h_R)/[g\ (1 - g)]$$

rounded up to the nearest integer. In this case, n_t would be

$$[2(1.827)(2.346)] / [0.0752(1 - 0.0752)] + 124$$

Operation of the Plan. For each value of n_{cum} that is less than the curtailment value, the acceptance number A is found by rounding the quantity $gn_{cum} - h_A$ down to the nearest integer. The rejection number R is found by rounding the quantity $gn_{cum} + h_R$ up to the nearest integer.

The acceptance number corresponding to the curtailed sample size A_t is determined by rounding the quantity gn_t down to the nearest integer. The corresponding rejection number R_t is $A_t + 1$.

If the acceptance number A is negative, the cumulative sample size is too small to allow acceptance of the lot. Conversely, if the rejection number R is larger than the cumulative sample size, the cumulative sample size is too small to reject the lot. The smallest cumulative sample size permitting lot acceptance can be found by rounding h_A/g up to the nearest integer. The smallest cumulative sample size permitting lot rejection is found by rounding $h_R / (1 - g)$ up to the nearest integer.

The example in Sec. 8.10.3 has the following parameters:

$$h_A = 1.827$$
$$h_R = 2.346$$
$$g = 0.0752$$
$$n_t = 128$$

The curtailment acceptance number, A_t, is $gn_t = 9.63$, rounded down to 9. The curtailment rejection number is 10. The formula for the acceptance number, A, is $0.0752\ n_{cum} - 1.827$ rounded down to the nearest integer. The rejection number, R, is found by rounding $0.0752gn_{cum} + 2.346$ up to the nearest integer. The procedure can be tabulated, as in Table 8.3.

The acceptance and rejection numbers corresponding to the cumulative sample sizes 1,2,3, …,127 are determined by successively inserting the values of n_{cum} in the above formulas and rounding appropriately. The acceptance number cannot exceed 9, and the rejection number cannot be more than 10 (the curtailment values).

TABLE 8.3 Example Chart for Sequential Sampling Plan

Cumulative sample size n_{cum}	Aceptance number $gn_{cum} - h_R$	Rounded acceptance number A	Rejection number $gn_{cum} + h_g$	Rounded rejection number R
1	–1.752	*	2.421	†
2	–1.677	*	2.496	†
3	–1.912	*	2.572	3
4	–1.526	*	2.647	3
. . .	. . . .	. . .	. . . .	. . .
8	–1.225	*	2.948	3
9	–1.150	*	3.023	4
. . .	. . . .	. . .	. . . .	. . .
20	–0.323	*	3.850	4
21	–0.248	*	4.925	4
22	–0.173	*	4.000	4
23	–0.097	*	4.076	5
24	–0.022	*	4.151	5
25	0.053	0	4.226	5
. .	. . . .	. .	. . . .	. .
31	0.050	0	4.677	5
. .	. . . .	. .	. . . .	. .
35	0.805	0	4.978	5
36	0.880	0	5.053	6
37	0.955	0	5.128	6
38	1.031	1	5.204	6
. .	. . . .	. .	. . . .	. .
48	1.783	1	5.956	6
49	1.858	1	6.031	7
50	1.933	1	6.106	7
51	2.008	2	6.181	7
. .	. . . .	. .	. . . .	. .
61	2.760	2	6.933	7
62	2.835	2	7.008	8
63	2.911	2	7.084	8
64	2.986	2	7.159	8
65	3.061	3	7.234	8
. .	. . . .	. .	. . . .	. .
75	3.813	3	7.986	8
76	3.888	3	8.061	9
77	3.963	3	8.136	9
78	4.039	4	8.212	9
. .	. . . .	. .	. . . .	.
88	4.791	4	8.964	9
89	4.866	4	9.039	10
90	4.941	4	9.114	10
91	5.016	5	9.189	10
. .	. . . .	. .	. . . .	. .
104	5.994	5	10.167	10
105	6.069	6	—	10
. .	. . . .	. .	. . . .	. .
117	6.971	6	—	10
118	7.047	7	—	10
. .	. . . .	. .	. . . .	. .
127	7.723	7	—	10
128	—	9	—	10

*Cumulative sample size is too small to permit acceptance.
†Cumulative sample size is too small to permit rejection.

Two examples of the operation of this sequential sampling plan follow:

1. The first 20 items inspected were found to conform to specifications. The 21st item was nonconforming. If no other nonconforming items are found, the lot would be accepted after 38 items are inspected.

2. The 5th, 10th, 18th, 24th, and 31st items were found to be nonconforming. The lot would be rejected after inspection of the 31st item.

The earliest that a lot may be accepted is after 25 items if all are found to be conforming to specifications. If the first three items inspected are nonconforming, the lot may then be rejected.

Graphical Operation of the Plan. The example in the above section can be presented graphically as in Fig. 8.2, if desired. Figure 8.2 is really a graph of the information in Table 8.3. To use the graph, start at the origin and for each item inspected, move one unit to the right. If an item is nonconforming, move up one unit. If the lower line is crossed, accept the lot. If the upper line is crossed, reject the lot. Continue sampling as long as the inspection results are between the two lines.

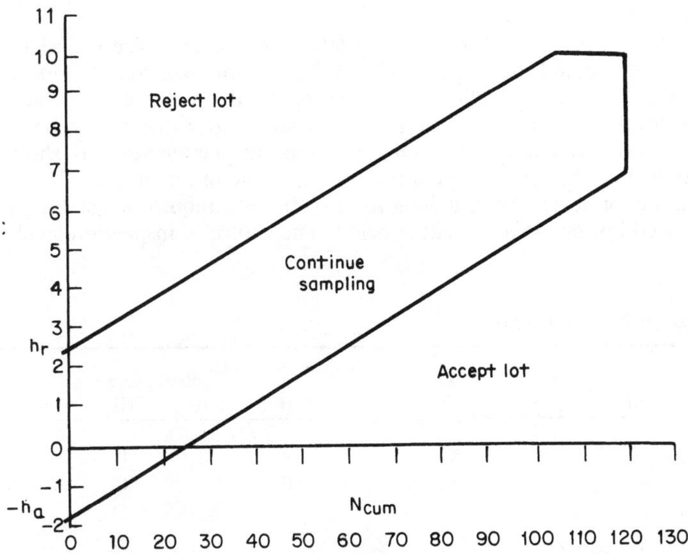

FIGURE 8.2 Graph for example sequential plan.

8.11 SAMPLING PLANS FOR A CONTINUING SERIES OF LOTS

8.11.1 Single Sampling Using ISO 2859-1 (ANSI/ASQC Z1.4)

ISO 2859-1 is an international standard that was updated and improved in 1989 and again in 1997 to use modern concepts of quality assurance. The 1987 edition will be used in this chapter. Earlier editions of the standard were similar to MIL-STD 105E, which has now been discontinued by the U.S. Department of Defense and replaced by American National Standard ANSI/ASQC Z1.4, which was last revised in 1993. In the near future, it is planned that the two standards (ANSI/ASQC Z1.4 and ISO 2859-1) will become one, with both designations in the United States. Most other countries use the ISO standard.

ISO 2859-1 uses sampling schemes designated by AQL and the sample size code letter. Each sampling scheme consists of three sampling plans, designated as normal, tightened, and reduced, together with rules for switching between the plans.

The standard contains single, double, and multiple sampling schemes that are matched as closely as possible; that is, they give approximately the same protection to the producer and the consumer as evidenced by their operating characteristics. Double sampling schemes have a smaller average sample size than single schemes. Multiple sampling schemes have a smaller average sample size than double schemes. Details on the use of double and multiple sampling schemes are to be found in the standards; however, they will not be discussed here. Sequential schemes with approximately the same protection as the single schemes in ISO 2859-1 can be found in ISO 8422. These are discussed in Sec. 8.11.2.

Sample Size Code Letters and Inspection Levels. The sample size code letter depends on the lot size and the inspection level. The sample size code letters can be found in Table 8.4. The inspection level is chosen by the user. The purpose of the inspection levels is to allow the user to maintain greater discrimination for some suppliers and lesser discrimination for others. Inspection level II should always be used, unless there is an urgent reason to use one of the others.

At each inspection level, the switching rules operate to require normal, tightened, and reduced inspection as specified below. The choice of inspection level is

TABLE 8.4 Sample Size Code Letters

	Inspection levels		
Lot or batch size	I	II	III
2 to 8	A	A	B
9 to 15	A	B	C
16 to 25	B	C	D
26 to 50	C	D	E
51 to 90	C	E	F
91 to 150	D	F	G
151 to 280	E	G	H
281 to 500	F	H	J
501 to 1,200	G	J	K
1,201 to 3,200	H	K	L
3,201 to 10,000	J	L	M
10,001 to 35,000	K	M	N
35,001 to 150,000	L	N	P
150,001 to 500,000	M	P	Q
500,001 and over	N	Q	R

quite separate from these three forms of inspection intensity. Thus the inspection level that has been specified shall be kept unchanged when switching between normal, tightened, and reduced inspection plans.

The amount of information about the quality of a lot obtained for a sample taken from the lot depends on the absolute size of the sample and not on the fraction of the lot inspected, provided the lot is large relative to the sample size. In spite of this, three reasons for varying the sample size with the lot size are given in the standard:

1. When the risk is high, it is more important to make the correct decision.
2. With a large lot a sample size can be afforded that would not be economical for a small lot.
3. A truly random selection is relatively more difficult if the sample is too small a proportion of the lot.

Obtaining a Sampling Plan. The AQL and the sample size code letter are used to obtain the sampling plans (normal, tightened, and reduced) from Tables 8.5, 8.6, and 8.7, respectively. When no sampling plan is available for a given combination of AQL and sample size code letter, the tables direct the user to a new letter by means of arrows. The sample size, acceptance, and rejection numbers to be used are the ones corresponding to the new code letter. If this procedure leads to different sample sizes for different classes of nonconformities, the sample size code letter corresponding to the largest sample size can be used for all classes of nonconformities.

Normal, Tightened, and Reduced Inspection. Normal inspection is to be used at the start of the sequence of lots submitted for inspection. Normal, tightened, or reduced inspection shall continue unchanged on successive lots until the switching rules signal a change. The switching rules shall be applied to each class of nonconformity independently. The following are the rules for switching between the three sampling intensities.

- *Normal to tightened:* When normal inspection is being carried out, tightened inspection shall be put into operation when two out of five or fewer consecutive lots have been rejected on original inspection, i.e., ignoring resubmitted lots.
- *Tightened to normal:* When tightened inspection is being carried out, normal inspection shall be reverted to when five consecutive lots have been considered acceptable on original inspection.
- *Tightened to discontinue:* If the cumulative number of lots not accepted on original tightened inspection reaches five, this acceptance sampling procedure shall be discontinued. It is not to be resumed until action has been taken by the supplier to improve the quality of the submitted product. Tightened inspection shall then be used as if switching from normal to tightened had just been invoked.
- *Normal to reduced:* Reduced inspection is an optional procedure which should be approved by both customer and supplier prior to its use. If approved, when normal inspection is being carried out, reduced inspection may be put into operation when the following condition is satisfied:

TABLE 8.5 Single Sampling Plans of Normal Inspection (Master Guide)

Acceptable quality levels (normal inspection)

Sample size code letter	Sample size	0.010	0.015	0.025	0.040	0.065	0.10	0.15	0.25	0.40	0.65	1.0	1.5	2.5	4.0	6.5	10	15	25	40	65	100	150	250	400	650	1000
		Ac Re	Ac Re	Ac Re	Ac Re	Ac Re	Ac Re	Ac Re	Ac Re	Ac Re	Ac Re	Ac Re	Ac Re	Ac Re	Ac Re	Ac Re	Ac Re	Ac Re	Ac Re	Ac Re	Ac Re	Ac Re	Ac Re	Ac Re	Ac Re	Ac Re	Ac Re
A	2	↓	↓	↓	↓	↓	↓	↓	↓	↓	↓	↓	↓	↓	↓	0 1	↓	↓	1 2	2 3	3 4	5 6	7 8	10 11	14 15	21 22	30 31
B	3	↓	↓	↓	↓	↓	↓	↓	↓	↓	↓	↓	↓	↓	0 1	↑	↓	1 2	2 3	3 4	5 6	7 8	10 11	14 15	21 22	30 31	44 45
C	5	↓	↓	↓	↓	↓	↓	↓	↓	↓	↓	↓	↓	0 1	↑	↓	1 2	2 3	3 4	5 6	7 8	10 11	14 15	21 22	30 31	44 45	↑
D	8	↓	↓	↓	↓	↓	↓	↓	↓	↓	↓	↓	0 1	↑	↓	1 2	2 3	3 4	5 6	7 8	10 11	14 15	21 22	30 31	44 45	↑	↑
E	13	↓	↓	↓	↓	↓	↓	↓	↓	↓	↓	0 1	↑	↓	1 2	2 3	3 4	5 6	7 8	10 11	14 15	21 22	30 31	44 45	↑	↑	↑
F	20	↓	↓	↓	↓	↓	↓	↓	↓	↓	0 1	↑	↓	1 2	2 3	3 4	5 6	7 8	10 11	14 15	21 22	↑	↑	↑	↑	↑	↑
G	32	↓	↓	↓	↓	↓	↓	↓	↓	0 1	↑	↓	1 2	2 3	3 4	5 6	7 8	10 11	14 15	21 22	↑	↑	↑	↑	↑	↑	↑
H	50	↓	↓	↓	↓	↓	↓	↓	0 1	↑	↓	1 2	2 3	3 4	5 6	7 8	10 11	14 15	21 22	↑	↑	↑	↑	↑	↑	↑	↑
J	80	↓	↓	↓	↓	↓	↓	0 1	↑	↓	1 2	2 3	3 4	5 6	7 8	10 11	14 15	21 22	↑	↑	↑	↑	↑	↑	↑	↑	↑
K	125	↓	↓	↓	↓	↓	0 1	↑	↓	1 2	2 3	3 4	5 6	7 8	10 11	14 15	21 22	↑	↑	↑	↑	↑	↑	↑	↑	↑	↑
L	200	↓	↓	↓	↓	0 1	↑	↓	1 2	2 3	3 4	5 6	7 8	10 11	14 15	21 22	↑	↑	↑	↑	↑	↑	↑	↑	↑	↑	↑
M	315	↓	↓	↓	0 1	↑	↓	1 2	2 3	3 4	5 6	7 8	10 11	14 15	21 22	↑	↑	↑	↑	↑	↑	↑	↑	↑	↑	↑	↑
N	500	↓	↓	0 1	↑	↓	1 2	2 3	3 4	5 6	7 8	10 11	14 15	21 22	↑	↑	↑	↑	↑	↑	↑	↑	↑	↑	↑	↑	↑
P	800	↓	0 1	↑	↓	1 2	2 3	3 4	5 6	7 8	10 11	14 15	21 22	↑	↑	↑	↑	↑	↑	↑	↑	↑	↑	↑	↑	↑	↑
Q	1250	0 1	↑	↓	1 2	2 3	3 4	5 6	7 8	10 11	14 15	21 22	↑	↑	↑	↑	↑	↑	↑	↑	↑	↑	↑	↑	↑	↑	↑
R	2000	↑	↑	1 2	2 3	3 4	5 6	7 8	10 15	14 15	21 22	↑	↑	↑	↑	↑	↑	↑	↑	↑	↑	↑	↑	↑	↑	↑	↑

↓ = use first sampling plan below arrow. If sample size equals, or exceeds, lot or batch size, carry out 100% inspection.

↑ = use first sampling plan arrow.

Ac = acceptance number.

Re = rejection number.

TABLE 8.6 Single Sampling Plans for Tightened Inspection (Master Table)

Acceptable quality levels tightened inspection)

Sample size code letter	Sample size	0.010	0.015	0.025	0.040	0.065	0.10	0.15	0.25	0.40	0.65	1.0	1.5	2.5	4.0	6.5	10	15	25	40	65	100	150	250	400	650	1000
		Ac Re	Ac Re	Ac Re	Ac Re	Ac Re	Ac Re	Ac Re	Ac Re	Ac Re	Ac Re	Ac Re	Ac Re	Ac Re	Ac Re	Ac Re	Ac Re	Ac Re	Ac Re	Ac Re	Ac Re	Ac Re	Ac Re	Ac Re	Ac Re	Ac Re	Ac Re
A	2	↓	↓	↓	↓	↓	↓	↓	↓	↓	↓	↓	↓	↓	↓	↓	↓	↓	↓	1 2	2 3	3 4	5 6	8 9	12 13	18 19	27 28
B	3	↓	↓	↓	↓	↓	↓	↓	↓	↓	↓	↓	↓	↓	↓	0 1	↓	↓	1 2	2 3	3 4	5 6	8 9	12 13	18 19	27 28	41 42
C	5	↓	↓	↓	↓	↓	↓	↓	↓	↓	↓	↓	↓	↓	0 1	↓	↓	1 2	2 3	3 4	5 6	8 9	12 13	18 19	27 28	41 43	↑
D	8	↓	↓	↓	↓	↓	↓	↓	↓	↓	↓	↓	↓	0 1	↓	↓	1 2	2 3	3 4	5 6	8 9	12 13	18 19	27 28	41 42	↑	↑
E	13	↓	↓	↓	↓	↓	↓	↓	↓	↓	↓	↓	0 1	↓	↓	1 2	2 3	3 4	5 6	8 9	12 13	18 19	27 28	41 42	↑	↑	↑
F	20	↓	↓	↓	↓	↓	↓	↓	↓	↓	↓	0 1	↓	↓	1 2	2 3	3 4	5 6	8 9	12 13	18 19	↑	↑	↑	↑	↑	↑
G	32	↓	↓	↓	↓	↓	↓	↓	↓	↓	0 1	↓	↓	1 2	2 3	3 4	5 6	8 9	12 12	18 19	↑	↑	↑	↑	↑	↑	↑
H	50	↓	↓	↓	↓	↓	↓	↓	↓	0 1	↓	↓	1 2	2 3	3 4	5 6	8 9	12 13	18 19	↑	↑	↑	↑	↑	↑	↑	↑
J	80	↓	↓	↓	↓	↓	↓	↓	0 1	↓	↓	1 2	2 3	3 4	5 6	8 9	12 13	18 19	↑	↑	↑	↑	↑	↑	↑	↑	↑
K	125	↓	↓	↓	↓	↓	↓	0 1	↓	↓	1 2	2 3	3 4	5 6	8 9	12 13	18 19	↑	↑	↑	↑	↑	↑	↑	↑	↑	↑
L	200	↓	↓	↓	↓	↓	0 1	↓	↓	1 2	2 3	3 4	5 6	8 9	12 13	18 19	↑	↑	↑	↑	↑	↑	↑	↑	↑	↑	↑
M	315	↓	↓	↓	↓	0 1	↓	↓	1 2	2 3	3 4	5 6	8 9	12 13	18 19	↑	↑	↑	↑	↑	↑	↑	↑	↑	↑	↑	↑
N	500	↓	↓	↓	0 1	↓	↓	1 2	2 3	3 4	5 6	8 9	12 13	18 19	↑	↑	↑	↑	↑	↑	↑	↑	↑	↑	↑	↑	↑
P	800	↓	↓	0 1	↓	↓	1 2	2 3	3 4	5 6	8 9	12 13	18 19	↑	↑	↑	↑	↑	↑	↑	↑	↑	↑	↑	↑	↑	↑
Q	1250		0 1	↓	↓	1 2	2 3	3 4	5 6	8 9	12 13	18 19	↑	↑	↑	↑	↑	↑	↑	↑	↑	↑	↑	↑	↑	↑	↑
R	2000	0 1	↑	↓	1 2	2 3	3 4	5 6	8 9	12 13	18 19	↑	↑	↑	↑	↑	↑	↑	↑	↑	↑	↑	↑	↑	↑	↑	↑
S	3150			1 2									↑	↑	↑	↑	↑	↑	↑	↑	↑	↑	↑	↑	↑	↑	↑

↓ = use first sampling plan below arrow. If sample size equals, or exceeds, lot or batch size, carry out 100% inspection.

↑ = use first sampling plan above arrow.

Ac = acceptance number.

Re = rejection number.

TABLE 8.7 Single Sampling Plans for Reduced Inspection (Master Table)

Sample size code letter	Sample size	Acceptable quality levels (reduced inspection)																									
		0.010	0.015	0.025	0.040	0.065	0.10	0.15	0.25	0.40	0.65	1.0	1.5	2.5	4.0	6.5	10	15	25	40	65	100	150	250	400	650	1000
		Ac Re	Ac Re	Ac Re	Ac Re	Ac Re	Ac Re	Ac Re	Ac Re	Ac Re	Ac Re	Ac Re	Ac Re	Ac Re	Ac Re	Ac Re	Ac Re	Ac Re	Ac Re	Ac Re	Ac Re	Ac Re	Ac Re	Ac Re	Ac Re	Ac Re	Ac Re
A	2	↓	↓	↓	↓	↓	↓	↓	↓	↓	↓	↓	↓	↓	↓	0 1	↓	↓	1 2	2 3	3 4	5 6	7 8	10 11	14 15	21 22	30 31
B	2	↓	↓	↓	↓	↓	↓	↓	↓	↓	↓	↓	↓	↓	0 1	↑	↓	↓	1 2	2 3	3 4	5 6	7 8	10 11	14 15	21 22	30 31
C	2	↓	↓	↓	↓	↓	↓	↓	↓	↓	↓	↓	↓	0 1	↑	↓	↓	1 2	2 3	3 4	4 5	6 7	8 9	10 11	14 15	21 22	↑
D	3	↓	↓	↓	↓	↓	↓	↓	↓	↓	↓	↓	0 1	↑	↓	↓	1 2	2 3	3 4	4 5	6 7	8 9	10 11	14 15	21 22	↑	↑
E	5	↓	↓	↓	↓	↓	↓	↓	↓	↓	↓	0 1	↑	↓	↓	1 2	2 3	3 4	4 5	6 7	8 9	10 11	14 15	21 22	↑	↑	↑
F	8	↓	↓	↓	↓	↓	↓	↓	↓	↓	0 1	↑	↓	↓	1 2	2 3	3 4	4 5	6 7	8 9	10 11	↑	↑	↑	↑	↑	↑
G	13	↓	↓	↓	↓	↓	↓	↓	↓	0 1	↑	↓	↓	1 2	2 3	3 4	4 5	6 7	8 9	10 11	↑	↑	↑	↑	↑	↑	↑
H	20	↓	↓	↓	↓	↓	↓	↓	0 1	↑	↓	↓	1 2	2 3	3 4	4 5	6 7	8 9	10 11	↑	↑	↑	↑	↑	↑	↑	↑
J	32	↓	↓	↓	↓	↓	↓	0 1	↑	↓	↓	1 2	2 3	3 4	4 5	6 7	8 9	10 11	↑	↑	↑	↑	↑	↑	↑	↑	↑
K	50	↓	↓	↓	↓	↓	0 1	↑	↓	↓	1 2	2 3	3 4	4 5	6 7	8 9	10 11	↑	↑	↑	↑	↑	↑	↑	↑	↑	↑
L	80	↓	↓	↓	↓	0 1	↑	↓	↓	1 2	2 3	3 4	4 5	6 7	8 9	10 11	↑	↑	↑	↑	↑	↑	↑	↑	↑	↑	↑
M	125	↓	↓	↓	0 1	↑	↓	↓	1 2	2 3	3 4	4 5	6 7	8 9	10 11	↑	↑	↑	↑	↑	↑	↑	↑	↑	↑	↑	↑
N	200	↓	↓	0 1	↑	↓	↓	1 2	2 3	3 4	4 5	6 7	8 9	10 11	↑	↑	↑	↑	↑	↑	↑	↑	↑	↑	↑	↑	↑
P	315	↓	0 1	↑	↓	↓	1 2	2 3	3 4	4 5	6 7	8 9	10 11	↑	↑	↑	↑	↑	↑	↑	↑	↑	↑	↑	↑	↑	↑
Q	500	0 1	↑	↑	↓	1 2	2 3	3 4	4 5	6 7	8 9	10 11	↑	↑	↑	↑	↑	↑	↑	↑	↑	↑	↑	↑	↑	↑	↑
R	800	↑	↑	↑	1 2	2 3	3 4	4 5	6 7	8 9	10 11	↑	↑	↑	↑	↑	↑	↑	↑	↑	↑	↑	↑	↑	↑	↑	↑

↓ = use first sampling plan below arrow. If sample size equal, or exceeds, lot size, carry out 100% inspection.

↑ = use first sampling plan above arrow.

Ac = acceptance number.

Re = rejection number.

When the acceptance number is 0 or 1, the preceding 15 lots have been submitted to normal inspection, and all have been accepted on original inspection.

When the acceptance number is 2 or higher, the preceding 10 lots have been submitted to normal inspection, and all would have been accepted if the *AQL* had been one step tighter.

- *Reduced to normal:* When reduced inspection is being carried out, normal inspection shall be reverted to if a lot on original inspection is not accepted.

Operating Characteristic Curves. Operating characteristic curves are given in the standard for the single sampling, normal, and tightened inspection plans. The tables and curves are organized by sample size code letter. The OC curves for *AQL*s of 10 or less are based on the binomial distribution for percent nonconforming inspection. The curves for *AQL*s greater than 10 and for number of nonconformities per 100 items for *AQL*s of 10 or less are based on the Poisson distribution. For these curves, the reader is referred to the standard.

Fractional Acceptance Number Plans. The 1998 edition of ISO 2859-1 includes an optional set of fractional acceptance number plans. These may be used when the user finds the plan needed for a sample size code letter and *AQL* combination corresponds to an arrow in Tables 8.5, 8.6, or 8.7 between the 0 and 1 acceptance number plans. For normal and tightened plans the fractions are $^1/_3$ and $^1/_2$. For reduced inspection plans the fractions are $^1/_5$, $^1/_3$, and $^1/_2$.

A fractional acceptance number of $^1/_2$ means that the lot may be accepted if there are no nonconforming items in the current sample or one nonconforming item in the current sample and none in the immediately preceding one. An acceptance number of $^1/_3$ means the lot may be accepted if there are no nonconforming items in the present sample or one nonconforming item in the present sample and none in the immediately preceding two samples. Similarly, an acceptance number of $^1/_5$ means the lot may be accepted if there are no nonconforming items in the current sample or one in the current sample and none in the immediately preceding four samples.

For details on the use of these plans, the reader is referred to the standard. This procedure is found in ISO 2859-1, but when this chapter was written, it was not in ANSI/ASQC Z1.4. It is anticipated that it will be included in later editions of that standard.

8.11.2 Sequential Sampling Using ISO 8422

This section presents sequential sampling plans that correspond to the single sampling plans of ISO 2859-1 (ANSI/ASQC Z1.4) discussed in Sec. 8.11.1. The sequential plans presented here and in ISO 8422 match, as closely as possible, the plans in Sec. 8.11.1. The sampling schemes are again indexed by sample size code and *AQL*. The procedure for selecting these indexes is the same as that described in Sec. 8.11.1.

The sampling plans to be described here again apply when there is a continuing series of lots and inspection is by attributes, for either percent nonconforming or number of nonconformities per 100 items.

The principles of sequential sampling by attributes discussed in Sec. 8.10.3 apply to this section as well. The only difference is in the selection of the plan parameters. The definitions and the symbols are the same as those used in the previous section. The parameters presented in this section are those that match as closely as possible the plans in Sec. 8.11.1 for the same sample size code letter and AQL.

Operation of the Procedure. For each value n_{cum} of the cumulative sample size that is less than the curtailment number n_t, the acceptance number A is found by rounding the quantity $gn_{cum} - h_A$ down to the nearest integer. The rejection number is found by rounding the quantity $gn_{cum} + h_R$ up to the nearest integer.

Values of h_A, h_R, and g for normal inspection are found in Table 8.8 and those values for tightened inspection are in Table 8.9. The parameter values in both tables are indexed by sample size code letter and AQL. Curtailment sample sizes for the normal and tightened plans of Tables 8.8 and 8.9 are found in Tables 8.10 and 8.11, respectively, along with the curtailment acceptance numbers A_t. Recall that the curtailment rejection numbers are one more than the curtailment acceptance numbers.

Example. For the sequential sampling scheme with $AQL = 2.5$ percent and sample size code letter J (from Table 8.4), Table 8.8 gives us the following parameter values for normal inspection:

$$h_A = 1.910$$
$$h_R = 1.650$$
$$g = 0.0706$$

The corresponding normal single sampling plan from Table 8.5 is $n = 80$, $Ac = 5$. The curtailment value from Table 8.10 is 120 with an acceptance number of 8 and rejection number of 9 for the curtailment sample size.

The value of the acceptance number for n_{cum} less than the curtailment value, 120, is

$$\begin{aligned} A &= gn_{cum} - h_A \\ &= 0.0706n_{cum} - 1.910 \end{aligned}$$

rounded down to the nearest integer. Similarly, the value of the rejection number is

$$\begin{aligned} R &= gn_{cum} + h_R \\ &= 0.0706n_{cum} + 1.650 \end{aligned}$$

rounded up to the nearest integer.

The rejection number cannot exceed the curtailment rejection number R_t, in this case 9, and the acceptance number cannot be less than zero. The acceptance and rejection numbers for this example are found in Table 8.12.

Figure 8.3 is the same plan shown graphically. Many persons prefer this method of operating a sequential sampling plan. The procedure for operating the plan is the same as that discussed in Sec. 8.10.3.

Switching Rules for Sequential Sampling Plans. With the exception that reduced inspection plans are not available in the sequential plans of ISO 8422, the switching rules are the same as those in Sec. 8.11.1. The plans are constructed in

TABLE 8.8 Sequential Sampling Plans for Normal Inspection for Percent Nonconforming (Master Table)

Code letter	Para-meters	Acceptable quality level in percent nonconforming (tightened inspection)															
		0.010	0.015	0.025	0.040	0.065	0.10	0.15	0.25	0.40	0.65	1.0	1.5	2.5	4.0	6.5	10
D	h_A h_R g	↓	↓	↓	↓	↓	↓	↓	↓	↓	↓	↓	*	↑	↓	0.824 0.798 0.2021	1.000 0.786 0.3214
E	h_A h_R g	↓	↓	↓	↓	↓	↓	↓	↓	↓	↓	*	↑	↓	1.011 0.753 0.1264	1.205 0.808 0.2009	1.244 1.173 0.2757
F	h_A h_R g	↓	↓	↓	↓	↓	↓	↓	↓	↓	*	↑	↓	0.998 0.829 0.0829	1.244 0.946 0.1318	1.446 1.205 0.1807	1.651 1.465 0.2790
G	h_A h_R g	↓	↓	↓	↓	↓	↓	↓	↓	*	↑	↓	1.043 0.802 0.0521	1.258 1.002 0.0830	1.500 1.227 0.1136	1.769 1.571 0.1759	1.917 1.801 0.2372
H	h_A h_R g	↓	↓	↓	↓	↓	↓	↓	*	↑	↓	1.060 0.826 0.0335	1.335 0.985 0.0533	1.561 1.268 0.0732	1.838 1.619 0.1128	2.114 1.872 0.1522	2.390 2.195 0.2119
J	h_A h_R g	↓	↓	↓	↓	↓	↓	*	↑	↓	1.080 0.824 0.0210	1.367 1.000 0.0333	1.570 1.271 0.0458	1.910 1.650 0.0706	2.238 1.952 0.0952	2.531 2.297 0.1328	2.865 2.683 0.1827
K	h_A h_R g	↓	↓	↓	↓	↓	*	↑	↓	1.085 0.824 0.0136	1.359 0.980 0.0215	1.575 1.262 0.0296	1.919 1.631 0.0456	2.198 1.894 0.0617	2.514 2.259 0.0857	2.840 2.585 0.1179	3.227 3.000 0.174 1

* Use corresponding single sampling plan in ISO 2859-1 with $Ac = 0$.

TABLE 8.8 (*Continued*)

Code letter	Parameters	Acceptable quality level in percent nonconforming (tightened inspection)															
		0.010	0.015	0.025	0.040	0.065	0.10	0.15	0.25	0.40	0.65	1.0	1.5	2.5	4.0	6.5	10
L	h_A h_R g	↓	↓	↓	↓	*	↑	↓	1.093 0.822 0.00848	1.372 0.987 0.0134	1.600 1.282 0.0184	1.965 1.661 0.0285	2.271 1.956 0.0385	2.623 2.328 0.0535	2.983 2.703 0.0737	3.500 3.239 0.1087	↑
M	h_A h_R g	↓	↓	↓	*	↑	↓	1.093 0.828 0.00534	1.377 0.991 0.00852	1.612 1.289 0.0117	1.992 1.679 0.0180	2.298 1.974 0.0244	2.684 2.382 0.0338	3.106 2.802 0.0467	3.690 3.414 0.0689	↑	↑
N	h_A h_R g	↓	↓	*	↑	↓	1.097 0.828 0.00338	1.390 0.995 0.00535	1.618 1.297 0.00737	2.002 1.682 0.0114	2.330 2.003 0.0153	2.716 2.401 0.0213	3.169 2.855 0.294	3.796 3.489 0.0434	↑	↑	↑
P	h_A h_R g	↓	*	↑	↓	1.102 0.831 0.00211	1.391 0.997 0.00335	1.628 1.303 0.00460	2.017 1.695 0.00710	2.345 2.015 0.00959	2.746 2.425 0.0134	3.212 2.891 0.0184	3.873 3.557 0.0271	↑	↑	↑	↑
Q	h_A h_R g	*	↑	↓	1.105 0.834 0.00135	1.393 0.998 0.00215	1.635 1.311 0.00294	2.024 1.700 0.00454	2.346 2.016 0.00614	2.778 2.454 0.00853	3.237 2.911 0.0117	3.927 3.607 0.0173	↑	↑	↑	↑	↑
R	h_A h_R g	↑	↑	1.102 0.831 0.000846	1.393 0.997 0.00134	1.634 1.309 0.00184	2.034 1.710 0.00284	2.358 2.025 0.00384	2.779 2.453 0.00534	3.233 2.907 0.00735	3.935 3.607 0.0108	↑	↑	↑	↑	↑	↑

* Use corresponding single sampling plan in ISO 2859-1 with Ac = 0.

TABLE 8.9 Sequential Sampling Plans for Tightened Inspection for Percent Nonconforming (Master Table)

Code letter	Para-meters	Acceptable quality level in percent nonconforming (normal inspection)															
		0.010	0.015	0.025	0.040	0.065	0.10	0.15	0.25	0.40	0.65	1.0	1.5	2.5	4.0	6.5	10
D	h_A h_R g	↓	↓	↓	↓	↓	↓	↓	↓	↓	↓	↓	↓	*	↓	↓	0.824 0.798 0.202 1
E	h_A h_R g	↓	↓	↓	↓	↓	↓	↓	↓	↓	↓	↓	*	↓	↓	1.011 0.753 0.126 4	1.205 0.808 0.200 9
F	h_A h_R g	↓	↓	↓	↓	↓	↓	↓	↓	↓	↓	*	↓	↓	0.998 0.829 0.082 9	1.244 0.946 0.131 8	1.446 1.205 0.180 7
G	h_A h_R g	↓	↓	↓	↓	↓	↓	↓	↓	↓	*	↓	↓	1.043 0.802 0.052 1	1.258 1.002 0.083 0	1.500 1.227 0.113 6	1.769 1.571 0.175 9
H	h_A h_R g	↓	↓	↓	↓	↓	↓	↓	↓	*	↓	↓	1.060 0.826 0.033 5	1.335 0.985 0.053 3	1.561 1.268 0.073 2	1.838 1.619 0.112 8	2.243 1.996 0.172 5
J	h_A h_R g	↓	↓	↓	↓	↓	↓	↓	*	↓	↓	1.080 0.824 0.021 0	1.367 1.000 0.033 3	1.570 1.271 0.045 8	1.910 1.650 0.070 6	2.342 2.073 0.107 8	2.710 2.483 0.157 3
K	h_A h_R g	↓	↓	↓	↓	↓	↓	*	↓	↓	1.085 0.824 0.013 6	1.359 0.980 0.021 5	1.575 1.262 0.029 6	1.919 1.631 0.045 6	2.303 2.025 0.069 7	2.690 2.438 0.101 8	3.046 2.813 0.149 9

* Use corresponding curtailed single sampling plan in ISO 2859-1 with Ac = 0.

TABLE 8.9 *(Continued)*

Code letter	Parameters	Acceptable quality level in percent nonconforming (normal inspection)															
		0.010	0.015	0.025	0.040	0.065	0.10	0.15	0.25	0.40	0.65	1.0	1.5	2.5	4.0	6.5	10
L	h_A h_R g	↓	↓	↓	↓		*	↓	↓	1.093 0.822 0.008 48	1.372 0.987 0.0134	1.600 1.282 0.0184	1.965 1.661 0.0285	2.395 2.085 0.0435	2.815 2.535 0.0635	3.313 3.031 0.0937	↑
M	h_A h_R g	↓	↓	↓	↓	*	↓	↓	1.03 0.828 0.00534	1.377 0.991 0.00852	1.612 1.289 0.0117	1.992 1.679 0.0180	2.425 2.114 0.0276	2.908 2.602 0.0403	3.453 3.159 0.0504	↑	↑
N	h_A h_R g	↓	↓	↓	*	↓	↓	1.097 0.828 0.00338	1.390 0.995 0.00535	1.618 1.297 0.00737	2.002 1.682 0.0114	2.477 2.157 0.0174	2.950 2.639 0.0254	3.537 3.229 0.0373	↑	↑	↑
P	h_A h_R g	↓	↓	*	↓	↓	1.102 0.831 0.00211	1.391 0.997 0.00335	1.628 1.303 0.00460	2.017 1.695 0.00710	2.494 2.171 0.0108	2.993 2.671 0.0158	3.68 3.298 0.0233	↑	↑	↑	↑
Q	h_A h_R g	↓	*	↓	↓	1.105 0.834 0.00135	1.393 0.998 0.00215	1.635 1.311 0.00294	2.024 1.700 0.00454	2.507 2.181 0.00694	3.021 2.694 0.0101	3.621 3.300 0.0149	↑	↑	↑	↑	↑
R	h_A h_R g	*	↑	↓	1.102 0.831 0.00846	1.393 0.997 0.00134	1.634 1.309 0.00184	2.034 1.710 0.00284	2.510 2,182 0.00434	3.023 2.694 0.00633	3.657 3.326 0.00935	↑	↑	↑	↑	↑	↑
S	h_A h_R g			1.103 0.832 0.000536													

* Use corresponding curtailed single sampling plan in ISO 2859-1 with Ac = 0.

TABLE 8.10 Curtailment Values for Sequential Sampling Plans for Normal Inspection for Percent Nonconforming

Code letter	Single sample size n_o	Curtailed sample size n_t	Acceptance number, A_t — Acceptable quality level in percent nonconforming: 0.025	0.040	0.065	0.10	0.15	0.25	0.40	0.65	1.0	1.5	2.5	4.0	6.5	10
D	8	12										*			2	3
E	13	20									*			2	4	5
F	20	30								*			2	3	5	8
G	32	48							*			2	3	5	8	11
H	50	75						*			2	3	5	8	11	15
J	80	120					*			2	3	5	8	11	15	21
K	125	188				*			2	4	5	8	11	16	22	32
L	200	300			*			2	4	5	8	11	16	22	32	
M	315	473		*			2	4	5	8	11	16	22	32		
N	500	750	*			2	4	5	8	11	15	22	32			
P	800	1200			2	4	5	8	11	16	22	32				
Q	1250	1875		2	4	5	8	11	15	21	32					
R	2000	3000	2	4	5	8	11	16	22	32						

* Use corresponding single sampling plan in ISO 2859-1 with $Ac = 0$.

TABLE 8.11 Curtailment Values for Sequential Sampling Plans for Tightened Inspection for Percent Nonconforming

Code letter	Single sample size	Curtailed sample size	Acceptance number, A_t													
			Acceptable quality level in percent nonconforming													
	n_o	n_t	0.025	0.040	0.065	0.10	0.15	0.25	0.40	0.65	1.0	1.5	2.5	4.0	6.5	10
D	8	12											*			2
E	13	20										*			2	4
F	20	30									*			2	3	5
G	32	48								*			2	3	5	8
H	50	75							*			2	3	5	8	12
J	80	120						*			2	3	5	8	12	18
K	125	188					*			2	4	5	8	13	19	28
L	200	300				*			2	4	5	8	13	19	28	
M	315	473			*			2	4	5	8	13	19	28		
N	500	750		*			2	4	5	8	13	19	28			
P	800	1200	*			2	4	5	8	12	18	27				
Q	1250	1875			2	4	5	8	13	18	27					
R	2000	3000		2	4	5	8	13	18	28						
S	3150	4725	2													

* Use corresponding single sampling plan in ISO 2859-1 with $Ac = 0$.

TABLE 8.12 Example Sequential Plan

Cumulative sample size n_{cum}	Acceptance number $gn_{cum} - h_A$	Rounded acceptance number A	Rejection number $gn_{cum} + h_R$	Rounded rejection number R
1	–1.839	*	1.721	†
2	–1.769	*	1.791	2
3	–1.698	*	1.862	2
4	–1.628	*	1.932	2
5	–1.557	*	2.003	3
…	…	…	…	…
20	–0.498	*	3.062	4
…	…	…	…	…
28	0.067	0	3.627	4
…	…	…	…	…
34	0.490	0	4.050	5
…	…	…	…	…
42	1.055	1	4.615	5
…	…	…	…	…
48	1.479	1	5.039	6
…	…	…	…	…
56	2.044	2	5.604	6
…	…	…	…	…
63	2.538	2	6.098	7
…	…	…	…	…
70	3.032	3	6.592	7
…	…	…	…	…
76	3.456	3	7.016	8
…	…	…	…	…
84	4.020	4	7.580	8
…	…	…	…	…
90	4.444	4	8.004	9
…	…	…	…	…
98	5.009	5		9
…	…	…		…
113	6.068	6		9
…		…		…
119		6		…
120		8		9

* Cumulative sample size is too small to permit acceptance.
† Cumulative sample size is too small to permit rejection.

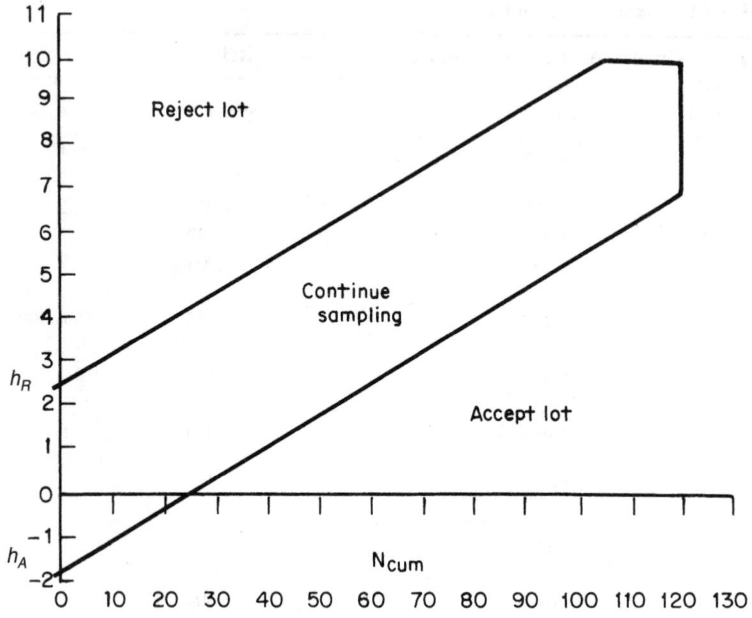

FIGURE 8.3 Example graph for sequential sampling using ISO 8422.

such a way that the operating characteristic curves match, as closely as possible, those found in ISO 2859-1 and ANSI/ASQC Z1.4.

8.12 CONCLUSIONS

Many other acceptance sampling procedures are available for the user that have not been discussed in this chapter. It is believed by the author that the ones discussed here are the ones that are most widely used for industrial applications. For a discussion of these procedures the reader is referred to texts such as Schilling (1982), Duncan (1986), or Wadsworth, Stephens, and Godfrey (1998).

A partial list of other procedures and appropriate standards, if they exist, that might be of interest to the reader is

- Variables sampling: ISO 3951, ANSI/ASQC Z1.9, MIL STD 414-1957, ISO 8423
- Skip-lot sampling procedures: ISO 2859-3, ANSI/ASQC S1, Stephens (1995*b*)
- Continuous sampling procedures: MIL STD 1235, Stephens (1995*a*)
- Attributes sampling procedures: ISO 2859-2, ANSI/ASQC S2, ANSI/ASQC Q3, ISO 2859-4

Most of the above references are available through the American Society for Quality.

REFERENCES

ANSI/ASQC S1: *An Attribute Skip-Lot Sampling Program,* American Society for Quality, Milwaukee, 1996.

ANSI/ASQC S2: *Introduction to Attribute Sampling,* American Society for Quality, Milwaukee, 1995.

ANSI/ASQC Q3: *Sampling Procedures and Tables for Inspection of Isolated Lots by Attributes,* American Society for Quality, Milwaukee, 1988.

ANSI/ASQC Z1.4: *Sampling Procedures and Tables for Inspection by Attributes,* American Society for Quality, Milwaukee, 1993.

ANSI/ASQC Z1.9: *Sampling Procedures and Tables for Inspection by Variables for Percent Nonconforming,* American Society for Quality, Milwaukee, 1993.

ANSI/ISO/ASQC A3534-2: *Statistics—Vocabulary and Symbols—Statistical Quality Control,* American Society for Quality, Milwaukee, 1993.

ANSI/ISO/ASQC A8402: *Quality Management and Quality Assurance—Vocabulary,* American Society for Quality, Milwaukee, 1994.

Duncan, A. J.: *Quality Control and Industrial Statistics,* 5th ed., Irwin, Homewood, IL, 1986.

ISO 2859-1: *Sampling Procedures for Inspection by Attributes—Part 1: Sampling Plans Indexed by Acceptable Quality Level (AQL) for Lot-by-Lot Inspection,* International Organization for Standardization, Geneva, 1998.

ISO 2859-2: *Sampling Procedures for Inspection by Attributes—Part 2: Sampling Plans Indexed by Limiting Quality (LQ) for Isolated Lots,* International Organization for Standardization, Geneva, 1987.

ISO 2859-3: *Sampling Procedures for Inspection by Attributes—Part 3: Skip Lot Sampling Procedures,* International Organization for Standardization, Geneva, 1988.

ISO 2859-4: *Sampling Procedures for Inspection by Attributes—Part 4: Sampling Plans for Assessment of Conformity to Stated Quality Levels,* International Organization for Standardization, Geneva, 1997.

ISO 3534-2: *Statistics—Vocabulary and Symbols—Part 2: Statistical Quality Control,* International Organization for Standardization, Geneva, 1993.

ISO 3951: *Sampling Procedures and Charts for Inspection by Variables for Percent Nonconforming,* International Organization for Standardization, Geneva, 1989.

ISO 8402: *Quality Management and Quality Assurance—Vocabulary,* International Organization for Standardization, Geneva, 1994.

ISO 8422: *Sequential Sampling Plans for Inspection by Attributes,* International Organization for Standardization, Geneva 1988.

ISO 8423: *Sequential Sampling Plans for Inspection by Variables for Percent Nonconforming (Known Standard Deviation)*, International Organization for Standardization, Geneva, 1988.

Schilling, E. G.: *Acceptance Sampling in Quality Control*, Marcel Dekker, New York, 1982.

Stephens, K. S.: *How to Perform Continuous Sampling*, 2d ed., American Society for Quality, Milwaukee, 1995*a*.

_______: *How to Perform Skip-Lot and Chain Sampling*, American Society for Quality, Milwaukee, 1995*b*.

U.S. Department of the Army: *Single and Multi-level Continuous Sampling Procedures and Tables for Inspection by Attributes,* Mil. Std. 1235 [ORD], Government Printing Office, Washington, DC, 1962.

———: *Sampling Procedures and Tables for Inspection by Variables for Percent Defective,* Mil. Std. 414, Government Printing Office, Washington, DC, 1957.

Wadsworth, H. M., K. S. Stephens, and A. B. Godfrey: *Modern Methods for Quality Control and Improvement, 2nd ed.,* Wiley, New York, 1998.

CHAPTER 9

ANALYSIS AND PRESENTATION OF RELIABILITY DATA

Jeffrey H. Hooper
Lucent Technologies, Warren, NJ

and

Sigmund J. Amster
Colts Neck, NJ

9.1 INTRODUCTION

Statistical methods for the analysis and presentation of reliability data play an important role in the design and manufacture of reliable products. These methods are an essential part of accelerated life testing of critical parts, prototype testing, process design and qualification, first office applications, and field-tracking studies.

This chapter is organized around the actions one can take to analyze and present reliability data for nonrepairable items effectively. We show where reliability questions and data occur within the context of a product reliability program, and we lay out a process for answering reliability questions and for classifying reliability data. We carefully describe the types of reliability data that can be handled by the analysis methods we present. The basic reliability models that are used in the simple situations with no explanatory variables and the more complex reliability models that are used with accelerated life testing or vendor comparisons are reviewed. A method for determining the sample size in some reliability studies is covered as well as the strategies for analyzing and presenting reliability data both with and without explanatory variables.

Because reliability data are often censored, simple analytical methods for analyzing reliability data do not exist. For this reason the chapter is written assuming that the reader has access to a software system for analyzing reliability data. Many such software systems exist. The excellent paper by Wagner and Meeker (1985) presents a careful comparison of the features of many of the software systems capable of analyzing reliability data. The examples in this chapter were analyzed using the STAR software system developed at AT&T Bell Laboratories.

9.2 SOURCES OF RELIABILITY QUESTIONS AND DATA

Reliability does not just happen in a product. It must be planned for and built into the product at each stage of the product realization process.

A product reliability program helps to ensure that the delivered product meets or exceeds its reliability objectives, and does so economically. Such a program is comprised of a coordinated set of activities that begin at the early stages of item design and continue through design and development, implementation, manufacturing, and sales and support. Figure 9.1 presents some reliability program activities that involve statistical methods for designing reliability studies and analyzing and presenting reliability data. A brief description of each of these reliability program activities and the important reliability questions they are designed to answer follows. These questions are stated in an imprecise manner, as they are typically asked of the reliability engineer or analyst. To provide a meaningful answer, the reliability engineer must then use his or her knowledge and judgment to interpret these questions precisely.

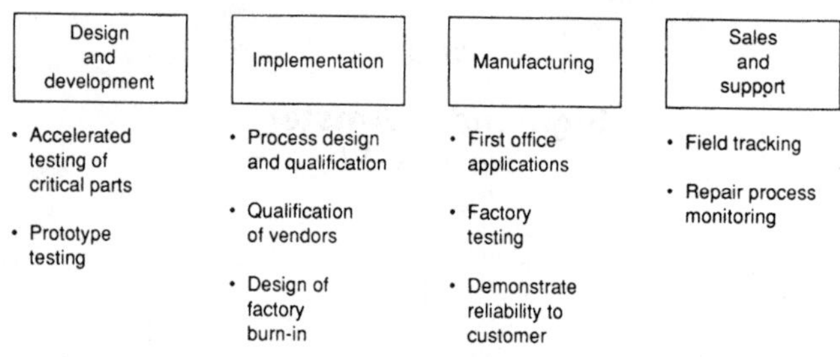

FIGURE 9.1 Reliability program activities involving the collection and analysis of reliability data.

Design and Development

1. *Accelerated testing of critical parts:* A small number of parts whose reliability is critical to the performance of the item are tested in the laboratory under high-stress conditions to induce early failures.

   Question: Do these critical parts meet their reliability objective at design stress and, if not, how can we improve the reliability of these parts?
2. *Prototype testing:* A small number of prototype items are tested in the laboratory either under accelerated conditions or under simulated use conditions for an extended period of time.

   Question: Can the current item design meet the item reliability objective?

Implementation

1. *Process design and qualification:* Items made on a pilot production line are tested at either design stress or accelerated stress conditions.

   Question: Can the manufacturing process as currently designed produce items that meet their reliability objective?

2. *Qualification of vendors:* Selected parts purchased from outside vendors are tested in the laboratory at either accelerated or design stress conditions.
   Question: Which vendor supplies the most reliable part?
3. *Design of factory burn-in:* Items manufactured on the final production line are tested in the laboratory at either accelerated or design stress conditions.
   Question: Is item operation in the factory or burn-in necessary for all production items to meet the reliability objective?

Manufacturing

1. *First office applications:* The first items installed in the field are monitored carefully and reliability data are collected on their performance.
   Question: Is the item performance in the hands of the customer meeting the item reliability objective, and what is an early estimate of the item warranty expense?
2. *Factory testing:* Samples of items are periodically taken from the production line and tested under simulated use conditions in the factory.
   Question: Is the production line continuing to produce reliable items, and can we offer evidence to our customers that our item meets its reliability objective?

Sales and Support

1. *Field tracking:* A large number of items are tracked carefully in the field, and their performance is monitored.
   Question: Is the item performance in the hands of the customers meeting its reliability objective?
2. *Repair process monitoring:* The item repair process is carefully monitored in order to collect reliability data on the item and its parts.
   Question: Do opportunities to improve item performance exist because certain parts are greatly exceeding their reliability objective, and how can we use this reliability information to improve our knowledge of part reliability?

9.3 ANSWERING RELIABILITY QUESTIONS

In this section we present a well-defined process for answering reliability questions. This requires not only an answer to the question, but also a measure of how precise that answer is. This process begins with the identification of the population or all items of interest, defines what is meant by a failure, selects a representative sample from the population, measures the time to failure of the items in the sample, analyzes the reliability data, and presents the results as the basis for action. Figure 9.2 illustrates the action steps in this process as well as the results or outputs of each action. In the development that follows, the process of answering reliability questions is presented in general, along with the specific process for answering questions such as, "What proportion of circuit packs will fail during the warranty period?".

Identify All Items of Interest. The items of interest are those items about which we want to be able to answer the question. In theory the items of interest should

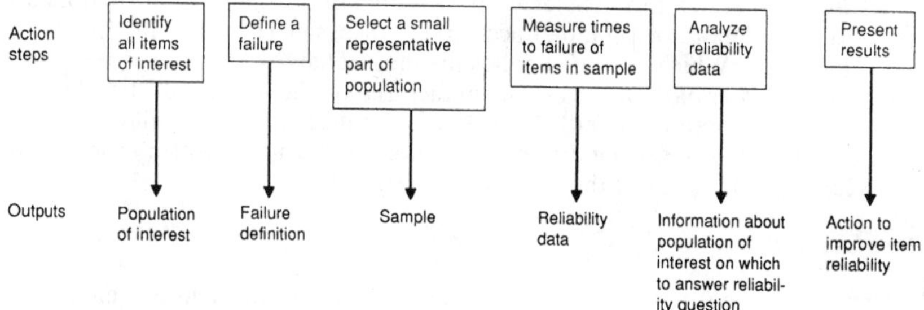

FIGURE 9.2 Process for answering reliability questions.

be carefully stated as part of the question, but in practice they almost never are. The collection of all items of interest is called the *population of interest*, or simply the population.

Example 9.1. All the circuit packs made in a specific manufacturing plant during the first month of production.

Define What Is Meant by a Failure. It is necessary to be able to distinguish between an item that has failed and one that has not. Often the precise definition of a failure is subjective, and it is very important that the failure definition be specified carefully before we have to measure the times to failure. The definition of failure may involve a degradation threshold below which the item is said to have failed. This is often the case with a large complex item where failure means the item can no longer deliver a specified proportion of its functionality. It is also important to decide in advance whether intermittent failures, no trouble founds, or dead on arrivals will be called failures.

Example 9.2. A circuit pack has failed when it can no longer pass the standard functional test that tests all of its functions. Intermittents, no trouble founds, and dead on arrivals will not be called failures. Thus every failed circuit pack must have passed an initial functionality test at the installation site and failed a final functionality test at the repair site.

Select a Representative Sample of the Population. Simple random sampling is the method of selecting a representative sample of a population when no prior information about the population is available. We always assume that the population is much larger than the sample. Simple random sampling means that every possible sample of size n from the population has an equal chance of being selected. In practice it is often neither possible nor desirable to select a simple random sample of either shipped parts or manufactured items. Instead it is usually preferable to ensure that our sample is spread out more uniformly over the population by either selecting some parts from each lot shipped or selecting some items from each day's or week's production run. See Chap. 10 for details on methods for selecting samples.

If we assume that the population from which we are sampling is very homogeneous, that is, the manufacturing process which was in a state of statistical control, then our more spread-out or stratified method of sampling will give the same results as simple random sampling. If the manufacturing process went out of control during the period in which we were drawing our sample, it would be necessary to treat the in-control and out-of-control portions of the population separately, and it may not be meaningful to analyze the out-of-control portion at all.

Sometimes it is not possible to sample from the entire population of interest. For example, we may be interested in the first 6 months of production, but be required to draw our sample after only 3 months of production. Our sample can only provide information about the part of the population from which it was drawn. Information about the unsampled part of the population must be based on engineering judgment and additional information about the unsampled part of the population.

Example 9.3. Using simple random sampling we select an equal number of circuit packs from each week's production. We monitor the process control information during this period carefully to make sure that the manufacturing process remains in a state of statistical control.

Measure Times to Failure. Reliability data are measurements of the times to failure and, possibly additional measurements, taken on all the items in our representative sample. It is often not possible to measure the times to failure very precisely, and many times all we know is that the time to failure of an item occurred in some time interval, which may be the unbounded time interval between some fixed time t and infinity. Such imprecise data are called *censored*, and this concept is covered in detail in Sec. 9.4. The units of failure time can be calendar units, such as hours or days, or count units, such as cycles or operations.

Example 9.4. The circuit packs in our sample were used in a first office application. They were not monitored continuously, but they were inspected periodically to see whether they were still functioning. Thus there were no exact failure times. For every failed circuit pack we know the interval in which it failed, and for all other circuit packs in the sample all we know is that their time to failure was greater than their last inspection time. The failure times of the circuit packs were measured in days.

Analyze Reliability Data. The process of statistical reliability analysis turns reliability data into information about the population of interest, which is used to answer the reliability question. The process of analyzing reliability data is covered in detail in Secs. 9.6 and 9.9.

Example 9.5. Using the measurements of the times to failure of our sample of circuit packs from the first month of production, we estimate the proportion of circuit packs from the first month of production which will fail during the warranty period. If the warranty period is 12 months, we estimate the probability of a circuit pack failing before 365 days. We also provide a 90 percent confidence interval for our estimate.

Present Results as a Basis for Action. A careful analysis of reliability data is of little use if a decision maker cannot readily understand the results. The results should be presented in a clear and uncluttered fashion so that a decision maker can quickly grasp the essential information. A measure of the precision of the results should always be included because the conclusiveness of the results is often as important as the results themselves in determining the best action.

Example 9.6. The proportion of circuit packs failing by time t is plotted as a function of time for $t = 0$ to 500 days. The 90 percent pointwise confidence bands and the nonparametric estimate of the cumulative distribution function (see Sec. 9.6) are included on the plot.

9.4 RELIABILITY DATA

9.4.1 Identifying Reliability Data

Reliability data are measurements of the time to failure and, possibly additional measurements on a sample of items. The following are important characteristics of reliability data.

Operating Time. For reliability data, time is the operating time of the item. This assumes that the item only ages when it is operating. Thus an item that had been in the field for 6 months but only operating for 3 months would be said to have aged 3 months.

Time Units. Time units are typically taken to be calendar time units such as hours, days, or months. However, reliability data can also be in terms of count units such as cycles or operations. We recommend that you put your data in the time units that make it easiest to answer the question. If the reliability question is in hours, then put your data in hours.

Dead on Arrivals. We assume that all items are initially functioning at time 0, so that dead on arrivals are not considered to be part of reliability data.

Time 0. Time 0 is the first time a failure can occur that is relevant to the person answering the reliability question. For example, suppose a customer wants to know the failure rate for a certain type of circuit pack at 10,000 hours. Then time 0 is when the circuit pack first begins operation on the customer's site, not when it first began operating in the factory test.

While our definition of reliability data is quite broad, it does not cover all types of reliability data. The following data types are *not* covered in this chapter.

Degradation Data. Degradation data are measurements over time of a product performance characteristic. Thus instead of a single time-to-failure measurement on an item we could have hundreds or even thousands of measurements of a performance characteristic as it changed over time. The analysis of this type of data is very complex, and the statistical methods for dealing with this are not well developed (see Carey and Grize, 1986).

Repairable Systems. Repairable items can experience a large number of failures over their economic lives since they are repaired and restored to operation at failure. Thus instead of a single time-to-failure measurement on an item we have a sequence of times between failure. The models and methods for analyzing these kinds of items are very different from the models and methods presented in this chapter for analyzing nonrepairable items. Today most reliability data are collected on nonrepairable items.

Count Data. Count data are the simplest types of reliability data. For this type of data the time interval of interest is fixed, as is typically the case with a military mission time, and all that is recorded is whether or not an item failed or passed its mission. These types of data are very easy to analyze using the simple binomial or Poisson model, and the appropriate types of analyses are covered in Chaps. 4 and 7 of this handbook.

9.4.2 Classifying Reliability Data

The appropriate statistical analysis for a set of reliability data depends on the characteristics of the data. In this section we classify reliability data along two different dimensions. The first dimension is the precision with which the times to failure have been measured. Lack of precision in measurement is called censoring, and we will classify reliability data according to the type of censoring. The second dimension is the type of explanatory variable that is included in the data. Explanatory variables explain some of the variability in the observed times to failure. For simplicity, we only consider the case of a single explanatory variable.

Types of Censoring. There are four basic types of censoring: exact failure, right censoring, left censoring, and interval censoring. These are shown in Fig. 9.3.

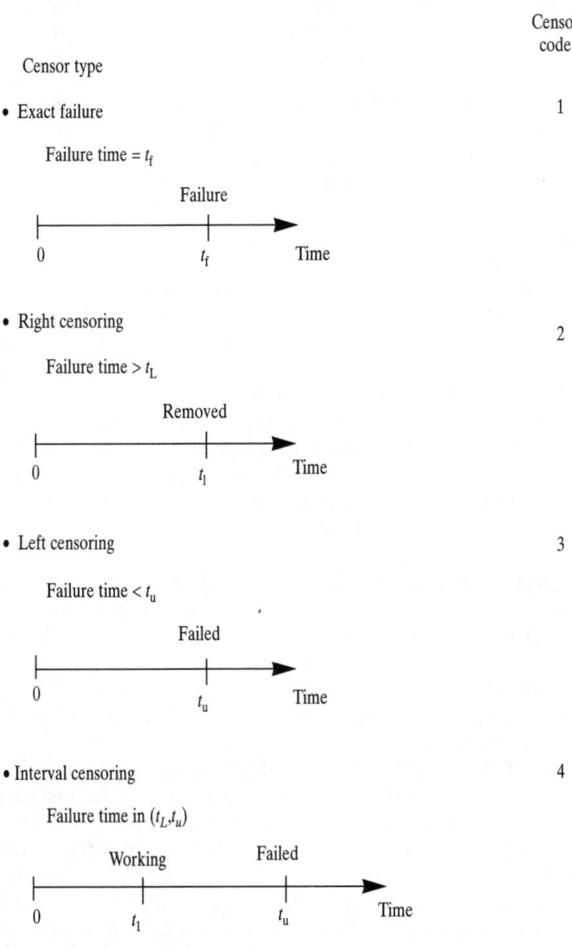

FIGURE 9.3 Types of censoring.

1. *Exact failure:* For an exact failure an item is put on test at time 0 and monitored continuously until it fails at time t_f. Time is assumed to be measured without error. In this case we know the failure time of the item exactly. We use a censor code of 1 to denote an exact failure.

2. *Right censoring:* For a right-censored observation an item is put on test at time 0 and then removed from test at time t_l while the item is still working. In this case all we know is that the item would have failed at some time greater than $t_l \equiv$ lower time. Clearly this is a very imprecise measurement of the failure time. We use a censor code of 2 to denote a right-censored observation.

3. *Left censoring:* For a left-censored observation an item is put on test at time 0 and then not monitored again until time t_u, at which time it is inspected and found to be no longer functioning. In this case all we know is that the item failed at some time greater than zero and less than $t_u \equiv$ upper time. Since there are no dead on arrivals in our data, all failure times are strictly greater than zero. We use a censor code of 3 to denote a left-censored observation.

4. *Interval censoring:* For an interval-censored observation an item is put on test at time 0, inspected at time t_l and found to be functioning, and then inspected again at time t_u and found to have failed. In this case all we know is that the item failed at some time greater than t_l and less than t_u, but we do not know where in the interval it failed. We use a censor code of 4 to denote an interval-censored observation.

Basic Censoring Assumption. In order to analyze censored reliability data it is necessary to make a basic censoring assumption. The assumption is that items removed from test at time *t* should be representative of the items on test at time *t*. The implication of this assumption is that items cannot be removed from test because they have an unusually high or low risk of failure. Censoring schemes which satisfy this assumption are called *uninformative*, because the censoring scheme tells us nothing about the time-to-failure distribution. When this assumption is violated, the censoring scheme is called *informative*, and in this case the censoring scheme must be included in the statistical analysis of the data. This is generally very complicated, and we do not deal with this situation.

Types of Explanatory Variables. We consider three types of explanatory variables: no explanatory variables, a single qualitative explanatory variable, and a single quantitative explanatory variable.

1. *None:* With no explanatory variables a single time-to-failure distribution will explain all the variability in the observed times to failure. This is the simplest and most common situation.

   Example 9.7 Circuit pack first office application: A sample of 2275 circuit packs was selected from the first month's production and installed at the first customer location. Once in operation these circuit packs were not monitored continuously, but were periodically inspected to see whether they were still functioning. Thus there were no exact failure times in the data. For every failed circuit pack we know the interval in which it failed, and for all other circuit packs in the sample all that we know is that their time to failure was greater than the last inspection time. Table 9.1 contains this data set with time measured in days. The "weights" column specifies the number of circuit packs that have the same times and censoring codes. We need to answer the question, "What proportion of circuit packs will fail during the warranty period of 12 months?" The process for answering this question is described in detail in Sec. 9.3

2. *Qualitative:* A qualitative explanatory variable denotes the category in which an item is placed. For example, in a vendor comparison study some items were

TABLE 9.1 Circuit Pack First Office Application

Row	Lower time, days	Upper time, days	Censoring code	Weight
1	1	1	3	10
2	1	2	4	2
3	2	3	4	3
4	3	6	4	2
5	1	8	4	5
6	7	11	4	2
7	8	13	4	3
8	11	24	4	4
9	11	25	4	1
10	7	27	4	1
11	28	28	2	10
12	27	28	4	1
13	26	29	4	1
14	30	31	4	1
15	24	52	4	2
16	25	53	4	1
17	26	54	4	1
18	32	55	4	2
19	31	55	4	2
20	52	65	4	1
21	93	93	2	1
22	55	96	4	1
23	101	101	2	1
24	104	104	2	2
25	107	107	2	6
26	52	108	4	1
27	111	111	2	4
28	52	123	4	5
29	53	126	4	3
30	54	126	4	8
31	55	127	4	1
32	188	188	2	669
33	126	188	4	1
34	127	191	4	5
35	191	191	2	1048
36	126	191	4	1
37	192	192	2	460
38	126	192	4	3

Censoring summary	
Total number of units	2275
Number of units with exact failure times	0
Number of units right-censored	2201
Number of units left-censored	10
Number of units interval-censored	64

produced by vendor 1 and some items were produced by vendor 2. In this case vendor is a qualitative explanatory variable, and some of the variability in the observed times to failure will be explained by the vendor category. Because vendor is a qualitative explanatory variable, knowing something about the reliability of the items produced by vendors 1 and 2 does not tell us anything about the reliability of items produced by vendor 3.

Example 9.8 Vendor comparison study: Two vendors (1 and 2) presently supply circuit packs used in a system. A new system is being developed, and it is necessary to choose one of these two suppliers as the sole source on the basis of the more reliable product. A sample of 1041 circuit packs from vendor 1 and 1245 circuit packs from vendor 2 are put on life test and continuously monitored for 1 year, at which time the test is terminated. Table 9.2 contains this data set with time measured in days. Can a definitive choice between vendors be made on reliability considerations alone, or must other factors be given importance since there is little to choose on the basis of reliability?

3. *Quantitative:* A quantitative explanatory variable can take on a continuum of values and denotes the explanatory level of the variable at which the item was tested. The most common situation is accelerated life testing in which items are tested at different stress levels so that we can model the relationship between the level of stress and the time-to-failure distribution. This allows us to determine the time-to-failure distribution at any level of stress for which the model holds.

Example 9.9 Device accelerated life test: A new device is to be used in a high-reliability application. It must be determined whether or not the device can meet its failure rate objective of 1000 FITs at 10,000 hours at the design temperature of 10°C, where 1 FIT ≡ $1/(10^9$ hours). A sample of 165 devices was selected for a 5000-hour accelerated life test. 30 devices were tested at 10°C, 100 devices were tested at 40°C, 20 devices were tested at 60°C, and 15 devices were tested at 80°C. Table 9.3 contains these data.

9.5 RELIABILITY MODELS: NO EXPLANATORY VARIABLES

This section presents the basic measures for describing populations of times to failure and also the parametric models most frequently used to describe them. Even though no model is ever completely "correct," they can be extremely useful for summarizing reliability data. The population descriptors are often used to specify a reliability objective.

9.5.1 Population Descriptors

Cumulative Distribution Function. The cumulative distribution function (cdf) is denoted by $F(t)$, the probability of failure before time t as a function of t.

Reliability Function. The reliability function is denoted by $R(t)$, the probability of failure after time t as a function of t. $R(t) = 1 - F(t)$.

Probability Density Function. The probability density function (pdf) is denoted by $f(t)$,

TABLE 9.2 Circuit Pack Vendor Comparison Study

Row	Failure time, days	Censoring code	Weight	Vendor
1	1.3	1	1	1
2	1.7	1	1	1
3	2.4	1	1	1
4	3.2	1	1	1
5	7.5	1	1	1
6	11.1	1	1	1
7	28.7	1	1	1
8	34.6	1	1	1
9	38.4	1	1	1
10	45.8	1	1	1
11	52.9	1	1	1
12	58.4	1	1	1
13	66.1	1	1	1
14	68.2	1	1	1
15	94.7	1	1	1
16	128.7	1	1	1
17	143.6	1	1	1
18	162.3	1	1	1
19	183.6	1	1	1
20	245.9	1	1	1
21	331.6	1	1	1
22	365.0	2	1020	1
23	1.4	1	1	2
24	2.6	1	1	2
25	2.7	1	1	2
26	4.7	1	1	2
27	12.8	1	1	2
28	18.2	1	1	2
29	33.5	1	1	2
30	45.6	1	1	2
31	50.0	1	1	2
32	60.3	1	1	2
33	73.6	1	1	2
34	79.1	1	1	2
35	98.1	1	1	2
36	150.6	1	1	2
37	174.1	1	1	2
38	230.5	1	1	2
39	311.2	1	1	2
40	326.4	1	1	2
41	365.0	2	1227	2

Censoring summary	
Total number of units	2286
Number of units with exact failure times	39
Number of units right-censored	2247
Number of units left-censored	0

TABLE 9.3 Device Accelerated Life Test

Row	FITs	Censoring code	Weight	Temperature, °C
1	5000.000	2	30	10
2	1298.363	1	1	40
3	1390.389	1	1	40
4	3186.731	1	1	40
5	3240.641	1	1	40
6	3261.180	1	1	40
7	3313.471	1	1	40
8	4501.063	1	1	40
9	4567.932	1	1	40
10	4840.625	1	1	40
11	4982.190	1	1	40
12	5000.000	2	90	40
13	581.2401	1	1	60
14	924.7739	1	1	60
15	1431.521	1	1	60
16	1586.338	1	1	60
17	2452.262	1	1	60
18	2733.527	1	1	60
19	2771.710	1	1	60
20	4105.611	1	1	60
21	4674.027	1	1	60
22	5000.000	2	11	60
23	283.0105	1	1	80
24	360.7869	1	1	80
25	514.6884	1	1	80
26	637.7716	1	1	80
27	853.8023	1	1	80
28	1024.483	1	1	80
29	1030.349	1	1	80
30	1044.687	1	1	80
31	1767.128	1	1	80
32	1776.572	1	1	80
33	1855.928	1	1	80
34	1950.633	1	1	80
35	1963.522	1	1	80
36	2884.282	1	1	80
37	5000.000	2	1	80

Censoring summary

Total number of units	165
Number of units with exact failure times	33
Number of units right-censored	132
Number of units left-censored	0
Number of units interval-censored	0

$$f(t) \frac{d}{dt} \{F(t)\}$$

Failure Rate (Hazard Rate). The failure, or hazard, rate is denoted by $h(t)$,

$$h(t) = \frac{f(t)}{R(t)}.$$

It is approximated by

$$h(t)\ dt \approx \text{Prob(failing in } (t, t + dt) \text{ given that unit survived to } t)$$

Note: $h(t)$ is a population characteristic. For example, a misture of two populations with constant but different failure rates will exhibit a decreasing failure rate. This is true even though none of the individual items is improving. Also, since failure rates per hour can be very small numbers, the use of FITs is frequently convenient, where 1 FIT $\equiv 1/(10^9$ hours).

Quantile. A quantile is denoted by $t(p)$,

$$P[T < t(p)] = p$$

that is, a fraction p of the population will fail before the time $t(p)$.

9.5.2 Parametric Reliability Models

Exponential Model

$$f(t) = \lambda \exp(-\lambda t)$$
$$R(t) = \exp(-\lambda t)$$
$$h(t) = \lambda$$

where λ is the failure rate. Since the failure rate for this distribution is constant, an item that had been operating for t hours would have the same reliability as a brand new item from the same population. Stated more precisely, this model has the "lack of memory property." For any $s > 0$,

$$P(T < t + s \text{ given that } T > t) = P(T < s).$$

Weibull Model

$$f(t) = \left(\frac{\beta}{\alpha}\right)\left(\frac{t}{\alpha}\right)^{\beta-1} \exp\left[-\left(\frac{t}{\alpha}\right)^{\beta}\right]$$
$$R(t) = \exp\left[-\left(\frac{t}{\alpha}\right)^{\beta}\right]$$
$$h(t) = \left(\frac{\beta}{\alpha}\right)\left(\frac{t}{\alpha}\right)^{\beta-1}$$

where α is the scale parameter and β is the shape parameter. Therefore, the failure rate is monotonically increasing (IFR) or decreasing (DFR), depending on the shape parameter β. In the special case where $\beta = 1$, the Weibull distribution is an exponential distribution. Figure 9.4 is a plot of the Weibull failure rates.

Lognormal Model. *T* follows the lognormal distribution if and only if

$$\ln T \sim N(\mu, \sigma^2)$$

(see Chap. 2). The lognormal failure rate satisfies $h(0) = 0$, and $h(t)$ increases to a maximum and then decreases to zero. Even though the failure rate for any lognormal distribution is not monotonic, if "very early" failures are not observed, it may be difficult or impossible to distinguish a DFR Weibull distribution from a lognormal distribution. However, for *t* large, the cdf of the Weibull distribution will be higher than that of the lognormal model. Therefore, the Weibull distribution will give a more conservative extrapolation. Figure 9.5 is a plot of lognormal failure rates.

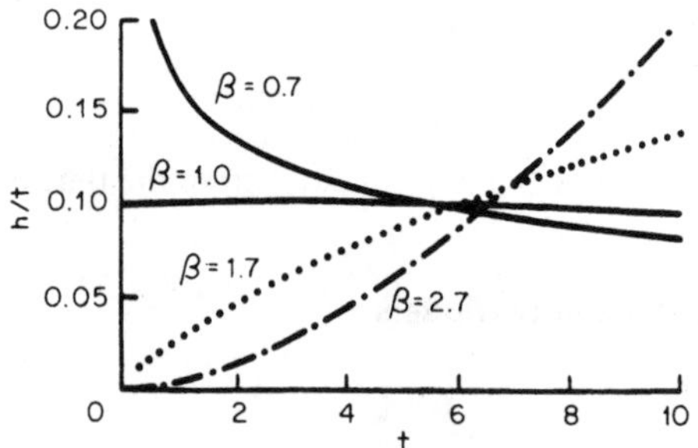

FIGURE 9.4 Weibull failure rates.

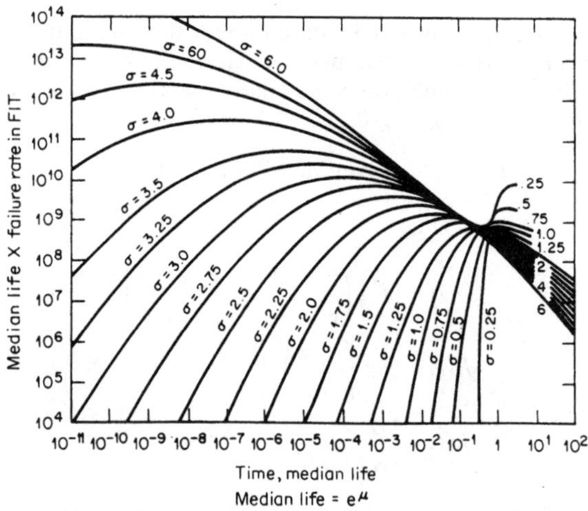

FIGURE 9.5 Lognormal failure rates.

9.6 ANALYZING RELIABILITY DATA: NO EXPLANATORY VARIABLES

In this case it is assumed that all the data come from a single population and that the reliability question will concern some characteristic of this population.

9.6.1 State Problem Precisely

Example 9.10. Determine a point estimate and a 95 percent confidence interval for the probability of circuit pack failure before 1000 days. The data found in Table 9.1 are used for the analysis. The "weight" column gives the number of circuit packs having the same failure experience. For example, the first row means that 10 circuit packs worked properly at time zero, were placed in operation, and failed in the first day of operation.

9.6.2 Construct a Nonparametric Estimate of the cdf

1. Estimate the cdf without making any parametric model assumptions. This estimate can be used for quantiles but not to estimate the pdf or the failure rate.

2. For a confidence interval on $F(t)$ for a single t, use a pointwise confidence interval at a selected confidence level. Otherwise, for more than one $F(t)$ or one or more quantiles, use simultaneous confidence intervals (see Nair, 1984).

3. A nonparametric estimate of the cdf is required to obtain a probability plot. Table 9.4 illustrates this for the data in Table 9.1.

4. If the nonparametric estimate is "reasonably" smooth and the reliability question does not involve an extrapolation, it may be possible to answer the question without any additional analysis. If so, the inference will not depend on a particular parametric distribution assumption, and may therefore be more easily accepted. However, the confidence interval or intervals in such a case will be wider than those obtained from a parametric analysis.

9.6.3 Obtain an Exponential Probability Plot

1. If the plot is convex (concave), this indicates that the population has an increasing (decreasing) failure rate.

2. For a DFR distribution, no preventive replacement policy can improve the reliability.

3. An exponential probability plot is not a good decision tool for determining an exponential distribution. If a "straight" line is obtained on a Weibull probability plot, fit a Weibull distribution. Then see if the confidence interval on β includes 1. If so, an exponential distribution may be a simple useful model of the population.

Example 9.11. Figure 9.6 contains the exponential probability plot for this set of data.

TABLE 9.4 Nonparametric Estimate of cdf with 90% Pointwise Confidence Bands for Circuit Pack First Office Application

Lower time, days	Upper time, days	Estimated cdf	Standard error	Lower confidence limit	Upper confidence limit
1	1	0.0044	0.0014	0.0021	0.0067
2	2	0.0058	0.0017	0.0030	0.0085
3	3	0.0078	0.0019	0.0046	0.0110
6	7	0.0091	0.0021	0.0057	0.0125
8	8	0.0107	0.0022	0.0071	0.0142
11	11	0.0107	0.0022	0.0071	0.0142
13	24	0.0145	0.0025	0.0104	0.0186
25	26	0.0145	0.0025	0.0104	0.0186
27	27	0.0145	0.0025	0.0104	0.0186
28	28	0.0166	0.0028	0.0120	0.0211
29	30	0.0166	0.0028	0.0120	0.0211
31	32	0.0176	0.0028	0.0131	0.0221
52	52	0.0176	0.0028	0.0131	0.0221
53	53	0.0176	0.0028	0.0131	0.0221
54	54	0.0176	0.0028	0.0131	0.0221
55	55	0.0247	0.0037	0.0186	0.0307
65	93	0.0282	0.0035	0.0225	0.0339
96	107	0.0282	0.0035	0.0225	0.0339
108	111	0.0282	0.0035	0.0225	0.0339
123	126	0.0282	0.0035	0.0225	0.0339
127	127	0.0282	0.0035	0.0225	0.0339
188	188	0.0296	0.0038	0.0235	0.0358
191	191	0.0337	0.0041	0.0269	0.0404
192	192	0.0346	0.0050	0.0263	0.0429

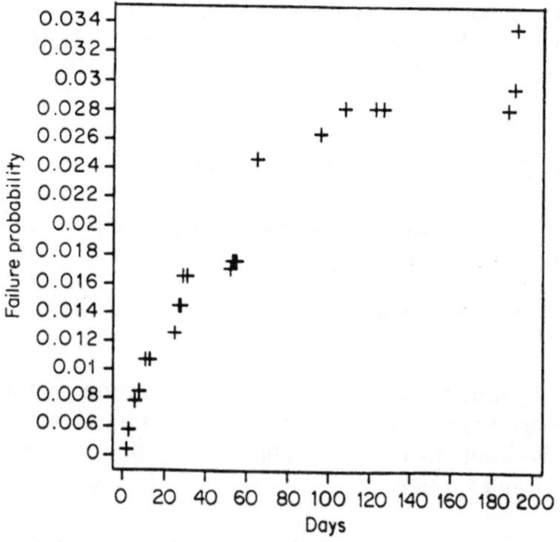

FIGURE 9.6 Exponential probability plot for circuit pack first office application.

9.6.4 Select a Parametric Distribution Using a Probability Plot

1. Select a parametric distribution and a confidence level.
2. Obtain a probability plot for that distribution and simultaneous confidence bands at that confidence level.
3. If the plot is "reasonably" straight and a straight line can be placed within the bands without touching the bands, that distribution is a candidate for parametric fitting. If not, try another distribution.
4. If several distributions satisfy the criteria of step 3, a parametric analysis can be applied using all of the candidate distributions and the results compared. If no distribution can be found which satisfies the criteria of step 3, then procedure 9.6.5 should not be used. This case is very complex. Possible explanations include mixtures of distributions, data that have been written down incorrectly, or other problems in the conduct of the experiment which generated the data.

Example 9.12. Figure 9.7 contains a Weibull probability plot for this set of data. Although simultaneous bands are not possible, the plot is extremely linear, and therefore a parametric model fit to this data would be in order. Figure 9.8 is a similar plot using the lognormal distribution.

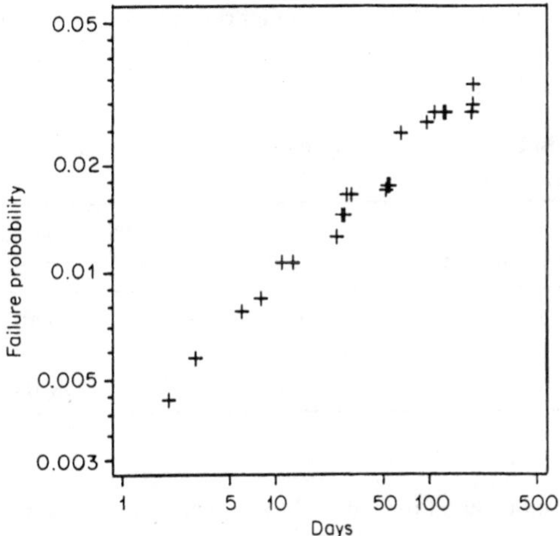

FIGURE 9.7 Weibull probability plot for circuit pack first office application.

9.6.5 Obtain a Parametric Model Fit to the Data

1. Using maximum-likelihood or other procedures, the data can be used to obtain statistical estimates of the unknown parameters in the selected parametric distribution.
2. Confidence intervals for these parameters can also be calculated.

Example 9.13. For these data we may use maximum-likelihood methods to obtain the estimates for the Weibull parameters listed in Table 9.5. Table 9.6 includes the data with 90% confidence bands and Fig. 9.9 is a plot of these data.

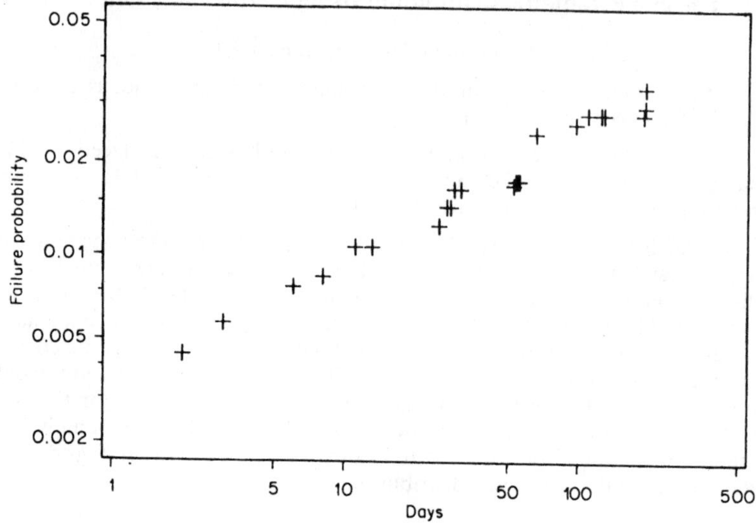

FIGURE 9.8 Lognormal probability plot for circuit pack first office application.

TABLE 9.5 Weibull Parameters for Circuit Pack First Office Application

Parameter	Estimate	95% Confidence interval
α	$1.462(10)^6$	$[1.50(10)^5, 1.43(10)^7]$
β	0.381	(0.298, 0.486)

TABLE 9.6 Weibull Failure Probability Table with 90 percent Pointwise Confidence Bands for Circuit Pack First Office Application

Failure time, days	Probability of failure	Standard error	Lower confidence limit	Upper confidence limit
1	0.0045	0.0012	0.0029	0.0070
5	0.0083	0.0017	0.0059	0.0116
25	0.0152	0.0023	0.0119	0.0194
50	0.0198	0.0026	0.0159	0.0245
100	0.0256	0.0030	0.0211	0.0311
250	0.0362	0.0042	0.0299	0.0436
365	0.0416	0.0049	0.0343	0.0505
500	0.0468	0.0057	0.0382	0.0572
1000	0.0605	0.0083	0.0482	0.0757

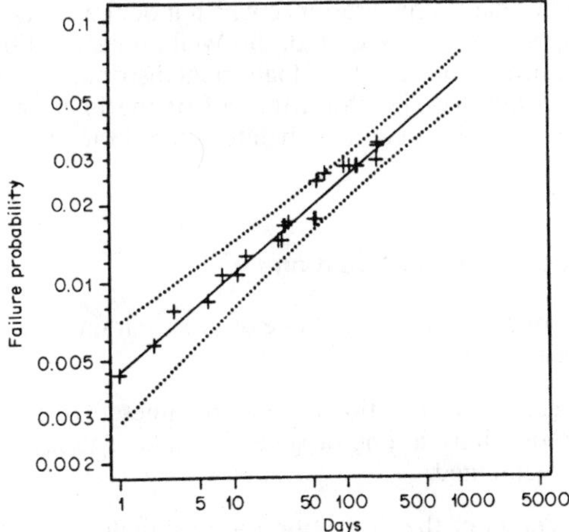

FIGURE 9.9 Weibull failure probability plot with 90% pointwise confidence bands for circuit pack first office application.

3. Compare lognormal and Weibull distributions.

Example 9.14. Figure 9.10 shows the result of fitting the circuit pack data to both the lognormal and the Weibull distributions. Both distributions "fit" the data very well. If the reliability question involved an interpolation, within the range of

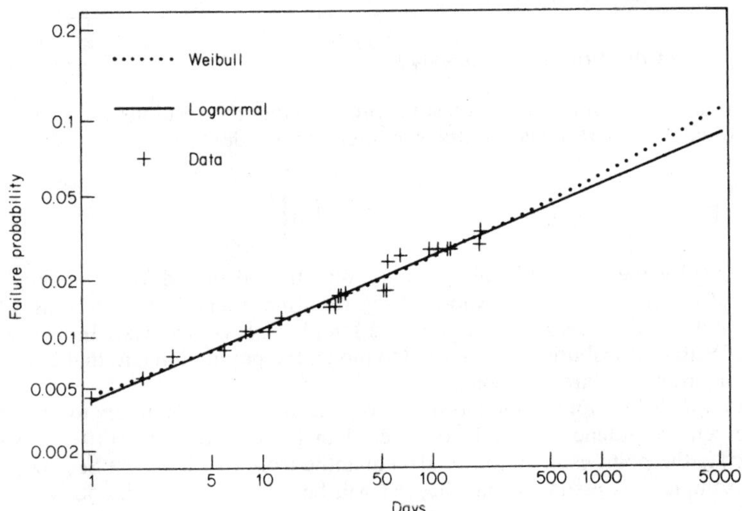

FIGURE 9.10 Comparison of Weibull and lognormal distributions for circuit pack first office application.

the data, it would not make any difference which model were chosen. However, if a large time extrapolation were required, the Weibull distribution would be more conservative. Technically, the tail of the lognormal distribution is heavier than that of the Weibull distribution. Therefore, the cdf of the lognormal will always be below that for the Weibull for sufficiently later time if they are similar for earlier time.

9.6.6 Answer the Reliability Question

1. If the question concerns the value of one or more parameters, the results of Sec. 9.6.5 can be used directly.

2. If the question concerns a function of these parameters (e.g., failure rate, quantile, probability of failure at one or several points of time), point and interval estimates can be obtained.

3. Once the parameters of the distribution are estimated, quantities beyond the range of the data can be estimated if the same distribution can be assumed to hold. But the physical question as to whether the same model holds beyond the actual data can only be answered by engineering considerations. If that is valid, the measurement of the extrapolation error is a statistical problem.

Example 9.15. Since the study was discontinued after 192 days, only a parametric analysis can be used to answer the question. For these data the point estimate of failure probability before 1000 days is 0.061 with a 95 percent confidence interval of (0.043, 0.084).

9.7 SAMPLE SIZE DETERMINATION WITH THE WEIBULL DISTRIBUTION*

9.7.1 State the Problem Precisely

Find a sample size n to place on test for time t so that the resulting reliability data will provide a $(1 - \alpha/2)$ level confidence interval no wider than

$$\left\{\frac{t'(p)}{1+r}, (1+r)t'(p)\right\}$$

where $t'(p)$ is the point estimate of the pth quantile calculated from the data. The value of r will determine the width of the resulting interval. As r decreases, the interval width decreases and the required sample size will increase. It is assumed that a Weibull distribution can be used to model the population and that estimates of both parameters are available.

Example 9.16. Find a sample size n to place on test for 60 hours to provide a 99 percent confidence interval no wider than $(t'(0.05)/1.25,\ 1.25t'(0.05)$, where $t'(0.05)$ is the point estimate of the 0.05 quantile from the data (see Sec. 9.6). For this example it is assumed that the test will be monitored continuously so that

*The discussion in this section is based on Meeker and Nelson (1976).

"exact" failure times will be obtained and failed items will not be replaced. A Weibull model with prior parameter estimates

$$a = 1045 \text{ hours}$$

$$b = 0.70$$

is also assumed, where a and b are estimates of α and β, respectively. (These could be obtained from previous tests on similar devices, literature searches, or elsewhere.)

9.7.2 Compute Standardized Censoring Time

$$Z(t) = b \ln \left(\frac{t}{a} \right)$$

Example 9.17

$$Z(60) = 0.70 \ln \left(\frac{60}{1045} \right) = -2.0.$$

9.7.3 Determine $V(p)$ from Fig. 9.11

Figure 9.11 is used by entering the horizontal scale at $100p$ [assuming there is interest in estimating $t(p)$], finding the intersection of that vertical line with the curve $Z(t)$, and reading $V(p)$ from the vertical scale corresponding to that intersection.

Example 9.18. Since $t(0.05)$ is being estimated, use 5 percent and $Z(60) = -2.0$, and $V(0.05)$ is read from Fig. 9.11 as 14.5.

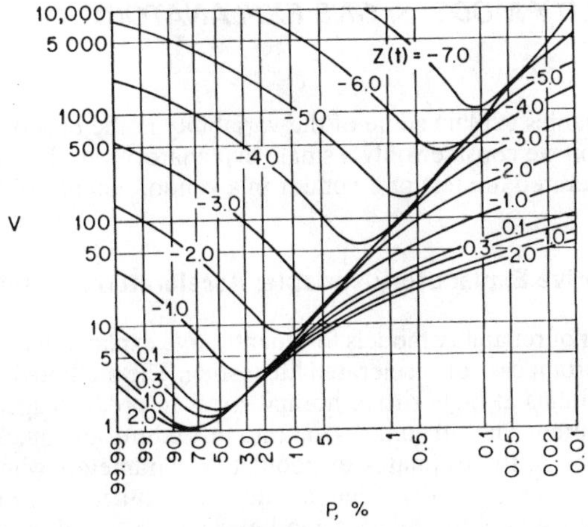

FIGURE 9.11 Determination of sample size for Weibull model.

9.7.4 Calculate the Sample Size *n*

$$n = V(p)\left[\frac{K(\alpha)}{b \ln (1 + r)}\right]^2$$

where $K(\alpha)$ is the $(1 - \alpha)$ quantile of the standard normal distribution. For example,

Example 9.19

α	0.1	0.05	0.01	0.005	0.001
$K(\alpha)$	1.28	1.645	2.327	2.575	3.08

$$n = 14.5\left[\frac{2.575}{0.70 \ln (1.25)}\right]^2 = 3941$$

A sample of 3941 units should be placed on test for a period of 60 hours and continuously monitored for failure if the assumptions are valid. In practice a plot of n versus r is frequently obtained to see the sensitivity of n to changes in r. Also, since prior information is usually very uncertain, it is recommended that the calculation of n be performed with various combinations of a and b. A choice of n can then be made by comparing the desire for conservatism and the strength of the prior information.

9.7.5 Sample Size in Accelerated Life Testing

See Meeker and Hahn (1985) for an excellent discussion.

9.8 RELIABILITY MODELS: ONE EXPLANATORY VARIABLE

Explanatory variables explain some of the variability in the observed times to failure. In this section we consider only a single explanatory variable, such as temperature in an accelerated life test or a vendor in a vendor comparison study.

9.8.1 Quantitative Explanatory Variable: Accelerated Life Testing

The development of reliability models for quantitative explanatory variables is done only for the important case of accelerated life testing. The method of accelerated life testing is to test items at higher than normal stress in order to speed up time and induce early failures. The advantages of this procedure are quicker results and increased precision in the estimates of population parameters when the test time and the sample size are tightly constrained. The disadvantage of accelerated life testing is that we must model the relationship between stress and the time to failure. This involves understanding the physics of failure. It is also difficult to verify the reasonableness of this model, and an incorrect model can give very misleading results.

Accelerated life tests are based on the physical assumption that all failure mechanisms are affected by accelerating stress in the same way. This assumption is approximately correct when there is a single or predominant failure mechanism. When this assumption is seriously violated, we may see failure mechanisms at high stress which are very different from the failure mechanisms we would see at design stress. Since most of our data will come from high-stress testing, we are really modeling the effect that stress has on the high-stress failure mechanisms, and then extrapolating this effect down to design stress. But the high-stress failure mechanisms do not occur at design stress, and the extrapolation will give erroneous results. To protect yourself against the phenomenon of masked failure mechanisms, failure mode analysis should be attempted. Also, some testing should always be done near design stress.

Accelerated Failure Time Model. The accelerated failure time model relates the time-to-failure distribution to the stress level. The basic idea of this model is that the level of stress is only compressing or expanding time (changing scale), and *not* changing the shape of the time-to-failure distribution. Let S, S_0 be two stress levels with $S > S_0$. Then testing for a time t at stress level S is equivalent to testing for time $a(S, S_0) \times t$ at stress level S_0, where $a(S, S_0) > 1$ is called the *acceleration factor.*

Let $T(S)$ be the time to failure at stress level S. Then for $S > S_0$,

$$T(S_0) = a(S, S_0) \times T(S), \qquad a(S, S_0) > 1 \tag{9.1}$$

or, equivalently,

$$\ln [T(S_0)] = \ln [a(S, S_0)] + \ln [T(S)]. \tag{9.2}$$

This means that stress affects the time-to-failure distribution by changing scale or, equivalently, stress affects the ln (time to failure) distribution by changing location.

If we assume

$$\ln [T(S)] \sim F(t; \mu(S), \sigma(S)) \tag{9.3}$$

where $\mu(S)$ = location parameter at stress level S

$\sigma(S)$ = scale parameter at stress level S

then the accelerated failure time model implies

$$\mu(S_0) = \ln [a(S, S_0)] + \mu(S)$$

$$\sigma(S_0) = \sigma(S) = \sigma$$

so that

$$\ln [T(S)] \sim F(t; \mu(S), \sigma). \tag{9.4}$$

Figure 9.12 illustrates the accelerated failure time model assuming $\mu(S)$ is a linear function of transformed stress. The probability density functions for the ln (time to failure) distribution at stress levels S_D, S_1, and S_2 are shown plotted around the location parameter line

$$\mu(S) = \beta_0 + \beta_1 X(S)$$

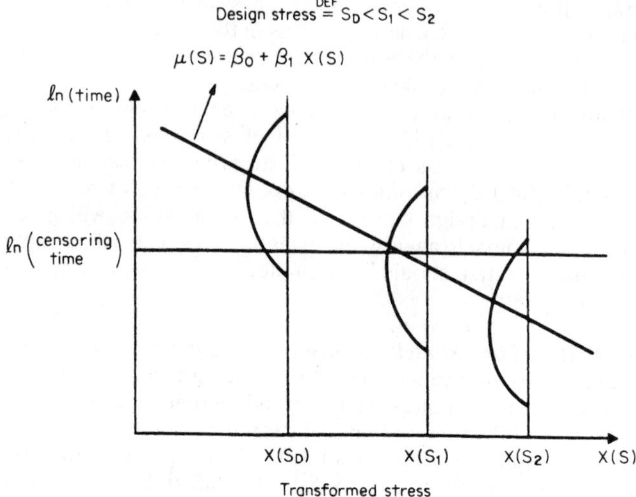

FIGURE 9.12 Accelerated failure time model.

where $X(S)$ is an increasing function of stress. The ln(time to failure) distribution for any stress level for which this model holds is obtained by sliding one of these probability density functions along this line.

We now develop the functional form for $\mu(S)$ when temperature and voltage are the accelerating stresses.

Arrhenius Stress Function. When temperature is the accelerating stress, we assume exp $[\mu(S)]$ follows the Arrhenius relationship. This means

$$\exp[\mu(S)] = R_0 \exp\left[\frac{E_a}{C(S+273)}\right] \tag{9.5}$$

where S = temperature, °C
C = Boltzmann's constant, = $8.6(10)^{-5}$ eV/K
E_a = activation energy, eV

Then the Arrehenius stress function is

$$\mu(S) = \ln R_0 + \frac{E_a}{C(S+273)} \tag{9.6}$$

$$= \beta_0 + \frac{\beta_1}{S+273}$$

where R_0 = rate constant, = exp (β_0)
E_a = activation energy, = $\beta_1 C$

so that

$$\ln [T(S)] \sim F\left(t;\, \beta_0 + \beta_1\left(\frac{1}{S + 273}\right),\, \sigma\right) \tag{9.7}$$

is the ln(time to failure) distribution. By introducing different temperature levels into the model we now must estimate three parameters, β_0, β_1, and σ. Equation (9.7) says that temperature affects the ln (time to failure) distribution by changing the location parameter according to Eq. (9.6), but leaving the scale parameter σ unchanged.

The acceleration factor when temperature is the accelerating stress and the accelerated failure time model holds is found by solving the following two equations for $a(S, S_0)$. For temperatures S, S_0 with $S > S_0$,

$$\mu(S_0) = \ln[a(S, S_0)] + \mu(S)$$

$$\mu(S) = \ln R_0 + \frac{E_a}{C(S + 273)}$$

which implies

$$a(S, S_0) = \exp\left[\left(\frac{E_a}{C}\right)\left(\frac{1}{S_0 + 273} - \frac{1}{S + 273}\right)\right] > 1$$

Inverse Power Stress Function. When voltage is the accelerating stress, we assume $\exp[\mu(S)]$ follows the inverse power relationship. This means the inverse power stress function is

$$e^{\mu(S)} = e^{\beta_0} S^{\beta_1} \tag{9.8}$$

$$\mu(S) = \beta_0 + \beta_1 \ln S$$

where $\beta_1 < 0$ and S is measured in a unit of voltage, so that

$$\ln [T(S)] \sim F(t;\, \beta_0 + \beta_1 \ln S,\, \sigma) \tag{9.9}$$

The location parameter of the ln (time to failure) distribution depends on voltage, and the scale parameter does not. The acceleration factor for the inverse power stress function is easily shown to be

$$a(S, S_0) = \left(\frac{S_0}{S}\right)^{\beta_1} > 1 \tag{9.10}$$

9.8.2 Qualitative Explanatory Variable: Vendor Comparison

A qualitative explanatory variable indicates in which of several groups or categories an item belongs. For example, if we are testing circuit packs from several vendors, the qualitative explanatory variable indicates which vendor supplied the circuit pack. If the qualitative explanatory variable is denoted by g, then the reliability model that includes this explanatory variable is

$$\ln [T(g)] \sim F(t;\, \mu(g),\, \sigma) \tag{9.11}$$

This model implies that the location parameter of the ln (time to failure) distribution depends on g but the scale parameter does not. This model is similar to the quantitative explanatory variable model, except that there is no function that relates the value of $\mu(g)$ for one group to the value of $\mu(g)$ for another. So if there are G groups, the qualitative variable reliability model has $G + 1$ parameters, the location parameter for each group, and σ.

9.9 ANALYZING RELIABILITY DATA: ONE EXPLANATORY VARIABLE

Reliability models that incorporate an explanatory variable model the time-to-failure distribution for each value of the explanatory variable. The typical way this is done is to assume the same parametric time-to-failure model for each value of the explanatory variable, and then to specify the parameters of this model as a function of the explanatory variable.

The reliability data analysis strategy with explanatory variables closely follows this modeling strategy. The first stage of the data analysis strategy is to verify that the same parametric time-to-failure model is appropriate for each value of the explanatory variable, and that the relationship between the parameters of these models approximately follows the assumed function of the explanatory variables. The next stage is to fit the explanatory variable reliability model, and to use the residuals from this fit as another verification of the appropriateness of the model. The third stage of the data analysis strategy is to use the fitted model to answer the reliability questions.

9.9.1 Qualitative Explanatory Variable: Vendor Comparison

The analysis of reliability data with a single qualitative (categorical) variable will be illustrated with a vendor comparison example. Here the categorical variable is the vendor who is supplying the item. The problem discussed in Sec. 9.4.2 is used as a specific example. A single-factor analysis of variance model is the basis for the analysis.

State the Problem Precisely. For the vendor comparison study in Sec. 9.4.2 we wish to determine whether the reliability for one vendor is always less than that for the other. A convenient way to proceed is to analyze the data in terms of a location-scale model with a common scale parameter. Then a "significant" difference between the location parameters of the two vendors would directly imply that the vendor with the larger location parameter would have the higher reliability for all time t.

Nonparametric Individual Group Analysis. Do a separate nonparametric analysis of the data from each group. This has two purposes:

1. To determine whether all groups can be modeled by a member of the same parametric family (checked by observing straight lines on a multiple-probability plot)
2. To help determine whether these distributions have the same scale parameter (checked by observing parallel straight lines on the multiple-probability plot)

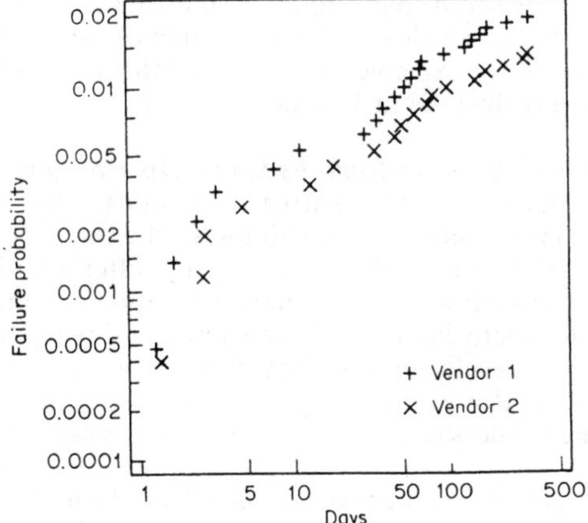

FIGURE 9.13 Lognormal multiple-probability plot for vendor comparison study.

Example 9.20. Separate nonparametric cdf estimates for the two vendors are shown on the lognormal probability plot of Fig. 9.13. These lines are reasonably straight and fairly parallel.

Parametric Individual Group Analysis. To obtain a quantitative check on an apparent parallelism of the lines in a probability plot, individual parametric distributions can be fit to each group. If the confidence limits for the scale parameters overlap, the groups may be modeled by a common scale parameter. Then the comparisons among the groups can be made solely in terms of their location parameters.

Example 9.21. A lognormal distribution was fit to the data of each vendor separately with the results given in Table 9.7 for the scale parameters σ of the underlying normal distributions.

TABLE 9.7 Data for Vendor Comparison Study

Vendor	Parameter estimate σ	95% confidence limits
1	6.20	(4.19, 9.18)
2	6.27	(4.10, 9.59)

Since the confidence limits for σ overlap, a common scale for the underlying distributions will be assumed in the analysis of the data from this example.

Fitting an Analysis of Variance Model to All Data. When a common scale factor can be used for all of the groups, a categorical model can be fit to all of the data. The location parameters directly reflect the reliability of each vendor, and a single scale parameter is used. Larger location parameters correspond to vendors with higher reliability. Therefore if the confidence interval on the difference between the location parameter for vendor 2 and the location parameter for vendor 1 does not include zero and is, for example, positive, then the second vendor has significantly higher reliability than the first vendor.

Checking Residuals of All Data from the Single Distribution. If the model assumptions are reasonably valid for a particular set of data, the differences (residuals) of the data from the model should all follow the same location-scale distribution, where the location parameter is zero. Any differences between vendors should be removed by subtracting the estimated location parameter for each vendor from the data for that vendor. Therefore a probability plot of these residuals should look like the results of sampling from a single distribution without explanatory variables. If not, the model assumptions are not confirmed. If so, comparisons among vendors can be made entirely in terms of their location parameters.

Example 9.22. An analysis of variance model was fit to the vendor data, and standardized residuals were calculated. Here each residual was divided by an estimate of the common scale parameter so that the result should follow a standardized normal distribution if the assumptions are valid. These results are found in the residual probability plot of Fig. 9.14. The linearity of this plot supports all the assumptions of the analysis of variance model for this data set.

Answering the Reliability Question. When the assumptions of the analysis of variance have been confirmed, vendor reliability can be compared by examining a confidence interval on the difference between location parameters. If the interval contains zero, the observed difference in reliability could be due to chance. For the vendor data, the 90% confidence interval on the difference between the two location parameters is (– 0.73, 2.41) with a point estimate of 0.84. Therefore, even though vendor 2 appears to have a more reliable product, the apparent superiority could be due to chance.

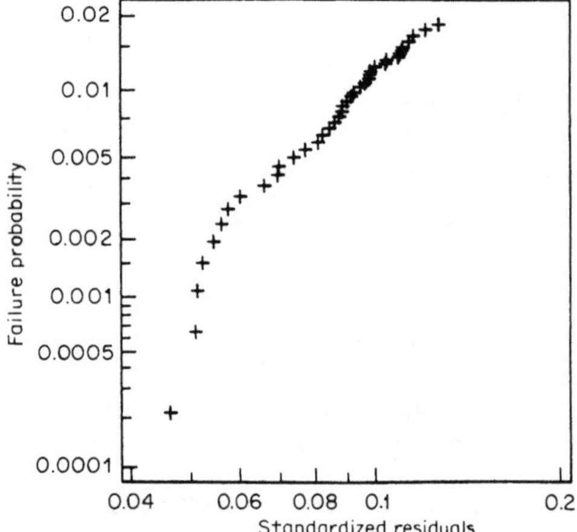

FIGURE 9.14 Lognormal residual probability plot for vendor comparison study.

9.9.2 Quantitative Explanatory Variable: Accelerating Stress

Analyzing reliability data with a single quantitative explanatory variable is presented in the important context of accelerated life testing. In this context the explanatory variable is the level of accelerating stress. All the items tested at the same level of stress are called a group. The device accelerated life test discussed in Sec. 9.4.2 and presented in Table 9.3 is used as an example.

State the Problem Precisely. The reliability question must be made precise before it can be answered. This often means a very careful interpretation of what the reliability objective is and what constitutes proof that we are meeting it.

Example 9.23. For the device accelerated life test in Sec. 9.4.2 we must determine whether the device can meet its failure rate objective at 10,000 hours. We interpret this to mean that the device must have a decreasing failure rate which is below the 10,000-hour objective for all times greater than 10,000 hours. We will ensure that the device can meet its failure rate objective if the 90 percent confidence interval on the failure rate at 10,000 hours is below the failure rate objective.

Nonparametric Individual Group Analysis. Do a separate nonparametric analysis of the items in each group. The interest is in the nonparametric estimate of the cdf for each group. Do not do the nonparametric analysis on any group with less than five failures. These individual nonparametric cdf estimates are displayed on a single plot called a *multiple cdf plot*. Ideally these cdf estimates should not cross, and the estimate associated with the highest stress group should be on top, followed by the next highest stress group below it, and so on.

After creating the multiple cdf plot, a parametric time-to-failure model is selected such as Weibull or lognormal. We then create a multiple-probability plot by rescaling the x and y axes of the multiple cdf plot.

If the accelerated failure time model, Eq. (9.4), holds and the common time-to-failure distribution is, say, Weibull, then the Weibull multiple-probability plot will appear approximately as parallel straight lines. The straightness of the lines indicates the goodness of fit of the Weibull model to the individual group time-to-failure distributions, and the approximate parallel nature of the lines indicates the common scale parameter of the ln (time to failure) distribution.

If the lines are all straight but not parallel, this indicates that the chosen time-to-failure distribution is appropriate for all stress levels, but perhaps the scale parameter is a function of the stress level, and this must be added to the model. In this case the time-to-failure model can be individually fit to each group to check more carefully whether the scale parameter is a function of stress (see next section).

If the probability plots for the high-stress groups are neither straight nor parallel, this indicates that we have tested at too high a stress level and these high-stress groups should be discarded. If the probability plots for the low-stress groups are neither straight nor parallel, this indicates that the accelerated failure time model is not valid near design stress, which is the main region of interest. In this case the entire accelerated life test is suspect. If the probability plot of an intermediate stress group is neither straight nor parallel, then the analysis should be done both with and without this group and the more conservative answer chosen.

Example 9.24. Separate nonparametric cdf estimates are constructed for the 40°C, 60°C, and 80°C groups. These estimates are displayed in the multiple cdf plot in Fig. 9.15. Notice that these estimates do not cross and are ordered from high to low stress. Figure 9.16 is the Weibull multiple-probability plot. Based on this plot,

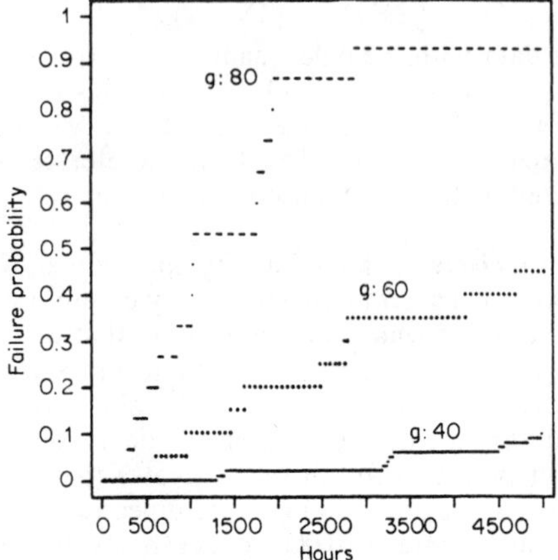

FIGURE 9.15 Multiple cdf plot for device accelerated life test.

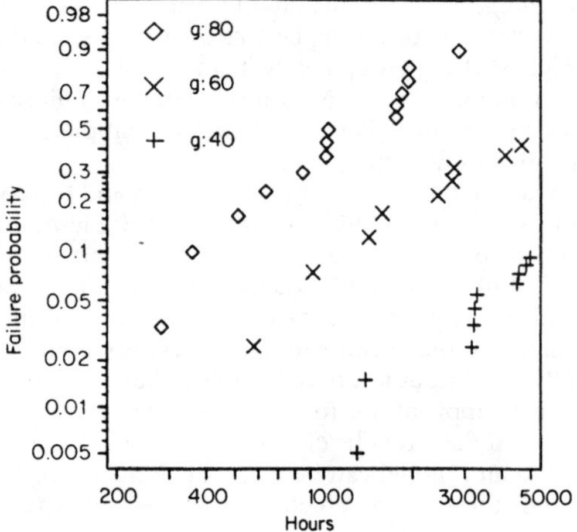

FIGURE 9.16 Weibull multiple-probability plot for device accelerated life test.

the accelerated failure time model with the Weibull time-to-failure distribution is a very good fit to these data.

Parametric Individual Group Analysis. If the individual lines on the multiple-probability plot are straight but their parallelism is questionable, then the assumed time-to-failure model should be fit to each group individually. The confidence intervals for the scale parameters of the ln (time to failure) distributions can be compared to see whether they overlap. If so, the scale parameter is assumed constant with stress. If the confidence intervals do not overlap, then it will be necessary to model the effect of stress on the scale parameter. Usually the scale parameter will be a monotone function of stress.

Example 9.25. The Weibull multiple-probability plot with the Weibull model fit individually to each group is shown in Fig. 9.17. The individual group Weibull models are indicated by the straight lines. The confidence intervals for the scale parameters associated with this plot are given in Table 9.8. All diagnostics indicate that the accelerated failure time model is a good fit to these data.

Fitting the Accelerated Failure Time Model. In order to fit the accelerated failure time model to the data we must first choose the appropriate stress function. If temperature is the accelerating stress, then use the Arrhenius stress function, Eq. (9.6), and if voltage is the accelerating stress, use the inverse power stress function, Eq. (9.8). The method of maximum likelihood is used to estimate the model parameters and to construct their confidence intervals.

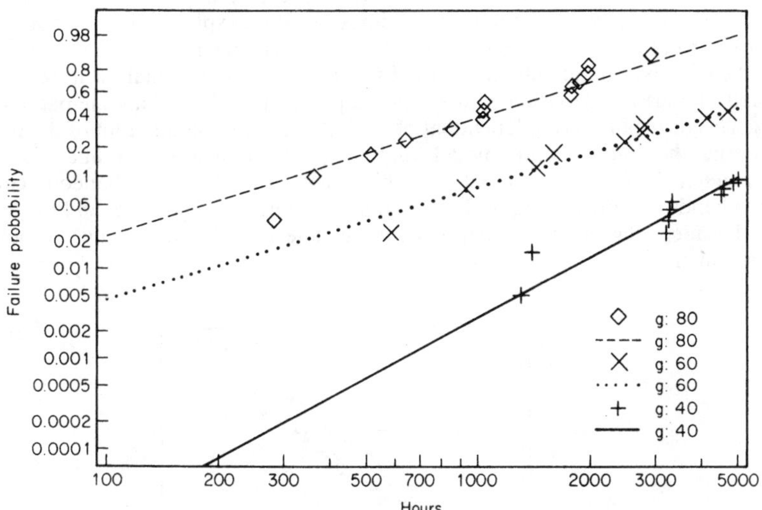

FIGURE 9.17 Weibull multiple-failure probability plot for device accelerated life test.

TABLE 9.8 Checking Equality of Individual Scale Parameters for Device Accelerated Life Test

Group	$\hat{\sigma}$	95% confidence interval on σ
40	0.448	(0.243, 0.826)
60	0.801	(0.438, 1.46)
80	0.762	(0.511, 1.14)

TABLE 9.9 Accelerated Failure Time Model Parameter Estimates

Parameter	Point estimate	95% confidence interval
β_0	– 13.32	(– 19.81, –6.82)
β_1	$7.36(10)^3$	$(5.15(10)^3, 9.56(10)^3)$
σ	0.707	(0.531, 0.940)

Example 9.26. The accelerated failure time model with the inverse power stress function, Eq. (9.7), is fit to these data. The estimates and confidence intervals for the parameters β_0, β_1, and σ are given in Table 9.9.

Checking the Residuals from the Fit of the Accelerated Failure Time Model. At this point in the analysis we have verified the accelerated failure time model, but we have not checked whether the stress function adequately models the dependence of the location parameter of the ln (time to failure) distribution on stress. This is done by an analysis of the residuals from the model fit. The residuals are what remains after the variability in the times to failure explained by the model are removed from the data. If we have modeled the effect of stress on the time-to-failure distribution adequately, then the distribution of the residuals will be the ln (time to failure) distribution with a location parameter of 0 and a scale parameter of σ. This distribution is independent of stress. Thus a probability plot of the residuals using the time-to-failure model will be approximately a straight line.

Example 9.27. The residual probability plot with 90% simultaneous confidence bands is shown in Fig. 9.18. It is approximately a straight line, and this verifies the adequacy of the accelerated failure time model and the inverse power stress function.

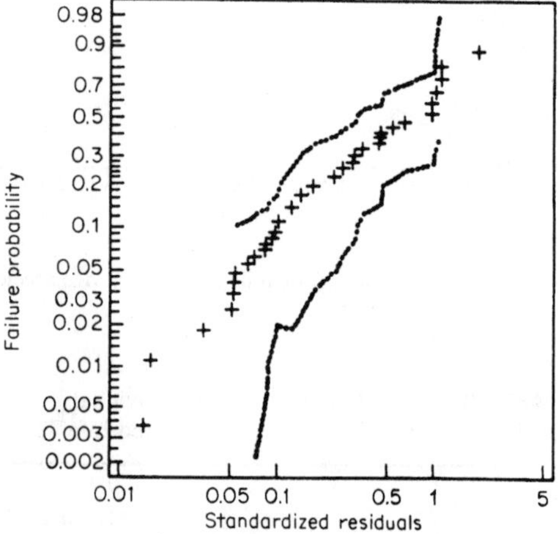

FIGURE 9.18 Weibull residual probability plot with 90% simultaneous confidence bands for device accelerated life test.

Using the Accelerated Failure Time Model. Once we have fit the accelerated failure time model we can use the model fit to answer reliability questions involving the time-to-failure distribution at any stress level for which the model holds. This means that we have the ability to extrapolate in both time and stress. We can estimate quantiles, failure probabilities, and failure rates, with pointwise or simultaneous confidence bands, at any stress level and time for which the model holds. We can present these estimates and confidence bands in a variety of ways. We can plot the failure probability estimates as a function of time for design stress and the stress levels for which the nonparametric individual group estimates exist, and superimpose these nonparametric estimates on the plot. This multiple failure probability plot shows the overall accelerated failure time model fit as well as the effect of extrapolation to design stress. We can also present multiple failure rate plots in the same manner, but without any nonparametric estimates superimposed. Finally we can present quantiles as a function of stress. We can plot quantiles at several stress levels estimated from the accelerated failure time model. This gives us a picture of the accelerated failure time model. If we add to this plot the quantiles estimated from the individual group parametric analyses and their confidence intervals, we get a visual representation of the goodness of fit of the accelerated failure time model.

Example 9.28. The multiple failure probability plot with the individual group nonparametric estimates is shown in Fig. 9.19. This plot shows an excellent model fit as well as the extrapolation to 10°C design stress. Table 9.10 shows the failure rate at 10°C design stress and the associated confidence intervals. From this table and the associated plot it is clear that we are not meeting the reliability objective since the failure rate is increasing. Figure 9.20 gives a plot of the quantiles from the accelerated failure time model and the quantiles from the individual group parametric analyses, which again shows the excellent fit of the accelerated failure time model.

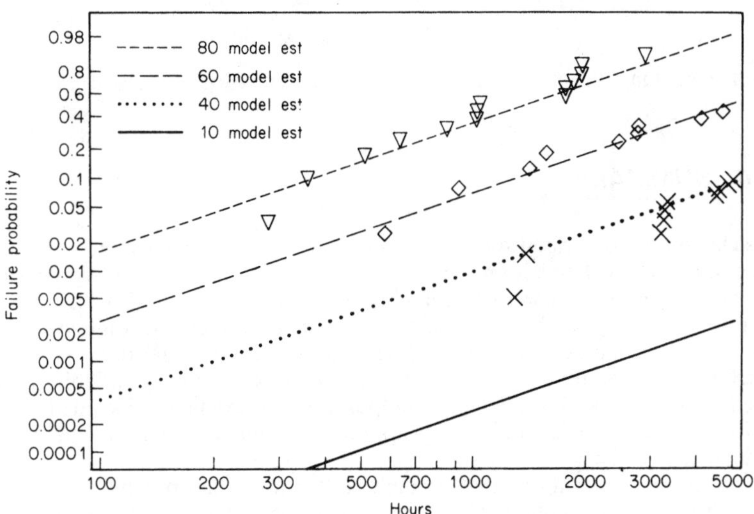

FIGURE 9.19 Weibull failure probability plot for device accelerated life test.

TABLE 9.10 Weibull Failure Rate at 10°C for Device Accelerated Life Test

Test hours	FITs	Standard error	Lower confi-dence limit	Upper confi-dence limit
500	310.9	253.3	81.4	1187.6
1000	414.3	305.9	123.0	1395.9
5000	807.2	527.3	275.5	2364.6
10000	1075.7	716.4	359.6	3217.6
25000	1572.6	1142.7	475.8	5197.3
50000	2095.8	1675.5	562.5	7808.7

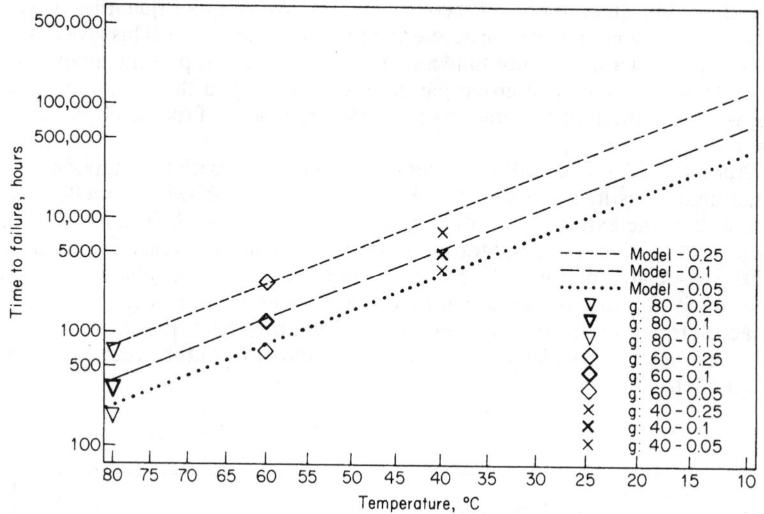

FIGURE 9.20 Weibull quantiles as a function of stress for device accelerated life test.

9.10 SUMMARY

This chapter has described some of the sources of reliability questions and certain types of reliability data within the framework of a product reliability program. A general strategy for answering reliability questions is presented, with a specific example to illustrate each point. Some of the more useful reliability models with no explanatory variables are presented as well as several important population descriptors. The analysis of reliability data without explanatory variables is illustrated by some circuit pack data, and output from this example is included. A brief section on the determination of an appropriate sample size using the Weibull model is then given.

Two specific cases are used as examples of the important reliability problem of reliability models with one explanatory variable: using temperature in an accelerated life test and a vendor comparison study. In both cases data are used to illustrate the techniques of data analysis appropriate to the particular problem and to exhibit output in a real-life situation. Also, specific questions are asked and answered for both examples.

REFERENCES

Carey, M. C., and Y. L. Grize: "Models for the Analysis of Degradation Data," *Proc. 5th Int. Conf. on Reliability and Maintainability*, pp. 321–321e, 1986.

Goldthwaite, L.: "Failure Rate Study for the Lognormal Lifetime Model," *Proc. 7th Nat. Symp. on Reliability and Quality Control*, pp. 208–213, 1961.

Lawless, J. F.: *Statistical Models and Methods for Lifetime Data*, Wiley, New York, 1982.

Meeker, W. Q., and G. J. Hahn: *How to Plan an Accelerated Life Test*, ASQC Basic References in Quality Control: Statistical Techniques, vol. 10, Am. Soc. Quality Control, Milwaukee, WI, 1985.

_____ and W. Nelson: "Weibull Percentile Estimates and Confidence Limits, from Singly Censored Data by Maximum Likelihood," *IEEE Trans. Reliability*, vol. R-25, pp. 20–25, Apr. 1976.

Nair, V. N.: "Confidence Bands for Survival Functions with Censored Data: A Comparative Study," *Technometrics*, vol. 26, pp. 265–275, 1984.

Nelson, W.: *How to Analyze Reliability Data*, ASQC Basic References in Quality Control: Statistical Techniques, vol. 6, Am. Soc. Quality Control, Milwaukee, WI, 1983.

Wagner, A. E., and W. Q. Meeker: "A Survey of Statistical Software for Life Data Analysis," *Proc. Statistical Computing Section, 1985 Am. Statistical Assoc. Nat. Meetings* (Las Vegas, NV).

CHAPTER 10
SURVEY SAMPLING METHODS*

Keith R. Eberhardt
Statistical Engineering Division
National Institute of Standards and Technology
Gaithersburg, MD

10.1 INTRODUCTION

Survey sampling methods are statistical methods for collecting data and making inferences about the characteristics of defined finite populations. The term "population" is used broadly to designate a group to be studied. It may consist of people, or it may consist of elements such as machines in a factory, cartons in a warehouse, bottles of a chemical reagent, or purchase orders for new equipment. The survey sampling approach consists of observing, or making measurements on, a subset (i.e., a sample) of the population in order to make inferences about the population as a whole.

It is useful to contrast survey sampling methods with methods for designing and analyzing experiments. In designed experiments, the investigator can decide which "treatments" will be applied to which experimental units in order to obtain information on how the treatments affect a response variable. In survey problems, the treatments will have already been applied to the population units before the sampler comes on the scene. Broadly speaking, designed experiments are focused on *why* the data came to be, whereas survey sampling methods are more aimed at discovering *what is*. Partly to emphasize the distinction between the passive nature of survey sampling investigations and the active nature of studies of cause-and-effect relations, Deming (1950, chap. 7) used the term "enumerative studies" for the former and "analytic studies" for the latter.

Briefly, the methods described in this chapter address the following issues:

- How large a sample to take
- How to choose the sample

- How to estimate a population mean, total, or proportion
- How to characterize the uncertainty in the estimate

Methods for practical situations and questions that go beyond those presented here may be found in some of the many fine textbooks that have been written about survey sampling theory and methods. Some of these are Cochran (1977), Thompson (1992), Barnett (1986), Deming (1950, 1960), Foreman (1991), Hansen, Hurwitz, and Madow (1953), Kish (1995), Levy and Lemeshow (1991), Raj (1968, 1972), and Sukhatme et al. (1984).

10.1.1 Considerations in Planning a Sample Survey

Some important practical considerations that should be taken into account when planning a sample survey are listed hereafter. The list is given in roughly chronological order, which does not necessarily coincide with the order of importance.

1. *Objectives:* As with any investigation or experiment, better results will be obtained when the objectives of the project are clearly stated in advance. In survey sampling problems, special attention must be given to the population about which inference is to be made. In the familiar example of preelection polling, the population consists of voters (or, more specifically, the votes to be cast by the voters) who will take part in the coming election. A common situation in practical surveys is that the population of ultimate interest is not identical to the population from which the sample will be drawn. Thus we differentiate between the *sampled population* and the *target population*.

If an opinion poll is conducted by telephone between the hours of 6 P.M. and 9 P.M., the sampled population will consist of people who have telephone service and who would be available to answer a call during the time period of the survey. Opinions of people who work during the 6 P.M. to 9 P.M. time period would be systematically excluded from the survey, and so the statistical inferences possible from such a survey would not apply to them. Similar statements apply to people who are uncooperative in answering survey questions over the telephone and others that a telephone survey would exclude. In practice, the sampled population is defined operationally as the collection of all individuals or items for which the probability of inclusion in the sample is nonzero.

The target population is the ideal population defined by the objectives of the survey. Conceptually, one might say that if the data of interest were known for every individual in the target population, then all the questions of interest could be unequivocally answered by simply summarizing the available data.

Typically, the target population will be somewhat more extensive than the best practically realizable sampled population. Statistical inference from the survey sample will only be possible from the sample to the sampled population. Further extrapolations beyond the sampled population to the target population will rest on the judgment of those making the extrapolation. When the difference between sampled and target populations is large, these considerations may lead to redefining the goals of the survey or, possibly, abandoning the project altogether.

2. *Sampling units:* An *element* of a population is the item, or individual, to which the survey measurement, or observation, applies. In the execution of a survey, data are often most conveniently collected for groups of one or more elements, called *sampling units*. For example, a door-to-door interview survey may be conducted by selecting a sample of blocks in a city and then conducting an interview at every household in the selected blocks. In this case, the block is the sampling unit, while the household is the element. It is critical that the chosen sampling units be well defined, nonoverlapping, and easily identified by the persons collecting the data.

A physical, or conceptual, list of all of the sampling units in the population is the *frame* for the survey. Consideration of the collection of elements contained in the sampling units that are included in the frame will reveal the operational definition of the sampled population for the survey.

As an example, a survey of the condition of items stored in a warehouse could proceed by working from a list of the shelves in the warehouse. A sample of shelves might be selected at random and the surveyors could be instructed to gather data for each item on each selected shelf. In this case, the items on the shelves would be the population elements, and the sampling units would be shelves. Note that any items on shelves not included on the original list would not be included in the sampled population.

3. *Method of observation:* The method by which the basic data are collected will have much to do with the success or failure of a survey. For example, one must recognize that mail questionnaires, telephone interviews, and personal interviews all typically produce different results. Whenever several surveyors are used, it is critical to ensure that they use consistent measurement methods. The questionnaire, or other data recording form, to be used should be carefully designed for ease of use, clarity, and ease of summarization. Provision should be made for handling nonresponse or other cases where the data cannot be obtained for a sample element. Adequate field supervision will be needed so that quick decisions can be made when (not "if" but "when") unforeseen circumstances arise during the data collection.

4. *Sample design:* The *sample* is the set of elements or sampling units from the population that are ultimately measured or observed. Many aspects of sample design are dealt with in detail later in this chapter. Perhaps the most easily recognized consideration is choosing an appropriate sample size, that is, the number of sampling units or elements that will be observed. When the population to be studied may be readily subdivided into recognizable subgroups, this structure may be incorporated in the sample design by stratification (Sec. 10.3) or by cluster sampling (Secs. 10.5 and 10.6). It may be appropriate to conduct the study in phases, perhaps collecting information about gross features of the population by an initial general purpose survey that will be followed up by one or more specialized studies.

5. *Data analysis:* In practice, this aspect of a sample survey involves much more than choice of proper estimation and uncertainty analysis procedures. As described in the following sections, these will be dictated by the sample design and the structure of the population under study. In addition, practical procedures for entering the data into a computer, including building-in appropriate ways to handle missing data, will be needed. After that some amount of programming will be needed, including the important need to build-in checks of the raw data for reasonableness, internal consistency, recording errors, consistency with similar studies, and so on.

6. *Reporting of results:* This obvious step of the survey process is listed primarily because the cost and time required for adequate reporting are often underestimated when planning a survey. An important part of a good report should be a discussion of what was learned in terms of how a similar survey might be done better in the future.

10.1.2 Theories of Finite Population Sampling

Practical survey sampling typically involves the use of artificial randomization in the process of selecting a sample of units. Classical finite population sampling theory is based on studying the properties of statistical procedures strictly as they relate to the probability distribution generated by random sampling. In particular, the bias and variance of estimators with respect to the random sampling distribution are often used to characterize the estimators.

A complimentary theory of finite population sampling has been developed in which simple probability models are used to describe the finite population itself. Then the properties of estimation procedures are studied in terms of the models. This model-based theory provides additional insight into the behavior of finite population estimators and has led to the development of useful modifications of classical estimation procedures.

Statistical properties of the sampling and estimation procedures discussed in this chapter will be described in terms of both the classical randomization-based theory and the model-based theory. In both cases, the theoretical properties to be discussed implicitly assume that any measurement error incurred in making observations on the sample units is negligible. Thus the standard error estimates and confidence intervals presented here reflect only uncertainty due to *sampling error*, that is, the error associated with the fact that only a sample, rather than the whole population, was observed.

Deming (1960, p. 50) gives an especially clear operational definition of sampling error along the following lines. The "equal complete count" is defined as the result that would be obtained by applying the actual survey procedures to all of the elements in the frame (same field workers, using the same definitions, data acquisition, and summarization procedures, exercising the same care as they exercised on the sample, etc.). Then the sampling error is simply the difference between the estimate obtained from the sample and the value that would be obtained from an equal complete count of the whole population.

10.1.3 Bulk Sampling

The sampling methods discussed in this chapter take for granted that the finite population of interest consists of discrete elements, or individuals, that are predefined before sampling. Another kind of population consists of material in bulk form. Bicking (1967, p. 96; quoted in Schilling, 1982, p. 192) describes bulk sampling as follows:

> Bulk materials are essentially continuous and do not consist of populations of discrete, constant, identifiable, unique units or items that may be drawn into the sample. Rather, the ultimate sampling units must be created, at the time of sampling, by means of some sampling device.

Thus bulk sampling deals with problems such as the following: sampling a drum of bulk chemical, sampling a train load of ore, or sampling a tank truckload of milk or fuel.

The unique aspects of bulk sampling problems, which include construction of the sample units and the use of composited samples, are beyond the scope of this chapter. Some useful references to this subject are Schilling (1982), Bicking (1978), Duncan (1962), Gy (1992), Ishikawa (1973), Pitard (1993), and ASTM (1951).

10.2 SIMPLE RANDOM SAMPLING

This section describes methods of estimation that are associated with the simple random sampling design and are based on the unweighted sample mean or sample proportion. The methods of this section are foundational in the sense that the material in later sections builds on the concepts presented here. Most practical sampling problems will have additional structure (strata, clustering, auxiliary variables, etc.) and so will require use of the methods described in Secs. 10.3 through 10.6.

10.2.1 Drawing a Simple Random Sample

Simple random sampling without replacement (SRS) is a procedure for drawing a sample of n units from a population in such a way that each and every subset consisting of n distinct units has the same probability of being selected. In a population of size N there are

$$\binom{N}{n} = \frac{N!}{n!(N-n)!}$$

possible samples of size n without replacement, so the probability that any one of these is selected must be $1/\binom{N}{n}$.

Simple random sampling may be implemented sequentially by drawing units one at a time from the population, as follows. Draw the first unit from the population in such a way that each unit has equal probability $1/N$ of being selected. Remove the selected unit from consideration for the next steps of sampling. Draw the second unit from the remaining $N-1$ units with probability of selection $1/(N-1)$ for each unit. Again, remove the selected unit from the population. Continue by this process, drawing the $(k+1)$th selected unit with probability $1/(N-k)$ from the remaining $N-k$ units in the population, until the desired sample size of n units is obtained.

The physical procedure for equal probability sampling at each stage of this process may be accomplished using random integers, either computer generated or from a published table, such as Table A.23 in the Appendix. [A very extensive table of random integers is published in RAND (1955).] The practical procedure for use is most readily described by an example; the generalization will be obvious.

Example 10.1. A research project requires drawing a simple random sample of $n = 25$ file cards from a drawer. A preliminary estimate states that the population represented in the drawer consists of approximately $N = 3000$ cards, and surely no more than 3500 cards. How can the desired simple random sample be drawn?

Assume that the most convenient source of random numbers is a computer routine. Calls to a uniform random number generator yield a sequence of decimal numbers on the interval from 0.0 to 1.0 like the following:

0.24375, 0.59747, 0.17328, 0.89746, ...

Using the first four digits to the right of the decimal yields a corresponding sequence of random integers, chosen with equal probability from 0000 to 9999:

2437, 5974, 1732, 8974, ...

This sequence of random integers can be used to select sample numbers in the range of 1 to 3500 by simply discarding numbers outside the desired range. In addition, any integer that appears twice in the sequence would be discarded to achieve sampling without replacement.

In describing this example, it was assumed that the total number of items in the population N is not known exactly. The exact value of N is not crucial to problems involving estimation of a population mean or proportion, as will be seen later. However, some provision must be made to ensure that the sample is indeed chosen from the whole population. In the present example, choosing random numbers between 1 and the maximum conceivable population size (3500), rather than simply using the rough estimate of N(3000 in the example), ensures that any items in the population with labels higher than 3000 will indeed have an equal chance of being selected. However, there is some risk that some of the selected labels will not correspond to any unit in the population. For example, suppose that the population in the example in fact has exactly 3100 units, and that one of the random integers selected happens to be greater than 3100. Then the sample size obtained by using the set of random integers would only be 24—smaller than intended. To guard against this possibility, a few "extra" random integers could be generated in the original randomization, so that they would be available in case the actual value of N turns out to be smaller than 3500.

The procedure outlined in Example 10.1 is inefficient in that random integers in the range from 3501 to 9999 will be "wasted." If a computer is being used to generate the random integers, the efficiency can be improved quite simply by ensuring that the computer generates random integers in exactly the desired range. In the example, this could be accomplished by the following steps:

1. Generate a sequence of uniform random numbers U.
2. Then transform U to random integers, denoted by RINT, in the range from 1 to 3500 by the substitution RINT = INT(3500 * U) + 1, where the function INT(X) returns the greatest integer less than, or equal to, the argument X.

Example 10.2: Some Sampling Methods that Are Not *Equivalent to Simple Random Sampling.* Recall that simple random sampling implies that every possible subset of n units from the population is given the same chance of selection. This implies that every individual unit has the same chance of being selected, but it is more than that. In addition, all possible pairs of units must have the same probability of selection, and also all triples, all quadruples, and so on. The following examples illustrate some sampling procedures that are *not* true simple random sampling.

1. *Cluster sampling:* A sample of employees in a large firm is selected as follows. First a simple random sample of departments is selected, and then two employees are selected at random from each selected department. This sampling procedure is actually an example of two-stage cluster sampling, with clusters defined by departments. Sections 10.5 and 10.6 discuss how to implement cluster sampling and how to analyze the resulting data. One readily recognized difference from simple random sampling is that *triples* of employees are not all equally likely to be selected into the sample. For example, the probability of selection is zero for any triple of employees belonging to the same department. In addition, unless the number of employees per department is the same for all departments, individual employees do not all have the same chance of being selected. (Employees from smaller departments have a relatively higher chance of being selected.)

2. *Stratified sampling:* In the previous example, another way to obtain a sample of employees is as follows. Two employees are selected (by simple random sampling) from each and every department in the firm. In this approach, the departments are treated as strata, and the sampling procedure is called *stratified random sampling.* Appropriate methods for stratified sampling are discussed in Sec. 10.3. Some properties of this procedure that are different from simple random sampling are: triples of employees are not all equally likely to be selected into the sample, and individual employees are not all equally likely to be selected if the departments are not all the same size.

3. *Systematic sampling:* Suppose that a frame, consisting of an alphabetical list of all employees of the firm, were available and that every 100th employee were chosen for the sample (after a randomly chosen starting point). This procedure, called systematic sampling, is sometimes convenient and may represent a practical approximation to true simple random sampling (see Cochran, 1977, chap. 8). One clear difference with simple random sampling is that all pairs of employees do not have the same chance of being selected into the sample. In fact, any pair of employees that do not happen to fall exactly 100 positions apart on the list has zero probability of being chosen.

4. *Haphazard selection and judgment sampling:* The formal use of random numbers in the implementation of simple random sampling differentiates that procedure in important ways from informal methods that are sometimes used to select "representative" samples. For instance, a file card drawn "with one's eyes closed" from a drawer is more likely to come from the middle of the drawer than from either end. An assistant asked to randomly draw a specimen bottle from a refrigerator full of bottles is more likely to choose one near the front of the shelf than one in the back, owing to convenience. In some cases, such haphazard methods of sample selection may in fact reasonably represent the population from which they are drawn, but it is easy to imagine problems: for instance refrigerators tend to be warmer near the front than near the back.

To try to improve the representativeness of a haphazard sample, additional requirements are sometimes imposed on the set of sample units to be selected. An assistant might be instructed to select one bottle from the front of a shelf and one from the back. A structured approach of this kind may indeed produce better results than pure haphazard selection, but ultimately, the inference from a judgment sample to the population will be based on the judgment that went into defining the sample, rather than on statistical theory. Most situations where it is desirable to ensure that the sample include units from a number of subsets of the population can be handled statistically by the method of stratified sampling (Sec.10.3).

10.2.2 Estimation of a Population Mean

We will denote by y_i the numerical value of a measured characteristic associated with the ith unit in the population. The population values of y are then $y_1, y_2, \ldots, y_N$.

The population mean of the characteristic y is

$$\overline{Y} = \frac{1}{N}\sum_{i=1}^{N} y_i.$$

The corresponding sample mean is

$$\bar{y} = \frac{1}{n} \sum_{s}^{n} y_i$$

where Σ_s^n denotes summation over the n units that were drawn into the sample s.

The classical randomization based theory supporting the use of $\bar{y}$ as an estimator of $\bar{Y}$ is based on the probability distribution of $\bar{y}$ generated by simple random sampling. An important property is that $\bar{y}$ is unbiased as an estimator of $\bar{Y}$,

$$\text{ave}_{\text{SRS}}(\bar{y}) = \bar{Y}$$

where $\text{ave}_{\text{SRS}}(\bar{y})$ denotes the average value of $\bar{y}$ over all of the (equally likely) $\binom{N}{n}$ possible samples of size n from the population of size N. The variance of $\bar{y}$ with respect to the simple random sampling distribution over samples is

$$\text{ave}_{\text{SRS}}(\bar{y} - \bar{Y})^2 = \frac{S_y^2}{n} \left(1 - \frac{n}{N}\right) \tag{10.1}$$

where

$$S_y^2 = \frac{1}{N-1} \sum_{i=1}^{N} (y_i - \bar{Y})^2$$

denotes the finite population variance of y. The square root of Eq. (10.1) is called the *standard error* of $\bar{y}$. If the sample size n is large, then the distribution of $\bar{y}$ with respect to simple random sampling is approximately normal, with mean $\bar{Y}$ and variance given by Eq. (10.1). A discussion of this finite population version of the central limit theorem and its limitations is given in Cochran (1977, sec. 2.15). See also Thompson (1992, secs. 3.2 and 7.4).

In practical problems, the population variance in Eq. (10.1) is generally unknown, and so it is natural to consider estimating S_y^2 by the corresponding sample variance,

$$s_y^2 = \frac{1}{n-1} \sum_{s}^{n} (y_i - \bar{y})^2.$$

As is true in the infinite population case, the sample variance is an unbiased estimator for the population variance,

$$\text{ave}_{\text{SRS}}(s_y^2) = S_y^2. \tag{10.2}$$

Substituting the sample variance into Eq. (10.1) and taking the square root yields the *estimated standard error* of $\bar{y}$,

$$s(\bar{y}) = \left[\frac{s_y^2}{n}\left(1 - \frac{n}{N}\right)\right]^{1/2}. \tag{10.3}$$

In practice, $s(\bar{y})$ is used to form approximate confidence intervals for the population mean as a way of summarizing the probable sampling uncertainty in an observed value of $\bar{y}$. If n is about 30 or more, the interval

$$\bar{y} \pm z s(\bar{y})$$

may be interpreted as an approximate $100(1 - \alpha)\%$ confidence interval for $\overline{Y}$, where $z = z_{1-\alpha/2}$ denotes the $(1 - \alpha/2)$ quantile of the standard normal distribution. If n is small, it is good practice to replace the multiplier z with the corresponding quantile of the Student's t distribution with $n - 1$ degrees of freedom (see Chap. 4, Sec. 4.3.2).

10.2.3 The Finite Population Correction Factor

In sampling from an infinite population it is a familiar fact that the estimated standard error of the sample mean $\overline{y}$ is $s_y/\sqrt{n}$ (see, for example, Chap. 2, Sec. 2.7.1). Equation (10.3) shows that the corresponding result for a finite population involves the extra factor $1 - n/N$. This factor and its square root are called *finite population correction factors* (fpc). As the population size increases, the fpc approaches 1, so the finite population result agrees with the infinite population result in the limit as $N \to \infty$. In practice, the approximation obtained by replacing the fpc by 1 is usually insignificant whenever the *sampling fraction* n/N is less than about 0.05.

Note that the standard error of $\overline{y}$ is relatively insensitive to the population size N, compared to its dependence on the sample size n. Thus if N is only known approximately, using its approximate value in $s(\overline{y})$ affects only the fpc and will have little effect on the resulting inference about the population mean.

10.2.4 Estimation of a Population Total

Many practical problems call for estimating a population total, such as the total value of items in a warehouse, or the total cost of repairs for machines in a plant. The *population total* for the characteristic y in a finite population is

$$Y = \sum_{1}^{N} y_i = N\overline{Y}.$$

When the population mean is estimated by $\overline{y}$, a natural estimator of Y is the so-called *expansion estimator*,

$$\hat{Y}_E = N\overline{y} = \frac{N}{n}\sum_{s}^{n} y_i. \tag{10.4}$$

The property that $\hat{Y}_E$ is unbiased as an estimator of Y with respect to simple random sampling follows immediately from the unbiasedness of $\overline{y}$ for $\overline{Y}$,

$$\text{ave}_{\text{SRS}}(N\overline{y}) = N\,x\,\text{ave}_{\text{SRS}}(\overline{y}) = N\overline{Y} = Y.$$

Note that estimation of the population total requires accurate knowledge of the population size N. The factor N/n in $\hat{Y}_E$, called the *expansion factor*, is the multiple by which the sample size n must be "scaled up" to equal the population size N. This factor is used to scale up the sample total $\Sigma_s\, y_i$ to form the expansion estimator of the population total Y.

The estimated standard error of $N\overline{y}$ is obtained simply by multiplying the standard error of $\overline{y}$ by N,

$$s(N\bar{y}) = Ns(\bar{y}) = N\left[\frac{s_y^2}{n}\left(1 - \frac{n}{N}\right)\right]^{1/2}. \qquad (10.5)$$

An approximate confidence interval for Y is given by $N\bar{y} \pm zs(N\bar{y})$, where z is the appropriate quantile of the standard normal distribution.

Example 10.3: Estimation of a Population Mean and Total with SRS. A manufacturer conducted a survey to study downtime of machinery during the past quarter. A simple random sample of $n = 20$ machines was selected from the population of $N = 90$ machines at the plant. An initial summary of the data showed that the mean downtime was $\bar{y} = 29$ hours, with a standard deviation of $s_y = 12$ hours. The standard error of $\bar{y}$, as an estimate of the population mean, is $s(\bar{y}) = (12/\sqrt{20})(1 - 20/90)^{1/2} = 2.37$ hours. Since n is not large, we form an approximate 95% confidence interval for the population mean by using the 0.975 quantile of the t distribution with 19 degrees of freedom, $t_{0.975}(19) = 2.093$. The resulting interval is $29 \pm 2.093 \times 2.37 = 29 \pm 5$ hours per machine. The corresponding 95% confidence interval for the total hours of downtime during the quarter is $90 \times (29 \pm 5) = 2610 \pm 450$ hours.

10.2.5 Model-Based Development

This section outlines a model-based theory of $\bar{y}$ as an estimator of $\bar{Y}$ that yields results which parallel the classical theory described in the previous sections. See Royall (1988) for an overview of this approach. One advantage of the model-based approach is that it explicitly connects survey sampling theory with the more familiar theory of linear models and regression. Another value of the model-based approach comes into play in the following sections on ratio estimation and cluster sampling where it is used to develop certain refinements of classical estimation and sampling methods.

We will assume that the N numbers y_i in the population under study can be well represented as *realized values* of random variables described by a working model of the form

$$y_i = \mu + \epsilon_i \qquad (10.6)$$

where the errors ϵ_i have common mean zero, $E(\epsilon_i) = 0$, and variance $\text{var}(\epsilon_i) = E(\epsilon_i^2) = \sigma^2$.

It is then instructive to consider the problem of estimating the total Y as a prediction problem. After the sample is drawn, the y values for sample units are known and only those y values associated with nonsample units need to be estimated. Thus, after the sample is drawn we can write

$$Y = \sum_{i=1}^{N} y_i = \sum_{s}^{n} y_i + \sum_{r}^{N-n} y_i \qquad (10.7)$$

where, in the last term, Σ_r^{N-n} denotes summation over the $N - n$ units in the remainder r of the population after the sample s has been removed. (Hereafter, the summation over r'will be denoted by Σ_r in text.) The first of the two sums in the rightmost equality of Eq. (10.7) is $n\bar{y}$. Since $\bar{y}$ is *known* after the sample is observed, it remains to estimate the second term. Under the working model, the expected value of $\Sigma_r y_i$ is $(N - n)\mu$, so it seems natural to estimate $\Sigma_r y_i$ by $(N - n)\bar{y}$. In fact, it can be shown that $(N - n)\bar{y}$ is the *best linear unbiased predictor* of $\Sigma_r y_i$ (Royall, 1970). Substituting the estimator $(N - n)\bar{y}$ for the second term in Eq. (10.7) yields

$$\hat{Y}_E = n\bar{y} + (N - n)\bar{y} = N\bar{y}.$$

This is again the expansion estimator described previously in Eq. (10.4).

We now develop the formula for the standard error of the expansion estimator as an estimate of Y. The error in estimating Y by $\hat{Y}_E$ is

$$\hat{Y}_E - Y = (N - n)\bar{y} - \sum_r^{N-n} y_i.$$

Under the working model of Eq. (10.6), $\hat{Y}_E$ is unbiased, $E(\hat{Y}_E - Y) = 0$, and the variance of the error of estimation (which is equal to the mean squared error since we are working with an unbiased estimator) is

$$\begin{aligned}\text{var}(\hat{Y}_E - Y) &= E(\hat{Y}_E - Y)^2 = \text{var}\,[(N - n)\bar{y}] + \text{var}\left(\sum_r^{N-n} y_i\right)\\ &= (N - n)^2 \frac{\sigma^2}{n} + (N - n)\sigma^2 = N^2 \frac{\sigma^2}{n}\left(1 - \frac{n}{N}\right).\end{aligned} \tag{10.8}$$

The corresponding error variance for estimating $\bar{Y}$ by $\bar{y}$ is obtained by dividing Eq. (10.8) by N^2. The result is quite analogous to Eq. (10.1), which is based on classical randomization theory. One distinction between the developments leading to Eqs. (10.1) and (10.8) is the following. Equation (10.1) is logically based on considering the average squared error over all possible samples that *might* be drawn by simple random sampling. By contrast, Eq. (10.8) is based on whatever (fixed) sample happens to be obtained. The mathematical expectation in Eq. (10.8) is defined as the average with respect to the superpopulation model of Eq. (10.6), which specifies a probabilistic structure that describes how all the y_i in the population (including those selected for the sample) came to be.

Under the model of Eq. (10.6), the sample variance s_y^2 is an unbiased estimator for σ^2, and so we are again led to using Eq. (10.5) to estimate the standard error of $\hat{Y}_E$, this time on the basis that the square of Eq. (10.5) is an unbiased estimator for Eq. (10.8) [as opposed to Eq. (10.1)].

Model Failure. The preceding developments show that the model-based properties of $\hat{Y}_E$ obtained from the simple working model of Eq. (10.6) completely parallel the properties obtained from classical randomization-based theory. An important question arises as to how these properties would be affected if the model of Eq. (10.6) failed.

If the mean of the y_i is not constant, as assumed by Eq. (10.6), then $\hat{Y}_E$ is generally biased as an estimator of Y (and $\bar{y}$ is biased as an estimator of $\bar{Y}$). In many practical situations the population under study is made up of several distinct subgroups, and it may be expected that the value of $E(y_i)$ varies from group to group. Such situations may be handled by appropriate stratification, the methodology for which is presented in Sec. 10.3. Alternatively, it is sometimes reasonable to model $E(y_i)$ as a simple function of an auxiliary variable x_i. This possibility is taken up in Sec. 10.4.

If the variance of y_i is not constant, say, $\text{var}(y_i) = \text{var}(\epsilon_i) = \sigma_i^2$, then $\hat{Y}_E$ remains unbiased for Y, but estimation of the error variance $E(\hat{Y}_E - Y)^2$ is affected. If a model relating the variance σ_i^2 to a stratification or other auxiliary variable is appropriate, the methods of Secs. 10.3 and 10.4 apply. Failing that, the following shows that simple random sampling tends to produce samples for which unequal variances will not greatly affect the validity of computed standard errors.

If $\mathrm{var}(y_i) = \sigma_i^2$, then the square of the usual estimate of standard error, Eq. (10.5) or (10.3), estimates a quantity proportional to the average variance of the sampled observations,

$$E[Ns(\bar{y})]^2 = \frac{N^2}{n}\left(1 - \frac{n}{N}\right)\left(\frac{\Sigma_s \sigma_i^2}{n}\right). \tag{10.9}$$

By comparison, the actual error variance in this case is

$$\mathrm{var}(\hat{Y}_E - Y) = \frac{(N-n)^2}{n}\frac{1}{n}\sum_s \sigma_i^2 + \sum_r \sigma_i^2. \tag{10.10}$$

Straightforward algebra shows that Eqs. (10.9) and (10.10) are equal for any sample in which the average variance for sample units is equal to the average for nonsample units,

$$\frac{\Sigma_s \sigma_i^2}{n} = \frac{\Sigma_r \sigma_i^2}{N-n}. \tag{10.11}$$

When the sample is chosen by simple random sampling, the condition in Eq. (10.11) will hold *on the average*. Thus although the model-based properties of $\bar{y}$ and $s(\bar{y})$ make no formal reference to probability sampling methods, the above property suggests that simple random sampling is a good practical choice for use with these estimates. Samples chosen by simple random sampling tend to satisfy Eq. (10.11) approximately, and therefore such samples may be expected to yield inferences about $\bar{y}$ and $s(\bar{y})$ that are insensitive to the assumption of equal variances.

10.2.6 Estimation of a Population Proportion

Many surveys involve the estimation of a population proportion, such as the proportion of voters in favor of a tax levy, the proportion of expense vouchers containing errors, or the proportion of office equipment needing repair. Clearly, if the population size is known, an estimate of a population proportion also provides information about the *number* of units that have the property of interest.

We denote by N_κ the number of units in the population (i.e., the frame) that belong to a category κ of interest. Then the *population proportion* of units in κ is $P = N_\kappa/N$. Machine-readable data describing membership in category κ in a population can be conveniently coded by defining a variable y_i as follows:

$$y_i = \begin{cases} 1 & \text{if unit } i \text{ belongs to category } \kappa \\ 0 & \text{if unit } i \text{ does not belong to } \kappa. \end{cases} \tag{10.12}$$

With this coding system, the population total of y_i is

$$Y = \sum_{i=1}^{N} y_i = N_\kappa$$

and the population mean becomes

$$\overline{Y} = \frac{Y}{N} = \frac{N_\kappa}{N} = P.$$

It follows that the methods for estimating $\overline{Y}$ and Y reduce to corresponding methods for the estimation of P and N_κ, respectively.

Estimation of P. The estimate of P (the population mean) is the sample mean, which reduces to the sample proportion, denoted by p,

$$\begin{aligned}\bar{y} &= \frac{\Sigma_s y_i}{n} \\ &= \frac{\text{number of sample units in category } \kappa}{n} \\ &= p.\end{aligned}$$

Similar simplification results when the formulas for variance and standard error are applied to the 0–1 data y_i defined in Eq. (10.12). For example, the sample variance of the y_i becomes

$$\begin{aligned}s_y^2 &= \frac{1}{n-1}\sum_s^n (y_i - \bar{y})^2 = \frac{1}{n-1}\left[\sum_s^n y_i^2 - n(\bar{y})^2\right] \\ &= \frac{1}{n-1}[np - np^2] = \frac{n}{n-1}p(1-p)\end{aligned} \tag{10.13}$$

since $y_i^2 = y_i$ for $y_i = 0$ or 1. In view of Eq. (10.13), Eq. (10.3) for the estimated standard error of $\bar{y} = p$ becomes

$$s(p) = \left[\frac{p(1-p)}{n-1}\left(1 - \frac{n}{N}\right)\right]^{1/2}. \tag{10.14}$$

The term $n - 1$ in the denominator of $s(p)$ may look out of place when compared to the more familiar formula $p(1-p)/n$ for the variance of p in infinite population sampling [see Eq. (4.11) in Sec. 4.2.2]. This factor, which arises from the term $n/(n-1)$ in Eq. (10.13), results in making $s^2(p)$ an unbiased estimator for the variance of p in the sense that

$$\text{ave}_{\text{SRS}}[s^2(p)] = \text{ave}_{\text{SRS}}(p - P)^2.$$

The more familiar formula, $p(1-p)/n$, is not unbiased even in the infinite population case. However, the difference between dividing by n and by $n - 1$ rarely makes a significant difference in practice. The main use of Eq. (10.14) is for forming approximate confidence intervals when n is large, in which case $n/(n-1) \approx 1$.

When the sample size is sufficiently large, the standard error can be used to form a confidence interval for P by the formula

$$p \pm \left[z\, s(p) + \frac{1}{2n}\right] \tag{10.15}$$

where z is the normal deviate corresponding to the desired confidence probability, and the term $1/2n$ is a continuity correction. The accuracy of the normal approximation used in Eq. (10.15) depends on n and P, being better when n is large and P is not too close to 0 or 1. Cochran (1977, Sec. 3.6) gives the criteria listed in Table 10.1 for deciding when to use Eq. (10.15) in practice. The table is constructed so that a nominal 95% confidence interval based on Eq. (10.15) (i.e., with $z = 1.96$) will cover the true value of P, at least 94.5% of the time.

TABLE 10.1 Smallest Values of n and np for Use with Confidence Interval, Eq. (10.15), Based on Normal Approximation

$\min[p, (1-p)]$	$\min[np, n(1-p)]$	Sample size n
0.5	15	30
0.4	20	50
0.3	24	80
0.2	40	200
0.1	60	600
0.05	70	1400

Adapted, with permission, from W. G. Cochran, *Sampling Techniques*, 3d ed., Table 3.3, p. 58. Copyright © 1977 by John Wiley & Sons, Inc., New York.

Small Sample Methods. When the conditions of Table 10.1 are not satisfied, exact methods based on the hypergeometric distribution (Chap. 4, Sec. 4.2.1) may be used. Chung and DeLury (1950) give charts for confidence limits for P at confidence levels 90, 95, and 99% and for population sizes $N = 500$, 2500, and 10,000. Intermediate cases may be obtained by interpolation.

A simpler approach, though approximate, makes use of the Clopper-Pearson charts given in the Appendix, Table A.6, and discussed in Chap. 4, Sec. 4.4.1. The method, which approximates the hypergeometric distribution by the binomial, is most easily described through a numerical example.

Example 10.4: Confidence Interval for P Using Clopper-Pearson Charts. A simple random sample of $n = 50$ cartons is obtained from a warehouse containing $N = 300$ sealed cartons. Inspection reveals that 5 of the cartons have faulty seals. We wish to obtain a 95% confidence interval for the proportion P of faulty seals in the warehouse. Entering the Clopper-Pearson charts at the values $n = 50$ and $p = 5/50 = 0.10$ gives 95% confidence limits (for the infinite population case) of 0.03 to 0.225. Since the sampling fraction 50/300 is substantial, the fpc may be used to shrink the interval toward p by the factor $(1 - 50/300)^{1/2} = 0.913$ as follows:

$$\text{Upper limit} = 0.10 + 0.913(0.225 - 0.10) = 0.214$$
$$\text{Lower limit} = 0.10 - 0.913(0.10 - 0.03) = 0.036$$

In this example, the standard error of p is $s(p) = 0.039$, so the approximate 95% confidence interval for the proportion of faulty seals, obtained from Eq. (10.15), would be $0.10 \pm [1.96(0.039) + 1/100] = 0.10 \pm 0.087$, or 0.013 to 0.187. The main difference between the normal and the binomial approximations for this example appears as a shift of the interval away from zero, a difference that is due to the asymmetry of the binomial limits. In extreme cases, the normal approximation can yield negative values for the lower limit.

Estimation of $\boldsymbol{N_\kappa}$. If N is a accurately known, the relation $N_\kappa = NP$ can be used to obtain estimates of the number of units in a population that belong to the category κ. The estimate of N is Np, with estimated standard error $s(Np) = Ns(p)$, where $s(p)$ is given by Eq. (10.14). Similarly, confidence limits for N_κ can be obtained by multiplying confidence limits for p [from Eq. (10.15) or the method of Example 10.4] by the population size N.

10.2.7 Determination of Sample Size

The formulas for standard errors of the estimators specify a relationship between sample size n and the uncertainty of estimation. This relationship can be exploited to determine the sample size required to obtain an estimate with a desired level of precision.

Sample Size for Means and Totals. First we consider the sample size for estimating a population mean. In planning a survey, it is often of interest to limit the uncertainty in estimating $\overline{Y}$ by $\overline{y}$ to some tolerable error specified numerically as Δ. The uncertainty of estimation is given by $zs(\overline{y})$, where z is the quantile from a normal distribution to achieve a specified confidence probability. Thus the sample size can be determined by solving the equation $\Delta = zs(\overline{y})$ for n,

$$n = \frac{n_0}{1 + n_0/N} \tag{10.16}$$

where

$$n_0 = \left(\frac{zS_y}{\Delta}\right)^2 \tag{10.17}$$

the population standard deviation S_y having been substituted for the sample value in the solution. The calculation of n is presented in two steps partly for convenience and simplicity, but partly because the quantity n_0 is of interest in itself. For large populations, as $N \to \infty$, Eq, (10.16) reduces to simply $n = n_0$, reflecting the fact that n_0 is the solution for the infinite population case. If, after n_0 has been calculated, the ratio n_0/N indicates that a consequential fraction of the population is to be sampled, the required sample size can be reduced to take into account the finiteness of the population by applying Eq. (10.16).

In the case of estimating a population *total*, the desired error limit Δ is set equal to $zs(N\overline{y})$. This equation leads to the solution $n_0 = (NzS_y/\Delta)^2$, with the same refinement, Eq. (10.16), to be used when n_0/N is not small.

Presurvey Estimates of Standard Deviation. The sample size formulas depend on the population standard deviation S_y, which is unknown when the survey is being planned. This leads to some unavoidable uncertainty in specifying the required sample size. Sometimes previous experience or studies by others with populations similar to the population to be sampled will provide a rough estimate to go on. In other cases, the results from a small pilot survey may be used. A disadvantage of this method is that pilot surveys tend, for convenience, to be restricted to readily accessible, and therefore possibly unrepresentative, portions of the population. This could cause the standard deviation from a pilot study to seriously understate the variation that will be found in the population in the full survey.

In some cases, knowledge about the range (minimum to maximum) of likely variation in the population can be put to use. Deming (1960, p. 260) gives the relation between the range, say R, and the standard deviation for some simply shaped distributions. If the distribution is rectangular, $S_y = R/\sqrt{12}$; if it is wedge-shaped, $S_y = R/\sqrt{18}$; if it is shaped like an isosceles triangle, $S_y = R/\sqrt{24}$. The most extreme case is a distribution with proportion P of the values at one extreme and $1 - P$ at the other. In that case $S_y = R[P(1 - P)]^{1/2}$, which achieves its maximum value of $R/2$ when $P = {}^1/_2$.

Finally, Cochran (1977, sec. 4.7) describes formal sequential procedures that guarantee, under certain conditions, achieving a prespecified bound on the sampling error. In these procedures, the sample is drawn in two phases, with the sample size in the second phase determined from analysis of the results of the first-phase sample. More generally along these lines, Thompson (1992, chaps. 23–26) describes a number of adaptive sampling design techniques in which the selection procedure depends on observations made during the survey.

Specification of Relative Error. In some applications, the magnitude of tolerable sampling error is most naturally expressed in *relative* terms as a fraction (or percent) of the quantity to be estimated. When estimating the mean $\overline{Y}$, this would imply that the error limit Δ is given as $\gamma\overline{Y}$, where the fraction γ expresses the limit to the relative error. The required value of n may be obtained immediately by substituting $\gamma\overline{Y}$ for Δ in Eq. (10.17), with the "infinite population" result $n_0 = (zS_y/\gamma\overline{Y})^2$ and, as usual, n given by Eq. (10.16). Note that this expression for n_0 depends on the ratio $S_y/\overline{Y}$, that is, the *relative standard deviation* of y in the population.

The case of specifying the relative error for estimation of the total, Y, works out similarly: solving the equation $\gamma N\overline{Y} = zs(N\overline{y})$ for n and ignoring the fpc (i.e., terms in n/N) yields the solution $n_0 = (zS_y/\gamma\overline{Y})^2$. Again Eq. (10.16) applies to give the finite population solution when needed. These results are summarized in Table 10.2.

TABLE 10.2 Sample Size Required to Obtain Specified Sampling Error in Simple Random Sampling

Quantity estimated	Estimate	Standard error	Sample size* n_0 to obtain: Absolute error Δ†	Relative error Δ†
$\overline{Y}$	$\overline{y}$	$\dfrac{s_y}{\sqrt{n}}\left(1-\dfrac{n}{N}\right)^{1/2}$	$\left(\dfrac{zS_y}{\Delta}\right)^2$	$\left(\dfrac{zS_y}{\gamma\overline{Y}}\right)^2$
Y	$N\overline{y}$	$\dfrac{Ns_y}{\sqrt{n}}\left(1-\dfrac{n}{N}\right)^{1/2}$	$\left(\dfrac{NzS_y}{\Delta}\right)^2$	$\left(\dfrac{zS_y}{\gamma\overline{Y}}\right)^2$
P	p	$\left[\dfrac{p(1-p)}{n-1}\left(1-\dfrac{n}{N}\right)\right]^{1\backslash 2}$	$\dfrac{z^2P(1-P)}{\Delta^2}$	$\dfrac{z^2(1-P)}{\gamma^2P}$
N_κ	Np	$N\left[\dfrac{p(1-p)}{n-1}\left(1-\dfrac{n}{N}\right)\right]^{1/2}$	$\dfrac{N^2z^2P(1-P)}{\Delta^2}$	$\dfrac{z^2(1-P)}{\gamma^2P}$

*The value n_0 computed from the table may be adjusted to take into account the finite population size N by the relation $n = n_0/(1 + n_0/N)$.

†The quantity z refers to the $(1 - \alpha/2)$ quantile of the normal distribution.

Sample Size for Proportions. The approximate sample size required to estimate a proportion P within sampling error Δ can be found by solving the equation $\Delta = zs(p)$ for n. To make the ensuing development parallel the case for $\bar{y}$, we temporarily replace $n - 1$ in the denominator of Eq. (10.14) by n. Then the "infinite population" solution n_0 is the same as that obtained for $\bar{y}$ upon substituting the quantity $[P(1 - P)]^{1/2}$ for S_y in Eq. (10.17). With this same substitution, the sample size n_0 for estimating N_κ with error Δ is identical to the solution for estimating Y. In both cases, the correction for finite population size in Eq. (10.16) applies as well. These results are given explicitly in Table 10.2.

The fact that the formulas for n depend on an unknown quantity, here $P(1 - P)$, affects the determination of the sample size for proportions, as it does for means. However, some advantage can be taken of the simple shape of the function $P(1 - P)$ as P varies from 0 to 1. As is well known, the maximum occurs at $P = 0.5$, and the value decreases monotonically as P departs from 0.5 toward either 0 or 1. Thus whenever an interval of plausible values for P can be specified in advance of sampling, an upper bound for the required sample size can be obtained by using the value in the interval nearest 0.5.

When the desired sampling error for estimating P or N_κ is specified as *relative* error, the value $P = 0.5$ is not the worst case. To estimate P within the relative error γ, the equation to be solved is [again replacing $n - 1$ by n in the denominator of Eq. (10.14) and p by P]

$$\gamma P = zs(p) \approx z \left[\frac{P(1 - P)}{n}\right]^{1/2} \left(1 - \frac{n}{N}\right)^{1/2}. \tag{10.18}$$

The solution is

$$n_0 = \left(\frac{z}{\gamma}\right)^2 \frac{1 - P}{P} \tag{10.19}$$

with n given by Eq. (10.16). In Eq. (10.19) n_0 increases as P approaches 0. Therefore the worst case for estimating a proportion within a given *relative* error occurs at the plausible value of P nearest 0.

Equations (10.18) and (10.19) also apply to estimating N_κ with specified relative error γ, since this problem requires the solution of the equation $\gamma NP = zs(Np)$, which reduces to Eq. (10.18) on dividing both sides by N. The result is listed in Table 10.2.

Example 10.5: Sample Size for Estimating P *with Specified Precision.* An opinion poll in a small town is being planned as a simple random sample to be conducted by telephone interviews. Note first that the frame for this survey (and therefore the population to which inference can be drawn, based on sampling theory) consists only of those residents who can be reached by telephone. Suppose we wish to determine the sample size so that P, the proportion of residents in favor of a tax levy, will be estimated with sampling error less than 0.05 at the 95% confidence level. Suppose further that the population size is 20,000, and it is expected that the town may be about evenly divided on the issue (so that $P = 0.5$ is plausible). Taking $z \approx 2$, $\Delta = 0.05$, $P = 0.5$, and using the appropriate formula from the third row of Table 10.2, we find

$$n_0 = \frac{2^2(0.5)(1 - 0.5)}{(0.05)^2} = 400.$$

Now applying Eq. (10.16) to take the fpc into account, gives

$$n = \frac{400}{1 + 400/20{,}000} = 392.$$

The small difference between n_0, the "infinite population" solution, and n often comes as a surprise when the phenomenon is first encountered. In fact, it makes little difference in this example whether the population size is 20 thousand or 20 million. The insensitivity of Eq. (10.16) to the population size is a manifestation of the fact, noted in Sec. 10.2.3, that the uncertainty of an estimate is not strongly affected by the fraction of the population sampled, unless that fraction is quite substantial.

10.3 STRATIFIED SAMPLING

Many real populations are made up of a number of nonoverlapping subgroups, called *strata*, for which important statistical properties vary systematically from stratum to stratum. Examples are geographic regions in a national opinion poll, union and nonunion employees in a study of company policy on health benefits, or manufacturing lots in a study of product performance. Formally, a stratified population consists of nonoverlapping subpopulations, that is, strata, such that every unit in the population may be identified uniquely with a single stratum. Normally the strata are defined through the sampling frame, and thus the stratum membership of every population unit is known before sampling.

In essence, a sample survey of a stratified population can be thought of as a collection of independent surveys conducted within each stratum. The population information provided by the stratification structure can be incorporated into the sample design in a number of ways, and for a number of reasons. With stratification, one can control the precision attained within each stratum, strata may be administratively convenient for the conduct of the field work, and different sampling methods are sometimes appropriate for different strata. Ordinarily, the accuracy of population estimates is increased by proper use of stratification.

10.3.1 Drawing a Stratified Random Sample

A *stratified random sample* (StRS) is a sample drawn from a stratified population in such a way that a (nonempty) simple random sample is drawn independently within each stratum. We denote the population variate values by

$$y_{hi} = \text{observation for unit } i \text{ of stratum } h$$

where $h = 1, 2, \ldots, H$ and $i = 1, 2, \ldots, N_h$, so that there are H strata with N_h units in the hth stratum. The collection of sample units selected in the hth stratum will be denoted by $s(h)$. Each sample $s(h)$ is obtained by simple random sampling (Sec. 10.2.1) of n_h units drawn from the N_h units in stratum h, $1 \le n_h \le N_h$.

10.3.2 Estimation of a Population Mean or Total

The y total in a stratified population is

$$Y = \sum_{h=1}^{H} \sum_{i=1}^{N_h} y_{hi} = \sum_{h=1}^{H} N_h \overline{Y}_h$$

where $\overline{Y}_h = \Sigma y_{hi}/N_h$ is the mean of the N_h units in stratum h. Dividing this expression for Y by the population size $N = N_1 + N_2 + \cdots + N_H$ yields an expression for the population mean as a weighted average of stratum means, weighted by the stratum sizes,

$$\overline{Y} = \sum_{h=1}^{H} \frac{N_h}{N} \overline{Y}_h.$$

Assuming the weights N_h/N are known, the problem of estimating $\overline{Y}$ is thus reduced to that of estimating the $\overline{Y}_h$ stratum by stratum. Applying the methods of Sec. 10.2 within each stratum, we estimate $\overline{Y}_h$ by the corresponding sample mean,

$$\overline{y}_h = \frac{1}{n_h} \sum_{s(h)}^{n} y_{hi}$$

and form the estimator for $\overline{Y}$ as the stratum-size-weighted average of the $\overline{y}_h$,

$$\overline{y}_{\text{St}} = \sum_{h=1}^{H} \frac{N_h}{N} \overline{y}_h.$$

Since each $\overline{y}_h$ is unbiased for $\overline{Y}_h$ with respect to SRS, $\overline{y}_{\text{St}}$ is unbiased for $\overline{Y}$ with respect to StRS in the sense that

$$\text{ave}_{\text{StRS}}(\overline{y}_{St}) = \overline{Y}.$$

We emphasize that this property, along with the properties described below, treat the stratum weights N_h/N as being known without error.

The sampling error of $\overline{y}_{\text{St}}$ may be described in terms of the population and sample variances of the y variable within the strata. To that end let

$$S_h^2 = \frac{1}{N_h - 1} \sum_{i=1}^{N_h} (y_{hi} - \overline{Y}_h)^2$$

and

$$s_h^2 = \frac{1}{n_h - 1} \sum_{s(h)}^{n_h} (y_{hi} - \overline{y}_h)^2.$$

Since each $\overline{y}_h$ is based on an independent SRS, the sampling variance of $\overline{y}_{\text{St}}$ with respect to StRS is the sum of variances,

$$\text{ave}_{\text{StRS}}(\overline{y}_{St} - \overline{Y})^2 = \sum_{h=1}^{H} \left(\frac{N_h}{N}\right)^2 \text{ave}_{\text{SRS}}(\overline{y}_h - \overline{Y}_{\text{h}})^2$$

$$= \sum_{h=1}^{H} \left(\frac{N_h}{N}\right)^2 \frac{S_h^2}{n_h} \left(1 - \frac{n_h}{N_h}\right). \qquad (10.20)$$

Substituting s_h for S_h in Eq. (10.20) yields an unbiased estimator for the variance, the square root of which is the *estimated standard error* of $\bar{y}_{St}$,

$$s(\bar{y}_{St}) = \left[\sum_{h=1}^{H} \left(\frac{N_h}{N}\right)^2 \frac{s_h^2}{n_h}\left(1 - \frac{n_h}{N_h}\right)\right]^{1/2}. \tag{10.21}$$

Estimation of the total Y follows the usual pattern: the unbiased estimator is $N\bar{y}_{St}$ with estimated standard error $s(N\bar{y}_{St}) = Ns(\bar{y}_{St})$.

If the sample size is reasonably large, approximate confidence intervals for $\bar{Y}$ and Y may be obtained by treating $\bar{y}_{St}$ as a normal random variable with mean $\bar{Y}$ and standard deviation $s(\bar{y}_{St})$.

Example 10.6: Estimation of a Population Mean with StRS. Returning to the problem described in Example 10.3, suppose that a stratified random sample was selected to study machine downtime with four machine types used to define four strata. The data are given in Table 10.3. The calculation of the estimate of the population mean is

$$\bar{y}_{St} = (0.18 \times 38 + 0.33 \times 19 + 0.27 \times 36 + 0.22 \times 18) = 26.7.$$

Using Eq. (10.21), the estimated standard error is $s(\bar{y}_{St}) = 1.65$. Since the sample sizes in the strata are not large, the usual practice is to construct an approximate confidence interval using an appropriate quantile from the t distribution, rather than the normal. But what should we use for the degrees of freedom? While the distribution of $s(\bar{y}_{St})$ is too complex to allow assignment of exact degrees of freedom in general, a useful approximate method is as follows (Satterthwaite, 1946). The *effective degrees of freedom* for $s(\bar{y}_{St})$, ν, is

$$\hat{\nu} = \frac{[s(\bar{y}_{St})]^4}{\Sigma Q_h^2/(n_h - 1)}$$

where

$$Q_h = \left(\frac{N_h}{N}\right)^2 \frac{s_h^2}{n_h}\left(1 - \frac{n_h}{N_h}\right).$$

TABLE 10.3 Machine Dountime with Four Machines (Ex. 10.6)

			Downtime		
Machine type h	Number in sample n_h	Number in stratum N_h	Sample mean $\bar{y}_h$	Sample standard deviation s_h	Stratum weight $N_h/N = W_h$
1	4	16	38	12.5	0.18
2	5	30	19	7.2	0.33
3	8	24	36	7.5	0.27
4	3	20	18	6.0	0.22
	20	$N = 90$			1.00

Note that Q_h is just the hth term in the sum in Eq. (10.21), that is, $s(\bar{y}_{St}) = (\Sigma Q_h)^{1/2}$. For the example data, calculation of the effective degrees of freedom yields $\hat{\nu} = (1.65)^4/0.659 = 11.25$. Thus to form an approximate 95% confidence interval, we use $t_{0.975}(11.25) \approx 2.2$ by eyeball interpolation in the t table. The resulting interval is $26.7 \pm 2.2 \times 1.65 = 26.7 \pm 3.6$ hours per machine.

10.3.3 Model-Based Development

A simple description of a stratified population is given by the familiar one-way fixed effects model (See Sec. 15.2 for a detailed discussion of this model)

$$y_{hi} = \mu_h + \epsilon_{hi}, \qquad \text{with } E(\epsilon_{hi}) = 0, \qquad \text{var}(\epsilon_{hi}) = \sigma_h^2 \tag{10.22}$$

The properties of $\bar{y}_{St}$ with respect to this model parallel those obtained from the randomization-based theory of StRS. The unbiasedness of $\bar{y}_{St}$ is expressed as $E(\bar{y}_{St} - \bar{Y}) = 0$. The error variance of $\bar{y}_{St}$ under the model of Eq. (10.22) is directly analogous to Eq. (10.20),

$$\text{var}(\bar{y}_{St} - \bar{Y}) = \sum_{h=1}^{H} \left(\frac{N_h}{N}\right)^2 \frac{\sigma_h^2}{n_h} \left(1 - \frac{n_h}{N_h}\right). \tag{10.23}$$

Finally, the square of the estimated standard error, from Eq. (10.21), is an unbiased estimator of Eq. (10.23), $E[s(\bar{y}_{St})]^2 = \text{var}(\bar{y}_{St} - \bar{Y})$.

10.3.4 Determination of Sample Size per Stratum and Total Sample Size

As shown by Eq. (10.21), the sampling error in $\bar{y}_{st}$ is related to overall sample size through the individual sample sizes n_h in each stratum. It is convenient to separate the problem of determining sample sizes into two parts:

1. Determination of *relative* sample sizes $w_h = n_h/n$
2. Determination of *overall* sample size $n = \Sigma n_h$

We begin with the second part, supposing that the relative sample sizes in part 1 are specified, and show how to determine the total sample size n to achieve a desired sampling error Δ.

For given w_h, the relation between n and the sampling error may be expressed by substituting nw_h for n_h in Eq. (10.21) and setting up the equation $\Delta = zs(\bar{y}_{St})$, where Δ represents the required limit to sampling error and the multiplier z is the normal distribution quantile corresponding to the desired confidence probability. Solving this equation for n yields

$$n_0 = \left(\frac{z}{\Delta}\right)^2 \sum_{h=1}^{H} \frac{W_h^2 S_h^2}{w_h} \tag{10.24}$$

and

$$n = \frac{n_0}{1 + (z^2/\Delta^2 N) \sum_{h=1}^{H} W_h S_h^2} \tag{10.25}$$

where

$$W_h = \frac{N_h}{N}.$$

The quantity W_h represents the fraction of the population in the hth stratum. The "infinite population" approximation to the sample size, n_0, is obtained by ignoring the within-stratum fpc's, $(1 - n_h/N_h)$.

The sample size required for estimating a population *total* with sampling error Δ is the solution to the equation $\Delta = zNs(\bar{y}_{St})$. The solution may be obtained from Eqs. (10.24) and (10.25) by substituting the quantity Δ/N for Δ in the formulas.

Optimum Allocation. We now return to part 1 of our problem, which deals with how to determine efficient relative sample sizes w_h. This problem is referred to as the *optimum allocation* problem in survey sampling literature. The criterion for determining optimum allocation accounts for possibly differing costs of sampling in different strata by assuming the simple cost function

$$C = C(n_1, n_2, \ldots, n_H) = c_0 + \Sigma c_h n_h$$

where c_0 represents a fixed cost and c_h the cost of each additional observation in stratum h. The goal of efficient allocation can then be stated in either of two ways:

1. Minimize the sampling cost C required to achieve a specified sampling error Δ.
2. Minimize the sampling error Δ subject to a fixed sampling cost C.

It turns out that both criteria lead to the same *relative* sample sizes (i.e., allocation scheme) and differ only in how the total sample size n is determined. The optimum allocation is given by

$$w_h = \frac{\lambda W_h S_h}{\sqrt{c_h}} \tag{10.26}$$

where the proportionality constant λ is chosen to make the w_h sum to unity:

$$\lambda = \frac{1}{\Sigma W_h S_h / \sqrt{c_h}}.$$

A simple derivation of this result may be found in Cochran (1977, sec. 5.5) or Thompson (1992, pp. 109–111). The qualitative features of the optimum allocation formula accord with intuition in that Eq. (10.26) leads to allocating a larger portion of sampling effort to a given stratum if the stratum size W_h is larger, or if the variability in the stratum S_h is larger, or if the cost of sampling c_h is lower.

If the total sample size to achieve a specified sampling error Δ is to be determined, then Eq. (10.26) can be used with Eqs. (10.24) and (10.25) to determine n. This satisfies the first stated optimization criterion. In the second case, where a fixed *cost* constraint must be satisfied, the total sample size n is the solution of the equation $C = c_0 + \Sigma c_h w_h n$, which gives

$$n = \frac{C - c_0}{\Sigma c_h w_h}. \tag{10.23}$$

Some important special cases of allocation schemes deserve mention. If the cost of sampling is taken to be equal in all strata, then the dependence on cost drops out of Eq. (10.26), leading to

$$w_h \propto W_h S_h. \tag{10.27}$$

This is usually called *Neyman allocation*. Like Eq. (10.26), allocation according to Eq. (10.27) satisfies one of two optimization criteria: either it minimizes the total sample size n required to attain a given sampling error [using Eqs. (10.24) and (10.25)], or it provides the most precise estimation possible, minimizing Δ subject to a fixed total sample size n.

In practice the allocation used must satisfy additional constraints that were not taken into account in Eqs. (10.24) and (10.27). For instance, the sample sizes $n_h = w_h n$ must be integers. In addition, the sample size cannot exceed the size of the whole stratum, $n_h \le N_h$. In practice, if an optimum allocation formula leads to $n_h > N_h$ in some stratum, n_h is set equal to N_h, that is, a census (100% sampling) is conducted in that stratum. Then the optimum allocation formula is applied again, this time leaving out the stratum that will be 100% sampled.

Example 10.7: Optimum Allocation in Stratified Sampling. Returning once again to the survey for estimating machine downtime, consider using the stratum sizes and standard deviations shown in Example 10.6 to determine the allocation scheme for a future stratified sample of the population. The Neyman allocation is computed in Table 10.4 for a total sample size of $n = 24$.

The final column, which was obtained by rounding the values of $24w_h$ to the nearest integer, gives the actual sample sizes that would be used in the survey.

TABLE 10.4 Computation of Neyman Allocation

Machine type h	Stratum weight $W_h = N_h/N$	Estimated standard deviation s_h	$W_h s_h$	$w_h = \lambda W_h s_h$	$24w_h$	Rounded optimum n_h
1	0.178	12.5	2.22	0.279	6.70	7
2	0.333	7.2	2.40	0.302	7.24	7
3	0.267	7.5	2.00	0.251	6.04	6
4	0.222	6.0	1.33	0.168	4.02	4
	1.000		7.96	1.000	24.00	24

Proportional Allocation. The optimum allocation formulas depend on having prior information on the stratum standard deviations S_h. If this information is not available, a simple expedient is to assume equal standard deviations, in which case Eq. (10.27) becomes $w_h = W_h$, or $n_h/n = N_h/N$. This sampling design, called *proportional allocation*, has the appealing property that every stratum is sampled at the same rate—the sampling fraction n_h/N_h is constant. The relation of proportional allocation to Eqs. (10.24) and (10.27) shows that it will not be far from optimum unless the sampling cost or within-stratum variability differs greatly from stratum to stratum. An additional convenience of proportional allocation is that $\bar{y}_{St}$ reduces to the unweighted sample mean when $n_h/n = N_h/N$,

$$\bar{y}_{St} = \sum_{h=1}^{H} \frac{N_h}{N}\bar{y}_h = \sum_{h=1}^{H} \frac{n_h\bar{y}_h}{n} = \sum_{h=1}^{H}\sum_{s(h)}^{nh} \frac{y_{hi}}{n} = \bar{y}. \tag{10.28}$$

Optimality aside, the combined benefits of computational simplicity afforded by Eq. (10.28), plus the "public relations" value of taking an equal fraction of every stratum into the sample, are often enough to make proportional allocation the practical method of choice.

The computational simplicity in Eq. (10.28), that reduces the estimator of $\overline{Y}$ in a stratified population to that used in the unstratified case, does not extend to the formula for estimated standard error. That is, Eq. (10.3) for $s(\bar{y})$ is *not* a valid estimate of the standard error in stratified sampling with proportional allocation, even though $\bar{y}_{St} = \bar{y}$. However, the relevant standard error, Eq. (10.21), does simplify to some extent. If $w_h = W_h$, Eq. (10.21) becomes

$$s(\bar{y}_{St}) = \left[\frac{1}{n}\left(1 - \frac{n}{N}\right)\sum_{h=1}^{H} W_h s_h^2\right]^{1/2}. \tag{10.29}$$

Another simplification that is sometimes useful applies to the formulas for sample size. Under the assumption of proportional allocation, Eqs. (10.24) and (10.25), for the sample size required to achieve sampling error Δ, reduce to

$$n_0 = \left(\frac{z}{\Delta}\right)^2 \sum_{h=1}^{H} W_h S_h^2 \quad \text{and} \quad n = \frac{n_0}{1 + n_0/N}.$$

10.3.5 Estimation of a Population Proportion

As described in Sec. 10.2.6, the proportion P of units in a population belonging to some category κ of interest is mathematically equivalent to the population mean of a variable that takes the value 1 for units in the category and 0 otherwise. Furthermore, the sample mean $\bar{y}_h$ in stratum h reduces to the proportion, say p_h, of units in $s(h)$ that belong to the category, and the sample standard deviation in stratum h becomes $s_h = [p_h(1 - p_h)n_h/(n_h - 1)]^{1/2}$. With these substitutions, the formulas and methods described in Sec. 10.3.2 apply to the estimation of a proportion in a stratified population. In particular, the population proportion P is estimated by the analogue of $\bar{y}_{St}$, which is

$$p_{St} = \sum_{h=1}^{H} W_h p_h \tag{10.30}$$

and the estimated standard error is given by the analogue of Eq. (10.21),

$$s(p_{St}) = \left[\sum_{h=1}^{H} \frac{W_h^2 p_h(1 - p_h)}{n_h - 1}\left(1 - \frac{n_h}{N_h}\right)\right]^{1/2}. \tag{10.31}$$

Analogous to estimating the population total Y with continuous data is estimation of N_κ, the total number of units in the population that belong to the category of interest. Since $N_\kappa = NP$, the appropriate estimate is Np_{St} with estimated standard error $Ns(p_{St})$.

The methods given in Sec. 10.3.4, for the determination of sample size, also carry over to the discrete data case. The only change required is to substitute the quantity $P_h(1 - P_h)$ for S_h^2 wherever the latter appears in the sample size formulas. Here P_h refers to the proportion of units that belong to the category κ in stratum h. As usual, it is necessary to obtain some kinds of prior estimates of the unknown quantities P_h in order to use the sample size formulas.

10.3.6 Stratification with More General Estimators

The estimator of the y total of a stratified population can be expressed as the sum over strata of estimators of the individual stratum totals,

$$N\bar{y}_{St} = \sum_{h=1}^{H} N_h\bar{y}_h = \sum_{h=1}^{H} \hat{Y}_h.$$

This representation suggests the valid generalization that the estimator of the y total in stratum h need not be limited to the expansion estimator $N_h\bar{y}_h$, but could be any appropriate estimator of Y_h in stratum h, including a ratio estimator or the estimators for cluster sampling described in Secs. 10.4 to 10.6. It is not even necessary to use the same estimation method in every stratum. Generally, then, an estimate of the total in a stratified population may be constructed as a sum over strata,

$$\hat{Y}_{St} = \sum_{h=1}^{H} \hat{Y}_h$$

where $\hat{Y}_h$ now represents a general estimator of Y_h based on the data from stratum h.

When estimation is independent across strata, the estimated standard error of $\hat{Y}_{St}$ can be obtained by combining the standard errors for each component estimator. The generalization of Eq. (10.21) is

$$s(\hat{Y}_{St}) = \left\{\sum_{h=1}^{H} [s(\hat{Y}_h)]^2\right\}^{1/2}$$

where $s(\hat{Y}_h)$ represents an appropriate estimator for the standard error of $\hat{Y}_h$ as an estimator of Y_h in stratum h.

Similarly, a general estimator of the population mean and the corresponding standard error may be obtained by dividing $\hat{Y}_{St}$ and $s(\hat{Y}_{St})$, respectively, by N.

10.3.7 Poststratification

For some desirable stratification criteria, the stratum to which a unit belongs cannot be determined until the data have been collected. Some examples might be personal characteristics such as age, sex, race, or educational level in a survey of a human population. However, some of the benefits of using stratification in estimation can be obtained by the technique called *poststratification*.

The method of poststratification amounts to taking a single simple random sample of total size n from a stratified population and then classifying the sample units into the appropriate (predefined) strata after sampling. As with ordinary stratification, the stratum sizes N_h or W_h must be accurately known. Also, it is critical that the criteria used for classifying sample units into strata be identical with those used to define the strata in the source of information that provides stratum sizes. The estimation methods of Secs. 10.3.2, 10.3.5, and 10.3.6 are then used with the resulting sample. A careful theoretical discussion of the benefits and trade-offs of post-stratification is given in Holt and Smith (1979).

A disadvantage of this method, compared with true stratified random sampling, is that the sample size per stratum is not under control of the sampler. For example, if $n_h = 0$ in some stratum, the estimation methods of stratified sampling cannot be used. However, if the total sample size is large enough and there are not too many strata, the likely result is that at least one or two units will be obtained in each stratum. Recall that the formulas for estimated standard errors in Sec. 10.3.2 actually require $n_h \geq 2$ in every stratum.

10.4 RATIO ESTIMATION

Like stratification, ratio estimation methods seek to improve the accuracy of estimation by applying weighting factors to the sample mean. The methods associated with stratified sampling use information about stratum sizes to define appropriate weights. In a sense, this weighting corrects for "unrepresentativeness" of the sample as measured by differences between the known stratum weights $W_h = N_h/N$ and the sample weights $w_h = n_h/n$. [Recall, from Eq. (10.28), that when $w_h = W_h$ the stratified estimate of the population mean $\bar{y}_{St}$ reduces to the unweighted sample mean $\bar{y}$.] The method of ratio estimation adjusts the sample mean, in a different way, to correct for "unrepresentative" sample values of an auxiliary variable x.

Example 10.8: Applications Where an Auxiliary Variable Can Be Useful for Ratio Estimation. In the following situations the x variable is a measure of the size (in some sense) of the ith unit.

1. A survey of repair costs for test instruments, with y_i being total annual repair cost for instruments at plant i and x_i the number of units in service at plant i (= "size" of plant i)

2. A survey of farm productivity, with y_i being the yield of wheat from farm i and x_i the acreage under wheat at farm i

3. A survey of hospital usage, with y_i being the number of patient discharges from hospital i during a given month and x_i the in-patient capacity (number of beds) of hospital i

More formally, conditions where ratio estimates are useful may be described as follows. Each unit in the population is associated with a pair of observables (x_i, y_i) such that the x, y pairs in the population are positively correlated and cluster about a straight line going through the origin $x = 0$, $y = 0$. Something must be known about the population of x values. Ratio estimates of the mean or total of y require a knowledge of the population mean for x, $\bar{X}$, or the total $X = N\bar{X}$. If the x values are known before sampling for every unit in the population, the method of balanced sampling can be used to make the estimation procedures relatively insensitive to the exact nature of the relationship between y and x in the population.

10.4.1 Ratio Estimation of a Population Mean or Total

An intuitive understanding of the use of ratio estimation is perhaps best gained by thinking through an example.

Example 10.9: Ratio Estimation of a Total. A survey of hospital usage was conducted in a region containing $N = 1000$ hospitals by a simple random sample of $n = 100$ hospitals. The number of patient discharges during the month of June was recorded for each sample hospital, with an average of $\bar{y} = 500$ discharges per hospital. With no further information, it would seem reasonable to use $N\bar{y} = 500{,}000$ as an estimate of the total number of discharges during June in the region. However, our attitude toward this estimate would naturally change if we learned that the average size of the hospitals in the region is $\bar{X} = 270$ beds, while the sample hospitals had an average of only $\bar{x} = 180$ beds. Since $\bar{x} < \bar{X}$, it would seem likely that $N\bar{y} < Y$. A simple adjustment to the estimate would be to multiply $N\bar{y}$ by the ratio $\bar{X}/\bar{x} = 270/180 = 1.5$ to compensate for the fact that smaller-than-average hospitals were chosen for the sample. This adjusted estimate is called the *ratio estimator* of the total number of discharges. The resulting estimate, in this example, is $N\bar{y}(\bar{X}/\bar{x}) = 1000(500)(270/180) = 750{,}000$ patient discharges.

Example 10.9 motivates the formula for the ratio estimator of the y total Y in a population, which is denoted by

$$\hat{Y}_R = X\frac{\bar{y}}{\bar{x}} = N\bar{X}\frac{\bar{y}}{\bar{x}}. \tag{10.32}$$

The randomization-based theoretical properties of $\hat{Y}_R$ with respect to simple random sampling are somewhat awkward mathematically, compared with the theory for the estimators presented in the previous sections; see Cochran (1977, chap. 6), Thompson (1992, chap. 7), or Raj (1968, chap. 5). However, useful approximate results are available. Under SRS, the ratio estimator is approximately unbiased, $\text{ave}_{\text{SRS}}(\hat{Y}_R) \approx Y$, and an approximate confidence interval for the mean may be obtained as

$$\hat{Y}_R \pm z s(\hat{Y}_R) \tag{10.33}$$

where z is the appropriate normal distribution quantile corresponding to the desired confidence level, and $s(\hat{Y}_R)$, the *estimated standard error* of $\hat{Y}_R$, is given by

$$s(\hat{Y}_R) = \frac{X\, s_R}{\bar{x}\sqrt{n}}\left(1 - \frac{n}{N}\right)^{1/2} \tag{10.34}$$

where

$$s_R = \left\{\sum_s^n \frac{[y_i - (\bar{y}/\bar{x})x_i]^2}{n-1}\right\}^{1/2}. \tag{10.35}$$

The quantity s'_R is a measure of the average of the deviations of y_i about a straight line drawn through the origin and the point $(\bar{x}, \bar{y})$. When s_R must be calculated by hand, the following algebraically equivalent formula is convenient:

$$s_R^2 = \frac{\Sigma_s\, y_i^2 - 2(\bar{y}/\bar{x})\Sigma_s\, x_i y_i + (\bar{y}/\bar{x})^2 \Sigma_s\, x_i^2}{n-1}$$

If, however, the calculation is made on a computer, Eq. (10.35) should be used directly because it is less sensitive to possible round-off error.

Note that the population size N does not enter the calculation of the estimate $\hat{Y}_R = X\bar{y}/\bar{x}$. Except for the fpc (which can often be ignored without significant effect), the same can be said of the standard error in Eq. (10.34). Occasionally, practical use can be made of these facts when the x total X is known but the population size N is not.

Example 10.10: Ratio Estimation of a Total with N Unknown. Suppose it is of interest to estimate the total sucrose content in a truckload of sugar beets. Let (x_i, y_i) denote the weight and the sucrose content, respectively, of the ith sugar beet (or ith bucket full of beets) selected at random from the truckload. Laboratory analyses can determine the proportion by weight of sucrose $\bar{y}/\bar{x}$, which can then be multiplied by the total net weight of the truckload X to obtain a ratio estimate $X\bar{y}/\bar{x}$ of the total sucrose content of the truckload. The *number* of sugar beets in the truckload N is not required for the estimate, and (except for the fpc) is not needed for estimating the standard error of the estimate.

The ratio estimator of the population mean $\bar{Y}$ is $\hat{Y}_R/N = \bar{X}\bar{y}/\bar{x}$. Expressions for the approximate confidence interval and the standard error for $\hat{Y}_R/N$ are obtained by dividing Eqs. (10.33) and (10.34), respectively, by N. In particular,

$$s\left(\frac{\hat{Y}_R}{N}\right) = \frac{1}{N}\, s(\hat{Y}_R) = \frac{\bar{X}\, s_R}{\bar{x}\sqrt{n}}\left(1 - \frac{n}{N}\right)^{1/2}. \tag{10.36}$$

In some applications, the object of estimation is simply the population ratio $R = \bar{Y}/\bar{X} = Y/X$. Examples might be the ratio of machine downtime in the current year to downtime in the previous year, or the ratio of salary cost to total budget obtained from a sample of administrative divisions in a large company. Since the population ratio is obtained by dividing the population total Y by X, the estimator of R is obtained from $\hat{Y}_R$ in the same way,

$$\hat{R} = \frac{\hat{Y}_R}{X} = \frac{\bar{y}}{\bar{x}} \tag{10.37}$$

with estimated standard error

$$s(\hat{R}) = \frac{1}{X}\, s(\hat{Y}_R) = \frac{s_R}{\bar{x}\sqrt{n}}\left(1 - \frac{n}{N}\right)^{1/2}. \tag{10.38}$$

Example 10.11: Ratio Estimation of a Total and a Ratio of Totals. Returning to the hospital discharge survey of Example 10.9, suppose that further calculations yield $s_R = 245$. Then an approximate 90% confidence interval for the total number of discharges in the region is given by

$$\hat{Y}_R \pm 1.645 s(\hat{Y}_R) = 750{,}000 \pm 1.645(1000)\left(\frac{270}{180}\right)\left(\frac{245}{\sqrt{100}}\right)\left(1 - \frac{100}{1000}\right)^{1/2}$$

$$= 750{,}000 \pm 57{,}351.$$

The five-digit uncertainty ± 57,351 belies the approximate nature of the theory underlying the calculation. A better way to summarize the numerical result, for public reporting, would be "750 thousand ± 57 thousand." An estimate of the ratio R, the number of discharges per hospital bed in the population, can be obtained from the above result by dividing the estimate and the confidence limits by the total number of hospital beds in the population, $X = N\overline{X} = 1000(270) = 270{,}000$. The resulting approximate 90% confidence interval is $\hat{R} \pm 1.645s(\hat{R}) = 2.78 \pm 0.21$ discharges per bed.

10.4.2 Model-Based Development and Balanced Sampling

A simple linear model that represents the situation where ratio estimation is appropriate is, for $i = 1, 2, \ldots, N$

$$y_i = \beta x_i + \epsilon_i$$

$$E(\epsilon_i) = 0 \quad \text{and} \quad \text{var}(\epsilon_i) = \sigma_i^2. \tag{10.39}$$

It is easy to show (Royall, 1970) that the ratio estimator is unbiased under this model; $E(\hat{Y}_R - Y) = 0$ for any sample s. It turns out that the nature of the error variance $E(\hat{Y}_R - Y)^2$ under this model is sensitive to assumptions about the variance function σ_i^2. Also affected strongly by the variance function are properties of variance estimators that might be considered for forming confidence intervals. Detailed investigations into this problem carried out by Royall and Cumberland (1978, 1981), Royall and Eberhardt (1975), and others have resulted in the development of some refinements to the standard error and confidence interval formulas given in Sec. 10.4.1. However, for present purposes, the following very simple summary of these studies will suffice. The methods given in Sec. 10.4.1 are reasonably robust in the following sense. In large samples from large populations (n large and n/N small), the confidence intervals given in Sec. 10.4.1 cover the true population values with approximately the nominal probabilities for a wide range of assumptions about the σ_i. For additional discussion of the model-based approach, see Royall (1988) and Särndall, Swensson, and Wretman (1991).

Balanced Sampling. Another concept derived from model-based theory, of perhaps more immediate practical consequence for ratio estimation, is the sample design called *balanced sampling*. The usefulness of balanced sampling arises when possible failure of the model of Eq. (10.39) is considered and more general regression functions are allowed. One simple generalization is a straight-line model with a nonzero intercept. In that case, we have

$$y_i = \beta_0 + \beta_1 x_i + \epsilon_i. \tag{10.40}$$

Under the model of Eq. (10.40), the ratio estimator is generally biased, with bias

$$E(\hat{Y}_R - Y) = N\beta_0\left(\frac{\overline{X}}{\overline{x}} - 1\right).$$

Note that the bias becomes zero if the sample is chosen so that $\overline{x} = \overline{X}$. This condition, which defines a (first-order) balanced sample, simply states that the average size of the sample units is the same as the average size of units in the population.

The next more general model is the quadratic model

$$y_i = \beta_0 + \beta_1 x_i + \beta_1 x_i^2 + \epsilon_i. \tag{10.41}$$

Under the model of Eq. (10.41), the ratio estimator has zero bias for any sample which is balanced on the first two moments of x,

$$\bar{x} = \bar{X} \quad \text{and} \quad \sum_{s}^{n} \frac{x_i^2}{n} = \sum_{i=1}^{N} \frac{x_i^2}{N} \tag{10.42}$$

(see Royall and Herson, 1973). Higher-order polynomial models lead to requirements for the balance of correspondingly higher-order moments of x in order to achieve unbiasedness for $\hat{Y}_R$.

A computational convenience which accompanies the use of balanced samples is that, when $\bar{x} = \bar{X}$, $\hat{Y}_R = N\bar{y}$, and the ratio estimate of the population mean $\hat{Y}_R/N = \bar{X}\bar{y}/\bar{x}$ reduces to the unweighted sample mean $\bar{y}$. This kind of simplification, in which a weighted mean reduces to the unweighted mean, is analogous to what happens in stratified sampling when the sample is "representative" in the sense that proportional allocation is used. Roughly speaking, a balanced sample is "representative" with respect to x, and so the ratio adjustment $\bar{X}/\bar{x}$ cancels out when the sample is balanced.

The desirability of obtaining balanced samples that satisfy, or nearly satisfy, Eq. (10.42) suggests that all of the equally likely samples that might arise by simple random sampling are not all equally desirable with ratio estimation. Many practical situations exist in which the x_i are known in advance of sampling for all units in the population. In such cases it is feasible to control the sample selection procedure so that a well-balanced sample will be obtained. While balanced sampling methods have not yet come into widespread use, practical methods for achieving balance have been developed. The most sophisticated implementation is probably the "basket method" developed by Wallenius (1973). This method creates N/n "baskets" (possible samples of size n) all of which are well balanced according to the criteria in Eq. (10.42). The survey sample is then specified by choosing one of these baskets at random. A less efficient, but possibly useful, method of achieving a reasonably balanced sample is to select a sample by simple random sampling, and then *reject* the sample (and try again) if the resulting sample does not turn out to be reasonably well balanced.

10.4.3 Determination of Sample Size

Assuming a balanced sample, the sample size required to obtain a $(1 - \alpha/2)$ confidence interval for Y of half-width Δ using the ratio estimator $\hat{Y}_R$ is the solution to the equation

$$\Delta = zs(\hat{Y}_R) = z\frac{Ns_R}{\sqrt{n}}\left(1 - \frac{n}{N}\right)^{1/2}$$

where $z = z_{1-\alpha/2}$ is the $(1 - \alpha/2)$ quantile of the standard normal distribution. The solution is

$$n_0 = \left(\frac{zNs_R}{\Delta}\right)^2 \quad \text{with} \quad n = \frac{n_0}{1 + n_0/N}. \tag{10.43}$$

If the sampling method to be used is SRS, so that attainment of a balanced sample is not guaranteed, the following line of reasoning indicates that Eq. (10.43) is still the best one can do. The ratio $\bar{X}/\bar{x}$, which affects the magnitude of the sampling error in Eq. (10.34), will not be known in advance of taking an SRS, and so the best practical prediction is that the sample will be an "average" one. Since $\text{ave}_{\text{SRS}}(\bar{x}) = \bar{X}$, setting $\bar{x}$ equal to its average value over SRS in Eq. (10.34) leads again to Eq. (10.43).

The sample size required to estimate $\bar{Y}$ by $\hat{Y}_R/N$ with sampling error Δ may be obtained from Eq. (10.43) by replacing Δ with $(N\Delta)$. Similarly, to estimate the population ratio R with sampling error Δ, the required sample size is given by Eq. (10.43) with Δ replaced by $(X\Delta) = (N\bar{X}\Delta)$.

As always with sample size determination formulas, advance information on an appropriate standard deviation is required. In this case, the needed information is a prior value of s_R, defined by Eq. (10.35), which is a measure of the variation of y values about a straight line through the origin.

10.5 SINGLE-STAGE CLUSTER SAMPLING

Cluster sampling is appropriate in populations for which the most convenient sampling frame is a list of *clusters*. The clusters are nonoverlapping subgroups, within which the individuals (or elements) of the population are physically or administratively located. Thus, the sampling unit becomes the cluster. We will denote the number of population elements in the ith cluster by M_i and the total number of clusters by N. The total number of *elements* in the population will thus be $M = \Sigma_{i=1}^{N} M_i$.

10.5.1 Drawing a Single-Stage Cluster Sample

The simplest method of cluster sampling is based on simple random sampling. In particular, a simple random sample of n clusters is taken from a frame listing the N clusters in the population. Then the survey data are obtained for *each* of the M_i elements in each of the sampled clusters. The following applications will help fix the ideas and notation.

Example 10.12: Applications of Single-Stage Cluster Sampling

1. A company wishes to estimate the average number of hours of training per employee in the past year. A survey is conducted by selecting a sample of n departments from a list of the N departments in the company. The required data on amount of training are then obtained for each of the M_i employees in each selected department.

2. An institution wishes to estimate the amount (in megabytes, say) of confidential or proprietary data stored on floppy disks. A list of the N computer workstations at the institution is available for use as a frame. A sample of n workstations is selected, and the amount of sensitive data is recorded for each of the M_i data diskettes associated with each selected workstation.

Comparison with Stratified Sampling and Two-Stage Cluster Sampling. Populations consisting of nonoverlapping subgroups have been discussed in Sec. 10.3 on stratified sampling. The essential difference between cluster sampling and stratified sampling reduces to whether or not all of the subgroups are represented in the sample. If at least one element is selected from *every* subgroup, then the subgroups are treated as *strata* and the methods of Sec. 10.3 on stratified sampling apply. If some, but not all, of the subgroups are selected into the sample, then the subgroups are treated as *clusters*.

In single-stage cluster sampling, data are obtained from all elements of the selected clusters. If data are obtained for only a sample of elements from selected clusters, methods appropriate for two-stage (or, generally, multistage) sampling are required. Two-stage sampling is discussed in Sec. 10.6.

10.5.2 Estimation of a Population Mean or Total

In cluster sampling, the observed datum for the jth element within the ith cluster is denoted by y_{ij}. The y total in the population is then

$$Y = \sum_{i=1}^{N} \sum_{j=1}^{M_i} y_{ij} = \sum_{i=1}^{N} y_i$$

where

$$y_i = \sum_{j=1}^{M_i} y_{ij}$$

denotes the y *total* in the ith cluster.

The population *mean per element*, or grand mean, which we denote by $\overline{\overline{Y}}$, is the ratio of the y total to the total number of elements among the N clusters,

$$\overline{\overline{Y}} = \frac{Y}{M} = \frac{\sum_{i=1}^{N} \sum_{j=1}^{M_i} y_{ij}}{\sum_{i=1}^{N} M_i}.$$

The *estimator of the population mean per element* is the sample mean per element,

$$\bar{y}_c = \frac{\Sigma_s y_i}{\Sigma_s M_i} = \frac{\bar{y}}{\overline{M}_s}. \qquad (10.44)$$

In the last expression of Eq. (10.44) $\bar{y}$ denotes the mean of the y *totals* per cluster,

$$\bar{y} = \frac{1}{n} \sum_{s}^{n} y_i = \frac{1}{n} \sum_{s}^{n} \sum_{j=1}^{M_i} y_{ij}$$

and $\overline{M}_s$ denotes the mean number of elements per sample cluster, $\overline{M}_s = \Sigma_s M_i/n$.

The estimator $\bar{y}_c$ has the properties of a ratio estimator since it is the ratio of two sample means, $\bar{y}$ and $\overline{M}_s$. Comparison of Eq. (10.44) with Eq. (10.37) shows that $\bar{y}_c$ is mathematically equivalent to the estimator of the population ratio R upon substituting the cluster size M_i for the x variable of Eq. (10.37). Building on this equivalence, the *estimated standard error* of $\bar{y}_c$ is obtained by substituting M_i for x_i in Eq. (10.38) with the result

$$s(\bar{y}_c) = \frac{S_c}{\overline{M}_s \sqrt{n}} \left(1 - \frac{n}{N}\right)^{1/2} \qquad (10.45)$$

where

$$s_c = \left[\sum_s^n \frac{(y_i - \bar{y}_c M_i)^2}{n-1} \right]^{1/2}. \tag{10.46}$$

The ratio estimator of the y total in single-stage cluster sampling is obtained by multiplying the estimator of the mean per element by the number of elements in the population,

$$\hat{Y}_c = M\bar{y}_c = \frac{M\bar{y}}{\overline{M}_s}.$$

The estimated standard error of $\hat{Y}_c$ is given by

$$s(\hat{Y}_c) = s(M\bar{y}_c) = Ms(\bar{y}_c)$$

where $s(\bar{y}_c)$ is from Eq. (10.45).

Model-Based Development and Balanced Sampling. As has been pointed out, the estimators $\hat{Y}_c$ and $\bar{y}_c$ for single-stage cluster sampling are essentially equivalent to the ratio estimators discussed in Sec. 10.4. As such, the model-based theory sketched for ratio estimators carries over to the present case. In brief, the basic working model for clustered populations represents the cluster total y_i as proportional to cluster size M_i plus error. If the relationship between y_i and M_i is better represented by a regression line with nonzero intercept or by a quadratic function, then $\bar{y}_c$ is biased unless the sample is balanced on M_i. This implies that a controlled sample selection procedure that results in

$$\overline{M}_s = \frac{M}{N} \quad \text{and} \quad \sum_s^n \frac{M_i^2}{n} = \sum_1^N \frac{M_i^2}{N}$$

is desirable for protection against such biases. See Sec. 10.4.2 for more discussion of balanced sampling.

10.5.3 Determination of Sample Size

Sample size formulas are obtainable from those given for the ratio estimators in Sec. 10.4.3 with the substitution of M_i for x_i. The results are as follows.

To estimate $\overline{\overline{Y}}$ by $\bar{y}_c$ with a $(1 - \alpha/2)$ confidence interval of half-width no more than some tolerable sampling error Δ, the required sample size is

$$n_0 = \left(\frac{zs_c}{\Delta \overline{\overline{M}}} \right)^2 \quad \text{with} \quad n = \frac{n_0}{1 + n_0/N} \tag{10.47}$$

where $z = z_{1-\alpha/2}$ is the $(1 - \alpha/2)$ quantile of the standard normal distribution and $\overline{\overline{M}} = M/N$.

To estimate Y by $\hat{Y}_c$ with sampling error Δ, the required sample size is given by Eq. (10.47) with Δ replaced by Δ/NM.

10.5.4 Estimation of a Population Proportion

Here we describe the estimation of the proportion of elements in a clustered population that belong to some category of interest κ. As has been seen previously, methods for estimating a population proportion are typically built on the corresponding methods for estimating a mean. In the case of cluster sampling, the correspondence is specified by interpreting the data y_{ij} as

$$y_{ij} = \begin{cases} 1 & \text{if element } j \text{ in cluster } i \text{ belongs to category } \kappa \\ 0 & \text{if not.} \end{cases}$$

With this interpretation, the ith cluster total is

$$y_i = (\text{number of elements in cluster } i \text{ that belong to category } \kappa)$$

and the population proportion is

$$P = \sum_{i=1}^{N} \frac{y_i}{M}.$$

The estimator of P is the analogue of $\overline{y}_c$, which becomes the overall sample proportion of elements in κ,

$$p_c = \frac{\Sigma_s\, y_i}{\Sigma_s\, M_i}$$

From Eqs. (10.45) and (10.46) the estimated standard error of p_c is

$$s(p_c) = \frac{s_c}{\overline{M}_s \sqrt{n}} \left(1 - \frac{n}{N}\right)^{1/2} \tag{10.48}$$

where

$$s_c = \left[\sum_{s}^{n} \frac{(y_i - p_c M_i)^2}{n - 1} \right]^{1/2}. \tag{10.49}$$

Notice that the formula for $s(p_c)$ does not involve the quantity $p_c(1 - p_c)$. These calculations are illustrated in the following example.

Example 10.13: Estimation of P *in Single-Stage Cluster Sampling.* A survey of computer workstations was conducted by an organization to estimate the proportion of data diskettes in use that contain sensitive data. From a frame listing all $N = 100$ workstations in the organization, a simple random sample of $n = 10$ was selected. The survey results are given in Table 10.5. Calculations give

$$p_c = \frac{82}{146} = 0.56$$

$$\Sigma_s (y_i - p_c M_i)^2 = 214.33$$

$$s_c = \left(\frac{214.33}{9}\right)^{1/2} = 4.88$$

$$\overline{M}_s = \frac{146}{10} = 14.6$$

$$s(p_c) = \frac{4.88}{14.6\sqrt{10}} \left(1 - \frac{10}{100}\right)^{1/2} = 0.10.$$

TABLE 10.5 Survey Results of Computer Workstations

Observation	Number of data diskettes M_i	Number of diskettes with sensitive data y_i
1	17	15
2	25	6
3	5	5
4	8	3
5	19	4
6	21	18
7	16	11
8	9	3
9	12	11
10	14	6
	$\Sigma_s M_i = 146$	$\Sigma_s y_i = 82$

Taking $z \approx 2$, an approximate 95% confidence interval for the population proportion of diskettes containing sensitive information is $0.56 \pm 2 \times 0.10 = 0.56 \pm 0.20$, or 0.36 to 0.76.

It is important to account for clustering in the analysis of survey results. Ignoring the clustered structure of this sample could wrongly lead to treating it as a random sample of 146 diskettes from the population. In that case, the standard error formula of Eq. (10.14) would give the value (ignoring the fpc) $s(p) = [0.56(1 - 0.56)/145]^{1/2} = 0.04$, a result substantially smaller than the correct standard error, 0.10.

10.6 TWO-STAGE CLUSTER SAMPLING

The structure of a clustered population was described in Sec. 10.5. The "second stage" of two-stage cluster sampling refers to subsampling the elements contained in selected clusters, as opposed to sampling the entire cluster, as is done in single-stage cluster sampling.

10.6.1 Drawing a Two-Stage Cluster Sample

In parallel with the discussion of single-stage cluster sampling, the sampling method described is based on simple random sampling at both stages. The first step is to obtain a frame that lists all N clusters in the population. Note that a list of all population *elements* is not required, just clusters. A simple random sample of n clusters is drawn at the first stage. Next, n second-stage frames listing all of the elements in each of the n selected clusters must be obtained or constructed. From each sample cluster, a simple random sample of elements is then drawn in the second stage of sampling. A practical situation illustrating the use of two-stage sampling was given as part 1 of Example 10.2.

The second-stage sample in cluster i will be denoted by $s(i)$, and the number of elements in $s(i)$ will be denoted by m_i. As was the case in Sec. 10.5, M_i denotes the total number of elements in cluster i and y_i is the y total for the M_i elements in cluster i.

10.6.2 Estimation of a Population Mean or Total

The population mean (per element) $\overline{\overline{Y}}$ and y total Y were defined at the beginning of Sec. 10.5.2. The estimator of $\overline{\overline{Y}}$ for single-stage cluster sampling $\bar{y}_c$, defined in Eq. (10.44), depends on the cluster totals y_i in the sample clusters. The y_i are not known in the two-stage case, but can be estimated from the mean of the second-stage samples using the analogue of Eq. (10.4), the expansion estimator of a total, as follows:

$$\text{Mean of second-stage sample, } \bar{y}_i = \sum_{s(i)}^{m_i} \frac{y_{ij}}{m_i}$$

$$\text{Estimator of cluster total } y_i \text{ in cluster } i, M_i\bar{y}_i.$$

Replacing y_i by $M_i\bar{y}_i$ in the formula for $\bar{y}_c$ yields the *ratio-type estimator* of the mean $\overline{\overline{Y}}$ for two-stage sampling,

$$\bar{y}_{\text{RT}} = \frac{\Sigma_s M_i \bar{y}_i}{\Sigma_s M_i} = \frac{\Sigma_s M_i \, (\Sigma_{s(i)}\, y_{ij}/m_i)}{\Sigma_s\, M_i}\,. \tag{10.50}$$

An estimate of the standard error of $\bar{y}_{\text{RT}}$ is given by

$$s(\bar{y}_{\text{RT}}) = \frac{1}{M_s}\left[\left(1-\frac{n}{N}\right)\frac{s_B^2}{n} + \frac{1}{Nn}\sum_s^n M_i^2\left(1-\frac{m_i}{M_i}\right)\frac{s_i^2}{m_i}\right]^{1/2} \tag{10.51}$$

where

$$s_B^2 = \sum_s^n \frac{(M_i\bar{y}_i - M_i\bar{y}_{\text{RT}})^2}{n-1}$$

and

$$s_i^2 = \sum_{s(i)}^{m_i} \frac{(y_{ij} - \bar{y}_i)^2}{m_i - 1}$$

denotes the within-cluster variance. Note that $s(\bar{y}_{\text{RT}})$ has terms that reflect both variability *between* clusters, through s_B, and also *within* clusters, through the s_i. Qualitatively, we can see that the between-cluster term estimates uncertainty in $\bar{y}_{\text{RT}}$ due to differences between cluster totals, while the within-cluster term estimates uncertainty due to the need to estimate the totals in sample clusters from the sample means $\bar{y}_i$. Furthermore, the two terms contain appropriate finite population correction factors: the between-cluster term contains the factor $1 - n/N$ which represents the population fraction of *clusters* not sampled, and the within-cluster term has fpc equal to $1 - m_i/M_i$ for cluster i, which represents the fraction of *elements* in cluster i not sampled.

An approximate $100(1-\alpha)\%$ confidence interval for $\overline{\overline{Y}}$ is given by the interval $\bar{y}_{\text{RT}} \pm zs(\bar{y}_{\text{RT}})$, where $z = z_{1-\alpha/2}$ is the $(1-\alpha/2)$ quantile of the standard normal distribution. If n is not large (say, less than 30), it is good practice to replace z with the corresponding quantile of the t distribution with $n-1$ degrees of freedom.

If the total number of elements in the population $M = \Sigma_{i=1}^{N} M_i$ is known, then the population total can be estimated using $\bar{y}_{\text{RT}}$ by multiplying by M,

$$\hat{Y}_{\text{RT}} = M\bar{y}_{\text{RT}}.$$

The estimated standard error for $\hat{Y}_{RT}$ may be obtained from Eq. (10.51) by multiplying by M,

$$s(\hat{Y}_{RT}) = Ms(\bar{y}_{\text{RT}}).$$

The following example illustrates the estimation of a population total in two-stage cluster sampling.

Example 10.14: Estimation of a Total in Two-Stage Cluster Sampling. An employee survey in a 1200-employee company was conducted to determine the number of hours of training that were undertaken in the past year. A first-stage sample of $n = 7$ departments, of the $N = 42$ departments in the firm, was obtained. Training records of 2 to 4 employees were examined within each sampled department. The results are summarized in Table 10.6.

TABLE 10.6 Results of Training Survey

Sampled department	Employees in department M_i	Employees in sample m_i	Mean hours of training y_i	Standard deviation s_i	$M_i\bar{y}_i$
1	27	3	4	6.93	108
2	38	2	18	8.49	684
3	21	3	28	10.58	588
4	45	4	26	15.49	1170
5	12	2	15	21.21	180
6	53	3	14	16.37	742
7	14	2	36	5.66	504
	210				3976

The ratio-type estimate of the population mean is $\bar{y}_{\text{RT}} = 3976/210 = 18.933$ hours per employee, and the corresponding estimate of the total is $\hat{Y}_{\text{RT}} = 1200(3976/210) = 22{,}720$ hours of training companywide. To obtain a 95% confidence interval for the estimated total, we compute

$$M = 1200 \quad \bar{M}_s = \frac{210}{7} = 30$$

$$s_B^2 = \sum_s^n \frac{(M_i\bar{y}_i - M_i\bar{y}_{\text{RT}})^2}{n-1} = 71{,}481.0$$

$$\sum_s^n M_i^2\left(1 - \frac{m_i}{M_i}\right)\frac{s_i^2}{m_i} = 450{,}849.3$$

$$s(\hat{Y}_{\text{RT}}) = \left(\frac{1200}{30}\right)\left[\left(1 - \frac{7}{42}\right)\left(\frac{71{,}481.0}{7}\right) + \left(\frac{1}{42 \times 7}\right)450{,}849.3\right]^{1/2} = 4009.$$

Thus, an approximate 95% confidence interval for total training hours is 22,720 ± 2 × 4009, or 22,720 ± 8018 hours.

10.6.3 Further Remarks on Two-Stage Sampling

Estimation of a Proportion. As is the case for other sampling designs, the estimation of the population proportion of elements that belong to some category κ is a special case of estimation of the mean. Specifically, the y variables are coded as $y_{ij} = 1$ if element (i, j) belongs to κ otherwise $y_{ij} = 0$. The population proportion of elements in κ is then equal to $\overline{\overline{Y}}$, and the appropriate estimate and standard error may be obtained by applying Eqs. (10.50) and (10.51), respectively, to the 0–1 coded data.

Randomization-Based Theoretical Properties. The estimators $\overline{y}_{RT}$ and $\hat{Y}_{RT}$ share with the ordinary ratio estimator of Sec. 10.4 the characteristic that they involve ratios of quantities that vary from sample to sample. Thus they also share with the ordinary ratio estimator some awkwardness in the mathematical description of their statistical behavior with respect to random sampling. Cochran (1977, Sec. 11.8) shows that the ratio-type estimators are approximately unbiased and gives arguments supporting the reasonableness of Eq. (10.51) as an estimate of standard error.

Model-Based Properties. Royall(1976, 1985) has conducted an extensive model-based theoretical study of robust estimation in two-stage sampling. Some simple conclusions from these studies are the following. Balanced sampling is advantageous in the same way, and for essentially the same reasons, as in single-stage cluster sampling and ordinary ratio estimation (see Secs. 10.4.2 and 10.5.2). The standard error estimator of Eq. (10.51) is a simplification of, and an approximation to (for n large and n/N small), the robust standard error estimator developed in Royall (1985).

Allocation. The sampler has control over the allocation of second-stage sample sizes m_i to clusters. Two popular and practical choices are *equal allocation*, in which m_i is a chosen constant (≥ 2) in each cluster, and *proportional allocation*, in which m_i is chosen as a constant fraction of M_i, resulting in m_i/M_i constant. Royall (1976) shows that proportional allocation is the optimal allocation method for use with $\overline{y}_{RT}$ under certain simple conditions.

10.7 ACKNOWLEDGMENT

I am grateful to several individuals for contributions to the preparation of this chapter. Robert Elder, of the U.S. Department of Agriculture, provided useful references to the literature of bulk sampling. Richard Royall, of The Johns Hopkins University, Raghu Kacker and Susannah Schiller, of the National Institute of Standards and Technology, and William Cumberland of UCLA read drafts of the manuscript at various stages and made many helpful comments. Harrison Wadsworth, editor of this handbook, gave helpful guidance, suggestions, and encouragement throughout the preparation of the manuscript. Finally, my wife Sue and my sons Kyle, Seth, and Kendall were very patient and encouraging while I was busy working on "that chapter."

REFERENCES

ASTM: "Symposium on Bulk Sampling," STP 114, ASTM Committee E-11, Am. Soc. Testing and Materials, Philadelphia, 1951.

Barnett, V.: *Elements of Sampling Theory*, Chapman and Hall, London, 1986.

Bicking, C. A.: "The Sampling of Bulk Materials," *Mater. Res. and Stand.*, vol. 7, pp. 95–116, 1967.

_____: "Principles and Methods of Sampling," in *Treatise on Analytical Chemistry*, pt. I: *Theory and Practice*, 2d ed., vol. 1, I. M. Kolthoff and P. J. Elving, Eds., Wiley, New York, chap. 6, pp. 299–359, 1978 .

Chung, J. H., and D. B. De Lury: *Confidence Limits for the Hypergeometric Distribution*, Univ. of Toronto Press, Toronto, Ont., Canada, 1950.

Cochran, W. G.: *Sampling Techniques*, 3d ed., Wiley, New York, 1977.

Deming, W. E.: *Some Theory of Sampling*, Wiley, New York, 1950.

_____: *Sample Design in Business Research*, Wiley, New York, 1960.

Duncan, A. J.: "Bulk Sampling: Problems and Lines of Attack," *Technometrics*, vol. 4, pp. 319–344, 1962.

Foreman, E. K.: *Survey Sampling Principles*, Marcel Dekker, New York, 1991.

Gy, P. M.: *Sampling of Heterogeneous and Dynamic Material Systems: Theories of Heterogeneity, Sampling and Homogenizing*, Elsevier, Amsterdam, 1992.

Hansen, M. H., W. N. Hurwitz, and W. G. Madow: *Sample Survey Methods and Theory* (in two vols.), Wiley, New York, 1953.

Holt, D., and T. M. F. Smith: "Post Stratification," *J. R. Stat. Soc.,* ser A, vol. 142, pt. 1, pp. 33–46, 1979.

Ishikawa, K.: "How to Establish and Control the Sampling Procedure for Bulk Material," *Rep. Stat. Appl. Res.* (Japanese Union of Scientists and Engineers), vol. 20, pp. 89–99, 1973.

Kish, L.: *Survey Sampling,* Wiley, New York, 1995.

Levy, P., and P. Lemeshow: *Sampling Populations*, Wiley, New York, 1991.

Pitard, F.: *Pierre Gy's Sampling Theory and Sampling Practice*, CRC Press, Boca Raton, FL, 1993.

Raj, D.: *Sampling Theory,* McGraw-Hill, New York, 1968.

_____: *The Design of Sample Surveys*, McGraw-Hill, New York, 1972.

RAND Corp.: *A Million Random Digits with 100,000 Normal Deviates,* Free Press, Glencoe, IL., 1955.

Royall, R. M.: "On Finite Population Sampling Theory under Certain Linear Regression Models," *Biometrika*, vol. 57, pp. 377–387, 1970.

_____: "The Linear Least-Squares Prediction Approach to Two-Stage Sampling," *J. Am. Stat. Assoc.,* vol. 71, pp. 657–664, 1976.

_____: "The Prediction Approach to Robust Variance Estimation in Two-Stage Cluster Sampling," *J. Am. Stat. Assoc.,* vol. 81, pp. 119–123, 1985.

_____: "The Prediction Approach to Sampling Theory," in *Handbook of Statistics, Volume 6: Sampling*, P. R. Krishnaiah and C. R. Rao, Eds., North-Holland/Elsevier, New York, pp. 399–413, 1988.

_____ and W. G. Cumberland: "Variance Estimation in Finite Population Sampling." *J. Am. Stat. Assoc.,* vol. 73, pp. 351–358, 1978.

_____ and _____; "An Empirical Study of the Ratio Estimator and Estimators of Its Variance," *J. Am. Stat. Assoc.,* vol. 76, pp. 66–77, 1981.

_____ and K. R. Eberhardt: "Variance Estimates for the Ratio Estimator." *Sankhya,* ser. C, vol. 37, pp. 43–52, 1975.

_____ and J. H. Herson: "Robust Estimation in Finite Populations I," *J. Am. Stat. Assoc.*, vol. 68, pp. 880–889, 1973.

Särndal, C.-E., B. Swensson, and J. Wretman: *Model Assisted Survey Sampling,* Springer-Verlag, New York, 1991.

Satterthwaite, F. E.: "An Approximate Distribution of Estimates of Variance Components," *Biometrics*, vol. 2, pp. 110–114, 1946.

Schilling, E. G.: "Bulk Sampling," in *Acceptance Sampling in Quality Control*, Dekker, New York, chap. 9, pp. 192–221, 1982.

Sukhatme, P. V., and B. V. Sukhatme *Sampling Theory of Surveys with Applications*, 3d ed., Iowa State Univ. Press, Ames, S. Sukhatme, and C. Asok: 1984.

Thompson, S. K.: *Sampling*, Wiley, New York, 1992.

Wallenius, K. T.: "On Statistical Methods in Contract Negotiation—Part III," Rep. N45, Dept. of Mathematical Sciences, Clemson University, Clemson, SC., 1973.

Wolter, K. M.: *Introduction to Variance Estimation*, Springer-Verlag, New York, 1985.

CHAPTER 11
BAYESIAN ANALYSIS

Thomas W. Calvin
Consultant, Wappingers Falls, NY

11.1 INTRODUCTION

Many processes are treated as a series of units, such as lots, batches, days, shifts, or some other form of natural or rational subgroup. Since these discrete units are produced in a time sequence, the unit-to-unit variation is associated with time, while the within-unit variation is considered independent of time. It is common practice to designate unit-to-unit variation as due to assignable causes and within-unit variation as due to chance causes inherent in the process. Distributions based on the latter are frequently used for control charts and acceptance sampling. The goal is a process in a state of statistical control with only chance causes of variation, all assignable causes being isolated or removed. Implied in this goal is a process at a fixed level with the between-unit variation caused only by the inherent process fluctuations. Such fluctuations are due to many small variations in manufacturing, materials, machines, personnel, methods, measurements, and so on, which can cumulate to produce random within- and between-unit variations. The fraction nonconforming of such a process would be constant with no variation over time. Each unit selected from the process would then be a random sample from the process, encountering only sampling variation.

Realistically this goal is achieved infrequently since many assignable causes are retained. Some are unknown, particularly in a new and highly technical process, and must be tolerated. Others cannot be removed due to state-of-the-art limitations. Finally, there are those known causes that remain for economic reasons: the cost of elimination is not justified at that stage in the process or the quality of the product is not impacted. When these assignable causes remain, the fraction nonconforming will vary over time from unit to unit. Thus there is variation in the process fraction nonconforming, in addition to the sampling variation, which results from the selection of the unit from the process. The study and incorporation of this additional variation into statistical estimation constitutes bayesian analysis. When this process variation is random over time, the process distribution can be modeled and combined with the sampling distribution to obtain a more realistic assessment of the unit-to-unit variation and the state of the process.

All of the variation among lots is not random, but can drift, incur a process shift, or constitute a maverick lot, which are usually sources of between-unit variation that can be eliminated. The random variation is often more subtle and may require redesign of the process to reduce the variation. The bayesian approach provides a formal mechanism for incorporating past experience into the estimation and decision process. This past experience can be provided from existing sample and process data or from subjective prior knowledge and engineering judgment. Much of the controversy surrounding bayesian statistics has centered around the formalizing of this subjective input. However, few production lines would ever work without experienced personnel applying their cumulative knowledge and expertise to solve process problems. Many successful businesses are the result of the good instincts and business acumen of their management. Thus much of the controversy involves the methods for capturing this prior subjective information for use in the decision process instead of using observed data.

The available hard data from process measurements or product sampling often involve the assumption of a prior distribution, with some element of risk, for which parameters are estimated from the data. It is also possible to obtain bayesian estimates from hard data without specifying the prior distribution, but only assuming that a distribution exists. While there is no universal agreement on the definitions for these various approaches, the three discussed here could be referred to as (1) subjective, (2) empirical or objective, and (3) nonparametric empirical bayesian estimation. The latter two extract information from hard data to estimate the parameters of a specific distribution and acknowledge only that a distribution exists, respectively.

This chapter addresses mainly empirical or objective bayesian estimation, where prior knowledge and engineering judgment assume a prior distribution and hard data are used to estimate its parameters. This is not an unreasonable approach since many industrial processes and products can be represented realistically by some very flexible and known choices for prior distributions. As a result a weighted estimate of process or product performance combining present information with past data can be obtained to provide a more meaningful assessment of the process than that which would result from present data only. A famous paper published by the Reverend Thomas Bayes (1702-1761) provides the basis for the method known as bayesian statistical inference. The method is demonstrated initially by a short subjective input example, and then by using objective prior data.

11.2 SUBJECTIVE BAYESIAN ANALYSIS EXAMPLE

A company wishes to screen sales applicants with a "selling aptitude test." Information prior to aptitude testing was gleaned from the subjective opinions of area sales managers. A summary of this past experience with hiring practices led to the following prior distribution:

20% hired were good salespersons

45% hired were satisfactory salespersons

35% hired were unsatisfactory salespersons

The hope was that the test would reduce the 35% of unsatisfactory salespersons hired. All salespeople were required to take the test, and the following results were obtained:

90% of good salespersons passed the test

70% of satisfactory salespersons passed
30% of unsatisfactory salespersons passed

Certainly 30% unsatisfactory does not appear too different from the 35% hired prior to testing. In the future, however, if only those passing the test are hired, what is the probability of successful candidates being good, satisfactory, or unsatisfactory salespersons? First the overall probability of the applicants passing the test must be determined from the above information. This is done in Table 11.1.

TABLE 11.1

	Good	Satisfactory	Unsatisfactory	Total
Pass	0.9 × 0.2 = 0.18	0.7 × 0.45 = 0.315	0.3 × 0.35 = 0.105	0.6
Fail	0.1 × 0.2 = 0.02	0.3 × 0.45 = 0.135	0.7 × 0.35 = 0.245	0.4
	0.20	0.450	0.350	

Since 20% of all applicants hired were good and 90% of these passed, then 18% of all applicants were good and passed the test. The remaining categories are calculated in a similar manner. Adding the pass row yields 60% of all candidates passing the test and 40% failing. Dividing each category in the pass row by 0.6 produces the following results for successful candidates:

30% expected to be good salespeople
52.5% expected to be satisfactory salespeople
17.5% expected to be unsatisfactory salespeople

Thus the posterior results of hiring only those who passed the test resulted in 50% more good salespersons (30% to 20%) and 50% fewer poor salespersons (17.5% to 35%). The company decided to use the selling aptitude test to screen applicants for sales.

The prior distribution, which classified the salespersons, was based on the judgment of their sales managers. This example was chosen purposely since not all would agree that these data are truly subjective. Sales managers should know their people. But what criteria would they use? Some are experienced; others are new. Many territories are compact, as in a large city, while others are spread out over miles of countryside. Quotas may be set for each salesperson, and there may be valid reasons why some were not met. Performance is difficult to assess, and its measurement is not as easy as determining a product failure. Thus the degree of belief in subjective prior information is often in contention, contributing to the controversy surrounding bayesian analysis.

Mathematically this selling aptitude test study can be expressed as follows:

$$\underbrace{P(s;p)}_{\text{posterior probability}} = \frac{\overbrace{P(p;s)}^{\text{test result}} \times \overbrace{P(s)}^{\text{prior probability}}}{\underbrace{\Sigma P(p;s) \times P(s)}_{\text{probability of passing}}}$$

where s is the salesperson's classification and p represents a pass for a particular classification. For example, the probability of a satisfactory salesperson passing is

$$P(s; p) = \frac{0.7 \times 0.45}{0.9 \times 0.2 + 0.7 \times 0.45 + 0.3 \times 0.35} = 0.315$$

This is a statement of Bayes' theorem (see Appendix 11.1). The posttest results are essentially a weighted combination of the prior and the current information. The preceding illustration calculates the posterior probability for the satisfactory sales category as the ratio of the proportion of satisfactory salespeople passing the aptitude test to the total passing the test. Thus bayesian analysis is a method of incorporating prior or past information into present information to obtain a more meaningful and representative estimate.

As noted, this assessment of selling performance may indeed be quite subjective with respect to individual salespersons. However, cumulated input from several sales managers from across the country would average out much of the subjectivity and improve the reliability of the information. While there are some who may feel uncomfortable with such subjective information, the nature of any process that generates observed events is seldom certain, and assumptions must be made concerning the underlying process. There may be some question about the validity of these assumptions when applied to objective data from a process. However, some subjectivity enters into most statistical analyses. The amount of subjectivity affects the degree of belief in the prior information. Existing data frequently available from a process or product through sampling and testing will provide more objective prior information, but assumptions based on experience may still be required.

11.3 OBJECTIVE PRIOR INFORMATION

Often complete knowledge of past process performance data is unavailable, and some assumptions based on experience are necessary. An example of a known prior distribution would be flipping a coin or rolling a pair of dice. An objective bayesian analysis is presented in Appendix 11.2 for those familiar with the gambling game of craps. The prior probability distribution of the total on a pair of dice is known exactly. The probability of winning by means of each total in the prior distribution can be calculated according to the rules of the game. The posterior probability distribution, given a win, describes the probability of winning with the various totals on a pair of dice. The old, or prior, information for a pair of dice, common to many dice games, is combined with the current information, the rules for the game of craps, to derive the probability of winning with any throw of the dice. This procedure is expressed mathematically as Bayes' theorem:

posterior probability — game rules — prior probability for two dice

$$P(T; W) = \frac{P(W; T) \times P(T)}{\Sigma P(W; T) \times P(T)}$$

overall probability of winning

where T is a total on a pair of dice and W represents a win for that total.

This expression of Bayes' theorem has now been applied to a subjective prior probability distribution and to an objective prior distribution that was known exactly. It can also be applied to processes and products by assuming some familiar distributions from mathematical and applied statistics. Previous chapters have introduced the Poisson and binomial distributions. These usually represent nonconformities or other variants per unit and the number of nonconforming items in a unit. If the expected number of variants is fixed (such as a bowl of beads with 1% black beads) and only subject to sampling variations due to the random selection of a sample, then no prior distribution exists and bayesian procedures are not warranted. The estimate of the process average is the total number of variants or variant items divided by the total number of items.

Realistically many, if not most, processes vary about some process average with greater fluctuations than can be attributed to sampling alone. Not all items have an equal chance of being variant (for example, each flip of a coin might have a different probability of being a tail). A lot produced at one time will be different from a lot produced at another time. The items have a continually changing probability of being variant from lot to lot. This is because, in addition to chance variation, some large assignable causes may not have been removed because they are unknown, state-of-the-art limited, or uneconomic. The result is a separate distribution of the process parameters obtained from prior information, in addition to sampling variation. This leads to bayesian analysis of the objective data. (See Appendix 11.3 for descriptions of frequently encountered distributions.)

Bayes' theorem would now appear mathematically in this fashion:

$$\underbrace{P(p;x)}_{\text{posterior distribution}} = \frac{\overbrace{P(x;p)}^{\text{sampling distribution}} \times \overbrace{P(p)}^{\text{prior distribution}}}{\underbrace{\Sigma P(x;p) \times P(p)}_{P(x)\text{ distribution of variants over all possible fraction variants}}} \tag{11.1}$$

Typically the sampling distribution will be assumed to be a Poisson or binomial distribution and the process prior distribution a gamma or beta distribution (see Appendix 11.3). Bayes' theorem can then be presented as follows:

$$\text{Beta or gamma} = \frac{\text{binomial or Poisson} \times \text{beta or gamma}}{\text{Polya or negative binomial}}$$

Note that the denominator becomes a known distribution for either the fraction variant or the variants per unit. To simplify the remainder of the discussion, only the fraction variant is considered. Generally this involves the binomial–beta–Polya combination of distributions, although in practice, for small p, the calculations are facilitated by the use of the Poisson–gamma–negative binomial distributions. The latter set of distributions are used as an approximation to simplify complex computations. Bayes' theorem for fraction variants would be

$$\underbrace{P(p;a,b,x,n)}_{\text{posterior distributions of lot fraction variant (beta)}} = \frac{\overbrace{P(x;p,n)}^{\text{sample distribution (binomial)}} \times \overbrace{P(p;a,b)}^{\text{prior process distribution (beta)}}}{\underbrace{P(x;a,b,n)}_{\text{process plus sampling distributions (Polya)}}} \tag{11.2}$$

where x = number of variants in sample
n = sample size
a, b = process-determined β parameters
p = fraction variant

The prior distribution of p, $P(p; a, b)$, is weighted by the ratio of the probability of x for a particular p, $P(x; p, n)$, to that of x for all p, $P(x; a, b, n)$, to give the posterior distribution of p, from lot to lot after sampling results, $P(p; x, n, a, b)$. The posterior distribution depends on the properties of both the prior and the sampling distributions. The posterior estimate of p is a combination of sample observations x and the a and b obtained from process data,

$$\hat{p} = \frac{x + a}{n + a + b} \tag{11.3}$$

Sometimes there are no past data and no knowledge of the process, except that a prior distribution exists. In such a situation, every fraction nonconforming may be considered equally likely, and a uniform distribution, a special case of the beta distribution with $a = 1$ and $b = 1$, is assumed,

$$\hat{p} = \frac{x + 1}{n + 2} \tag{11.4}$$

This is not characteristic of most process situations, and past data are generally available. However, with past data, but no prior distribution, a and b go to infinity, only binomial sampling exists, and the average fraction variant is the sum of the variant items divided by the cumulative samples,

$$\bar{p} = \frac{\Sigma x}{kn} \tag{11.5}$$

where k is the number of samples of size n.

The above discussion indicates that all situations are considered by the beta prior distribution: no prior with a and b equal to infinity, no knowledge of the prior with a and b equal to 1, or a prior with a and b estimated from past process data. These two constants are called *shape parameters* of the beta distribution and are mathematically meaningful for describing the skewness of the distribution, but are of limited other practical use. The Polya distribution represents total variation, sampling plus process, for all p, and its average and variance can be used to replace a and b with more meaningful parameters,

$$\bar{p} = \frac{a}{a + b}$$

and

$$V(p) = \frac{ab(n + a + b)}{(a + b)(a + b)(1 + a + b)}$$

or

$$V(p) = \frac{\bar{p}(1 - \bar{p})\,(n + a + b)}{n(1 + a + b)}$$

Since the variance of the binomial is $p(1 - p)/n$, then

$$V(p) = \frac{\bar{p}(1 - \bar{p})r}{n}$$

where

$$r = \frac{n + a + b}{1 + a + b}$$

Thus a and b can be replaced by p and r as follows:

$$a = \frac{(n - r)\bar{p}}{r - 1} \qquad \text{and} \qquad b = \frac{(n - r)(1 - \bar{p})}{r - 1}$$

Physically these parameters are more meaningful because p is the overall process average and r is the variance ratio of the total variation to the sampling variation. With no prior distribution, $r = 1$ and a and b equal infinity, that is, we have binomial sampling. For $r > 1$ a prior beta distribution exists, the reciprocal of r represents the proportion of the total variance due to sampling, and $(r - 1)/r$ represents that due to process variation. The bayesian estimate of the fraction nonconforming now becomes

$$\hat{p} = \frac{x + a}{n + a + b} = \frac{(r - 1)x + (n - r)\bar{p}}{r(n - 1)} \tag{11.6}$$

Finally, for completeness, but with limited explanation, an unspecified prior distribution can be assumed for available past data (Martz and Waller, 1982), yielding

$$\hat{p} = \frac{(x + 1)\, f(x + 1)}{n\, f(x)} \tag{11.7}$$

where $f(x)$ is the frequency of x. If the relative frequencies of $x = 0, 1, 2, 3, \ldots$ differ greatly from the Poisson distribution, then this determination of p from two adjacent frequencies will yield a p very different from the overall average. Otherwise it will be essentially the same. The disadvantage to this approach is that only two frequencies are used at a time.

Some examples are now considered for various assumptions about past data to assess their impact on the estimation of the fraction nonconforming.

Example 11.1. To illustrate these concepts, consider the following data, referring to the number of nonconforming items in 25 samples, each containing 21 items:

0, 3, 0, 4, 1, 1, 0, 1, 1, 2, 2, 0, 0, 3, 0, 3, 1, 1, 2, 0, 1, 1, 0, 2, 0

The average fraction nonconforming, from Eq. (11.5), is $\bar{p} = 0.055$ for a binomial sampling assumption. With only the knowledge that $n = 21$ for each sample and ignoring the data completely, an estimate can be obtained using Eq. (11.4),

$$\hat{p} = \frac{x + 1}{n + 2} = \frac{0 + 1}{21 + 2} = 0.0435$$

This is the estimate of the fraction nonconforming given that the next sample contains no nonconforming items. Assuming that one nonconforming item would yield 0.087, and so on, and since there exist some data, this is not a sensible approach. The prior information from the 25 samples should be used. For an unspecified prior distribution the following would result from Eq. (11.7):

$$P(x = 0) = \frac{(x + 1)f(x + 1)}{n\,f(x)} = \frac{1 \times 8}{21 \times 9} = 0.042$$

There are eight 1's in the above set of data and nine 0's, and we have $n = 21$. For $P(x = 1)$ there are four 2's, giving a fraction nonconforming of $(2 \times 4)/(21 \times 8) = 0.0476$, and so on. The method uses only a portion of the data for each estimate, and as x gets larger, less data are used.

We may calculate the variance ratio r to be 1.3, from Appendix 11.4. Then, from Eq. (11.6),

$$\hat{p} = \frac{x + a}{n + b + a} = \frac{(r - 1)x + (n - r)\overline{p}}{r(n - 1)}$$

and

$$P(x = 0) = \frac{0.3 \times 0 + 19.7 \times 0.055}{1.3 \times 20} = 0.042$$

$$P(x = 1) = \frac{0.3 \times 1 + 19.7 \times 0.055}{1.3 \times 20} = 0.053$$

$$P(x = 2) = \frac{0.3 \times 2 + 19.7 \times 0.055}{1.3 \times 20} = 0.065$$

$$P(x = 3) = \frac{0.3 \times 3 + 19.7 \times 0.055}{1.3 \times 20} = 0.076$$

For $r = 1$, $\overline{p} = 0.055$ for no prior distribution and only binomial sampling variation. Usually, for $r < 1.5$, it is not very practical to assume a prior distribution since there is not much change in the estimate of the fraction nonconforming, except when a large number of variants occurs in a lot. Since the expected number of nonconforming items is 1 (0.055×21), a value of 4 or more is unlikely for a low r ratio, and such a lot may be a maverick lot not typical of production.

Example 11.2. Suppose the following number of nonconforming items were found in samples of 17 from each of 25 lots:

0, 3, 0, 2, 0, 4, 0, 2, 0, 3, 1, 0, 0, 1, 0, 5, 0, 0, 1, 0, 1, 2, 0, 1, 0

The fraction nonconforming is, from Eq. (11.5), $p = 0.0612$. Assuming complete ignorance as in the previous example, using Eq. (11.4),

$$P(x = 0) = \frac{0 + 1}{17 + 2} = 0.0526$$

$$P(x = 1) = \frac{1 + 1}{17 + 2} = 0.1053$$

For an unspecified prior distribution using Eq. (11.7), we find

$$P(x = 0) = \frac{1 \times 5}{17 \times 13} = 0.0226$$

$$P(x = 1) = \frac{2 \times 3}{17 \times 5} = 0.0706$$

$$P(x = 2) = \frac{3 \times 2}{17 \times 3} = 0.118$$

$$P(x = 3) = \frac{4 \times 1}{17 \times 2} = 0.118$$

$$P(x = 4) = \frac{5 \times 1}{17 \times 1} = 0.294$$

The higher the number of nonconforming items, the higher the estimated fraction nonconforming, as expected. However, each estimate uses less and less of the data, and none uses all of the data. Assuming a beta prior distribution and calculating the variance ratio as in Appendix 11.4, $r = 2.1$. The estimates of the fraction nonconforming then become, using Eq. (11.6),

$$P(x = 0) = \frac{(17 - 2.1) \times 0.0612}{2.1 \times 16} = 0.0271$$

$$P(x = 1) = \frac{1.1 \times 1 + 14.9 \times 0.0612}{2.1 \times 16} = 0.0599$$

$$P(x = 2) = \frac{1.1 \times 2 + 14.9 \times 0.0612}{2.1 \times 16} = 0.0926$$

$$P(x = 3) = \frac{1.1 \times 3 + 14.9 \times 0.0612}{2.1 \times 16} = 0.1254$$

$$P(x = 4) = \frac{1.1 \times 4 + 14.9 \times 0.0612}{2.1 \times 16} = 0.1581$$

All of the data were used to estimate $\overline{p}$ and r. Thus these estimates would be more efficient and have a smaller sampling variance. Experience has shown that in most practical situations a beta distribution is an adequate and, usually, good fit to the process variation around the fraction variant.

From these examples it is apparent that the higher the variants in the sample, the higher the estimated process fraction nonconforming when the process follows a prior frequency distribution. The existence of a prior probability distribution suggests that the process is varying continuously and samples from the lots reflect the state of the process at a particular time and vary about the process mean. If the process average is high, then both the lot and the sample will tend to be high, and vice versa. When no prior probability distribution exists and the fraction nonconforming is fixed, the mean of the sample variants is independent of the variants in the remainder of the lot. Thus the sample estimates the process average and not necessarily the lot, because the only variation affecting the lot is that due to sampling from the process. The sample statistics reflect sampling variation from the lot plus the between-lot variation due to the process. Some control charts for variables have followed bayesian principles for years.

11.4 BAYESIAN ANALYSIS USING THE VARIANCE RATIO

In the previous section, the variance ratio was introduced as a ratio of the total variation (process plus sampling) to the sampling variation. Rational subgroups such as sampled lots or adjacent items in a sequence provide a measure of the sampling variation. Between-subgroup variation is a measure of the process variation often called between-lot or between-sample variation. A variance ratio r greater than 1 suggests a prior process distribution. What size r is important? As discussed below this will depend upon the number of items in the sample or subgroup used to estimate r.

Consider a random sample from a lot or subgroup of adjacent items whose content is expected to be consistent with r very close to 1 and within a standard deviation σ_W. From several samples, a between-subgroup standard standard deviation σ_b can be determined. For a sample size n, the following result:

Total variance $= \sigma_b^2 + \sigma_w^2/n$ Sampling variance $= \sigma_w^2/n$

Variance ratio $r = (\sigma_b^2 + \sigma_w^2/n)/\sigma_w^2/n$ $(r - 1) = n\sigma_b^2/\sigma_w^2$

For an individual item $n = 1$

$$r_i = (\sigma_b^2 + \sigma_w^2)/\sigma_w^2 \qquad (r_i - 1) = \sigma_b^2/\sigma_w^2$$

and

$$(r - 1) = n(r_i - 1) \qquad r_i = (r - 1)/n + 1$$

Suppose $n = 50$ and $r = 3$. Then $r_i = 1 + (3 - 1)/50 = 1.04$, which is close to 1 and means process variation is almost negligible at the item level where sampling variation is 25 times the process variation. But at the sampling level, the process variance is twice the sampling variance of the sample mean. As n gets larger and larger, so does r. Since acceptance sampling is based on sample means, this author believes that a bayesian (Polya) distribution is more appropriate than the conventional binomial distribution assumption ($r = 1$). When binomial sampling is used in the presence of process variation, the actual probability of acceptance is often higher than expected and the average outgoing quality going to the customer is better than predicted.

The 0% or 100% inspection recommended by Deming, as more cost-effective than acceptance sampling, applies only to the binomial assumption which rarely exists. Generally sampling inspection will always be preferable and less expensive. Sampling costs can be separated into inspection costs K_i and escape costs K_e. The latter, which are detected in higher level assemblies, are often orders of magnitude higher than inspection costs. For binomial sampling, the criterion for 0% versus 100% inspection is K_i/K_e compared to the process fraction nonconforming p, and is called the Kp rule: 100% inspection if $(K_i/K_e) < p$ and 0% if $(K_i/K_e) > p$. At $K_i/K_e = p$, either 0% or 100% inspection is cost-effective. For this criterion to apply, the binomial assumption has to be exact: the process average is fixed and never varies,

$\sigma_b^2 = 0$ and $r = 1$. Realistically as major assignable causes are removed, this lower bound is approached but never achieved. Thus there will always be a slight process variation negligible at the item level but, as shown above, greatly amplified at the sample level. Now, the K ratio cannot be merely compared to p. Depending on the situation, it may be cost-effective to do 0%, 100%, or sampling inspection. Generally in a bayesian situation, $r > 1$, sampling inspection is justified for $K_i/K_e = p$ and for higher r values; otherwise the Kp rule applies. This will be illustrated in Sec. 11.6 for $c = 0$ acceptance sampling.

Suppose instead of a random sample from a lot or a sequential sample from a process, a variety of items from several sources are grouped into an assembly. Some assemblies, such as a jumbo jet or a large mainframe computer, become quite complex, with hundreds or thousands of items in many subassembly units. This is no longer a consistent random sample but a mixture of many individual simple and complex units. The variance ratio appears to be related to the logarithm of the complexity or number of items. The impact of the total number of individual items is reduced and more reflects the critical or major types and groups.

An example of a group could be a set of stacking dimensional specifications that must meet an overall subassembly dimension. This stacking situation has been addressed in the literature (Bender, 1962; Evans, 1975), where a rule of thumb was established with about 10 specifications:

$$\sigma = 1.5\sqrt{(\Sigma\sigma_i^2)}$$

This involves the standard procedure of adding the variances to obtain the overall variance. The standard deviation is the square root of this sum multiplied by 1.5 to allow for shifts and drifts or other process variations inherent in each of the individual specifications. The 1.5 reflects the 10 specifications and is 2.25 for the variance. This is approximately ln10 and represents the variance ratio r. If any item that is out of specification potentially causes an assembly failure, then this serial stacking philosophy can be generalized to n items with r proportional to ln n. One simple relationship this author has found useful is

$$r = \ln(1.72 + n)$$

where 1.72 allows r to be approximately 1 when $n = 1$. This should be treated as an empirical guideline and not a universal rule. If r_i is known at the item level, then a more complex relationship such as

$$r = \ln[e + k\,n\,(r_i - 1)/n_i]$$

could be developed. Unless r_i is very different from 1 or there is a unique assembly situation, the effort may not be justified over the simpler guideline. The sample size used to determine r_i at the item level is n_i and the constant k would be related to the nature of the assemblies. As alluded to previously, mainframe computer and jumbo jet assemblies are quite different, but their complexities would relate, empirically, to the logarithm of n.

Why all this concern over r for complex assemblies? For large assemblies to perform, individual items have to be virtually defect-free and extremely reliable. Consider the stacking of 10 specifications above. With no process distribution, shifts, or drifts, σ is determined by summing variances as above. The total specification process capability becomes

$$C_p = (S_U - S_L)/6\sigma$$

or incorporating the 1.5 rule of thumb for the 10 items

$$C_p = (S_U - S_L)/(6 \times 1.5\sigma)$$

Generalizing this procedure to all items is difficult because the item standard deviations may be all different dimensions. However, the C_p ratio is dimensionless and a bayesian capability index can be defined as

$$C_{pr} = C_p / \sqrt{r}$$

A popular capability index today is $C_p = 2$, which if applied to each assembly would result in the following table:

n = 1	10	100	1000	10,000	100,000
r = 1.00	2.46	4.62	6.91	9.21	11.51
C_{pr} = 2.00	1.28	0.93	0.76	0.66	0.59
ppm = 0.002	119	5,200	22,600	47,800	78,400
% def = 0.00	0.01	0.52	2.26	4.78	7.84

$r = \ln(1.72 + n)$;
ppm is parts per million.

This reflects the impact of r in complex assemblies. The "number of items," n could relate to major or critical specifications. Thus 100,000 is probably very unlikely even for a large complex assembly. Such specification definitions could relate to those that are incapable of meeting $C_p = 2$ or lack centering of a target producing a C_{pr}. Finally, the ppm and percent defective are low since assuming a $C_p = 2$ for each assembly before division by $\sqrt{r}$ may appear unrealistic as this criterion usually applies only to small subproducts and individual items. However, large subproduct assemblies typically undergo inspection, testing, and rework/repair, reducing ppm and r to make the table plausible.

11.5 CONTROL CHARTS AND BAYESIAN PRINCIPLES

Typically a control chart plots information from a sample from a rational subgroup, such as a batch, lot, or sequence of adjacent items, which are assumed to be very consistent, with any major differences occurring only between subgroups. Typical variables control charts are $\bar{X}$ and R charts, where a sample of three to five items is selected from a rational subgroup and used to obtain the average $\bar{X}$ and the range R. An attributes control chart involves counting the number of variants or variant items from a sample. A process in a state of statistical control assumes that assignable causes, the differences between these subgroups, have been eliminated to leave only random chance causes, which consist of many small disturbances responsible for the variation within subgroups. While this may be typical of some well-established processes of the fabrication industries such as appliances or automobiles, many industries, such as chemical or electronics industries, often do not eliminate all assignable causes. For such processes rational subgroups such as batches or lots continue to have large between-group variations compared to within-group variations. Here the ratio of the total variance to the within-group variance provides an estimate of the variance ratio r.

In the presence of systematic variations, such as cycles or trends, an adequate estimate of total variance can be obtained from a two-point between-group moving range. Limits derived from this moving range include both within- and between-group variations. Attributes charts can be treated as variables, using a fixed subgroup sample size, to determine the total variance for comparison to the within-group sampling variance, which is based on a binomial or Poisson assumption. When $r > 3$, varying the sample size has a minimal impact. Thus any moving-range chart for individuals or a chart for averages (see Chap. 7) is actually a form of bayesian control chart because it accommodates between-group variation in the estimate of the limits. For large r, subgroup averages can be treated as individuals, but cannot be assumed to be normally distributed.

If the process variability is very large, its reduction may require redesign of the process by such means as better equipment, different operating conditions, improved raw materials, or changes in specifications. A large assignable cause could appear as a bimodal distribution, suggesting not a redesign, but a discovery of the problem that can be pursued and eliminated. While with large r a process may not be considered to be in a state of statistical control, it could follow a stable probability distribution, indicating that the assignable causes occur at random and are not systematic. Thus when the between-subgroup variation is large, continuous efforts should be made to reduce the total process variation through problem solving or process redesign.

When samples from a process are collected over time, the individual sample characteristics are not necessarily the best estimates of the process characteristics at a specific time. When r is finite, the total error is reduced by shrinking the individual sample measurements toward the grand mean. The best estimate always lies between the estimated process average and the current sample observation. This best estimate is a shrinkage of the sample observation toward the estimated process average. Some of these ideas are explored in greater detail in a system for reporting audit results, called the *quality measurement plan* by Hoadley (1981). The items in the subgroup are not independent, but are collectively dependent on the process average at the time of sampling. The degree of shrinkage toward the average is related to a shrinkage factor which is the reciprocal of the variance ratio r. The best estimate is a weighted posterior average of the form

$$\begin{pmatrix}\text{best}\\\text{estimate}\end{pmatrix} = \frac{1}{r}\begin{pmatrix}\text{estimated}\\\text{process}\\\text{average}\end{pmatrix} + \begin{pmatrix}\text{current}\\\text{sample}\\\text{observation}\end{pmatrix}\left(\frac{r-1}{r}\right)$$

and, as noted previously,

$$r = \frac{\text{process variance} + \text{sampling variance}}{\text{sampling variance}}$$

The bigger the sampling variance relative to the process variance, the more weight is put on the process average, that is, as r goes to 1, the process variance approaches zero, the maximum shrinkage is obtained, and the best estimate is the process average.

Theoretically r can range from 1 to infinity. For $r = 1$ the process is defined as being in a "state of statistical control," with no assignable causes between subgroups. The best estimate of the process average is the mean of the prior 20 to 25 subgroup samples, and a classical Shewhart control chart is appropriate. However, when $r = \infty$ ($r > 10$ for practical purposes), a stable probability distribution probably does not exist. Each subgroup represents an independent process, and the subgroup sample average is the best estimate of the process. Here limits would be placed about each subgroup average and compared to a target or specification on an individual basis. Without a stable process, statistical control has no meaning. For r between 1 and 10, the best estimate is a weighted average of the estimated process average and the current sample observation as described. This implies between- and within-subgroup variation, and if this variation is random, a stable probability distribution can be assumed with a fixed process average and variance ratio r, which can be estimated from 20 to 25 subgroups as above. A classical Shewhart control chart could be utilized by multiplying total variance (process plus sampling) by $(r-1)^2/r^2$ to obtain the variance for the best estimate. Besides shrinking the sample observation toward the process average, the variance is also reduced.

If the process is not expected to be stable, then both the process average and r may vary. The variance of the posterior best estimate, particularly for attributes, becomes much more complex. A more rigorous approach, which is beyond the scope of this introduction, is in Hoadley (1981), who uses a moving average of recent process history to estimate the process average. A simplification, for variables following a normal distribution, is to view the varying process average from the perspective of an exponentially weighted moving average. This assumes a fixed variance ratio r as the reciprocal of the smoothing constant. The previous best estimate becomes the estimated process average, and the current best estimate becomes

$$\begin{pmatrix}\text{current}\\ \text{best}\\ \text{estimate}\end{pmatrix} = \frac{1}{r}\begin{pmatrix}\text{previous}\\ \text{best}\\ \text{estimate}\end{pmatrix} + \left(\frac{r-1}{r}\right)\begin{pmatrix}\text{current}\\ \text{sample}\\ \text{observation}\end{pmatrix}$$

An adequate estimate of the total random variation, in the presence of systematic assignable causes, can be obtained from a two-point between-subgroup moving range of sample averages. Comparing this to the within-group variation for 20 to 25 subgroups yields an estimate of r. The variance of the best estimate becomes $(r-1)/(r+1)$ multiplied by the total variance. Limits applied to each best estimate would allow comparison to a target or specification. Also this approach could be applied to a stable probability distribution situation using an exponentially weighted moving-average chart to detect systematic variation due to trends, cycles, or shifts. The process nonconforming target or variables specification would take the place of the process average.

Example 11.3. To demonstrate some of these ideas, consider samples drawn from Shewhart's normal bowl (Grant and Leavenworth, 1988), which is a bowl of numbered chips that have a normal distribution with a mean of 30 and a standard deviation of 10. Figure 11.1 plots the first 50 of 100 subgroups of four against $3\sigma_{\bar{x}}$ limits of 30 ± 15, where the standard deviation of subgroup averages is $10/\sqrt{4}$, or 5. The control chart shows a well-behaved normal process for $r = 1$. The second 50 subgroups were given irregular variation among their averages, with amount and duration determined at random, as given in Table 11.2, but with the same within-group standard deviation throughout all subgroups.

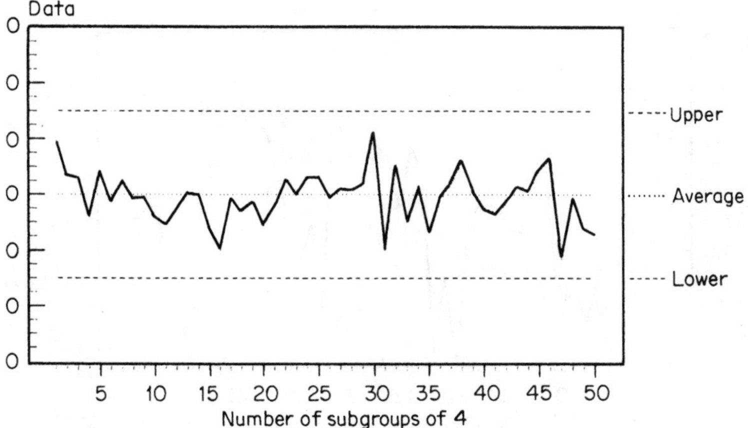

FIGURE 11.1 Drawing of samples from Shewhart's normal bowl. Mean = 30; standard deviation = 5.

TABLE 11.2 Variation in Subgroup Averages

Subgroup	Average	Run length
1-7	20	7
8-11	10	4
12	45	1
13-19	25	7
20-22	5	3
23-31	35	9
32-35	15	4
36-43	40	8
44-46	10	3
47-50	30	4

The result, plotted in Fig. 11.2, is obviously more erratic and no longer in control. The mean and the standard deviation of the subgroup averages are 26 and 13.4, respectively. This standard deviation calculation assumes that there are no large runs or random subgroup differences. We may then calculate the variance ratio r as the ratio of the observed variance of sample averages to the ratio that would be observed for a stable process, that is, 5,

$$r = \frac{(13.4)^2}{(5)^2} = 7.2$$

$$\frac{1}{r} = 0.14$$

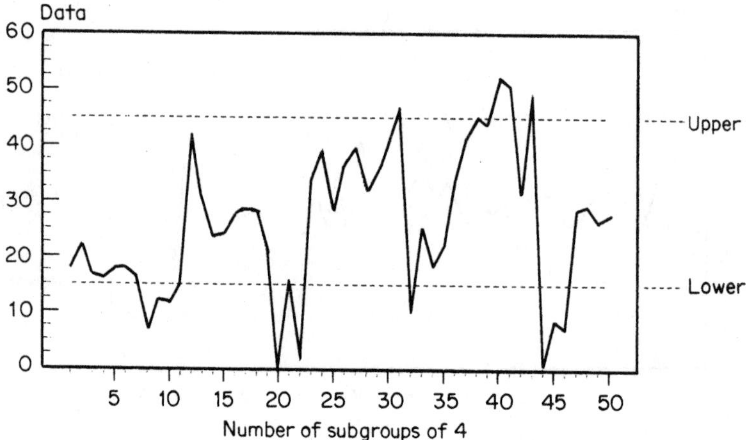

FIGURE 11.2 Drawing of samples from Shewhart's bowl. Mean = erratic; standard deviation = 5.

Then

$$\frac{r-1}{r} = 0.86$$

Thus the best estimate would be based on 86% of the current observations and on only 14% of the changes in the process average. Very little shrinkage results, and prior history would contribute little to the estimate. However, there are four large runs of 7, 7, 8, and 9, and the calculation of the total variance of $(13.4)^2$ includes these runs.

To remove this systematic variation from the estimate, use the two-point moving range to estimate the variation (Appendix 11.4). The estimated standard deviation then becomes 7.85, which is much smaller than the previous 13.4, and the resulting variance ratio r is $(7.85)^2/(5)^2 = 2.5$. The major effects of the prolonged shifts have been removed, and 7.85 represents the random between- and within-subgroup variation.

If past process history indicated a process average of 30 and $r = 2.5$, Fig. 11.3 shows a plot of the best estimates, where the current observation is weighted by $1.5/2.5 = 0.6$ and the process average by $1/2.5 = 0.4$. Because of the shrinking of the best estimate toward the process average, the data are less erratic compared to those in Fig. 11.2. Only three values are outside the limits, as compared to 15, and the standard deviation of averages has been reduced from 7.85 to $0.6 \times 7.85 = 4.7$. This strongly displays the stabilizing effect of combining past process performance with the current sample observation. If sustained shifts are the exception and can be diagnosed and removed, then a fixed process average assumption is reasonable. If erratic prolonged shifts occur frequently, lack of control is evident, and no long-term process average exists.

The approach in this situation is to combine an estimate of the current process average with the current sample observation. Figure 11.4 utilizes the geometric or exponentially weighted moving average, which is compared to a process target of

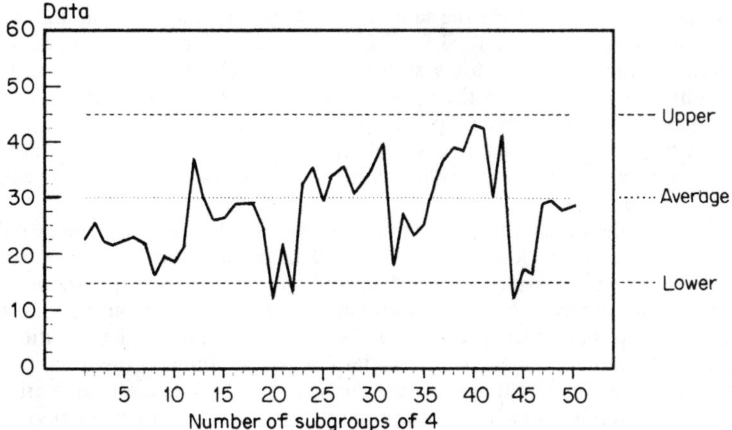

FIGURE 11.3 Bayesian estimate. $r = 2.5$; fixed process average = 30; standard deviation = 4.7.

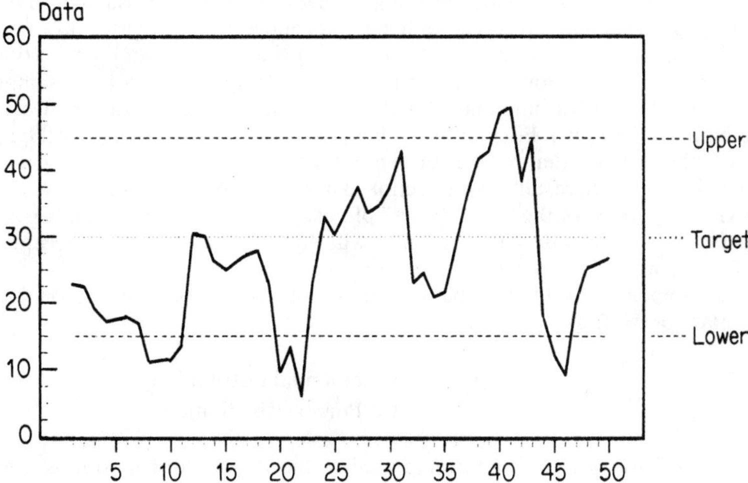

FIGURE 11.4 Bayesian geometric average. $r = 2.5$; process target = 30; standard deviation = 5.1.

30. Thus the limits about the target indicate when the process is not meeting target, not whether the process is in statistical control. Instead of limits about the target, limits could be placed about each best estimate as a box plot, and the process is not meeting its target if the target is outside the box. This moving-average method is more variable than a fixed-average method because of the continually changing

state of the process at the time the sample observation is made. The standard deviation is only slightly larger, $\sqrt{1.5/3.5} \times 7.85 = 5.1$, than the fixed standard deviation of sample averages. This is not a statistical control chart, but only a statistical assessment of departure from the target. Figure 11.4 shows how a Shewhart-style chart can assess an out-of-control process by comparison to a process target. A box chart may be preferred since it is less likely to be confused with a control chart, and it should be used for attributes where within-subgroup variation changes with the process average.

In all charts the limits were approximately the same because the standard deviations of averages were close: 5, 4.7, and 5.1. This permitted direct comparison of all charts. Erratic process shifts suggest a problem that, if eliminated, should bring the process into control. If the same spread were obtained without the sustained shifts, then a redesign of the process may be required to reduce the variation that is suggested by large r values. If the standard deviation found in the usual manner is substantially larger than that from the moving range, then systematic variations in the form of trends, shifts, or cycles probably exist and may be compensated for or eliminated.

11.6 BAYESIAN ACCEPTANCE SAMPLING

When items have a continually changing probability of being variant from lot to lot, acceptance sampling risks can only be determined from the prior process history of between-lot variation. Analysis of these prior data can supply an estimate of the process distribution, gamma or beta, which is combined with the sampling distribution, Poisson or binomial, to yield the distribution of the lot samples, negative binomial or Polya. For simplicity, bayesian acceptance sampling will be discussed only for plans that have a zero acceptance number c; that is, a single nonconforming item will result in lot rejection. These plans have become more important with the advent of the "zero defect" philosophy and the trend against passing lots with known defects. The discussion will, however, apply to any acceptance sampling plan.

From Appendix 11.5 the probability of acceptance P_a for a conventional sampling plan for $c = 0$ is

$$P_a = \begin{cases} (1-p)^n, & \text{for binomial distribution} \\ e^{-np}, & \text{for Poisson distribution} \end{cases}$$

For very low fraction nonconforming quality, the Poisson distribution is a good approximation, and combined with the gamma distribution it produces a negative binomial for $c = 0$ of

$$P_a = \left(\frac{\beta}{\beta+1}\right)^{\alpha} = \left(\frac{1}{r}\right)^{n\bar{p}/(r-1)}$$

for $\bar{\mu} = n\bar{p}$ and $\beta = 1/(r-1)$. As $r \to 1$, $P_a \to e^{-n\bar{p}}$. From the previous discussion the posterior average for $c = 0$ is

$$\hat{p} = \frac{(n-r)\bar{p}}{r(n-1)} \to \frac{\bar{p}}{r}$$

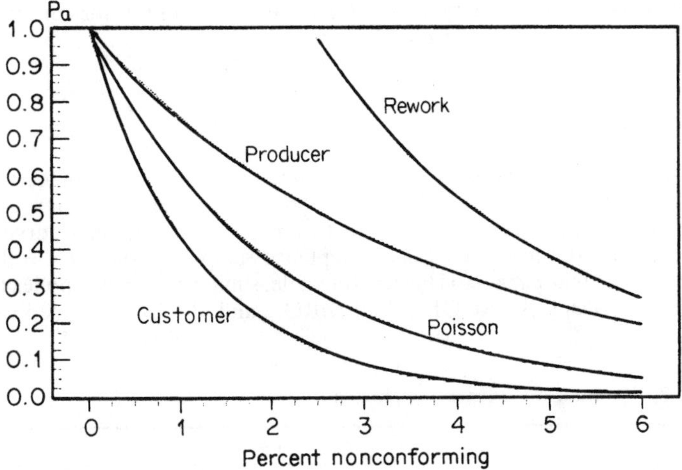

FIGURE 11.5 Operating-characteristic curve. $r = 3$; sample size = 50; acceptance number = 0.

Since the process average becomes $\bar{p} = r\hat{p}$, then the accepted lot operating characteristic (OC) curve becomes

$$P_a = \left(\frac{1}{r}\right)^{rn\hat{p}/(r-1)}$$

The average of the accepted lots is much better than the process average fraction nonconforming by a factor equal to the variance ratio r, and the consumer gets a better quality product than is produced by the process (Fig. 11.5). This is understandable if a distribution is considered to exist about the process average. Since the average of accepted and rejected lots must equal the process average, the average of the accepted lots will be better than the process average and the average of the rejected lots must be worse than the process average, resulting in greater rework and repair. With a lower sample size, the accepted lot curve can be made equal to the typically specified binomial curve, and thus the consumer gets equivalent protection with a smaller sample size due to bayesian acceptance sampling. For good process quality the producer does less inspection, but for poor quality the rework and the repair on rejected lots are very large as their quality is much worse than that for accepted lots. Compared to a binomial plan with a particular average outgoing quality limit (AOQL), the bayesian customer AOQL for the same sample size is

$$\frac{\text{AOQL}\ (r-1)}{r \ln (r)}$$

where $\ln (r)$ is the natural logarithm of r. The customer AOQL becomes smaller and smaller with larger and larger values of r. It is the limit of the average quality past an inspection process when rejected lots are 100% inspected. Similarly, for a given binomial limiting quality level (LQL), the bayesian LQL for consumer protection with the same sample size is

$$\frac{\text{LQL}\,(r-1)}{r\ln(r)}$$

The LQL is the worst lot quality that may be accepted. For conversion between the binomial and the producer's OC curve the factor is $(r-1)/\ln(r)$ times the binomial curve. The producer's curve divided by r gives the consumer curve. These values are reflected in the bayesian acceptance sampling table for $c = 0$ (Table 11.3). While the mathematics is far less tractable, similar reasoning applies to $c > 0$, and Table 11.3 displays nAOQL, nAQL, nRQL, and nLQL for selected values of

TABLE 11.3 Bayesian Acceptance Sampling Plans

	$R = 1$				$R = 2$				$R = 3$			
c	nAOQL	nAQL	nRQL	nLQL	nAOQL	nAQL	nRQL	nLQL	nAOQL	nAQL	nRQL	nLQL
0	37	5	230	230	27	8	329	165	22	9	421	140
1	84	36	389	389	63	23	523	298	53	23	624	253
2	137	82	532	532	104	50	682	417	89	42	805	364
3	194	137	668	668	151	87	825	529	129	68	969	471
4	254	197	799	799	200	131	971	644	172	101	1118	574
5	317	261	928	928	252	180	1106	754	218	139	1265	679
7	447	398	1177	1177	363	290	1384	981	316	229	1552	887
10	653	617	1541	1541	541	472	1770	1307	477	385	1955	1195
16	1088	1084	2240	2240	926	885	2515	1953	828	750	2739	1813
25	1776	1825	3263	3263	1547	1555	3593	2909	1405	1375	3854	2731
40	2977	3111	4930	4930	2653	2750	5321	4479	2443	2500	5645	4257

	$R = 4$				$R = 6$				$R = 10$			
c	nAOQL	nAQL	nRQL	nLQL	nAOQL	nAQL	nRQL	nLQL	nAOQL	nAQL	nRQL	nLQL
0	20	11	506	126	17	14	666	111	14	20	906	91
1	47	24	722	229	40	26	889	198	34	33	1246	175
2	79	40	920	333	68	40	1102	290	56	45	1477	254
3	115	61	1093	433	98	57	1303	384	81	59	1679	334
4	154	87	1244	529	132	76	1469	473	109	74	1874	416
5	196	118	1410	632	168	99	1652	569	138	91	2083	502
7	286	192	1707	830	246	154	1964	752	203	130	2424	670
10	433	328	2123	1124	375	258	2422	1034	311	202	2941	932
16	760	658	2931	1720	666	528	3280	1601	557	394	3880	1465
25	1302	1233	4088	2616	1156	1026	4480	2459	980	780	5181	2279
40	2288	2300	5921	4107	2064	1995	6390	3902	1780	1590	7209	3661

c up to 40. In today's very competitive environment, the use of values of $c > 5$ is highly unlikely. RQL is the process rejectable quality level.

Example 11.4. Consider an acceptance sampling plan with sample size $n = 50$ and acceptance number $c = 0$. The OC curves are illustrated in Fig. 11.5. To provide a comparison with traditional sampling, the OC curve using the Poisson distribution (as an approximation to the binomial distribution) is also included. This latter curve assumes no variation in the process percent nonconforming. Sampling variation only exists as the sample is randomly selected from the lot, which is also randomly selected from the fixed percent nonconforming process. A variance ratio of 1 would represent this situation. For $r = 3$, only 33% of the total variation is due to sampling and 67% is due to the process. Now if there is a distribution of the parameter about its average value, each lot sampled comes from a different process level, and three curves are required to describe the sampling plan completely.

Because the process distribution has a positive skew to the right, there will be more lots below the average than above. Thus the producer probability of acceptance P_a will be higher than the Poisson value for the same process average, that is, more lots will be accepted. While this is good news for the producer, it is even better for the customer: the average of the accepted lots will be lower than the process average. This can be seen in Fig. 11.5

Since the average of accepted and rejected lots must equal the process percent nonconforming, the average of the rejected lots will be higher than the process average. When these rejected lots are screened, the producer will encounter more rework and repair than with the Poisson curve, as indicated by the top curve in Fig. 11.5. Only the producer curve is a true probability of acceptance curve for this bayesian plan. The top and bottom curves merely reflect the quality of rejected and accepted lots, respectively, for a specific producer P_a. The AOQL and LQL for lots are determined from the customer curve, and the AQL and RQL for the process are obtained from the producer curve. Usually the Poisson and rework curves would not be displayed leaving the producer and customer curves as in a simplified Fig. 11.6. If the Poisson curve represents the quality desired by the customer, then a smaller sample would increase the customer curve to overlay the Poisson curve.

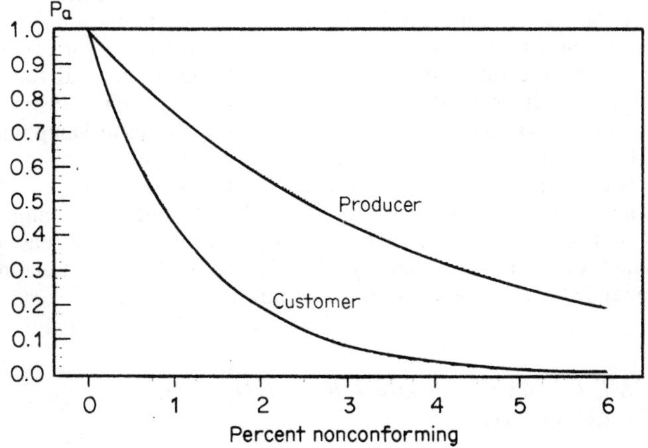

FIGURE 11.6 Operating-characteristic curve. $r = 3$; sample size = 50; acceptance number = 0.

TABLE 11.4 Bayesian Acceptance Sampling Costs in Dollars

Sample	R	R_i	K_i/K_e 0.005	0.01	0.02
100%			1010	1010	1010
0%			2010	1010	510
2.5%	1	1	1768	1010	631
2.5%	1.25	1.01	1478	854	542
2.5%	1.5	1.02	1276	745	479
2.5%	2	1.04	1010	600	395
2.5%	3	1.08	727	444	302
2.5%	5	1.16	482	305	217

Thus bayesian sampling for $r > 1$ can reduce the sample size. For this plan a bayesian sample size would be $n(r - 1)/[r \ln (r)] = 50 \times 2/[3 \ln (3)] = 30$, or only 60% of the equivalent Poisson sample.

Bayesian acceptance sampling has advantages over conventional acceptance sampling. Generally for the same criteria, the sample size is lower with bayesian sampling. But is acceptance sampling inspection worth performing at all? In Sec. 11.4, Deming's *Kp* rule was introduced which indicated that it was cost-effective to use 0% or 100% inspection for binomial-based sampling, suggesting acceptance sampling was not a less expensive alternative. In Table 11.4 a comparison is made between bayesian and conventional sampling which clearly demonstrates the utility of acceptance sampling when a process distribution and a sampling distribution exist. For Table 11.4, the process is assumed to be 1% nonconforming and a $c = 0$ plan is used under a negative binomial assumption as discussed above. The lot size is 1000 and the sample size is 25. The cost $K_i = \$1$ and cost $K_e = \$50, \$100, \$200$ for the cost formulas in Appendix 11.6.

The above costs assume the replaced items are inspected. Thus in the 0% and 100% inspection, the \$10 portion is for the 10 nonconforming items in 1000. For 100% inspection, \$1 an item is \$1000 plus \$10 for replacements equals \$1010. With 0% inspection, the cost 10(1 + 200, 100, 50) = 2010, 1010, 510 for cost of escapes and replacement inspection. The 2.5% sampling plan costs make up the remainder of the table. Each value should be compared to the 0% and 100% inspection costs. When the ratio is equal to the fraction nonconforming, bayesian sampling is cheapest except for $R = 1$ where all costs are equal. With a ratio less than 0.01, 100% inspection is best up to $R = 2$, after which it is better to sample. Finally a ratio greater than 0.01 suggests 0% inspection below $R = 1.5$ and then sampling inspection. Thus it is obvious bayesian sampling inspection can be the least expensive inspection method. To reiterate, the *Kp* rule of 0% or 100% inspection only applies for a binomial distribution which in practice rarely exists.

11.7 SOME COMMENTS ON THESE QUALITY CONTROL APPLICATIONS

The approach to control charts and acceptance sampling has been greatly simplified to reduce the complexity inherent in bayesian statistics. The attribute acceptance sampling formulas and the best estimate for attribute control charts assume a Poisson-gamma distribution system which approximates a binomial-beta system for low fraction nonconforming. This should not be very restrictive in the present atmosphere of "zero defects" and parts per million (ppm). The best-estimate formulas are assumed to apply to any process average and observed sample, attributes, or variables. The within-subgroup sample variances are specific to the sampling distribution, and the between-subgroup variances are determined from a moving range (Appendix 11.4). While this does not estimate pure random variation, this assumption is adequate for practical purposes for both attributes and variables. For calculating limits, count data could be transformed by a square-root transformation to compensate for the skewness. For a sample with zero variants, the best sample estimate is derived from $np = 0.693$ for Poisson samples, and from $(n + 1.5)$ $p = 0.693$ for small binomial samples. The variables calculations have assumed a normal distribution, and some skewed distributions may follow a lognormal distribution. While the simplified bayesian geometric moving-average approach may be useful for normal data, a more complex approach may be needed for attributes data where the variance is related to the mean.

REFERENCES

Bender, A.: "Benderizing Tolerances — A Simple Practical Probability Method of Handling Tolerances for Limit-Stack-ups," *Graphic Science*, p. 17, Dec. 1962.

Calvin, T. W.: "An Approach to Bayesian Quality Control," in *ASQC 33d Annual Technical Conf. Trans.* (Houston, Tex.), pp. 667-674, 1979.

_____: "How and When to Perform Bayesian Acceptance Sampling," *ASQC Basic References in Quality Control: Statistical Techniques*, vol. 7. Am. Soc. Quality Control, Milwaukee, Wis., 1984.

Case, K. E., and J. B. Keats: "On the Selection of a Prior Distribution in Bayesian Acceptance Sampling," *J. Qual. Technol.*, vol. 14, pp. 10-18, 1982.

_____, G. K. Bennett, and J. W. Schmidt: "The Effect of Inspection Error on Average Outgoing Quality," *J. Qual. Technol.*, vol. 7, pp. 28-33, 1975.

Dodge, H. F., and H. G. Romig: *Sampling Inspection Tables*. Wiley, New York, 1959.

Evans, D. H.: "Statistical Tolerancing: The State of the Art; Part III — Shifts and Drifts," *J. Qual. Technol.* vol. 7, no. 2, April 1975.

Grant, E. L., and R. S. Leavenworth: *Statistical Quality Control*, 6th ed. McGraw-Hill, New York, 1988.

Hald, A.: "The Compound Hypergeometric Distribution and a System of Single Sample Inspection Plans Based on Prior Distribution and Cost," *Technometrics*, vol. 2, pp. 275-340, 1960.

Hoadley, B.: "The Quality Measurement Plan (QMP)," *Bell Sys. Tech. J.*, vol. 60, pp. 215-273, Feb. 1981.

Jaraiedi, M., D. S. Kochhar, and S. C. Jaisingh: "Multiple Inspections to Meet Desired Outgoing Quality," *J. Qual. Technol.*, vol. 19, pp. 46-51, 1987.

Martz, H. F., and H. A. Waller: *Bayesian Reliability Analysis*. Wiley, New York, 1982.

MIL-STD-105D, "Sampling Procedures and Tables for Inspection by Attributes," U.S. Govt. Printing Office, Washington, D. C., 1963.

APPENDIX 11.1

BAYESIAN CONCEPTS

Realistically, the process varies about its average p, and each p has a different probability of occurrence. Other distributions could be used to describe this variation in p, but the most popular are the beta distribution and the gamma distribution approximation for low p.

Beta Distribution:

$$P(p; s, t) = \frac{(s + t - 1)!}{(s - 1)!(t - 1)!}\, p^{s-1}(1 - p)^{t-1}$$

Gamma Distribution:

$$P(np; \alpha, \beta) = \frac{\beta^{\alpha}(np)^{\alpha - 1}}{(\alpha - 1)!}\, e^{-\beta np}$$

where s, t, α, and β are parameters determined from the process past history. The mean and the variance are, for the beta distribution,

$$\text{Mean} = \bar{p} = \frac{s}{s + t}$$

$$\text{Variance} = \frac{st}{(s + t)^2(s + t + 1)}$$

$$= \frac{\bar{p}(1 - \bar{p})}{s + t + 1}$$

and for the gamma distribution,

$$\text{Mean} = n\bar{p} = \frac{\alpha}{\beta}$$

$$\text{Variance} = \frac{\alpha}{\beta^2}$$

$$= \frac{n\bar{p}}{\beta}$$

This description of the process variation is termed *prior distribution*. Since the sampling distribution and the prior distribution are independent, the distribution of p in the sample will be proportional to the product of the prior and the sampling distributions,

$$P(p; x, n, s, t) \approx P(x; p, n) \times P(p; s, t)$$

or

$$P(np; x, \alpha, \beta) \approx P(x; np) \times P(np; \alpha, \beta)$$

This is the essence of Bayes' theorem. The distributions $P(x; p, n)$ and $P(x; np)$ are binomial and Poisson, respectively (see Appendix 11.3).

The probability of x, independent of p, can be obtained by summing over all values of p between 0 and 1,

$$P(x; n, s, t) = \int_0^1 P(x; n, p) \times P(p; s, t)\, dp$$

which is the Polya distribution (see Appendix 11.3). Similarly, the negative-binomial distribution approximation to the Polya distribution can be derived for all np from 0 to ∞,

$$P(x; \alpha, \beta) = \int_0^\infty P(x; np) \times P(np; \alpha, \beta)\, dnp$$

These are the compound distributions of x for all p or np. The exact probability of occurrence of a particular p or np can be obtained. The relative probability of x at p for a specified np is the binomial or Poisson distribution divided by the appropriate compound distribution of x over all p or np. If the prior distribution is weighted by this relative probability, the exact distribution of p or np is called the *posterior distribution* of p or np,

$$P(p; x, n, s, t) = P(p; s, t) \times \frac{P(x; n, p)}{P(x; n, s, t)}$$

and

$$P(np; x, \alpha, \beta) = P(np; \alpha, \beta) \times \frac{P(x; np)}{P(x; \alpha, \beta)}$$

This is Bayes' rule. It permits prior process information to be utilized for current estimates. These posterior distributions are a beta distribution for a beta prior distribution and a gamma for a gamma prior.

The beta posterior distribution is

$$P(p; x, n, s, t) = \frac{(n + s + t + 1)!}{(s + x - 1)!(t + n - x - 1)!}\, p^{s + x - 1} (1 - p)^{t + n - x - 1}$$

The gamma posterior distribution is

$$P(np; x, \alpha, \beta) = \frac{(\beta + 1)^{x+\alpha}\, np^{x + \alpha - 1}\, e^{-(\beta + 1)np}}{(x + \alpha - 1)!}$$

By inspection it can be seen that these posterior distributions depend on the properties of both the prior distribution of the process and the sampling distribution. In this way bayesian estimation incorporates present and past information.

APPENDIX 11.2

AN ILLUSTRATION OF BAYESIAN ANALYSIS IN A FAMILIAR SITUATION

Consider the gambling game of craps. This involves a known prior distribution for the totals on a pair of dice. This distribution is easily obtained: each die has six sides, resulting in 11 totals from 36 possible combinations:

Total	2	3	4	5	6	7	8	9	10	11	12
Probability	$\frac{1}{36}$	$\frac{2}{36}$	$\frac{3}{36}$	$\frac{4}{36}$	$\frac{5}{36}$	$\frac{6}{36}$	$\frac{5}{36}$	$\frac{4}{36}$	$\frac{3}{36}$	$\frac{2}{36}$	$\frac{1}{36}$

The rules for the game of craps describe the criteria for winning or losing as follows. On the first roll a 7 or 11 wins and a 2, 3, or 12 loses. With a 4, 5, 6, 8, 9, or 10 obtained on the first roll and matched on a subsequent roll, the player wins, while a 7 at any time after the first roll loses. Given a total, the probability of a win can be determined. For example, there are three ways to get a 4 and six ways to get a 7 based on the prior distribution. The probability of winning given a 4 is 3/(3 + 6) = $^1/_3$ since 7 loses. Similarly:

Given	4	5	6	8	9	10
Probability of winning	$\frac{3}{3+6}$	$\frac{4}{4+6}$	$\frac{5}{5+6}$	$\frac{5}{5+6}$	$\frac{4}{4+6}$	$\frac{3}{3+6}$

These probabilities are conditional on having rolled one of these numbers on the first roll and are called *conditional probabilities*. Since the probability of a 4 on the first roll is $^3/_{36}$ and the probability of matching the 4 on a subsequent roll is $^1/_3$, the probability of winning with a 4 is $^3/_{36} \times {}^1/_3$, or $^1/_{36}$. The rules of the game, applied to the prior probability of getting a 4, yield the absolute probability of winning with a 4. Similarly:

Winning total	4	5	6	8	9	10
Cumulative probability of winning	$\frac{3}{36}\frac{3}{3+6}+$	$\frac{4}{36}\frac{4}{4+6}+$	$\frac{5}{36}\frac{5}{5+6}+$	$\frac{5}{36}\frac{5}{5+6}+$	$\frac{4}{36}\frac{4}{4+6}+$	$\frac{3}{36}\frac{3}{3+6}=0.2707$

The probability of winning with a 7 or 11 on the first throw of the dice is (6 + 2)/36 = 0.2222, and the total absolute probability of winning is 0.2707 + 0.2222 = 0.4929. This is a reasonably fair game since the probability of losing is 1 – 0.4929 = 0.5071.

Now after the fact, given a win, what is the probability that it was done with a 4? The probability of winning with a 4 over the probability of winning is $^1/_{36}$/0.4929 = 0.056. This is the posterior probability of winning with a 4.

Bayes' theorem is expressed as follows:

$$P(T; W) = \frac{P(W; T)P(T)}{\Sigma P(W; T)P(T)}$$

where T is a total on a pair of dice and W represents a win for that total.

APPENDIX 11.3

SUMMARY OF PROBABILITY DISTRIBUTIONS

Binomial Distribution. This distribution is used to describe the distribution of the current sample,

$$P(x; n, p) = \left[\frac{n!}{x!(n-x)!}\right] p^x(1-p)^{n-x}$$

$$\text{Mean} = np$$

$$\text{Variance} = np(1-p)$$

where n is the sample size and p is the proportion of variant items in the sample. To obtain the mean and the variance of p, divide by n and n^2, respectively.

Beta Distribution. This distribution is used to describe the prior distribution of the process parameters,

$$P(p; s, t) = \left[\frac{(s+t-1)!}{(s-1)!(t-1)!}\right] p^{s-1}(1-p)^{t-1}$$

$$\text{Mean} = \frac{s}{s+t} = \overline{p}$$

$$\text{Variance} = \frac{st}{(s+t)^2(s+t+1)} = \frac{\overline{p}(1-\overline{p})}{s+t+1}$$

where s and t are parameters estimated from the process.

Beta-Binomial or Polya Distribution. This distribution is used to measure the unconditional probability of x for all p,

$$P(x; n, s, t) = \frac{(x+s-1)!\,(n+t-x-1)!\,n!\,(s+t-1)!}{x!(s-1)!\,(n-x)!\,(t-1)!\,(n+s+t-1)!}$$

$$\text{Mean} = \frac{ns}{s+t} = n\overline{p}$$

$$\text{Variance} = \frac{nst(s+t+n)}{(s+t)^2(t+s+1)} = n\overline{p}(1-\overline{p})\left(\frac{s+t+n}{s+t+1}\right)$$

where s and t are the beta distribution parameters. To obtain the mean and the variance of p, divide by n and n^2, respectively.

The ratio of the variances of the Polya distribution to the binomial distribution is

$$r = \frac{s+t+n}{s+t+1}$$

Thus r and p can replace s and t as Polya parameters as follows:

$$s = \left(\frac{n-r}{r-1}\right)\bar{p} \qquad \text{and} \qquad t = \left(\frac{n-r}{r-1}\right)(1-\bar{p})$$

Poisson Distribution. This distribution is used to approximate the binomial distribution with $\mu = np$,

$$P(x; \mu) = \frac{\mu^x e^{-\mu}}{x!}$$

$$\text{Mean} = \mu$$

$$\text{Variance} = \mu$$

where x is the number of variants per item and μ the average variants per item.

Gamma Distribution. This distribution is used to approximate the beta distribution with $\mu = np$,

$$P(\mu; \alpha, \beta) = \frac{\beta^\alpha \mu^{\alpha-1}}{(\alpha-1)!} e^{-\beta\mu}$$

$$\text{Mean} = \frac{\alpha}{\beta} = \mu$$

$$\text{Variance} = \frac{\alpha}{\beta^2} = \frac{\mu}{\beta}$$

where α and β are parameters estimated from the process.

Negative-Binomial Distribution. This distribution is used to describe the number of trials to the first success in a binomial distribution,

$$P(x; \alpha, \beta) = \left[\frac{(x+\alpha-1)!}{(\alpha-1)!x!}\right]\left(\frac{\beta}{\beta+1}\right)^\alpha \left(\frac{1}{\beta+1}\right)^x$$

$$\text{Mean} = \frac{\alpha}{\beta} = \mu$$

$$\text{Variance} = \frac{\alpha(\beta+1)}{\beta^2} = \frac{\mu(\beta+1)}{\beta}$$

The ratio of the negative-binomial distribution to the Poisson distribution is

$$r = \frac{\beta+1}{\beta}$$

Thus r and μ can replace α and β as negative-binomial parameters,

$$\alpha = \frac{\mu}{r-1} \qquad \text{and} \qquad \beta = \frac{1}{r-1}$$

This allows these distributions to be described in terms of a process average p or μ and a variance ratio r, the ratio of total variation (process plus sampling) to sampling variation. These are the more meaningful process parameters.

APPENDIX 11.4

CALCULATION OF VARIANCE RATIO r

Attributes. For a stable process containing only random within- and between-subgroup variation and 20 or more samples of data,

$$\overline{\mu} \text{ or } \overline{p} = \frac{\text{total variants or variant items}}{\text{cumulative sample items}}$$

$$= \sum_{i=1}^{m} \frac{D_i}{mn} = \sum_{i=1}^{m} \frac{p_i}{m}$$

The sampling variance is

$$S_w^2 = \frac{\overline{p}(1-\overline{p})}{n} \qquad \text{or} \qquad \frac{\overline{\mu}}{n}$$

The total variance is

$$S_t^2 = \sum_{i=1}^{m} \frac{(p_i - p)^2}{m-1} = \sum_{i=1}^{m} \frac{p_i^2}{m-1} - \left(\frac{m}{m-1}\right)p^2$$

For the Poisson, μ is substituted for $\overline{p}$.

The variance ratio is

$$r = \frac{S_t^2}{S_w^2}$$

Here μ = variants per unit
p = fraction variant
D_i = number of variant items or variants
m = number of samples
n = sample size
S_w = within-subgroup standard deviation
S_t = total-subgroup standard deviation, which includes both the between- and the within-subgroup variation

Variables. For a stable process with only random between- and within-subgroup variation,

$$\overline{\overline{x}} = \sum_{i=1}^{m} \frac{\overline{x}_i}{m}$$

$$S_w = \sum_{i=1}^{m} \frac{R_i}{d_2(m-1)\sqrt{n}}$$

Here

d_2	1.128	1.693	2.059	2.326
n	2	3	4	5

and

$$\bar{x} = \frac{\Sigma x}{n} \text{ for each subgroup}$$

$$\bar{\bar{x}} = \text{overall grand average}$$

$$S_t^2 = \frac{\Sigma \bar{x}^2 - (\Sigma \bar{x})^2/m}{m-1}$$

The variance ratio is

$$r = \frac{S_t^2}{S_w^2}$$

For an unstable process with systematic variation we may use the moving range to estimate S_t,

$$S_t = \sum_{i=1}^{m-1} \frac{\bar{x}_i - \bar{x}_{i+1}}{1.128(m-1)}$$

APPENDIX 11.5

BAYESIAN ACCEPTANCE SAMPLING PLAN FORMULAS; c = 0

Binomial $P_a = (1 - p)^n$

Poisson $P_a = e^{-np}$

Producer $P_a = \left(\frac{1}{r}\right)^{n\bar{p}/(r-1)}$

Customer $P_a = \left(\frac{1}{r}\right)^{m\hat{p}/(r-1)}$

Rework $\tilde{p} = \hat{p}\left(\frac{r - P_a}{1 - P_a}\right)$, using the producer P_a

where P_a = probability of acceptance
$\bar{p}$ = process fraction nonconforming
n = sample size
r = variance ratio
$\hat{p}$ = customer fraction nonconforming
$\tilde{p}$ = rework/repair fraction nonconforming

APPENDIX 11.6

ACCEPTANCE SAMPLING COST FORMULAS: CONVENTIONAL AND BAYESIAN

100% inspection: $EC_{100} = N K_i/(1 - \bar{p})$

0% inspection: $EC_0 = N \bar{p} K_e + N \bar{p} K_i/(1 - \bar{p})$

Conventional binomial acceptance sampling inspection cost:

$$\text{Lot cost} \qquad \text{Sample cost} \qquad \text{Escape cost}$$

$$EC(N, n) = N K_i \frac{(1 - P_a)}{1 - \bar{p}} + \frac{n K_i P_a}{1 - \bar{p}} + (N - n) \bar{p} \left(K_e + \left(\frac{K_i}{1 - \bar{p}}\right)P_a\right.$$

$$= \frac{N K_i}{1 - \bar{p}} - (N - n) K_e P_a \left(\frac{K_i}{K_e} - \bar{p}\right) \quad \text{(latter is } Kp \text{ rule)}$$

Probability of acceptance ($c = 0$): $P_a = (1 - \bar{p})^n$

Bayesian Polya acceptance sampling inspection cost:

$$EC(N, n) = N K_i \frac{(1 - P_a)}{1 - \tilde{p}} + \frac{n K_i P_a}{1 - \hat{p}} + (N - n) \hat{p} \left(K_e + \frac{K_i}{1 - \hat{p}}\right)P_a$$

Probability of acceptance ($c = 0$, negative binomial approximation)

$$P_a = (1/r)^{n\bar{p}/(r-1)} \qquad \hat{p} = \bar{p}/r \qquad \tilde{p} = \hat{p}\,(r - P_a) / (1 - P_a)$$

Definitions of terms:

P_a = probability of acceptance
$\bar{p}$ = process fraction nonconforming
$\hat{p}$ = customer (accepted lot) fraction nonconforming
$\tilde{p}$ = rework/repair (rejected lot) fraction nonconforming
r = variance ratio
n = sample size
N = lot size
EC = expected cost
K_i = inspection cost per item
K_e = escape cost per item

CHAPTER 12
NONPARAMETRIC STATISTICS

Jean Dickinson Gibbons
North Hutchinson Island, FL

12.1 INTRODUCTION

Nonparametric statistics is a collective term given to methods of hypothesis testing and estimation that are valid under less restrictive assumptions than classical methods. For example, the classical analyses of variance tests require the assumption of normal distributions with equal variances, and the nonparametric counterparts require only the assumption that the samples come from continuous populations. Also, classical statistical methods are strictly valid only for data measured on at least an interval scale, while nonparametric statistics applies to data measured on a nominal scale or ordinal scale (rank data). Since interval-scale data can be reduced to ordinal or nominal scale, nonparametric methods can be used whenever classical methods are valid; the reverse is not true. Thus nonparametric statistics can legitimately be used in a very wide variety of practical situations, and the inferences reached need not be tempered by qualifying statements such as, "if the assumptions are true, then...."

Nonparametric methods are frequently called *distribution-free* methods, because the inferences are based on a test statistic whose sampling distribution does not depend on the specific distribution of the population from which the sample is drawn.

Several statistical methods regarded as classical techniques actually are nonparametric according to these definitions. For example, methods based on the central limit theorem, Chebyshev's inequality, and many chi square tests do not require specific distribution assumptions for large sample sizes.

Section 12.2 gives a brief description of the best-known and most frequently used nonparametric methods that are applicable for both small and large sample sizes. The methods are organized according to the types of data to be analyzed, such as one sample, paired samples, and two independent samples, and the type of hypothesis to be tested. An outline listing many of the well-developed methods is given in Sec. 12.3 together with numerous references.

Most of the methods covered here were originally proposed in the late 1940s and early 1950s. Many articles that have appeared later develop properties, modifications, refinements, and extensions of the basic procedures as well as new procedures. Considerable research is still being devoted to this topic.

12.2 DESCRIPTION OF SOME NONPARAMETRIC METHODS

The specific methodology and the logical rationale for the most useful nonparametric test and estimation procedures are given in this section. An assessment of the performance of each test is also given in terms of its asymptotic relative efficiency, which is loosely defined as follows. Suppose there are two tests A and B, either of which can be used to test a specific null hypothesis against a specific alternative. The *asymptotic relative efficiency* of test A relative to test B is the ratio of the sample size required by test B relative to test A to achieve the same power at the same level under the same distribution assumptions with large sample sizes. For example, suppose the asymptotic relative efficiency of a nonparametric test relative to a parametric test is computed to be 0.955 for a specific distribution. This means that the nonparametric test based on a sample of, say, 1000 observations is as efficient as the corresponding parametric test based on a sample of 955 observations from this specific distribution under asymptotic theory.

12.2.1 Location Tests for One Sample and Paired Samples

Assume that $x_1, x_2, \ldots, x_n$ are a random sample of n observations measured on at least an ordinal scale and drawn from a continuous population with unknown median M. We want to test the null hypothesis

$$H: \quad M = M_0.$$

If the null hypothesis is true, about half of the observations in the sample should be larger than M_0. The *sign test statistic* is defined as the number of plus signs S among the n differences $x_1 - M_0, x_2 - M_0, \ldots, x_n - M_0$. The null distribution of S is the binomial distribution with n and $\theta = 0.5$, which is given for $n \le 20$ in Table A.7 in the Appendix. For larger sample sizes we can use the normal approximation to the binomial distribution with the test statistic

$$z = \frac{S - 0.5n \pm 0.5}{\sqrt{0.25n}}. \tag{12.1}$$

The ± 0.5 term is a continuity correction introduced to improve the normal approximation to the binomial. If any of the differences $x_i - M_0$ are equal to zero, they should be ignored, and then n is reduced accordingly.

Suppose the alternative to H is one-sided in the positive direction as

$$A_+: \quad M > M_0.$$

If A_+ is true, much more than half of the sample observations will be larger than M_0, and S will be large. Therefore with A_+ the appropriate rejection region of S or

z has large values, that is, it is in the right tail of the null distribution. Similarly, if the alternative is one-sided in the negative direction as

$$A_-: \quad M < M_0$$

the appropriate rejection region of S or z has small values, that is, it is in the left tail of the null distribution.

Example 12.1. A large company with six manufacturing plants is concerned about whether the number of plant accidents has increased since last year when the median number of accidents per man-hour worked was equal to 0.025 per month, with small variations between months. The safety engineer collects data on the six plants for the current month and finds the accident rates are 0.036, 0.021, 0.028, 0.035, 0.040, and 0.032. Is there reason for concern?

The relevant hypotheses here are H: $M = 0.025$, A_+: $M > 0.025$. For the $n = 6$ observations, there are $S = 5$ plus signs, that is, observations larger than 0.025. Is this value large enough to reject H in favor of A_+? The classical approach would be to determine the appropriate critical value of S for, say, $\alpha = 0.05$ from Table A.7 with $n = 6$. However, Table A.7 shows that the only reasonable α values we can use are 0.0156 and 0.1094 because the distribution is discrete. The usual nonparametric approach is to report the P value, that is, the probability under H of obtaining a sample result as extreme as that observed in a particular direction, as indicated by the one-sided alternative. In our example, the alternative is in the positive direction, so the appropriate direction is large values of S. Since we observed $S = 5$, the P value is $P = \Pr(S \geq 5|H) = 0.1094$. If we want to use this P value to reach a decision at $\alpha = 0.05$ we cannot reject H in favor of A_+ because $P > \alpha$.

In general the decision rule using P values is to reject H for any $P \leq \alpha$ and not to reject H otherwise. Alternatively, we can simply report the P value and not choose a particular value of α. If the alternative is two-sided as

$$A: \quad M \neq M_0$$

the common practice is to report either twice the smaller one-tailed P value or 1, whichever is smaller.

For the sign test the appropriate P values for S are right-tail for the alternative A_+, left-tail for A_-, and twice the smaller tail probability for A; for larger samples the same tail probabilities are used as for z in Eq. (12.1).

For paired samples $(x_1, y_1), (x_2, y_2), \ldots, (x_n, y_n)$, the same procedure can be used to test the null hypothesis,

$$H: \quad M_D = M_0$$

where M_D represents the median of the population of differences $X - Y$. In this kind of application, S is defined as the number of plus signs among the n differences $x_1 - y_1 - M_0, x_2 - y_2 - M_0, \ldots, x_n - y_n - M_0$.

Example 12.2. Fifteen fabricated panels are divided into two equal pieces, and each half is treated by different products X and Y, which are both supposed to promote resistance to corrosion. Each half-panel is subjected to simulated moderate and extreme weather conditions, accelerated to be the equivalent of 10 years of typical exposure. The degree of corrosion is then rated in increasing order of magnitude as (1) below normal, (2) normal, and (3) above normal, as shown in Table 12.1. Which product, X or Y, seems to provide the greatest resistance?

The differences $X - Y$ have 10 minus signs, no plus signs, and five zeros. Thus the effective n is 10 with $S = 0$. The question asked in this example implies that a

TABLE 12.1 Degree of Corrosion

Panel	X	Y	Panel	X	Y
1	1	2	9	1	2
2	2	3	10	2	3
3	2	2	11	1	3
4	1	1	12	1	3
5	1	3	13	2	2
6	1	2	14	1	1
7	2	3	15	1	3
8	2	2			

one-sided alternative should be used. We select the direction of this alternative from the observations themselves. Since $S = 0$ is smaller than its expected value $n/2 = 5$, if any difference exists, it must be in the negative direction of $D = X - Y$. Thus the appropriate hypotheses are

$$H\colon\ M_D = 0 \qquad A_-\colon\ M_D < 0.$$

Table A.7 shows that the left-tail P value for $S = 0$ is $P = 0.0010$ with $n = 10$. Therefore we conclude at any $\alpha > 0.001$ that $M_D < 0$. This means that panels treated with X have a lower degree of corrosion than those treated with Y and that X provides the greater resistance.

Now we return to the same one-sample situation but suppose that we are willing to assume that the population is symmetric about its median. If the null hypothesis $H\colon\ M = M_0$ is true, then we expect not only that about half of the sample observations will exceed M_0, but also that the magnitudes of those larger than M_0 will about balance off the magnitudes of those smaller than M_0. The appropriate procedure here then is to rank the absolute values $|x_1 - M_0|, |x_2 - M_0|, \ldots, |x_n - M_0|$ from 1 to n, keeping track of the original signs. The sum of the positive ranks (ranks which are from originally positive differences $x_i - M_0$) should be about equal to the sum of the negative ranks under H. However, since these two sums add up to a constant, $1 + 2 + \cdots + n = n(n + 1)/2$, either one can be used as a test statistic by itself. The *Wilcoxon signed rank test statistic* is defined as either

$$T_+ = \text{sum of positive ranks}$$

or

$$T_- = \text{sum of negative ranks}$$

T_+ and T_- are identically distributed under the null hypothesis. This distribution is given here for $n \leq 15$ as Table A.8, where T is to be interpreted as either T_+ or T_-. The table makes use of the fact that T is symmetrically distributed about its mean $n(n + 1)/4$.

If the alternative $A_+\colon\ M > M_0$ is true, T_+ should be much larger than its expected value under H, and therefore the appropriate P value is right-tail for T_+. We get the same P value if we use a left-tail probability for T_-. For the alternative $A_-\colon\ M < M_0$, the appropriate P value is either left-tail for T_+ or right-tail for T_-. With a two-sided alternative $A\colon\ M \neq M_0$, the appropriate P value is twice the right-tail probability for the larger of T_+ or T_-.

The null distribution of T approaches the normal distribution, and this approximation is reasonably accurate for $n \geq 16$. The test statistics corresponding to T_+ and T_- are

$$z_{+,R} = \frac{T_+ - 0.5 - n(n+1)/4}{\sqrt{n(n+1)(2n+1)/24}} \qquad z_{-,R} = \frac{T_- - 0.5 - n(n+1)/4}{\sqrt{(n+1)(2n+1)/24}}. \tag{12.2}$$

The appropriate P values are right-tail for $z_{+,R}$ with A_+, right-tail for $z_{-,R}$ with A_-, and twice the right-tail probability for the larger of $z_{+,R}$ and $z_{-,R}$ with A.

Two problems that can arise in applications are zeros and ties. Since a zero is neither positive nor negative, each zero should be ignored and n reduced accordingly. A tie occurs whenever $|x_i - M_0| = |x_j - M_0|$. Ties should not be discarded. The accepted procedure is to assign each absolute difference in any tied set the average of the ranks they would be assigned if they were not tied. Zeros and ties should not be extensive when the assumption of a continuous distribution is met, and the preceding tests can be used in exactly the same way to give approximate P values. If the zeros or ties are extensive, more precise measurement techniques should be used.

The Wilcoxon signed rank test can be used in exactly the same way for paired samples $(x_1, y_1), (x_2, y_2), \ldots, (x_n, y_n)$ and H: $M_D = M_0$, where M_D is the median of the population of differences $D = X - Y$. The signed rank statistic is calculated in the same way, but for the absolute values $|x_i - y_i - M_0|$.

Example 12.3. The number of man-hours lost per month from plant accidents have been recorded for years at each of the six plants of a large company. A year ago, the company instituted an extensive industrial safety program at all plants. The number of hours lost this month and the number lost for the same month before instituting the safety program are shown in Table 12.2. Has the program been effective in reducing time lost from accidents?

TABLE 12.2 Man-Hours Lost before and after Safety Program

Plant	Before	After
1	51.2	45.8
2	46.5	41.3
3	24.1	15.8
4	10.2	11.1
5	65.3	58.5
6	92.1	61.2

Define D as the difference $A - B$, after minus before, and assume that the distribution of D is symmetric about M_D. The program is effective if $M_D < 0$, so the appropriate hypotheses are

$$H: \; M_D \geq 0 \qquad A_-: \; M_D < 0.$$

The differences for each plant and the signed ranks of their absolute values are shown in Table 12.3.

The Wilcoxon signed rank statistics are $T_+ = 1$, $T_- = 20$. Table A.8 gives the left-tail P value for $T_+ = 1$ as 0.031 for $n = 6$. The safety program has been effective at the $\alpha = 0.05$ level.

TABLE 12.3 Differences and Signed Ranks

Plant	Difference $A - B$	Signed rank of $\|A - B\|$
1	− 5.4	− 3
2	− 5.2	− 2
3	− 8.3	− 5
4	0.9	1
5	− 6.8	− 4
6	− 30.9	− 6

Notice that this example could have been carried out using the sign test. Then we do not need to assume symmetry and the test statistic is $S = 1$ for $n = 6$. Table A.7 gives the left-tail $P = 0.1094$, and we cannot conclude that the program is effective at the 0.05 level.

Which result should we believe? In this example the data are recorded on a higher than ordinal scale, and therefore the magnitudes of the differences can be ranked. The sign test does not use the information about relative magnitudes while the signed rank test does. Therefore, unless the symmetry assumption is suspect, the conclusion based on the Wilcoxon signed rank test is the better one in this example.

Note that the Wilcoxon signed rank test could also have been used in Example 12.1 instead of the sign test, but only the sign test is appropriate for Example 12.2 because the $X - Y$ differences cannot be ranked in any meaningful way.

In general, if either test can be used, the Wilcoxon signed rank test is preferred if the symmetry assumption appears reasonable. The asymptotic relative efficiency of the Wilcoxon signed rank test relative to Student's t test is 0.955 for the normal distribution and is at least 0.864 for any continuous, symmetric distribution, compared with only 0.333 for the sign test.

12.2.2 Location Test for Two Mutually Independent Samples

Assume that $x_1, x_2, \ldots, x_m$ and $y_1, y_2, \ldots, y_n$ are two mutually independent random samples measured on at least an ordinal scale and drawn from continuous distributions with medians M_X and M_Y, respectively. We want to test the null hypothesis that these unknown medians are equal, or

$$H: \quad M_X = M_Y.$$

If the distributions of X and Y are assumed to be the same except for possible differences in location, the distributions are identical under the null hypothesis.

The test procedure that is most appropriate for this situation is to pool the $m + n = N$ observations into a single array while keeping track of whether each is an X or a Y observation. Thus the array of a sample of $m = 3$ and $n = 4$ might be recorded as $XXXYYYY$. This means that the three smallest observations come from the X population and the four largest come from the Y population. This is the most extreme case to support the alternative A_-: $M_X < M_Y$. Thus we see that the relative positions of the Xs and Ys in the resulting array of the pooled observations give us information about the relative values of their population medians. One way to compare the relative positions is to assign the ranks $1, 2, \ldots, N$ to reflect the magnitudes. The test statistic is then

$$T_X = \text{sum of the } X \text{ ranks}$$

that is, the sum of the ranks assigned to the X observations. A large value of T_X supports the alternative A_+: $M_X > M_Y$ and calls for a right-tail P value. A small value of T_X supports A_-: $M_X < M_Y$ and calls for a left-tail P value. With the alternative A: $M_X \neq M_Y$ we double the smaller one-tail P value.

The null distribution of T_X is given in Table A.9 for $m \leq n \leq 10$. Since the table covers only $m \leq n$, the sample with fewer observations must be labeled X. For larger sample sizes, the normal approximation to the distribution of T_X can be used. The appropriate test statistics are

$$z_{X,L} = \frac{T_X + 0.5 - m(N+1)/2}{\sqrt{mn(N+1)/12}} \qquad z_{X,R} = \frac{T_X - 0.5 - m(N+1)/2}{\sqrt{mn(N+1)/12}} \tag{12.3}$$

where a left-tail probability is used with $z_{X,L}$ and a right-tail with $z_{X,R}$. With a two-sided alternative, the P value is twice the right-tail probability for z, where

$$z = \begin{cases} -z_{X,L} & \text{if } T_X < m(N+1)/2 \\ z_{X,R} & \text{if } T_X > m(N+1)/2 \end{cases}.$$

If ties occur between or within samples, each one should be assigned the midrank.

Example 12.4. A reliability engineer needs to obtain information on the times to failure of capacitors which are not known to follow any specific distribution. A sample of 18 capacitors is divided randomly into two groups. The capacitors in the first group are used in a normal operating environment (the control sample), while those in the second group are subjected to temperature stress (the exposure sample). The time to failure is noted for each capacitor, with the results shown in Table 12.4. Does it appear that the median time to failure is shorter for the exposure group than for the control group?

TABLE 12.4 Times to Failure for Two Capacitor Groups

Control sample		Exposure sample	
5.2	17.1	1.1	7.2
8.5	17.9	2.3	9.1
9.8	23.7	3.2	15.2
12.3	29.8	6.3	18.3
		7.0	21.1

Since the control sample has only eight observations while the exposure sample has 10, the control must be the X sample, so $m = 8$ and $n = 10$. The problem states that the hypotheses should be

$$H\colon\ M_X = M_Y \qquad A_+\colon\ M_X > M_Y.$$

The array of pooled observations and their ranks are shown in Table 12.5, with each X indicated by an asterisk.

TABLE 12.5 Pooled Observations and Ranks

Array	Rank	Array	Rank
1.1	1	9.8*	10*
2.3	2	12.3*	11*
3.2	3	15.2	12
5.2*	4*	17.1*	13*
6.3	5	17.9*	14*
7.0	6	18.3	15
7.2	7	21.1	16
8.5*	8*	23.7*	17*
9.1	9	29.8*	18*

**X* observations.

The test statistic is $T_X = 95$. Table A.9 gives the right-tail P value for $m = 8$, $n = 10$ as $P = 0.051$. At any $\alpha > 0.05$, we do not reject H and conclude that median times to failure are the same for each group.

This test, which is based on the sum of X ranks, is frequently called the *Wilcoxon rank sum test.* An equivalent test statistic was proposed independently by Mann and Whitney. Hence the test is also called the *Mann-Whitney-Wilcoxon test.* The parametric counter part is the two-sample t test, which requires the assumption of equal variances as well as normal distributions. The asymptotic relative efficiency of the Mann-Whitney-Wilcoxon test relative to the t test is at least 0.864 for any continuous distribution and 0.955 for the normal distribution.

12.2.3 Location Test for $k \geq 3$ Mutually Independent Samples

The nonparametric extension of the Mann-Whitney-Wilcoxon test to k mutually independent random samples is the *Kruskal-Wallis test,* or one-way analysis of variance (ANOVA) test. Assume we have k mutually independent random samples measured on at least an ordinal scale and drawn from continuous distributions with unknown medians. The null hypothesis is that these medians are equal, or

$$H\colon \quad M_1 = M_2 = \cdots = M_k.$$

The alternative here is always that the medians are not all the same, and this is a two-sided alternative A. If the k distributions are assumed the same except for possible differences in location, all distributions are identical under H.

The test procedure is to pool all N observations into a single array and rank them from 1 to N while keeping track of which sample produced which observation. The sample sizes need not be equal here, and we denote the sample sizes by $n_1, n_2, \ldots, n_k$ so that $N = n_1 + n_2 + \cdots + n_k$. Then the ranks assigned to each sample are summed to get the individual rank sums $R_1, R_2, \ldots, R_k$. Note that R_i is the sum of the n_i ranks assigned to sample i and that $R_1 + R_2 + \cdots + R_k = 1 + 2 + \cdots + N = N(N + 1)/2$. Under H, the ranks entering the sum R_i are a random sample of n_i of the possible ranks and we expect the average rank sums $\overline{R}_i = R_i/n_i$ to be about equal to each other and to the expected rank of any observation, which is $[N(N + 1)/2]/N = (N + 1)/2$.

The test statistic then is a function of the weighted sum of squares of deviations of the actual average rank sums from the expected average rank sum.

The *Kruskal-Wallis test statistic* is

$$Q = \frac{12\sum_{j=1}^{k} n_j\left[\bar{R}_j - \frac{(N+1)}{2}\right]^2}{N(N+1)} = \left(\frac{12}{N(N+1)}\sum_{j=1}^{k}\frac{R_j^2}{n_j}\right) - 3(N+1). \qquad (12.4)$$

The latter form of the test statistic is easiest for the purposes of calculation. Extensive tables of the null distribution of Q are given in Iman, Quade, and Alexander (1975) for small samples. A table for $k = 3$, each $n_j \le 5$, is given in Table A.10. For large sample sizes, Q is approximately chi square distributed with $k - 1$ degrees of freedom, given in Table A.3. Since small values of Q support H, the appropriate P value is right-tail from Table A.3 or Table A.10.

The parametric equivalent of this test is the one-way ANOVA test, which requires the assumption of equal variances and normal distributions. The asymptotic relative efficiency of the nonparametric to the parametric test is at least 0.864 for continuous distributions, and is 0.955 for normal distributions.

Example 12.5. Three different arrangements of instruments on a control panel of an airplane are being considered to improve the safety of the plane. An emergency condition is simulated, and pilot reaction time required to use instruments to correct the condition is measured in seconds. Fifteen different pilots are used to obtain five observations of the experiment on each of the three control panels. The pilots are assigned randomly to the arrangements and are a homogeneous group with respect to age and physical fitness (Table 12.6). Thus any differences in correction times should be attributed to differences in arrangements. Do the arrangements produce different median correction times?

TABLE 12.6 Pilot Reaction Times

Arrangement	Correction times, seconds
1	1.4, 0.9, 1.5, 1.3, 1.1
2	1.0, 0.8, 0.7, 1.2, 0.6
3	1.6, 1.7, 1.8, 2.1, 1.9

The null hypothesis is H: $M_1 = M_2 = M_3$. We pool all the observations and rank them from 1 to 15. For example, the ranks for arrangement 1 are 9, 4, 10, 8, and 6, respectively. The respective ranks and corresponding rank sums are shown in Table 12.7.

To use the second expression for Q in Eq. (12.4), we need to calculate $\Sigma R_j^2/n_j =$ 1183.6. Then since $k = 3$ and $n_1 = n_2 = n_3 = 5$, we have

$$Q = \frac{12}{15(16)}(1183.6) - 3(16) = 11.18.$$

The exact distribution of Q in Table A.10 gives right-tail $P < 0.001$. The chi square approximation with 2 degrees of freedom gives $0.001 < P < 0.01$ from Table A.3.

TABLE 12.7 Ranks and Rank Sums for Three Arrangements

1	2	3
9	5	11
4	3	12
10	2	13
8	7	15
6	1	14
37	18	65

12.2.4 Location Test for $k \geq 3$ Related Samples

The appropriate experimental design for a location test with $k \geq 3$ related samples is n treatments assigned randomly within each of k blocks, the randomized block design. The units within each block are as homogeneous as possible so that any differences that are produced in a block can be reasonably attributed to differences in the treatments. Comparisons of treatment effects within each block are made by ranking them from 1 to n. No comparisons are made between blocks.

The null hypothesis is that the n treatment effects $\mu_1, \mu_2, \ldots, \mu_n$ are all the same,

$$H: \quad \mu_1 = \mu_2 \cdots = \mu_n$$

against the alternative that they are not all the same. Under this null hypothesis the rankings in each block are a random permutation of the integers 1, 2, ..., n, and the sums of the ranks assigned to each treatment will tend to be equal to each other. This common value is $k(n + 1)/2$. Thus the test statistic is a function of the sums of squares of the deviations between the treatment rank sums $R_1, R_2, \ldots, R_n$ and $k(n + 1)/2$, or

$$S = \sum_{j=1}^{n} \left[R_j - \frac{k(n+1)}{2} \right]^2 = \sum_{j=1}^{n} R_j^2 - \frac{k^2 n(n+1)^2}{4}. \tag{12.5}$$

The null distribution of S is given in Table A.11 for $k \leq 8$ with $n = 3$ and for $k \leq 4$ with $n = 4$. A large value of S calls for rejection of H, so the appropriate P value is right-tail. For larger sample sizes, the distribution of

$$Q = \frac{12S}{kn(n+1)} \tag{12.6}$$

is approximately chi square with $n - 1$ degrees of freedom. The approximate P value is right-tail from Table A.3. The test based on S or Q is called the *Friedman test*.

The parametric analogue of the Friedman test is the two-way analysis of variance for the balanced randomized complete block design. Here the asymptotic relative efficiency is $0.955n/(n + 1)$ for normal distributions.

Example 12.6. An experiment is designed to study the effectiveness of four different detergents in cleaning normally soiled clothing. Since different models of washing machines might also have an effect, three different models are used as blocks. The data in Table 12.8 are measures of "whiteness" obtained using specially designed equipment for 12 loads of washing distributed randomly among the detergents and models. Larger measures of "whiteness" indicate greater effectiveness. Do the detergents have the same median effect?

The detergents are the treatments and the machine models are the blocks, so $k = 3$, $n = 4$. The null hypothesis is H: $M_1 = M_2 = M_3 = M_4$. We assign ranks 1 to 4 to each model and sum these ranks to get R_1, R_2, R_3, R_4 (Table 12.9).

Table A.11 gives the right-tail P value as $P = 0.002$. The detergents do not have the same median effect.

To illustrate the importance of blocking and making comparisons only within blocks, suppose that now we ignore the fact that the machines represent three different models and use the Kruskal-Wallis test on these data. Then we would pool the observations and rank from 1 to 12 as shown in Table 12.10. Then we have $\Sigma R_j^2/n_j = 563.17$ and, from Eq. (12.4), $Q = 4.32$, which is approximately chi square

TABLE 12.8 Measures of "Whiteness"

	Detergent			
Model	*A*	*B*	*C*	*D*
1	45	47	48	42
2	43	46	50	37
3	50	52	54	49

TABLE 12.9 Ranks and Rank Sums for Four Detergents

	Detergent			
Model	*A*	*B*	*C*	*D*
1	2	3	4	1
2	2	3	4	1
3	2	3	4	1
R	6	9	12	3

$\Sigma R^2 = 270$

$$S = 270 - \frac{3^2(4)(5^2)}{4} = 45$$

TABLE 12.10 Ranks and Rank Sums Ignoring Blocks

A	*B*	*C*	*D*
4	6	7	2
3	5	9.5	1
9.5	11	12	8
16.5	22	28.5	11

distributed with 3 degrees of freedom. Table A.3 gives $0.20 < P < 0.30$, an entirely different conclusion.

12.2.5 Measures of Association and Tests of Independence with Alternative of Association

For a random sample of n observations from a continuous bivariate distribution $(x_1, y_1), (x_2, y_2), \ldots, (x_n, y_n)$, we frequently need a descriptive measure of the relationship or association between the X and Y variables. The measure used in parametric statistics is the Pearson product moment correlation coefficient. Its value ranges between –1 and +1, with absolute values close to 1 indicating a strong linear relationship between the variables.

The two frequently used nonparametric descriptive measures of the relationship between X and Y are the Spearman rank correlation coefficient and the Kendall τ coefficient. Both require data measured on at least an ordinal scale.

The *Spearman rank correlation coefficient* is actually equal to the Pearson correlation coefficient with ranks substituted for variate values. The procedure is to rank the X sample from 1 to n and, independently, the Y sample from 1 to n, keeping track of the original pairs. Then the test statistic is a function of the sum of squares of the differences D between corresponding ranks of the X and Y variables. The easiest expression for calculation is

$$R = 1 - \frac{6 \sum_{i=1}^{n} D_i^2}{n(n^2 - 1)}. \tag{12.7}$$

The value $R = 1$ describes a perfect direct, or positive, relationship between ranks; $R = -1$ describes a perfect indirect, or negative, relationship between ranks; while $R = 0$ describes no relationship. Midranks are used for tied observations.

The *Kendall* τ *coefficient* describes the relationship between X and Y in a different way as the proportion of concordant pairs minus the proportion of discordant pairs in the samples. A pair of bivariate observations (x_i, y_i), (x_j, y_j) is called concordant whenever the product $(x_i - x_j)(y_i - y_j)$ is positive, that is, when the difference of the X pairs has the same sign as the difference of the Y pairs. A pair is called discordant when the same product is negative. If S is the number of concordant pairs minus the number of discordant pairs, the Kendall τ coefficient is

$$\tau = \frac{2S}{n(n-1)}. \tag{12.8}$$

This is a relative measure of association because there are $n(n-1)/2$ pairs. The easiest way to calculate S is to rearrange and list first the X observations in an array and then list the corresponding Y observations in the second column. Then for each Y observation on the list, count the number of Y observations that follow in the list and are larger than it. The total is U, the number of pairs of Ys for which $(y_i - y_j) < 0$ when $(x_i - x_j) < 0$ for $i < j$. Also for each Y, count the number of Y observations that follow in the list and are smaller than that Y. The total is V, the number of pairs of Ys for which $(y_i - y_j) > 0$ when $(x_i - x_j) < 0$. Then $U - V$ is the difference of the number of concordant and discordant pairs. Observations with a tie in either set are called neither concordant nor discordant and are not counted in calculating either U or V.

The values of τ are interpreted in exactly the same way as the values of R. Note that in either case, the existence of a relationship should not be interpreted as due to a cause and effect between the two variables. The relationship may exist because of many other factors.

Example 12.7. Thousands of rivets are used to join aluminum sheets and frames when airplane fuselage sections are manufactured. A study was carried out to determine whether there was a positive relationship between the number of rivet holes X and the number of minor repairs Y on a unit. A random sample of 10 units produced the following pairs:

X	38	56	65	43	72	69	64	47	62	44
Y	17	36	29	17	40	41	36	13	31	19

Calculate both R and τ to describe the relationship.

We first list the X observations in an array and the corresponding Y values, then assign ranks to each set, as shown in Table 12.11. For R we calculate D and D^2, while for τ we calculate U and V.

From Table 12.11, we compute

$$R = 1 - \frac{6(30)}{10(99)} = 0.18 \qquad \tau = \frac{2(27)}{10(9)} = 0.600.$$

By both measures we have a positive relationship.

Since R and τ are both descriptive measures of association, both can be used as test statistics for the null hypothesis that X and Y are independent or have no association. The corresponding alternatives are

A_+: positive association
A_-: negative association
A: association exists

TABLE 12.11 Computation of R and τ

X	Y	X rank	Y rank	D^2	U	V
38	17	1	2.5	2.25	7	1
43	17	2	2.5	0.25	7	1
44	19	3	4	1	6	1
47	13	4	1	9	6	0
56	36	5	7.5	6.25	2	2
62	31	6	6	0	3	1
64	36	7	7.5	0.25	2	1
65	29	8	5	9	2	0
69	41	9	10	1	0	1
72	40	10	9	1	–	–
				30	35	8

The null distribution of R is given in Table A.12, while the distribution of τ is given in Table A.13. In either case, a small right-tail P value leads to the rejection of H in favor of A_+ and left-tail for A_-. For larger samples, both R and τ are approximately normally distributed. The standard normal statistics are, respectively,

$$z = R\sqrt{n-1} \qquad \text{and} \qquad z = \frac{3\tau\sqrt{n(n-1)}}{\sqrt{2(2n+5)}}. \tag{12.9}$$

The τ statistic approaches normality much faster than the R statistic. It should be noted that generally when there are no ties, $|R| > |\tau|$ by about 50%, but the P values are usually of similar magnitude.

The one-tailed P values for Example 12.7 with the alternative A_+: positive association are $P = 0.003$ for R from Table A.12 and $P = 0.008$ for τ from Table A.13. In each case we conclude that there is a significant positive relationship between X and Y.

The natural extension of the Spearman rank correlation and the Kendall τ coefficients to measure the association between three or more variables is the Kendall coefficient of concordance. Suppose we have random samples of n observations from a continuous multivariate distribution with $k \ge 3$ variables. Then the procedure is to rank each sample separately from 1 to n, keeping track of the original k-tuples. Alternatively, the data may be collected in the form of $k \ge 3$ sets of rankings for n objects. The sum of the ranks given to the respective objects by the k judges are denoted by $R_1, R_2, \ldots, R_n$. If there were complete agreement between the judges about the overall ranking, these rank sums would be some permutation of the numbers $1k, 2k, 3k, \ldots, nk$. If there were no agreement between the judges, these rank sums would all tend to be equal to each other, with the average value of $k(n + 1)/2$. A relative measure of agreement is then the ratio of the actual sum of squares of deviations between the R_j and $k(n + 1)/2$, denoted by S, to the sum of squares of deviations under perfect agreement. This measure is called the *Kendall coefficient of concordance*. It is defined as

$$W = \frac{\sum_{j=1}^{n}\left[R_j - \frac{k(n+1)}{2}\right]^2}{\sum_{j=1}^{n}\left[jk - \frac{k(n+1)}{2}\right]^2} = \frac{12S}{k^2 n(n^2-1)} \tag{12.10}$$

where S can be calculated from Eq. (12.5). The sum of squares of deviations under perfect agreement is the maximum possible value of S, and therefore the value of W ranges between 0 and 1, with increasing values reflecting an increasing level of agreement.

Since W is a descriptive measure of agreement between rankings, it can be used as a test statistic for the null hypothesis of no association. The only possible alternative here is

$$A_+: \quad \text{positive association}$$

because perfect disagreement is not possible with three or more sets of rankings. The appropriate P value then is right-tail for W or S. Table A.11 gives the right-tail probabilities in the exact null distribution of S for $k \le 8$ with $n = 3$ and for $k \le 4$ with $n = 4$. For larger sample sizes the distribution of

$$Q = \frac{12S}{kn(n+1)} = k(n-1)W \tag{12.11}$$

is approximately chi square with $n - 1$ degrees of freedom. Thus the approximate P value is right-tail from Table A.3.

If we reject the null hypothesis of no association and conclude that agreement exists, we may want to estimate the consensus that is implied by the overall rankings of the n objects. The least squares estimate of the consensus is in accordance with the relative magnitudes of $R_1, R_2, \ldots, R_n$.

Example 12.8. Table 12.12 shows the number of mistakes made in a week of five successive days by three different technicians working in a photographic laboratory. Find a descriptive measure of the agreement between rankings during the week, test for no association, and estimate the consensus if there is one.

TABLE 12.12 Number of Mistakes Made by Three Technicians

	Technician		
Day	I	II	III
1	10	12	8
2	6	15	10
3	8	11	12
4	4	7	8
5	5	8	6

The technicians are the $n = 3$ objects and the days are the $k = 5$ judges. The technicians are ranked from 1 to 3 on each day, as shown in Table 12.13. Since $\Sigma R_j^2 =$ 326, we calculate S from Eq. (12.5) as

$$S = 326 - \frac{5^2(3)(4)^2}{4} = 26$$

and W from Eq. (12.10) as $W = 0.52$. The right-tail P value from Table A.11 for $S = 26$ with $n = 3$ and $k = 5$ is 0.093. The agreement is not especially strong, and thus we do not estimate a consensus.

TABLE 12.13 Ranks and Rank Sums for Three Technicians

	Technician		
Day	I	II	III
1	2	3	1
2	1	3	2
3	1	2	3
4	1	2	3
5	1	3	2
	6	13	11

12.2.6 Tests of Randomness with General Alternatives

One category of nonparametric tests has no parametric counterpart. These are generally called tests of randomness, and they are applied to one-sample data for which the order is meaningful, such as time-ordered or sequential observations. Then we test the null hypothesis that the order can be considered a random arrangement or, equivalently, that the observations are independent and identically distributed. The two tests we describe here are both based on the total number of runs. In each case we have either the original data or transformed data expressed as symbols of two types, keeping the original order. In a random arrangement the symbols should appear in a random order. We define a run as a succession of one or more identical symbols preceded and followed by a different symbol or no symbol at all. Both the total number of runs and their lengths can reflect nonrandomness as a tendency to mix or a tendency to cluster. However, these properties are interrelated and therefore only one need be used as a test statistic.

For the *ordinary-runs test* the data may originate as symbols of two types. For example, parts coming off an assembly line may be labeled as defective or good. If the data are measurements, they can be transformed to symbols according to whether the magnitude is A above (larger than) or B below some fixed value or the sample median. In either case the test statistic is the total number of runs U.

The null distribution of U is given in Table A.14 as a function of m and n, the number of symbols of each type, where $m \le n$. Since the distribution of U is not symmetric unless $m = n$, this table gives left-tail probabilities and right-tail probabilities in two separate parts. For larger m and n, and $m + n = N$, the distribution of

$$z = \frac{U \pm 0.5 - 1 - 2mn/N}{\sqrt{2mn(2mn - N)/[N^2(N-1)]}} \tag{12.12}$$

is approximately standard normal.

The one-tailed alternatives to the null hypothesis of randomness are A_+: tendency to mix and A_-: tendency to cluster. A right-tail P value is appropriate with A_+ and a left-tail value for A_-. With a two-sided alternative of nonrandomness the smaller-tail probability can be multiplied by 2.

The *runs up and down test* is applicable to measurement data on at least an ordinal scale. Here the symbols are produced by comparing the magnitude of each observation with the observation that immediately precedes it in the ordered sequence. A run up is started and recorded as a plus if the second observation is larger than the first and a run down as a minus if the second is smaller than the first. We assume a continuous distribution so that a tie has probability zero and we end up with a sequence of $N - 1$ plus and minus signs. The test statistic is the total number of runs V among these signs.

The null distribution of V is given in Table A.15. For larger sample sizes the normal approximation test statistic is

$$z = \frac{V \pm 0.5 - (2N - 1)/3}{\sqrt{(16N - 29)/90}}. \tag{12.13}$$

The null and alternative hypotheses for this test are the same as for the ordinary-runs test. The appropriate P values are also the same, but taken from Table A.15.

Example 12.9. A time and motion study was made in the permanent mold department of a foundry. The workers pour the molten metal into a die to form a casting, and rapid pouring times are very important to the efficiency of the process. Any pattern in the variation of pouring times that can be attributed to an assignable cause and corrected should be investigated. For example, perhaps the workers become more efficient as the day goes on. Sixteen observations on pouring time in seconds are taken at random times over the day from early morning to late afternoon, eight before lunch and eight after lunch. The time-ordered observations are 9.4, 9.8, 11.1, 11.2, 14.0, 11.8, 12.1, 12.5, 16.4, 15.8, 15.4, 14.0, 12.2, 11.8, 11.2, 11.4. The average pouring time is known to be 12.0 seconds. Test the sequence for randomness against the alternative of clustering.

For the ordinary-runs test, we can divide the data into symbols using magnitudes relative to the known average of 12.0 seconds or the sample median of 11.95. In either case, the data then become $B, B, B, B, A, B, A, A, A, A, A, A, A, B, B, B$, with $m = 8$, $n = 8$, and $U = 5$. The left-tail P value from Table A.14 is $P = 0.032$, so there is a tendency to cluster at $\alpha = 0.04$. The tendency seems to be more efficiency at the beginning and at the end of each workday.

For the runs up and down test, the signs of each observation relative to the immediately preceding one are +, +, +, +, −, +, +, +, −, −, −, −, −, −, + with $V = 5$. For $N = 16$, Table A.15 gives $P = 0.0009$, so there is a tendency to cluster at $\alpha = 0.01$. We get a somewhat different impression of the pattern here, however. The pouring times tend to increase mostly for the first eight differences, and then they tend to decrease. The first seven differences were taken before lunch and the last eight after lunch. Thus it appears here that the lunch break may increase efficiency.

The fact that the pattern appears somewhat different in the two test procedures should not be of concern because the two test statistics actually reflect different kinds of nonrandomness.

12.2.7 Tests of Randomness in Time Series Data

The ordinary-runs test with runs above and below the sample median or the runs up and down test can be used with measurement data on one variable taken over time, that is, time series data. The properties of interest in time series data usually are trends, cycles, and seasonal variations. A trend up or down would be reflected by A_-: tendency to cluster and a cyclical or seasonal pattern by A_+: tendency to mix.

Two other tests can be used with time series data and the alternative of a trend. The *Daniels test* uses the rank correlation coefficient with X representing time and Y representing the time-ordered observations. Here the alternatives are A_-: downward trend and A_+: upward trend. The *Mann test* uses the Kendall τ coefficient in exactly the same way as the Daniels test uses the rank correlation coefficient.

Example 12.10. A crew is making fuel tanks. The percentage of tanks with leaks is measured for 10 successive days of a randomly selected month as shown in Table. 12.14. Does there appear to be a trend during this period? Use $\alpha = 0.01$.

For the ordinary-runs test we compute the sample median as $(35.5 + 38.1)/2 = 36.8$, and the 10-day sequence is $B, B, B, B, B, A, A, A, A, A$. Thus $U = 2$ with $m = 5$ and $n = 5$. Too small a number of runs suggests a trend up or down, and Table A.14 gives the left-tail P value as $P = 0.008$. Thus we conclude that there is a trend for $\alpha = 0.01$. To determine the direction, we see that the pattern is that all Bs precede all As, so the trend is upward.

TABLE 12.14 Percentage of Tanks with Leaks

Day	Percentage
1	30.3
2	29.6
3	31.5
4	35.5
5	35.4
6	38.1
7	40.2
8	41.6
9	45.8
10	49.3

TABLE 12.15 Computation of R and τ

Day	Rank of percentage	D^2	U	V
1	2	1	8	1
2	1	1	8	0
3	3	0	7	0
4	5	1	5	1
5	4	1	5	0
6	6	0	4	0
7	7	0	3	0
8	8	0	2	0
9	9	0	1	0
10	10	0	–	–
		4	43	2

For the runs up and down test the sequence of plus and minus signs is –, +, +, –, +, +, +, +, + with $V = 4$. The left-tail P value from Table A.15 with $N = 10$ is $P = 0.0633$, so that no significant trend is exhibited at $\alpha = 0.01$ by this test. For the Daniels and Mann tests we use the calculations shown in Table 12.15.

With $\Sigma D^2 = 4$ and $n = 10$, we compute $R = 0.9758$. With $U = 43$ and $V = 2$, we have $S = 41$ and $\tau = 0.911$. The alternative for these tests is A_+: upward trend, which calls for a right-tail P value. Table A.12 for R gives $P = 0.000$ and Table A.13 for τ gives $P = 0.000$. With both tests we conclude that there is a positive trend at $\alpha = 0.01$.

12.2.8 Tests of Equality of $k \geq 2$ Distributions

If there are two mutually independent samples, we can test the null hypothesis that both come from the same unspecified population. This is like a goodness-of-fit test where the second population plays the role of the hypothesized distribution. Thus the hypotheses here are

$$H: \quad F_1(x) = F_2(x) \text{ for all } x$$

$$A: \quad F_1(x) \neq F_2(x) \text{ for some } x$$

We will cover three different tests, the chi square test, which applies when $F_1(x)$ and $F_2(x)$ are discrete, and the Kolmogorov-Smirnov test and the Wald-Wolfowitz test, both of which apply when $F_1(x)$ and $F_2(x)$ are continuous. In each case we have mutually independent random samples of n_1 from population 1 and n_2 from population 2.

In the case of hypothesized distributions which are discrete, assume that there are r categories of outcomes. The data to be analyzed are the frequencies of these outcomes, denoted by f_{ij} for the ith category with $i = 1, 2, \ldots, r$ and the jth sample where $j = 1, 2$. These observed frequencies are compared with the corresponding expected frequencies e_{ij} under H for the *chi square two-sample test statistic*

$$Q = \sum_{i=1}^{r} \sum_{j=1}^{2} \frac{(f_{ij} - e_{ij})^2}{e_{ij}} \tag{12.14}$$

The e_{ij} are not specified by H and must be estimated from the data by

$$e_{ij} = \frac{n_j f_{i\bullet}}{N} \tag{12.15}$$

where $f_{i\bullet} = \Sigma_{j=1}^{2} f_{ij}$ is the total observed frequency for category i in both samples combined and $N = n_1 + n_2$. The null distribution of Q is approximately chi square with $r - 1$ degrees of freedom. This approximation is good when $n_1, n_2 \geq 30$ and no e_{ij} is small, say, less than 3. If there are small e_{ij}, related categories of outcomes should be combined and the degrees of freedom reduced accordingly.

Example 12.11. Random samples of orders for refrigerator replacement parts are taken each week over a period of 50 weeks. The demand for two similar parts is shown in Table 12.16. Test the null hypothesis that the distribution of demand is the same for both parts.

TABLE 12.16 Orders for Refrigerator Parts

	Number of weeks		
Weekly demand	Part 1	Part 2	Total
0	28	22	50
1	15	21	36
2	6	7	13
3	1	0	1
4 or more	0	0	0
	50	50	100

The weekly demand is a discrete random variable for each part, and therefore the chi square two-sample test is appropriate for H. We have $n_1 = n_2 = 50$, $N = 100$, $f_{1\bullet} = 50, f_{2\bullet} = 36, f_{3\bullet} = 13, f_{4\bullet} = 1$, and $f_{5\bullet} = 0$. For 0 demand, the expected frequency for part 1 is $e_{11} = 50(50)/100 = 25$. For 1 demand, the expected frequency for part 1

TABLE 12.17 Computation of Q

	Part 1			Part 2		
Demand	f_1	e_1	$\frac{(f_1 - e_1)^2}{e_1}$	f_2	e_2	$\frac{(f_2 - e_2)^2}{e_2}$
0	28	25	0.36	22	25	0.36
1	15	18	0.50	21	18	0.50
2	6 } 7	6.5 } 7		7 } 7	6.5 } 7	
≥ 3	1	0.5	0.00	0	0.5	0.00
			0.86			0.86

is $e_{21} = 50(36)/100 = 18$. The remaining calculations are shown in Table 12.17. Note that demand ≥ 3 has expected frequency less than 1, so the final two categories are combined to represent a demand of 2 or more for each part. The test statistic is $Q = 0.86 + 0.86 = 1.72$ with $3 - 1 = 2$ degrees of freedom. The right-tail P value from Table A.3 is $0.30 < P < 0.50$.

The *Kolmogorov-Smirnov two-sample test* for H where $F_1(x)$ and $F_2(x)$ are continuous distributions requires ungrouped data measured on at least an interval scale. The two samples are pooled into a single array, and the two empirical distribution functions are calculated as

$$S_i(x) = \frac{\# \leq x \text{ in sample } i}{n_i} \tag{12.16}$$

for $i = 1, 2$ and each different observed value of x in either sample. Both $S_1(x)$ and $S_2(x)$ are step functions. The Kolmogorov-Smirnov two-sample test statistic is

$$D = \max |S_1(x) - S_2(x)|. \tag{12.17}$$

A small value of D supports H, so the P value is right-tail. The null distribution of $n_1 n_2 D$ is given in Table A.17. Note that this table covers only the cases where $n_1 \leq n_2$, so that sample 1 must be the one with fewer observations.

Example 12.12. Fatigue test data are recorded as millions of revolutions to failure for test specimens in samples from lots obtained from two different suppliers, as shown in Table 12.18. Test the null hypothesis that the distributions of fatigue are the same for the two suppliers.

TABLE 12.18 Fatigue Test Data

Supplier *A*	Supplier *B*
1.1	2.0
2.3	3.7
4.0	5.0
6.5	8.0
8.6	11.5
	13.0
	20.0
	23.5

TABLE 12.19 Computation of D

Array of x	$S_1(x)$	$S_2(x)$	$n_1n_2\|S_1(x) - S_2(x)\|$
1.1*	$^1/_5$	0	8
2.0	$^1/_5$	$^1/_8$	3
2.3*	$^2/_5$	$^1/_8$	11
3.7	$^2/_5$	$^2/_8$	6
4.0*	$^3/_5$	$^2/_8$	14
5.0	$^3/_5$	$^3/_8$	9
6.5*	$^4/_5$	$^3/_8$	17
8.0	$^4/_5$	$^4/_8$	12
8.6*	$^5/_5$	$^4/_8$	20
11.5	$^5/_5$	$^5/_8$	15
13.0	$^5/_5$	$^6/_8$	10
20.0	$^5/_5$	$^7/_8$	5
23.5	$^5/_5$	$^8/_8$	0

*Observations from lot 1.

Since supplier A has only five specimens while B has eight, supplier A must be labeled sample 1 so that $n_1 = 5 \le n_2 = 8$. Table 12.19 shows the calculation of $S_1(x)$, $S_2(x)$, and D. The observations from lot 1 are indicated by an asterisk. Note that $S_1(x)$ takes a jump of $^1/_5$ at each of these values, while $S_2(x)$ remains constant.

The largest entry in the last column is 20, so that $n_1n_2D = 20$. Table A.17 gives the right-tail probability $P > 0.126$ for $n_1 = 5$, $n_2 = 8$.

The *Wald-Wolfowitz two-sample runs test* can also be used to test the equality of two distributions. The procedure is simply to pool the $n_1 + n_2$ observations into a single array while keeping track of whether each is from sample 1 or from sample 2. If the distributions are the same, this array should be a random mixing of the 1's and 2's that indicate sample number. The Wald-Wolfowitz test statistic is the number of runs U in the arrangement of 1's and 2's. A small number of runs would imply that the null hypothesis is not true, and hence a left-tail P value from Table A.14 is appropriate for this test.

Example 12.13. The Wald-Wolfowitz runs test is illustrated by the data in Example 12.12. The array in Table 12.19 shows that the arrangement of the 1's and 2's is

1 2 1 2 1 2 1 2 1 2 2 2 2.

The runs test statistic is $U = 10$ with $n_1 = 5$, $n_2 = 8$. Table A.14 shows that the left-tail P value is $P > 0.5$, so that the null hypothesis of equality of distributions is accepted.

Both the chi square and the Kolmogorov-Smirnov tests can be extended for use with data on $k > 2$ samples and the null hypothesis of equal distributions. The chi square test statistic in Eq. (12.14) becomes

$$Q = \sum_{i=1}^{r} \sum_{j=1}^{k} \left(\frac{f_{ij} - e_{ij}}{e_{ij}} \right)^2 \qquad (12.18)$$

with the e_{ij} defined in Eq. (12.15). The distribution of Q is approximately chi square with $(r-1)(k-1)$ degrees of freedom.

The extension of the Kolmogorov-Smirnov test statistic in Eq. (12.17) is

$$D = \max_{x,\,i,j} \left| S_i(x) - S_j(x) \right| \tag{12.19}$$

where $S_i(x)$ is defined in Eq. (12.16) for the ith sample.

12.3 OUTLINE OF METHODS AVAILABLE

The outline given in this section lists most of the basic nonparametric hypothesis testing procedures, classified first according to the type of hypothesis to be tested (inference situation) and then according to the type of data, along with the original literature references and references to tables.

1. Location parameter hypothesis

- ***a.*** Data on one sample or two related samples (paired samples)
  - **(1)** Sign test (Arbuthnott, 1710)
  - **(2)** Wilcoxon signed rank test (Wilcoxon, 1945, 1947, 1949; Wilcoxon, Katti, and Wilcox, 1972)
- ***b.*** Data on two mutually independent samples
  - **(1)** Mann-Whitney U test (Mann and Whitney, 1947)
  - **(2)** Wilcoxon rank sum test (Wilcoxon, 1945, 1947, 1949; Wilcoxon, Katti, and Wilcox, 1972)
  - **(3)** Median test (Brown and Mood, 1951; Westenberg, 1948)
  - **(4)** Control median test (Mathisen, 1943)
  - **(5)** Tukey's (1959) quick test
  - **(6)** Normal scores tests (Terry, 1952; Hoeffding, 1951; van der Waerden, 1952, 1953; van der Waerden and Nievergelt, 1956)
  - **(7)** Percentile modified rank tests (Gastwirth, 1965; Gibbons and Gastwirth, 1970)
  - **(8)** Wald-Wolfowitz runs test (Wald and Wolfowitz, 1940)
- ***c.*** Data on k independent samples ($k \geq 3$)
  - **(1)** Kruskal-Wallis one-way analysis of variance test (Kruskal and Wallis, 1952; Iman, Quade, and Alexander, 1975)
  - **(2)** Extended median test (chi square test for equality of k proportions)
  - **(3)** Steel tests for comparison with a control (Steel, 1959a, b, 1960, 1961; Rhyne and Steel, 1965)
  - **(4)** Jonckheere-Terpstra test for ordered alternatives (Terpstra, 1952; Jonckheere, 1954)
- ***d.*** Data on k related samples ($k \geq 3$)
  - **(1)** Friedman two-way analysis of variance test (Friedman, 1937)
  - **(2)** Durbin (1951) test for balanced incomplete block designs
  - **(3)** Page (1963) test for ordered alternatives

2. Scale or dispersion parameter hypothesis (Duran, 1976)

- ***a.*** Data on two mutually independent samples
  - **(1)** Siegel-Tukey test (Siegel and Tukey, 1960)
  - **(2)** Mood test (Mood, 1954; Laubscher, Steffens, and deLange, 1968)

- **(3)** Freund-Ansari test (Freund and Ansari, 1957; Ansari and Bradley, 1960)
- **(4)** Barton-David test (David and Barton, 1958)
- **(5)** Normal-scores test (Klotz, 1962; Capon, 1961)
- **(6)** Sukhatme test (Sukhatme, 1957; Laubscher and Odeh, 1976)
- **(7)** Rosenbaum (1953) test
- **(8)** Kamat (1956) test
- **(9)** Percentile modified rank tests (Gastwirth, 1965; Gibbons and Gastwirth, 1970)
- **(10)** Moses (1963) ranklike tests

3. Tests of independence with alternative of association

a. Data on two related samples

- **(1)** Spearman rank correlation coefficient (Spearman, 1904; Kendall, 1962; Glasser and Winter, 1961; Franklin, 1988)
- **(2)** Kendall τ coefficient (Kendall, 1962; Kaarsemaker and van Wijngaarden, 1953)

b. Data on k related samples ($k \geq 3$)

- **(1)** Kendall coefficient of concordance for complete rankings (Kendall, 1962)
- **(2)** Kendall coefficient of concordance for balanced incomplete rankings (Durbin, 1951)
- **(3)** Partial correlation (Moran, 1951; Maghsoodloo, 1975; Maghsoodloo and Pallos, 1981)

c. Contingency table data

- **(1)** Chi square test of independence

4. Tests of randomness with general alternatives

a. Data on one sample

- **(1)** Number of runs test (Swed and Eisenhart, 1943)
- **(2)** Runs above and below the median
- **(3)** Runs up and down test (Olmstead, 1946; Edgington, 1961)
- **(4)** Rank von Neumann runs test (Bartels, 1982)

5. Tests of randomness with trend alternatives

a. Time series data

- **(1)** Daniels (1950) test based on rank correlation
- **(2)** Mann (1945) test based on Kendall τ coefficient
- **(3)** Cox-Stuart test (Cox and Stuart, 1955)
- **(4)** All four tests listed in 4a.

6. Tests about slope in linear regression models

a. Data on two related samples

- **(1)** Theil (1950) test based on Kendall's τ
- **(2)** Test based on Spearman's rank correlation

b. Data on two independent samples

- **(1)** Hollander (1970) test for parallelism

7. Tests of equality of k distributions ($k \geq 2$)

a. Data on two independent samples

- **(1)** Chi square test

    (2) Kolmogorov-Smirnov test (Kim and Jenrich, 1973)
    (3) Wald-Wolfowitz test (Wald and Wolfowitz, 1940)
  b. Data on k independent samples ($k \geq 3$)
    (1) Chi square test
    (2) Kolmogorov-Smirnov test (Birnbaum and Hall, 1960)

8. Tests of equality of proportions
  a. Data on one sample
    (1) Binomial test
  b. Data on two related samples
    (1) McNemar (1962) test
  c. Data on two independent samples
    (1) Fisher's exact test
    (2) Chi square test
  d. Data on k independent samples
    (1) Chi square test
  e. Data on k related samples ($k \geq 3$)
    (1) Cochran (1950) Q test

9. Tests of goodness of fit (see Chap. 6)

10. Multivariate analysis procedures (Puri and Sen, 1971)

Some publications give surveys of nonparametric statistics, but there are now many textbooks and handbooks that give thorough explanations of how to apply many of these methods. The best elementary books are Bradley (1968), Marascuilo and McSweeney (1977), Daniel (1990), Conover (1980), and Gibbons (1997). Gibbons (1993 a,b) provides a very elementary coverage of many of the techniques. Some of the more readable books that cover the theory of nonparametric statistics are Lehmann (1975), Pratt and Gibbons (1981), and Gibbons and Chakraborti (1992). All of these books include tables.

Many of the tests for hypotheses about parameter values have corresponding procedures for confidence interval estimation of the relevant parameters. A general reference here is Noether (1972), but these methods are also covered in several of the books referred to above. Several of the k-sample tests have corresponding nonparametric multiple comparison procedures. General references are Miller (1981), Hochberg and Tamhane (1987), or some of the other books listed.

REFERENCES

Ansari, A. R., and R. A. Bradley: "Rank Sum Tests for Dispersion," *Ann. Math. Stat.*, vol. 31, pp. 1174–1189, 1960.

Arbuthnott, J.: "An Argument for Divine Providence Taken from the Constant Regularity Observed in the Births of Both Sexes," *Phil. Trans.*, vol. 27, pp.186–190, 1710.

Bartels, R.: "The Rank Version of von Neumann's Ratio Test for Randomness," *J. Am. Stat. Assoc.*, vol. 77, pp. 40–46, 1982.

Birnbaum, Z. W., and R. A. Hall: "Small Sample Distributions for Multi-Sample Statistics of the Smirnov Type," *Ann. Math. Stat.*, vol. 22, pp. 592–596, 1960.

Bradley, J. V.: *Distribution-Free Statistical Tests*, Prentice-Hall, Englewood Cliffs, NJ, 1968.

Brown, G. W., and A. M. Mood: "On Median Tests for Linear Hypotheses," in *Proc. 2d Berkeley Symp. on Mathematical Statistics and Probability*, J. Neyman, Ed. Univ. of California Press, Berkeley, CA., 1951, pp. 159–166.

Capon, J.: "Asymptotic Efficiency of Certain Locally Most Powerful Rank Tests," *Ann. Math. Stat.*, vol. 32, pp. 88–100, 1961.

Cochran, W. G.: "The Comparison of Percentages in Matched Samples," *Biometrika*, vol. 37, pp. 256–266, 1950.

Conover, W. J.: *Practical Nonparametric Statistics*, 2d ed., Wiley, New York, 1980.

Cox, D. R., and A. Stuart: "Some Quick Tests for Trend in Location and Dispersion," *Biometrika*, vol. 42, pp. 80–95, 1955.

Daniel, W.: *Applied Nonparametric Statistics*, 2d ed., Houghton Mifflin, Boston, 1990.

Daniels, H. E.: "Rank Correlation and Population Models," *J. R. Stat. Soc. B*, vol. 12, pp. 171–181, 1950.

David, F. N., and D. E. Barton: "A Test for Birth-Order Effects," *Ann. Human Eugenics*, vol. 22, pp. 250–257, 1958.

Duran, B. J.: "A Survey of Nonparametric Tests for Scale," *Commun. Stat. — Theory and Methods*, vol. 14, pp. 1287–1312, 1976.

Durbin, J.: "Incomplete Blocks in Ranking Experiments," *Brit. J. Psychol.* (Stat. Sec.), vol. 4, pp. 85–90, 1951.

Edgington, E. S.: "Probability Table for Number of Runs of Signs or First Differences in Ordered Series," *J. Am. Stat. Assoc.*, vol. 56, pp. 156–159, 1961.

Franklin, L.A.: "The Complete Exact Null Distribution of Spearman's rho for n = 12(1)16," *J. of Stat. Comput. and Simulation*, vol. 29, pp. 255–269, 1988.

Freund, J. E., and A. R. Ansari: "Two-Way Rank Sum Tests for Variance," Tech. Rep. to Office of Ordnance Research and National Science Foundation 34, Virginia Polytechnic Institute, Blacksburg, VA, 1957.

Friedman, M.: "The Use of Ranks to Avoid the Assumption of Normality Implicit in the Analysis of Variance," *J. Am. Stat. Assoc.* vol. 32, pp. 675–701, 1937.

Gastwirth, J. L.: "Percentile Modifications of Two-Sample Rank Tests," *J. Am. Stat. Assoc.*, vol. 60, pp. 1127–1141, 1965.

Gibbons, J. D.: *Nonparametric Methods for Quantitative Analysis*, 3d ed. Am. Sciences Press, Syracuse, N. Y., 1997.

_____: *Nonparametric Measures of Association*, Sage University Paper Series on Quantitative Applications in the Social Sciences, 07–091, Newbury Park, CA, 1993*a*.

_____: *Nonparametric Statistics: An Introduction*, Sage University Paper Series on Quantitative Applications in the Social Sciences, 07–090. Newbury Park, CA, 1993*b*.

_____. and S. Chakraborti: *Nonparametric Statistical Inference*, 3d ed., Dekker, New York, 1992.

_____ and J. L. Gastwirth: "Properties of the Percentile Modified Rank Tests," *Ann. Inst. Stat. Math, Suppl.*, vol. 6, pp. 95–114, 1970.

Glasser, G. J., and R. F. Winter: "Critical Values of the Rank Correlation Coefficient for Testing the Hypothesis of Independence," *Biometrika*, vol. 48, pp. 444–448, 1961.

Hochberg, Y., and A. C. Tamhane: *Multiple Comparison Procedures*, Wiley, New York, 1987.

Hoeffding, W.: "Optimum Nonparametric Tests," in *Proc. 2d Berkeley Symp. on Mathematical Statistics and Probability*, J. Neyman, Ed., Univ. of California Press, Berkeley, CA, pp. 83–92, 1951.

Hollander, M.: "A Distribution-Free Test for Parallelism," *J. Am. Stat. Assoc.*, vol. 65, pp. 387–394, 1970.

Iman, R. L., D. Quade, and D. A. Alexander: "Exact Probability Levels for the Kruskal-Wallis Test," in *Selected Tables in Mathematical Statistics*, vol. 3, Inst. of Mathematical Statistics, Am. Math. Soc., Providence, RI, pp. 329–384, 1975.

Jonckheere, A. R.: "A Distribution-Free k-Sample Test Against Ordered Alternatives," *Biometrika*, vol. 41, pp. 133–145, 1954.

Kaarsemaker, L., and A. van Wijngaarden: "Tables for Use in Rank Correlation," *Statistica Neerlandica*, vol. 7, pp. 41–54, 1953.

Kamat, A. R.: "A Two-Sample Distribution-Free Test," *Biometrika*, vol. 43, pp. 377–387, 1956.

Kendall, M. G.: *Rank Correlation Methods*, 3d ed., Hafner, New York, 1962.

_____ and J. D. Gibbons: *Rank Correlation Methods*, 5th ed., Edward Arnold, London, 1990.

Kim, P. J., and R. J. Jennrich: "Tables of the Exact Sampling Distribution of the Two-Sample Kolmogorov-Smirnov Criterion D_{mn} $(m \leq n)$," in *Selected Tables in Mathematical Statistics*, vol. 1. Inst. of Mathematical Statistics, Am. Math. Soc., Providence, RI., pp. 79–170, 1973.

Klotz, J. H.: "Nonparametric Tests for Scale," *Ann. Math. Stat.*, vol. 33, pp. 495–512, 1962.

Kruskal, W. H., and W. A. Wallis: "Use of Ranks in One-Criterion Analysis of Variance," *J. Am. Stat. Assoc.*, vol. 47, pp. 583–621, 1952; errata, *ibid*, vol. 48, pp. 905–911, 1953.

Laubscher, N. F., and R. E. Odeh: "A Confidence Interval for the Scale Parameter Based on Sukhatme's Two-Sample Statistic," *Commun. Stat. — Theory and Methods*, vol. 14, pp. 1393–1407, 1976.

_____, F. E. Steffens, and E. M. deLange: "Exact Critical Values for Mood's Distribution-Free Test Statistic for Dispersion and Its Normal Approximation," *Technometrics*, vol. 10, pp. 497–508, 1968.

Lehmann, E. L.: *Nonparametrics: Statistical Methods Based on Ranks*. Holden Day, San Francisco, CA., 1975.

Maghsoodloo, S.: "Estimates of the Quantiles of Kendall's Partial Rank Correlation Coefficient and Additional Quantile Estimates," *J. Stat. Comput. Simulation*, vol. 4, pp. 155–164, 1975.

_____ and L. L. Pallos: "Asymptotic Behavior of Kendall's Partial Rank Correlation Coefficient and Additional Quantile Estimates," *J. Stat. Comput. Simulation*, vol. 13, pp. 41–48, 1981.

Mann, H. B.: "Nonparametric Tests against Trend," *Econometrica*, vol. 13, pp. 245–259, 1945.

_____ and D. R. Whitney: "On a Test whether One of Two Random Variables Is Stochastically Larger than the Other," *Ann. Math. Stat.*, vol. 18, pp. 50–60, 1947.

Marascuilo, L. A., and M. McSweeney: *Nonparametric and Distribution-Free Methods for the Social Sciences*, Brooks/Cole Publishing, Monterey, CA, 1977.

Mathisen, H. C.: "A Method of Testing the Hypothesis That Two Samples are from the Same Population," *Ann. Math. Stat.*, vol. 14, pp. 188–194, 1943.

McNemar, Q.: *Psychological Statistics*. Wiley, New York, 1962.

Miller, R. G.: *Simultaneous Statistical Inference*, 2d ed., Springer, New York, 1981.

Mood, A. M.: "On the Asymptotic Efficiency of Certain Nonparametric Two-Sample Tests," *Ann. Math. Stat.*, vol. 25, pp. 514–522, 1954.

Moran, P. A. P.: "Partial and Multiple Rank Correlation," *Biometrika*, vol. 38, pp. 26–32, 1951.

Moses, L. E.: "Rank Tests for Dispersion," *Ann. Math. Stat.*, vol. 34, pp. 973–983, 1963.

Noether, G. E.: "Distribution-Free Confidence Intervals," *Am. Statistician*, vol. 26, pp. 39–41, 1972.

_____: *Introduction to Statistics the Nonparametric Way*, Springer-Verlag, New York, 1991.

Olmstead, P. S.: "Distribution of Sample Arrangements for Runs Up and Down," *Ann. Math. Stat.*, vol.17, pp. 24–33, 1946.

Page, E. P.: "Ordered Hypotheses for Multiple Treatments: A Significance Test for Linear Ranks," *J. Am. Stat. Assoc.*, vol. 58, pp. 216–230, 1963.

Pratt, J. W., and J. D. Gibbons: *Concepts of Nonparametric Theory*, Springer-Verlag, New York, 1981.

Puri, M. L., and P. K. Sen: *Nonparametric Methods in Multivariate Analysis*, Wiley, New York, 1971.

Rhyne, A. L., and R. G. D. Steel: "Tables for a Treatment versus Control Multiple Comparisons Sign Test," *Technometrics*, vol. 7, pp. 293–306, 1965.

Rosenbaum, S.: "Tables for a Nonparametric Test of Dispersion," *Ann. Math. Stat.*, vol. 24, pp. 663–668, 1953.

Siegel, S., and N. J. Castellan: *Nonparametric Statistics for the Behavioral Sciences*, McGraw-Hill, New York, 1988.

_____ and J. W. Tukey: "A Nonparametric Sum of Ranks Procedure for Relative Spread in Unpaired Samples," *J. Am. Stat. Assoc.*, vol. 55, pp. 429–445, 1960; correction, *ibid.*, vol. 56, p. 1005, 1961.

Spearman, C.: "The Proof and Measurement of Association between Two Things," *Am. J. Psychol.*, vol. 15, pp. 73–101, 1904.

Sprent, P.: *Applied Nonparametric Statistical Methods*, 2d ed., Chapman & Hall, New York, 1993.

Steel, R. G. D.: "A Multiple Comparison Sign Test: Treatments versus Control," *J. Am. Stat. Assoc.*, vol. 54, pp. 767–775, 1959*a*.

_____: "A Multiple Comparison Rank Sum Test: Treatments versus Control," *Biometrics*, vol. 15, pp. 560–572, 1959*b*.

_____: "A Rank Sum Test for Comparing All Sets of Treatments," *Technometrics*, vol. 2, pp. 197–207, 1960.

_____: "Some Rank Sum Multiple Comparisons Tests," *Biometrics*, vol. 17, pp. 539–552, 1961.

Sukhatme, B. V.: "On Certain Two-Sample Nonparametric Tests for Variances," *Ann. Math. Stat.*, vol. 28, pp. 188–194, 1957.

Swed, F. S., and C. Eisenhart: "Tables for Testing Randomness of Grouping in a Sequence of Alternatives," *Ann. Math. Stat.*, vol. 14, pp. 66–87, 1943.

Terpstra, T. J.: "The Asymptotic Normality and Consistency of Kendall's Test Against Trend, When Ties Are Present in One Ranking," *Proc. K. Ned. Akad. Wet.*, vol. 55, pp. 327–333, 1952.

Terry, M. E.: "Some Rank Order Tests which are Most Powerful against Specific Parametric Alternatives," *Ann. Math. Stat*, vol. 23, pp. 346–366, 1952.

Theil, H.: "A Rank-Invariant Method of Linear and Polynomial Regression Analysis," *Indagationes Mathematicae*, vol. 12, pp. 85–91, 1950.

Tukey, J. W.: "A Quick, Compact, Two-Sample Test to Duckworth's Specifications," *Technometrics*, vol. 1, pp. 31–48, 1959.

van der Waerden, B. L.: "Order Tests for the Two-Sample Problem and Their Power: I, II, III," *Proc. K. Ned. Akad. Wet. A*, vol. 55 (*Indagationes Mathematicae*, vol. 14), pp. 453–458, 1952; *Indagationes Mathematicae*, vol. 15, pp. 303–310, 311–316, 1953; correction, *ibid.*, vol. 15, p. 80, 1953.

_____ and E. Nievergelt: *Tafeln zum Vergleich zweier Stichproben mittels X-test und Zeichentest.*, Springer-Verlag, Berlin-Göttingen-Heidelberg, West Germany, 1956.

Wald, A., and J. Wolfowitz: "On a Test Whether Two Samples Are from the Same Population," *Ann. Math. Stat.*, vol. 11, pp. 147–162, 1940.

Westenberg, J.: "Significance Test for Median and Interquartile Range in Samples from Continuous Populations of Any Form," *Proc. K. Ned. Akad. Wet.*, vol. 51, pp. 252–261, 1948.

Wilcoxon, F.: "Individual Comparisons by Ranking Methods," *Biometrics*, vol. 1, pp. 80–83, 1945.

_____: "Probability Tables for Individual Comparisons by Ranking Methods," *Biometrics*, vol. 3, pp. 119–122, 1947.

_____: *Some Rapid Approximate Statistical Procedures*. American Cyanamid Co., Stanford Research Lab., Stanford, CA, 1949.

_____, S. K. Katti, and R. A. Wilcox: "Critical Values and Probability Levels for the Wilcoxon Rank Sum Test and the Wilcoxon Signed Rank Test," in *Selected Tables in Mathematical Statistics*, vol., 1., Inst. Mathematical Statistics, Am. Math. Soc., Providence, RI, pp. 171–259, 1972.

CHAPTER 13
DISCRETE-EVENT COMPUTER SIMULATION

Jerry Banks and David Goldsman
Georgia Institute of Technology, Atlanta, GA

and

John S. Carson, II
AutoSimulations, Inc., Marietta, GA

13.1 INTRODUCTION

13.1.1 Simulating a Real-World Process

A simulation is the imitation of the operation of a real-world process or system over time. Whether performed by hand or on a computer, simulation involves the generation of an artificial history to draw inferences concerning the operating characteristics of the real system.

The behavior of a system as it evolves over time is studied by developing a simulation *model*. This model usually takes the form of a set of assumptions concerning the operation of the system. These assumptions are expressed in mathematical, logical, and symbolic relationships among the *entities*, or objects of interest, in the system. Once developed and validated, a model can be used to investigate a wide variety of "what if" questions about the real-world system. Simulation can also be used to study the system in the design stage before it is built. Further, potential changes to the system can first be simulated in order to predict their impact on system performance. Thus simulation modeling can be used both as an analysis tool for predicting the effects of changes to existing systems, and as a design tool to predict the performance of new systems under varying sets of circumstances.

In some instances a model can be developed that is simple enough to be "solved" by mathematical methods. Such solutions may be found by the use of differential calculus, probability theory, algebraic methods, or other mathematical techniques. However, many real world systems are so complex that models of these systems are virtually impossible to solve mathematically. In many of these instances, numerical, computer based simulation can be used to imitate the system over time. Data are collected from the simulation as if the real system were being observed. These simulation generated data are used to estimate the system performance measures.

13.1.2 Advantages and Disadvantages of Simulation

Although simulation is frequently an appropriate tool of analysis, its advantages and disadvantages should be considered before using the methodology in a particular instance. The primary advantages and disadvantages of simulation are discussed by Pegden, Shannon, and Sadowski (1995). The advantages include the following:

- A simulation study can help in understanding how the system operates.
- New policies, operating procedures, decision rules, information flows, and organizational procedures can be explored without disrupting the ongoing operations of the real system.
- New hardware designs, physical layouts, and transportation systems can be tested without committing resources for their acquisition.
- Hypotheses about how or why certain phenomena occur can be tested for feasibility.
- Time can be compressed or expanded allowing for a speed-up or slow-down of the phenomena under investigation.
- Insight can be obtained about the interaction of variables.
- Insight can be obtained about the importance of variables on the performance of the system.
- Bottleneck analysis can be performed indicating where work-in-process, information, and materials are being excessively delayed.
- "What if" questions can be answered. This is particularly useful in the design of new systems.
- Simulation data are usually much less costly to obtain than the analogous data from the real system.

The disadvantages include the following:

- Model building requires special training. It is an art that is learned over time and through experience. Furthermore, if models are constructed by two competent individuals, the models may have similarities, but it is highly unlikely that they will be the same.
- Simulation results may be difficult to interpret. Since most simulation outputs are essentially random variables (they are usually based on random inputs), it may be hard to determine whether an observation is the result of system interrelationships or randomness.
- Simulation modeling and analysis can be time-consuming and expensive. Skimping on resources for modeling and analysis may result in a simulation model and/or analysis that is not sufficient to the task.
- Simulation is used in some cases when an analytical solution is possible, or even preferable. This is particularly true in the simulation of some waiting lines where closed-form queuing models are available.

In defense of simulation, these four disadvantages, respectively, can be offset as follows:

- Vendors of simulation software have been actively developing packages that contain models that only need input data for their operation. Such models have the generic tag "simulators" or "templates."
- Some simulation software vendors have developed output analysis capabilities within their packages for performing very eloquent analyses.
- Computer simulation can be performed rapidly. This is attributable to the advances in software and hardware that permit rapid running of scenarios. For example, some software packages contain constructs for modeling material handling using transporters such as fork-lift trucks, conveyors, and automated guided vehicles.
- Closed-form models are not able to analyze most of the complex systems that are encountered in practice. In many years of consulting practice of one of the authors, not one problem has been encountered that could have been solved by a closed-form solution.

13.1.3 Areas of Application

As an indication of some of the broad areas of application of simulation, the contents of several recent *Proceedings of the Winter Simulation Conference* reveal the wide range of applications:

Manufacturing Applications

Automated material-handling systems for semiconductor manufacturing.
Interoperability for aircraft spares inventory planning.
Aircraft assembly operations.
Agile control of aerospace manufacturing.
Control of integrated manufacturing systems.
Agent-based flexible routing manufacturing control system.
Precise modeling and analysis of a large-scale AS/RS.
Inventory tracking of a Kanban production system.
Modeling strain of manual labor.

Public Systems Applications

Health systems
- Screening for abdominal aortic aneurysms.
- Lymphocyte development in immune compromised patients.
- Asthma dynamics and medical amelioration.
- Timing of liver transplants.
- Diabetic retinopathy.
- Evaluating nurse staffing and patient population scenarios.
- Evaluation of automated equipment for a clinical processing laboratory.
- Hospital surgical suite and critical care area.

Military systems
- Air Force resource allocation.

Analysis of material handling equipment for prepositioning ships.
Force allocation.
Airland compact modeling.
Theater airlift system productivity.
C-141 depot maintenance.
Air mobility command channel cargo system.

Natural resources

Nonpoint source pollution.
Weed scouting and weed control decision making.
Surface water quality data.

Public services

Emergency ambulance systems.
Flow of civil lawsuits.
Field offices within a government agency.

Service Systems Applications

Transportation

Analysis of plans for container port operations.
Demand-based toll plaza lane staffing.
Intelligent vehicle highway systems.
Traffic control procedures at highway work zones.
Taxi management and route control.
Rapid transit modeling with automatic and manual controls.

Computer systems performance

User transaction processing behavior.
Database transaction management protocols.
Evaluation of analytic models of memory queuing.

Chemical process industries

Simulation of fish processing.
Capacity expansion for the pistachio hulling process.
International competitiveness in broiler production.

Air transportation

Human behavior in aircraft evacuations.
Analysis of airport/airline operations.
Combination carrier air cargo hub.

Communications systems

Trunked radio network.
Telephone service provisioning process.
Picture archiving and communications systems.
Modeling of broadband telecommunication networks.
Virtual reality for telecommunication networks.

Restaurant and entertainment

Quick service restaurant traffic analysis.

Determining labor requirements at a quick service restaurant.
Opportunities in amusement parks.
Hazardous waste/environmental
Waste remediation system.
Solid waste management.
Operational efficiency at a nuclear power plant.
Predicting environmental restoration.
Oil spill modeling.

Construction Applications

Earthmoving/strip mining.
Cable-stayed bridges.
Strengthening the design/construction interface.

13.1.4 The Simulation Process

Figure 13.1 shows a set of steps to guide a model builder in a thorough and sound simulation study. Similar figures and discussions can be found in other sources (e.g., Law and Kelton, 1991). The number beside each symbol in Fig. 13.1 refers to the brief description that follows.

1. *Problem formulation:* Statement of the problem to be investigated.
2. *Setting of objectives and overall project plan*: The objectives indicate the questions to be answered by the simulation. The project plan should include a statement of the alternatives to be considered and methods to evaluate the effectiveness of the alternatives.
3. *Model conceptualization*: Abstraction of the system under study into mathematical-logical relationships.
4. *Data collection*: Identification, specification, and acquisition of data.
5. *Model translation*: Preparation of the model for computer processing.
6. *Verification*: Does the computer program execute as intended?
7. *Validation*: Does the desired accuracy or correspondence exist between the simulation model and the real system?
8. *Experimental design*: Determine the length of the initialization (warm-up) period, the length of the simulation runs, and the number of replications to be made of each run.
9. *Production runs and analysis*: Execution of the simulation runs to obtain output values.
10. *Additional runs*: Based on the analysis of runs that have been completed, are more runs required to obtain the accuracy and precision desired?
11. *Documentation and reporting*: Document the operation of the model and prepare a report for the decision maker.
12. *Implementation*: Decisions resulting from the simulation are put in place.

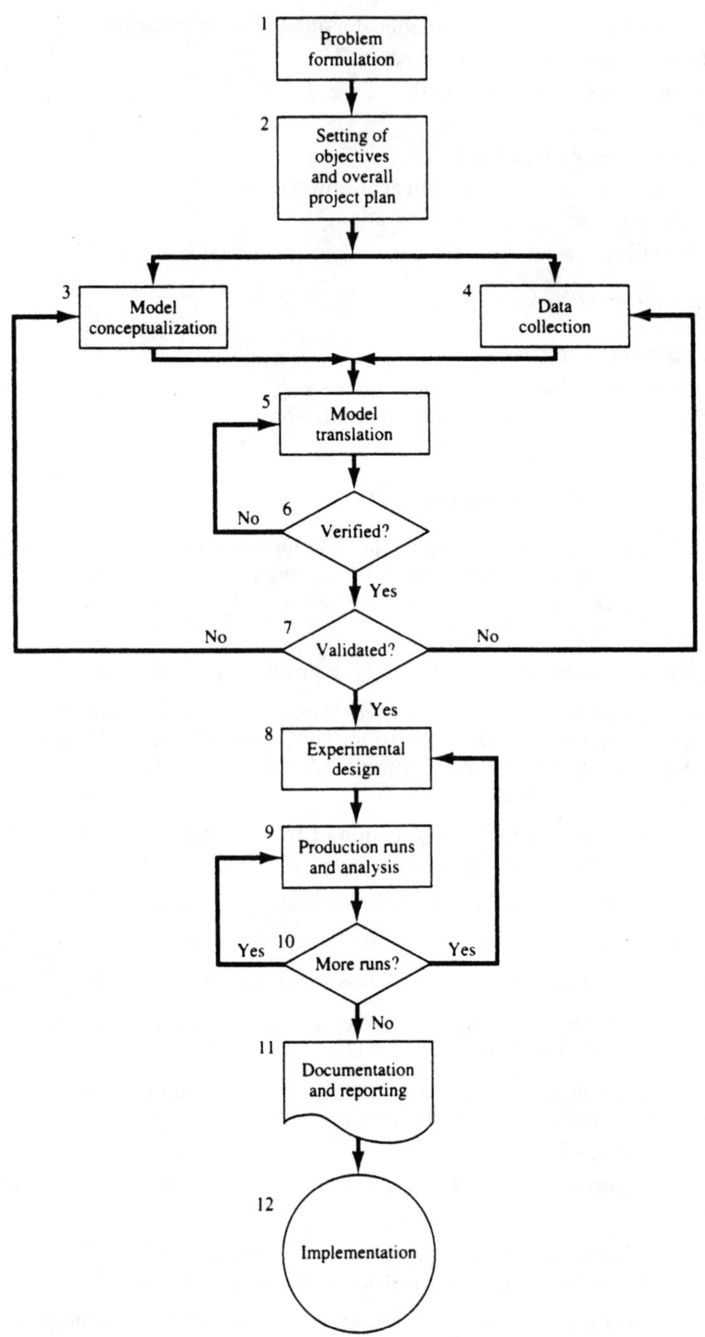

FIGURE 13.1 Steps in a simulation study. (From Banks, Carson, and Nelson, 1996, p. 14; reprinted by permission of Prentice-Hall, Upper Saddle River, NJ.)

13.1.5 Organization of the Chapter

Section 13.2 of this chapter introduces the reader to various simulation languages and their associated "world views." Section 13.3 deals with the use of random numbers and random variates. The topics of verification and validation are discussed in Sec. 13.4. Output analysis is the subject of Sec. 13.5.

13.2 SIMULATION SOFTWARE PRODUCTS

13.2.1 World Views

When developing a simulation model, a conceptual framework (or perspective) must be adopted. The framework contains a "world view" in which functional relationships between the entities can be described. Although there are others, we will discuss only the process interaction, event scheduling, and continuous world views. Various simulation languages implement these world views. We briefly describe several of these languages in this section and provide examples using each world view in Sec. 13.2.2.

The *process-interaction* approach employs statements to define the flow of entities (customers, jobs, transactions, etc.) through a network. In the *event scheduling* orientation, a system is modeled by defining the events that can change the state of the system (such as an arrival of a job to be processed or a machine breakdown), and by describing the logic associated with each type of event. In a *continuous* simulation model, the state of the system is represented by dependent variables whose values change over time in a continuous manner. The products described in this section were current as of 1994. The reader is advised to contact vendors for the latest product information.

Description of Several Simulation Languages. There are more than 100 simulation languages, manufacturing-oriented simulation products, or related software. Well known simulation languages include GPSS/H, SIMAN V, SIMSCRIPT II.5, and SLAMSYSTEM. The world views adopted by these languages are indicated in Table 13.1. We first describe these languages in general and then show how they can be used in terms of the world views described previously. We then describe seven manufacturing-oriented simulation products without showing examples.[1]

GPSS/H, a product of Wolverine Software Corp., is an implementation of GPSS, an acronym for *g*eneral-*p*urpose *s*imulation *s*ystem (see Crain and Smith, 1994). The language is used only for process-interaction simulation models. GPSS/H has a very understandable conceptual framework that provides building

[1]GPSS/H and PROOF Animation are trademarks of Wolverine Software Corporation; GPSS/PC is a trademark of Minuteman Software; SIMAN and ARENA are registered trademarks of Systems Modeling Corporation; SIMSCRIPT, SIMANIMATION, and SIMFACTORY are registered trademarks of CACI Products Company; SLAM, SLAMSYSTEM, and FACTOR are registered trademarks of Pritsker Corporation; AIM is a trademark of Pritsker Corporation; ProModel is a registered trademark of PROMODEL Corporation; WITNESS is a registered trademark of AT&T Istel; Windows is a trademark of Microsoft Corporation; OS/2 is a registered trademark of IBM Corporation.

TABLE 13.1 World Views Adopted by Selected Simulation Languages

Language	Process interaction	Event scheduling	Continuous	Combined
GPSS/H	×			
SIMAN V	×	×	×	×
SIMSCRIPT II.5	×	×	×	×
SLAMSYSTEM	×	×	×	×

blocks that are particularly applicable for the simulation of queuing systems. GPSS/H can be used for any situation in which entities (such as customers or jobs) can be viewed as passing through a system (e.g., a network of queues or a job shop). In using the GPSS/H language, a simulation programmer develops a "block diagram" describing the system under consideration. Animation for GPSS/H is available using PROOF Animation (Earle and Henriksen, 1994). Another implementation of GPSS is GPSS/PC from Minuteman Software.

SIMAN V, a product of Systems Modeling Corp., is a general-purpose simulation language that allows for the adoption of process interaction, event scheduling, or continuous approaches, or a mixture of approaches (see Banks, et al., 1995). SIMAN is an acronym for *sim*ulation *an*alysis. The language includes special features that facilitate the simulation of manufacturing and material-handling systems. These special features include statements to simulate conveyors, transporters, and manufacturing cells. SIMAN V can be animated within the ARENA environment. The ARENA environment (discussed in Sec. 13.2.2) includes other components such as an input data processor, an output data processor, and a template builder.

SIMSCRIPT II.5, from CACI Products Co., is a high level programming language that facilitates the development of a discrete event simulation model (see Russell, 1993). As a simulation language, it allows for the process interaction or event scheduling world views, or a mixture of the two. (A language extension allows continuous simulation.) In addition, it has powerful scientific computing and list processing capabilities. When used in this manner, it can be programmed quite efficiently. SIMSCRIPT II.5 is animated by SIMANIMATION.

SLAMSYSTEM, a product of Pritsker Corp., contains the simulation language SLAM II, an acronym for *s*imulation *l*anguage for *a*lternative *m*odeling (see O'Reilly, 1994). SLAM II supports the three different world views in a single integrated framework. SLAM II is the culmination of a family of languages developed by A. Alan B. Pritsker and others, starting with GASP, GERT, and Q-GERT. SLAMSYSTEM provides for output analysis and data management as well.

Manufacturing-Oriented Simulation Software. This section is taken, in part, from Banks (1994). The software discussed in this section is limited to those products associated with manufacturing and further to only seven within that category, including SIMFACTORY II.5, ProModel for Windows, AutoMod, Taylor II, WITNESS, AIM, and ARENA.

SIMFACTORY II.5, available from CACI Products Co., is a factory simulator for engineers who are not full time simulation analysts (see Goble, 1991). It operates on a PC under Windows 3.x and OS/2 2.0, or on many workstations. A system amenable to SIMFACTORY II.5 can be modeled rapidly. A model is best constructed in stages by first defining the layout consisting of processing stations, buffers, receiving areas, and transportation paths, then defining the products, then the resources, next the transporters, and finally any interruptions that might occur. The animation automatically follows from the definition of the model. These model elements are pulled from a pallet rather than a menu bar. The resulting model can be changed with a text editor. Flexible flow modeling is supported. For example, OR logic can be used (as in Request Part A OR Part B).

The layout is created by positioning icons, selected from a library, on the screen. As each icon is positioned, characteristics describing it are entered. The products are defined by process plans that define the operations performed on each part and the durations of those operations. Resources are added to the model in two steps. First, the resource is defined and its capacity is set. Second, the stations requiring specified resources are identified. While resources are moving, simulation time can elapse. Resource requirements are flexible, viz., one unit of Resource A and two units of Resource B can be requested. Transporters may be batch movers such as fork lifts or they may be conveyors. Characteristics of a transporter are specified (pickup speed, delivery speed, load time, unload time, and capacity of a fork lift, as an example). The transporter path is identified on the screen. Transporters can avoid each other by collision detection and they can carry resources. Any interruption, planned or unplanned, can be applied to any model element or group of elements (e.g., conveyors, queues, resources, and transporters). Interruptions can require any combination of resources.

Reports are available concerning equipment utilization, throughput, product makespan, and buffer utilization. Multiple business graphics (pie charts, histograms, and plots) can be compared at the same time. Data can be compared across multiple runs. Text reports can be customized, and specific statistics can be collected on the elements of interest. A summary report of all replications provides means, standard deviations, and confidence intervals on the model output.

ProModel for Windows, from PROMODEL Corp., has programming features within the environment, and the capability to add C or Pascal-type subroutines to a program (see Baird and Leavy, 1994). Some of the features of ProModel for Windows are as follows:

- Models are created using a point and click approach. Intuitive interfaces, interactive dialog, and online help are provided. An auto-build feature guides the user through the model building process. An online trainer is available.
- The software operates in the Windows and OS/2 environments, as a 32-bit application, taking advantage of memory management techniques, synchronized windowing, and data exchange. Windows fonts, printer drivers, cooperative multitasking, and the Dynamic Link Library are available.
- Virtually unlimited model size is possible.
- The simulator offers a 2-D graphics editor with scaling, rotating, and so on. Icons can be defined using either vector-based or pixel graphics. These icons are saved as bitmaps at runtime for fast animation during the simulation.
- CAD drawings as clip art can be imported as well as process information and schedules. Customized output reports and spreadsheet files can be produced. If

the data are another Windows application, cutting and pasting can be accomplished.

- The static and dynamic elements of the animation are developed while defining the model. That is, the simulation model and animation are integrated.
- Business output graphics are automatically provided and may be printed in color. Automatic statistics are available.
- Only standard hardware is required (IBM or compatible with VGA graphics). No special graphics cards, monitors, or math coprocessor chips are needed.
- Preprogrammed constructs are provided. This allows for fast modeling of multiunit and multicapacity locations, shared and mobile resources, downtimes, shifts, and so on.
- Submodels allow the creation of a library of templates of work steps, activities, or subprocesses that can be reused. This allows for model construction to be accomplished by a team with a later merger of submodels into one model.
- A free runtime, multiple-scenario capability is provided.

A model is constructed by defining a route for parts, defining the capacities of each of the locations along the route, defining additional resources such as operators or fixtures, defining the transporters, scheduling the part arrivals, and specifying the simulation parameters. The software then prompts the user to define the layout and the dynamic elements in the simulation.

AutoMod, a product of AutoSimulations, Inc., has general programming features including the specification of processes, resources, loads, queues, and variables (Rohrer, 1994). Processes are specified in terms of traffic limits, input and output connections, and itineraries. Resources are specified in terms of their capacity, processing time, mean time between failures, and mean time to repair. Loads are defined by their shape and size, attributes, generation rates, generation limits, and start times, as well as their priority.

The simulator is very powerful in its description of material handling systems. AGVs, conveyors, bridge cranes, AS/RSs, and power and free devices can be defined. The range of definition is extensive. For example, an AGV can be defined in terms of the following: multiple vehicle types, multiple-capacity vehicles, path options (unidirectional or bidirectional), variable speed paths, control points, flexible control and scheduling rules, arbitrary blocking geometries, automatic shortest distance routing, and vehicle procedures. Numerous control statements are available. For instance, process control statements include If-Then-Else, While-Do, Do-Until, Wait-Until, and Wait-For. Load control, resource control, and other statements are also available. C functions can be defined by the user. Attributes and variables may be specified. The animation capabilities include true 3-D graphics, rotation, and tilting, to mention a few. A CAD-like drawing utility is used to construct the model. Business graphics can be generated. The latest release contains a simulator within AutoMod. Features of the simulator include its spreadsheet interface. This eliminates the need for programming in building models. The spreadsheet interface also allows the definition of models outside of AutoMod. Another option as a separate utility is AutoStat. It provides simulation warmup capability, scenario management, confidence interval generation, and design of experiments capability.

Taylor II, developed by F&H Simulations, Inc., and Automation B.V., features menu-driven model building (see Nordgren, 1994). A model in Taylor II consists of four fundamental entities: elements, jobs, routings, and products. The element types are machine, buffer, conveyor, transport, path, aid, warehouse, and reservoir. One or more operations can take place at an element. The three basic operations are processing, transport, and storage.

Defining a layout is the first step when building a model. Layouts consist of element types. By selecting the elements in sequence, the product path or routing is defined. Routing descriptions can be provided from external files. The next step is detailing the model. In this step, the parameters are provided. In addition to a number of default values, Taylor II uses a macro language called TLI (Taylor Language Interface). TLI is a programming language that permits modifications of model behavior in combination with simulation specific predefined and user definable variables. TLI can also be used interactively during a simulation run to make queries and updates. An interface to C, Basic, and Pascal is also available. Local and global attributes are available.

During simulation, zoom, pan, rotate, and pause are possible. Modifications can be made on-the-fly. The time representation is fully user-definable (hours, days, seconds, and so on can be mixed). Output analysis possibilities include predefined graphics, user defined graphics, predefined tabular reports, and user defined reports. Examples of predefined graphics are queue histograms and utilization pies. User defined outputs include bar graphs, stacked bars, and other business graphics. Predefined tables include job, element, and cost reports. Animation capabilities include both 2-D and 3-D. The 3-D animation can be shaded. Standard indicators can be shown for elements. Icon libraries for both 2-D and 3-D animation are provided. Each of these libraries contains numerous icons. Additional features include online, context sensitive help with index and page search capability. Educational support materials are available.

WITNESS, from the Lanner Group, contains many elements for discrete-part manufacturing (see Thompson, 1994). For example, machines can be single, batch, production, assembly, multistation, or multicycle. Conveyors can be accumulating or nonaccumulating. Options exist for labor, vehicles, tracks, and shifts. WITNESS also contains elements for continuous processing including processors, tanks, and pipes.

Variables and attributes can be specified. Parts arrivals can be scheduled using a file. Distributions and functions can be used for specifying operation times and for other purposes. Machine downtimes can be scheduled on the basis of operations, busy time, or available time. Labor is a resource that can be preempted, use a priority system, and be scheduled based on current model conditions.

Track and vehicle logic allow requests for certain types of jobs, vehicle acceleration and deceleration, park when idle, and change destinations dynamically. Many types of routing logic are possible in addition to the standard push and pull. For example, If-Then-Else conditions may be specified.

Simulation actions, performed at the beginning and end of simulation events, may employ programming constructs such as For-Next, While-End, and GoTo-Label. The user can look at an element at any time and determine the status of a part. Reporting capabilities include dynamic on-screen information about machines and elements. Reports may be exported to spreadsheet software.

Tools within the language include access to the model data base through the C language, arithmetic and logical operators, save current status of model, built-in status functions, and many others. In addition, all of the above-mentioned features can be enhanced through the use of C. Debugging or brainstorming can be accomplished by stopping the model, changing desired parameters, and continuing with the model from the same point in simulation time. An animation is constructed along with the model definition. This animation and statistical feedback can be turned on or off during any run. Many changes to the model may be made at any time. Built-in experimentation capabilities are available from the menu bar.

The latest release of the software includes the following capabilities:

- The ability to store thousands of variables or attributes.
- Many distinct random number streams.
- Bitmap import/export with increased, icon sizes.
- "Module" elements for hierarchical modeling. One icon represents the detail existing in another portion of the model.
- "Selector Bar" for filling fields with rules, distributions, built-in functions, and so on, for defining elements on-the-fly.

AIM, a product of Pritsker Corp., stands for *A*nalyzer for *I*mproving *M*anufacturing, and is one of three applications of **FACTOR** (see Lilegdon, Martin, and Pritsker, 1994). Related packages are Factor Production Manager and Leitstand. Factor Production Manager performs detailed planning and scheduling of operations, order promising, order release, and supply-chain management. All of these applications use the same data base.

AIM models are OS/2-based and built graphically with icons that represent machines, operators, conveyors, and so on, placed directly on the screen. The animations are created in a virtual window. During a simulation, the model can be stopped to check its status or add other components, then continue the simulation.

Performance data are dynamically updated and displayed while the simulation is running. A dynamic Gantt chart is provided for tracking machine and operator status. Inventory levels and material handling utilization can also be graphed dynamically. Outputs include bar charts, pie charts, and plots of inventory levels. Alternatively, information can be transferred to other software for development of presentation graphics.

Features of AIM include the following:

- *Manufacturing representation.* Manufacturing-specific modeling components can represent a variety of discrete manufacturing processes. Standard rules provide choices that are interpreted with processes. Custom rules may be written to extend the logic.
- *Integration with scheduling applications.* Models written with AIM can be used with other FACTOR applications providing support for capacity, logistics, production scheduling, supply-chain analysis, and schedule management.
- *Manufacturing data.* AIM is built around a relational data base that stores the manufacturing operations and simulation output. Part descriptions, process plans, order release schedules, machine locations and schedules, shift schedules, and so on can be transferred from other data sources to the AIM data base.
- *Animation support.* AIM models are built graphically and are animated automatically during model construction.
- *Interactive model building and simulation.* Components are located on a scaled facility background. Intelligent defaults are provided for all components.

Components are customized by completing forms. During execution, the modeler can change the status of a component and observe the simulated impact on the manufacturing system.

- *Comparison of alternatives.* The AIM project framework organizes all aspects of a manufacturing simulation project. Alternative models of the manufacturing process are stored. Comparison reports show model performance data to identify differences between alternatives.
- *AIM Gantt charts.* AIM supports the creation of Gantt charts for the improved verification and validation of models. Model performance can be reviewed in the Gantt chart to follow a single load or the decisions of a resource.
- *Cost modeling methodology.* AIM includes a detailed cost modeling capability. AIM models can represent alternative costing philosophies such as standard cost and ABC.

ARENA, from Systems Modeling Corp., provides the tools and power of SIMAN to those for who find SIMAN burdensome to learn (Drevna and Kasales, 1994). A SIMAN programmer must understand the blocks used in the model and the elements used in the experiment frame to proceed (see Sec. 13.2.2). Under ARENA, instead, the user can extract a module, place it in its appropriate location, and parameterize it without having to learn the SIMAN language. For SIMAN language modelers, ARENA is intended to increase their functionality, eliminating the need to write similar code in different models.

SIMAN is the engine for ARENA, which can be used as an animation system. Other products included in ARENA are input and output processors. A shop-floor analysis capability is also supported by ARENA. The latter product is oriented toward scheduling and real-time shop-floor applications of simulation.

The term "module" is used to represent the building blocks available for creating models. The most fundamental feature of ARENA is that a simulation analyst can construct a module definition for use by others. These module definitions may be combined to create other modules. SIMAN Base Modules form the lowest possible level of modules. These correspond to basic SIMAN modeling constructs (blocks and elements). All other modules, called Derived Modules, are built from Base Modules or other Derived Modules. This increases the speed at which models can be built, and aids in understanding by those not familiar with SIMAN blocks and elements. Templates provide modelers with a domain-specific Module Definition set. For example, a Manufacturing Template could be sold by Systems Modeling or by third parties.

ARENA animation capability is integrated with other ARENA modules. For example, when adding a module to represent a manufacturing process, a modeler might get both the modeling logic to represent the process, as well as the animation components representing work-in-process and the status of the resource (busy, idle, in repair, etc.).

More Information about the Software. For more information about the software mentioned above, contact the following firms:

AutoMod
AutoSimulations, Inc.
P. O. Box 307
Bountiful, UT 84011-0307
(801)298-1398

GPSS/H, PROOF Animation
Wolverine Software Corporation
Little River Turnpike
Suite 900
Annandale, VA 22003-2603
(800)456-5671

GPSS/PC
Minuteman Software
Box 131
Holly Springs, NC 27540
(800)223-1430

ProModel for Windows
PROMODEL Corp.
1875 State St.
Suite 3400
Orem, UT 84058
(801)226-6036

SIMAN V, ARENA
Systems Modeling Corp.
504 Beaver Street
Sewickley, PA 15143
(412)741-3727

SIMSCRIPT II.5, SIMANIMATION, SIMFACTORY
CACI Products Co.
3333 North Torrey Pines Ct.
La Jolla, CA 92037
(619)457-9681

SLAM II, SLAMSYSTEM, AIM
Pritsker Corp.
8910 Purdue Rd.
Suite 500
Indianapolis, IN 46268
(800)428-7636

Taylor II
F&H Simulations, Inc.
P. O. Box 658
Orem, UT 84059-0658
(801)224-6914

WITNESS
Lanner Group, Inc.
31225 Bainbridge Road
Solon, OH 44138
(216) 519-1200

13.2.2 Language Examples

In this section, we pose a very simple problem and then give programs for solving the problem using the process-interaction world view of GPSS/H and SIMSCRIPT II.5. Next we show the solution of the same problem using the event scheduling approach of SIMAN. Finally, we use SLAM II to solve a problem that requires a continuous world view. The syntax will be briefly explained in every instance.

First Problem Statement. There are two identical machines in a center. They can process a part in a time that is distributed uniformly from 4 to 12 min. There is a single queue in front of the machine center, and parts are worked on a first-come, first-served basis. The parts arrive in a uniform fashion from 0 to 10 min apart. Find the average number in the queue.

Solution in GPSS. First we show the code and explain the solution in GPSS/H. (Other variations of GPSS yield similar solutions.) The reader should realize that a very straightforward problem has been posed, and virtually no realistic problem solution will be as simple as that indicated in Fig. 13.2.

The rightmost column consists of comments that are not processed by the program. These comments indicate the flow of the parts through the machine center. There are four *control* statements in the program and seven *blocks*. A control statement specifies an operating condition and a block specifies an action to be taken. The SIMULATE control statement tells the program that an execution is desired. Omitting this statement would result only in a compilation. The STORAGE control statement specifies the number of machines that are in the machine center. The START control statement specifies the number of parts that are to finish processing in the simulation. The END control statement tells the processor that there are no more statements.

The verbs associated with the blocks describe the actions that are taken. Note that QUEUE and DEPART blocks operate as pairs, as do ENTER and LEAVE blocks. The QUEUE-DEPART blocks provide for the collection and display of statistics about waiting lines that form. The ENTER-LEAVE combination provides for entry to and exit from a storage. Lastly, the notation A, B is used in GPSS to represent a uniform $(A - B, A + B)$ random variable. Many types of probability distributions can be accommodated in GPSS/H by using internal functions, by calling outside. Fortran subroutines, or by defining empirical cumulative distribution functions.

Solution in SIMSCRIPT. A SIMSCRIPT program is shown in Figs. 13.3 to 13.6. The first section in the program, the PREAMBLE, is displayed in Fig. 13.3.

```
SIMULATE                   RUN A SIMULATION
STORAGE    S(CENTER),2     SET CAPACITY OF CENTER TO 2
GENERATE   5,5             CREATE PARTS
QUEUE      LINE            PARTS JOIN A QUEUE
ENTER      CENTER          PARTS ENTER THE CENTER
DEPART     LINE            PARTS DEPART THE QUEUE
ADVANCE    8,4             PARTS ARE PROCESSED
LEAVE      CENTER          PARTS LEAVE THE CENTER
START      100             RUN SIMULATION FOR 100 PARTS
END                        STOP THE SIMULATION
```

FIGURE 13.2 Solution in GPSS/H.

```
PREAMBLE
   PROCESSES INCLUDE GENERATOR AND ARRIVAL
   RESOURCES INCLUDE MACHINE
   ACCUMULATE AVE.Q.LENGTH AS THE AVERAGE
      AND MAX.Q.LENGTH AS THE MAXIMUM OF
      N.Q.MACHINE
END
```

FIGURE 13.3 SIMSCRIPT PREAMBLE.

The PREAMBLE identifies the various modeling elements (e. g., processes and resources) that will be used. The first statement in the PREAMBLE indicates that there are two processes, called process GENERATOR and process ARRIVAL. The second lists the different resources that will be used. In this program, we need only one type of resource, and we name it MACHINE. In the following statement, we tell the computer to collect the average and the maximum length of the queue that will be formed in front of the machine.

The next section of a SIMSCRIPT program is called MAIN (Fig. 13.4). The first statement indicates that the resource to be used is MACHINE. The following statement says that there are two units of this resource. The next statements begin execution of the process GENERATOR at the present time, tell the computer to begin the simulation run, and have the computer print the average and the maximum queue length statistics.

```
MAIN
      CREATE EVERY MACHINE(1)
      LET U.MACHINE(1) = 2
      ACTIVATE A GENERATOR NOW
      START SIMULATION
      PRINT 2 LINES WITH AVE.Q.LENGTH(1) AND MAX.Q.LENGTH(1) THUS
AVERAGE QUEUE LENGTH IS ***.***
MAXIMUM QUEUE LENGTH IS ***
END
```

FIGURE 13.4 SIMSCRIPT MAIN program.

The subsequent section of a SIMSCRIPT program includes the various PROCESS routines. A PROCESS routine contains a sequence of actions an object experiences throughout its life in the model. In this SIMSCRIPT program, we have processes named GENERATOR and ARRIVAL. In our process GENERATOR. shown in Fig. 13.5, 1000 objects are created. The time between creations follows a uniform (0,10) distribution. After an object has been created, the object goes through the sequence of actions specified in the ARRIVAL routine. As an object enters the ARRIVAL process shown in Fig. 13.6, it will attempt to find a free unit of the resource MACHINE. If it finds one, then the object will make the unit unavailable for other objects until the object has been serviced. If an object cannot find a free unit of the resource, it will have to wait in a queue until a unit of the resource becomes free. The service time of a unit of the resource MACHINE follows a uniform (4,12) distribution.

Solution in SIMAN. The SIMAN solution employs the event scheduling approach. SIMAN programs have a *model* frame and an *experiment* frame. These

```
PROCESS GENERATOR
   FOR I = 1 TO 1000
   DO
      ACTIVATE AN ARRIVAL NOW
      WAIT UNIFORM.F(0.0,10.0,1) MINUTES
   LOOP
END
```

FIGURE 13.5 SIMSCRIPT PROCESS GENERATOR.

```
PROCESS ARRIVAL
   REQUEST 1 MACHINE(1)
   WORK UNIFORM.F(4.0,12.0,2) MINUTES
   RELINQUISH 1 MACHINE(1)
END
```

FIGURE 13.6 SIMSCRIPT PROCESS ARRIVAL.

frames for solving the previously stated problem are shown in Figs. 13.7 and 13.8. (We note that a SIMAN solution using the process interaction approach could be accomplished with only seven statements.) The model frame describes the system to be simulated. The experiment frame defines the experimental conditions under which the model is run to generate specific output data. One advantage of separating the two is that recompilation of just a part of the model may be necessary when an error occurs or when a model condition is to be changed.

In order to fully appreciate the event scheduling approach in SIMAN, the simulationist must be familiar with standard programming languages such as Fortran. In the next few paragraphs, we describe the subroutines in Fig. 13.7. A similar model is presented in Pegden, Shannon, and Sadowski (1995).

The events that can change the state of the system are the arrival of a new part, the start of processing of a part on a machine, and the completion of processing of a part on a machine. The state is represented by the status of each machine (in use or not in use) and the number of parts waiting to use a machine.

The subroutines in Fig. 13.7 are EVENT, PRIME, ARRIVE, and DEPART. One file is used and is denoted as NQ(1). (SIMAN returns the current number of entities in file J when NQ(J) appears in a program.) Attribute 1, denoted by A(PART,1), specifies the machine where a PART is being processed. We define two X variables such that $X(N) = 0.0$ if machine N is idle and $X(N) = 1.0$ if machine N is busy, $N = 1,2$.

The first subroutine is named EVENT. If an event is an arrival, subroutine ARRIVE is called. If the event is a departure, DEPART is called. Subroutine PRIME establishes the initial conditions. (In this example, the system is initialized empty and idle.) An entity is then scheduled to arrive to the system by a call to subroutine SCHED that schedules a PART to be processed following a delay of 0 minutes.

The logic for the arrival event is shown in subroutine ARRIVE. NEWPT is the arriving part. The first action taken is to reschedule the arrival of NEWPT to occur at the current time plus a sample value from a uniform distribution with endpoints 0 and 10. [UN(1,1) in the program indicates a uniform distribution given by parameter set 1 utilizing random number stream 1. The parameter sets are given in the experiment frame.] Since NEWPT is rescheduled as the next arrival, subroutine

```
      SUBROUTINE EVENT(PART,N)
      GOTO (1,2), N
1     CALL ARRIVE(PART)
      RETURN
2     CALL DEPART(PART)
      RETURN
      END

      SUBROUTINE PRIME
      CALL CREATE(PART)
      CALL SCHED(PART,1,0.0)
      RETURN
      END

      SUBROUTINE ARRIVE(NEWPT)
      COMMON/SIM/D(50),DL(50),S(50),SL(50),X(50),DTNOW,TNOW,TFIN,J,NRUN
      CALL SCHED(NEWPT,1,UN(1,1))
      CALL CREATE(PART)
      IF(X(1)+X(2).EQ.2.0) THEN
          CALL INSERT(PART,1)
          RETURN
      ENDIF
      IF(X(1).EQ.0.0) THEN
          CALL SETA(PART,1,1.0)
          N=1
      ELSE
          CALL SETA(PART,1,2.0)
          N=2
      ENDIF
          X(N) = 1.0
          CALL SCHED(PART,1,UN(2,1))
          RETURN
      END

      SUBROUTINE DEPART(PART)
      COMMON/SIM/D(50),DL(50),S(50),SL(50),X(50),DTNOW,TNOW,TFIN,J,NRUN
      N = A(PART,1)
      CALL DISPOS(PART)
      CALL COUNT(1,1)
      IF(NQ(1).GE.1) THEN
          NXTPT = LFR(1)
          CALL REMOVE(NXTPT,1)
          CALL SETA(NXTPT,1,FLOAT(N))
          CALL SCHED(NXTPT,2,UN(2,1))
      ELSE
          X(N) = 0.0
      ENDIF
      RETURN
      END
```

FIGURE 13.7 SIMAN model frame.

CREATE is called to create an entity record representing the part that enters the system at this time. A test is made to determine whether both machines are busy. If they are, the PART is inserted into file (queue) 1. Otherwise the value of $X(1)$ is

```
   BEGIN;
10 PROJECT,MACHINE CENTER,BANKS,5/16/96;
20 DISCRETE,20,,1;
30 PARAMETERS:1,0,10:2,4,12;
40 COUNTERS,1,PARTS COMPLETED,100;
50 DSTAT:1,NQ(1),PARTS QUEUE:2,X(1),MACH.#1 UTILIZ.:3,X(2),MACH.#2 UTILIZ.;
60 END;
```

FIGURE 13.8 SIMAN experiment frame.

tested to determine the status of machine 1. If machine 1 is available, attribute 1 of PART is set to 1.0, as is the local variale N. If machine 1 is busy, attribute 1 is set to 2.0 (i.e., machine 2 is free), as is N.

Subroutine DEPART contains the logic for the departure event. The value of N will be the machine number on which PART just completed service. The entity PART is then DISPOSed. The counter (used to terminate the run) is incremented by 1 with a call to subroutine COUNT. Then a test is conducted to see whether there are additional PARTs waiting in the queue. If one or more PARTs are waiting, the variable NXTPT is set to the location of the first record in file 1. NXTPT is removed from file 1, attribute 1 of NXTPT is set to the current machine number, and a departure event is scheduled for NXTPT. If no PARTs are waiting, $X(N)$ is set to zero to indicate that machine N is idle.

Finally, we display the experiment frame in Fig. 13.8. The PROJECT element gives a name to the model and specifies the analyst and the date. The DISCRETE element indicates that at most 20 parts will be in the system at one time and that we have specified one file (queue). There are two PARAMETER sets identified. The first is associated with the time between arrivals, and the second is associated with the time required for parts to go through a machine. (In line 30, the element "1,0,10" means that parameter set 1 has values 0 and 10. The 0 and 10 are the endpoints of the uniform distribution of arrivals.) Line 40 limits the number of PARTs completed to 100. The desired time-persistent statistics are specified in the DSTAT element. In this example we are requesting the average number in the queue and the utilization of each machine.

Second Problem Statement. To further facilitate comparison among various language implementations, consider the following simple exercise from Pritsker (1986). A bank has a drive-in teller whose service time is distributed exponentially with a mean time of 0.5 min. Customer interarrival times are also exponential with a mean of 0.4 min. Only one car can wait for the drive-in teller; that is, customers that arrive when one car is waiting balk from the system and do not return. Letting $p_n(t)$ denote the probability that n customers are in the system (queue plus service) at time t, we can derive the following:

$$\frac{dp_0(t)}{dt} = -2.5 \times p_0(t) + 2.0 \times p_1(t)$$

$$\frac{dp_1(t)}{dt} = 2.5 \times p_0(t) - 4.5 \times p_1(t) + 2.0 \times p_2(t)$$

$$\frac{dp_2(t)}{dt} = 2.5 \times p_1(t) - 2.0 \times p_2(t)$$

Assume that at $t = 0$ there are no customers in the system so that the teller is idle. Let $N(t)$ be the number of customers in the system at time $t = 0,2,4, \ldots,100$. The problem is to prepare a table and plot estimates of $p_i(t)$ and $E[N(t)]$ using DTSAV = DTPLT = 0.05 min. (DTSAV is the frequency at which data values are recorded, and DTPLT is the printing increment.)

Solution in SLAM. It is not possible to explain all of the prerequisite material and generally prepare the reader to solve a continuous simulation problem in such limited space as is available. However, we will provide the code for solving the above problem and explain the purpose of the various statements in the solution. There are two portions of the code. The first consists of the subroutine STATE (Fig. 13.9). STATE is called to obtain updates for the state variables that define the condition of the system over time. When specified accuracy requirements are met, STATE is no longer called. STATE is written in Fortran. Thus the user must understand such standard programming languages in order to use the continuous (and event scheduling, but not process interaction) approach of SLAM.

The values of SLAM variables are transferred between the SLAM processor and user-written subroutines through the COMMON block. Therefore, any user-written continuous (or event scheduling) subroutine that references a SLAM variable must contain the COMMON block shown in Fig. 13.9. The SLAM variable SS(I) is used to represent state variable I. The derivative of state variable I is defined by the SLAM variable DD(I), that is, DD(I) $\equiv d$SS(I)/dt. In the code of Fig. 13.9, SS(I) represents the estimated probability of $I - 1$ customers in the system. The translation of the differential equations given in the problem statement into SLAM differential equations is direct as indicated by DD(1), DD(2), and DD(3). SS(4) is the estimated expected number of customers in the system at the time subroutine STATE is called.

The input statements for the program are given in Fig. 13.10. (Note that all SLAM statements end with a semicolon.) The GEN statement in Fig. 13.10 identifies the analyst, names the project, gives a date, and specifies the number of runs to be made. (The generic version of these control statements allows for much more complexity.) The CONTINUOUS statement indicates that there are three differential equations; there is one state variable [SS(4)] that can be defined by state equations; the minimum and maximum step sizes are 0.005 and 0.05, respectively; the error checking is defaulted (indicated by two commas) to a warning message; and the absolute and relative truncation errors are 0.001. The INTLC statements provide initial (starting) conditions. We see that SS(1) is set to 1, which prescribes a probability of 1 that there are zero customers initially in the system; that is, the system is starting empty and idle.

```
      SUBROUTINE STATE
      COMMON/SCOM1/ATRIB(100),DD(100),DDL(100),DTNOW,II,MFA,MSTOP
     1,NCLNR,NCRDR,NPRNT,NNRUN,NNSET,NTAPE,SS(100),SSL(100),TNEXT
     2,TNOW,XX(100)
      DD(1) = - 2.5*SS(1) + 2.*SS(2)
      DD(2) = 2.5*SS(1) - 4.5*SS(2) + 2.*SS(3)
      DD(3) = 2.5*SS(2) - 2.*SS(3)
      SS(4) = SS(2) + 2.*SS(3)
      RETURN
      END
```

FIGURE 13.9 SLAM subroutine STATE.

```
GEN,TEAM,SLAM CONTINUOUS,5/16/96,1;
CONTINUOUS,3,1,.005,.05,,.001,.001;
INTLC,SS(1)=1,SS(2)=0,SS(3)=0;
RECORD,TNOW,TIME,0,B,.05;
VAR,SS(1),0,PR OF 0,0,1;
VAR,SS(2),0,PR OF 1,0,1;
VAR,SS(3),0,PR OF 2,0,1;
VAR,SS(4),N,NUMBER IN SYS,0,2;
INIT,0,5;
FIN;
```

FIGURE 13.10 SLAM input statements for continuous example.

The RECORD statement provides general information concerning the values to be recorded at save times. In Fig. 13.10, we see that the independent variable is TNOW (the current time), and its identification is TIME. The 0 indicates that the values are to be stored in SLAM reserved storage areas (NSET/QSET). The B indicates that both a plot and a table are to be printed. Lastly, the printing increment is 0.05.

There are four VAR statements. These are used in conjunction with the RECORD statement to define the dependent variables that are to be recorded for each value of the independent variable. As an example, consider the last VAR statement. SS(4) is the dependent variable whose symbol is N and whose identification is NUMBER IN SYS. The statement indicates that it is not possible to have less than zero or more than two in the system at any time.

The INIT, or INITIALIZE, statement specifies that the beginning time of the simulation is zero and the ending time of the simulation is 5 (minutes). The FIN statement denotes an end to all SLAM input statements.

13.3 USE OF RANDOM NUMBERS

13.3.1 Random Number and Random Variate Generation

Stochastic simulations require a source of random numbers with which to generate sample values representing random quantities in the model. These random quantities are called random variates and typically are defined in terms of a probability distribution. Examples of *random variates* include random interarrival times, service times, processing times, repair times, and time until failure of an item.

All major simulation languages have built-in random number generators (r.n.g.'s) that produce upon demand a sequence of (supposedly) independent and identically distributed (i.i.d.) uniform (0,1) random numbers. The r.n.g.'s in the major languages have been extensively tested and no evidence has been found to reject the presumed uniformity and independence of their resulting random numbers. Before using a language, a modeler should ascertain whether the built-in r.n.g. has been adequately tested. Unfortunately, defective generators have not been removed from all computer systems. One example is the infamous RANDU, which still resides on some old IBM systems. Another is the r.n.g. in interpreted BASIC on some PCs using DOS 2.x or lower (Hultquist, 1984). Both of those generators are known to produce poor sequences that exhibit nonuniformity when r-tuples $(R_{i+1}, R_{i+2}, \ldots, R_{i+r})$ are "plotted" in r space.

Many r.n.g.'s in use today are based on the *linear congruential method*, for which $R_i = X_i/m$, where

$$X_i = (aX_{i-1} + c)\text{mod}\{m\}$$

a, c, m are integers, and the "seed" X_0 is a positive integer. (See Knuth, 1969, for a detailed explanation.) The number sequence produced by a given r.n.g. only simulates randomness; in fact, it is deterministic. The modeler specifies a seed value (usually a positive integer, sometimes odd, depending on the type of generator) that completely determines the sequence. For example, one popular, easy-to-use linear congruential generator takes $a = 16807$, $c = 0$, and $m = 2^{31} - 1$. (See Bratley, Fox, and Schrage, 1987, for implementation details.)

Every r.n.g. has a period P that gives the number of random numbers generated before the sequence repeats itself. Periods for good generators are at least on the order of $P = 2^b$, where b (or $b + 1$) is the number of bits in one computer word. 32-bit computers (e.g., IBM mainframes) yield periods on the order of $P = 2^{31} > 2$ billion. In most applications that require fewer than 2^{31} random numbers, the ultimate repeatability of a sequence is of no practical consequence.

The importance of a good source of uniform random numbers lies in the fact that all procedures for generating nonuniform random variates involve a mathematical transformation of the uniforms. The most popular random variate generation procedure is the *inverse X transform* technique; this method produces a realization of a random variable X by setting $X = F^{-1}(R)$, where R is uniform (0,1) and $F^{-1}(\cdot)$ is the inverse cumulative distribution function (c.d.f.) of X. (For details, see Banks, Carson, and Nelson, 1996.)

For example, to generate an exponential random variable with mean λ, we note that the c.d.f. is $F(x) = 1 - e^{-x/\lambda}$, $x \geq 0$, and then set

$$X = -\lambda \ell\text{n}(1 - R) \; [\text{or} - \lambda \ell\text{n}(R)]$$

As another example, if $\Phi(x)$ is the standard normal c.d.f., then

$$X = \Phi^{-1}(R)$$

can be used to generate samples from the standard normal distribution. Of course, one must be able to evaluate the standard normal's inverse c.d.f. (either numerically or via table lookup); but a reasonable approximation is given by Schmeiser (1979):

$$\Phi^{-1}(R) \approx \frac{R^{0.135} - (1 - R)^{0.135}}{0.1975}$$

Other techniques for generating nonuniform random variates include convolution, acceptance-rejection, and composition (see Law and Kelton, 1991, and Bratley, Fox, and Schrage, 1987).

The major simulation languages have built-in random variate generators for the most widely used distributions, including the exponential, normal, gamma, Weibull, Poisson, geometric, binomial, and negative binomial. Most also provide a facility for generating samples from empirical distributions, or distributions defined by a table. Fortran scientific libraries such as IMSL (1987) contain numerous random variate generation routines.

In all but rare instances, a modeler needs to know how to use r.n.g.'s and random variate generators properly, but not how to develop them. The use of r.n.g.'s to duplicate experimental conditions in a technique called common random numbers is explained in Sec. 13.5.4.

13.3.2 Choosing a Statistical Distribution to Represent an Input Variable

For each random element in a system being modeled, the modeler must decide upon a way to represent the associated random variables. The techniques used may vary depending on:

- The amount of available data.
- Whether the data are "hard" or someone's best guess.
- Whether each variable is independent of other input random variables, or related in some way to other inputs.

In the case of a univariate variable independent of other variables, the choices to the modeler range over the following:

- Assume that the variable is deterministic.
- Fit a probability distribution to the available data.
- Use the empirical distribution of the data.

The choices are discussed further in the following subsections.

Assuming Randomness Away. Some modelers may be tempted to assume that a variable is deterministic, that is, to use a constant value in the model, possibly obtained by averaging the few (if any) available observations or by taking someone's best guess. If the variable does vary randomly, this is usually a poor choice and can result in an invalid model.

As an example, consider a single machine whose nominal processing rate is $\mu =$ 1000 parts per hour. However, the machine is subject to stoppages and requires an adjustment before production can proceed. The mean time to failure (or run time until a stoppage occurs) is 11 min, while the average time to adjust the machine is 1 min. Thus, on the average, the machine will experience five 12-min cycles during each hour, each cycle consisting of a runtime (on average 11 min) and a downtime (on average 1 min). Some modelers may be tempted to model the machine at a constant processing rate of

$$\mu_e = 1000\left(\frac{11}{12}\right) + 0\left(\frac{1}{12}\right) = 916.7 \text{ parts per hour}$$

with no random downtimes occurring in the model. This approach is certainly simple, but it does not result in a valid model. The proper approach is to model the machine with random downtimes; that is, the machine should run at 1000 parts per hour during run times and zero parts per hour during downtimes. This on-off behavior will have different consequences for overall system performance than will the simplistic assumption of a constant process rate of 916.7 parts per hour. To model the random downtimes properly, one of the approaches discussed in the next three subsections should be followed.

Fitting a Distribution to Data. Fitting a probability distribution to available data may be appropriate when there is a large amount of data, say at least 100 independent observations, or when the nature of the data suggests a certain probabilistic form. For instance, if we wish to model random arrivals to a facility, or random times to failure of a large class of machines or components, then theoretical considerations plus much past experience suggest that a Poisson model might be appropriate. In these cases, the times between arrivals and times between failures tentatively are modeled as arising from exponential distributions.

Nevertheless, a great deal of data is usually needed to discriminate among several distributional forms. With a small amount of data, it is not unusual for a goodness-of-fit test to fail to reject several candidate distributions. In such cases, the goodness-of-fit test offers no guidance in choosing an appropriate distribution. If it is decided to fit a distribution to data, we recommend the following three-step procedure:

1. *Hypothesize a candidate distribution.* First ascertain whether the variable is discrete or continuous. If discrete, possible candidate distributions include binomial, geometric, Poisson, and negative binomial. If continuous, candidates include uniform, exponential, normal, lognormal, gamma, and Weibull. The use of histograms and probability plots may be helpful when hypothesizing an appropriate distributional form for the data.
2. *Estimate the parameters of the hypothesized distribution.* For example, if the hypothesis is

$$H_0\text{: data are normal } (\mu, \sigma^2)$$

then μ and σ^2 must be estimated from the data.
3. *Perform a goodness-of-fit test such as the chi square test.* If the test rejects the hypothesis, that is a strong indication that the hypothesis is false. Return to step 1, or conclude that none of the standard distributions provide a good fit to the data. In the latter case, use of the empirical distribution is recommended.

Recently, a number of vendors have introduced software that is capable of fitting a wide variety of data to appropriate distributions. The interested reader should peruse Vincent and Law (1995) and Jankauskas and McLafferty (1995) for more details on the software products ExpertFit and BestFit, respectively.

Empirical Distribution of the Data. If only a small amount of data are available, then an attempt to fit a distribution is not appropriate because there is insufficient data to justify assuming one distributional form over another. In fact, with a small amount of data, a goodness-of-fit test might "accept" (i.e., fail to reject) *several* candidate distributions. It appears, under these circumstances, that use of the empirical distribution of the data may well be the only justifiable approach. For additional arguments supporting the use of empirical distributions (possibly with a fitted exponential tail), see Bratley, Fox, and Schrage (1987).

When No Data Are Available. Quite often hard data may not be available, especially in the early stages of a study, or when it is too expensive to gather data, or when the system being modeled simply does not exist in complete form. In some cases, personnel familiar with the real system may be able to provide subjective estimates of the minimum, maximum, mean, or most likely value of the random variable. For example, if operating personnel estimate that the time to repair a minor machine jam ranges from 1 to 8 min, then a crude but reasonable assumption is to make this repair time uniform (1,8). If further questioning reveals that the most repair times are about 2.5 min in duration, then perhaps a better assumption would be that the repair times are triangularly distributed with a minimum of 1 min, a mode of 2.5, and a maximum of 8. If personnel estimate that this same machine will jam at random about twice an hour but no hard data are available, a justifiable assumption would be that the time to failure is exponentially distributed with a mean time to failure (MTTF) of 30 min.

It is important for modelers to remember the tentative nature of the input data assumptions. If and when data (or more data) become available, the modeler should update input parameters such as the MTTF, and assumed distributional forms (e.g., uniform, triangular) should be changed to reflect the data more accurately. In addition, if experimentation with a model through sensitivity analysis reveals that the performance measures (outputs) of greatest importance are highly sensitive to the value of an input parameter such as the MTTF of a machine, then every effort should be made to collect more data so that the input parameter can be estimated as accurately as possible.

13.4 VERIFICATION AND VALIDATION

In the application of simulation, the real-world system under investigation is abstracted by a *conceptual model*—a series of mathematical and logical relationships concerning the components and structure of the system. The conceptual model is then coded into a computer recognizable form (i.e., an *operational model*), which the simulationist hopes is an accurate imitation of the real world system. The accuracy of the simulation must be checked before the simulationist can make valid conclusions based on the results from a number of runs. This checking process consists of two main components:

1. *Verification:* Here we determine whether the computer implementation of the conceptual model is correct. Does the computer code represent the model that has been formulated?

2. *Validation:* Here we determine whether the conceptual model is a reasonable representation of the real world system.

This checking process is iterative. If there are discrepancies among the operational and conceptual models and the real world system, the relevant computer code must be examined for errors, or the conceptual model must be modified in order to better represent the real world system. The verification and validation process should then be repeated.

13.4.1 Verification

The verification process involves the debugging of the simulation code so that the conceptual model is accurately reflected by the operational model. The following are a number of commonsense suggestions that can be used in the verification process:

1. Write the simulation program in a logical, well ordered manner. Make use of detailed flowcharts when writing the code.
2. Make the code as self documenting as possible. Define all variables and state the purpose of each section of the program.
3. Have the computer code checked by more than one person.
4. Check to see that the values of the input parameters have not been changed inadvertently during the course of a simulation run.
5. For a variety of input parameter values, examine the output of simulation runs for reasonableness.
6. Use traces to check that the program performs as intended.

13.4.2 Validation

A variety of subjective and objective techniques can be used to validate the conceptual model. Naylor and Finger (1967) and Sargent (1984) describe a validation process that includes the following procedure.

Face Validity. A conceptual model of a real world system must appear to be reasonable "on its face" to those who are knowledgeable about the real world system. Thus it is sound practice to have experts examine the assumptions or the mathematical relationships of the conceptual model for correctness. Such a critique by experts would be of aid in identifying any deficiencies or errors in the conceptual model. The credibility of the conceptual model would then be enhanced as these deficiencies or errors are corrected during the iterative verification and validation process.

If the conceptual model is not overly complicated, additional methods can be used to check face validity. Sargent (1984) suggests that a manual trace of the conceptual model be conducted. Banks, Carson, and Nelson (1996) point out that the modeler may be able to perform elementary sensitivity analysis by varying selected "critical" input parameters and observing whether the model behaves as expected.

Validation of Model Assumptions. We consider two types of model assumptions—structural assumptions (i.e., assumptions concerning the operation of the real-world system) and data assumptions. Structural assumptions can be validated by observing the real world system and by discussing the system with the appropriate personnel. For instance, we could make the following structural assumptions about the queues that form in the customer service area at a bank. Patrons form one long line, with the person at the front of the line receiving service as soon as one of the tellers becomes idle; a customer might leave the line if the customers in line are moving too slowly; a customer seeing 10 or more patrons in the system may decide not to join the line.

Assumptions concerning the data that are collected may also be necessary. For example, consider the interarrival times at the above bank during peak banking periods. We might suppose that these interarrivals are i.i.d. exponential random variables. In order to validate these assumptions, we should proceed as follows.

1. Consult with bank personnel to determine when peak banking periods occur.
2. Collect interarrival data from these periods.

3. Conduct a statistical test to check that the assumption of independent interarrivals is reasonable.

4. Estimate the parameter of the (supposedly) exponential distribution.

5. Conduct a statistical goodness-of-fit test to check that the assumption of exponential interarrivals is reasonable.

Validating Input-Output Transformations. We can treat the conceptual model as a function that *transforms* certain input parameters into output performance measures. In the banking example, input parameters could include the distributional forms of the patron interarrival times and teller service times, the number of tellers present, and the customer queuing discipline. The average customer waiting time and server utilization might be the output performance measures of interest.

The basic principle of input-output validation is the comparison of output from the verified operating model to data from the real world system. Input-output validation requires that the real world system currently exist.

One method of comparison uses the familiar t test. (Nonparametric tests might also be appropriate.) To illustrate, suppose we collected data from the bank under study, and the average customer service time during a particular peak banking period was 2.50 min. Further suppose that five independent simulation runs of this banking period were conducted (and that the simulations were all initialized under the same conditions). The average customer service times from the five simulations were 1.60, 1.75, 2.12, 1.94, and 1.89 min, respectively. We would expect the simulated average service times to be consistent with the observed average service time. Therefore, the hypothesis to be tested is

$$H_0: \quad \text{E}[X_i] = 2.50 \text{ min}$$

versus

$$H_1: \quad \text{E}[X_i] \neq 2.50 \text{ min}$$

where X_i is the random variable corresponding to the average customer service time from the ith simulation run.

Define $\mu_0 \equiv 2.50$ (= $\text{E}[X_i]$ under H_0), $n = 5$ (the number of independent simulation runs), $\bar{X} \equiv \Sigma_{i=1}^{n} X_i/n$, and $S^2 \equiv \Sigma_{i=1}^{n}(X_i - \bar{X})^2/(n-1)$. By design and a central limit theorem, the X_i's are approximately i.i.d. normal random variables. So $t_0 \equiv (\bar{X} - \mu_0)/(S/\sqrt{n})$ is approximately a t random variable with $n - 1$ degrees of freedom.

We reject H_0 if $|t_0| > t_{\alpha/2,n-1}$, where $t_{\alpha/2,n-1}$ is the $(1 - \alpha/2)$ quantile of the t distribution with $n - 1$ degrees of freedom and where the experimenter must specify $\alpha = \Pr(\text{reject } H_0 | H_0 \text{ true})$. For the example at hand, $\bar{X} = 1.86$, $S^2 = 0.0387$, and $t_0 = -7.28$. Taking $\alpha = 0.05$, t tables give $t_{0.025,4} = 2.78$. Hence H_0 is rejected. Since α is very small, the rejection of H_0 is a "strong" conclusion. This suggests that our operational model does not produce realistic customer service times and that changes in the conceptual model or computer code may be necessary (in which case another iteration of the verification-validation process would be required). The reader should refer to Balci and Sargent (1981, 1982) for additional discussion of the use of hypothesis testing in validation.

Suppose we have validated the conceptual model (and verified the associated simulation code) of the existing real world system, so we can say that the simulation adequately mimics the real world system. If we wish to compare the real world system to nonexisting systems with alternative designs or with different input parameters, it is necessary that the conceptual model (and associated code) be somewhat *robust*. That is, assume that some nonexisting system of interest and our conceptual model have only *minor* differences. We should be able to make small modifications in our operational model and then use this new version of the code to generate valid output performance values for the nonexisting system. Such minor changes might involve certain numerical input parameters (e.g., the customer interarrival rate) or the form of a certain statistical distribution (e.g., the service time distribution). Unfortunately, it may be quite difficult to validate the model of a nonexisting system if it differs substantially from the conceptual model of the real world system.

Validation Using Historical Input Data. Instead of running the operational model with artificial input data, we could drive the operational model with the actual historical record. It would then be reasonable to expect the simulation to yield output results very close to those observed from the real-world system.

As an example, suppose that we have collected interarrival and service time data from the bank during n independent peak periods. Let W_j denote the observed average customer waiting time from the jth peak period, $j = 1, \ldots, n$. For fixed j, we can drive the operational model with the actual interarrival and service times to get the (simulated) average customer waiting time Y_j. We would hope that $D_j \equiv W_j - Y_j \approx 0$ for all j. If we are willing to assume that the D_j's are approximately i.i.d. normal random variables, then a paired t test can be used to test H_0: $E[D_j] = 0$.

For instance, say that we collect the following i.i.d. normal observations:

W_j	16.5	17.3	16.9	14.3	17.2
Y_j	17.1	17.6	17.0	14.9	16.6
D_j	−0.6	−0.3	−0.1	−0.6	0.6

We define $\mu_0 \equiv 0$ (= $E[D_j]$ under H_0), $n = 5$ (the number of pairs of observations), $\bar{D} \equiv \Sigma_{j=1}^{n} D_j/n$, $S_D^2 \equiv \Sigma_{j=1}^{n} (D_j - \bar{D})^2/(n-1)$, and $t_0 \equiv (\bar{D} - \mu_0)/(S_D/\sqrt{n})$. The hypothesis H_0 is rejected if $|t_0| > t_{\alpha/2,n-1}$. In the current example, $\bar{D} = -0.2$, $S_D^2 = 0.2450$, and $t_0 = -1.825$. For $\alpha = 0.05$, recall that $t_{0.025,4} = 2.78$. So we fail to reject H_0 at level 0.05.

13.5 OUTPUT ANALYSIS

13.5.1 Introduction

If the input processes driving a simulation model are specified as being random, then the output produced from running the simulation should be considered as a statistical sample. Therefore, a run (or even many runs) of the model does not directly yield the desired measure of system performance, but rather an estimate of the measure of system performance. This estimate is subject to sampling error. Any inferences made concerning system behavior should be given with some indication of their reliability. The purpose of this section is to give a broad overview of the design and analysis of simulation experiments and to outline some of the available output analysis techniques.

To facilitate our discussion, we distinguish between two types of simulations with respect to output analysis:

1. *Terminating, or transient, simulations:* In this case, the nature of the problem defines the simulation run length.
2. *Nonterminating, or steady-state, simulations:* Of interest here is the long-run or steady-state behavior of the system. It is presumed that the probability law governing system behavior approaches a "steady state," independent of initial conditions.

Examples of terminating simulations include any facility that closes down after a fixed period of time, such as a bank or other commercial facility. In addition, a terminating simulation may be used to study the startup of a factory or the effects of some extraordinary event. In such cases, the simulation run length is determined by the nature of the problem and is not a matter of the modeler's choice. Examples of nonterminating simulations include continuously running production lines and hospitals for which the modeler is interested in some long-run measure of system performance.

Numerous techniques have been developed, or adapted from classical statistics, for dealing with statistical problems inherent in stochastic simulations. For terminating simulations, all techniques are based on the method of *independent replications*; see Secs. 13.5.3 and 13.5.4. Additional tactical problems arise with respect to nonterminating simulations. For instance, in Sec. 13.5.2, we discuss the simulation *startup* policy (i.e., initialization of the system state at time zero), the *truncation* policy (i.e., determination of the point at which data are retained for analysis), and the specification of the run length needed to reach approximate steady state. If startup and truncation are unwittingly ignored by the modeler, then initialization bias is introduced into statistical estimators of steady-state system performance. Finally, Sec. 13.5.3 gives methods for estimating confidence intervals for simulation output parameters, and Sec. 13.5.4 concerns the important problem of comparing two or more alternative system designs.

13.5.2 Initialization Bias

A difficult simulation problem concerns the treatment of *initialization bias*. Before the simulation can be run, one must provide initial values for all state variables. Since the simulationist typically does not know what initial values are appropriate for the state variables, these values are usually chosen somewhat arbitrarily. Such a choice of initial conditions can have a significant but unrecognized influence on the outcome of the simulation experiment. Thus the initialization bias problem can lead to errors.

Although a great deal of the simulation literature is devoted to the topic of initialization bias, there is no definitive solution to the problem. In fact, there are a number of different levels on which to consider the initialization bias problem. On one hand, suppose that the simulated process of interest $Y_1, \ldots, Y_n$ has cumulative distribution function F_t at time t and that the steady-state distribution function is $F \equiv \lim_{t \to \infty} F_t$. The goal of the simulation study might be to estimate some parameter of F (e.g., the expected value). It is clearly desirable that any such estimate be based on observations from F. So we could say that initialization bias exists in the process if the F_t's do not equal the limiting distribution function F. On a more practical level, we might wish to use a realization of $Y_1, \ldots, Y_n$ in order to make certain decisions. Suppose that we could also observe a realization of $Y_{1+T}, \ldots, Y_{n+T}$ (where T is very large). If the corresponding decisions arising from the former and latter samples are significantly different, then we could again say that initialization bias exists.

Some Examples. Schruben and Goldsman (1985) give examples of problems (several of which are paraphrased below) that can arise from simulation initialization.

1. *Visual detection of initialization effects*: It is sometimes quite difficult to detect the presence of initialization effects simply by looking at process output. Figure 13.11 plots successive customer waiting times from a simulated queuing system initialized with no customers in the system. The initial portion of the output does not appear to be greatly different from the latter portion of the output. However, the average waiting time for the first 50 customers is 2.9 min, whereas the average waiting time for the first 500 customers is fully 4.4 min. (The discrepancy is due to the fact that the empty queue initialization allowed early customers to proceed relatively rapidly through the system.)
2. *Initial demand*: Suppose a machine shop closes at exactly 5:00 P.M. each afternoon, even if there are jobs waiting in line to be served. These unserved jobs are

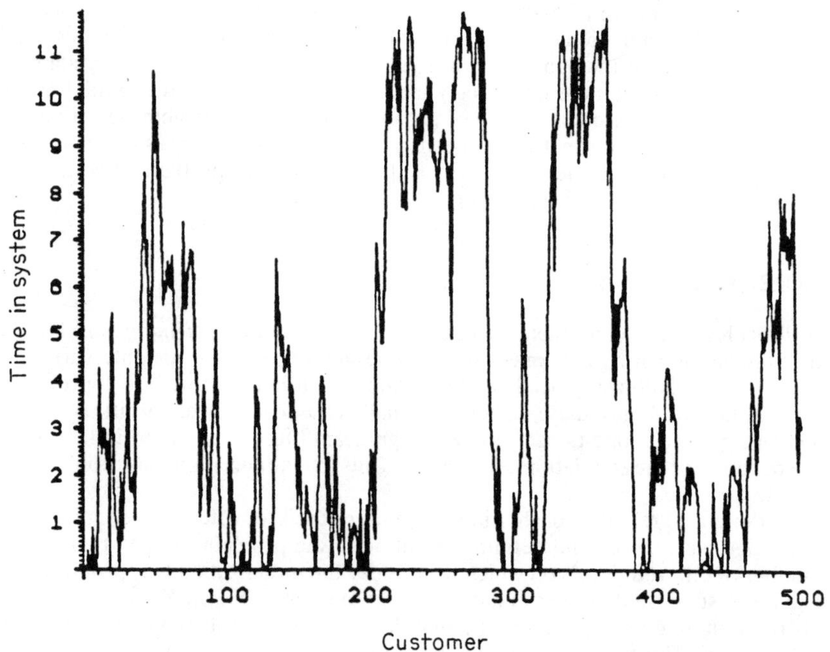

FIGURE 13.11 Initialization bias is sometimes difficult to detect visually.

still in line when the shop reopens the next morning. A simulation model of the shop should start each day with an initial demand that is dependent on the number of unserved jobs from the end of the previous day.

3. *Estimator bias*: Consider the simple first-order autoregressive process

$$Y_k = \mu + \alpha(Y_{k-1} - \mu) + \epsilon_k \quad k = 1, 2, \ldots$$

where $\mu \equiv \lim_{k \to \infty} E[Y_k]$ is the steady-state process mean, $|\alpha| < 1$, and $\{\epsilon_k\}$ is a sequence of i.i.d. normal $(0,1 - \alpha^2)$ random variables. Suppose that we wish to estimate μ and that we initialize $Y_0 = 0$. Then it can be shown that $E[Y_k] = \mu(1 - \alpha^k)$; the early Y_k's are rather biased as estimators of μ.

4. *Estimator mean squared error*: Say that $\hat{\theta}$ is a point estimator for some parameter θ. The mean squared error (MSE) of $\hat{\theta}$ combines the bias and variance of $\hat{\theta}$ into one measure:

$$\text{MSE} \equiv (E[\hat{\theta}] - \theta)^2 + \text{Var}(\hat{\theta})$$

It is desirable for an estimator to have low MSE. Fishman (1972) investigates the effects of initialization on the MSE of the sample mean (as an estimator of μ) for the autoregressive process described previously. He concludes that the use of *data truncation* (i.e., allowing the simulation to "warm up" before data are retained for analysis) is sometimes detrimental from an MSE point of view. Other relevant papers concerning MSE are Turnquist and Sussman (1977), Kelton and Law (1984), and Snell and Schruben (1985).

5. *Confidence interval estimators*: There are a number of methods used to calculate confidence intervals for the steady-state mean $\mu \equiv \lim_{k \to \infty} E[Y_k]$ of an arbitrary process $Y_1, \ldots, Y_n$. Specific methods are discussed in the next subsection, and we shall see that such approximate $100(1 - \alpha)\%$ confidence intervals are usually of the form

$$\bar{Y} \pm t_{\alpha/2} \sqrt{\frac{V}{n}}$$

where $\bar{Y} \equiv \sum_{k=1}^{n} Y_k/n$ is a point estimator for μ, V is a point estimator for the *variance parameter* $\sigma^2 \equiv \lim_{n \to \infty} n\text{Var}(\bar{Y})$, and $t_{\alpha/2}$ is an appropriate t distribution quantile.

Initialization bias can cause these confidence interval estimators to perform poorly. For instance, if $\bar{Y}$ is a significantly biased estimator of μ, then the above confidence interval estimator is incorrectly centered. Bias of V as an estimator of σ^2 may also be detrimental. If $E[V] \ll \sigma^2$, the probability that the confidence interval estimator actually covers μ might be considerably less than the nominal value $1 - \alpha$. Confidence intervals that are too wide to be of practical use might result in the case that $E[V] \gg \sigma^2$.

Detection of Initialization Bias. A number of methods to detect the presence of (significant) initialization bias in a simulated process have been proposed in the literature. We survey some of these methods below.

Visual Methods. The simplest means of detecting initialization bias involves visually scanning a raw realization of the simulated process under scrutiny. Of course, as illustrated previously, a naive visual analysis can often allow bias to escape notice. Further, a simulation study may require the experimenter to make a large number of simulation runs; it can then be excessively tedious for the simulationist to methodically scan such voluminous output for bias. Suggestions for facilitating visual analysis include:

- Transform the output (e.g., take square roots or logs).
- Smooth the output (see Sargent, 1979; Welch, 1981, 1983).
- Average observations across several simulation runs (i.e., average the first observation from each run, average the second from each run, etc.).
- Construct CUSUM plots (see Schruben, 1982).

Statistical Tests. Various tests for initialization bias have been developed in the literature. For instance, Kelton and Law (1983) give a sequential test that proceeds roughly as follows:

1. A number of independent simulation runs are made, and the output is averaged across the runs. The across-averaged output should be smoother than that from an individual run.
2. The across-averaged output is then divided into nonoverlapping, contiguous batches (see Sec. 13.5.3), and the last few *batch means* (the sample means from each batch) are tested for stationarity, that is, are the batch means identically distributed? If the test does not indicate stationarity, we take additional observations from the original independent simulation runs (i.e., we extend the original simulation runs) and repeat step 2.
3. If all goes well, this test will eventually indicate stationarity of the last few batch means. We then sequentially test to see whether any of the earlier batches can be included in the *stationary* group.

When the testing procedure terminates, we retain for further analysis only those observations from the independent simulation runs that correspond to the stationary group of batches.

The Kelton and Law procedure is intuitively appealing. Indeed, empirical results concerning this testing procedure have been encouraging. However, it should be noted that the user of the procedure is required to prespecify several parameters before the procedure can be started. (Note that Kelton and Law suggest setting the level of each test at 0.5, a seemingly high value.) Lastly, we point out that this procedure is not designed for use in certain situations; one should consult Kelton and Law (1983). A more sophisticated procedure is given in Fishman (1996).

Schruben (1982), Schruben, Singh and Tierney (1983), and Goldsman, Schruben, and Swain (1994) propose other tests for initialization bias. The tests all operate under the assumption that the simulated process $Y_1, \ldots, Y_n$ can be modeled as

$$Y_i = \mu_i + X_i \qquad \text{for all } i$$

where the X_i's are stationary random variables with mean 0. Specifically, we test H_0: $\mu_1 = \mu_2 = \cdots = \mu_n$. If H_0 is rejected, we shall say that there is initialization bias present.

The idea behind these tests can be motivated as follows. Partition a realization of $Y_1, \ldots, Y_n$ into two nonoverlapping intervals. Then calculate a variance parameter estimate based solely on the first portion of the output and a variance parameter estimate based solely on the latter portion. We use a variant of the standard F test to see whether the two variance parameter estimates are statistically significantly different; if so, H_0 is rejected. We caution that the tests require large sample sizes in order to be valid, but empirical results to date reveal that the procedures perform as intended.

Common Treatment of Initialization Bias. Once the presence of initialization bias is detected, the simulationist may wish to ameliorate its effects. Two general methods for dealing with initialization bias are suggested in the literature: namely, truncation and long runs.

Output Truncation. Many experimenters allow their simulations to "warm up" before data are collected. This practice is called *output truncation*. It is hoped that the data retained for analysis are not affected significantly by any initialization bias and are representative of the steady-state system under study. Because of its simplicity, output truncation is probably the most popular method for dealing with the initialization bias problem. In fact, many simulation languages have built-in functions to facilitate truncation. For example, the RESET command in GPSS allows the collection of data only after a certain number of transactions have passed through the system or only after a fixed period of simulated time.

The main difficulty concerning output truncation lies in determining the data truncation point. If the output is truncated "too early," then significant initialization bias might still be present in the remaining data; if the output is truncated "too late," then "good" observations are wasted. A number of simple truncation rules have been proposed (see the surveys in Gafarian, Ancker, and Morisaku, 1978, and Wilson and Pritsker, 1978*a*,*b*). Unfortunately, these rules are often naive and do not perform well in general. Current practice seems to call for the experimenter to conduct several independent runs of the simulation, average the observations across these runs, and then visually choose a truncation point based on the averaged run. An explicit visual/graphic approach is described in Welch (1983).

As alluded to previously, Kelton and Law (1983) and Fishman (1996) propose more sophisticated approaches for controlling the effects of initialization bias. One could also adapt the hypothesis testing procedures from Schruben (1982), Schruben, Singh, and Tierney (1983), or Goldsman, Schruben, and Swain (1994) to *sequentially* truncate data until the null hypothesis of no initialization bias is accepted (see Heidelberger and Welch, 1983).

Making a Long Run. We might attempt to overwhelm any initialization bias effects by conducting an extremely long simulation run. In this case, we retain *all* of the data for analysis and hope that the influence of the bias will wash out as the run becomes of great length. This method of bias control has various advantages. Clearly, it is simple to carry out. Further, Fishman (1972) and others show that under certain specific conditions, a nontruncated process yields point estimators for the steady-state mean that have lower MSEs than do corresponding estimators from the truncated process. Nevertheless, the disadvantages of making one long run militate against its use as a bias control technique. This method can obviously be very wasteful with observations. It can also be demonstrated that for some systems a prohibitively excessive simulation run length might be required before the initialization effects become negligible.

13.5.3 Confidence Interval Estimators

When a simulation is run, we garner information concerning certain system characteristics. Since a simulation is usually driven by "random" numbers, the simulation will most likely give a different point estimate for the parameters of interest (e.g., the process mean or variance parameter) each time the simulation is run. Because of this variability, an experimenter might want to calculate *confidence intervals* (CIs) for the parameters of interest.

Over the last 30 years, several general methodologies for conducting discrete-event simulation output analysis have been proposed. We shall examine the problem of CI estimation for steady-state parameters by discussing the two simplest methodologies: *batch means* and *independent replications*. CIs for parameters from *terminating* simulations are usually calculated via the method of independent replications.

Batch Means. The batch means methodology is frequently used to calculate CIs for the mean μ of a stationary process $Y_1, \ldots, Y_n$. Suppose $Y_1, \ldots, Y_n$ is partitioned into b nonoverlapping, contiguous batches of m Y_k's each (assume that $n = bm$). So the ith batch consists of the random variables $Y_{(i-1)m+1}, Y_{(i-1)m+2}, \ldots, Y_{im}$, $i = 1, \ldots, b$. The ith batch mean is defined as the average of the m Y_k's from the ith batch,

$$\bar{Y}_i \equiv \frac{1}{m} \sum_{k=1}^{m} Y_{(i-1)m+k} \qquad i = 1, \ldots, b$$

The batches and associated batch means are as follows:

$$\underbrace{Y_1, Y_2, \ldots, Y_m}_{\text{Batch 1}} \quad \underbrace{Y_{m+1}, Y_{m+2}, \ldots, Y_{2m}}_{\text{Batch 2}} \quad \cdots \quad \underbrace{Y_{(b-1)m+1}, Y_{(b-1)m+2}, \ldots, Y_{bm}}_{\text{Batch } b}$$

$$\downarrow \qquad\qquad \downarrow \qquad\qquad \downarrow$$

$$\bar{Y}_1 \qquad\qquad \bar{Y}_2 \qquad \cdots \qquad \bar{Y}_b$$

In order to use the method of batch means, we must implicitly assume that the batch means are approximately i.i.d. normal random variables with unknown mean μ and variance. We further define

$$\bar{Y} \equiv \frac{1}{n} \sum_{k=1}^{n} Y_k = \frac{1}{b} \sum_{i=1}^{b} \bar{Y}_i$$

and

$$V \equiv \frac{m}{b-1} \sum_{i=1}^{b} (\bar{Y}_i - \bar{Y})^2$$

as point estimators for the mean μ and the variance parameter $\sigma^2 \equiv \lim_{n \to \infty} n\,\mathrm{Var}(\bar{Y})$, respectively. Then standard statistical manipulations yield the following approximate $100(1 - \alpha)\%$ CI estimator for μ:

$$\bar{Y} \pm t_{\alpha/2,b-1} \sqrt{\frac{V}{n}} \tag{13.1}$$

Example. Suppose that we divide a simulation run of 10,000 observations into five contiguous batches, each consisting of 2000 observations. Also suppose that we calculate the following batch means:

i	1	2	3	4	5
$\overline{Y}_i$	3.2	4.3	5.1	4.2	4.6

Then $\overline{Y} = 4.28$ and $V = 974.0$. Setting $\alpha = 0.05$, we have $t_{0.025,4} = 2.78$. This yields [3.41, 5.15] as a 95% CI for μ.

Although the batch means method is intuitively appealing and is easy to implement, problems can arise if

- The Y_k's are not stationary (e.g., if initialization bias is present).
- The batch means (or $\overline{Y}$ and V) are not independent.
- The batch means are not normally distributed.

If any of these problems exist, poor CI coverage may result. Recall from our discussion of initialization bias that there are two widely used techniques to alleviate the effects of any such bias: (1) output truncation and (2) making an extremely long run. In addition, the absence of independence or normality of the batch means can be counteracted somewhat by increasing the batch size m. For additional information concerning the method of batch means (including the choice of batch size), the reader should see Schmeiser (1982), Song and Schmeiser (1995), Fishman (1996), and Chien, Goldsman, and Melamed (1997).

Independent Replications. Of the various difficulties encountered when using the method of batch means, Law (1977) maintains that the possibility of correlation among the batch means appears to be the most deleterious. This problem is avoided by the use of independent replications, another popular method of calculating CIs for μ. We sample b independent runs or *replications* (of m observations each, say) from the simulation process, each of the replications being initialized by a different random number seed. We denote the random variable corresponding to the jth observation from the ith replication by $Y_{(i-1)m+j}$, $i = 1, \ldots, b$ and $j = 1, \ldots, m$:

$$\begin{array}{ll} \text{Replication 1:} & Y_1, Y_2, \ldots, Y_m \to \overline{Y}_1 \\ \text{Replication 2:} & Y_{m+1}, Y_{m+2}, \ldots, Y_{2m} \to \overline{Y}_2 \\ & \vdots \\ \text{Replication } b\text{:} & Y_{(b-1)m+1}, Y_{(b-1)m+2}, \ldots, Y_{bm} \to \overline{Y}_b \end{array}$$

The quantity $\overline{Y}_i$ denotes the sample mean from the ith replication, $i = 1, \ldots, b$. Note that the Y_k's are independent *between* replications but not necessarily *within* replications. CIs for μ are obtained by performing the calculations given in Eq. (13.1), with the batch means now replaced by the replicate means.

The replicate means are independent by construction, so the use of independent replications has considerable appeal; but this method is not without possible pitfalls. Initialization bias is more of a problem when using independent replications than when batch means are used since *each* of the b replication runs can contain bias (whereas only the single batch means run can have any bias). Further, the normality of the independent replicate means is not guaranteed.

Notice that as the replicate size $m \to \infty$, the methods of batch means and independent replications become approximately equivalent (assuming that the initialization and independence problems wash out). For "moderate" m, Law and Kelton (1984) regard the possibility of initialization bias in each of the replications as a serious problem and therefore recommend the use of batch means over replications.

Other Confidence Interval Methods. We briefly mention several other methods for obtaining CIs for the process mean μ.

ARMA Time Series Modeling. This method calls for the experimenter to fit a finite autoregressive moving average (ARMA) time series model to the output from a simulation run (see Fuller, 1976, for a discussion about ARMA processes). We then use properties of the fitted ARMA model in order to form CIs for μ. Fishman (1971, 1973, 1978) first developed such a time series modeling technique, and Schriber and Andrews (1984) and Yuan and Nelson (1994) later proposed more general procedures. Yuan and Nelson report promising results on a variety of simulation processes.

Spectral Estimation. The CI estimation methods outlined here all rely on obtaining good estimates of $\text{Var}(\bar{Y})$. The spectral method estimates this quantity in a completely different fashion than the aforementioned methods. Namely, this approach works in the so-called *frequency domain*, while those previously discussed work in the so-called *time domain*. The spectral estimation method is somewhat complicated to conduct, but it works well enough to suggest that the reader see Heidelberger and Welch (1981, 1983).

Overlapping Batch Means. Meketon and Schmeiser (1984) study the method of over lapping batch means. This method intentionally overlaps batches, and then uses spectral analysis to obtain variance estimators with about 50% more degrees of freedom than "regular" batch means. See Song and Schmeiser (1995) for additional results.

Regeneration. Crane and Iglehart (1975) and Crane and Lemoine (1977) comment that the output from many simulations is derived from i.i.d. blocks or groups; that is, the simulation "starts from scratch" probabilistically at certain regeneration points. For example, consider an M/M/1 queue's waiting time process that periodically returns to zero; the blocks whose endpoints are defined by waiting times equal to zero are i.i.d. (see Fig. 13.12). The regeneration method exploits this i.i.d. structure and, under certain conditions, produces excellent CIs.

Advantages of this method include:

- After some study, it is easy to understand and implement.
- The initialization bias problem is virtually nonexistent. (One might simply choose the regeneration points to be those for which the simulation returns to its initial state.)

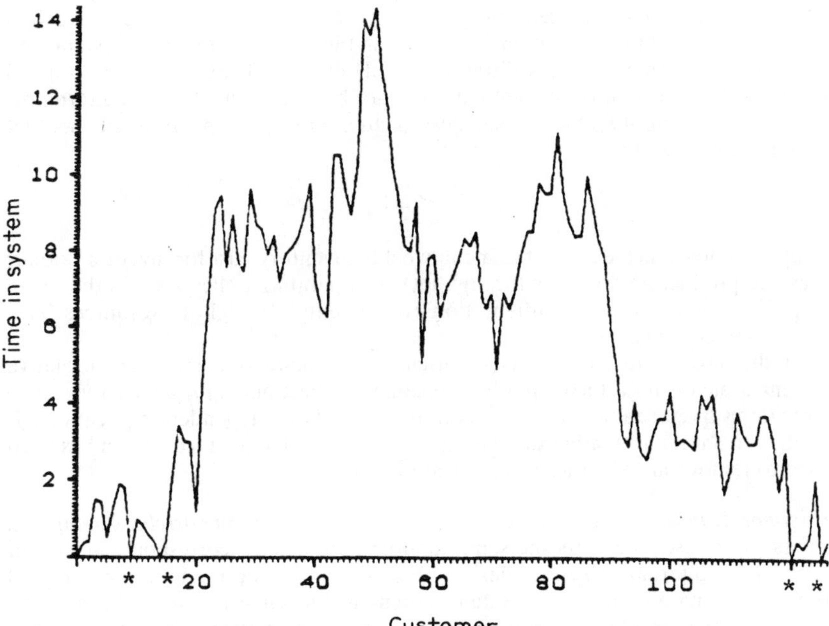

FIGURE 13.12 Regeneration points, denoted by *.

Disadvantages include:

- It is sometimes difficult to define regeneration points.
- *Extremely long* simulation runs might be required in order to obtain a sufficient number of i.i.d. blocks; enormous computing resources might be needed.

Standardized Time Series. It is common practice to use the central limit theorem to standardize i.i.d. random variables into a standard normal random variable. Schruben (1983) uses a *process* central limit theorem to standardize an arbitrary simulation process into a *Brownian bridge* process. (This Brownian bridge process is essentially a generalization of the standard normal random variable.) Various properties of Brownian bridges are then used to calculate CIs for μ.

The standardized time series method is easy to apply and has various asymptotic advantages over batch means and independent replications (see Goldsman and Schruben, 1984). However, long simulation runs may be required before the necessary asymptotics come into play (see Sargent, Kang, and Goldsman, 1992).

13.5.4 Comparison of Alternative System Designs

Suppose that we are comparing d alternative system designs on the basis of a specified measure of system performance θ_i, $i = 1, 2, \ldots, d$. Here the subscript i refers to the ith alternative design. For example, θ_1 might be the mean number of parts in the queue for the machine center of the first problem statement of Sec. 13.2.2, and θ_2 could represent the mean number of parts in the queue for the same system with three machines (perhaps of a different speed) instead of the two in the original problem. The purpose of the simulation might be to determine which alternative system design, call it i^*, has the smallest mean number of parts in the queue, that is, determine i^* so that

$$\theta_{i^*} \leq \theta_i \qquad i = 1, 2, \ldots, d$$

Other examples include θ_i as the mean weekly inventory cost for inventory policy i, as the production level (or throughput) for operating policy i, or as the mean response time to fires for configuration i of a county's firefighting resources (personnel and equipment).

In the subsequent subsections, we discuss a technique for comparing alternative system designs and a technique for selecting the best among d alternatives. The techniques discussed here are based on the method of independent replications. If applied to a nonterminating simulation, it is assumed that data truncation has been used to remove any significant initialization bias.

Common Random Numbers. *Common random numbers*, or *correlated sampling*, refers to the use of exactly the same random numbers for corresponding runs of each of the alternative system designs. By using the same random numbers and thus subjecting all alternative system designs to the same random "shocks," the modeler hopes to obtain a better comparison of these alternatives. It is argued that if the alternatives can be compared under identical experimental conditions (or as nearly identical as possible), then it will be easier to distinguish which systems are "better" on the basis of estimators that are subject to sampling error. Indeed, this argument is correct for many types of models provided that the use of the random number streams in any two alternative models can be properly synchronized. Synchronization means that a given random number used for a certain purpose in one model should be used for the same purpose in all other models. If synchronization is ignored or implemented carelessly, then the potential benefits of common random numbers can be lost. For examples and some guidelines, see, for example, Banks, Carson, and Nelson (1996) or Bratley, Fox, and Schrage (1987).

Let $\hat{\theta}_{ir}$ be the estimator of θ_i from replication r of the simulation $r = 1, 2, \ldots, R$. By choice of the random number seeds, the replications should be statistically independent. Thus, for system alternative i,

$$\hat{\theta}_{i1}, \hat{\theta}_{i2}, \ldots, \hat{\theta}_{iR}$$

constitute a set of i.i.d. random variables, all having mean θ_i and variance σ_i^2, say. On the other hand, *within* corresponding replications, say, replication r,

$$\hat{\theta}_{1r}, \hat{\theta}_{2r}, \ldots, \hat{\theta}_{dr}$$

are correlated because of the use of common random numbers. Correct implementation should, in most cases, lead to a positive correlation,

$$\rho_{ij} \equiv \text{Corr}(\hat{\theta}_{ir}, \hat{\theta}_{jr}) > 0$$

If, in fact, ρ_{ij} is positive, then the variance of the difference estimator $\hat{\theta}_{ir} - \hat{\theta}_{jr}$ between system designs i and j will be smaller than if independent sampling had been used. Mathematically,

$$\text{Var}(\hat{\theta}_{ir} - \hat{\theta}_{jr}) = \sigma_i^2 + \sigma_j^2 - 2\rho_{ij}\sigma_i\sigma_j \tag{13.2}$$

If independent sampling were used, then ρ_{ij} would be zero. If common random numbers work as desired, then ρ_{ij} is positive and the left-hand side of Eq. (13.2) is smaller than it would be otherwise. This reduction in estimator variability provides the statistical rationale for common random numbers.

To obtain an overall point estimator for system design i, let

$$\hat{\theta}_i \equiv \frac{1}{R}\sum_{r=1}^{R} \hat{\theta}_{ir} \qquad i = 1, 2, \ldots, d$$

Notice that $\text{Var}(\hat{\theta}_i) = \sigma_i^2/R$.

Initially assume that $d = 2$ system designs are being compared. Compute for each replication $r = 1, 2, \ldots, R$,

$$D_r \equiv \hat{\theta}_{1r} - \hat{\theta}_{2r}$$

Then compute the average difference,

$$\bar{D}_{12} \equiv \frac{1}{R}\sum_{r=1}^{R} D_r = \hat{\theta}_1 - \hat{\theta}_2 \tag{13.3}$$

and the sample variance of the differences,

$$S_{12}^2 \equiv \frac{1}{R-1}\sum_{r=1}^{R} (D_r - \bar{D}_{12})^2 \tag{13.4}$$

Then an approximate $100(1 - \alpha)\%$ confidence interval for $\theta_1 - \theta_2$ is given by

$$\bar{D}_{12} \pm t_{\alpha/2, R-1} \frac{S_{12}}{\sqrt{R}} \tag{13.5}$$

Since $D_1, D_2, \ldots, D_r$ are i.i.d. with mean $\theta_1 - \theta_2$, the procedure outlined here is nothing but the method of independent replications applied to the estimation of $\theta_1 - \theta_2$.

We interpret the CI in Eq. (13.5). If it is totally positive (i.e., to the right of zero), then we have strong evidence that $\theta_1 - \theta_2 > 0$, or $\theta_1 > \theta_2$; if a small θ value is desirable, then system 2 is better than system 1. If the CI is totally negative, then we conclude that $\theta_1 < \theta_2$. The third possible conclusion occurs if the CI contains zero, in which case the result of the simulation experiment based on these R pairs of runs is inconclusive with respect to which system is better. The width of the CI is a measure of the reliability of the difference estimator $\bar{D}_{12}$. Thus, if the CI contains zero and the length of the CI in Eq. (13.5) is "small," then we can conclude that the difference $\theta_1 - \theta_2$ between the two designs is of no practical significance. On the other hand, if the CI contains zero and the length of the CI in Eq. (13.5) is "large," then more replications are needed before a sound conclusion can be drawn.

The following example (from Banks, Carson, and Nelson, 1996) illustrates the logic of the preceding discussion. A vehicle safety inspection station has three attendants, each of whom performs three jobs: brake, headlight, and steering checks. The present system (model 1) has one queue feeding three stalls in parallel; that is, when its turn comes, a vehicle enters a stall, where one of the attendants makes all three inspections. Based on the data from the existing system, it has been assumed that arrivals occur according to a Poisson process at an average rate of 9.5 per hour. The times for brake, headlight, and steering checks are assumed to be normally distributed with means 6.5, 6.0, and 5.5 min, respectively, and a common standard deviation of approximately 0.5 min. There is no limit on the length of the queue of waiting vehicles.

An alternative system design (model 2) is being considered. Each attendant will specialize in a single task, and each vehicle will pass through three workstations in series. The workstations, in order, are for brake, headlight, and steering checks. No space is allowed for vehicles to wait between the workstations. Owing to the increased specialization of the inspectors, it is anticipated that the mean inspection times for each type of check will decrease by 10% to 5.85, 5.4, and 4.95 min, respectively, for the brake, headlight, and steering inspections. We will compare models 1 and 2 on the basis of mean response time per vehicle, where a response time is defined as the total time from the arrival of a vehicle until its departure from the system. In order to demonstrate the greater accuracy of common random numbers, we shall compare models 1 and 2 using both independent sampling and common random numbers.

Each vehicle arriving to be inspected has four input random variates associated with it:

$$A_n = \text{interarrival time between vehicles } n \text{ and } n+1$$

$$B_n^{(i)} = \text{brake inspection time for vehicle } n \text{ in model } i = 1, 2$$

$$H_n^{(i)} = \text{headlight inspection time for vehicle } n \text{ in model } i = 1, 2$$

$$S_n^{(i)} = \text{steering inspection time for vehicle } n \text{ in model } i = 1, 2$$

For model 2, all mean service times are decreased by 10%. When using independent sampling, completely different values of service and interarrival times would be generated for model 2. But when conducting correlated sampling, the random number generator must be used in such a way that exactly the same values are generated for the interarrival times $A_1, A_2, \ldots$. Service times are *not identical* in the two models, so the same values cannot be used. However, the service times should be generated so that they are as highly correlated (between models) as possible. Taking brake inspections as an example, service times with correlation (between models) of 1 could be generated by

$$B_n^{(i)} = E[B_n^{(i)}] + \sigma Z_n^B \qquad i = 1, 2$$

where Z_n^B is a standard normal variate, $\sigma = 0.5$ min, $E[B_n^{(1)}] = 6.5$ min for model 1 and $E[B_n^{(2)}] = 5.85$ min for model 2; the other two types of inspection times would be generated in a similar fashion.

Table 13.2 gives the average response time for each of $R = 10$ replications, each of run length 16 h. We assumed that two cars were present at time 0, waiting to be inspected. Model 2 was run using independent random numbers (column 2I) and common random numbers (column 2C). The purpose of the simulation is to estimate the mean difference in response times for the two systems, $\theta_1 - \theta_2$. A difference of 1.5 min or less is regarded as being of no practical signifi-

TABLE 13.2 Comparison of System Designs for the Vehicle Safety Inspection System

	Average response time for model			Differences
Replication r	1	2I	2C	$1 - 2C\ (D_r)$
1	29.59	51.62	29.55	0.04
2	23.49	51.91	24.26	−0.77
3	25.68	45.27	26.03	−0.35
4	41.09	30.85	42.64	−1.55
5	33.84	56.15	32.45	1.39
6	39.57	28.82	37.91	1.66
7	37.04	41.30	36.48	0.56
8	40.20	73.06	41.24	−1.04
9	61.82	23.00	60.59	1.23
10	44.00	28.44	41.49	2.51
Sample mean	37.63	43.04		0.37
Sample variance	118.90	244.33		1.74

cance. For the two independent runs (1 and 2I), the 95% CI for the mean difference using the standard formulas in Sec. 13.1.3 of Banks, Carson, and Nelson (1996) turns out to be

$$-5.4 \pm 2.11 \times 6.03$$

or

$$-18.1 \leq \theta_1 - \theta_2 \leq 7.3 \qquad (13.6)$$

The 95% CI in this expression contains zero. This indicates that there is no strong evidence that the observed difference of −5.4 min is due to anything other than random variation in the output data. Thus, if the simulator had decided to use independent sampling, no strong conclusion would have been possible. In addition, the estimate of $\theta_1 - \theta_2$ is quite imprecise. Indeed, it is impossible to conclude whether the difference in mean response times is greater or less than 1.5 min. The only options open to the modeler are to make more replications to reduce the length of the CI or to abandon the experiment as inconclusive.

When complete synchronization of the random numbers was used in run 2C, the point estimate of the mean difference in response times was $D_{12} = 0.37$ min and the sample variance was $S_{12}^2 = 1.74$ (with $R - 1 = 9$ degrees of freedom). The resulting 95% CI for the mean difference is

$$0.37 \pm 2.26 \times 0.42$$

or

$$-0.58 \leq \theta_1 - \theta_2 \leq 1.32 \qquad (13.7)$$

This CI again contains zero, but it is considerably shorter than the previous interval. This greater accuracy in the estimation of $\theta_1 - \theta_2$ is due to the use of synchronized common random numbers. The short length of the interval suggests that the true difference $\theta_1 - \theta_2$ is close to zero. In fact, it can be concluded that there is no practical difference between the two alternatives since the entire CI lies between –1.5 and 1.5 min.

By comparing the two CIs given by Eqs. (13.6) and (13.7), we see that the width of the CI is reduced by 93% when using common random numbers. To achieve accuracy comparable to that of common random numbers, over 2000 independent replications would have had to have been made. So considerable computer time is saved by using common random numbers, and a statistically sound conclusion is reached based on only $R = 10$ replications.

When the number of alternatives is $d > 2$, a slight modification of Eq. (13.5) is needed, depending on the number of CIs to be computed. If overall confidence of $100(1 - \alpha)\%$ is desired and c intervals are to be computed, replace α by α/c in Eq. (13.5). For example, if system 1 is to be compared to all other system alternatives, then $c = d - 1$. In a manner similar to Eqs. (13.3) and (13.4), compute $\overline{D}_{1j}$ and S_{1j}^2 for each pair $(1, j)$, and finally a CI analogous to Eq. (13.5) for each difference $\theta_1 - \theta_j$. The interpretation is similar to the case $d = 2$. For an extension of the vehicle safety example to $d = 4$ alternative systems, see Banks, Carson, and Nelson (1996), Example 13.4.

When implemented correctly, common random numbers can reduce the sampling effort required to reach sound conclusions. Unfortunately, for certain systems, it is possible for a variance *increase* to occur rather than a variance *decrease.* Therefore, we recommend that pilot studies be used to evaluate the technique for the particular systems being simulated.

Selecting the Best Alternative System. *Ranking and selection* procedures are statistical techniques for selecting the best (or nearly the best) of several alternative systems. The user of such a procedure must usually specify two constants P^* and δ^*. The quantity P^* is a lower bound on the desired probability of correct selection (CS) given that the parameter of interest from the best system and that from the second best system differ by at least an amount δ^*. If the difference is less than δ^*, then we are "indifferent" as to which system is selected as best — for all practical purposes, the first and second best are equivalent. An introduction to these topics for simulation use is provided by Law and Kelton (1991). A more detailed treatment is presented in Bechhofer, Santner, and Goldsman (1995).

Our goal here is to select the best of d alternative system designs, where "best" is defined to be that system having the largest mean. Suppose that $Y_{i1}, Y_{i2}, \ldots$ are independent replication sample means from alternative $i = 1, 2, \ldots, d$; these are our "observations." Thanks to a central limit theorem, we will assume that $Y_{i1}, Y_{i2}, \ldots,$ are approximately i.i.d. normal with unknown mean θ_i and variance σ_i^2. If we denote the unknown ordered means by $\theta_{[1]} \le \theta_{[2]} \le \ldots \le \theta_{[d]}$, then our goal is to select the alternative associated with the largest mean $\theta_{[d]}$. We impose the probability requirement that

$$P\{\text{CS}\} \ge P^* \text{ whenever } \theta_{[d]} - \theta_{[d-1]} \ge \delta^*$$

The following is an extension of a *two-stage* procedure due to Rinott (1978) for finding the best normal population. The first stage of the Rinott procedure uses the observations to estimate marginal variances. These estimates establish the number of observations to be taken in the second stage of sampling in order to meet the probability requirement. See Goldsman and Nelson (1998) for additional motivation.

1. Specify P^*, δ^*, and the first-stage sample size R_0.
2. Take an i.i.d. sample $Y_{i1}, Y_{i2}, \ldots, Y_{iR_0}$ from each of the d systems simulated *independently*.
3. Compute the marginal sample means

$$\bar{Y}_i \equiv \frac{1}{R_0} \sum_{j=1}^{R_0} Y_{ij}$$

and sample variances

$$S_i^2 \equiv \frac{\sum_{j=1}^{R_0} (Y_{ij} - \bar{Y}_i)^2}{R_0 - 1} = \frac{\sum_{j=1}^{R_0} Y_{ij}^2 - R_0 \bar{Y}_i^2}{R_0 - 1}$$

for $i = 1, 2, \ldots, d$.

4. Find h from Table 13.3 for $P^* = 0.90$ and 0.95 (or from the more detailed tables in Bechhofer, Santner, and Goldsman 1995). Compute the final sample sizes

$$R_i \equiv \max \left\{R_0, \left\lceil (hS_i/\delta^*)^2 \right\rceil \right\}$$

for $i = 1, 2, \ldots, d$, where $\lceil \cdot \rceil$ means to round up.

5. Take $R_i - R_0$ additional i.i.d. observations from system i, independently of the first-stage sample and the other systems, for $i = 1, 2, \ldots, d$.
6. Compute the overall sample means

$$\bar{\bar{Y}}_i \equiv \frac{1}{R_i} \sum_{j=1}^{R_i} Y_{ij}$$

for $i = 1, 2, \ldots, d$.

7. Select the system with the largest $\bar{\bar{Y}}_i$ as best.

TABLE 13.3 Values of h Required by Rinott's Procedure

		d			
P^*	R_0	2	3	4	5
	5	2.291	3.058	3.511	3.837
	10	1.999	2.587	2.913	3.137
0.90	15	1.928	2.479	2.779	2.983
	20	1.896	2.431	2.720	2.916
	30	1.866	2.387	2.666	2.855
	5	3.107	3.905	4.390	4.744
	10	2.614	3.166	3.476	3.693
0.95	15	2.502	3.006	3.285	3.477
	20	2.452	2.937	3.203	3.385
	30	2.407	2.874	3.129	3.303

Additional extensions related to Rinott's procedure are given by Matejcik and Nelson (1995) and Nelson and Matejcik (1995). Another popular normal means procedure is discussed in Law and Kelton (1991) and Dudewicz and Dalal (1975). Many other classes of ranking and selection procedures have been developed in the simulation and statistics literature. For instance, procedures exist for screening out inferior systems, so that further experimentation and consideration can be given to those alternatives most likely to be the best. Procedures have also been developed for more specific problems: find the best exponential population; find the Bernoulli population having the largest success parameter; and find the most probable multinomial cell. Most of these procedures are very easy to use, usually requiring only one or two tabled constants to implement.

13.6 A FINAL WORD

One of the most common, and most dangerous mistakes made by managers who desire to use simulation to gain insight into their operations is to regard simulation as an exercise in computer programming. The task of developing the model might be assigned to a programmer and the output (sometimes only one replication!) might be accepted in the same manner as the output from a payroll or database system. One of the main purposes of this chapter is to point to the wide variety of skills required to carry out a valid study based on a simulation model. Computer programming, although important, is only one of the aspects of simulation; modeling, data collection and analysis, statistical tools, validation, and output analysis are all essential parts of a simulation project.

This chapter was supported in part by National Science Foundation grant DMI-9622269.

REFERENCES

Baird, S. P., and J. J. Leavy: "Simulation Modeling Using ProModel for Windows," *Proceedings of the 1994 Winter Simulation Conference* (ed. J. Tew, M. S. Manivannan, D. A. Sadowski, and A. F. Seila), pp. 527–532, The Institute of Electrical and Electronics Engineers, Piscataway, NJ, 1994.

Balci, O., and R. G. Sargent: "A Methodology for Cost-Risk Analysis in the Statistical Validation of Simulation Models," *Communications of the ACM*, vol. 24, pp. 190–197, 1981.

_____ and _____ "Some Examples of Simulation Model Validation Using Hypothesis Testing," *Proceedings of the 1982 Winter Simulation Conference* (ed. H. Highland, Y. Chao, and O. Madrigal), pp. 620–629, The Institute of Electrical and Electronics Engineers, Piscataway, NJ, 1982.

Banks, J.: "Software for Simulation," *Proceedings of the 1994 Winter Simulation Conference* (ed. J. Tew, M. S. Manivannan, D. A. Sadowski, and A. F. Seila), pp. 26–33, The Institute of Electrical and Electronics Engineers, Piscataway, NJ, 1994.

_____, B. Burnett, J. D. Rose, and H. Kozloski: *SIMAN V* and *CINEMA V*, Wiley, New York, 1995.

_____, J. S. Carson, II, and B. L. Nelson: *Discrete-Event System Simulation*, 2d ed., Prentice-Hall, Upper Saddle River, NJ, 1996.

Bechhofer, R. E., T. J. Santner, and D. Goldsman: *Design and Analysis for Statistical Selection, Screening and Multiple Comparisons*, Wiley, New York, 1995.

Bratley, P., B. L. Fox, and L. E. Schrage: *A Guide to Simulation*, 2d ed., Springer-Verlag, New York, 1987.

Chien, C., D. Goldsman, and B. Melamed: "Large-Sample Results for Batch Means," to appear in *Management Science*, 1997.

Crain, R. C., and D. S. Smith: "Industrial Strength Simulation Using GPSS/H," *Proceedings of the 1994 Winter Simulation Conference* (ed. J. Tew, M. S. Manivannan, D. A. Sadowski, and A. F. Seila), pp. 502–508, The Institute of Electrical and Electronics Engineers, Piscataway, NJ, 1994.

Crane, M. A., and D. L. Iglehart: "Simulating Stable Stochastic Systems: III. Regenerative Processes and Discrete-Event Simulations," *Operations Research*, vol. 23, pp. 33–45, 1975.

_____ and A. J. Lemoine: *An Introduction to the Regenerative Method for Simulation Analysis*, Springer-Verlag, Berlin, 1977.

Drevna, M. J., and C. J. Kasales: "Introduction to ARENA," *Proceedings of the 1994 Winter Simulation Conference* (ed. J. Tew, M. S. Manivannan, D. A. Sadowski, and A. F. Seila), pp. 431–436, The Institute of Electrical and Electronics Engineers, Piscataway, NJ, 1994.

Dudewicz, E. J., and S. R. Dalal: "Allocation of Observations in Ranking and Selection with Unequal Variances," *Sankhya-B*, vol. 37, pp. 28–78, 1975.

Earle, N. J., and J. O. Henriksen: "PROOF Animation: Reaching New Heights in Animation," *Proceedings of the 1994 Winter Simulation Conference* (ed. J. Tew, M. S. Manivannan, D. A. Sadowski, and A. F. Seila), pp. 509–516, The Institute of Electrical and Electronics Engineers, Piscataway, NJ, 1994.

Fishman, G. S.: "Estimating Sample Size in Computer Simulation Experiments," *Management Science*, vol. 18, pp. 21–38, 1971.

_____: "Bias Considerations in Simulation Experiments," *Operations Research*, vol. 20, pp. 785–790, 1972.

_____: *Concepts and Methods in Discrete Event Digital Simulation*, Wiley, New York, 1973.

_____: *Principles of Discrete Event Simulation*, Wiley, New York, 1978.

_____: *Monte Carlo Concepts, Algorithms, and Applications*, Springer-Verlag, New York, 1996.

Fuller, W. A.: *Introduction to Statistical Time Series*, Wiley, New York, 1976.

Gafarian, A. V., C. J. Ancker, Jr., and T. Morisaku: "Evaluation of Commonly Used Rules for Detecting 'Steady State' in Computer Simulation," *Naval Research Logistics Quarterly*, vol. 25, pp. 511–529, 1978.

Goble, J.: "Introduction to SIMFACTORY II.5," *Proceedings of the 1991 Winter Simulation Conference* (ed. B. L. Nelson, W. D. Kelton, and G. M. Clark), pp. 77–80, The Institute of Electrical and Electronics Engineers, Piscataway, NJ, 1991.

Goldsman, D., and B. L. Nelson: "Comparing Systems Via Simulation," to appear in *Handbook of Simulation* (ed. J. Banks), chap. 8, Wiley, New York, 1998.

_____ and L. W. Schruben: "Asymptotic Properties of Some Confidence Interval Estimators for Simulation Output," *Management Science*, vol. 30, pp. 1217–1225, 1984.

_____, _____, and J. J. Swain: "Tests for Transient Means in Simulated Time Series," *Naval Research Logistics*, vol. 41, pp. 171–187, 1994.

Heidelberger, P., and P. D. Welch: "A Spectral Method for Confidence Interval Generation and Run Length Control in Simulations," *Communications of the ACM*, vol. 24, pp. 233–245, 1981.

_____ and _____: "Simulation Run Length Control in the Presence of an Initial Transient," *Operations Research*, vol. 31, pp. 1109–1144, 1983.

Hultquist, P. F.: "Random Number Generators," *PC Tech. J.*, pp. 86–103, August 1984.

IMSL: STAT/LIBRARY, Version 1.0, IMSL, Houston, TX, April 1987.

Jankauskas, L., and S. McLafferty: "BestFit, Distribution Fitting Software by Palisade Corporation," *Proceedings of the 1995 Winter Simulation Conference* (ed. C. Alexopoulos, K. Kang, W. R. Lilegdon, and D. Goldsman), pp. 457–461, The Institute of Electrical and Electronics Engineers, Piscataway, NJ, 1995.

Kelton, W. D., and A. M. Law: "A New Approach for Dealing with the Startup Problem in Discrete-Event Simulation," *Naval Research Logistics Quarterly*, vol. 30, pp. 641–658, 1983.

_____ and _____: "An Analytical Evaluation of Alternative Strategies in Steady-State Simulation," *Operations Research*, vol. 32, pp. 169–184, 1984.

Knuth, D. W.: *The Art of Computer Programming*, vol. 2: Semi-Numerical Algorithms, Addison-Wesley, Reading, MA, 1969.

Law, A. M.: "Confidence Intervals in Discrete Event Simulation: A Comparison of Replication and Batch Means," *Naval Research Logistics Quarterly*, vol. 24, pp. 667–678, 1977.

_____ and W. D. Kelton: "Confidence Intervals for Steady-State Simulations, I: A Survey of Fixed Sample Size Procedures," *Operations Research*, vol. 32, pp. 1221–1239, 1984.

_____ and _____: *Simulation Modeling and Analysis*, 2d ed., McGraw-Hill, New York, 1991.

Lilegdon, W. R., D. L. Martin, and A. A. B. Pritsker: "FACTOR/AIM: A Manufacturing Simulation System," *Simulation*, vol. 62, pp. 367–372, 1994.

Matejcik, F. J., and B. L. Nelson: "Two-Stage Multiple Comparisons with the Best for Computer Simulation," *Operations Research*, vol. 43, pp. 633–640, 1995.

Meketon, M. S., and B. W. Schmeiser: "Overlapping Batch Means: Something for Nothing?" *Proceedings of the 1984 Winter Simulation Conference* (ed. S. Sheppard, U. Pooch, and D. Pegden), pp. 227–230, The Institute of Electrical and Electronics Engineers, Piscataway, NJ, 1984.

Naylor, T. H., and J. M. Finger: "Verification of Computer Simulation Models," *Management Science*, vol. 14, pp. 92–101, 1967.

Nelson, B. L., and F. J. Matejcik: "Using Common Random Numbers for Indifference-Zone Selection and Multiple Comparisons in Simulation," *Management Science*, vol. 41, pp. 1935–1945, 1995.

Nordgren, B.: "Taylor II Manufacturing Simulation Software," *Proceedings of the 1994 Winter Simulation Conference* (ed. J. Tew, M. S. Manivannan, D. A. Sadowski, and A. F. Seila), pp. 446–449, The Institute of Electrical and Electronics Engineers, Piscataway, NJ, 1994.

O'Reilly, J. J.: "Introduction to SLAM II and SLAMSYSTEM," *Proceedings of the 1994 Winter Simulation Conference* (ed. J. Tew, M. S. Manivannan, D. A. Sadowski, and A. F. Seila), pp. 415–419, The Institute of Electrical and Electronics Engineers, Piscataway, NJ, 1994.

Pegden, C. D., R. E. Shannon, and R. P. Sadowski: *Introduction to Simulation Using SIMAN*, 2d ed., McGraw-Hill, New York, 1995.

Pritsker, A. A. B.: *Introduction to Simulation and SLAM II*, Wiley, New York, 1986.

Rinott, Y.: "On Two-Stage Selection Procedures and Related Probability-Inequalities," *Communications in Statistics – Theory and Methods*, vol. A8, pp. 799–811, 1978.

Rohrer, M.: "AutoMod," *Proceedings of the 1994 Winter Simulation Conference* (ed. J. Tew, M. S. Manivannan, D. A. Sadowski, and A. F. Seila), pp. 487–492, The Institute of Electrical and Electronics Engineers, Piscataway, NJ, 1994.

Russell, E. C.: "SIMSCRIPT II.5 and SIMGRAPHICS Tutorial," *Proceedings of the 1993 Winter Simulation Conference* (ed. G. W. Evans, M. Mollaghasemi, E. C. Russell, and W. E. Biles), pp. 223–227, The Institute of Electrical and Electronics Engineers, Piscataway, NJ, 1993.

Sargent, R. G.: "An Introduction to Statistical Analysis of Simulation Output Data," *Proc. AGARD Symp.*, Paris, 1979.

_____: "A Tutorial on Verification and Validation of Simulation Models," *Proceedings of the 1984 Winter Simulation Conference* (ed. S. Sheppard, U. Pooch, and D. Pegden), pp. 115–121, The Institute of Electrical and Electronics Engineers, Piscataway, NJ, 1984.

_____, K. Kang, and D. Goldsman: "An Investigation of Finite-Sample Behavior of Confidence Interval Estimators," *Operations Research*, vol. 40, pp. 898–913, 1992.

Schmeiser, B. W.: "Approximations to the Inverse Cumulative Normal Function for Use on Hand Calculators," *Applied Statistics*, vol. 28, pp. 175–176, 1979.

_____: "Batch Size Effects in the Analysis of Simulation Output," *Operations Research*, vol. 30, pp. 556–568, 1982.

Schriber, T. J., and R. W. Andrews: "ARMA-Based Confidence Intervals for Simulation Output Analysis," *Am. J. Math. Manage. Sci.*, vol. 4, pp. 345–373, 1984.

Schruben, L. W.: "Detecting Initialization Bias in Simulation Output," *Operations Research*, vol. 30, pp. 569–590, 1982.

_____: "Confidence Interval Estimation Using Standardized Time Series," *Operations Research*, vol. 31, pp. 1090–1108, 1983.

_____ and D. Goldsman: "Initialization Effects in Computer Simulation Experiments," Tech. Report #594, School of OR&IE, Cornell University, Ithaca, NY, 1985.

_____ H. Singh, and L. Tierney: "Optimal Tests for Initialization Bias in Simulation Output," *Operations Research*, vol. 31, pp. 1167–1178, 1983.

Snell, M. K., and L. W. Schruben: "Weighting Simulation Data to Reduce Initialization Effects," *IIE Transactions*, vol. 17, pp. 354–362, 1985.

Song, W.-M. T., and B. W. Schmeiser: "Optimal Mean-Squared-Error Batch Sizes," *Management Science*, vol. 41, pp. 110–123, 1995.

Thompson, W. B.: "A Tutorial for Modeling with the WITNESS Visual Interactive Simulator," *Proceedings of the 1994 Winter Simulation Conference* (ed. J. Tew, M. S. Manivannan, D. A. Sadowski, and A. F. Seila), pp. 517–521, The Institute of Electrical and Electronics Engineers, Piscataway, NJ, 1994.

Turnquist, M. A., and J. M. Sussman: "Toward Guidelines for Designing Experiments in Queuing Simulation," *Simulation*, vol. 28, pp. 137–144, 1977.

Vincent, S., and A. M. Law: "ExpertFit: Total Support for Simulation Input Modeling," *Proceedings of the 1995 Winter Simulation Conference* (ed. C. Alexopoulos, K. Kang, W. R. Lilegdon, and D. Goldsman), pp. 395–400, The Institute of Electrical and Electronics Engineers, Piscataway, NJ, 1995.

Welch, P. D.: "On the Problem of the Initial Transient in Steady State Simulations," IBM Watson Research Center, Yorktown Heights, NY, 1981.

_____: "The Statistical Analysis of Simulation Results," in *The Computer Performance Handbook* (ed. S. Lavenberg), Academic Press, Orlando, FL, 1983.

Wilson, J. R., and A. A. B. Pritsker: "A Survey of Research on the Simulation Startup Problem," *Simulation*, vol. 31, pp. 55–58, 1978*a*.

_____ and _____: "Evaluation of Startup Policies in Simulation Experiments," *Simulation*, vol. 31, pp. 79–89, 1978*b*.

Yuan, M., and B. L. Nelson: "Autoregressive Output Analysis Methods Revisited," *Annals of Operations Research*, vol. 53, pp. 391–418, 1994.

CHAPTER 14
LINEAR REGRESSION

Thomas P. Ryan
Case Western Reserve University Cleveland, OH

14.1 INTRODUCTION

It is often said that regression is probably the most widely used statistical tool. Unfortunately it is also generally believed that it may be the most misused statistical tool. The word "regression," in statistics, was coined by Sir Francis Galton (1822–1911), who observed that children's heights regressed toward the average height for the population rather than digressing from it.

This is essentially unrelated to modern-day linear regression. The uses of regression are (1) description, (2) prediction, (3) estimation, and (4) control. Each of these uses is discussed in this chapter. In regression there are one or more "independent" variables and (in univariate regression) a single "dependent" variable. The term "dependent variable" does not imply an exact functional relationship, however, as would be the case with the equation $F = {}^9/_5\, C + 32$, which expresses temperature in degrees Fahrenheit as a function of temperature measured on the Celsius (centigrade) scale.

Such exact relationships do not exist in regression analysis. The dependent variable may actually be part of a cause-and-effect relationship, or it may simply be "correlated" with one or more independent variables without any evidence of a cause-and-effect relationship.

14.2 SIMPLE LINEAR REGRESSION

The word "simple" means that there is a single independent variable, but the word "linear" does not have the interpretation that might seem self-evident. Specifically, it does not (necessarily) mean that the relationship between the dependent variable and the single independent variable will be portrayed graphically as a straight line. Rather, it means that the regression equation is linear in the parameters. For example, the equation

$$Y = \beta_0 + \beta_1 X + \epsilon \tag{14.1}$$

is a linear regression equation, but so is the equation

$$Y = \beta_0' + \beta_1'X + \beta_{11}X^2 + \epsilon' \tag{14.2}$$

Even though Eq. (14.2) contains the quadratic term X^2, the equation is still linear in the parameters (the betas), so it is a linear regression equation.

The general form for simple linear regression is given by Eq. (14.1) which is also in the general form of the equation for a straight line (usually given in algebra books as $y = mx + b$) except for the error ϵ. This is illustrated in Fig. 14.1.

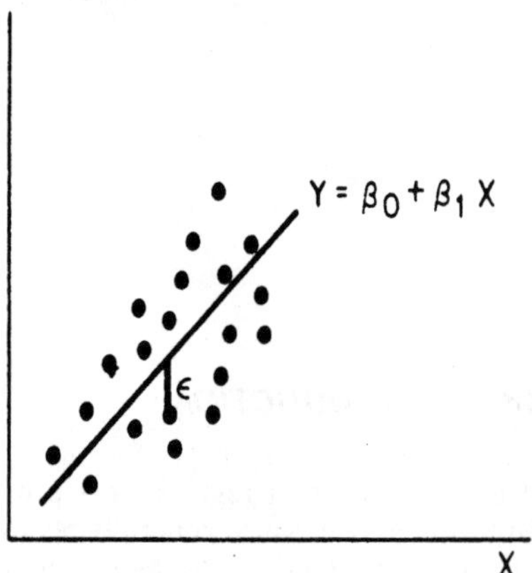

FIGURE 14.1 Scatter plot.

If all of the points are close to the line, there is a strong linear relationship between Y and X. This is of paramount importance since such a relationship must exist for a linear regression equation to be of any value.

In Fig. 14.1 the regression line $Y = \beta_0 + \beta_1 X$ is drawn with the implicit assumption that β_0 and β_1 are known. They will generally be unknown, however, so they must be estimated. There are various methods for estimating the parameters in a regression equation; the most widely used method is the *method of least squares*. The use of the least-squares estimators $\hat{\beta}_0$ and $\hat{\beta}_1$ will produce a regression line such that the sum of the squared vertical deviations from each point to the line is minimized, and the sum of the vertical deviations is zero.

For the ith observation in a sample of size n,

$$Y_i = \beta_0 + \beta_1 X_i + \epsilon_i$$

so that

$$\epsilon_i = Y_i - \beta_0 - \beta_1 X_i$$

It is $\Sigma_{i=1}^{n} \epsilon_i^2$ that is to be minimized, and

$$\Sigma \epsilon_i^2 = \Sigma (Y_i - \beta_0 - \beta_1 X_i)^2. \tag{14.3}$$

Using calculus, the right-hand side of Eq. (14.3) would be differentiated first with respect to β_0 and then with respect to β_1. This will produce two equations in β_0 and β_1, and the solution for each will be the estimators $\hat{\beta}_0$ and $\hat{\beta}_1$. The results are

$$\hat{\beta}_1 = \frac{\Sigma (X - \overline{X})(Y - \overline{Y})}{\Sigma (X - \overline{X})^2}$$

$$= \frac{\Sigma XY - (\Sigma X)(\Sigma Y)/n}{\Sigma X^2 - (\Sigma X)^2/n}$$

$$\hat{\beta}_0 = \overline{Y} - \hat{\beta}_1 \overline{X}$$

TABLE 14.1 Least-Squares Computations

	Temperature X, °F	Process yield Y	XY	X^2
	310	17.3	5363.0	96100
	290	14.2	4118.0	84100
	300	15.5	4650.0	90000
	315	18.1	5701.5	99225
	320	20.1	6432.0	102400
	305	16.7	5093.5	93025
	325	20.2	6565.0	105625
	295	15.3	4513.5	87025
	285	14.1	4018.5	81225
Totals	2745	151.5	46455.0	838725

$$\hat{\beta}_1 = \frac{46455 - 2745(151.5)/9}{838725 - (2745)^2/9} = \frac{247.5}{1500} = 0.165$$

$$\hat{\beta}_0 = \frac{151.5}{9} - 0.165\left(\frac{2745}{9}\right) = -33.4917$$

$$\hat{Y} = -33.4917 + 0.165X$$

where $\overline{Y}$ and $\overline{X}$ are the averages of the Y and X values, respectively, for a sample of size n. Obviously $\hat{\beta}_1$ would have to be computed first since it is used in computing $\hat{\beta}_0$. The necessary calculations are illustrated for the data in Table 14.1.

The regression equation is thus $\hat{Y} = -33.4917 + 0.165X$. In practice it would be desirable to first plot the data, as in Fig. 14.1, to see whether there is evidence of a linear relationship. For the data in Table 14.1 a scatter plot of the X and Y values would have indicated such a relationship.

After the regression equation has been determined, the strength of the linear relationship can be found by comparing the Y and $\hat{Y}$ values to see how well the observed values are predicted by the equation. The results are given in Table 14.2.

It can be observed that $\Sigma(Y - \hat{Y})$ is approximately zero. (Theoretically it is zero. The displayed sum would have been zero if more decimal places had been displayed for the $\hat{Y}$ values.)

Although Table 14.2 indicates that the predicted values are fairly close to the observed values, it would be difficult to judge the worth of a regression equation solely from the values of $Y - \hat{Y}$. We would expect the latter to be larger for Y values that are larger than those in Table 14.2, and smaller for smaller Y values.

A more objective measure is the percentage of the total variability in Y that is accounted for by using X to predict Y. The total variability in Y can be represented by $\Sigma(Y - \overline{Y})^2$, which is the numerator of the sample variance of Y. It can be shown that

$$\Sigma(Y - \overline{Y})^2 = \Sigma[(Y - \hat{Y}) + (\hat{Y} - \overline{Y})]^2$$
$$= \Sigma(Y - \hat{Y})^2 + \Sigma(\hat{Y} - \overline{Y})^2. \quad (14.4)$$

The second sum on the right-hand side of Eq. (14.4) is also a measure of the value of the regression equation. Specifically, it indicates the superiority of $\hat{Y}$ as a predictor of Y as compared with $\overline{Y}$. Since $\hat{\beta}_0 = \overline{Y} - \hat{\beta}_1\overline{X}$, it follows that

TABLE 14.2 Predicted and Observed Values

Observed values Y	Predicted values $\hat{Y}$	$Y - \hat{Y}$
17.3	17.66	–0.36
14.2	14.36	–0.16
15.5	16.01	–0.51
18.1	18.48	–0.38
20.1	19.31	0.79
16.7	16.83	–0.13
20.2	20.13	0.07
15.3	15.18	0.12
14.1	13.53	0.57
		0.01

$$\hat{Y} = \hat{\beta}_0 + \hat{\beta}_1 X$$

$$= \overline{Y} + \hat{\beta}_1(X - \overline{X}).$$

Thus if $\hat{\beta}_1 = 0$ (which would be a logical estimate of β_1 if $\beta_1 = 0$), then $\hat{Y} = \overline{Y}$. Therefore we would hope that $\Sigma(\hat{Y} - \overline{Y})^2$ is large (relatively speaking). If so, this would mean that $\Sigma(Y - \hat{Y})^2$ is small since the sum of the two components, $\Sigma(Y - \overline{Y})^2$, is fixed in that it does not depend upon X.

Accordingly,

$$R^2 = \frac{\Sigma(\hat{Y} - \overline{Y})^2}{\Sigma(Y - \overline{Y})^2} \tag{14.5}$$

is an objective measure of the value of the regression equation where, by definition, $0 \leq R^2 \leq 1$. It is the percentage of the total variability in Y that is accounted for by using X to predict Y. Therefore,

$$1 - R^2 = \frac{\Sigma(Y - \hat{Y})^2}{\Sigma(Y - \overline{Y})^2}$$

is the "residual" variability expressed as a percentage of the total variability.

If all of the points fell on the regression line, then R^2 would equal 1. Ideally, we would like to have an R^2 value very close to 1, but there is no dividing line that separates "acceptable" R^2 values from "unacceptable" values. R^2 could be computed using Eq. (14.5), but it would be preferable to compute it as

$$R^2 = \hat{\beta}_1 \frac{S_{xy}}{S_{yy}}$$

where $S_{xy} = \Sigma XY - (\Sigma X)(\Sigma Y)/n$ and $S_{yy} = \Sigma Y^2 - (\Sigma Y)^2/n$. Also, $S_{xx} = \Sigma X^2 - (\Sigma X)^2/n$, which will be used later. (Notice that only S_{yy} will not have been obtained by the time the regression equation has been produced.) For the data in Table 14.1 it can be shown that $R^2 = 0.9636$. Thus 96.36% of the variability in Y is explained by the regression equation. Since $\hat{\beta}_1 = 0.165$ we would predict that Y would increase by 0.165 units per unit increase in X. We can see from Table 14.1 that the actual increases are close to the predicted increases. For example, as X goes from 300 to 310 we would predict that Y would increase by 10(0.165) = 1.65 units. This compares favorably with the actual increase of 1.80 units.

Thus the regression equation provides insight into the relationship between X and Y. This may or may not be a causal relationship, however. If the data in Table 14.1 had come from an experiment designed to investigate the relationship between temperature and yield, a decision might have been made to vary the temperature in 5-degree increments from 285 to 325. Because of the strength of the linear relationship between X and Y, it would be reasonable to infer that an increase in temperature does cause an increase in yield, *within the range of temperatures used in the experiment.* The italicized qualifier is important since we should not assume that the relationship between Y and X is the same for values of X not included in the study. The experimenter who does so is extrapolating beyond the range of the data. For example, it would be unwise to attempt to predict Y for values of X much greater than 325; the slope might be considerably different in that region, or the relationship might not even be linear. Although some extrapolation is often unavoidable, one should not extrapolate very far beyond the range of X values that are used in obtaining the regression equation.

A regression equation can also be produced when the values of X are not preselected. For example, we might investigate the relationship between the number of nonconforming units produced by plant workers before and those produced after a training program. If X represents before and Y represents after, we would hope that $\hat{\beta}_0$ would be "significantly" below zero. (The significance of regression coefficients is discussed later.) Information from the regression analysis might then be used as an aid in predicting how much money would be saved as a result of the continuation of the training program.

Obviously we cannot "fix" the number of nonconforming units produced by a plant worker, so the question arises as to whether the same calculations could be performed on such data as were used in Table 14.1. Sampson (1974) demonstrated that the "fixed X" and the "random X" cases can be treated the same. The point estimators $\hat{\beta}_0$ and $\hat{\beta}_1$ will be different for the two cases, but the *values* of the estimators will be the same, so there is no need to distinguish between the two cases. [See also Neter and Wasserman (1974, p. 76) for further discussion.] Mandel (1984) draws a different conclusion, however, and presents a different approach. The problem has also been addressed by Madansky (1959) and others. Nevertheless, there is general agreement that the classical regression approach can be used safely when the variation in the X values is small relative to the variation in the Y values.

In addition to this "random error," measurement errors often are a problem, especially when regression analysis is used in the physical sciences. If measurement errors were always small and random with an expected value of zero, they would not be a major problem. If the errors are systematic, however, there is definitely a problem, although the problem has not been given much attention in the literature. Seber (1977, sec. 6.4) summarizes the results concerning both random measurement error and round-off error and also discusses the random regressor case in his sec. 6.5. Measurement error is discussed in Neter and Wasserman (1974, sec. 5.7); see also Draper and Smith (1981, sec. 2.14). In general, the consequences of measurement error can vary greatly, depending on whether Y or X is measured with error. If Y is measured with error and the error is random, this simply becomes a part of the random error term ϵ in the model, and the conventional approach can be used. The measurement error for X can also be effectively ignored when the variance of the error term is small compared to the error for Y (which might be comprised of both random and measurement errors).

The round-off error can be a serious error in regression, even when a computer is used. If a hand calculator is used, only the final result should be rounded. Rounding intermediate calculations to, say, two decimal places can cause a serious loss of accuracy. The extent of the loss will depend on the characteristics of the data. If a computer is being used, it will be necessary for certain data sets to have the calculations performed in double precision. Rounding off data when they are recorded can also cause a loss of accuracy. See Freund (1963) for additional information concerning round-off error in regression.

14.2.1 Assumptions

Various assumptions are usually made in (linear) regression analysis. The only one that has been alluded to thus far is the implicit assumption that the relationship between X and Y is linear. Assumptions should always be checked, and this assumption can be checked by a simple scatter plot, as was indicated earlier.

The other assumptions relate to the error term ϵ in the model $Y = \beta_0 + \beta_1 X + \epsilon$. These assumptions are represented by the expression

$$\epsilon_i \sim N(0, \sigma^2).$$

This indicates that the errors have a normal distribution with mean zero and are independent with the same variance, σ^2, for each X_i.

It should be understood, however, that these last assumptions are unnecessary when regression is used simply to examine the relationship between Y and X and to develop a regression equation. A regression user will typically want to go further than this, however, and construct prediction intervals, confidence intervals, hypothesis tests, and a test for the adequacy of the model.

14.2.2 Confidence Intervals

A confidence interval is constructed for a parameter, and there are two parameters in the simple linear regression model, β_0 and β_1. The confidence interval for each one is of the form

$$\hat{\beta}_i \pm t_{\alpha/2,n-2}\, S_{\hat{\beta}_i}$$

where $t_{\alpha/2,n-2}$ designates the $(1 - \alpha/2)$ percentage point of a t distribution with $n - 2$ degrees of freedom and $S_{\hat{\beta}_i}$ is the estimated standard deviation of $\hat{\beta}_i$.

A $100(1 - \alpha)\%$ confidence interval for β_1 would be constructed from

$$\hat{\beta}_1 \pm t_{\alpha/2,n-2} \frac{\hat{\sigma}}{\sqrt{S_{xx}}} \tag{14.6}$$

where

$$\hat{\sigma} = \left(\frac{S_{yy} - \hat{\beta}_1 S_{xy}}{n-2}\right)^{1/2}$$

and S_{xx}, S_{xy}, and S_{yy} are as defined previously. Thus with $\alpha = 0.05$, Eq. (14.6) would be used to obtain a 95% confidence interval for β_1. Therefore, before the data are collected and the computations performed, there is a 95% chance that the confidence interval that is to be constructed will contain β_1.

For the data in Table 14.1, $n = 9$, $S_{xx} = 1500$, $S_{xy} = 247.5$, $S_{yy} = 42.38$, and $\hat{\beta}_1 = 0.165$, so that $\hat{\sigma} = 0.4694$. It can be seen from Table A.4 in the Appendix that $t_{0.025,7} = 2.365$, so the 95% confidence interval would be

$$0.165 \pm 2.365 \frac{0.4694}{\sqrt{1500}} = 0.165 \pm 0.0287. \tag{14.7}$$

The lower limit is 0.1363 and the upper limit is 0.1937. The fact that this interval does not include zero indicates that the corresponding hypothesis test of H_0: $\beta_1 = 0$ versus H_1: $\beta_1 \neq 0$, with $\alpha = 0.05$, would lead to rejection of H_0 and acceptance of H_1. The hypothesis test is of the form

$$t = \frac{\hat{\beta}_1}{S_{\hat{\beta}_1}} = \frac{\hat{\beta}_1}{\hat{\sigma}/\sqrt{S_{xx}}} \tag{14.8}$$

where t is the test statistic.

There are other confidence intervals that can be constructed, notably for β_0 and for the mean of Y for a given X (usually denoted by $\mu_{Y|X}$). The former can be obtained from

$$\hat{\beta}_0 \pm t_{\alpha/2,n-2}\left(\frac{\Sigma X^2}{nS_{xx}}\right)^{1/2}\hat{\sigma}$$

and the latter from

$$\hat{Y}_0 \pm t_{\alpha/2,n-2}\left[\frac{1}{n} + \frac{(X_0 - \overline{X})^2}{S_{xx}}\right]^{1/2}\hat{\sigma} \tag{14.9}$$

where the confidence interval for $\mu_{Y|X}$ is seen to depend on X (here denoted by X_0), and $\hat{Y}_0$ is the predicted value of Y using X_0.

14.2.3 Prediction Interval

A prediction interval for Y given X may be more useful than a confidence interval for $\mu_{Y|X}$ since there is often more interest in individual predicted values than in estimating the mean. Such a prediction interval is obtained from

$$\hat{Y}_0 \pm t_{\alpha/2,n-2}\left[1 + \frac{1}{n} + \frac{(X_0 - \overline{X})^2}{S_{xx}}\right]^{1/2}\hat{\sigma}. \tag{14.10}$$

It is called a prediction interval rather than a confidence interval since the latter is for parameters and a future value of Y is not a parameter. A prediction interval is a measure of the precision of the regression equation in predicting future values of Y. Notice the difference between Eqs. (14.9) and (14.10). In particular, the prediction interval will be wider than the confidence interval since there will naturally be less precision in predicting a particular value of Y than in estimating the average value of Y.

The three confidence intervals are robust in the presence of slight to moderate departures from normality, but this is not true of the prediction interval. "Robust" means that a statistical procedure is not sensitive to violations of the prerequisite assumptions.

14.2.4 Checking Assumptions

There is an important point concerning confidence intervals, prediction intervals, and hypothesis tests in regression which cannot be overemphasized. The confidence interval given by Eq. (14.7) was constructed under the assumption that the error terms have a normal distribution and are independent (with constant variance). How are these assumptions checked, and what if the appropriate checks indicate that the assumptions are apparently not valid? The assumption of normality is not as critical as the other two assumptions, however, since it is well known that confidence intervals and hypothesis tests using the t distribution are relatively robust (insensitive) to slight to moderate departures from normality. It is still a good idea to check the normality assumption. There are a number of ways to do this. One approach is to produce a normal probability plot of the residuals and see whether the points form approximately a straight line. (The residuals are $e_i = Y_i - \hat{Y}_i$, $i = 1, 2, \ldots, n$.)

The assumptions of independent errors and constant error variance can be checked by plotting e_i against X_i. Figure 14.2 indicates a desirable plot of the residuals. The spread of the residuals is fairly constant over X, and the residuals do not exhibit an obvious pattern. Compare this with the residual plot given in Fig. 14.3. Such a plot can easily occur when regression is (inappropriately) used for time series data, rather than time series regression-type methods. The potential hazards of such misuse are clearly demonstrated in Box and Newbold (1971), who give an example where a study by Coen, Gomme, and Kendall (1969) showed that car sales (seven quarters earlier) could be used to predict stock prices. This resulted from the improper use of ordinary least squares. Specifically, $\hat{\beta}_1$ was 14 times its standard error so that the regression was deemed highly significant. Unfortunately, the errors were also correlated, and when Box and Newbold fit an appropriate model that accounted for the error structure, they showed that no significant relationship

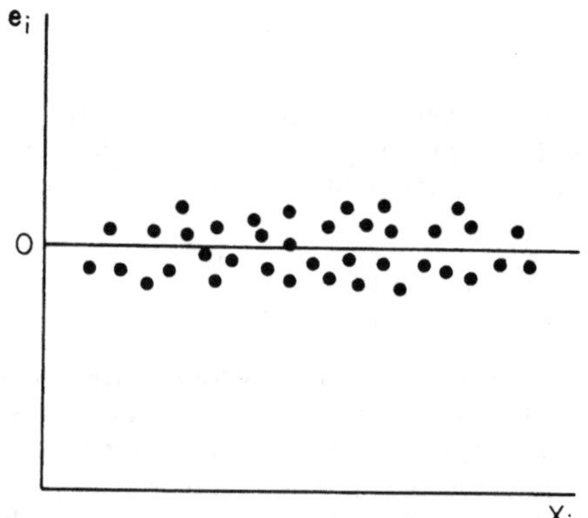

FIGURE 14.2 Residual plot.

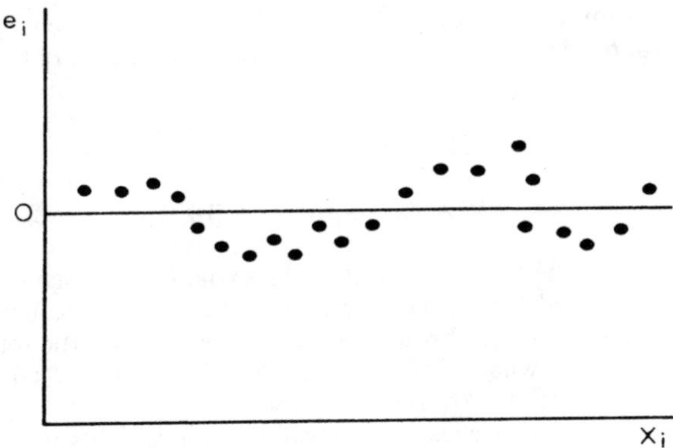

FIGURE 14.3 Plot illustrating nonrandom residuals.

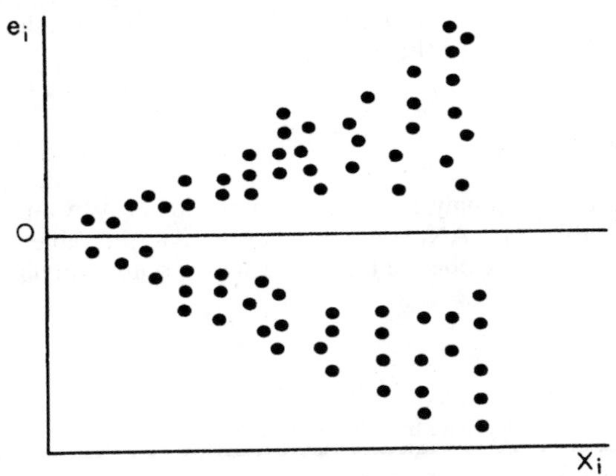

FIGURE 14.4 Plot illustrating nonconstant residual variance.

existed between the two variables. This example is also discussed in Box, Hunter, and Hunter (1978, p. 496).

Another residual plot that indicates a problem is shown in Fig. 14.4. This plot provides evidence that the assumption of a constant error variance has been violated. One way to correct for this problem would be to transform the data (discussed later in this chapter). Another possibility would be to use *weighted least squares*.

14.2.5 Weighted Least Squares

Figure 14.4 indicates that Var(e) increases as X increases, where Var denotes variance. Roughly, Var(ϵ_i) = kX_i, where k could be estimated from a plot such as Fig. 14.4. We could obtain $\hat{\beta}_0$ and $\hat{\beta}_1$ without having to estimate k, however, since the appropriate formulas do not involve k. Specifically, as given by Neter and Wasserman (1974, p. 136), the solutions for $\hat{\beta}_0$ and $\hat{\beta}_1$ would be obtained by solving the equations

$$\Sigma W_i Y_i = \hat{\beta}_0 \Sigma W_i + \hat{\beta}_1 \Sigma W_i X_i$$

$$\Sigma W_i X_i Y_i = \hat{\beta}_0 \Sigma W_i X_i + \hat{\beta}_1 \Sigma W_i X_i^2$$

with $W_i = 1/kX_i$. Since the constant $1/k$ in $1/kX_i$ is common to each of the summations, it can be extracted from each summation and will not be involved in the expression for either $\hat{\beta}_0$ or $\hat{\beta}_1$. This would also obviously hold true for the general expression $W_i' = 1/kf(x_i)$, where $f(x_i)$ is a function of X_i[such as $f(x_i) = X_i^2$]. [This would correspond to $\text{Var}(\epsilon_i) = kf(x_i)$.]

More sophisticated approaches for determining the weights may be found in Draper and Smith (1981, p. 115) and in Carroll and Ruppert (1988). There will usually not be enough repeats of each data point to allow a standard weighted least-squares approach to be used, as when only a few repeats are available each W_i cannot be efficiently determined (Carroll and Cline, 1988). Accordingly it will often be better to model the variance by using all of the data rather than determining each W_i by using only the repeats of the ith point.

14.2.6 ANOVA Tables

The results of a regression analysis can be summarized partly by an analysis of variance (ANOVA) table. The ANOVA table for the data in Table 14.1 is given in Table 14.3. These values are obtained from the appropriate formulas given in Table 14.4.

TABLE 14.3 ANOVA Table, Data from Table 14.1

Source	Degrees of freedom	Sum of squares	Mean square	F	p
Regression	1	40.8375	40.8375	185.32	2.7×10^{-6}
Residual	7	1.5425	0.2204		
Total	8	42.3800			

TABLE 14.4 ANOVA Table, Formulas for Simple Linear Regression

Source	Degrees of freedom	Sum of squares SS	Mean square MS	F
Regression	1	$\hat{\beta}_1 S_{xy}$	$\hat{\beta}_1 S_{xy}$	$\frac{MS_{reg}}{MS_{res}}$
Residual	$n-2$	$S_{yy} - \hat{\beta}_1 S_{xy}$	$\frac{S_{yy} - \hat{\beta}_1 S_{xy}}{n-2}$	
Total	$n-1$	S_{yy}		

Although the p value in Table 14.3 is not usually found in statistics books, it is given here because it is frequently part of the output from computer programs. (See Sec. 14.5 for information on software for regression.) The p value is the probability of obtaining an F value as large as or larger than the calculated value (in this case 185.32) when $\beta_1 = 0$. It is thus a test of the hypothesis H_0: $\beta_1 = 0$ against the alternative hypothesis H_1: $\beta_1 \neq 0$. Also, it can be shown easily that $t^2 = F$, where t^2 is obtained from Eq. (14.8). [In general, for the two statistical distributions t and F, $(t_{\alpha/2, n-2})^2 = F_{\alpha, 1, n-2}$.] In order to conclude that $\beta_1 \neq 0$, we require p values that are quite small, which is the same as saying that we require large F values. How large should F be? If we use a significance level of 0.05, we would reject the hypothesis that $\beta_1 = 0$ if, for $n = 9$, $F > 5.59$. A value of F not much larger than 5.59 would, however, mean that the regression equation is of very little value for predicting Y.

To illustrate, if we had used the Y in Table 14.1 but had selected another X and found that $F = 5.59$, then R^2 would equal 0.4440. Thus less than half of the variation in Y would have been explained by that particular X, and if the X was random instead of fixed, one could then say that the (Pearson) correlation coefficient between X and Y was (only) $\sqrt{0.4440} = 0.6663$.

This correlation coefficient measures the strength of the linear relationship between two random variables. For random variables X and Y, the coefficient would be denoted by r_{xy} and defined as

$$r_{xy} = \frac{S_{xy}}{\sqrt{S_{xx}S_{yy}}}$$

where S_{xy}, S_{xx}, and S_{yy} are as defined previously. The value of r_{xy} must be between −1 and +1, inclusive. The latter would indicate a perfect positive correlation, as all of the (x, y) points in a scatter plot could be connected by a single straight line, which has a slope of +1. Similarly, a coefficient of −1 would mean that all of the points could be connected by a single straight line which has a slope of −1. If $r_{xy} = 0$, there would be no evidence of a linear relationship between X and Y. For example, if r_{xy} is calculated using certain points which form a circle, r_{xy} can be zero even though there is a definite relationship between X and Y (but not a linear relationship).

The F value should be considerably larger than the minimum value needed to reject H_0: $\beta_1 = 0$. But how much larger? This question is addressed in Draper and Smith (1981, pp. 129–133) who conclude that, based upon the work of Wetz (1964), the value of F should be at least four or five times this minimum value.

Using five as the multiple in conjunction with this same Y would imply that X should be such that $R^2 \geq 0.7997$. Obviously this would be much more reasonable than $R^2 \geq 0.4440$. This requirement would also correspond to $p \leq 0.0011$.

14.2.7 Lack of Fit

The F value in Table 14.3 was calculated under the implicit assumption that the residual mean square provides an unbiased estimate of σ^2. This will be true, however, only if the model $Y = \beta_0 + \beta_1 X + \epsilon$ is the "true" model. Since we will never know the true model, we simply use the best model that we can find, where best might include "simplicity."

A "lack-of-fit" test is used to determine whether the model is correct. For example, higher-order terms in X might be needed in the model.

For the data in Table 14.1 let us assume that an additional reading was made for each of the following temperatures: 290, 300, 310, and 320°F, and the corresponding process yields were recorded. These would then be added to the set of existing values. The augmented data set is given in Table 14.5.

The difference in process yields for the same temperature is an indication of the pure error in the data, which cannot be reduced by trying to improve upon the existing model. The use of these repeat readings enables us to separate pure error from error due to lack of fit. Evidence of lack of fit might mean that nonlinear terms in X (such as X^2) should be added to the model, or that some nonlinear model such as an exponential one might be more appropriate. The numerical results of this analysis are given in Table 14.6, using the formulas in Table 14.7. In the latter, k represents the number of different X values, n_j is the number of Y values corresponding to X_j, and $\overline{Y}_j$ is the average of those values.

The data in Table 14.6 give strong evidence of lack of fit since $F = 47.8$ ($= 0.3585/0.0075$) results in a small p value of 0.0011. As indicated previously, this indicates that the model $Y = \beta_0 + \beta_1 X + \epsilon$ could be improved by adding higher-order terms in X, such as X^2 or X^3. The selection of one or more terms to add to an existing model (or to replace one or more terms in an existing model) can be aided by the use of residual plots. These were discussed previously in regard to checking the assumptions of (1) constant error variance and (2) independence of the errors. They can also be used to check the adequacy of a model.

In this instance, however, we cannot improve the existing model very much since $R^2 = 0.96$ (that is, there is not much room for improvement). Thus although

TABLE 14.5 Data with Repeat Observations

Temperature X	Process yield Y
310	17.3
310	17.4
290	14.2
290	14.0
300	15.5
300	15.4
315	18.1
320	20.1
320	20.1
305	16.7
325	20.2
295	15.3
285	14.1

TABLE 14.6 ANOVA Table, Data from Table 14.5

Source	Degrees of freedom		Sum of squares SS		Mean square MS	F	p
Regression	1		60.9005		60.9005	263.79	4.9×10^{-9}
Residual	11		2.5395		0.2309		
Lack of fit		7		2.5095	0.3585	47.8	0.0011
Pure error		4		0.0300	0.0075		
Total	12		63.4400				

TABLE 14.7 Lack-of-Fit Computation

Source	Degrees of freedom	Sum of squares SS
Residual	$n-2$	$S_{yy} - \hat{\beta}_1 S_{xy}$
Lack of fit	$k-2$	By subtraction; $SS_{residual} - SS_{pure\ error}$
Pure error	$n-k$	$\sum_{j=1}^{k} \sum_{i=1}^{n_j} (Y_{ij} - \overline{Y}_j)^2$

we could plot the residuals $Y_i - \hat{Y}_i$ against X_i and consider adding an X^2 term if the plot exhibited quadrature, such an addition might be ill-advised. The reason for this is that whenever a term is added to a regression model, the variance of $\hat{Y}$ can never decrease and will almost always increase. This is something that must be considered when multiple regression (more than one X) is used. Since the addition of terms to the current model falls within the realm of multiple regression, consideration of the addition of other terms will be deferred until later.

14.2.8 Regression through the Origin

One possible modification of the basic model that can be discussed in the context of simple linear regression is the possible deletion of the constant term β_0 from the model. As indicated earlier, the significance of β_0 in the model can be assessed by testing H_0: $\beta_0 = 0$ versus H_1: $\beta_0 \neq 0$, or by constructing a confidence interval for β_0 and seeing whether the interval includes zero. Such tests would be appropriate, only if $X = 0$ is included in the range of data. If we *knew* that $Y = 0$ when $X = 0$, we might wish to consider the model $Y = \beta_1' X + \epsilon'$ (that is, a no-intercept model), but the intercept model would be generally used if the range of values of X to be used in the regression equation does not come close to including zero, even though $Y = 0$ when $X = 0$.

14.2.9 Regression for Control

Until now the following uses of regression have been illustrated: (1) description, (2) prediction, and (3) estimation. The relationship between Y and X can be described by using the sign and the magnitude of the coefficients $\hat{\beta}_0$ and $\hat{\beta}_1$, and by the size of R^2, and the regression equation can be used to predict future values of Y. Finally, if the model $Y = \beta_0 + \beta_1 X + \epsilon$ is appropriate, the parameters β_0 and β_1 are estimated by the least squares coefficients $\hat{\beta}_0$ and $\hat{\beta}_1$.

The other use of regression, for control, is not mentioned prominently in textbooks, but it is a potentially important use. Specifically, if a cause-and-effect relationship can be inferred, Y could conceivably be controlled by varying X. For example, if Y is some measure of water pollution and X is a known contributor to water pollution, X could be varied in an effort to keep Y at a tolerable level. Of course, for this to be feasible X must be controllable by the experimenter. See Draper and Smith (1981, p. 413) and Hahn (1974) for more information concerning the use of regression for control.

14.2.10 Transformations

If it appears that the true relationship between Y and X can be represented by an exponential model such as

$$Y = \exp(X\beta + \epsilon) \tag{14.11}$$

linear regression methods as presented previously can still be used if this nonlinear model can be transformed into a linear model. In this case, this can be accomplished by taking the logarithm of both sides of the equation. Specifically,

$$\ln Y = X\beta + \epsilon \tag{14.12}$$

which is in the general form of a linear regression model, where ln Y represents the natural logarithm of Y and is thus the dependent variable.

On the other hand, the model

$$Y = X \ln\left(\frac{X}{\beta_1}\right)^{\beta_0} + \epsilon \tag{14.13}$$

cannot be transformed into a linear regression model. Thus, a model such as Eq. (14.11) would generally be called *intrinsically linear* since it can easily be transformed into a linear model, whereas Eq. (14.13) would be called *intrinsically nonlinear*. Nonlinear regression methods would have to be used to estimate the parameters in Eq. (14.13). These methods are discussed in Chap. 18; see also Draper and Smith (1981, chap. 10).

These are model transformations; transformations on X and/or Y may sometimes be beneficial, particularly to better satisfy the regression assumptions discussed earlier. For example, a weighted least-squares analysis with $W_i = 1/kX_i^2$ will produce the same results as use of the model

$$Y^* = \beta_0^* + \beta_1^* X^* + \epsilon^* \tag{14.14}$$

where $Y^* = Y/X$, $X^* = 1/X$, $\beta_0^* = \beta_1$, $\beta_1^* = \beta_0$, and $\epsilon^* = \epsilon/X$.

Thus Eq. (14.14) involves the transformation of both X and Y for the purpose of stabilizing the error variance, where $\text{Var}(\epsilon_i) = kX_i^2$. Transformations that stabilize the variance of the error term may simultaneously correct for nonnormality of the error term. Therefore it is generally advisable to correct for the nonconstant variance first, and then check to see whether the transformation has produced error terms that appear to be approximately normally distributed.

Transformations are also performed for the purpose of improving the fit to the data. For example, regressing Y on $\log(X)$ might produce a better regression equation than that obtained by regressing Y on X. Assume that a model, with possibly nonlinear regressors, has been identified that provides a good fit to the data but the error assumptions appear to be grossly violated. If Y is subsequently transformed in an effort to correct for this problem, the quality of the fit may be greatly affected. To prevent this from happening, the right side of the equation would be transformed in the same manner. For example, if we start from the simple linear regression model of Eq. (14.1) and Y is transformed to Y^λ, the new model would be $Y^\lambda = (\beta_0 + \beta_1 X)^\lambda + \epsilon$. Notice that this is a nonlinear regression model (Chap. 18). This transformation approach is the "transform both sides" approach popularized by Carroll and Ruppert (1984, 1988).

Many useful transformations are members of the Box and Cox (1964) family of transformations. To determine whether a transformation of Y has improved the model fit, it is necessary to convert the predicted values back to the original scale and then compute R^2. If Y has not been transformed, it makes no difference whether R^2 is computed using Eq. (14.5) or by using one of the equivalent expressions. Equivalence will generally not occur when Y has been transformed and the predicted values converted back to the original scale, and the R^2 formula recommended by Kvålseth (1985) will generally be negative when the model fit is not good. Nevertheless, the need to transform back to the original scale has been demonstrated by Scott and Wild (1991). See also the discussions in Anderson-Sprecher (1994) and Ryan (1997).

14.2.11 Inverse Regression

One area of application of simple linear regression is in calibration work, but the way in which it is used is much different from customary usage of regression.

Suppose we have two different methods of measuring the transconductance of a certain electronic device. One method employs a "standard" measuring instrument, which is highly accurate but also slow and expensive to use. The other method uses a measuring device which is much faster than the standard instrument, but has virtually a constant bias. If there is a linear relationship between the measurements obtained using the two instruments (as would be implied by the words "constant bias"), the measurement obtained using the second device could then be used to estimate the true measurement that would presumably be obtained using the standard instrument.

In the context of regression, the measurement obtained using the second device would have to be the dependent variable since it is measured with error, and the measurement obtained with the standard instrument would be the independent variable.

The objective would then be to obtain the regression equation

$$\hat{Y} = \hat{\beta}_0 + \hat{\beta}_1 X$$

and then "solve" for X to obtain

$$X = \frac{\hat{Y} - \hat{\beta}_0}{\hat{\beta}_1}.$$

Here X is being estimated rather than Y, so for a known value of Y, say Y_k, the equation is

$$\hat{X}_k = \frac{Y_k - \hat{\beta}_0}{\hat{\beta}_1}. \tag{14.15}$$

Thus the standard measurement is estimated (predicted) using a known measurement from the second device.

This approach has been termed the classical estimation approach to calibration. Krutchkoff (1967) reintroduced the concept of "inverse" regression to estimate X, where X is regressed on Y. Thus,

$$\hat{X}_k^* = \hat{\beta}_0^* + \hat{\beta}_1^* Y_k \tag{14.16}$$

where the asterisks are used to signify that the estimated value of X_k and the parameter estimates will be different from the estimates obtained using the classical approach in Eq. (14.15).

We shall use the data in Hunter and Lamboy (1981, table 1) to illustrate the two methods. The data are given here in Table 14.8.

Regressing Y on X produces the equation

$$\hat{Y} = 0.76 + 0.9873X$$

so that X would be estimated as

$$\hat{X} = \frac{Y - 0.76}{0.9873}$$

$$= -0.7698 + 1.0129Y.$$

Regressing X on Y produces the inverse regression estimator

$$\hat{X}^* = -0.7199 + 1.0048Y.$$

Thus for a measured amount of molybdenum of 8.6 the classical estimator would lead us to estimate the actual amount as 7.94, whereas the inverse regression estimator would produce an estimate of 7.92. Thus there is hardly any difference in the two estimates. Note that at $Y = 1.7$ the two estimates would differ somewhat more (0.95 versus 0.99). In general, the two methods will produce estimates that are very

TABLE 14.8 Calibration Data

Known amount of molybdenum X	Measured amount of molybdenum Y
1	1.8
1	1.6
2	3.1
2	2.6
3	3.6
3	3.4
4	4.9
4	4.2
5	6.0
5	5.9
6	6.8
6	6.9
7	8.2
7	7.3
8	8.8
8	8.5
9	9.5
9	9.5
10	10.6
10	10.6

The data were originally provided by G. E. P. Box and are given (without X and Y designated) as table 1 in Hunter and Lamboy (1981). Reproduced, with permission, from *Technometrics*, vol. 23, p. 326, 1981.

close whenever the correlation coefficient r_{xy} is almost 1.0. (Such a value for r_{xy} would result if the second device did have almost a constant bias.)

14.3 MULTIPLE LINEAR REGRESSION

There is often more than one independent variable in a typical application of linear regression. For p independent variables the model for multiple linear regression is

$$Y = \beta_0 + \beta_1 X_1 + \beta_2 X_2 + \cdots + \beta_p X_p + \epsilon. \tag{14.17}$$

It should be noted that this is a natural extension of the simple linear regression model to the case of more than one X. As such, the principles of simple regression also apply, in general, to multiple regression. The use of more than one X does, however, create additional complexities. As Snedecor and Cochran (1980, p. 334) state: "Multiple linear regression is a complex subject."

In particular, hand calculation is inadvisable when p is greater than 2 or 3 because of the amount of work involved. This is more drudgery than complexity. The real complexity involves (1) determining how many and which variables to use, including the form of each variable (such as linear or nonlinear), (2) interpreting the results, especially the regression coefficients, and (3) determining whether an alternative to ordinary least squares should be used.

In what follows it is assumed that the reader is familiar with matrix algebra. Readers without this requisite knowledge are referred to Appendix 14.1 at the end of this chapter.

For the forthcoming example we shall assume that the appropriate model is

$$Y = \beta_0 + \beta_1 X_1 + \beta_2 X_2 + \epsilon \tag{14.18}$$

that is, there are two regression variables (also called *regressors*) and one response variable. This can be written in matrix notation as

$$\mathbf{Y} = \mathbf{X}\boldsymbol{\beta} + \boldsymbol{\epsilon}$$

for n observations on Y, X_1, and X_2. Specifically,

$$\mathbf{Y} = \begin{bmatrix} Y_1 \\ Y_2 \\ \cdot \\ \cdot \\ \cdot \\ Y_n \end{bmatrix}, \quad \mathbf{X} = \begin{bmatrix} 1 & X_{11} & X_{12} \\ 1 & X_{21} & X_{22} \\ \cdot & \cdot & \cdot \\ \cdot & \cdot & \cdot \\ \cdot & \cdot & \cdot \\ 1 & X_{n1} & X_{n2} \end{bmatrix}, \quad \boldsymbol{\beta} = \begin{bmatrix} \beta_0 \\ \beta_1 \\ \beta_2 \end{bmatrix}, \quad \boldsymbol{\epsilon} = \begin{bmatrix} \epsilon_1 \\ \epsilon_2 \\ \cdot \\ \cdot \\ \cdot \\ \epsilon_n \end{bmatrix}$$

where X_{ij} denotes the ith observation on the jth regression variable. Thus,

$$\begin{aligned} Y_1 &= \beta_0 + \beta_1 X_{11} + \beta_2 X_{12} + \epsilon_1 \\ Y_2 &= \beta_0 + \beta_1 X_{21} + \beta_2 X_{22} + \epsilon_2 \\ &\cdot \\ &\cdot \\ &\cdot \\ Y_n &= \beta_0 + \beta_1 X_{n1} + \beta_2 X_{n2} + \epsilon_n \end{aligned}$$

As with simple linear regression, the method of least squares stipulates that $\Sigma_{i=1}^{n} \epsilon_i^2$ is to be minimized. For the model in Eq. (14.18)

$$\Sigma\epsilon_i^2 = \Sigma(Y_i - \beta_0 - \beta_1 X_{i1} - \beta_2 X_{i2})^2$$

and the differentiation of the right-hand side of the equation with respect to β_0, β_1, and β_2 (separately) produces three equations in the three unknown parameters. These so-called normal equations can be written in matrix notation as

$$\mathbf{X'X}\beta = \mathbf{X'Y}$$

the solution of which is

$$\hat{\beta} = (\mathbf{X'X})^{-1}\mathbf{X'Y}. \tag{14.19}$$

The use of Eq. (14.19) to obtain the regression equation by hand calculator is relatively simple for $p = 2$, and the confidence intervals, prediction intervals, and hypothesis tests can also be obtained using matrix products. Many hand-held calculators have a built-in subroutine for multiple regression (with small p), which further simplifies the task. The use of a larger computing device is desirable when p is much greater than 2.

Large values of p can still create problems, even when large computers are used. Round-off error was discussed previously, and numerical problems are also alluded to in Sec. 14.5.

The following hypothetical example will be used to demonstrate not only the computational aspects of multiple regression, but also some of the nuances and interpretational difficulties that are frequently encountered. The data are given in Table 14.9.

TABLE 14.9 Regression Example ($p = 2$)

Y	X_1	X_2
30.1	8	22
32.2	9	19
34.3	11	19
35.4	12	22
34.4	10	18
30.0	8	22
31.4	9	19
32.3	10	18
34.0	12	22
33.1	11	19
33.2	11	19
34.3	12	22
31.1	9	19
30.0	8	22
32.3	10	18
34.4	12	22
30.3	8	22
31.6	9	19
33.3	11	19
32.0	10	18

Using the model $Y = \beta_0 + \beta_1 X_1 + \beta_2 X_2 + \epsilon$, the regression equation using the data in Table 14.9 is obtained in the following manner.

If we "center" the data by subtracting from each variable value the mean of that variable, we then need to invert only a 2×2 matrix rather than a 3×3 matrix, and the resultant prediction equation is of the form

$$\hat{Y} - \overline{Y} = \hat{\beta}_1(X_1 - \overline{X}_1) + \hat{\beta}_2(X_2 - \overline{X}_2)$$

which can then be rearranged to produce

$$\hat{Y} = \hat{\beta}_0 + \hat{\beta}_1 X_1 + \hat{\beta}_2 X_2$$

where $\hat{\beta}_0 = \overline{Y} - \hat{\beta}_1 \overline{X}_1 - \hat{\beta}_2 \overline{X}_2$. Specifically,

$$\mathbf{X} = \begin{bmatrix} -2 & 2 \\ -1 & -1 \\ 1 & -1 \\ 2 & 2 \\ \cdot & \cdot \\ \cdot & \cdot \\ \cdot & \cdot \end{bmatrix}, \qquad \mathbf{Y} = \begin{bmatrix} -2.385 \\ -0.285 \\ 1.815 \\ 2.915 \\ \cdot \\ \cdot \\ \cdot \end{bmatrix}$$

$$\mathbf{X'X} = \begin{bmatrix} 40 & 0 \\ 0 & 56 \end{bmatrix}, \qquad \mathbf{X'Y} = \begin{bmatrix} 43.0 \\ -5.2 \end{bmatrix}$$

so that

$$\hat{\boldsymbol{\beta}} = \mathbf{(X'X)^{-1}X'Y}$$

$$= \begin{bmatrix} \frac{1}{40} & 0 \\ 0 & \frac{1}{56} \end{bmatrix} \begin{bmatrix} 43.0 \\ -5.2 \end{bmatrix}$$

$$= \begin{bmatrix} 1.075 \\ -0.093 \end{bmatrix}$$

Thus, with $\overline{Y} = 32.485$, $\overline{X}_1 = 10$, and $\overline{X}_2 = 20$,

$$\hat{Y} - 32.485 = 1.075(X_1 - 10) + (-0.093)(X_2 - 20)$$

which leads to

$$\hat{Y} = 23.592 + 1.075X_1 - 0.093X_2. \tag{14.20}$$

If X_1 and X_2 represented pressure and temperature, respectively, and Y is process yield, there would appear to be a positive relationship between Y and X_1 and a negative relationship between Y and X_2 based on the signs of the coefficients of X_1 and X_2.

Can we infer that yield should increase when pressure is increased (and temperature is held constant), and decrease when temperature is increased (and pressure is held constant)? In this instance the answer is "yes" if the data in Table 14.9 can be thought of as having come from a designed experiment where pressure was varied from 8 to 12 and temperature from 18 to 22, and these are the regions of future interest.

Looking at the data in Table 14.9 for $X_2 = 22$, the difference between the average values of Y for $X_1 = 8$ and $X_1 = 12$ is 4.425. This compares favorably with $4(1.075) = 4.3$, which is the predicted difference from Eq. (14.20).

Such interpretation of the regression coefficients is possible for these data because X_1 and X_2 are uncorrelated, as evidenced by the fact that the off-diagonal elements of $\mathbf{X'X}$ are zero. This means that the correlation coefficients for Y and X_1 and for Y and X_2 will have the same sign as the regression coefficients for X_1 and X_2 when the regression equation contains both variables.

However, this will not be true in general. If the last column in Table 14.9 is changed so that $X_2 = 18$ when $X_1 = 8$, 22 when $X_1 = 11$, and 19 when $X_1 = 10$, the regression equation will be

$$\hat{Y} = 23.39 + 1.278X_1 - 0.184X_2. \tag{14.21}$$

Notice that there is very little difference between this equation and Eq. (14.20), but Eq. (14.21) does not lend itself to similar interpretation. This is due primarily to the fact that X_1 and X_2 are now highly correlated ($r = 0.93$). This, coupled with the fact that X_1 is more highly correlated with Y than X_2 (0.931 versus 0.840), causes the coefficient for X_2 to be negative. Thus an engineer could *not* decrease temperature (holding pressure constant) and expect yield to increase as Eq. (14.21) might seem to suggest since there is a *positive* correlation between Y and X_2. Accordingly, the yield could be expected to decrease rather than increase. Recall that the correlation was negative for the data in Table 14.9. Thus the correlation coefficient has changed from negative to positive, even though the regression coefficient has remained negative.

This illustrates the value of designing an experiment to be analyzed by regression methods in such a way that the variables are uncorrelated, or at least have very small correlation coefficients. (Here the term "correlation" is being used loosely since a correlation coefficient is defined, strictly speaking, only when both variables are random.) See Chap. 15 for additional information concerning the design of experiments.

For additional information concerning difficulties in interpreting regression coefficients the reader is referred to Mullet (1976).

14.3.1 ANOVA Tables

The general form of the ANOVA table for multiple regression, using matrix notation, is given in Table 14.10. (Here it is assumed that the data are centered.)

TABLE 14.10 ANOVA Table, Centered Data, p Regressors

Source	Degrees of freedom	Sum of squares SS	Mean square MS	F
Regression	p	$\hat{\boldsymbol{\beta}}'\mathbf{X}'\mathbf{Y}$	$\left(\frac{1}{p}\right)\hat{\boldsymbol{\beta}}'\mathbf{X}'\mathbf{Y}$	$\frac{MS_{reg}}{MS_{res}}$
Residual	$n - p - 1$	$\mathbf{Y}'\mathbf{Y} - \boldsymbol{\beta}'\mathbf{X}'\mathbf{Y}$	$\frac{\mathbf{Y}'\mathbf{Y} - \hat{\boldsymbol{\beta}}'\mathbf{X}'\mathbf{Y}}{n - p - 1}$	
Total	$n - 1$	$\mathbf{Y}'\mathbf{Y}$		

This table corresponds to Table 14.4 for simple regression. If the data were not centered, $\mathbf{Y'Y}$ would be replaced by $\mathbf{Y'Y} - (\Sigma Y)^2/n$ and $\hat{\boldsymbol{\beta}}'\mathbf{X'Y}$ would be replaced by $\hat{\boldsymbol{\beta}}'\mathbf{X'Y} - (\Sigma Y)^2/n$.

The data in Table 14.9 produce the ANOVA table presented in Table 14.11. The small p value in Table 14.11 indicates that the regression equation with both variables is of value in predicting Y. The same message is received when R^2 is calculated. The latter can be obtained from an ANOVA table as

$$R^2 = \frac{\text{SS}_{\text{reg}}}{\text{SS}_{\text{total}}}$$

Thus,

$$R^2 = \frac{46.7079}{53.3455} = 0.876.$$

Neither this value nor the p value give any indication of the contribution of each individual regressor. When the regressors are orthogonal, as in Table 14.9 (that is, $\mathbf{X'X}$ is a diagonal matrix), the respective contributions can be determined from the individual components of the product $\hat{\boldsymbol{\beta}}'\mathbf{X'Y}$. Table 14.12 is a more detailed ANOVA table for these data, showing not only the contributions of each regressor but also lack of fit. In multiple regression the sum of squares for lack of fit is computed in the same manner as in simple regression, namely, by computing the pure error sum of squares from repeats and subtracting that from the residual sum of squares. Here a "repeat" means that the values for both regressors are repeated.

Table 14.12 indicates that X_1 is the variable that causes the regression equation to be significant; X_2 contributes very little. There is also no evidence of lack of fit,

TABLE 14.11 ANOVA Table, Centered Data from Table 14.9

Source	Degrees of freedom	Sum of squares SS	Mean square MS	F	p
Regression	2	46.7079	23.3540	59.821	2×10^{-8}
Residual	17	6.6376	0.3904		
Total	19	53.3455			

TABLE 14.12 Detailed ANOVA Table, Centered Data from Table 14.9

Source	Degrees of freedom		Sum of squares SS		Mean square MS	F	p
Regression	2		46.7079		23.3540	59.821	2×10^{-8}
Due to X_1		1		46.2250	46.2250	118.404	4×10^{-9}
Due to X_2		1		0.4829	0.4829	1.237	0.2812
Residual	17		6.6376		0.3904		
Lack of fit		2		0.2051	0.1026	0.2393	0.7951
Pure error		15		6.4325	0.4288		
Total	19		53.3455				

where lack of fit is tested as in simple regression. Specifically, F is obtained as the ratio of the lack-of-fit mean square to the pure error mean square, and the resultant value is compared to a value from the F table. Alternatively, as in Table 14.12, the p value can be assessed for its "smallness." With these data there is no evidence of lack of fit. Accordingly, the residual mean square is used as the divisor in testing for the individual contributions of X_1 and X_2. If lack of fit appeared to be significant, the pure error mean square would be used for this purpose.

It should be noted that it is not necessary to have replicated data points in order to obtain an estimate of the error variance. Indeed, when data are obtained from an observational study, there may not be any replicated points. When this is the case, the "method of near neighbors" discussed in Daniel and Wood (1980, p. 134) can be used to provide a measure of pure error. The latter is estimated by using Y values whose corresponding X values $(X_1, X_2, \ldots, X_p)$ are reasonably close.

The question of what to do with variable X_2 must now be addressed. Should it be retained or dropped from the regression equation? Although R^2 is 0.876 with X_2 included in the equation, and 0.867 with X_2 excluded, this alone is not sufficient justification for including X_2 since R^2 will virtually always increase (and can never decrease) when variables are added to a model. A better measure is the adjusted R^2, which is defined as

$$R^2_{\text{adj}} = 1 - (1 - R^2)\left(\frac{n-1}{n-p}\right)$$

where R^2 is the unadjusted value. Since $n - p$ becomes smaller as p increases, $1 - R^2$ must decrease at a faster rate (that is, R^2 must increase at a faster rate) in order for R^2_{adj} to increase.

For the present set of data, with only X_1 in the equation, $R^2_{\text{adj}} = 0.867$.(This is the same as the unadjusted R^2 when $p = 1$.) However, when X_2 is added to the equation, $R^2_{\text{adj}} = 0.869$ so that there is hardly any change. (This is not to suggest that R^2_{adj} should be used in selecting which variables to use in a regression equation; there are better methods for that purpose, which are discussed later. The intent here is simply to show that the adjusted R^2 gives us a somewhat better picture than the unadjusted R^2 when variables are added to a model.)

Whether we use the F-test result in Table 14.12 or the adjusted R^2 value, it would seem desirable to exclude X_2 from the regression equation. One important justification for doing so is the following result. Walls and Weeks (1969) showed that the variance of $\hat{Y}$ can never be decreased (and will almost always be increased) when new regressors are added. Specifically,

$$\text{Var}(\hat{Y}_0) = \sigma^2 \left[\mathbf{X}_0' \, (\mathbf{X}'\mathbf{X})^{-1} \, \mathbf{X}_0 + \frac{1}{n} \right] \tag{14.22}$$

where $\mathbf{X}_0$ contains a single input value for each regressor and $\hat{Y}_0$ is the predicted value of Y using those input values. Thus the variance depends on the values of the regressors. [Note: We are still assuming centered data at this point. If the data are not centered, there would not be a $1/n$ term inside the brackets in Eq. (14.22). Also, the first column of $\mathbf{X}$ would contain a 1 in each position, and $\mathbf{X}_0$ would contain a 1 in the first position.]

To illustrate the computation for Eq. (14.22) we shall obtain $\text{Var}(\hat{Y}_0)$ for $X_i = 2$ and $X_2 = 2$, using the MS_{res} from Table 14.12 as an estimate of σ^2. Thus,

$$\text{Est. Var}(\hat{Y}_0) = \text{MS}_{\text{res}} \left[\mathbf{X}_0' \, (\mathbf{X}'\mathbf{X})^{-1}\mathbf{X}_0 + \frac{1}{n} \right]$$

$$= 0.3904\left[(2\ \ 2)\begin{bmatrix}\frac{1}{40} & 0\\ 0 & \frac{1}{56}\end{bmatrix}\begin{pmatrix}2\\2\end{pmatrix}+\frac{1}{20}\right]$$

$$= 0.0864$$

where Est. Var indicates that the variance is estimated. (Since the data are centered, $X_1 = 2$ represents 12 in Table 14.9 and $X_2 = 2$ represents 22.) This same number, 0.0864, would have been obtained if the raw data had been used, but the computations would have been a bit more tedious.

14.3.2 Confidence and Prediction Intervals

The estimated variance of Y_0 is used to obtain a confidence interval for the mean of Y given $\boldsymbol{X}_0$, and in obtaining a prediction interval for an individual value of Y given $\boldsymbol{X}_0$. (These two types of intervals were illustrated in Secs. 14.2.2 and 14.2.3.)

The general form of a $100(1-\alpha)\%$ confidence interval for the mean of Y given $\boldsymbol{X}_0$ is

$$\hat{Y}_0 \pm t_{\alpha/2,n-p-1}\sqrt{\text{MS}_{\text{res}}}\left[\mathbf{X}_0'(\mathbf{X}'\mathbf{X})^{-1}\mathbf{X}_0+\frac{1}{n}\right]^{1/2}.$$

Using the (centered) values of $X_1 = 2$ and $X_2 = 2$ to produce $\boldsymbol{X}_0' = (2, 2)$, and plugging the uncoded values $X_1 = 12$ and $X_2 = 22$ into Eq. (14.20) to obtain $\hat{Y}_0$, a 95% confidence interval with the present data is

$$34.449 \pm 2.110\sqrt{0.3904}\left(\frac{1}{10}+\frac{1}{14}+\frac{1}{20}\right)^{1/2} = 34.449 \pm 0.620.$$

Thus, the lower limit is 33.829 and the upper limit is 35.069.

The general form of a $100(1-\alpha)\%$ prediction interval for Y given $\mathbf{X}_0$ is

$$\hat{Y}_0 \pm t_{\alpha/2,n-p-1}\sqrt{\text{MS}_{\text{res}}}\left[1+\frac{1}{n}+\mathbf{X}_0'(\mathbf{X}'\mathbf{X})^{-1}\mathbf{X}_0\right]^{1/2}.$$

Again using the centered values of $X_1 = 2$ and $X_2 = 2$, a 95% prediction interval with the present data is

$$34.449 \pm 2.110\sqrt{0.3904}\left(1+\frac{1}{20}+\frac{1}{10}+\frac{1}{14}\right)^{1/2} = 34.449 \pm 1.457$$

so that the lower limit is 32.992 and the upper limit is 35.906.

Individual confidence intervals can also be obtained for $\beta_1, \beta_2, \ldots, \beta_p$. These are of the form

$$\hat{\beta}_i \pm t_{\alpha/2,n-p-1}\sqrt{\text{MS}_{\text{res}}}\,(c_{ii})^{1/2}$$

where c_{ii} is the ith diagonal element of $(\mathbf{X}'\mathbf{X})^{-1}$. For example, the 95% confidence interval on β_1 would be obtained as

$$1.075 \pm 2.110\sqrt{0.3904}\left(\frac{1}{40}\right)^{1/2} = 1.075 \pm 0.208$$

That is, the lower limit is 0.867 and the upper limit is 1.283. The interval for β_2 may be obtained similarly, with 1.075 replaced by –0.093 and 1/40 replaced by 1/56. (*Note*: There is no confidence interval for β_0 with centered data because there is no $\hat{\beta}_0$, which is obtained when raw data are used.)

14.3.3 Hypothesis Tests

Hypothesis tests on the β_i ($i = 1, 2, \ldots, p$) can also be performed, but these are unnecessary when a detailed ANOVA table such as Table 14.12 is available. This is because these t tests will lead to the same conclusions as the F-test results given in Table 14.12. For example, the test statistic for testing $\beta_1 = 0$ is of the form

$$t = \frac{\hat{\beta}_1}{s_{\hat{\beta}_1}} = \frac{\hat{\beta}_1}{\sqrt{\text{MS}_{\text{res}}\, c_{11}}}$$

which for the present data produces

$$t = \frac{1.075}{\sqrt{0.3904(1/40)}} = 10.8814.$$

This would lead to rejection of the hypothesis that $\beta_1 = 0$. The square of this t value is equal to the F value in Table 14.12 (within rounding), so there is no need to use both. (The fact that the confidence interval for β_1 does not include zero would also dictate rejection of the hypothesis.)

Thus when the data are orthogonal, as in the present example, the value of the F statistic for testing the contribution of a particular regressor will be the same, regardless of the order in which the regressors are entered in the prediction equation. That is, the marginal contribution of each regressor is the same as the total contribution, and each F value will be the square of the corresponding t value.

This relationship does not hold when the data are nonorthogonal since the value of the F statistics will depend on the order in which the regressors are entered in the prediction equation. Order is irrelevant with t tests, however, since with a t test one is testing the marginal contribution of a regressor with the implicit assumption that all of the other regressors are *already in* the prediction equation.

The t tests and the confidence intervals will agree, regardless of whether the data are orthogonal. That is, if a confidence interval on β_i does not include zero, the t test would lead to rejection of $\beta_i = 0$, and the latter would not be rejected when the interval does include zero.

As mentioned in Sec. 14.2.3, these t tests and confidence intervals are insensitive to slight to moderate departures from normality, but the prediction interval is somewhat sensitive.

14.3.4 Extrapolation

This topic was touched upon in Sec. 14.2. It is easy to see when extrapolation occurs in simple regression, but in multiple regression it is less obvious. Extrapolation can be determined easily, however, when there are only two regressors. Assume that a

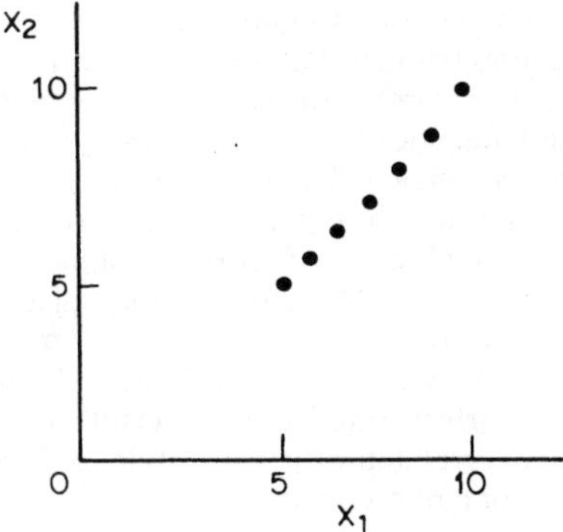

FIGURE 14.5 Plot of X_1 versus X_2.

regression equation has been developed using points (X_1, X_2), which are graphed in Fig. 14.5.

If a user of the resultant regression equation were to attempt to predict Y using $X_1 = 5$, $X_2 = 10$ or $X_1 = 10$, $X_2 = 5$, the user would obviously be guilty of (extreme) extrapolation, since neither point is even close to the points that were used in developing the equation. On the other hand, it would be safe to use a point that lies between any two of the plotted points, or a point that lies slightly to either side of the line that would be formed by connecting these points.

The problem of determining when extrapolation has occurred is much more difficult when there are more than two regressors, however, because the region cannot be displayed easily. Weisberg (1980, p. 216) presents two methods for determining the ellipsoid that encloses the experimental region, and the means for determining when extrapolation has occurred for each method.

14.4 PROBLEMS AND DIAGNOSTICS

The fundamental concepts of multiple regression have been presented in the preceding sections. There are, however, certain conditions that can preclude a routine analysis of regression data. Problems that can often occur are presented in the following sections, along with the suggested remedies.

14.4.1 Multicollinear Data

Multicollinear (that is, nonorthogonal) data present special problems in regression analysis. The orthogonal data in Table 14.9 were used previously to illustrate ANOVA tables, F tests, t tests, and confidence and prediction intervals, and the interpretation was straightforward. Such is not the case with multicollinear data. The term "multicollinearity" is typically used to represent an extreme departure from orthogonality. Such a departure can take several different forms. One possibility would be a high pairwise correlation between at least one pair of regressors.

As indicated previously, regression coefficients are difficult to interpret with such data, and other problems also ensue. For example, t tests can indicate that neither of two highly correlated regressors is significant when, in fact, both are highly correlated with Y. This is due to the fact that multicollinearity causes the standard deviations (also called standard errors) of the coefficients to be relatively high. This leads to t tests which are not statistically significant, as well as to unstable coefficients. Here the term "unstable" means that a slight change in one or more values in the data set that is used to develop the equation will cause a considerable change in the coefficients. Fortunately, multicollinearity does not have a similarly deleterious effect on the predicted response. Therefore, if the objective is to predict Y, multicollinearity will not be a serious problem.

Because of the relationship between confidence intervals and hypothesis tests, the intervals for the β_i are similarly affected; that is, the intervals could include zero when each regressor is highly correlated with Y. Confidence intervals for the mean of Y and prediction intervals for a particular value of Y are not substantially affected by multicollinearity, however.

The results of F tests will not generally agree with t test results for multicollinear data, but F tests are affected somewhat differently. Specifically, the relative value of each regressor will depend on the order in which the regressors are assumed to be entered into the equation. For example, for two highly correlated regressors, each of which is also highly correlated with Y, the variable that is entered first will be judged significant by the appropriate F test, while the regressor that is entered second will be deemed not significant; that is the results will be similar to those displayed in Table 14.12 for orthogonal data.

Possible remedies for multicollinear data include augmenting the existing data set with additional data, collected in such a way as to reduce the multicollinearity. Other solutions include deleting variables and using alternatives to least squares. These are briefly discussed in later sections, and are also covered in Chap. 15.

The effects of multicollinearity just summarized will be illustrated using the data in Table 14.13. It should be observed that these are the same data as those shown in Table 14.9, except that the values for X_2 have been rearranged.

As noted previously, the regression equation for the data in Table 14.13 is

$$\hat{Y} = 23.39 + 1.278X_1 - 0.184X_2.$$

Thus, in this instance the multicollinearity (the correlation between X_1 and X_2 is 0.93) partly causes the coefficient of X_2 to be negative, even though the correlation between X_2 and Y is positive ($r = 0.84$).

Table 14.14 illustrates what was discussed earlier, namely, that X_2 has considerable value when it enters the equation before X_1, but not when it enters after X_1. With orthogonal data it makes no difference how the regressors are entered; the individual contributions are the same, regardless of the order.

The effect of multicollinearity on the t tests and confidence intervals can be seen by first comparing the $(\mathbf{X'X})^{-1}$ matrix from the data of Table 14.13 with those from Table 14.9. For Table 14.13,

$$\mathbf{X'X} = \begin{bmatrix} 40 & 44 \\ 44 & 56 \end{bmatrix}$$

so

$$(\mathbf{X'X})^{-1} = \frac{1}{304}\begin{bmatrix} 56 & -44 \\ -44 & 40 \end{bmatrix}$$

TABLE 14.13 Multicollinear Data

Y	X_1	X_2
30.1	8	18
32.2	9	19
34.3	11	22
35.4	12	22
34.4	10	19
30.0	8	18
31.4	9	19
32.3	10	19
34.0	12	22
33.1	11	22
33.2	11	22
34.3	12	22
31.1	9	19
30.0	8	18
32.3	10	19
34.4	12	22
30.3	8	18
31.6	9	19
33.3	11	22
32.0	10	19

TABLE 14.14 Detailed ANOVA Table, Centered Data from Table 14.13

Source	Degrees of freedom		Sum of squares SS		Mean square MS	F	p
Regresion	2		46.4829		23.2414	57.573	3×10^{-8}
Due to X_1		1		46.2250	46.2250	114.508	6×10^{-9}
Due to $X_2 \| X_1$*		1		0.2579	0.2579	0.639	0.435
Due to X_2		1		37.6216	37.6216	93.196	3×10^{-8}
Due to $X_1 \| X_2$		1		8.8613	8.8613	21.951	0.0002
Residual	17		6.8626		0.4037		
Total	19		53.3455				

*(Marginal) contribution of X_2 after X_1 has entered equation.

Recall that the standard error of $\hat{\beta}_i$ can be estimated by $\sqrt{\text{MS}_{\text{res}}\, c_{ii}}$, where c_{ii} is the ith diagonal element of $(\mathbf{X}'\mathbf{X})^{-1}$. Here $\text{MS}_{\text{res}} = 0.4037$, and the estimated standard error of $\hat{\beta}_1$ is $\sqrt{0.4037(56/304)}$, or 0.2727. With the data from Table 14.9 the estimated standard error was $\sqrt{0.3904(1/40)}$, or 0.0988. Thus the standard error of $\hat{\beta}_1$ has almost tripled, even though neither X_1 nor Y was changed. This means that the confidence interval for β_1 will have a width that is almost three times the width of the interval that was obtained by using the orthogonal data. Specifically, the 95% confidence interval for β_1 with the nonorthogonal data is obtained as

$$1.278 \pm 2.110 \sqrt{0.4037\left(\frac{56}{304}\right)} = 1.278 \pm 0.575$$

so that the lower limit is 0.703 and the upper limit is 1.853.

Since this interval does not include zero, the t test for testing the hypothesis that $\beta_1 = 0$ will be rejected, but the value of the t statistic will be considerably smaller than the value obtained with the orthogonal data. Specifically,

$$t = \frac{1.278}{\sqrt{0.4037\left(\frac{56}{304}\right)}} = 4.686$$

as compared to 10.8814 with the orthogonal data. Thus although the t statistic is considerably smaller, it is still "significant" ($p = 0.0001$). See Neter and Wasserman (1974, p. 343) for a two regressor example where the test of the regression equation has a p value of 2×10^{-6}, but the two t tests each yield a p value in excess of 0.05.

As noted previously, confidence intervals for the mean of Y and prediction intervals for Y are generally not adversely affected by multicollinearity. This is because the estimated variance of $\hat{Y}$ is not inflated by multicollinearity. The estimated variance was shown earlier to be 0.0864 for $X_1 = 12$ and $X_2 = 22$. Using this same point for the nonorthogonal data,

$$\begin{aligned}\text{Est. Var}(\hat{Y}_0) &= \text{MS}_{\text{res}}\left[\mathbf{X}_0'(\mathbf{X}'\mathbf{X})^{-1}\mathbf{X}_0 + \frac{1}{n}\right] \\ &= 0.4037\left[(2 \quad 2)\left(\frac{1}{304}\right)\begin{bmatrix} 56 & -44 \\ -44 & 40 \end{bmatrix}\begin{pmatrix} 2 \\ 2 \end{pmatrix} + \frac{1}{20}\right] = 0.0733\end{aligned}$$

Thus the estimated variance is actually slightly smaller than with the orthogonal data. Consequently, the confidence and prediction intervals will be narrower than before.

Inadvertent extrapolation can easily result from multicollinearity, however. For example, $X_1 = 12$ and $X_2 = 18$ are not within the range of the data (although each one is individually), and the estimated variance of $\hat{Y}_0$ for this combination is much larger (0.9976).

In summary, it is preferable to use orthogonal data whenever possible, as the analysis and interpretation are more straightforward. When this is not possible, the use of centered data can help alleviate some of the problems caused by nonorthogonal data, especially when polynomial terms in at least one of the regressors are to be used in the equation. Scaling, such as dividing each regressor by its standard deviation, can also be helpful. The importance of centering and scaling has been stressed in a number of papers by Marquardt and Snee. In particular, see Snee (1983, p. 235) and Marquardt (1980).

14.4.2 Variable Selection

The problem of multicollinearity discussed in the preceding section can be avoided by simply not using all of the available regressors. The question then arises as to which variables should be used and what method or methods should be employed to obtain a good subset.

Various methods have been used for many years for the purpose of obtaining a good subset. One such method is called *forward selection.* With this method a subset is obtained by sequentially adding variables one at a time until a "point of diminishing returns" is reached. This point is determined through the use of partial-F tests of the type illustrated in Table 14.14. Specifically, the point occurs where the marginal contribution of a regressor is deemed insignificant by the F test. (In Table 14.14 this point occurred when X_2 was tested after X_1 had been added.) The selection of which variable to test for inclusion at each stage is determined by the one that has the largest partial correlation coefficient in absolute value of those that are remaining. The opposite of forward selection is *backward elimination.* With this method the starting point is the full equation with all of the available variables. A variable is then deleted if the partial-F test indicates that its marginal contribution is not significant, acting as if the variable were the last one that entered the equation.

Stepwise regression is essentially a combination of these two methods in that it allows for both addition and deletion of variables. After an F test indicates that a variable should be added to the equation, subsequent F tests are performed to determine whether any of the other variables in the equation have become redundant and should thus be removed.

One shortcoming of these three methods is that the nominal significance level α of each F test is not the true significance level because the tests are correlated. In fact, the true levels are generally unknown. Also, the "best" model (by some criterion) may not be selected since not all of the possible subset models are examined. This does not preclude the use of these methods, but it should be kept in mind.

One method that does not rely on significance tests is the method of *all possible regressions.* There are $2^p - 1$ possible subsets with p variables, and with this method all of these subsets are examined, relative to some criterion, either explicitly or implicitly. The latter is certainly preferable if p is of at least moderate size, since many of the subsets will be of little value. The Furnival and Wilson (1974) algorithm is probably the best-known algorithm for implicitly considering all possible subsets. The "best" k subsets of each possible size can be determined for a specified value of k as well as the best k subsets overall. The subsets that are best are determined by specifying one of the following criteria: (1) R^2, (2) adjusted R^2, or (3) Mallows' C_p statistic. The latter has not been discussed previously. It was originally proposed by Mallows as a means of determining a reasonable range for the number of variables that should be used, not for selecting a particular subset (see, for example, Mallows, 1973). The statistic is defined as

$$C_r = \frac{\text{RSS}_r}{\hat{\sigma}^2} - (n - 2r)$$

where r is the number of parameters in the model (it equals the number of variables when the data are centered.) RSS_r is the residual sum of squares for a given subset of size r (if the data are centered; $r - 1$ if not centered), and $\hat{\sigma}^2$ is an estimate of σ^2 that may be obtained from the full equation with all p available regressors. Good subsets have a C_r value that is small and close to r. (*Note*: C_r is used here rather than C_p since p has been used previously to denote the number of available regressors.)

Which of these methods should be used? None is so compelling that it can always be used to the total exclusion of the other methods. Berk (1978) compared forward selection, backward elimination, and all possible regressions and concluded, in particular, that there was very little difference in the three methods when the subset size was fixed in each comparison and the data were generated from known populations.

In virtually any statistical application it is wise to analyze the data in more than one way; the same applies to variable selection. There is usually no one best subset. An experimenter may wish to include a particular variable in the regression equation if it is "known" that the variable is important, regardless of what the selection procedure specifies. Therefore, a good strategy would be to use the methods described in this section to identify a modest number of good subsets, and then allow for the possibility of nonstatistical considerations influencing the final subset selection.

The methods discussed in this section need to be used with a computer. Accordingly, they are discussed in Sec. 14.5. The reader is also referred to Draper and Smith (1981, secs. 6.1–6.6) for additional reading.

14.4.3 Alternatives to Least Squares

One alternative to variable selection is to use most, if not all, of the available regressors, but to use something other than least squares. Ridge regression is one such alternative. It produces regression coefficients that are more stable than those produced by least squares, and it is also of (lesser) value when the objective is prediction rather than estimation. (Recall that multicollinearity is not a serious problem when regression is used for prediction.) Another alternative is robust regression, which has been proposed for use when the assumptions on the error term, especially the normality assumption, are not at least approximately met. Ridge regression and robust regression are discussed in detail in Chaps. 16 and 17, respectively.

14.4.4 Plots and Model Building

The types of plots that should be displayed in a regression analysis were introduced in the section on simple regression but, of course, other plots are needed for multiple regression. Probably the most extensive collection of multiple-regression plots to be found in any one source are those in Daniel and Wood (1980). They do not advocate the use of every conceivable type of plot. For example, even though a plot of Y versus X is a logical starting point in simple regression, plots of Y against each regressor are not recommended in multiple regression, as they can be misleading (see Daniel and Wood, 1980, p. 406). Plots of the *residuals* against each regressor are recommended (including potential regressors not in the equation), as well as plots against $\hat{Y}$. The residuals should not be plotted against Y, however, as the two are generally correlated.

As was discussed in Sec. 14.2.4, residual plots are used to check the assumptions on the error term that must be met at least approximately before confidence intervals, hypothesis tests, and, in particular, prediction intervals can be constructed.

In simple regression it makes very little difference whether the residuals are plotted against $\hat{Y}$ or against X. The plots will have the same general configuration when $\hat{\beta}_1$ is positive since $\hat{Y}$ and X then have perfect positive correlation. When $\hat{\beta}_1$ is negative, one plot will be the mirror image of the other. For multiple regression, however, the residuals should be plotted against $\hat{Y}$ and against the individual regressors. If, for example, the plot against $\hat{Y}$ shows that the spread of the residuals is not constant, plots against the individual regressors might indicate how weighted least squares could be used to remove the nonconstant residual variance.

Residual plots against the regressors can also indicate the need for nonlinear terms in one or more regressors (such as X_1^2). Plotting the residuals against variables that are not in the equation can also be important. In particular, a plot of the residuals against time (assuming that the time order has been recorded) could indicate that time should be included as a variable in the equation. A linear trend in the plot would indicate the need for a linear term, a quadratic trend would indicate the need for a quadratic term, and so on. This would also apply to plots against other omitted variables, including cross products of variables already in the equation.

When the constant error variance is checked, the standardized residuals, not the raw residuals, should be plotted against each regressor. This is because the raw residuals do not all have the same variance, whereas the standardized residuals do have a constant variance. Therefore, a plot using the raw residuals could exhibit some evidence of a nonconstant variance for this reason alone. The ith standardized residual is given by $e_i^* = e_i/(s\sqrt{1 - h_{ii}})$, where e_i is the raw residual, $s = \sqrt{\text{MS}_{\text{res}}}$, and h_{ii} is the ith diagonal element of the matrix $\mathbf{I}–\mathbf{X}(\mathbf{X}'\mathbf{X})^{-1}\mathbf{X}'$.

Partial residual plots are often more informative in indicating the specific form for a nonlinear term. (Partial residuals to be plotted against regressor X_j are defined as $e_i + \hat{\beta}_j X_{ij}$, $i = 1, 2, \ldots, n$, where the e_i are the residuals.) Gunst and Mason (1980, p. 251) provide an example where a partial residual plot is superior to a residual plot in indicating the specific form of a nonlinear term to be added to the model. Mansfield and Conerly (1987) provide insight into partial residual plots as compared to regular residual plots.

As in simple linear regression, a normal probability plot of the standardized residuals can be helpful in assessing possibly severe nonnormality of the error term, although interpretation of the plot might be difficult for inexperienced users. Accordingly, many analysts may find simulation envelopes to be helpful. These provide boundaries within which each standardized residual (or some other type of residual) should fall when the errors have a normal distribution. The n pairs of boundaries form the envelope. See Atkinson (1981, 1985) and Flack and Flores (1989) for additional details. A histogram of the standardized residuals will also frequently be useful.

There are other plots that sometimes work better than residual plots and partial residual plots, including CERES plots (Cook, 1993, 1994). The relative merits of these and other plots are discussed by Berk and Booth (1995). The reader is referred to the latter for additional reading. In general, no one plot can always be relied upon, so the use of different types of plots is preferable to the selection of one type of plot. Dynamic graphics can also be of value, including 3-dimension graphs that can be rotated (Cook and Weisberg, 1994).

All of these plots are generally constructed by computer. The computer packages discussed in Sec. 14.5 can be used to construct a wide variety of regression plots, and some will display the regression diagnostics that are alluded to in Sec. 14.4.5. It is important to bear in mind, however, that intelligent use of computer packages requires that the user play an active role. This is true in general, and especially when it comes to building regression models. What has been termed "data snooping" is as important in regression analysis as it is in other areas of statistics. Henderson and Velleman (1981) illustrate the importance of interacting with a computer in developing a regression model rather than relying on some automated procedure.

The reader is referred to Berk and Booth (1995), Draper and Smith (1981, chap. 3), and Montgomery and Peck (1992, chap. 3) for a discussion of residual plots, to Daniel and Wood (1980) for numerous examples of such plots, and to Mosteller and Tukey (1977, chap. 16 and appendix) for discussions of residual analysis and model building through reexpression of the regressor terms. Box and Draper (1986, chap. 8) is also an excellent source of information on transformations for regression.

14.4.5 Regression Diagnostics

Of fairly recent origin, regression diagnostics should now play an important role in any regression analysis. Books such as Belsley, Kuh, and Welsch (1980) and Cook and Weisberg (1982) represent different schools of thought. There is some disagreement as to what regression "diagnostics" represents. Weisberg (1983) views diagnostic methods as a collection of procedures that are used for "model criticism" (including residual plots). Belsley, Kuh, and Welsch (1980), however, present diagnostics as also including methods for detecting multicollinearity (and its cause), outliers (observations far removed from the others), and influential data points. Weisberg (1983) refers to the latter as "influence analysis."

Hampel, Ronchetti, Rousseeuw, and Stahel (1986) claim that all data sets have 1 to 10% gross errors, so it is important that good diagnostics be used to detect bad data points. Bad data points will frequently be outliers, and many outliers will also be influential data points. There are three categories of outliers that can occur in regression: (1) coordinate outliers, (2) residual outliers, and (3) regression outliers. A data point could be in error because one or more coordinates are wrong, such as a misplaced decimal point. A data point might also not be well-fit by a particular model, thus resulting in a residual outlier. The term *regression outlier* has apparently appeared in only two books, Rousseeuw and Leroy (1987) and Ryan (1997), but regression outliers are the most important outliers to detect. As defined by Rousseeuw and Leroy (1987), a regression outlier is a point that does not conform to the line or plane that would be fit to the majority of the data. If such a point is highly influential, it might have a small standardized residual and thus escape detection. It is important to recognize that least-squares diagnostics cannot be relied upon to detect regression outliers. More sophisticated techniques are needed (see Chap. 17).

The distinction between outliers and influential data points can be seen by comparing Fig. 14.6*a* and *b*. In Fig. 14.6*a* the scatter plot (for simple regression) shows that the circled point is far removed from the other points. It is an outlier, but it is not an influential data point. The regression line is essentially determined by the (strong) configuration of the other points; the circled point could be expected to lie on or close to that line. Even if the point were moved up or down somewhat, it would hardly cause any change in the position of the line. The reverse is true in Fig. 14.6*b*, however, as the position of the circled point will essentially determine the position of the line. Thus the point is both an outlying observation and an influential observation.

These two graphs depict extreme cases, which are easily detected because there is only one regressor. With more regressors such detection is not generally possible with simple plots. Partial regression (leverage) plots can be useful in detecting influential data, especially when multiple data points are influential [see Belsley, Kuh, and Welsch (1980, p. 30)]. A list of suggested diagnostic statistics for detecting influential data points can be found in the review paper on regression by Hocking (1983, table 1). As the author indicates, however, there has been insufficient evidence to judge the relative merits of the suggested statistics.

Once an influential data point is detected, the question arises as to what should be done with it. Should the point be retained or deleted? If we have n data points,

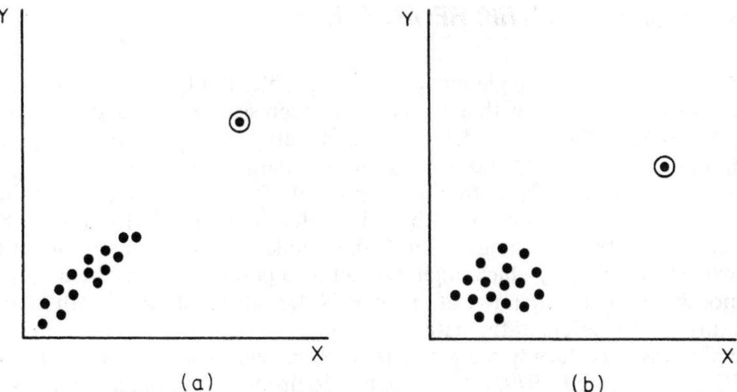

FIGURE 14.6 Scatter plots. (*a*) Possible outlier. (*b*) Influential data point and possible outlier.

we would ideally want each point to have approximately equal influence on the resultant regression equation. An alternative to deleting an influential, but valid, data point is to downweight the influence of such a point. Such an approach is called *bounded influence regression*, as described, for example, in Krasker and Welsch (1982).

14.5 LOGISTIC REGRESSION

Linear logistic regression is used when Y is binary. If Y can assume only two possible values (say, 0 and 1), it is easy to show that the error term could not possibly have a normal distribution. Accordingly, ordinary least squares would not be used in obtaining the parameter estimates. The estimates are determined iteratively with one or more convergence criteria used to determine when the iterations should cease. One obvious choice for a criterion would be the change in the parameter estimates. As discussed by McCullagh and Nelder (1989, p. 117), however, the parameter estimates could converge to a bad solution, so a criterion that is a function of the quality of the fit should also be used.

Logistic regression presents various other problems and some of these problems have not been solved. In particular, the diagnostics that are used are not independent of the predicted response (usually denoted by $\hat{\pi}$), unlike linear regression where residuals are independent of $\hat{Y}$. Jennings (1986) showed that the square of one type of residual, the deviance residual, has an expected value that varies considerably over π. There is also the question of which type of residual to use for diagnostic purposes.

Another question that has not been answered in specific terms is when exact logistic regression (Cox, 1970, p. 46) should be used instead of the more frequently used maximum likelihood approach. In general, exact logistic regression should be used when there is a considerable imbalance in the relationship between the number of zeros and the number of ones, and also when the sample size is small.

Books on logistic regression include Hosmer and Lemeshow (1989) and Kleinbaum (1994), the latter being especially well-suited for self-study.

14.6 NONPARAMETRIC REGRESSION

Assume that we have a single independent variable. If a linear relationship seems apparent, we might start with a simple linear regression model and see if the fit is adequate. Since whatever model is fit almost certainly is not the true model, one might argue that we should not attempt to fit a parametric model.

Consider Fig. 14.7. The dotted line represents the least squares fit and the solid line gives a nonparametric "smooth." The latter consists of fitted values that are obtained without using a model. Notice that the least-squares line is influenced by the two extreme points, which might be bad data points, whereas the nonparametric smooth follows the majority of the data. Notice also that the smooth follows the curvature in the middle of the data.

In this instance locally weighted regression, which is usually referred to as LOWESS (Cleveland, 1979, 1981), was used to produce the nonparametric smooth. The general idea is to estimate the mean of Y given X, for each value of X, say x_0, in the data set. We would expect to have virtually none of the X values repeated when the data have come from an observational study, however, so rather than attempting to perform averaging in the vertical direction, the averaging (weighting) is done horizontally. That is, neighboring points are used to estimate the mean of Y at $X = x_0$, with points that are the closest to x_0 receiving the most weight when LOWESS is used.

A neighborhood size must first be selected and a frequently used value is half of the data points. Assume that we have 50 data points (x_i, y_i) and we first consider data points for which the x-coordinate is near the average X value. Consider the

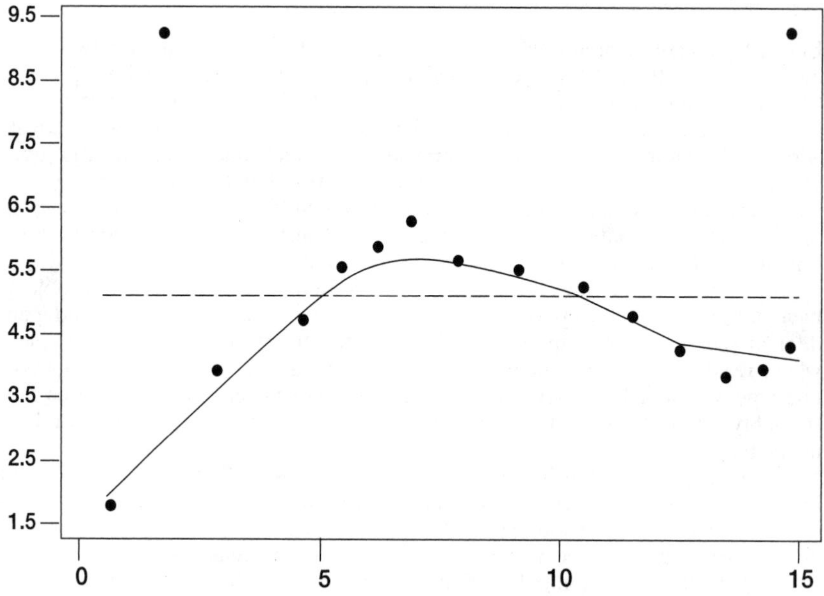

FIGURE 14.7 LOWESS smooth (solid line) and ordinary least squares fit (dotted line).

data point (x_0, y_0) and assume that a neighborhood size of 25 is being used. To obtain the fitted value for x_0, the 12 points that are on each side of x_0 would be used in addition to x_0. The weight, W_i^*, for each point (x_i^*, y_i^*) in the neighborhood of points is given by

$$W_i^* = \left\{1 - \left(\frac{|x_i^* - x_0|}{\max_i |x_i^* - x_0|}\right)^3\right\}^3 \tag{14.23}$$

with $W_0 = 1$.

For points (x_i, y_i) with x_i close to its minimum or close to its maximum value, it will obviously not be possible to construct a symmetric neighborhood about such points. For such points the preferred approach is to use a variable neighborhood size with, in this example, less than 12 points on one side of x_0. This is preferred over using a fixed neighborhood size and using as many as 24 points on one side of x_0. The rationale is that points that are far removed from x_0 should not be used in determining the fit at x_0.

Weighted simple linear regression is used for each point (x_0, y_0), with the weights given by equation (14.23). It is also possible to define the weights in such a way that the LOWESS is robust to outliers. Readers are referred to Härdle (1990) for additional reading on nonparametric regression.

14.7 COMPUTER PROGRAMS

Efficient handling of multiple regression data requires the use of some type of computer. In this section the regression capabilities of several well-known statistical packages that run on large computers are discussed. In addition, some statistical packages for microcomputers which have notable regression capabilities are also described.

There are many statistical packages which provide substantial regression output, including most of the features presented in this chapter. Unfortunately, many of these regression programs are not thoroughly tested or do not use the best numerical algorithms. There are instances in which a "test" data set has been used with each of several regression programs, and different numerical results were obtained. The general form of regression output can also vary some, what from package to package, but this is not generally a problem when there is good documentation.

In this section four mainframe packages and three microcomputer packages are discussed, although the mainframe packages have been modified to run also on smaller computers. The selection of a particular set of such packages for discussion is not meant to imply that these are the best ones available for regression. Rather, they are simply among the best known and are widely available.

The base SAS* software product is a general-purpose statistical package which contains several programs for regression analysis. The SAS system can also be used for information storage and retrieval, data modification and programming, report writing, file handling, and graphics. There are two programs for variable selection, PROC RSQUARE and PROC STEPWISE. The first procedure computes all possible regressions, and the output consists solely of the (unadjusted) R^2 and C_p values for each subset. The second procedure incorporates forward selection, back-

*SAS is a registered trademark of SAS Institute, Inc., Cary, NC, USA.

ward elimination, stepwise regression, and two other methods. More extensive output is produced with this procedure than with the former. The output includes an ANOVA table in addition to the regression coefficients and F-test results on each coefficient. PROC REG is the general-purpose regression procedure and provides the most extensive output. In addition to output of the same general type as that provided by PROC STEPWISE, the output includes many of the multicollinearity diagnostics and influential data diagnostics described in Belsley, Kuh, and Welsch (1980).

Since these diagnostics are generally obtained from matrix expressions, additional diagnostics, if desired, can be obtained by using PROC MATRIX as a programming language and writing the necessary program. (An example of such a program can be found in Velleman and Welsch, 1981.) PROC MATRIX can be used for creating other specialized analyses not provided for explicitly by the SAS system.

Linear regression can also be performed using PROC GLM. For a complete description of these procedures, the reader is referred to Ray (1985).

Another well-known statistical package with considerable regression capabilities is BMDP (Biomedical Computer Programs—P series). Like the SAS system, BMDP can also be used to obtain all possible regressions; the appropriate program is P9R. The latter is based on the Furnival and Wilson (1974) algorithm, which provides a moderate amount of output for each subset. The stepwise regression program is P2R and the general regression program is P1R. These programs provide a considerable number of options, which include the capability to produce various types of residual plots. Diagnostics can also be produced with each program, but most of them are in P2R. As with the SAS system, BMDP allows the user to write small programs to be used with the BMDP programs for accomplishing specific objectives. BMDP also contains several other programs for regression, including nonlinear regression. The reader is referred to Dixon (1985) for details.

Minitab is another popular package with good regression capabilities, although not quite in the same class with the SAS system and BMDP. However, Minitab does have the important feature of being essentially an interactive package, whereas the other two are not. Specifically, additional output can be obtained without having to start from the beginning. Minitab does not consist of a set of programs, but rather is comprised of a set of commands which are applied to a worksheet. All possible regressions can be performed using the BREG command and the STEPWISE command can be used for forward selection, backward elimination, and stepwise regression. The general-purpose multiple-regression command is REGRESS. This command can be used to obtain standard regression output as well as residual plots, multicollinearity diagnostics, and influential data diagnostics. Minitab also allows individual programming to be used in conjunction with the commands, so that output can be produced to accommodate special needs (see Ryan, 1985).

SPSS-X is another well-known general-purpose statistical package which can be used for regression analyses (SPSS, 1986). It has one program, REGRESSION, for general regression analysis. This can be used for forward selection, backward elimination, and stepwise regression, but not for all possible regressions. Some multicollinearity and influential data diagnostics can be produced in addition to the basic output but, like Minitab, its total regression capabilities are somewhat less than those of the SAS system and BMDP.

Although all four of these packages were originally developed for mainframes, conversions have allowed them to also run on smaller computers, including microcomputers. This is especially true for Minitab, which runs on a wide variety of mainframes and smaller computers and is not as machine-dependent as the other three packages.

The popularity of microcomputers has led to the development of a considerable number of statistical packages, some of which have regression capabilities that are generally comparable to those of the mainframe packages discussed previously. Three packages which fall into this category are, in alphabetical order, STATGRAPHICS, STATPRO, and SYSTAT.

STATGRAPHICS is essentially an interactive statistical package (originally developed for mainframes) consisting of over 350 functions organized into 26 chapters. The chapter on regression provides for stepwise regression, but not all possible regressions. General regression capabilities include residual plots and influential data diagnostics. Other forms of regression can also be handled, including nonlinear regression. In addition user-written functions (using the APL language) can be used for special needs. (STATGRAPHICS operates independently of APL, in the absence of user-written functions.)

STATPRO has regression capabilities similar to those of STATGRAPHICS. Specifically, it will not produce all possible regressions, but does provide for stepwise regression. The general regression capabilities include residual plots, but multicollinearity and influential data diagnostics are not available.

Both of these packages run on the IBM-PC, and STATPRO also runs on all versions of the Apple II.

SYSTAT is one of the best-known and most powerful statistical packages for microcomputers, and it also runs on larger computers. It has extensive regression capabilities, which include stepwise, polynomial, and weighted regression, in addition to other procedures such as nonlinear regression. Regression diagnostics are also provided.

Lesage and Simon (1985) compared a number of microcomputer statistical packages for accuracy in performing least squares calculations with multicollinear data. The study included STATPRO and SYSTAT, and both were found to be very accurate, as was another package, STATPAC.

Carpenter, Deloria, and Morganstein (1984) surveyed two dozen general-purpose statistical packages including STAN and others which were viewed as having excellent regression capabilities.

The software packages presented at the beginning of this section (SAS system, BMDP, Minitab, and SPSS-X) are also available on microcomputers (under slightly different names), so a user of regression analysis has a variety of excellent software from which to choose.

14.8 SUMMARY: STEPS TO BE PERFORMED IN ANALYZING REGRESSION DATA

Many facets of linear regression analysis have been presented. The logical steps that an experimenter should take in a regression study might be the following:

1. Are the data to come from an observational study (that is, "happenstance" data) or from a designed experiment? If they are from the former, recognize that the analysis may be difficult; if they are from a designed experiment, an orthogonal or near-orthogonal design should be used.
2. Determine the purpose for which the study is to be performed and identify the single dependent variable and the one or more independent variables. *Note*: Multivariate regression (more than one dependent variable) can be used, but it was not covered in this chapter.

3. If there is only one independent variable, construct a scatter plot of Y versus X to see whether there is evidence of a linear relationship. When there is more than one independent variable, a display of the correlation matrix is one possible starting point. This will show the linear correlation of Y with each X, as well as the pairwise correlations for the independent variables. It will show which regressors are most highly correlated with Y and which regressors, if any, are highly correlated with other regressors (as in an observational study). The objective here would be to simply subject the candidate variables to an initial screening. This would be particularly advantageous when there is a large number of potential regressors.
4. Using perhaps a subset of the potential regressors obtained from the previous step, the remaining candidate regressors could be examined through all possible regressions, provided that the number of such regressors is not exceedingly large (at most 27 with BMDP 9R, unless the program is modified).
5. A particular subset might be selected at this point, using both statistical and nonstatistical considerations. The other variable selection procedures (forward, backward, and stepwise) might also be used and the results compared.
6. A thorough regression analysis would then be performed for the chosen subset. This would include residual plots, a lack-of-fit test, and influential data diagnostics. Multicollinearity might not be a problem at this point, as all redundant variables could have been eliminated in steps 4 and 5. *Note*: Multicollinearity may result when an experimenter insists on using regressors that he or she "knows" to be important, but may be correlated with other regressors. A solution to this problem is not to use correlated regressors. Another solution is to allow correlated regressors in the regression equation, but to use some alternative to least squares. The choice between the two solutions has been termed the *regression dilemma* by at least one pair of authors (Hocking and Pendelton, 1983).
7. Once the form of the regression equation has been decided upon, it will generally be used for an indefinite period of time, with care being exercised to avoid inadvertent (extreme) extrapolation. When conditions change (for example, a change in the correlation between Y and a particular X), a new regression equation may have to be developed. A reduction in R^2 over time would be a sign of the need for a new equation.

The reader is referred to Draper and Smith (1981, chap. 8) for additional information concerning the development and maintenance of a regression equation.

REFERENCES

Anderson-Sprecher, R.: "Model Comparisons and R^2," *Am. Stat.*, vol. 48, pp. 113–117, 1994.

Atkinson, A. C.: "Two Graphical Displays for Outlying and Influential Observations in Regression," *Biometrika*, vol. 68, pp. 13–20, 1981.

_____: *Plots, Transformations, and Regression*. Clarendon Press, Oxford, England, 1985.

Belsley, D. A., E. Kuh, and R. E. Welsch: *Regression Diagnostics: Identifying Influential Data and Sources of Collinearity*. Wiley, New York, 1980.

Berk, K. N.: "Comparing Subset Regression Procedures," *Technometrics*, vol. 20, pp. 1–6, Feb. 1978.

_____ and D. E. Booth: "Seeing a Curve in Multiple Regression," *Technometrics*, vol. 37, 385–398, 1995.

Box, G. E. P., and D. R. Cox: "An Analysis of Transformations," *J. R. Stat. Soc. B*, vol. 26, pp. 211–243, 1964; discussion, *ibid*., pp. 244–252.

_____ and N. R. Draper: *Empirical Model Building and Response Surfaces*. Wiley, New York, 1986.

_____ W. G. Hunter, and J. S. Hunter: *Statistics for Experimenters*. Wiley, New York, 1978.

_____ and P. Newbold: "Some Comments on a Paper by Coen, Gomme, and Kendall," *J. R. Stat. Soc. A*, vol. 134, pp. 229–240, 1971.

Carpenter, J., D. Deloria, and D. Morganstein: "Statistical Software for Microcomputers," *Byte*, vol. 9, pp. 234–264, Apr. 1984.

Carroll, R. J., and D. B. H. Cline: "An Asymptotic Theory for Weighted Least Squares with Weights Estimated by Replication," *Biometrika*, vol. 75, pp. 35–43, 1988.

_____ and D. Ruppert: "Power Transformations When Fitting Theoretical Models to Data." *Journal of the Am. Stat. Assoc.*, vol. 79, pp. 321–328, 1984.

_____ and _____: *Transformation and Weighting in Regression,* Chapman and Hall, New York, 1988.

Cleveland, W. S.: "Robust Locally Weighted Regression and Smoothing Scatterplots." *Journal of the Am. Stat. Assoc.*, vol. 74, pp. 829–836, 1979.

_____: "LOWESS: A Program for Smoothing Scatterplots by Robust Locally Weighted Regression." *The Am. Stat.*, vol. 35, p. 54, 1981.

Coen, P. J., E. E. Gomme, and M. G. Kendall: "Lagged Relationships in Economic Forecasting," *J. R. Stat. Soc. A*, vol. 132, pp. 133–152, 1969.

Cook, R. D.: "Exploring Partial Residual Plots." *Technometrics*, vol. 35, pp. 351–362, 1993.

_____ "On the Interpretation of Regression Plots." *J. Am. Stat. Association*, vol. 89, pp. 177–189, 1994.

_____ and S. Weisberg: *Residuals and Influence in Regression*, Chapman and Hall, New York, 1982.

_____ and _____: *An Introduction to Regression Graphics*, Wiley, New York, 1994.

Cox, D. R.: *Analysis of Binary Data*, Chapman and Hall, London, 1970.

Daniel, C., and F. S. Wood: *Fitting Equations to Data*, 2d ed., Wiley, New York, 1980.

Dixon, W. J., Ed.: *BMDP Statistical Software 1985*, Univ. of California Press, Berkeley, 1985.

Draper, N. R., and H. Smith: *Applied Regression Analysis*, 2d ed., Wiley, New York, 1981.

Flack, V. F., and R. A. Flores: "Using Simulated Envelopes in the Evaluation of Normal Probability Plots of Regression Residuals," *Technometrics*, vol. 31, pp. 219–225, 1989.

Freund, R. J.: "A Warning of Round-off Errors in Regression," *Am. Stat.*, vol. 17, pp. 13–15, Dec. 1963.

Furnival, G. M., and R. W. Wilson: "Regression by Leaps and Bounds," *Technometrics*, vol. 16, pp. 499–511, Nov. 1974.

Gunst, R. F., and R. L. Mason: *Regression Analysis and Its Application*. Dekker, New York, 1980.

Hahn, G. J.: "Regression for Prediction versus Regression for Control," *Chemtech*, pp. 574–576, Sept. 1974.

Hampel, F. R., E. M. Ronchetti, P. J. Rousseeuw, and W. A. Stahel: *Robust Statistics: The Approach Based on Influence Functions*, Wiley, New York, 1986.

Härdle, W.: *Applied Nonparametric Regression*, Cambridge University Press, London, 1990.

Henderson, H. V., and P. F. Velleman: "Building Multiple Regression Models Interactively," *Biometrics*, vol. 37, pp. 391–411, June 1981.

Hocking, R. R.: "Developments in Linear Regression Methodology: 1959-1982," *Technometrics*, vol. 25, pp. 219-230, Aug. 1983; discussion, *ibid.*, pp. 230–249.

_____ and O. J. Pendelton: "The Regression Dilemma," *Commun. Stat., Theory and Methods*, pt. *A*, vol. 12, pp. 497–527, 1983.

Hosmer, D. W., Jr., and S. Lemeshow: *Applied Logistic Regression*. Wiley, New York, 1989.

Hunter, W. G., and W. F. Lamboy: "A Bayesian Analysis of the Linear Calibration Problem," *Technometrics*, vol. 23, pp. 323–328, Nov. 1981; discussion, *ibid.*, pp. 329–350.

Jennings, D. E.: "Outliers and Residual Distributions in Logistic Regression." *Journal of the American Statistical Association*, vol. 81, pp. 987–990, 1986.

Kleinbaum, D. G.: *Logistic Regression: A Self-Learning Text*, Springer-Verlag, New York, 1994.

Krasker, W. S., and R. E. Welsch: "Efficient Bounded Influence Regression Estimation," *J. Am. Stat. Assoc.*, vol. 77, pp. 595–604, Sept. 1982.

Krutchkoff, R. G.: "Classical and Inverse Regression Methods of Calibration," *Technometrics*, vol. 9, pp. 425–439, Aug. 1967.

Kvalseth, T. O.: "Cautionary Note about R^2." *Am. Stat.*, vol. 39, pp. 279–285, 1985.

Lesage, J. P., and S. D. Simon: "Numerical Accuracy of Statistical Algorithms for Microcomputers," *Comput. Stat. Data Anal.*, vol. 3, pp. 47–57, May 1985.

Madansky, A.: "The Fitting of Straight Lines when Both Variables Are Subject to Error," *J. Am. Stat. Assoc.*, vol. 54, pp. 173–205, Mar. 1959.

Mallows, C. L.: "Some Comments on *Cp*," *Technometrics*, vol. 15, pp. 661–675, Nov. 1973.

Mandel, J.: "Fitting Straight Lines when Both Variables Are Subject to Error," *J. Qual.Technol.*, vol. 16, pp. 1–14, Jan. 1984.

Mansfield, E. R., and M. D. Conerly: "Diagnostic Value of Residual and Partial Residual Plots," *Am. Stat.*, vol. 41, pp. 107–116, May 1987.

Marquardt, D. W.: "You should Standardize the Predictor Variables in Your Regression Models" (discussion of a paper by G. Smith and F. Campbell), *J. Am. Stat. Assoc.*, vol. 75, pp. 87–91, Mar. 1980.

McCullagh, P., and J. A. Nelder: *Generalized Linear Models,* 2d ed., Chapman and Hall, London, 1989.

Montgomery, D. C. and E. A. Peck: *Introduction to Linear Regression Analysis,* 2d ed., Wiley, New York, 1992.

Mosteller, F., and J. W. Tukey: *Data Analysis and Regression*, Addison-Wesley, Reading, MA, 1977.

Mullet, G. M.: "Why Regression Coefficients Have the Wrong Sign," *J. Qual. Technol.,* vol. 8, pp. 121–126, July 1976.

Neter, J., and W. Wasserman: *Applied Linear Statistical Models*, Irwin, Homewood, I, 1974.

Ray, A., Ed.: *SAS User's Guide: Statistics*, SAS Inst., Cary, NC, 1985.

Rousseeuw, P. J., and A. Leroy: *Robust Regression and Outlier Detection*, Wiley, New York, 1987.

Ryan, T. A., B. L. Joiner, and B. F. Ryan: *Minitab Student Handbook*, Duxbury, North Scituate, MA, 1985.

Ryan, T. P.: *Modern Regression Methods*, Wiley, New York, 1997.

Sampson, A. R.:" A Tale of Two Regressions," *J. Am. Stat. Assoc.*, vol. 69, pp. 682–689, Sept. 1974.

Scott A., and C. Wild: "Transformations and R^2," *Am. Stat.*, 48, pp. 127–129. 1991.

Seber, G. A. F.: *Linear Regression Analysis*, Wiley, New York, 1977.

Snedecor, G. W., and W. G. Cochran: *Statistical Methods,* 7th ed., Iowa State Univ. Press, Ames, 1980.

Snee, R. D.: "Discussion" (of an invited paper by R. R. Hocking), *Technometrics*, vol. 25, pp. 230–237, Aug. 1983.

SPSS, Inc.: *SPSS-X User's Guide*, 2d ed., SPSS, Inc., Chicago, 1986.

Velleman, P. F., and R. E. Welsch: "Efficient Computing of Regression Diagnostics," *Am. Stat.*, vol. 35, pp. 234–242, Nov. 1981.

Walls, R. C., and D. L. Weeks: "A Note on the Variance of a Predicted Response in Regression," *Am. Stat.*, vol. 23, pp. 24–26, June 1969.

Weisberg, S.: *Applied Linear Regression*, Wiley, New York, 1980.

_____ : "Some Principles for Regression Diagnostics and Influence Analysis," *Technometrics*, vol. 25, pp. 240–244, Aug. 1983.

Wetz, J. M.: "Criteria for Judging Adequacy of Estimation by an Approximating Response Function," Ph.D. dissertation, Univ. of Wisconsin, Madison, 1964.

APPENDIX 14.1

MATRIX ALGEBRA

Some matrix algebra expressions were used in the section on multiple regression. A matrix is a rectangular array of numbers arranged in rows and columns. Matrices are generally represented by capital letters. For example,

$$\mathbf{X} = \begin{bmatrix} -7 & -10 \\ 6 & 11 \\ -10 & -13 \\ 11 & 12 \end{bmatrix}$$

is a 4×2 matrix (4 rows and 2 columns) which could be the centered data matrix in a multiple regression problem with four data points and two regressors. The transpose of a matrix (denoted by $\mathbf{X}'$) is obtained by writing each column of the matrix as a row. Thus,

$$\mathbf{X}' = \begin{bmatrix} -7 & 6 & -10 & 11 \\ -10 & 11 & -13 & 12 \end{bmatrix}$$

Matrix multiplication (of the type used in this chapter) is "row times column," and can be performed whenever the number of rows in the first matrix equals the number of columns in the second matrix. The multiplication is accomplished by multiplying corresponding numbers and adding the individual products:

$$\mathbf{X}'\mathbf{X} = \begin{bmatrix} 306 & 398 \\ 398 & 534 \end{bmatrix}$$

where 398 is obtained as $-7(-10) + 6(11) + (-10)(-13) + 11(12)$. In general, the (i, j) element of a product is obtained by multiplying the ith row of the first matrix times the jth column of the second matrix. Thus the (1, 2) element of $\mathbf{X}'\mathbf{X}$ is obtained by multiplying the first row of $\mathbf{X}'$ by the second column of $\mathbf{X}$. In regression, $\mathbf{X}'\mathbf{X}$ will always be symmetric, that is, the (i, j) element will always equal the (j, i) element. It will also always be a square matrix, where the number of rows equals the number of columns.

A 2×2 $\mathbf{X}'\mathbf{X}$ matrix can easily be inverted by hand. If the elements of $\mathbf{X}'\mathbf{X}$ are denoted by

$$\mathbf{X}'\mathbf{X} = \begin{bmatrix} a & b \\ b & c \end{bmatrix}$$

then

$$(\mathbf{X}'\mathbf{X})^{-1} = \frac{1}{ac - b^2} \begin{bmatrix} c & -b \\ -b & a \end{bmatrix}$$

A vector is a single row or column of numbers. Specifically,

$$\mathbf{Y}' = [3 \quad 1 \quad 6 \quad 2]$$

is a row vector which, if written as a column, would be denoted by **Y** (that is, without the prime). Multiplication of a string of vectors and matrices is permissible as long as the criterion for matrix multiplication given previously (which also applies to vectors) is satisfied.

CHAPTER 15
DESIGN AND ANALYSIS OF EXPERIMENTS

Peter R. Nelson
Clemson University, Clemson, SC

15.1 INTRODUCTION

For the purpose of this chapter we will define an experiment as an investigation where the parameters of the system being studied are under the control of the investigator. More specifically, we will assume that treatments (materials, actions, etc.), experimental units (batches, lots, plots, etc.), and the nature of the observations are all decided on by the experimenter. It is very important in this situation that treatments and experimental units be chosen and observations made so that, as much as possible, the effects of any outside influences, either systematic or non-systematic, are eliminated or reduced.

To this end we introduce the idea of "good" data versus "bad" data. Generally speaking, good data come from well designed experiments that have been carefully carried out. While most of this chapter deals with the design and analysis aspects of experimentation, it is important not to overlook the actual conduct of a well designed experiment. The best designed experiment will be of little value if the experimenter or technician conducting it is not careful to carry it out as prescribed. As an example, imagine an experiment where one is interested in measuring the yield from a particular chemical reaction. The product is in powdered form and yield is to be measured by weight. Unfortunately, a lab technician drops a container of the resultant powder and it spills on the lab bench before it can be weighed. There are then several possible courses of action. One would hope that if the result of a particular trial were inadvertently destroyed, then that information would be recorded in the lab book along with the results from the other trials of the experiment. The experimenter could then decide how it would be best to proceed. On the other hand, attempts to hide the accident, say by trying to recover the spilled resultant without making any notes to that effect or by simply guessing at what it might have weighed, lead to bad data. Unfortunately, particularly in situations where experiments are being conducted by people not actually interested in the results, bad data do not occur as infrequently as one would hope.

There are two other concepts that are fundamental to the design of experiments and help to eliminate systematic biases. These are the ideas of random sampling and randomization. *Random sampling*, where each set of elements of the population being sampled has an equal chance of being chosen, ensures that, on the average, the samples are independent and representative, that is, they are not related to one another other than coming from the same population. For example, in an experiment to compare the rate of wear of different brands of automobile tires, samples from each brand should be obtained randomly rather than, say, using tires manufactured consecutively on the same machine. While truly random sampling is not possible in all situations, samples should be chosen as much as possible to be independent and representative of their populations. A more complete discussion of random sampling can be found in Wright and Tsao (1985).

Fisher (1935) introduced the concept of *randomization*, that is, assigning treatments to experimental units at random and, when applicable, performing the experimental trials in a random order, as a technique to ensure the validity of an experiment by providing an appropriate reference distribution. Randomization averages over the treatments any biases due to the experimental units not behaving independently because of some unknown or uncontrollable factor. As an example, consider again the experiment to compare rates of tire wear and suppose one is interested in only two brands. The tires are to be used on cars for, say, 10,000 miles of driving, and then the tread wear is to be measured and compared. Since it is desirable to subject the two brands of tires to the same conditions affecting wear, it would be reasonable to conduct a paired experiment (see the discussion on pairing in Chap. 4) and assign one of each brand of tire to each car used. Also, since it is known that front and rear tires do not necessarily wear in the same way, the two tires should both be mounted on either the front or the rear. Having chosen, say, the rear, all that remains is to decide which brand goes on which side, and this is where randomization comes into the picture. In order to average over both brands any effect on wear due to the side on which a tire is mounted, the brands should be assigned to the sides at random. This could be done, for example, by using a table of random numbers and associating odd numbers with the assignment of brand A to the driver's side. Randomization can also be used as a justification for the normal theory t and F tests, even when the assumptions of normality are not necessarily met since the t and F distributions serve as approximations to the appropriate randomization (reference) distributions. A more complete discussion of randomization can be found in Box, Hunter, and Hunter (1978).

15.2 ONE-FACTOR EXPERIMENTS AT MORE THAN TWO LEVELS

The techniques of analysis of single-factor experiments at more than two levels (that is, with more than two treatments) are fundamental to the analyses of all designed experiments. These techniques and their extensions permeate the subject of the design of experiments. Therefore, we commence with a discussion of these analyses.

In a single-factor experiment the factor can be one of two kinds: fixed or random. A factor is said to be *fixed* or to have a *fixed effect* if the levels of that factor constitute the entire population of interest. For example, if one step in a particular chemical production process requires the use of an alcohol, and three different alcohols are known to work, one might be interested in determining whether the alcohols behaved differently with respect to average yield. The three alcohols are then three levels of a fixed factor. A second example, where the factor is not discrete, would be comparing three reaction temperatures rather than three alcohols. On the other hand, a factor is said to be *random* or to have a *random effect* if the levels of that factor constitute a random sample from the population of interest rather than the entire population itself. For example, in an experiment to study the time required to teach someone to operate a particular machine, each person taught is a different level of a single factor, and that factor would be random if the operators in the experiment were a random sample from a population of potential operators. With a random factor the interest is in estimating the variability of the population rather than the effects of specific levels of the factor.

The first part of this chapter deals exclusively with fixed effects. Random effects are discussed in the last section. The single-factor fixed effects model is an extension of the two-sample problems discussed in Chap. 4. In general, one is interested in comparing I treatments, which will be referred to as I levels of a factor A. We will assume that the observations are all independent quantitative values. More specifically, observations taken at the ith level of factor A ($i = 1, \ldots, I$) are assumed to be normally distributed with mean μ_i and variance σ^2 [denoted by $N(\mu_i, \sigma^2)$]. That is, there are I normal populations each having the same (unknown) variance but potentially different means. The problem is to determine whether the I populations are all the same, and if not, which ones are different. The question of whether the I populations are all the same can be dealt with by testing the null hypothesis

$$H_0\colon\ \mu_1 = \mu_2 = \cdots = \mu_I. \tag{15.1}$$

In order to keep the notation as simple as possible, particularly in the later sections, dot notation will be used to indicate averaging, that is, a subscript replaced by a dot indicates that the values have been averaged over that subscript. For the case of a single factor with possibly different sample sizes from each population let

X_{ij} = jth observation from the ith population

n_i = number of observations from the ith population

$X_{i\bullet}$ = sample mean for the ith population,.

$$X_{i\bullet} = \frac{1}{n_i}\sum_{j=1}^{n_i} X_{ij} \tag{15.2}$$

N = total sample size,

$$N = \sum_{i=1}^{I} n_i$$

$X_\bullet$ = grand mean,

$$X_{\bullet\bullet} = \frac{1}{N}\sum_{i=1}^{I}\sum_{j=1}^{n_i} X_{ij} \tag{15.3}$$

$$= \frac{1}{N}\sum_{i=1}^{I} n_i X_{i\cdot} \tag{15.4}$$

S_i^2 = sample variance for ith population,

$$S_i^2 = \frac{1}{n_i - 1} \sum_{j=1}^{n_i} (X_{ij} - X_{i\bullet})^2 \tag{15.5}$$

$$\mu = \mu_{\bullet} = \frac{1}{N} \sum_{i=1}^{I} n_i \mu_i \tag{15.6}$$

α_i = main effect of ith level of factor A,

$$\alpha_i = \mu_i - \mu. \tag{15.7}$$

Our model can now be written as

$$X_{ij} = \mu_i + \epsilon_{ij} \tag{15.8}$$

$$= \mu + \alpha_i + \epsilon_{ij} \tag{15.9}$$

where the assumption that the X_{ij} are distributed $N(\mu_i, \sigma^2)$ [denoted by $X_{ij} \sim N(\mu_i, \sigma^2)$] and are independent implies that

$$\epsilon_{ij} \sim N(0, \sigma^2)$$

and (15.10)

$$\epsilon_{ij} \text{ are independent.}$$

Estimators for the various parameters (denoted by carets) are given by

$$\hat{\mu}_i = X_{i\bullet} \tag{15.11}$$

$$\hat{\mu} = X_{\bullet\bullet} \tag{15.12}$$

$$\hat{\alpha}_i = X_{i\bullet} - X_{\bullet\bullet} \tag{15.13}$$

$$\hat{\sigma}^2 = \text{MS}_e = \frac{\text{SS}_e}{N - I} \tag{15.14}$$

$$\text{SS}_e = \sum_{i=1}^{I} (n_i - 1) S_i^2 \tag{15.15}$$

$$= \sum_{i=1}^{I} \sum_{j=1}^{n_i} (X_{ij} - X_{i\bullet})^2. \tag{15.16}$$

In general, terms of the form $\Sigma(\cdot)^2$ are referred to as *sums of squares* (SS), and sums of squares divided by their degrees of freedom (df) are called *mean squares* (MS). In particular, Eqs. (15.15) and (15.16) are formulas for the error sum of square SS_e, and the estimate of σ^2 given in Eq. (15.14) is the mean square error MS_e. Under assumptions (15.10), SS_e/σ^2 is distributed as a chi square random variable with $N - I$ degrees of freedom [denoted by $\chi^2(N - I)$].

15.2.1 One-Way Analysis of Variance

Hypothesis (15.1), which will subsequently be referred to as H_0, is equivalent to

$$H_A: \quad \alpha_i = 0 \qquad \text{for all } i \tag{15.17}$$

and can be tested by comparing the $\hat{\alpha}_i$ with the MS_e using a one-way *analysis of variance* (ANOVA). An ANOVA using a model with only fixed effects is sometimes referred to as a model I (one) ANOVA. The notation H_A used in hypothesis (15.17) should not be confused with a similar notation sometimes used by other authors to indicate an alternative hypothesis.

Unequal Sample Sizes. The $\hat{\alpha}_i$ are combined to form the sum of squares for factor A

$$\text{SS}_A = \sum_{i=1}^{I} n_i \hat{\alpha}_i^2 \tag{15.18}$$

$$= \sum_{i=1}^{I} n_i (X_{i\bullet} - X_{\bullet\bullet})^2. \tag{15.19}$$

Under hypothesis H_A [with assumptions (15.10)], SS_A/σ^2 is distributed as $\chi^2(I-1)$ and is independent of SS_e. Therefore, under H_A

$$F_A = \frac{\text{MS}_A}{\text{MS}_e} = \frac{\text{SS}_A/(I-1)}{\text{SS}_e/(N-I)} \tag{15.20}$$

is distributed as an F distribution with numerator degrees of freedom $I - 1$ and denominator degrees of freedom $N - I$ [denoted by $F(I-1, N-I)$]. Large values of the $\hat{\alpha}_i$ (either positive or negative) tend to discredit H_A, as do large values of SS_A, and hence large values of F_A. One would, therefore, reject H_A at level of significance α if

$$F_A > F(\alpha; I-1, N-I) \tag{15.21}$$

where $F(\alpha; \nu_1, \nu_2)$ is the upper α quantile of an $F(\nu_1, \nu_2)$ distribution. Alternatively, one would reject H_A at level α if

$$\Pr[F(I-1, N-I) > F_A] < \alpha. \tag{15.22}$$

The probability in inequality (15.22) is called the *descriptive* (or *attained*) *level of significance* (DLS). It can be obtained easily using a computer program for the F distribution, such as the one given in Craig (1984).

The relationship

$$\text{SS}_{\text{tot}} = \sum_{i=1}^{I} \sum_{j=1}^{n_i} (X_{ij} - X_{\bullet\bullet})^2 \tag{15.23}$$

$$= \text{SS}_A + \text{SS}_e \tag{15.24}$$

is sometimes convenient for computing either SS_A or SS_e. Care, however, must be taken to ensure sufficient accuracy for the (corrected) total sum of squares SS_{tot}

TABLE 15.1 ANOVA Table for a One-Way Layout with Fixed Effects

Source of variation	Sum of squares	Degrees of freedom	Mean square	F	Pr > F
Factor A	$SS_A = \sum_{i=1}^{I} n_i(X_{i\bullet} - X_{\bullet\bullet})^2$	$I - 1$	$MS_A = \frac{SS_A}{I-1}$	$F_A = \frac{MS_A}{MS_e}$	$\Pr[F(I-1, N-I) > F_A]$
Error	$SS_e = \sum_{i=1}^{I} \sum_{j=1}^{n_i} (X_{ij} - X_{i\bullet})^2$	$N - I$	$MS_e = \frac{SS_e}{N-1}$		
	$= \sum_{i=1}^{I} (n_i - 1)S_i^2$				
	$= SS_{tot} - SS_A$				
Total	$SS_{tot} = \sum_{i=1}^{I} \sum_{j=1}^{n_j} (X_{ij} - X_{\bullet\bullet})^2$	$N - 1$			

TABLE 15.2 Maximum Output Rates (fl oz/h) for Three Brands of Ultrasonic Humidifiers (Example 15.1)

	Brand	
1	2	3
13	14	13
15	17	12
14		
$X_{1\bullet} = 14$	$X_{2\bullet} = 15.5$	$X_{3\bullet} = 12.5$
$S_1^2 = 1$	$S_2^2 = 4.5$	$S_3^2 = 0.5$

and SS_A (or SS_e). Otherwise, using Eq. (15.24) can result in substantial errors. Information from a one-way ANOVA is usually summarized in a table similar to Table 15.1. Note that not only are the sums of squares additive [Eq. (15.24)], but so are the degrees of freedom [that is, $(I - 1) + (N - I) = N - 1$].

Example 15.1. Three brands of ultrasonic humidifiers were compared with respect to the rate at which they output moisture. Tests were done in a chamber controlled at 70° F with a relative humidity of 30%. The data from the experiment are given in Table 15.2 together with the sample mean and the variance for each brand. Using Eqs. (15.4), (15.15), (15.19), and (15.20), one computes

$$X_{\bullet\bullet} = \frac{1}{7}[3(14) + 2(15.5) + 2(12.5)] = 14$$

$$SS_e = 2(1) + 1(4.5) + 1(0.5) = 7$$

$$SS_A = 3(14 - 14)^2 + 2(15.5 - 14)^2 + 2(12.5 - 14)^2 = 9$$

$$F_A = \frac{9/2}{7/4} = 2.57.$$

From Table A.5 in the Appendix one finds $F(0.1; 2, 4) = 4.32$, and therefore since $2.57 < 4.32$, one would not reject H_0 and would conclude that the sample means are not statistically significantly different at level $\alpha = 0.1$ (at the 10% level). The ANOVA table for this example is given in Table 15.3, from which one sees that the attained level of significance is 0.19 (that is, DLS = 0.19).

TABLE 15.3 ANOVA Table for Example 15.1

Source	SS	df	MS	F	$\Pr > F$
Humidifiers	9	2	4.5	2.57	0.19
Error	7	4	1.75		
Total	16	6			

Equal Sample Sizes. In the case where equal samples of size n are taken from each population, the formulas used in a one-way ANOVA simplify somewhat. Namely,

$$X_{\bullet\bullet} = \frac{1}{I}\sum_{i=1}^{I} X_{i\bullet} \tag{15.25}$$

$$\text{MS}_e = \frac{1}{I}\sum_{i=1}^{I} S_i^2 \tag{15.26}$$

$$\text{SS}_e = (N - I)\text{MS}_e \tag{15.27}$$

$$\text{SS}_A = n\sum_{i=1}^{I} (X_{i\bullet} - X_{\bullet\bullet})^2. \tag{15.28}$$

Example 15.2. The warm-up time in seconds was determined for three different types of cathode-ray tubes using four observations on each type of tube. The results are given in Table 15.4 together with the sample mean and the variance for each tube type. Using Eqs. (15.25), (15.26), (15.28), and (15.20), one computes

$$X_{\bullet\bullet} = \frac{1}{3}(20 + 26 + 19) = 21.67$$

$$\text{MS}_e = \frac{1}{3}(6 + 72 + 10) = 29.33$$

$$\text{SS}_A = 4[(20 - 21.67)^2 + (26 - 21.67)^2 + (19 - 21.67)^2] = 114.67$$

$$F_A = \frac{114.67/2}{29.33} = 1.95.$$

TABLE 15.4 Warm-Up Times (seconds) for Three Different Types of Cathode-Ray Tubes (Example 15.2)

Tube type		
1	2	3
17	20	23
20	20	20
20	38	16
23	26	17
$X_{1\bullet} = 20$	$X_{2\bullet} = 26$	$X_{3\bullet} = 19$
$S_1^2 = 6$	$S_2^2 = 72$	$S_3^2 = 10$

Comparing 1.95 with $F(0.1; 2, 9) = 3.01$ from Table A.5, one would not reject H_0 and would conclude that the sample means are not statistically significantly different at level $\alpha = 0.1$. The ANOVA table for this example is given in Table 15.5.

TABLE 15.5 ANOVA Table for Example 15.2

Source	SS	df	MS	F	Pr > F
Tube type	114.67	2	57.34	1.95	0.20
Error	264	9	29.33		
Total	378.67	11			

15.2.2 One-Way Analysis of Means

An alternative method for testing H_A is the *analysis of means* (ANOM), which rejects H_A if any of the $\hat{\alpha}_i$ are too large in magnitude. The ANOM has two advantages over the ANOVA: (1) if any of the sample means are statistically different, the ANOM indicates exactly which ones are different; and (2) the ANOM allows one to easily evaluate the practical significance of the differences. For the case of equal sample sizes from each of the I populations, exact critical values are available (Appendix 15.1), and for the case of unequal sample sizes approximate (conservative) critical values have been calculated (Appendix 15.2).

Equal Sample Sizes. When equal samples of size n are taken from each of the I populations, one rejects H_A if for any i

$$|\hat{\alpha}_i| > h(\alpha; I, N-I)\sqrt{\mathrm{MS}_e}\sqrt{\frac{I-1}{nI}} \tag{15.29}$$

where the critical values h(α; L, ν) are given in Appendix 15.1. Actually, rather than making the comparison (15.29), the ANOM is usually carried out graphically by computing decision lines

$$X_{\bullet\bullet} \pm h(\alpha; I, N-I)\sqrt{\mathrm{MS}_e}\sqrt{\frac{I-1}{nI}} \tag{15.30}$$

and checking to see whether any of the $X_{i\bullet}$ fall outside these lines.

Example 15.3. Analyzing the data in Table 15.4 using the ANOM decision lines (15.30) with $\alpha = 0.1$, one computes

$$X_{\bullet\bullet} \pm h(0.1; 3, 9)\sqrt{\mathrm{MS}_e}\sqrt{\frac{2}{12}} = 21.67 \pm (2.34)\sqrt{29.33}\sqrt{\frac{1}{6}}$$

$$= 21.67 \pm 5.17$$

where $X_{\bullet\bullet}$ and MS_e were computed in Example 15.2. Each $X_{i\bullet}$ is compared with these decision lines in Fig. 15.1, from which one finds that the sample means are not significantly different at the $\alpha = 0.1$ level.

Unequal Sample Sizes. When unequal samples of size n_i are taken from the I populations, one rejects H_A if for any i

$$|\hat{\alpha}_i| > m(\alpha; I, N-I)\sqrt{\mathrm{MS}_e}\sqrt{\frac{N-n_i}{Nn_i}} \tag{15.31}$$

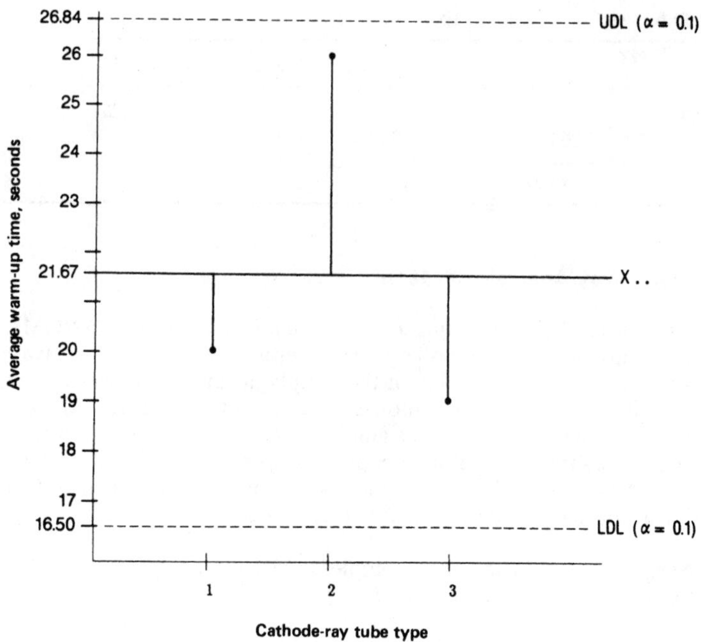

FIGURE 15.1 ANOM chart for Example 15.3.

where values of $m(\alpha; I, \nu)$ are given in Appendix 15.2. For parameter combinations not given in Appendix 15.2, values of $h^*(\alpha; I, \nu)$, which are slightly more conservative than $m(\alpha; I, \nu)$, can be obtained using the nomograph for the t distribution published by L. S. Nelson (1975) and the identity

$$h^*(\alpha; I, \nu) = t(\alpha^*/2; \nu) \tag{15.32}$$

where $t(\alpha; \nu)$ is the upper alpha quantile of a t distribution with ν degrees of freedom (Table A.4 in the Appendix), and

$$\alpha^* = 1 - (1 - \alpha)^{1/I}. \tag{15.33}$$

The graphical equivalent of comparison (15.31) is to plot each $X_{i\bullet}$ with decision lines

$$X_{\bullet\bullet} \pm m(\alpha; I, N - I)\sqrt{\mathrm{MS}_e}\sqrt{\frac{N - n_i}{Nn_i}}. \tag{15.34}$$

Note that different decision lines are needed for each different sample size n_i.

Example 15.4. Four brands of extended service light bulbs were compared for average lifetime. The results (artificial data to simplify the analyses) are given in Table 15.6 together with the sample mean and the variance for each brand. Using Eqs. (15.4) and (15.15), one obtains

$$X_{\bullet\bullet} = \frac{1}{14}\,[3(2500) + 3(2500) + 4(2500) + 4(2999)] = 2642.6$$

$$SS_e = 2(75{,}625) + \cdots + 3(46{,}030) = 720{,}292.$$

The decision lines (14.34) for $\alpha = 0.05$ and $n_i = 4$ are therefore

$$X_{\bullet\bullet} \pm m(0.05; 4, 10)\sqrt{MS_e}\sqrt{\frac{10}{14(4)}} = 2642.6 \pm 2.98\sqrt{72{,}029.2}\,(0.4226)$$

$$= 2642.6 \pm 338.0.$$

The ANOM chart for this example is given in Fig. 15.2, from which one sees that Brand 4 extended service light bulbs last significantly ($\alpha = 0.05$) longer. From this chart one also sees that the difference between the average Brand 4 bulb's lifetime and the grand average is approximately 360 hours, which can be used to easily assess the practical significance. It is interesting to note that using the ANOVA on this example one would compute $F_A = 3.29$, which is not significant at the $\alpha = 0.05$ level.

15.2.3 Choice of Sample Size

Specifying the level of significance and the power of a test determines the necessary sample size. More specifically, one must specify the power for some particular configuration of parameters such that H_0 is not true. For the one-way layout one usually specifies the power for detecting any two μ_i that are Δ/σ units apart. Sample size tables of this kind are given in Appendixes 15.3 and 15.4, respectively, for the ANOVA and the ANOM. Appendix 15.4 for the ANOM does not contain sample sizes for the case $I = 2$. However, since the ANOM and the ANOVA are equivalent in that case, Appendix 15.3 can be used. The sample sizes given in these appendixes are the number of observations (replicates) that must be taken at each treatment level (that is, n). A comparison of the tables shows that the sample sizes are essentially the same for small numbers of treatment levels, and that for large num-

TABLE 15.6. Lifetimes (hours) for Four Brands of Extended Service Light Bulbs (Example 15.4)

Brand			
1	2	3	4
2775	2535	2136	3121
2500	2210	2505	3238
2225	2755	2884	2797
		2475	2840
$X_{1\bullet} = 2500$	$X_{2\bullet} = 2500$	$X_{3\bullet} = 2500$	$X_{4\bullet} = 2999$
$S_1^2 = 75{,}625$	$S_2^2 = 75{,}175$	$S_3^2 = 93{,}534$	$S_4^2 = 46{,}030$

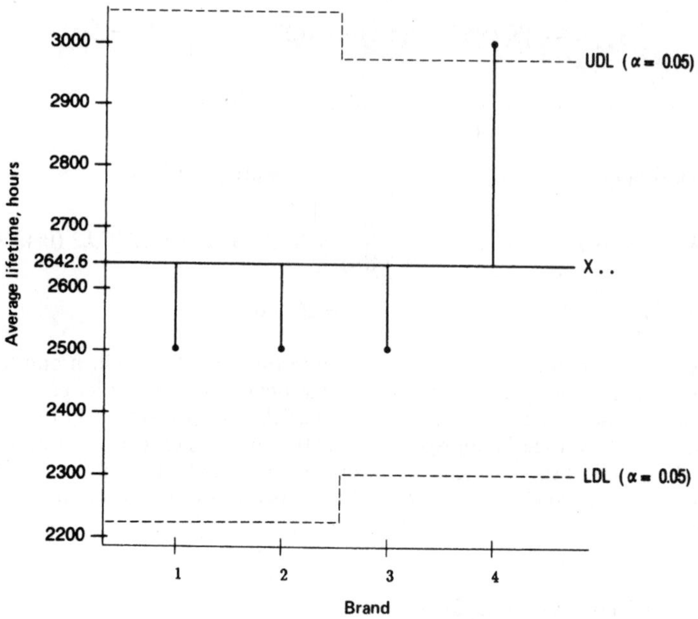

FIGURE 15.2 ANOM chart for Example 14.4.

bers of treatment levels the sample sizes for the ANOM are uniformly smaller, with the difference increasing as the number of treatment levels increases.

Example 15.5. In the cathode-ray-tube warm-up time experiment (Example 15.2), how many observations with each type of tube would be necessary to detect with probability 0.9 a difference of 2σ for any two tube means when using a level of significance of 0.05? If the data are to be analyzed using the ANOM, Appendix 15.4 is entered with $\alpha = 0.05$, power $= 0.9$, and $\Delta/\sigma = 2$. One finds that $n = 8$ samples are needed for each tube type. Using Appendix 15.3 for the ANOVA, one also obtains $n = 8$.

Even in situations where cost or time limits the possible number of replicates, Appendixes 15.3 and 15.4 can be used to determine whether differences of practical importance can be detected with reasonably high probability.

Example 15.6. If in Example 15.2 the analysis is to be done using $\alpha = 0.05$ and the ANOM (ANOVA), is it worthwhile conducting the experiment with only $n = 4$ replicates when differences of 2σ are of practical importance? From Appendix 15.4 (15.3) with $\alpha = 0.05$, $I = 3$, $\Delta/\sigma = 2$, and $n = 4$ one finds a power of 0.5. Thus one would expect to be able to detect a 2σ difference only half the time, and there seems to be little justification for conducting the experiment.

15.2.4 Checking the Assumptions

Both the ANOVA and the ANOM depend on the assumptions that $X_{ij} \sim N(\mu_i, \sigma^2)$ and the X_{ij} are independent. Therefore, the first part of any analysis should consist of checking for obvious violations of these assumptions.

Testing for Equal Variances. The assumption of equal variance can be easily tested using the maximum F ratio. The largest and the smallest sample variances are used to form the ratio

$$F_{max} = \frac{S^2_{max}}{S^2_{min}} \tag{15.35}$$

which is compared with a critical value $F_{max}(\alpha; k, \nu)$ from Appendix 15.5. If

$$F_{max} > F_{max}(\alpha; k, \nu) \tag{15.36}$$

then the hypothesis of equal variances is rejected at level α. These critical values are indexed by the number k of sample variances being compared and the degrees of freedom ν associated with each sample variance. In the one-way ANOVA (ANOM) situation with equal sample sizes one would use $F_{max}(\alpha; I, n - 1)$. In the case of unequal sample sizes a conservative test is obtained by using $F_{max}(\alpha; I, n_{min} - 1)$, where $n_{min} = \min_i n_i$. For cases where the n_i are not at least approximately equal, Bartlett's test may be more appropriate. A discussion of Bartlett's test can be found in Montgomery (1984). Neither of these tests is robust against nonnormality; however, since in this case one is also interested in testing the assumption of normality, this is not a practical difficulty.

Example 15.7. For Example 15.2 $F_{max} = 72/6 = 12$, $I = 3$, $n = 4$, and from Appendix 15.5, $F_{max}(0.1; 3, 3) = 16.77$. Since $12 < 16.77$, there is no evidence of unequal variances at the $\alpha = 0.1$ level.

Example 15.8. For Example 15.4 $F_{max} = 93{,}534/46{,}030 = 2.03$, $I = 4$, $n_{min} = 3$, and from Appendix 15.5, $F_{max}(0.1; 4, 2) = 69.13$. Since $2.03 < 69.13$, there is no evidence of unequal variances at the $\alpha = 0.1$ level.

Residual Analysis. The assumptions (15.10) of normality and independence can be checked using the estimates of the ϵ_{ij}. From Eq. (15.8) one has

$$\epsilon_{ij} = X_{ij} - \mu_i \tag{15.37}$$

which leads to the estimator

$$\hat{\epsilon}_{ij} = X_{ij} - X_{i\cdot} \tag{15.38}$$

The quantities $\hat{\epsilon}_{ij}$ in Eq. (15.38) are called *residuals*. If the order in which the observations X_{ij} were obtained is known, a plot of the residuals in that time sequence can be used to check the assumption of independence. Any nonrandomness in the plot (such as too many or too few runs) would indicate a lack of independence of the observations. Proper randomization when conducting the experiment, however, will ensure the required independence.

A plot of the residuals versus the fitted values (in this case the $X_{i\bullet}$) is another way to check for nonconstant variances. If the variances are equal, the residuals plotted in this way should fall within a horizontal band.

Finally, the assumption of normality can be checked by plotting the residuals on normal probability paper. The fact that the residuals are not independent is not a practical difficulty. Normal probability paper is a special kind of graph paper having a uniform scale for the horizontal axis and a scale on the vertical axis constructed so that normally distributed data will plot as a straight line. One orders the residuals and plots the ith smallest residual versus the modified cumulative percentage

$$P_i = \frac{100(i - 0.5)}{N}. \tag{15.39}$$

Possible alternatives to the P_i values given by Eq. (15.39) are discussed in Barnett (1975). For small values of N the plotted points may appear quite nonlinear without indicating nonnormality. Daniel and Wood (1980) have several examples of probability plots of random normal deviates that show how nonlinear such plots can be without necessarily indicating nonnormality. Some care and experience is therefore necessary in order to evaluate normal probability plots correctly.

If normal probability paper is not available, the same type of plot can be obtained using ordinary graph paper (both axes having a uniform scale) and plotting the ordered residuals versus the expected values of the order statistics for a standard normal distribution. These expected values can be obtained from Harter (1961), from L. S. Nelson (1983*b*), or approximated by

$$E(X_{(i)}) \doteq z\left(\frac{P_i}{100}\right) = z\left(\frac{i - 0.5}{N}\right) \tag{15.40}$$

where $z(\alpha)$ is the α quantile from a standard normal distribution, which can be obtained from Table A.1 in the Appendix.

Example 15.9. The ordered residuals for Example 15.4 together with the P_i values computed using Eq. (15.39) are given in Table 15.7 and plotted on normal probability paper in Fig. 15.3. Examination of Fig. 15.3 reveals nothing to indicate nonnormality. Table 15.7 also contains the approximate expected values of the normal order statistics obtained using Eq. (15.40). The ordered residuals are plotted versus these values on ordinary graph paper in Fig. 15.4.

Testing for Normality. Since the examination of normal probability plots is somewhat subjective, it is sometimes helpful to perform a statistical test for normality of the residuals in addition to constructing the normal probability plot. The

TABLE 15.7 Ordered Residuals and Normal Probability Plotting Positions for Example 15.4 (see Example 15.9)

Rank i	$\hat{\epsilon}_{ij}$	P_i	$z(P_i/100)$
1	–364	3.57	–1.80
2	–290	10.7	–1.24
3	–275	17.9	–0.92
4	–202	25.0	–0.67
5	–159	32.1	–0.46
6	–25	39.3	–0.27
7	0	46.4	–0.09
8	5	53.6	0.09
9	35	60.7	0.27
10	122	67.9	0.46
11	239	75.0	0.67
12	255	82.1	0.92
13	275	89.3	1.24
14	384	96.4	1.80

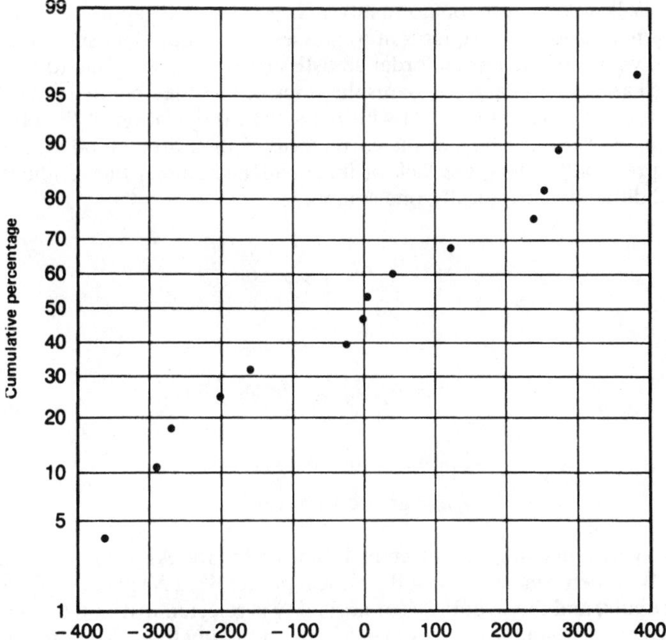

FIGURE 15.3 Normal probability plot of the residuals in Example 15.4 (see Example 15.9).

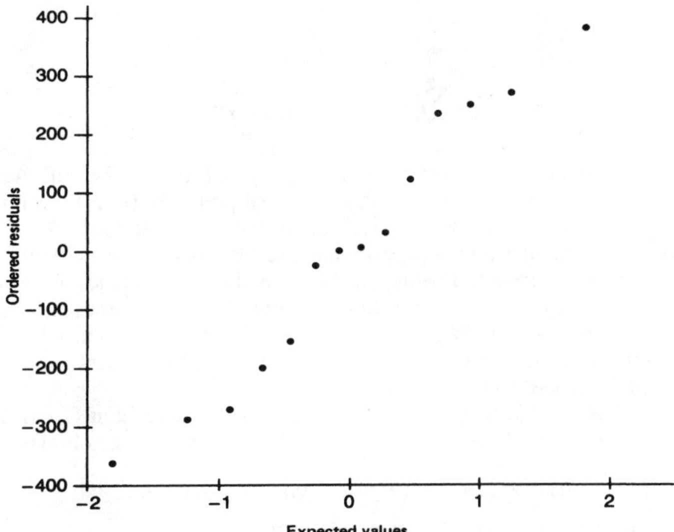

FIGURE 15.4 Plot of ordered residuals versus expected values of normal order statistics for Example 15.4 (see Example 15.9).

Shapiro-Wilk (1965) test for normality is a good one to apply in this situation. Using generalized least squares it regresses the ordered observations on the expected values of the normal order statistics (that is, it fits a line to the points in plots such as Fig. 15.4) and compares the slope of the line obtained (an estimate of σ if the normal model is correct) with the standard deviation of the observations (an estimate of σ regardless of the correctness of the normal model). As with the normal probability plots, the lack of independence among the residuals is not a practical difficulty. Specifically, one computes

$$W = b^2 \Big/ \sum_{i=1}^{N} e_{(i)}^2 \tag{15.41}$$

where

$$b = \sum_{i=1}^{u} a_{N-i+1}\left[e_{(N-i+1)} - e_{(i)}\right] \tag{15.42}$$

$$e_{(i)} = i\text{th smallest residual}$$

$$u = \text{greatest integer in } N/2$$

and the coefficients a_{N-i+1} are given in Table A.20 in the Appendix for $N \le 50$. The statistic W is then compared with the critical points $W(\alpha; N)$ given in Table A.18 in the Appendix, and the hypothesis of normality is rejected if $W < W(\alpha; N)$. For values of N greater than 50 and less than 99, an approximate test of normality due to Shapiro and Francia (1972) can be performed by computing

$$W' = (b')^2 \Big/ \sum_{i=1}^{N} e_{(i)}^2 \tag{15.43}$$

where

$$b' = \sum_{i=1}^{u} a'_{N-i+1}\left[e_{(N-i+1)} - e_{(i)}\right] \tag{15.44}$$

and the coefficients a'_{N-i+1} are given as b_{n-i+1} in Table A.19 in the Appendix. The statistic W' is then compared with the critical points $W'(\alpha; N)$ given in Table A.21 in the Appendix, and the hypothesis of normality is rejected if $W' < W'(\alpha; N)$. The statistic W' provides only an approximate test because it regresses the ordered observations on the expected values of the normal order statistics using ordinary least squares, which does not take into account the correlations introduced by ordering the observations. When applied to residuals, which themselves are not independent, both the test using W and the test using W' are only approximate. They are, nonetheless, still extremely useful.

Example 15.10. Applying the Shapiro-Wilk test to the residuals from Example 15.4 (given in Table 15.7) and using Eqs. (15.41) and (15.42), one obtains

$$\begin{aligned} b &= 0.5251(384 + 364) + 0.3318(275 + 290) + 0.2460(255 + 275) \\ &\quad + 0.1802(239 + 202) + 0.1240(122 + 159) \\ &\quad + 0.0727(35 + 25) + 0.0240(5 - 0) \\ &= 829.416 \end{aligned}$$

$$\sum_{i=1}^{N} e_{(i)}^2 = 720{,}292$$

$$W = \frac{(829.416)^2}{720{,}292} = 0.955.$$

Since 0.955 is greater than $W(0.5; 14) = 0.947$, there is no significant nonnormality at the 50% level.

Transformation to Stabilize the Variances. When analysis indicates either nonhomogeneity of variances (that is, nonconstant variances) or nonnormality, these problems can sometimes be eliminated by a nonlinear transformation of the data. (Linear transformations have no effect on the homogeneity of the variances or the normality of the observations, but can be used for computational convenience.) We will discuss transformation to stabilize the variances, which often also has the desirable effect of making the observations more nearly normal. In general, however, even a nonlinear transformation will have little effect unless the observations vary greatly in magnitude. As a model for the nonhomogeneity of variances we assume that the variance is a function of the mean. Specifically, when X_{ij} is not $N(\mu_i, \sigma^2)$ but has mean μ_i and variance σ_i^2 and is not necessarily normally distributed, we assume

$$\sigma_i = c\mu_i^p. \tag{15.45}$$

This would be true, for example, if the observations had a Poisson distribution, in which case $\sigma_i^2 = \mu_i$. When the variances are related to the mean as in Eq. (15.45), they can be stabilized using the transformation

$$X_{ij} \rightarrow X_{ij}^{1-p}. \tag{15.46}$$

When $p = 1$, one uses $X_{ij} \rightarrow \ln(X_{ij})$. Thus for observations having a Poisson distribution one would use the transformation

$$X_{ij} \rightarrow \sqrt{X_{ij}}. \tag{15.47}$$

Employing a more general form of Eq. (15.45) leads to the transformation

$$X_{ij} \rightarrow \arcsin(\sqrt{X_{ij}}) \tag{15.48}$$

for the binomial distribution. Two slightly different transformations for these distributions are discussed in L. S. Nelson (1983*a*) in connection with the ANOM for attribute data.

When μ_i and σ_i are unknown (and the necessary replicates are present), it is possible to estimate p. Taking logs in Eq. (15.45) and replacing the unknown distribution parameters with their sample values gives

$$\ln(S_i) \doteq \ln(c) + p\ln(X_{i\bullet}). \tag{15.49}$$

Therefore, p can be estimated as the slope of the line fitted to a plot of $\ln(S_i)$ versus $\ln(X_{i\cdot})$. Further discussion of transformation can be found in Box, Hunter, and Hunter (1978).

Example 15.11. Three brands of gloss white aerosol paints were compared for hiding capability by repeatedly spraying a hiding chart (panels of light to dark gray stripes on a white background) until the pattern was completely obliterated. The amount of paint used was determined by weighing the paint cans both before and after spraying. Four samples of each brand were used and the results are given in Table 15.8 together with the sample mean and the variance for each brand. Checking for nonhomogeneity of variance using the maximum F ratio, one obtains $F_{max} = 1.14/0.0333 = 34.23$, which is significant at the 5% level [$F_{max}(0.05; 3, 3) =$

TABLE 15.8 Covering Capabilities (oz/ft^2) for Three Brands of Gloss White Aerosol Paints (Example 15.11)

Brand 1	Brand 2	Brand 3
2.1	4.7	6.4
1.9	3.6	8.5
1.8	3.9	7.9
2.2	3.8	8.8
$X_{1\bullet} = 2.0$	$X_{2\bullet} = 4.0$	$X_{3\bullet} = 7.9$
$S_1^2 = 0.0333$	$S_2^2 = 0.233$	$S_3^2 = 1.14$
$\ln(X_{1\bullet}) = 0.693$	$\ln(X_{2\bullet}) = 1.39$	$\ln(X_{3\bullet}) = 2.07$
$\ln(S_1) = -1.70$	$\ln(S_2) = -0.728$	$\ln(S_3) = 0.0655$

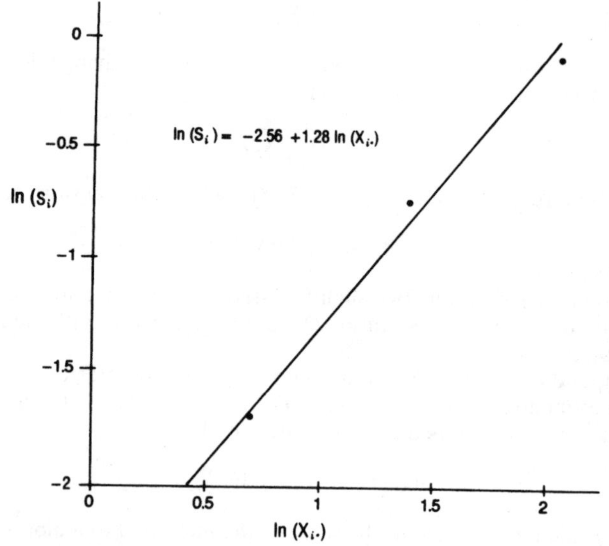

FIGURE 15.5 Plot of $\ln(S_i)$ versus $\ln(X_{i\bullet})$ for Example 15.11.

27.76]. The $\ln(S_i)$ and $\ln(X_{i.})$ are given in Table 15.8 and are plotted in Fig. 15.5 together with the least squares line

$$\ln(S_i) = -2.56 + 1.28 \ln(X_{i.}). \tag{15.50}$$

Therefore, $\hat{p} = 1.28 \doteq 1.3$ and one would employ the transformation (15.46)

$$X_{ij} \to X_{ij}^{1-1.3} = X_{ij}^{-0.3}. \tag{15.51}$$

The transformed (and scaled) values $X'_{ij} = 10X_{ij}^{-0.3}$ are given in Table 15.9 together with the sample means and the variances. Analysis of these transformed observations results in the following. Testing for nonhomogeneity of variances, one obtains $F_{max} = 1.10$, which is not significant at the 10% level. A normal probability plot (Fig. 15.6) reveals nothing to indicate nonnormality. For the Shapiro-Wilk test of

TABLE 15.9 Transformed Observations $X'_{ij} = 10X_{ij}^{-0.3}$ for Example 15.11

	Brand	
1	2	3
8.005	6.286	5.730
8.248	6.809	5.379
8.383	6.648	5.262
7.894	6.700	5.208
$X_{1\bullet} = 8.132$	$X_{2\bullet} = 6.611$	$X_{3\bullet} = 5.395$
$S_1^2 = 0.0500$	$S_2^2 = 0.0514$	$S_3^2 = 0.0550$

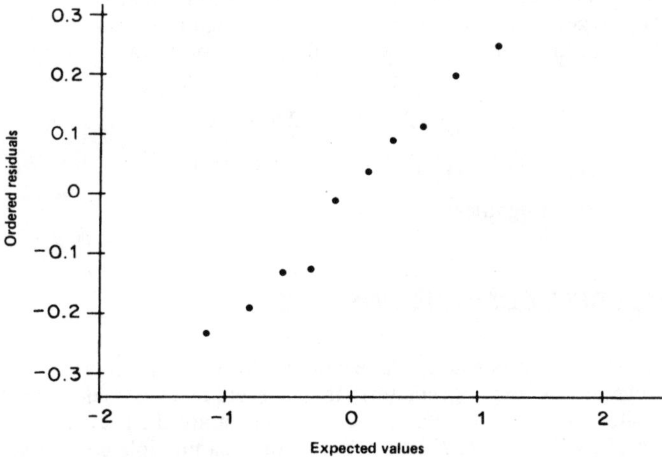

FIGURE 15.6 Normal probability plot of residuals for the transformed observations in Example 15.11.

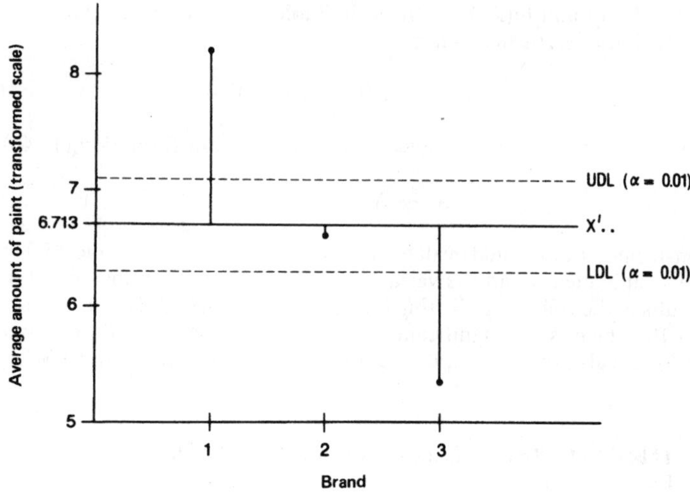

FIGURE 15.7 ANOM chart for Example 15.11 using transformed data.

normality one computes $b = 0.6753$, $\Sigma e^2_{(i)} = 0.4685$, and $W = 0.973$, which is not significant at the 50% level. The $\alpha = 0.01$ ANOM decision lines for the transformed values are [Eq. (15.30)]

$$X'_{\bullet\bullet} \pm h(0.01; 3, 9)\sqrt{\text{MS}_e}\sqrt{\frac{2}{12}} = 6.713 \pm (3.84)\sqrt{0.0521}\sqrt{\frac{1}{6}}$$

$$= 6.713 \pm 0.358$$

and examination of the ANOM chart in Fig. 15.7 reveals that Brand 1 values are significantly large (good) and Brand 3 values are significantly small (bad) at the $\alpha = 0.01$ level. In this case the practical significance is best evaluated in the original metric, where

$$X_{1\bullet} - X_{\bullet\bullet} = 2 - 4.63 = -2.63 \text{ oz/ft}^2$$

$$X_3 - X_{\bullet\bullet} = 7.9 - 4.633 = 3.27 \text{ oz/ft}^2$$

are both of practical significance.

15.3 MULTIPLE COMPARISONS

Multiple comparison procedures are techniques for making simultaneous comparisons of several unknown parameters. In this section we consider the ANOVA (ANOM) situation where the parameters of interest are the means of the normal populations. (The maximum F ratio test is actually a multiple comparison procedure for the variances of normal populations.) The need for these procedures arises, for example, when after rejecting H_0 using the ANOVA, one is interested in exactly which means are different. Only Fisher's protected least significant difference procedure and procedures for constructing simultaneous confidence intervals that are the most useful in practice will be considered. The multiple-range tests of Newman-Keuls and Duncan are not discussed and cannot be recommended even in situations that do not require confidence intervals.

Specifically of interest are simultaneous confidence intervals for contrasts. A contrast is a linear combination of the means,

$$\ell = \sum_i c_i \mu_i \tag{15.52}$$

for which $\Sigma_i c_i = 0$. Thus, contrasts are a generalization of pairwise differences of means (that is, $\mu_i - \mu_j$) that includes the differences of weighted averages such as $\mu_1 - (\mu_2 + \mu_3)/2$. The ANOM is actually a multiple comparison procedure for the contrasts α_i, where $c_i = 1 - n_i/N$ and $c_j = -n_j/N$ for $j \neq i$. In the most common situation where one is interested in simultaneous confidence intervals for pairwise contrasts of I means,

$$\binom{I}{2} = \frac{I(I-1)}{2}$$

intervals are needed.

15.3.1 Fisher's Protected Least Significant Difference

Fisher's (1935) (protected) least significant difference (LSD) procedure was designed to indicate which means are different from other means after the ANOVA has rejected H_0. If (and only if) H_0 is rejected at level α, then each of the $\binom{I}{2}$ pairwise contrasts $\mu_i - \mu_j$ can be tested to see if it is different from zero by checking whether

$$|X_{i\bullet} - X_{j\bullet}| > t\left(\frac{\alpha}{2}; \nu_e\right) \sqrt{\text{MS}_e} \sqrt{\left(\frac{1}{n_i} + \frac{1}{n_j}\right)} \tag{15.53}$$

where ν_e is the number of degrees of freedom associated with the MS_e. For a one-way ANOVA, $\nu_e = N - I$. Two sample means $X_{i\bullet}$ and $X_{j\bullet}$ are declared to be significantly different (and μ_i is declared to be different from μ_j) if inequality (15.53) holds. The overall type 1 error rate for this procedure (that is, the probability of declaring some of the means to be different when they are actually all the same) is assured to be no greater than α because of the requirement that H_0 be rejected at level α (hence the adjective "protected"). Simulation results indicate that the overall error rate is actually quite close to α, so the procedure is not overly conservative. It is possible that the ANOVA will reject H_0 at level α, but none of the inequalities (15.53) will hold, and thus no two sample means will be declared significantly different.

15.3.2 The Tukey-Kramer Procedure

Another multiple comparison technique for pairwise contrasts is the Tukey-Kramer procedure. It controls the experimentwise error rate without relying on an initial F test by using the intervals

$$X_{i\bullet} - X_{j\bullet} \pm q(\alpha; k, \nu_e)\sqrt{\text{MS}_e}\sqrt{\frac{1}{2}\left(\frac{1}{n_i}+\frac{1}{n_j}\right)} \tag{15.54}$$

where $q(\alpha; k, \nu)$ is the upper α quantile of a studentized range distribution. The parameter k is the total number of means, and ν_e is the number of degrees of freedom associated with the MS_e. Values of $q(\alpha; k, \nu)$ are given in Appendix 15.6. In the one-way ANOVA situation one would use $q(\alpha; I, N - I)$. Tukey (1953) proposed this procedure and proved that in the equal sample size case it has overall error exactly α. He also conjectured that it is conservative (that is, the overall error is less than α) for unequal sample sizes. Kramer (1956) independently proposed the unequal sample size modification used in the intervals (15.54). Hayter (1984) proved the conjecture that the procedure is conservative for unequal sample sizes, and simulations done by Dunnett (1980) suggest that it is not too conservative.

Example 15.12. In Example 15.4 the ANOVA did not reject hypothesis (15.1) at level $\alpha = 0.05$, and therefore Fisher's LSD procedure is not applicable. Simultaneous 95% confidence intervals for the $\binom{I}{2} = \binom{4}{2} = 6$ pairwise contrasts can, however, be obtained using the Tukey-Kramer procedure. The four sample means are given in Table 15.6, and in Example 15.4 we found $\text{MS}_e = 72{,}029.2$ with 10 degrees of freedom. Using Eq. (15.54) to compute the interval for $\mu_4 - \mu_3$, one obtains

$$2999 - 2500 \pm q(0.05; 4, 10)\sqrt{\text{MS}_e}\sqrt{\frac{1}{2}\left(\frac{1}{4}+\frac{1}{4}\right)} = 499 \pm (4.33)\sqrt{72{,}029.2}\left(\frac{1}{2}\right)$$
$$= 499 \pm 581.$$

The remaining intervals are computed in a similar fashion, and all six intervals are given in Table 15.10. Since all six intervals contain zero, no two sample means are significantly different at $\alpha = 0.05$.

TABLE 15.10 Simultaneous 95% Confidence Intervals for the Pairwise Contrasts in Example 15.4 Using the Tukey-Kramer Procedure (see Example (15.12)

Pair	Interval
$\mu_4 - \mu_3$	499 ± 581.0
$\mu_4 - \mu_2$	499 ± 627.6
$\mu_4 - \mu_1$	499 ± 627.6
$\mu_3 - \mu_2$	0 ± 627.6
$\mu_3 - \mu_1$	0 ± 627.6
$\mu_2 - \mu_1$	0 ± 670.9

15.3.3 Scheffé's Procedure

Scheffé (1953) proposed a method for constructing simultaneous confidence intervals for all possible contrasts in the ANOVA. He showed that with probability $1 - \alpha$ all contrasts ℓ [Eq. (15.52)] will be contained simultaneously in the intervals

$$\hat{\ell} \pm [(I-1)\, F(\alpha; I-1, \nu_e)]^{1/2} \sqrt{\text{MS}_e} \sqrt{\sum_i \frac{c_i^2}{n_i}} \tag{15.55}$$

where

$$\hat{\ell} = \sum_{i=1}^{I} c_i X_{i\cdot} \tag{15.56}$$

For a one-way ANOVA, $\nu_e = N - I$.

He also showed that the intervals (15.55) are equivalent to the ANOVA test of H_0 in the sense that H_0 is rejected (at level α) if and only if at least one $\hat{\ell}$ is significantly different from zero (at level α). While Scheffé's procedure could be used for pairwise contrasts (since they are a subset of all possible contrasts), it is not recommended because the intervals obtained will be longer than those obtained using the Tukey-Kramer procedure. Scheffé's procedure is useful when one does not know in advance which contrasts will be of interest (data snooping).

15.3.4 Šidák's Studentized Maximum Modulus Procedure

In situations where one is interested in contrasts other than pairwise and the number of contrasts k is known in advance, conservative simultaneous $(1 - \alpha)$ 100% confidence intervals are given by

$$\hat{\ell} \pm m(\alpha; k, \nu_e) \sqrt{\text{MS}_e} \sqrt{\sum_i \frac{c_i^2}{n_i}} \tag{15.57}$$

where $\hat{\ell}$ is defined in Eq. (15.56) and $m(\alpha; k, \nu)$ is the upper α quantile of a studentized maximum modulus distribution (Appendix 15.2). Note that the ANOM decision lines (15.34) for unequal sample sizes are a special case of Eq. (15.57). Dunn (1958) proved that the intervals (15.57) are conservative for $k = 2, 3$ and conjectured it was true in general. Šidák (1967) provided the general proof. The intervals (15.57) are uniformly less conservative (that is, shorter) than those produced by the Bonferroni procedure, which would replace $m(\alpha; k, \nu)$ with $t(\alpha/(2k); \nu)$. For small values of k the studentized maximum modulus intervals (15.57) tend to be shorter than Scheffé's intervals (15.55), and for larger values of k the reverse is true.

Example 15.13. Suppose that in Example 15.1 (data given in Table 15.2) one is interested in simultaneous 90% confidence intervals for the three contrasts

$$\ell_1 = \mu_1 - \mu_3$$

$$\ell_2 = \mu_1 - \frac{1}{2}(\mu_2 + \mu_3)$$

$$\ell_3 = \mu_1 - \frac{1}{3}(\mu_1 + \mu_2 + \mu_3).$$

Comparing

$$[(2)F(0.1; 2, 4)]^{1/2} = \sqrt{2(4.32)} = 2.94$$

with $m(0.1; 3, 4) = 2.98$, one finds that in this case Scheffé's procedure will give shorter intervals. For ℓ_2 one computes [using Eq. (15.56)]

$$\hat{\ell}_2 = 14 - \frac{1}{2}(15.5 + 12.5) = 0$$

$$\sum_i \frac{c_i^2}{n_i} = (1)^2\left(\frac{1}{3}\right) + \left(-\frac{1}{2}\right)^2\left(\frac{1}{2}\right) + \left(-\frac{1}{2}\right)^2\left(\frac{1}{2}\right) = 0.5833$$

and formula (15.55) gives

$$0 \pm (2.94)\sqrt{\frac{7}{4}}\sqrt{0.5833} = 0 \pm 2.97.$$

Similarly for ℓ_1 and ℓ_3 one obtains 1.5 ± 3.55 and 0 ± 1.98, respectively.

15.4 MULTIFACTOR EXPERIMENTS

The motivation for designing experiments is to obtain more (or better) information with less work. The following simple example is due to Hotelling (1944). If one is interested in determining the weights of two objects using only two weighings, rather than weigh them separately, it is possible to determine their weights with the same precision as if they had each been weighed twice. One first weighs both objects together and then obtains the difference in their weights (say by putting one object in each pan of a two-pan balance and adding weights until it balances). The average of the two weights obtained is the weight of one object, and the average (half of the) difference is the weight of the other.

In general, studying more than one factor at a time leads to much greater efficiency. The simplest experimental designs for studying two or more factors are complete layouts. Suppose the chemical production process mentioned in the Introduction calls for first a reaction with an alcohol (factor A) and then a reaction with a base (factor B). There are I possible alcohols and J possible bases that can be used, and one is interested in studying their effects on the average yield. The I alcohols are I levels of factor A, and the J bases are J levels of factor B. Observations for this experiment could be arranged in an $I \times J$ table consisting of one row for each level of factor A and one column for each level of factor B, where the observations corresponding to the ith level of A and the jth level of B are recorded in the (i, j) cell. This is called a *two-way layout*, and if every cell in the table contains at least one observation, then it is a *complete layout*. Table 15.11 is an example of this with $I = 3$, $J = 2$, and $n = 4$ observations per cell.

In a complete layout the factors are said to be *crossed*, and in an incomplete layout two factors can be either crossed or nested. The levels of a factor B are said to be *nested* within the levels of a factor A (B is nested within A) if in the set of observations each level of B appears with only a single level of A. For example, suppose one is interested in comparing I machines that manufacture computer disks for uniformity of their coating process. Several disks, say J, are chosen at random from the output of each machine, and the uniformity of each disk's coating is determined in two areas chosen at random. The disks are nested within the machines since the jth disk from machine i has nothing to do with the jth disk from machine i'. In an

TABLE 15.11 Percent Yields for Alcohol/Base Combinations

		Base			
		1		2	
Alcohol	1	91.3 90.7	89.9 91.4	87.3 91.5	89.4 88.3
	2	89.3 90.4	88.1 91.4	92.3 90.6	91.5 94.7
	3	89.5 88.3	87.6 90.3	93.1 91.5	90.7 89.8

incomplete layout, two factors are crossed if neither is nested within the other. Designs where all the factors are crossed are called *factorial* designs.

We assume in this section that all factors are crossed and the layout is complete. We also restrict our attention to balanced designs, that is, designs having the same number of observations in each cell. Incomplete layouts are discussed in Sec. 15.5 and nested designs are discussed in Sec. 15.11.

15.4.1 Interaction

In addition to the improved efficiency obtained from the use of multifactor designs, one is also afforded the opportunity to study possible *interaction* between the factors. Two factors are said to interact if the behavior of one factor depends on the particular level of the other factor. Thus, it is not possible to study interaction with one factor at a time experiments, and the unidentified presence of an interaction can lead to erroneous conclusions. Factorial designs, on the other hand, can estimate interactions, and if none are present, they allow the effect of every factor to be evaluated as if the entire experiment were devoted entirely to that factor.

The possible presence of an interaction can easily be detected graphically with an interaction plot. The average cell response is plotted versus the level of one of the factors, and points corresponding to the same level of the second factor are connected with lines. If the line segments for the levels of the second factor are parallel, then there is no interaction. An interaction plot for the average yield data in Table 15.12*a* is shown in Fig. 15.8. Average yield is plotted versus the level of factor *B*, and points corresponding to the same level of factor *A* are connected with

TABLE 15.12*a* Average Yields for Alcohol/Base Combinations together with Row Means, Column Means, and the Grand Mean

		Base		
		1	2	
Alcohol	1	90.825	89.125	$89.975 = X_{1\bullet\bullet}$
	2	89.8	92.275	$91.037 = X_{2\bullet\bullet}$
	3	88.925	91.275	$90.1 = X_{3\bullet\bullet}$
		89.85	90.892	$90.371 = X_{\bullet\bullet\bullet}$
		$X_{\bullet 1\bullet}$	$X_{\bullet 2\bullet}$	

TABLE 15.12*b* Sample Variances for Alcohol/Base Combinations

		Base	
		1	2
Alcohol	1	0.4758	3.243
	2	2.02	3.096
	3	1.456	1.963

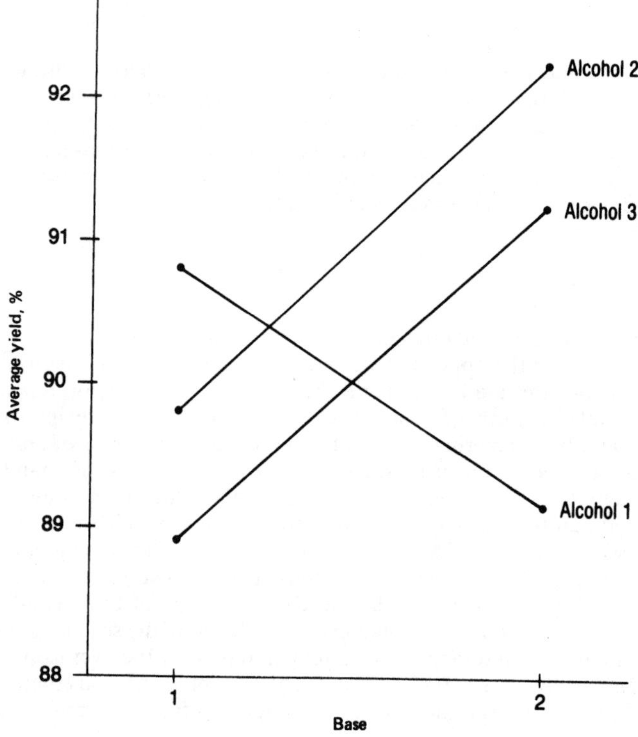

FIGURE 15.8 Interaction plot for the average yield data in Table 15.12*a*.

lines. The three line segments are clearly not parallel, indicating a possible Alcohol × Base interaction. The roles of the two factors could be reversed in the plotting procedure, leading to the interaction plot given in Fig. 15.9.

Interaction plots, however, can only indicate the possible presence of an interaction. Whether the differences in the slopes of the line segments are statistically significant depends on the experimental error, which is not taken into account in an interaction plot. Statistical tests for interactions as well as for the main effects of both factors can be performed using a two-way ANOVA.

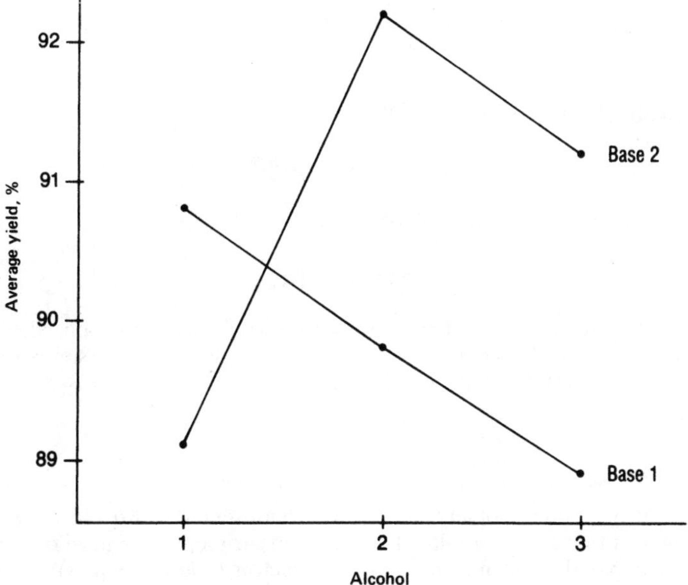

FIGURE 15.9 Interaction plot for the average yield data in Table 15.12*a*.

15.4.2 Two-Way Layouts

In a balanced complete two-way layout with factor A at I levels and factor B at J levels, n replicate observations are taken at each of the IJ treatment combinations. We assume for the moment that $n > 1$. Also, the observations are still assumed to be normally distributed and independent. Specifically, the observations in cell (i, j) are assumed to be $N(\mu_{ij}, \sigma^2)$. Thus, there are IJ normal populations, each having the same (unknown) variance but potentially different means.

Let

X_{ijk} = kth observation from cell (i, j)

S_{ij}^2 = sample variance for cell (i, j),

$$S_{ij}^2 = \frac{1}{n-1} \sum_{k=1}^{n} (X_{ijk} - X_{ij\bullet})^2 \tag{15.58}$$

$$\mu = \mu_{\bullet\bullet} = \frac{1}{IJ} \sum_{i=1}^{I} \sum_{j=1}^{J} \mu_{ij} \tag{15.59}$$

$\mu_{i\bullet}$ = mean of the ith level of factor A,

$$\mu_{i\bullet} = \frac{1}{J} \sum_{j=1}^{J} \mu_{ij} \tag{15.60}$$

$\mu_{\bullet j}$ = mean of the jth level of factor B,

$$\mu_{\bullet j} = \frac{1}{I} \sum_{i=1}^{I} \mu_{ij} \tag{15.61}$$

α_i^A = main effect of the *i*th level of factor *A*,

$$\alpha_i^A = \mu_{i\bullet} - \mu_{\bullet\bullet} \tag{15.62}$$

α_j^B = main effect of the *j*th level of factor *B*,

$$\alpha_j^B = \mu_{\bullet j} - \mu_{\bullet\bullet} \tag{15.63}$$

To simplify the notation, when there is no possible confusion, a particular subscript that is replaced by a dot on every term in an equation will be suppressed. For example, Eq. (15.62) would be written as

$$\begin{aligned} \alpha_i^A &= \mu_{i\bullet} - \mu_{\bullet\bullet} \\ &= \mu_i - \mu_{\bullet} \end{aligned}$$

Thus, the main effects of any factor can be represented by Eq. (15.7), regardless of how many other factors are involved in the experiment. One need only remember, for example, that when dealing with two factors (and hence μ_{ij}'s),

$$\begin{aligned} \alpha_j^B &= \mu_j - \mu_{\bullet} \\ &= \mu_{\bullet j} - \mu_{\bullet\bullet} \end{aligned}$$

The advantage of this notation is that different formulas are not needed when an additional factor is added to the experimental design. The convention will also be used that if more subscripts than superscripts appear on α, then the extra subscript is added to all the terms in the definition (or estimator) of that effect. For example,

$$\alpha_{ij}^A = \mu_{ij} - \mu_{\bullet j} \tag{15.64}$$

which is the main effect of the *i*th level of factor *A* specific to the *j*th level of factor *B*.

The concept of interaction, which was presented in a graphical form, can now be defined algebraically. The interaction between the *i*th level of factor *A* and the *j*th level of factor *B* (denoted by α_{ij}^{AB}) is the difference between the main effect of the *i*th level of factor *A* specific to the *j*th level of factor *B* and the main effect of the *i*th level of factor *A*. Namely,

$$\alpha_{ij}^{AB} = \alpha_{ij}^A - \alpha_i^A \tag{15.65}$$

$$= \mu_{ij} - \mu_{i\bullet} - \mu_{\bullet j} + \mu_{\bullet\bullet}. \tag{15.66}$$

Using Eqs. (15.59), (15.62), (15.63), and (15.66), our model can be written as

$$X_{ijk} = \mu_{ij} + \epsilon_{ijk} \tag{15.67}$$

$$= \mu + \alpha_i^A + \alpha_j^B + \alpha_{ij}^{AB} + \epsilon_{ijk} \tag{15.68}$$

where the assumption that the X_{ijk} are distributed $N(\mu_{ij}, \sigma^2)$ and are independent implies that

$$\epsilon_{ijk} \sim N(0, \sigma^2)$$

and (15.69)

$$\epsilon_{ijk} \text{ are independent.}$$

Estimators for the various parameters are given by

$$\hat{\mu}_{ij} = X_{ij\bullet} \tag{15.70}$$

$$\hat{\mu}_{i\bullet} = X_{i\bullet\bullet} = \frac{1}{J}\sum_{i=1}^{J} X_{ij\bullet} \tag{15.71}$$

$$\hat{\mu}_{\bullet j} = X_{\bullet j\bullet} = \frac{1}{I}\sum_{i=1}^{I} X_{ij\bullet} \tag{15.72}$$

$$\hat{\mu} = X_{\bullet\bullet\bullet} = \frac{1}{IJ}\sum_{i=1}^{I}\sum_{j=1}^{J} X_{ij\bullet} \tag{15.73}$$

$$\hat{\alpha}_i^A = X_{i\bullet\bullet} - X_{\bullet\bullet\bullet} \tag{15.74}$$

$$\hat{\alpha}_j^B = X_{\bullet j\bullet} - X_{\bullet\bullet\bullet} \tag{15.75}$$

$$\hat{\alpha}_{ij}^{AB} = X_{ij\bullet} - X_{i\bullet\bullet} - X_{\bullet j\bullet} + X_{\bullet\bullet\bullet} \tag{15.76}$$

$$\hat{\sigma}^2 = \text{MS}_e = \frac{I}{IJ}\sum_{i=1}^{I}\sum_{j=1}^{J} S_{ij}^2 \tag{15.77}$$

$$\text{SS}_e = IJ(n-1)\text{MS}_e. \tag{15.78}$$

Under assumptions (15.69), SS_e/σ^2 is distributed as $\chi^2(IJ(n-1))$. Note that Eq. (15.70) is just Eq. (15.11) with the i subscript replaced by an ij subscript (a population mean is estimated by the average of all the observations from that population), and applying the convention for removing extra dots reduces Eqs. (15.71) and (15.74) to Eqs. (15.11) and (15.13), respectively. Specifically, this means one computes the estimates of the main effects of a factor in the same way regardless of how many other factors are in the experimental design.

In a two-way layout there are three hypotheses of potential interest, namely,

$$H_A\colon \quad \alpha_i^A = 0 \qquad \text{for all } i \tag{15.79}$$

$$H_B\colon \quad \alpha_j^B = 0 \qquad \text{for all } j \tag{15.80}$$

$$H_{AB}\colon \quad \alpha_{ij}^{AB} = 0 \qquad \text{for all } (i, j) \tag{15.81}$$

These hypotheses correspond, respectively, to factor A having no effect, factor B having no effect, and no interaction occurring between factors A and B. It is necessary to consider hypothesis H_{AB} first, since if there is an interaction effect, it makes no sense to average the estimates of the effect of the ith level of factor A over different levels of factor B [as in Eq. (15.74)], and even if the $\hat{\alpha}_i^A$ were all zero, it would not imply that factor A had no effect.

Two-Way Analysis of Variance. Hypothesis H_{AB} is best tested using a two-way ANOVA. The $\hat{\alpha}_{ij}^{AB}$ are combined to form the sum of squares for the interaction

$$SS_{AB} = n \sum_{i=1}^{I} \sum_{j=1}^{J} (\hat{\alpha}_{ij}^{AB})^2 \tag{15.82}$$

$$= n \sum_{i=1}^{I} \sum_{j=1}^{J} (X_{ij\bullet} - X_{i\bullet\bullet} - X_{\bullet j\bullet} + X_{\bullet\bullet\bullet})^2 \tag{15.83}$$

and under H_{AB} [with assumptions (15.69)], SS_{AB}/σ^2 is distributed as $\chi^2((I-1)(J-1))$ and is independent of SS_e. Therefore, under H_{AB}

$$F_{AB} = \frac{MS_{AB}}{MS_e} = \frac{SS_{AB}/[(I-1)(J-1)]}{SS_e/[IJ(n-1)]} \tag{15.84}$$

is distributed as $F((I-1)(J-1), IJ(n-1))$, and one would reject H_{AB} at level of significance α if

$$F_{AB} > F(\alpha; (I-1)(J-1), IJ(n-1)) \tag{15.85}$$

or, alternatively, if

$$\Pr[F((I-1)(J-1), IJ(n-1)) > F_{AB}] < \alpha. \tag{15.86}$$

If H_{AB} is not rejected, then one can test H_A using the sum of squares for factor A

$$SS_A = Jn \sum_{i=1}^{I} (\hat{\alpha}_i^A)^2 \tag{15.87}$$

$$= Jn \sum_{i=1}^{I} (X_{i\bullet\bullet} - X_{\bullet\bullet\bullet})^2 \tag{15.88}$$

to compute

$$F_A = \frac{MS_A}{MS_e} = \frac{SS_A/(I-1)}{SS_e/[IJ(n-1)]} \tag{15.89}$$

and rejecting H_A at level of significance α if

$$F_A > F(\alpha; I-1, IJ(n-1)). \tag{15.90}$$

Similarly, one would reject H_B if

$$F_B > F(\alpha; J-1, IJ(n-1)) \tag{15.91}$$

where

$$F_B = \frac{MS_B}{MS_e} = \frac{SS_B/(J-1)}{MS_e} \tag{15.92}$$

$$SS_B = In \sum_{j=1}^{J} (\hat{\alpha}_j^B)^2 \tag{15.93}$$

$$= In \sum_{j=1}^{J} (X_{\bullet j \bullet} - X_{\bullet\bullet\bullet})^2. \tag{15.94}$$

The relationship

$$SS_{tot} = \sum_{i=1}^{I} \sum_{j=1}^{J} \sum_{k=1}^{n} (X_{ijk} - X_{\bullet\bullet\bullet})^2 \tag{15.95}$$

$$= SS_A + SS_B + SS_{AB} + SS_e \tag{15.96}$$

is sometimes useful for computing SS_{AB}. As with the use of Eq. (15.24), however, sufficient accuracy is necessary for all the other sums of squares. Information from a two-way ANOVA is usually summarized in a table similar to Table 15.13.

Example 15.14. Analyzing the percent yield data in Table 15.11 using the equations given in Table 15.13 and the means and variances given in Tables 15.12, one obtains

$$MS_e = \frac{1}{6}(0.4758 + \cdots + 1.963) = 2.042.$$

$$SS_e = 6(3)(2.042) = 36.756$$

$$SS_A = 2(4)\,[(89.975 - 90.371)^2 + \cdots + (90.1 - 90.371)^2] = 5.391$$

$$SS_B = 3(4)\,[(89.85 - 90.371)^2 + (90.892 - 90.371)^2] = 6.515$$

$$SS_{tot} = [(91.3 - 90.371)^2 + \cdots + (89.8 - 90.371)^2] = 71.230$$

$$SS_{AB} = 71.230 - 5.391 - 6.515 - 36.756 = 22.568$$

$$F_{AB} = \frac{22.568/2}{2.042} = 5.53.$$

This information is summarized in Table 15.14, from which one sees that the AB interaction is significant at the 5% level. Since there is a significant interaction, the significance of the main effects of factors A and B cannot be determined from Table 15.14. Factor A must be studied separately for each level of factor B, or vice versa. With factor B fixed, say, at level 1, the effect of factor A can be determined using a one-way ANOVA or, alternatively, a one-way ANOM. From Tables 15.12 (using data for Base 1 only) one computes

$$MS_e = \frac{1}{3}(0.4758 + 2.02 + 1.456) = 1.317$$

$$SS_A = 4\,[(90.825 - 89.85)^2 + \cdots + (88.925 - 89.85)^2] = 7.235$$

$$F_A = \frac{7.235/2}{1.317} = 2.75.$$

TABLE 15.13 ANOVA Table for a Two-Way Layout with Fixed Effects and $n > 1$ Observations per Cell

Source of variation	Sum of squares	Degrees of freedom	Mean square	F	$\Pr > F$
A	$SS_A = Jn \sum_{i=1}^{I} (X_{i\bullet\bullet} - X_{\bullet\bullet\bullet})^2$	$\nu_A = I - 1$	$MS_A = \frac{SS_A}{\nu_A}$	$F_A = \frac{MS_A}{MS_e}$	$\Pr[F(\nu_A, \nu_e) > F_A]$
B	$SS_B = In \sum_{j=1}^{J} (X_{\bullet j\bullet} - X_{\bullet\bullet\bullet})^2$	$\nu_B = J - 1$	$MS_B = \frac{SS_B}{\nu_B}$	$F_B = \frac{MS_B}{MS_e}$	$\Pr[F(\nu_B, \nu_e) > F_B]$
AB	$SS_{AB} = n \sum_{i=1}^{I} \sum_{j=1}^{J} (X_{ij\bullet} - X_{i\bullet\bullet} - X_{\bullet j\bullet} + X_{\bullet\bullet\bullet})^2$	$\nu_{AB} = (I - 1)(J - 1)$	$MS_{AB} = \frac{SS_{AB}}{\nu_{AB}}$	$F_{AB} = \frac{MS_{AB}}{MS_e}$	$\Pr[F(\nu_{AB}, \nu_e) > F_{AB}]$
	$= SS_{tot} - SS_e - SS_A - SS_B$				
Error	$SS_e = \sum_{i=1}^{I} \sum_{j=1}^{J} \sum_{k=1}^{n} (X_{ijk} - X_{ij\bullet})^2$	$\nu_e = IJ(n - 1)$	$MSe = \frac{SS_e}{\nu_e}$		
			$= \frac{1}{IJ} \sum_{i=1}^{I} \sum_{j=1}^{J} S_{ij}^2$		
Total	$SS_{tot} = \sum_{i=1}^{I} \sum_{j=1}^{J} \sum_{k=1}^{n} (X_{ijk} - X_{...})^2$	$IJn - 1$			

TABLE 15.14 ANOVA Table for Example 15.14

Source	SS	df	MS	F	Pr > F
A	5.391	2	2.696	1.32	0.292
B	6.515	1	6.515	3.19	0.091
AB	22.568	2	11.284	5.53	0.013
Error	36.756	18	2.042		
Total	71.230	23			

Since $2.75 < F(0.1; 2, 9) = 3.01$, the effect of Alcohol is not significant at $\alpha = 0.1$ when using Base 1.

For the purpose of illustrating different methods of analysis, we will study the main effects of factor A when factor B is fixed at level 2 using a one-way ANOM, although one would not normally analyze the data for factor A using different techniques for the different levels of factor B. From Tables 15.12 (using data for Base 2 only) one computes

$$\text{MS}_e = \frac{1}{3}(3.243 + 3.096 + 1.963) = 2.767$$

and hence $\alpha = 0.1$ ANOM decision lines

$$90.892 \pm h(0.1; 3, 9)\sqrt{2.767}\sqrt{\frac{2}{3(4)}} = 90.892 \pm 2.34\sqrt{2.767}\sqrt{\frac{1}{6}}$$

$$= 90.892 \pm 1.59.$$

From the ANOM chart in Fig. 15.10 one sees that the effect of Alcohol is significant at $\alpha = 0.1$ when using Base 2, and this is due to Alcohol 1 producing a significantly low yield.

If in this experiment interest centered on finding the Alcohol/Base combination that produces the highest yields, after finding a significant interaction, the following alternative analysis would be in order. The six Alcohol/Base combinations are considered to be six populations ($I = 6$) in a one-way layout. Using $\text{MS}_e = 2.042$ from Table 15.14, one computes $\alpha = 0.05$ ANOM decision lines

$$90.371 \pm h(0.05; 6, 18)\sqrt{2.042}\sqrt{\frac{5}{6(4)}} = 90.371 \pm 2.91\sqrt{2.042}\sqrt{\frac{5}{24}}$$

$$= 90.371 \pm 1.90.$$

From the ANOM chart in Fig. 15.11 one sees that the Alcohol 2 and Base 2 combination produced a significantly ($\alpha = 0.05$) higher average yield.

This example also serves to illustrate that the type of analysis done will depend on the question(s) one is interested in answering. Even in specific designs, such as a two-way layout, there is no one sequence of steps in the analysis.

Two-Way Analysis of Means. In a two-way layout one must first test for interactions using the ANOVA. If interaction is present, then the analysis must be performed using one-way layouts, as was done in Example 15.14. If interaction is not

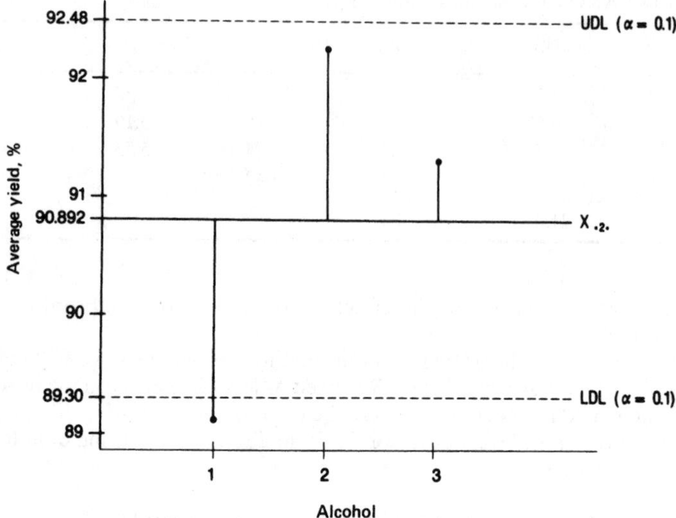

FIGURE 15.10 ANOM chart for Example 15.14 using data for Base 2 only.

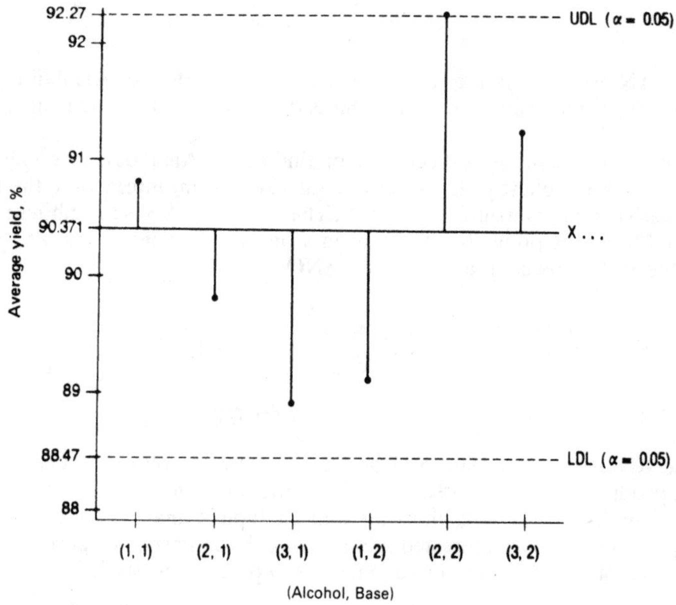

FIGURE 15.11 ANOM chart for Alcohol/Base combinations in Example 15.14.

present, then the ANOM can be used to test for main effects. For two-way (and higher-order) balanced designs the ANOM decision lines (15.30) generalize to

$$\hat{\mu} \pm h(\alpha; f, \nu_e)\sqrt{\text{MS}_e}\sqrt{\frac{f-1}{N}} \tag{15.97}$$

where f = number of levels of the factor

ν_e = degrees of freedom for MS_e

N = total number of observations.

Thus, in a two-way layout the ANOM decision lines for testing the main effects of factor A are

$$X_{\bullet\bullet\bullet} \pm h(\alpha; I, IJ(n-1))\sqrt{\text{MS}_e}\sqrt{\frac{I-1}{Jn}}. \tag{15.98}$$

Similarly for factor B one uses

$$X_{\bullet\bullet\bullet} \pm h(\alpha; J, IJ(n-1))\sqrt{\text{MS}_e}\sqrt{\frac{J-1}{IJn}}. \tag{15.99}$$

Example 15.15. The output in milliamperes of a newly designed electronic component was measured at three temperatures and three humidity levels. Two replicate measurements (different components) were made at each temperature/humidity combination, and the results are given in Table 15.15. Using the means and the variances given in Tables 15.16, one obtains the ANOVA table given in

TABLE 15.15 Outputs (mA) of an Electronic Component (Example 15.15)

		Temperature, °C		
		10	20	30
Relative humidity, %	20	174 164	179 168	182 188
	30	192 186	198 204	199 208
	40	198 202	205 199	206 210

TABLE 15.16a Average Outputs for Temperature/Humidity Combinations Together with Row Means, Column Means, and the Grand Mean (Example 15.15)

		Temperature, °C			Row means
		10	20	30	
Relative humidity, %	20	169	173.5	185	175.83
	30	189	201	203.5	197.83
	40	200	202	208	203.33
Column means		186	192.17	198.83	192.33

TABLE 15.16b Sample Variances for Temperature/Humidity Combinations (Example 15.15)

		Temperature, °C		
		10	20	30
Relative humidity, %	20	50	60.5	18
	30	18	18	40.5
	40	8	18	8

Table 15.17. Since there is no significant interaction, the main effects of the two factors can be studied using the ANOM decision lines (15.98) and (15.99). For Humidity (factor A) one computes

$$192.33 \pm h(0.001; 3, 9)\sqrt{26.56}\sqrt{\frac{2}{18}} = 192.33 \pm 9.43$$

and from the ANOM chart in Fig. 15.12 one sees that low humidity (20%) causes significantly low output, and high humidity (40%) causes significantly high output

TABLE 15.17 ANOVA Table for Example 15.15

Source	SS	df	MS	F	Pr > F
Humidity	2541	2	1270.5	47.8	0.00002
Temperature	494.33	2	247.2	9.31	0.006
Interaction	87.67	4	21.9	0.82	0.544
Error	239	9	26.56		
Total	3362	17			

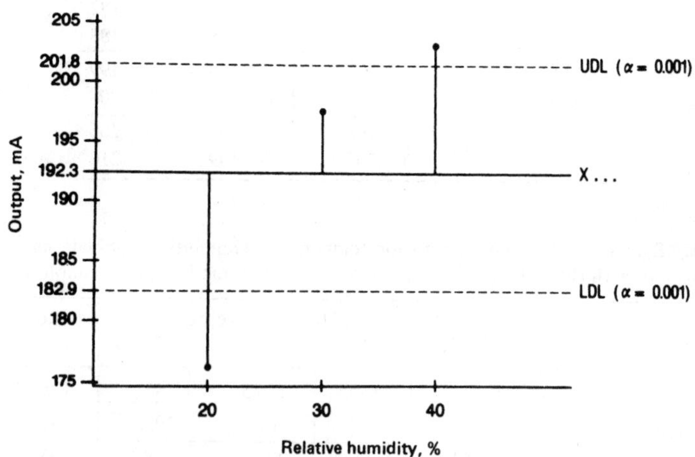

FIGURE 15.12 ANOM chart for Humidity in Example 15.15.

at $\alpha = 0.001$. The effect of Temperature can be studied in a similar fashion. (In this particular case the decision lines would be exactly the same as those for Humidity.) Also, note that since there is no significant interaction, the descriptive levels of significance given in Table 15.17 for the main effects of Temperature and Humidity could have been used to determine the significance of those factors.

One Observation per Cell. When there is only one observation in each cell ($n = 1$), it is not possible to obtain an estimate of the experimental error using Eqs. (15.58) and (15.77). Instead, one assumes that no interaction is present and uses the interaction sum of squares [Eq. (15.83)] as a measure of error. (Obviously, this technique should not be used when an interaction is known to exist or when one is suspected.) The ANOVA table for this situation is given in Table 15.18. Note that Eq. (15.96) reduces to

$$SS_{tot} = SS_A + SS_B + SS_e. \qquad (15.100)$$

Example 15.16. Silicon crystals were grown using the Czochralski process and treatment conditions consisting of combinations of type of impurity used (factor A) and rate of cooling (factor B). A two-way layout was used with one observation in each cell. The (coded) electrical properties of the resulting crystals are given in Table 15.19 together with row means, column means, and the grand mean. Using the equations in Table 15.18, one obtains

$$SS_A = 3[(6 - 18)^2 + (30 - 18)^2] = 864$$

$$SS_B = 2[(15 - 18)^2 + \cdots + (17 - 18)^2] = 52$$

$$SS_{tot} = (2 - 18)^2 + \cdots + (30 - 18)^2 = 928$$

$$SS_e = 928 - 864 - 52 = 12.$$

The complete ANOVA table is given in Table 15.20, from which one sees that Impurity Type (DLS = 0.007) is significant at level $\alpha = 0.01$, but Cooling Rate (DLS = 0.188) does not have a significant effect.

Since when there is only one observation per cell one assumes that there is no interaction, the ANOM can be applied without a preliminary test. In this situation the decision lines (15.97) become

$$X_{\bullet\bullet} \pm h(\alpha; f, (I-1)(J-1))\sqrt{MS_e}\sqrt{\frac{f-1}{IJ}} \qquad (15.101)$$

where $f = I$ for factor A and $f = J$ for factor B, and the MS_e is computed using the formulas in Table 15.18.

Example 15.17. Analyzing the data in Table 15.19 using the ANOM decision lines (15.101) for factor B with $\alpha = 0.1$, one computes

$$X_{\bullet\bullet} \pm h(0.1; 3, 2)\sqrt{6}\sqrt{\frac{2}{6}} = 18 \pm 5.76.$$

Each $X_{\bullet j}$ is compared with these decision lines in Fig. 15.13, from which one sees that the Cooling Rate does not have a significant effect at $\alpha = 0.1$.

TABLE 15.18 ANOVA Table for a Two-Way Layout with Fixed Effects and One Observation per Cell

Source of variation	Sum of squares	Degrees of freedom	Mean square	F	$\Pr > F$
A	$SS_A = J \sum_{i=1}^{I} (X_{i\bullet} - X_{\bullet\bullet})^2$	$\nu_A = I - 1$	$MS_A = \frac{SS_A}{\nu_A}$	$F_A = \frac{MS_A}{MS_e}$	$\Pr[F(\nu_A, \nu_e) > F_A]$
B	$SS_B = I \sum_{j=1}^{J} (X_{\bullet j} - X_{\bullet\bullet})^2$	$\nu_B = J - 1$	$MS_B = \frac{SS_B}{\nu_B}$	$F_B = \frac{MS_B}{MS_e}$	$\Pr[F(\nu_B, \nu_e) > F_B]$
Error	$SS_e = \sum_{i=1}^{I} \sum_{j=1}^{J} (X_{ij} - X_{i\bullet} - X_{\bullet j} + X_{\bullet\bullet})^2$ $= SS_{tot} - SS_A - SS_B$	$\nu_e = (I-1)(J-1)$	$MS_e = \frac{SS_e}{\nu_e}$		
Total	$SS_{tot} = \sum_{j=1}^{I} \sum_{j=1}^{J} (X_{ij} - X_{\bullet\bullet})^2$	$IJ - 1$			

TABLE 15.19 Electrical Properties (Coded) of Silicon Crystals for Various Impurity/Cooling Rate Combinations together with Row Means, Column Means, and the Grand Mean (Example 15.16)

		Rate of cooling 1	2	3	Row means
Impurity type	1	2	12	4	6
	2	28	32	30	30
Column means		15	22	17	18

TABLE 15.20 ANOVA Table for Example 15.16

Source	SS	df	MS	F	Pr > F
Impurity	864	1	864	144	0.007
Cooling	52	2	26	4.33	0.188
Error	12	2	6		
Total	928	5			

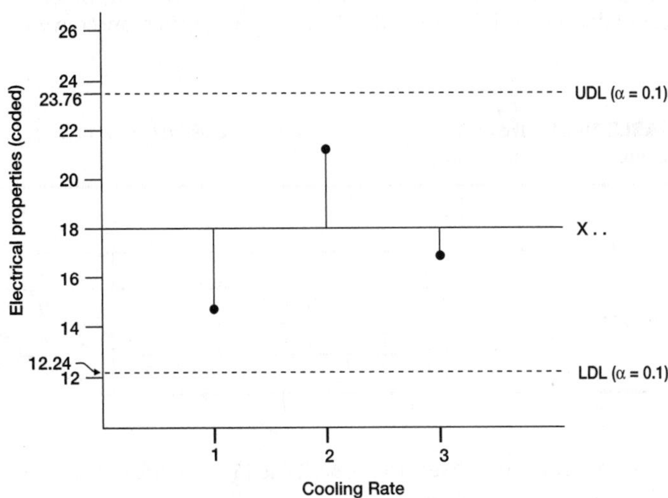

FIGURE 15.13 ANOM chart for Cooling Rate in Example 15.17.

Blocking Factors. In order to decrease the variability associated with the factors of interest in an experiment, the experimental trials can sometimes be split into subgroups that are more homogeneous than the entire experiment. These subgroups are called *blocks*, and a factor used to accomplish this subdivision is called a *blocking factor*. The simplest example of blocking is a paired experiment such as the tire example in the Introduction, where the blocks are cars. Other examples of blocking factors are batches of raw material, technicians, and days on which experimental trials are performed. In each instance the experimenter is not specifically interested in the effect of the blocking factor, but in separating the variability associated with it from the variability associated with the factors of interest. It is possible to have more than one blocking factor in an experiment, such as the situation where experimental trials are performed by different technicians over several days. The analysis associated with a blocking factor is the same as for any other factor. The only difference is that because of their nature blocking factors are assumed not to interact with the other factors.

Consider an experiment to compare four different brands of tires with regard to tread wear. The tires are to be mounted on cars, driven for 10,000 miles, and then compared for tread wear. In order to remove any effect due to differences in cars (such as being driven over different routes), cars are used as blocks, and one of each brand of tire is mounted on each car. The positions of the different brands are chosen at random for each car. This is called a *randomized* (complete) *block design*, since the randomization is restricted to being done within the blocks. The adjective "complete," which will be suppressed but assumed unless "incomplete" is specified, signifies that all the levels of the factor of interest appear in each block. In this case all four brands of tires appear on each car. Designs that are randomized without any restrictions are called *completely randomized*.

Example 15.18. Four brands of tires (labeled *a*, *b*, *c*, and *d*) were compared for tread wear using four cars and the randomized block design shown in Table 15.21. The letters in the table indicate the positions assigned to the brands on each car (the design), and the numbers are the resulting tread wear (in units of 0.1 mm) after 10,000 miles of driving. In Table 15.22 the data are rearranged more conveniently for

TABLE 15.21 Tread Wear (0.1 mm) for Four Brands (*a*, *b*, *c*, *d*) of Tires Using a Randomized Block Design (Example 15.18)

		Car							
		1		2		3		4	
Tire position	1	*a*	8	*c*	11	*c*	9	*b*	8
	2	*c*	7	*b*	9	*a*	7	*c*	9
	3	*d*	3	*a*	5	*d*	2	*d*	4
	4	*b*	4	*d*	5	*b*	2	*a*	3

TABLE 15.22 Tread Wear Data from Table 15.21 Rearranged More Conveniently for Analysis

		Car				Row
		1	2	3	4	means
Brand	*a*	8	5	7	3	5.75
	b	4	9	2	8	5.75
	c	7	11	9	9	9
	d	3	5	2	4	3.5
Column means		5.5	7.5	5	6	6

analysis. Using the formulas in Table 15.18, one obtains the ANOVA table given in Table 15.23. The ANOM decision lines (15.101) for Brands with $\alpha = 0.05$ are

$$X_{\bullet\bullet} \pm h(0.05; 4, 9)\sqrt{5.17}\sqrt{\frac{3}{16}} = 6 \pm 2.94$$

and from the ANOM chart given in Fig. 15.14 one sees that Brands are significantly different at $\alpha = 0.05$ due to Brand c being significantly bad.

If, as in Example 15.18, the blocking variable does not have a significant effect (from Table 15.23 DLS = 0.478 for Cars), then the sum of squares for blocks (SS_{blocks}) can be pooled with SS_e to obtain an improved estimator of σ^2. Specifically, one could use

$$\hat{\sigma}^2 = \frac{SS_e + SS_{blocks}}{\nu_e + \nu_{blocks}} \tag{15.102}$$

where ν_e and ν_{blocks} are the degrees of freedom for error and blocks, respectively. This is the MS_e that would be obtained if the data in Table 15.22 were analyzed as

TABLE 15.23 ANOVA Table for Example 15.18

Source	SS	df	MS	F	Pr > F
Brands	61.5	3	20.5	3.97	0.047
Cars (blocks)	14	3	4.67	0.90	0.478
Error	46.5	9	5.17		
Total	122	15			

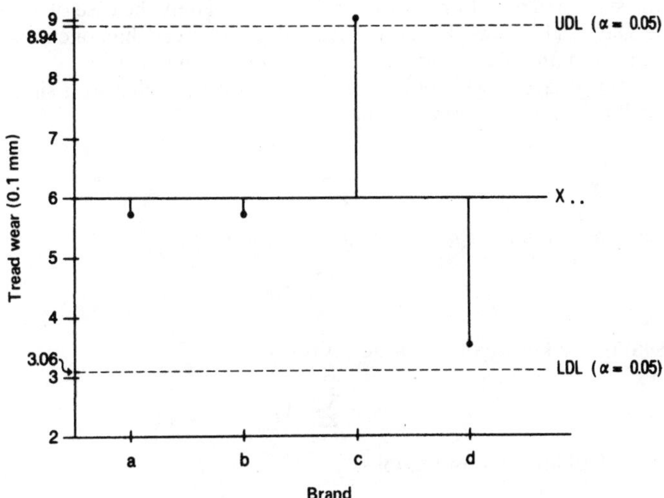

FIGURE 15.14 ANOM chart for brand in Example 15.18.

a one-way layout without taking Cars into account. For Example 15.18, however, this would have no practical effect. The $MS_e = 5.17$ (from Table 15.23) would be replaced [using Eq. (15.102)] by

$$\hat{\sigma}^2 = \frac{46.5 + 14}{9 + 3} = 5.04$$

and the $\alpha = 0.05$ ANOM decision lines would become 6 ± 2.77.

Missing Values. A missing observation (missing value) is the result of a planned experimental trial that for some reason, such as a sample being spilled or a tire blowing out, was not obtained. Designed experiments should always be conducted so as to minimize the chances of any missing observations. When there are missing values in a two-way (or higher-order) layout, the design is no longer balanced, the estimates of the effects cease to be independent, and the design is less efficient. That is, for a given number of trials the effects are estimated with less precision. If there are several missing values, then it may even be impossible to estimate some effects.

It is possible to deal with missing values using the general theory of linear models. That, however, is cumbersome and does not exploit the structure of a designed experiment. Instead, the following techniques are suggested. The missing values are replaced with estimates chosen so as to minimize the SS_e, and the associated degrees of freedom are reduced by the number of estimated values. The usual ANOVA then results in the same value for the MS_e as the linear model approach. Specifically, when a cell containing a missing value or values is not empty, the SS_e is minimized by using the average of the remaining observations in that cell as an estimate of the missing value(s). This adds nothing to the SS_e since the estimates are exactly the predicted values for each cell. For empty cells each missing value is replaced with a variable, and the SS_e is computed in terms of these variables. The values of the variables (estimates of the missing values) are obtained by differentiating the SS_e with respect to each variable, setting the results equal to zero, and solving the system of equations. We consider in detail only the case of a single missing value. Davies (1978) works several examples with more than one missing value, and more information can be found in Little and Rubin (1987).

In a two-way layout with only one observation per cell and a single missing value in cell (i', j') the resulting estimator is

$$\hat{X}_{i'j'} = \frac{IR + JC - T}{(I-1)(J-1)} \tag{15.103}$$

where C = sum of the known observations in column j',

$$C = \sum_{i \neq i'} X_{ij'} \tag{15.104}$$

R = sum of the known observations in row i',

$$R = \sum_{j \neq j'} X_{i'j} \tag{15.105}$$

T = sum of all known observations,

$$T = \sum_{(i,j) \neq (i',j')} X_{ij}. \tag{15.106}$$

Using the estimator (15.103), one can conduct approximate tests for main effects with either the ANOVA or the ANOM. (Recall that for a two-way layout with only one observation per cell, in order to test for main effects one must assume that there is no interaction.) The ANOVA test for main effects is performed in the usual way using the estimate of the missing value and then reducing the degrees of freedom for the SS_e (and thus the SS_{tot}) by 1. The ANOM test for main effects is performed using the decision lines (15.97), and $\hat{X}_{i'j'}$ is used instead of the missing $X_{i'j'}$ in all computations. Specifically,

$$\hat{\mu} = \frac{T + \hat{X}_{i'j'}}{IJ} \qquad (15.107)$$

(that is, $\hat{\mu}$ is the overall average, including the estimated value) and N includes the estimated value.

Example 15.19. Suppose that in Example 15.18 Tire Brand b on Car 3 blew out before going 10,000 miles. The (b, 3) cell value of 2 in Table 15.22 would then be missing. Using Eq. (15.103), one would compute an estimate of this missing value to be

$$\hat{X}_{b3} = \frac{4(21) + 4(18) - 94}{3(3)} = 6.89.$$

The ANOVA table computed using the value 6.89 in cell (b, 3) is given in Table 15.24, from which one finds that the Brands are significantly different (approximate DLS = 0.029). Using Eq. (15.107), one obtains

$$\hat{\mu} = \frac{94 + 6.89}{16} = 6.31$$

and the ANOM decision lines (15.97) with $\alpha = 0.05$ are ($MS_e = 4.133$ from Table 15.24)

$$\hat{\mu} \pm h(0.05; 4, 8)\sqrt{4.133}\sqrt{\frac{4-1}{16}} = 6.31 \pm (3.07)\sqrt{4.133}\sqrt{\frac{3}{16}}$$
$$= 6.31 \pm 2.70.$$

The ANOM chart for this example is given in Fig. 15.15, from which one sees that Brand d is significantly good at (approximately) $\alpha = 0.05$.

When there are replicate values in the cells, one can (and should) first test for possible interactions. Missing values should be estimated as previously described and used to produce the usual ANOVA table, which provides approximate tests for

TABLE 15.24 ANOVA Table for Example 15.19 (One Missing Value)

Source	SS	df	MS	F	Pr > F
Brands	63.54	3	21.18	5.12	0.029
Cars	8.70	3	2.90	0.70	0.578
Error	33.06	8	4.133		
Total	105.30	14			

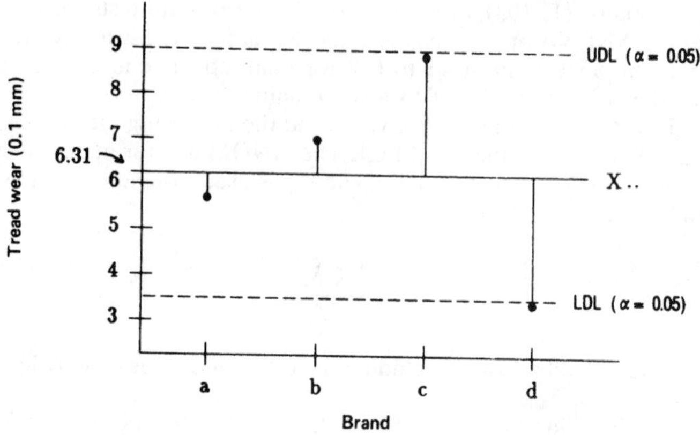

FIGURE 15.15 ANOM chart for Brand in Example 15.19.

all the usual hypotheses, in particular H_{AB}. If there is not a significant interaction, then the ANOM decision lines (15.97) can be used as an approximate test for main effects. The estimates are used in place of the missing values in all computations. If there is a significant interaction, then factor A must be studied separately for each level of factor B, and in this one-way situation missing values complicate the analysis only to the extent that one must then deal with unequal sample sizes.

Example 15.20. Suppose that in Example 15.15 the observation $X_{121} = 179$ were missing due to a power surge destroying the component for that particular trial. The missing value would be estimated as $\hat{X}_{121} = 168$ (the average of the remaining values in that cell). The ANOVA with $\hat{X}_{121} = 168$ used in cell (1, 2) and the degrees of freedom for SS_e (and SS_{tot}) reduced by 1 is given in Table 15.25. The approximate DLS for H_{AB} from this table is 0.219, indicating no apparent interaction. Using $MS_e = 22.31$ from Table 15.25, the ANOM decision lines (15.97) for Temperature (also for Humidity) with $\alpha = 0.01$ are

$$\hat{\mu} \pm h(0.01; 3, 8)\sqrt{22.31}\sqrt{\frac{3-1}{18}} = 191.72 \pm (3.98)\sqrt{22.31}\sqrt{\frac{1}{9}}$$

$$= 191.72 \pm 6.27$$

The ANOM chart for Temperature is given in Fig. 15.16, from which one sees that the output is significantly ($\alpha = 0.01$) high at 30°.

TABLE 15.25 ANOVA Table for Example 15.20 (One Missing Value)

Source	SS	df	MS	F	Pr > F
Humidity	2917	2	1459	65.40	0.00001
Temperature	511.4	2	255.7	11.46	0.004
Interaction	162.2	4	40.56	1.818	0.219
Error	178.5	8	22.31		
Total	3769.1	16			

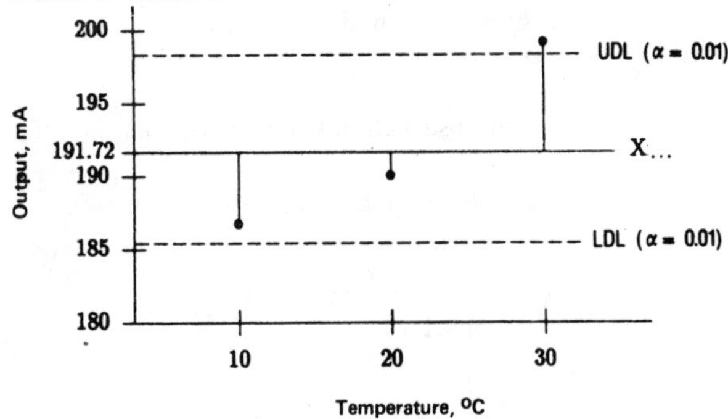

FIGURE 15.16 ANOM chart for Temperature in Example 15.20.

15.4.3 Three-Way Layouts

All the definitions and techniques associated with studying two factors simultaneously extend to the study of three factors. For three factors A, B, and C, with numbers of levels I, J, and K, respectively, when observations corresponding to the ith level of A, the jth level of B, and the kth level of C are recorded in cell (i, j, k) of a table, it is called a three-way layout. Data of this form are conveniently organized as K $I \times J$ two-way tables, one for each level of factor C. We will for the present restrict our attention to designs having $n > 1$ observations per cell. Imagine, as an example, that for the chemical production process discussed previously one is interested in studying not only the effects of Alcohol (factor A) and Base (factor B), but also the effect on yield of the Temperature at which the reaction is carried out (factor C). Table 15.26 is an example of this with $I = 3$, $J = 2$, $K = 2$, and $n = 2$ observations per cell.

For a balanced complete three-way layout there are IJK treatment combinations (populations) that are assumed to be normally distributed with common variance σ^2 and means depending on the particular treatment combination. That is, the

TABLE 15.26 Percent Yields for Alcohol/Base/Temperature Combinations

		Temperature, °C			
		20		40	
		Base		Base	
		1	2	1	2
Alcohol	1	89.9 91.5	88.3 89.4	90.6 89.1	88.2 87.1
	2	88.4 90.1	92.3 93.7	85.7 87.6	90.4 91.8
	3	89.5 87.6	93.1 91.4	87.5 85.8	89.7 91.5

observations from cell (i, j, k) are assumed to be $N(\mu_{ijk}, \sigma^2)$. All the observations are still assumed to be independent. Let

$$X_{ijk\ell} = \ell\text{th observation from cell } (i, j, k)$$

$$S_{ijk}^2 = \text{sample variance from cell } (i, j, k),$$

$$S_{ijk}^2 = \frac{1}{n-1} \sum_{\ell=1}^{n} (X_{ijk\ell} - X_{ijk\bullet})^2 \tag{15.108}$$

and

$$\mu = \mu_{\bullet\bullet\bullet} = \frac{1}{IJK} \sum_{i=1}^{I} \sum_{j=1}^{J} \sum_{k=1}^{K} \mu_{ijk}. \tag{15.109}$$

As was mentioned in Sec. 15.4.2, the main effects for any factor can be obtained from Eq. (15.7) by recalling the convention for removing dots and keeping track of which subscript represents which factor. In the case of three factors one obtains

$$\alpha_i^A = \mu_{i\bullet\bullet} - \mu_{\bullet\bullet\bullet} \tag{15.110}$$

$$\alpha_j^B = \mu_{\bullet j\bullet} - \mu_{\bullet\bullet\bullet} \tag{15.111}$$

$$\alpha_k^C = \mu_{\bullet\bullet k} - \mu_{\bullet\bullet\bullet} \tag{15.112}$$

Similarly, all two-way interaction effects can be obtained from Eq. (15.66). Thus, with three factors,

$$\alpha_{ij}^{AB} = \mu_{ij\bullet} - \mu_{i\bullet\bullet} - \mu_{\bullet j\bullet} + \mu_{\bullet\bullet\bullet} \tag{15.113}$$

$$\alpha_{jk}^{BC} = \mu_{\bullet jk} - \mu_{\bullet j\bullet} - \mu_{\bullet\bullet k} + \mu_{\bullet\bullet\bullet} \tag{15.114}$$

$$\alpha_{ik}^{AC} = \mu_{i\bullet k} - \mu_{i\bullet\bullet} - \mu_{\bullet\bullet k} + \mu_{\bullet\bullet\bullet}. \tag{15.115}$$

Estimators of the various parameters are easily obtained from Eqs. (15.11) to (15.13) and (15.76). For example, using Eq. (15.11), one obtains

$$\hat{\mu}_{ijk} = X_{ijk}. \tag{15.116}$$

using Eq. (15.13), one obtains

$$\hat{\alpha}_k^C = X_{\bullet\bullet k\bullet} - X_{\bullet\bullet\bullet}. \tag{15.117}$$

and using Eq. (15.76), one obtains

$$\hat{\alpha}_{ik}^{AC} = X_{i\bullet k\bullet} - X_{i\bullet\bullet\bullet} - X_{\bullet\bullet k\bullet} + X_{\bullet\bullet\bullet}. \tag{15.118}$$

Note that the estimators given in Eqs. (15.117) and (15.118) could also have been obtained using definitions (15.112) and (15.115), respectively, and replacing each μ_* with its estimator $X_{*\bullet}$ ($* = i\bullet\bullet, ij\cdot, i\cdot k$, etc.). Also, the estimator for σ^2 is the usual pooled sample variance

$$\hat{\sigma}^2 = \text{MS}_e = \frac{1}{IJK} \sum_{i=1}^{I} \sum_{j=1}^{J} \sum_{k=1}^{K} S_{ijk}^2 \tag{15.119}$$

and

$$\text{SS}_e = IJK(n-1)\text{MS}_e \tag{15.120}$$

where SS_e/σ^2 is distributed as $\chi^2(IJK(n-1))$.

Three-Way Interaction. Introduction of a third factor into an experiment (with $n > 1$) allows one to test for a three-way interaction between the factors. The three-way interaction effect between the ith level of factor A, the jth level of factor B, and the kth level of factor C (denoted by α_{ijk}^{ABC}) is the difference between the two-way interaction effect α_{ij}^{AB} specific to the kth level of factor C and the α_{ij}^{AB} effects averaged over all the levels of factor C. Algebraically,

$$\alpha_{ijk}^{ABC} = \alpha_{ijk}^{AB} - \alpha_{ij}^{AB} \tag{15.121}$$

$$= \mu_{ijk} - \mu_{ij\bullet} - \mu_{\bullet jk} - \mu_{i\bullet k} + \mu_{i\bullet\bullet} + \mu_{\bullet j\bullet} + \mu_{\bullet\bullet k} - \mu_{\bullet\bullet\bullet} \tag{15.122}$$

and replacing each μ_* in Eq. (15.122) with its estimator X_{*}. , one obtains

$$\hat{\alpha}_{ijk}^{ABC} = X_{ijk\bullet} - X_{ij\bullet\bullet} - X_{\bullet jk\bullet} - X_{i\bullet k\bullet} + X_{i\bullet\bullet\bullet} + X_{\bullet j\bullet\bullet} + X_{\bullet\bullet k.} - X_{\bullet\bullet\bullet}. \tag{15.123}$$

Using Eqs. (15.109) to (15.115) and (15.122), our model can be written as

$$X_{ijk\ell} = \mu_{ijk} + \epsilon_{ijk\ell} \tag{15.124}$$

$$= \mu + \alpha_i^A + \alpha_j^B + \alpha_k^C + \alpha_{ij}^{AB} + \alpha_{jk}^{BC} + \alpha_{ik}^{AC} + \alpha_{ijk}^{ABC} + \epsilon_{ijk\ell} \tag{15.125}$$

where

$$\epsilon_{ijk\ell} \sim N(0, \sigma^2)$$

and (15.126)

$$\epsilon_{ijk\ell} \text{ are independent.}$$

Rarely, if ever, are there physically interpretable three-way (or higher-order) interactions. Therefore, little is gained by using graphical methods to try to identify the source of any three-way interaction. Instead, it is reasonable to simply test the hypothesis

$$H_{ABC}\colon\ \alpha_{ijk}^{ABC} = 0 \quad \text{for all } (i, j, k) \tag{15.127}$$

which corresponds to no three-way interaction. This is, in fact, the first thing that must be done since in the presence of a three-way interaction it makes no sense to average two-way interaction effects over different levels of the third factor or to average the main effects for a single factor over different levels of the other two factors. As with the estimates of main effects in the presence of a two-way interaction, having all zeros for the estimates of main effects or two-way-interaction effects would not imply lack of effects in the presence of a three-way interaction.

Three-Way Analysis of Variance. Hypothesis H_{ABC} is best tested using a three-way ANOVA. The $\hat{\alpha}_{ijk}^{ABC}$ [Eq. (15.123)] are combined to form the sum of squares for the ABC interaction

$$SS_{ABC} = n \sum_{i=1}^{I} \sum_{j=1}^{J} \sum_{k=1}^{K} (\hat{\alpha}_{ijk}^{ABC})^2 \tag{15.128}$$

and under H_{ABC} [with assumptions (15.126)], SS_{ABC}/σ^2 is distributed as $\chi^2(I-1)(J-1)(K-1))$ and is independent of SS_e [Eq. (15.120)]. Therefore, under H_{ABC}

$$F_{ABC} = \frac{MS_{ABC}}{MS_e} = \frac{SS_{ABC}/[(I-1)(J-1)(K-1)]}{SS_e/[IJK(n-1)]} \tag{15.129}$$

is distributed as $F((I-1)\,(J-1)\,(K-1), IJK(n-1))$, and one would reject H_{ABC} at level α if

$$F_{ABC} > F(\alpha; (I-1)(J-1)(K-1), IJK(n-1)) \tag{15.130}$$

or, alternatively, if

$$\Pr[F((I-1)(J-1)(K-1), IJK(n-1)) > F_{ABC}] < \alpha. \tag{15.131}$$

All the other hypotheses of potential interest (H_{AB}, H_{BC}, H_{AC}, H_A, H_B, and H_C) are tested in a fashion similar to both the test of H_{ABC} and the tests described for a two-way ANOVA by using the estimates for the effects obtained from equations such as Eqs. (15.117) and (15.118). All the necessary equations for a three-way ANOVA with $n > 1$ are summarized in Table 15.27. Recall that one should start by testing H_{ABC} using the DLS at the bottom of the table and then work up. If H_{ABC} is not rejected, two-way interactions can be tested, and if none of them are significant, then main effects can be tested.

How to proceed in the face of significant interactions depends on which interactions are significant. If a three-way interaction or at most two pairwise interactions exist, then the analysis must be performed using two-way layouts (for example, study $A \times B$ separately for each level of C). Exactly which two-way layouts can be used depends on the situation. In some cases, such as when a three-way interaction exists, several choices are acceptable. In other cases, however, only one choice is reasonable. For example, significant AB and AC interactions imply that one cannot average B over A or C over A, and therefore one must consider $B \times C$ separately for different levels of A. If all three pairwise interactions exist, then the analysis must be performed using one-way layouts [for example, study A separately for each (j, k) level of $B \times C$)].

Example 15.21. Analyzing the percent yield data in Table 15.26, one obtains the ANOVA table given in Table 15.28. The various sums of squares are most easily computed by constructing two-way tables such as the one given in Table 15.29 for Alcohol/Base combinations (averaged over Temperature and Replicates). Using these values, one obtains

TABLE 15.27 ANOVA Table for a Three-Way Layout with Fixed Effects and $n > 1$ Observations per Cell

Source of variation	Sum of squares	Degrees freedom	Mean square	F	$\Pr > F$
A	$SS_A = JKn \sum_{i=1}^{I} (X_{iZ\bullet\bullet\bullet} - X_{\bullet\bullet\bullet\bullet})^2$	$\nu_A = I - 1$	$MS_A = \frac{SS_A}{\nu_A}$	$F_A = \frac{MS_A}{MS_e}$	$\Pr[F(\nu_A, \nu_e) > F_A]$
B	$SS_B = IKn \sum_{i=1}^{I} (X_{\bullet j\bullet\bullet} - X_{\bullet\bullet\bullet\bullet})^2$	$\nu_B = J - 1$	$MS_B = \frac{SS_B}{\nu_B}$	$F_B = \frac{MS_B}{MS_e}$	$\Pr[F(\nu_B, \nu_e) > F_B]$
C	$SS_C = IJn \sum_{k=1}^{K} (X_{\bullet\bullet k\bullet} - X_{\bullet\bullet\bullet\bullet})^2$	$\nu_C = K - 1$	$MS_C = \frac{SS_c}{\nu_C}$	$F_C = \frac{MS_C}{MS_e}$	$\Pr[F(\nu_C, \nu_e) > F_C]$
AB	$SS_{AB} = Kn \sum_{i=1}^{I} \sum_{j=1}^{J} (X_{ij\bullet\bullet} - X_{i\bullet\bullet\bullet} - X_{\bullet j\bullet\bullet} + X_{\bullet\bullet\bullet\bullet})^2$	$\nu_{AB} = (I-1)(J-1)$	$MS_{AB} = \frac{SS_{AB}}{\nu_{AB}}$	$F_{AB} = \frac{MS_{AB}}{MS_e}$	$\Pr[F(\nu_{AB}, \nu_e) > F_{AB}]$
BC	$SS_{BC} = In \sum_{j=1}^{J} \sum_{k=1}^{K} (X_{\bullet jk\bullet} - X_{\bullet j\bullet\bullet} - X_{\bullet\bullet k\bullet} + X_{\bullet\bullet\bullet\bullet})^2$	$\nu_{BC} = (J-1)(K-1)$	$MS_{BC} = \frac{SS_{BC}}{\nu_{BC}}$	$F_{BC} = \frac{MS_{BC}}{MS_e}$	$\Pr[F(\nu_{BC}, \nu_e) > F_{BC}]$
AC	$SS_{AC} = Jn \sum_{i=1}^{I} \sum_{k=1}^{K} (X_{i\bullet k\bullet} - X_{i\bullet\bullet\bullet} - X_{\bullet\bullet k\bullet} + X_{\bullet\bullet\bullet\bullet})^2$	$\nu_{AC} = (I-1)(K-1)$	$MS_A = \frac{SS_{AC}}{\nu_{AC}}$	$F_{AC} = \frac{MS_{AC}}{MS_e}$	$\Pr[F(\nu_{AC}, \nu_e) > F_{AC}]$
ABC	$SS_{ABC} = n \sum_{i=1}^{I} \sum_{j=1}^{J} \sum_{k=1}^{K} (X_{ijk\bullet} - X_{ij\bullet\bullet} - X_{i\bullet k\bullet} - X_{\bullet jk\bullet}$ $+ X_{i\bullet\bullet\bullet} + X_{\bullet j\bullet\bullet} + X_{\bullet\bullet k\bullet} - X_{\bullet\bullet\bullet\bullet})^2$ $= SS_{tot} - SS_e - SS_A - SS_B - SS_C - SS_{AB} - SS_{BC} - SS_{AC}$	$\nu_{ABC} = (I-1)(J-1)(K-1)$	$MS_{ABC} = \frac{SS_{ABC}}{\nu_{ABC}}$	$F_{ABC} = \frac{MS_{ABC}}{MS_e}$	$\Pr[(F(\nu_{ABC}, \nu_e) > F_{ABC}]$
Error	$SS_e = \sum_{i=1}^{I} \sum_{j=1}^{J} \sum_{k=1}^{K} \sum_{\ell=1}^{n} (X_{ijk} - X_{ijk\bullet})^2$	$\nu_e = IJK(n-1)$	$MS_e = \frac{SS_e}{\nu_e} = \frac{1}{IJK} \sum_{i=1}^{I} \sum_{j=1}^{J} \sum_{k=1}^{K} S_{ijk}^2$		
Total	$SS_{tot} = \sum_{i=1}^{I} \sum_{j=1}^{J} \sum_{k=1}^{K} \sum_{\ell=1}^{n} (X_{ijk\ell} - X_{\bullet\bullet\bullet\bullet})^2$	$IJKn - 1$			

TABLE 15.28 ANOVA Table for Example 15.21

Source	SS	df	MS	F	Pr > F
A (Alcohol)	2.2508	2	1.1254	0.892	0.435
B (Base)	23.207	1	23.207	18.39	0.001
C (Temperature)	17.002	1	17.002	13.48	0.003
AB	47.876	2	23.938	18.97	0.0002
BC	0.06	1	0.06	0.0476	0.831
AC	1.5258	2	0.76292	0.605	0.562
ABC	0.2775	2	0.13875	0.110	0.897
Error	15.140	12	1.2617		
Total	107.3391	23			

TABLE 15.29 Average Yields for Alcohol/Base Combinations together with Row Means, Column Means, and the Grand Mean (Example 15.21)

		Base		
		1	2	
Alcohol	1	90.275	88.25	$89.2625 = X_{1\bullet\bullet\bullet}$
	2	87.95	92.05	$90 = X_{2\bullet\bullet\bullet}$
	3	87.6	91.425	$89.5125 = X_{3\bullet\bullet\bullet}$
		88.6083	90.575	$89.5917 = X_{\bullet\bullet\bullet\bullet}$
		$X_{\bullet 1\bullet\bullet}$	$X_{\bullet 2\bullet\bullet}$	

$$\begin{aligned} SS_A &= 2(2)(2)[(89.2625 - 89.5917)^2 + \cdots + (89.5125 - 89.5917)^2] \\ &= 2.2508 \end{aligned}$$

$$\begin{aligned} SS_B &= 3(2)(2)[(88.6083 - 89.5917)^2 + (90.575 - 89.5917)^2] \\ &= 23.207 \end{aligned}$$

$$\begin{aligned} SS_{AB} &= 2(2)[(90.275 - 88.6083 - 89.2625 + 89.5917)^2 + \cdots \\ &\quad + (91.425 - 90.575 - 89.5125 + 89.5917)^2] \\ &= 47.876. \end{aligned}$$

Clearly, for three-way and higher-order layouts the computations can be quite tedious. There are, however, a variety of computer programs available for this task.

From Table 15.28 one sees that the ABC interaction is not significant (DLS = 0.897), but that the AB interaction is (DLS = 0.0002). Therefore, different levels of B (or A) must be considered separately. (Note that if A and B had both interacted with C, further subdivision would be necessary.) Using the data for Base 1 only (Table 15.30), one obtains the ANOVA table given in Table 15.31. Since the AC interaction is not significant (DLS = 0.618), the significance of the main effects can

TABLE 15.30 Percent Yields for Alcohol/Temperature Combinations with Base 1 (from Table 15.26)

		Temperature, °C	
		20	40
Alcohol	1	89.9	90.6
		91.5	89.1
	2	88.4	85.7
		90.1	87.6
	3	89.5	87.5
		87.6	85.8

TABLE 15.31 ANOVA Table for the Data in Table 15.30 (Example 15.21)

Source	SS	df	MS	F	$\Pr > F$
A	16.91	2	8.46	5.70	0.041
C	9.54	1	9.54	6.43	0.044
AC	1.55	2	0.775	0.522	0.618
Error	8.91	6	1.485		
Total	36.91	11			

be read directly from Table 15.31. Alternatively, the ANOM decision lines (15.97) can be used to study the main effects. The ANOM chart for Alcohol is shown in Fig. 15.17, from which one sees that when using Base 1, Alcohol 1 produces a significantly ($\alpha = 0.05$) high yield.

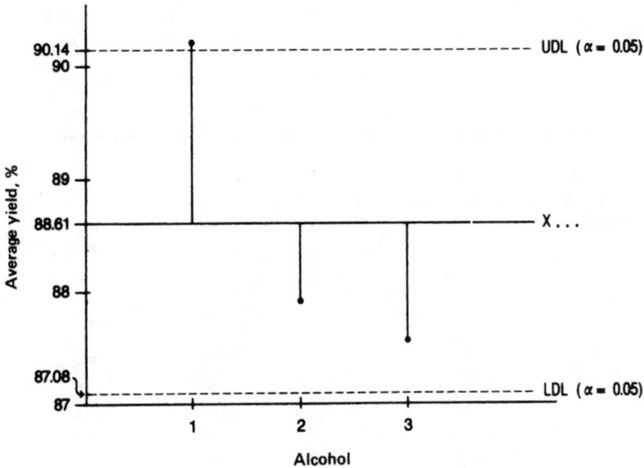

FIGURE 15.17 ANOM chart for Alcohol in Example 15.21, using data for Base 1 only.

Three-Way Analysis of Means. In a three-way layout one must first test for interactions using the ANOVA. Recall that if a three-way interaction or at most two pairwise interactions are present, then the analysis must be performed using two-way layouts (as was done in Example 15.21); and if all three pairwise interactions exist, then the analysis must be performed using one-way layouts. Example 15.3 shows the analysis for a one-way layout. If no interaction is present, then the ANOM decision lines (15.97) can be used to test for significant main effects.

Example 15.22. Suppose that in Table 15.26 the percent yield values obtained with Alcohol 1 and Base 2 (four values) were modified as shown in Table 15.32. Analyzing these Alcohol/Base/Temperature data, one obtains the ANOVA Table given in Table 15.33. Since there are no significant interactions, the ANOM decision lines (15.97) can be used to test for significant main effects. For Alcohol these decision lines with $\alpha = 0.05$ are

$$90.28 \pm h(0.05; 3, 12)\sqrt{1.245}\sqrt{\frac{2}{24}} = 90.28 \pm 0.86$$

and the ANOM chart is shown in Fig. 15.18.

One Observation per Cell. In a k-way layout with only one observation per cell ($n = 1$), one cannot obtain an estimate of σ^2 by pooling cell sample variances [that

TABLE 15.32 Percent Yields for Alcohol/Base/Temperature Combinations (Example 15.22)

		Temperature, °C			
		20		40	
		Base		Base	
		1	2	1	2
Alcohol	1	89.9	94.1	90.6	91.7
		91.5	93.2	89.1	90.6
	2	88.4	92.3	85.7	90.4
		90.1	93.7	87.6	91.8
	3	89.5	93.1	87.5	89.7
		87.6	91.4	85.8	91.5

TABLE 15.33 ANOVA Table for Example 15.22

Source	SS	df	MS	F	Pr > F
A (Alcohol)	14.29	2	7.145	5.74	0.018
B (Base)	67.33	1	67.33	54.08	0.0000
C (Temperature)	21.66	1	21.66	17.40	0.0013
AB	4.58	2	2.29	1.84	0.201
BC	0.082	1	0.082	0.066	0.802
AC	0.378	2	0.189	0.152	0.861
ABC	1.556	2	0.778	0.625	0.552
Error	14.94	12	1.245		
Total	124.816	23			

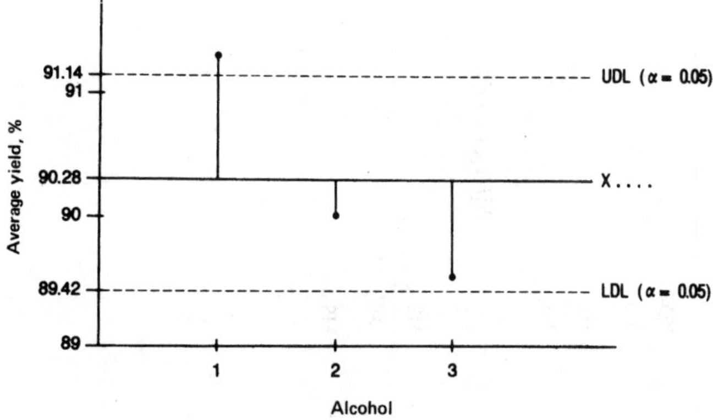

FIGURE 15.18 ANOM chart for Alcohol in Example 15.22.

is, by using equations such as Eq. (15.119)]. Instead one assumes that the highest-order (k-way) interaction does not actually exist and uses its sum of squares for the SS_e. All the necessary formulas for a three-way ANOVA with only one observation per cell are given in Table 15.34. Decision lines for the ANOM are computed exactly as previously described, using the MS_e from Table 15.34.

Example 15.23. An experiment was conducted to study the effects of temperature, pressure, and stirring time on the thickening time of a certain type of cement. The coded results are given in Table 15.35, and the ANOVA using the formulas from Table 15.34 is given in Table 15.36. Since none of the interactions are significant, the ANOM decision lines (15.97) can be used to test the significance of the main effects. In this case all three factors have the same number of levels and would, therefore, use the same decision lines. For $\alpha = 0.05$ they would be

$$X_{\bullet\bullet\bullet} \pm h(0.05; 2, 1)\sqrt{2}\sqrt{\frac{1}{8}} = 0 \pm 6.36$$

and only Temperature is seen to have a significant ($\alpha = 0.05$) effect. However, since in this example many effects are being estimated relative to the number of observations, and since clearly none of the interactions are significant, one could pool all the interaction effects to obtain $MS_e = 4$ (with 4 degrees of freedom) and $\alpha = 0.05$ ANOM decision lines of 0 ± 1.96. One would then conclude that all three factors have significant ($\alpha = 0.05$) effects.

15.5 INCOMPLETE LAYOUTS

While the complete layouts discussed in the last section allow one to estimate all possible interactions, a heavy price is paid in terms of the number of observations required. Even, for example, with only three levels for each of three factors and $n = 2$ replicates, 3(3)(3)(2) = 54 experimental runs are required. Therefore, the question naturally arises as to whether something can be done with fewer observations. We will consider four specific kinds of incomplete designs: Latin squares, Graeco-

TABLE 15.34 ANOVA Table for a Three-Way Layout with Fixed Effects and One Observation per Cell

Source of variation	Sum of squares	Degrees freedom	Mean square	F	$\Pr > F$
A	$SS_A = JK \sum_{i=1}^{I} (X_{i\bullet\bullet} - X_{\bullet\bullet\bullet})^2$	$\nu_A = I - 1$	$MS_A = \frac{SS_A}{\nu_A}$	$F_A = \frac{MS_A}{MS_e}$	$\Pr[F(\nu_A, \nu_e) > F_A]$
B	$SS_B = IK \sum_{j=1}^{J} (X_{\bullet j\bullet} - X_{\bullet\bullet\bullet})^2$	$\nu_B = J - 1$	$MS_B = \frac{SS_B}{\nu_B}$	$F_B = \frac{MS_B}{MS_e}$	$\Pr[F(\nu_B, \nu_e) > F_B]$
C	$SS_C = IJ \sum_{k=1}^{K} (X_{\bullet\bullet k} - X_{\bullet\bullet\bullet})^2$	$\nu_C = K - 1$	$MS_C = \frac{SS_C}{\nu_C}$	$F_C = \frac{MS_C}{MS_e}$	$\Pr[F(\nu_C, \nu_e) > F_C]$
AB	$SS_{AB} = K \sum_{i=1}^{I} \sum_{j=1}^{J} (X_{ij\bullet} - X_{i\bullet\bullet} - X_{\bullet j\bullet} + X_{\bullet\bullet\bullet})^2$	$\nu_{AB} = (I-1)(J-1)$	$MS_A = \frac{SS_{AB}}{\nu_{AB}}$	$F_{AB} = \frac{MS_{AB}}{MS_e}$	$\Pr[F(\nu_{AB}, \nu_e) > F_{AB}]$
BC	$SS_{BC} = I \sum_{j=1}^{J} \sum_{k=1}^{K} (X_{\bullet jk} - X_{\bullet j\bullet} - X_{\bullet\bullet k} + X_{\bullet\bullet\bullet})^2$	$\nu_{BC} = (J-1)(K-1)$	$MS_{BC} = \frac{SS_{BC}}{\nu_{BC}}$	$F_{BC} = \frac{MS_{BC}}{MS_e}$	$\Pr[F(\nu_{BC}, \nu_e) > F_{BC}]$
AC	$SS_{AC} = J \sum_{i=1}^{I} \sum_{k=1}^{K} (X_{i\bullet k\bullet} - X_{i\bullet\bullet} - X_{\bullet\bullet k} + X_{\bullet\bullet\bullet})^2$	$\nu_{AC} = (I-1)(K-1)$	$MS_{AC} = \frac{SS_{AC}}{\nu_{AC}}$	$F_{AC} = \frac{MS_{AC}}{MS_e}$	$\Pr[F(\nu_{AC}, \nu_e) > F_{AC}]$
Error	$SS_e = \sum_{i=1}^{I} \sum_{j=1}^{J} \sum_{k=1}^{K} (X_{ijk} - X_{ij\bullet} - X_{i\bullet k} - X_{\bullet jk}$ $+ X_{i\bullet\bullet} + X_{\bullet j\bullet} + X_{\bullet\bullet k} - X_{\bullet\bullet\bullet})^2$ $= SS_{tot} - SS_A - SS_B - SS_C - SS_{AB} - SS_{BC} - SS_{AC}$	$\nu_e = (I-1)(J-1)(K-1)$	$MS_e = \frac{SS_e}{\nu_e}$		
Total	$SS_{tot} = \sum_{i=1}^{I} \sum_{j=1}^{J} \sum_{k=1}^{K} (X_{ijk} - X_{\bullet\bullet\bullet})^2$	$IJK - 1$			

TABLE 15.35 Thickening Times of Cement (Coded) for Temperature/Pressure/Stirring Time Combinations (Example 15.23)

		Stirring Time			
		1		2	
		Pressure		Pressure	
		1	2	1	2
Temperature	1	5	17	–3	13
	2	–11	–1	–15	–5

TABLE 15.36 ANOVA Table for Example 15.23

Source	SS	df	MS	F	Pr > F
A (Temperature)	512	1	512	256	0.040
B (Pressure)	288	1	288	144	0.053
C (Stirring Time)	50	1	50	25	0.126
AB	8	1	8	4	0.295
BC	2	1	2	1	0.500
AC	2	1	2	1	0.500
Error	2	1	2		
Total	864	7			

Latin squares, balanced incomplete block designs, and Youden squares. Each is potentially useful in a different experimental situation.

15.5.1 Latin Squares

Let us reconsider the tire tread wear problem of Example 15.18. Note that in the randomized block design given in Table 15.21, Brand c never appears in Positions 3 or 4, and Brand d never appears in Positions 1 or 2. If Positions 1 and 2 were the front and 3 and 4 were the rear, then any difference found between Brands c and d would be confounded with use on the front or rear. In fact, since one would expect tread wear to be different from front to rear, any real differences between Brands c and d would be either at least partially hidden or accentuated. Therefore, instead of using the randomized block design given in Table 15.21, it would be better to consider Position as a third factor (a second blocking factor) and balance the Positions of the different Brands, that is, use a design such as the one given in Table 15.37, where each Brand (that is, each level of the factor of interest) appears once and only once in each Position (row) and once and only once on each Car (in each column). Such a design is called a *Latin square* and is a special incomplete three-way layout. This type of design originated in agricultural experiments, where the rows and columns were actually positions of different plots within a field.

What is gained by using a Latin square design is that one can study three factors each at M levels with only M^2 rather than M^3 observations. There are, however, two restrictions. First, the factors must all have the same number of levels, and second,

TABLE 15.37 *A* 4 × 4 Latin Square Design for Comparing Four Brands (*a*, *b*, *c*, *d*) of Tires with Regard to Tread Wear

		Car			
		1	2	3	4
Position	1	*a*	*b*	*c*	*d*
	2	*b*	*c*	*d*	*a*
	3	*c*	*d*	*a*	*b*
	4	*d*	*a*	*b*	*c*

one must assume that there is no interaction between the factors. Specifically, one assumes the model

$$X_{ijk} = \mu + \alpha_i^A + \alpha_j^B + \alpha_k^C + \epsilon_{ijk} \tag{15.132}$$

where the ϵ_{ijk} are independent $N(0, \sigma^2)$ and the (i, j, k) combinations belong to a set $\mathfrak{D}$ [denoted by $(i, j, k) \in \mathfrak{D}$] which depends on the particular Latin square. The particular Latin square should be chosen at random. This is accomplished by first choosing a standard square at random (like the one given in Table 15.37 where the first row and the first column are in order) and then randomizing the columns and all but the first row. Tables of standard squares can be found in Fisher and Yates (1963). The ANOVA table for an $M \times M$ Latin square is given in Table 15.38, and the ANOM decision lines (15.97) can be used to test for main effects.

In the tread wear example the rows and columns were both blocking factors, but any three factors (with the same number of levels) could be studied using a Latin square design if the assumption of no interactions were reasonable. Unfortunately, in most physical science and engineering experiments that is not the case, and Latin square designs should be used with great caution since the presence of interactions can invalidate the results. Hunter (1985) discusses this issue in detail.

Example 15.24. Table 15.39 contains the results of a tire tread wear experiment designed using a Latin square with Cars and Tire Position as blocks. The design (the letters), row means, column means, and the grand mean are also included in the table. For convenience the numbers have been chosen so that SS_{tot} and the sums of squares for Cars and Brands are the same as in Example 15.18 (see Table 15.22), and one need only compute (using equations from Table 15.38)

$$SS_{position} = 4[(7.25 - 6)^2 + (4.5 - 6)^2 + (7 - 6)^2 + (5.25 - 6)^2] = 21.5$$

$$SS_e = 122 - 61.5 - 14 - 21.5 = 25.$$

The complete ANOVA table is given in Table 15.40. Using $MS_e = 4.17$ from Table 15.40, one could compute $\alpha = 0.05$ ANOM decision lines

$$6 \pm h(0.05; 4, 6)\sqrt{4.17}\sqrt{\frac{3}{16}} = 6 \pm 2.93.$$

Comparing the Brand means in Table 15.22 with these limits, one finds that Brand *c* is significantly bad at $\alpha = 0.05$.

TABLE 15.38 ANOVA Table for an $M \times M$ Latin Square

Source of variation	Sum of squares	Degrees freedom	Mean square	F	Pr $> F$
A	$SS_A = M \sum_{i=1}^{M} (X_{i\bullet\bullet} - X_{\bullet\bullet\bullet})^2$	$\nu_A = M - 1$	$MS_A = \frac{SS_A}{\nu_A}$	$F_A = \frac{MS_A}{MS_e}$	$\Pr[F(\nu_A, \nu_e) > F_A]$
B	$SS_B = M \sum_{j=1}^{M} (X_{\bullet j\bullet} - X_{\bullet\bullet\bullet})^2$	$\nu_B = M - 1$	$MS_B = \frac{SS_B}{\nu_B}$	$F_B = \frac{MS_B}{MS_e}$	$\Pr[F(\nu_B, \nu_e) > F_B]$
C	$SS_C = M \sum_{k=1}^{M} (X_{\bullet\bullet k} - X_{\bullet\bullet\bullet})^2$	$\nu_C = M - 1$	$MS_C = \frac{SS_C}{\nu_C}$	$F_C = \frac{MS_C}{MS_e}$	$\Pr[F(\nu_C, \nu_e) > F_C]$
Error	$SS_e = \sum_{(i,j,k) \in \mathcal{D}} (X_{ijk} - X_{i\bullet\bullet} - X_{\bullet j\bullet} - X_{\bullet\bullet k} + 2X_{\bullet\bullet\bullet})^2$ $= SS_{tot} - SS_A - SS_B - SS_C$	$\nu_e = (M-1)(M-2)$	$MS_e = \frac{SS_e}{\nu_e}$		
Total	$SS_{tot} = \sum_{(i,j,k) \in \mathcal{D}} (X_{ijk} - X_{\bullet\bullet\bullet})^2$	$M^2 - 1$			

15.5.2 Graeco-Latin Squares

Two $M \times M$ Latin squares (one denoted by Roman letters and one denoted by Greek letters) are said to be *orthogonal* if when they are superimposed, as in Table 15.41, every one of the M^2 letter combinations appears only once. Such a design is a special incomplete four-way layout and is called a *Graeco-Latin square*. It can be used to study four factors each at M levels with only M^2 observations, if, as with the Latin square, none of the factors interact. Specifically, one assumes the model

$$X_{ijk\ell} = \mu + \alpha_i^A + \alpha_j^B + \alpha_k^C + \alpha_\ell^D + \epsilon_{ijk\ell} \tag{15.133}$$

where the $\epsilon_{ijk\ell}$ are independent $N(0, \sigma^2)$ and $(i, j, k, \ell) \in \mathfrak{D}$. Obviously, all the cautions regarding Latin squares also apply to Graeco-Latin squares. Fisher and Yates (1963) provided sets of $M - 1$ orthogonal Latin squares of size M (a set of Latin squares is called orthogonal if every pair in the set is orthogonal) for $M = 3$, 4, 5, 7, 8, 9. (No 6×6 orthogonal Latin squares exist.) One should choose a Graeco-Latin square by choosing two orthogonal Latin squares at random, superimposing them, and then randomizing the columns and all but the first row. The ANOVA for a Graeco-Latin square is a straightforward extension of the ANOVA for a Latin square. In addition to the sums of squares in Table 15.38, one simply computes

$$SS_D = M \sum_{\ell=1}^{M} (X_{\bullet\bullet\bullet\ell} - X_{\bullet\bullet\bullet})^2 \tag{15.134}$$

and subtracts that amount from the previously computed SS_e to obtain a new SS_e with degrees of freedom reduced by $M - 1$.

Example 15.25. Suppose that in the tire tread wear experiment considered in Example 15.24 four mechanics participated and the Graeco-Latin square design given in Table 15.41 was used (the Greek letters representing the four mechanics). In order to minimize the amount of additional computation assume that the results of the experiment are the same as those given in Table 15.39. One need then only compute the sum of squares for Mechanics

$$SS_D = 4[(6 - 6)^2 + (6.25 - 6)^2 + (6.5 - 6)^2 + (5.25 - 6)^2] = 3.5$$

TABLE 15.39 Tread Wear (0.1 mm) for Four Brands (a, b, c, d) of Tires Using a Latin Square Design together with Column Means, Row Means, and the Grand Mean (Example 15.24)

		Car				
		1	2	3	4	
Tire position	1	*a* 8	*c* 11	*d* 2	*b* 8	7.25
	2	*c* 7	*a* 5	*b* 2	*d* 4	4.5
	3	*d* 3	*b* 9	*a* 7	*c* 9	7
	4	*b* 4	*d* 5	*c* 9	*a* 3	5.25
		5.5	7.5	5	6	6

TABLE 15.40 ANOVA Table for Example 15.24

Source	SS	df	MS	F	Pr > F
Brands	61.5	3	20.5	4.92	0.047
Cars	14	3	4.67	1.12	0.412
Positions	21.5	3	7.17	1.72	0.262
Error	25	6	4.17		
Total	122	15			

TABLE 15.41 A 4 × 4 Graeco-Latin Square

$a\ \alpha$	$c\ \gamma$	$d\ \delta$	$b\ \beta$
$c\ \delta$	$a\ \beta$	$b\ \alpha$	$d\ \gamma$
$d\ \beta$	$b\ \delta$	$a\ \gamma$	$c\ \alpha$
$b\ \gamma$	$d\ \alpha$	$c\ \beta$	$a\ \delta$

and $SS_e = 25 - 3.5 = 21.5$. The complete ANOVA table is given in Table 15.42, and $\alpha = 0.1$ ANOM decision lines (for all four factors) would be 6 ± 4.07.

15.5.3 Balanced Incomplete Block Designs

In a randomized block design the size of the block must equal the number of treatments (the number of levels of the factor of interest), as in Example 15.18 where the blocks were of size 4 (four tires per car) and four brands of tires were being compared. Sometimes it is convenient (or even necessary) to use block sizes that are smaller than the number of treatments as, for example, if one were interested in comparing five brands of tires. For that situation *balanced incomplete block* (BIB) *designs* are useful. A BIB design is a design where

1. Each treatment is replicated the same number of times.
2. All blocks are the same size.
3. No treatment appears twice in the same block.

TABLE 15.42 ANOVA Table for Example 15.25

Source	SS	df	MS	F	Pr > F
Brands	61.5	3	20.5	2.86	0.206
Cars	14	3	4.67	0.65	0.634
Positions	21.5	3	7.17	1.00	0.500
Mechanics	3.5	3	1.17	0.16	0.917
Error	21.5	3	7.17		
Total	122	15			

4. Each pair of treatments appears in the same number of blocks. (This is the "balanced" part of the design.)

Let

I = number of treatments
J = number of blocks
n = number of replicates
b = block size.

A BIB design with $I = 5$, $J = 5$, $n = 4$, and $b = 4$ is shown in Table 15.43. It is easier to see that this is a BIB design by looking at the design as a two-way layout and indicating cells where observations are to be taken with an ×. This is done in Table 15.44. Tables of BIB designs can be found in Fisher and Yates (1963), Davies (1978); and Box, Hunter, and Hunter (1978). Once a design has been chosen, everything possible should be randomized. Namely, numbers should be assigned to the blocks at random, letters should be assigned to the treatments at random, and the treatments within the blocks should be randomized.

With a BIB design one assumes the model

$$X_{ij} = \mu + \alpha_i^A + \alpha_j^B + \epsilon_{ij} \tag{15.135}$$

for the $N = In = Jb$ pairs $(i, j) \in \mathfrak{D}$, and that the ϵ_{ij} are independent $N(0, \sigma^2)$. Factor A represents the treatments, and factor B represents the blocks. This is actually a design with many missing values for which, because of its structure, it is possible to

TABLE 15.43 *A* BIB Design for Five Treatments (*a, b, c, d, e*) in Blocks of Size 4

Block				
1	2	3	4	5
a	*a*	*b*	*a*	*a*
b	*b*	*c*	*c*	*b*
c	*d*	*d*	*d*	*c*
d	*e*	*e*	*e*	*e*

TABLE 15.44 *A* Different Representation for the BIB Design in Table 15.43

		Block				
		1	2	3	4	5
Treatment	*a*	×	×		×	×
	b	×	×	×		×
	c	×		×	×	×
	d	×	×	×	×	
	e		×	×	×	×

compute directly the sums of squares needed for an exact test of H_A. The sum of squares for treatments must be adjusted (often referred to as "adjusted for blocks" or "eliminating blocks") because of the missing values, and hence is computed slightly differently. Let

g_i = ith treatment total
T = sum of all the observations
T_i = sum of block totals in which the ith treatment appears
T'_i = sum of block totals in which the ith treatment is absent,

$$T'_i = T - T_i$$

G_i = ith adjusted treatment total,

$$G_i = g_i - \frac{T_i}{b} \tag{15.136}$$

$$= g_i - \frac{T - T'_i}{b}. \tag{15.137}$$

Then the adjusted sum of squares for treatments is given by

$$\text{SS}_{A(adj)} = \frac{I-1}{J(b-1)} \sum_{i=1}^{I} G_i^2 \tag{15.138}$$

where the G_i are most conveniently computed using Eq. (15.137) if $2n > J$, and using Eq. (15.136) otherwise. The complete ANOVA for a BIB design is given in Table 15.45. Note that SS_B cannot be used to test for block effects without first adjusting it for treatments. Davies (1978) discusses this in more detail.

For BIB designs the usual ANOM procedure is not applicable because for these designs the estimators of the treatment effects α_i^A are

$$\hat{\alpha}_i^A = \frac{I-1}{J(b-1)} G_i \tag{15.139}$$

rather than being of the form $X_i - X$. One can, however, compare the $\hat{\alpha}_i^A$ with ANOM type decision line

$$0 \pm h(\alpha; I, \nu_e)\,(I-1)\,\frac{\sqrt{\text{MS}_e}}{\sqrt{IJ(b-1)}}. \tag{15.140}$$

Example 15.26. Five different brands of half-height fixed disk drives were to be compared with regard to their average access time. In order to eliminate any differences due to the computer in which they were used, it was desired to install samples of all the brands on the same machine. Unfortunately, the machines available would accommodate no more than four half-height drives. Therefore, it was decided to use the BIB design given in Table 15.44. The resulting average access times in milliseconds are given in Table 15.46. Using Eq. (15.137), one computes

$$G_a = 148 - \frac{736 - 154}{4} = 2.5$$

TABLE 15.45 ANOVA Table for a BIB Design

Source of variation	Sum of squares	Degrees of freedom	Mean square	F	$\Pr > F$
Treatments (adjusted)	$SS_{A(adj)} = \frac{I-1}{J(b-1)} \sum_{i=1}^{I} G_i^2$	$\nu_A = I - 1$	$MS_{A(adj)} = \frac{SS_{A(adj)}}{\nu_A}$	$F_{A(adj)} = \frac{MS_{A(adj)}}{MS_e}$	$\Pr[F(\nu_A, \nu_e) > F_{A(adj)}]$
Blocks	$SS_B = b \sum_{j=1}^{J} (X_{\bullet j} - X_{\bullet\bullet})^2$	$\nu_B = J - 1$	*		
Error	$SS_e = SS_{tot} - SS_{A(adj)} - SS_B$	$\nu_e = N - I - J + 1$	$MS_e = \frac{SS_e}{\nu_e}$		
Total	$SS_{tot} = \sum_{(i,j) \in \mathcal{D}} (X_{ij} - X_{\bullet\bullet})^2$	$N - 1$			

*SS_B cannot be used to test for block effects.

TABLE 15.46 Average Access Times (ms) for Five Brands of Half-Height Fixed Disk Drives together with Row Totals, Column Totals, and the Grand Total (Example 15.26)

		Block 1	2	3	4	5	Totals
Brand	*a*	35	41	×	32	40	148
	b	42	45	40	×	38	165
	c	31	×	42	33	35	141
	d	30	32	33	35	×	130
	e	×	40	39	36	37	152
Totals		138	158	154	136	150	736

$$G_b = 165 - \frac{736 - 136}{4} = 15$$

$$G_c = 141 - \frac{736 - 158}{4} = -3.5$$

$$G_d = 130 - \frac{736 - 150}{4} = -16.5$$

$$G_e = 152 - \frac{736 - 138}{4} = 2.5.$$

(Note that the G_i must sum to zero.) Using these results in Eq. (15.138), one obtains

$$\text{SS}_{A(adj)} = \frac{4}{15}[(2.5)^2 + \bullet\bullet\bullet + (2.5)^2] = 139.2.$$

The remaining sums of squares are computed in the usual fashion, and the complete ANOVA table is given in Table 15.47. The $\alpha = 0.05$ ANOM type decision lines (15.140) are

$$0 \pm h(0.05; 5, 11)(4)\frac{\sqrt{9.71}}{\sqrt{5(5)(3)}} = 0 \pm 4.33.$$

TABLE 15.47 ANOVA Table for Example 15.26

Source	SS	df	MS	*F*	Pr > *F*
Brands (adj)	139.2	4	34.8	3.58	0.042
Blocks	95.2	4	—	—	—
Error	106.8	11	9.71		
Total	341.2	19			

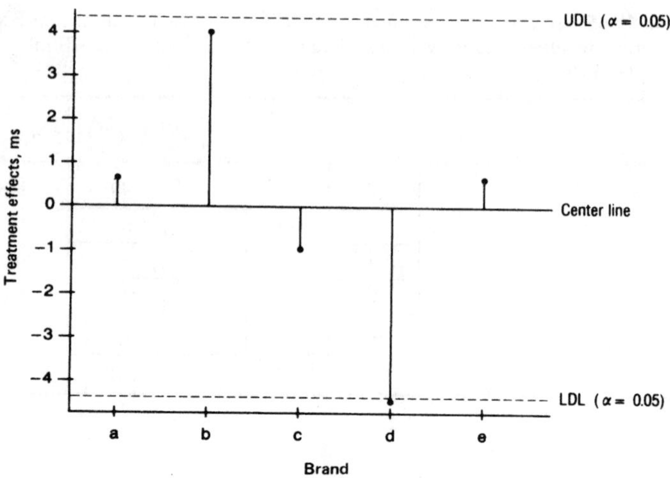

FIGURE 15.19 ANOM chart for Example 15.26.

Comparing the $\hat{\alpha}_i^A$ with these decision lines (Fig. 15.19), one finds that Brand *d* is significantly ($\alpha = 0.05$) faster.

15.5.4 Youden Squares

Suppose that in Example 15.26 one were concerned that the positions of the disk drives in the machine would have an effect on their performance, perhaps because of temperature variation within the machine. In a situation like this, one would like a design where each drive appeared once in each position. Designs having the same properties as BIB designs, and where in addition each treatment appears once in each position within the block, are called *Youden squares. A* Youden square is not actually a square, but an incomplete Latin square where position within the block is the third factor. If $b(b - 1)/(J - 1)$ is an integer, then a Youden square with J blocks can be obtained from a $J \times J$ Latin square by discarding $J - b$ rows. Table 15.48 is an example of this with $J = 5$ and $b = 4$. Analysis for a Youden square design is the same as that for a BIB design except that one has a sum of squares for posi-

TABLE 15.48 A Youden Square with $J = 5$ Blocks and $b = 4$ Positions within Blocks Obtained from a 5 × 5 Latin Square

		Block 1	2	3	4	5	
Position	1	*a*	*d*	*b*	*c*	*e*	
	2	*b*	*e*	*c*	*d*	*a*	
	3	*c*	*a*	*d*	*e*	*b*	
	4	*d*	*b*	*e*	*a*	*c*	
		e	*c*	*a*	*b*	*d*	← Discard

tion with $b - 1$ degrees of freedom. The ANOVA table for a Youden square is given in Table 15.49, and the ANOM type decision lines (15.140) are also appropriate for this design.

Example 15.27. Suppose that the data on average access times for fixed disk drives given in Table 15.46 had actually been collected using the Youden square design given in Table 15.48. The sums of squares $SS_{A(\text{adj})}$, SS_B, and SS_{tot} would still be computed in exactly the same way. But one would then compute

$$SS_C = 5[(35.4 - 36.8)^2 + \cdots + (36.2 - 36.8)^2] = 61.6$$

$$SS_e = 341.2 - 139.2 - 95.2 - 61.6 = 45.2.$$

The complete ANOVA table is given in Table 15.50, and the $\alpha = 0.05$ ANOM type decision lines (15.140) would be

$$0 \pm h(0.05; 5, 8)(4)\frac{\sqrt{5.65}}{\sqrt{5(5)(3)}} = 0 \pm 3.52$$

indicating that at $\alpha = 0.05$ Brand b is significantly slow and Brand d is significantly fast. If Position had not had a significant effect, then SS_C and SS_e could be pooled, and one would obtain exactly the same analysis as in Example 15.26.

15.6 FACTORIAL DESIGNS

As mentioned earlier, factorial designs (designs where all the factors are crossed) are very efficient ways to study multiple factors in the same experiment. They allow one to assess any possible interactions, and if no interactions are present, then the effect of each factor can be evaluated with the same efficiency (that is, the same precision with a fixed number of runs) as if the whole experiment had been devoted entirely to that factor. However, factorial designs with even a moderate number of factors can lead to large numbers of experimental trials, and industrial-type experiments are almost always restricted as to the number of trials that can be run due to time or expense. Therefore, two extremely useful factorial designs are those with all the factors at only two levels or all the factors at only three levels. These are referred to as 2^k and 3^k factorial designs, respectively.

15.6.1 2^k Factorial Designs

While 2^k designs can be analyzed using the general procedures for complete layouts (as was done, for example, with the 2^3 design of Example (15.23), different notation and a different (although equivalent) technique of analysis are usually applied to this special case. For each factor the two levels are denoted by 0 and 1 and are referred to as low level and high level, respectively. Different treatment combinations are then represented using lowercase letters for the different factors and the 0's and 1's as exponents. For example, with two factors A and B, both factors at the low level would be represented by $a^0b^0 = (1)$, and factor A at the high level with factor B at the low level would be represented by $a^1b^0 = a$. Thus, in a 2^2 experiment the four possible treatment combinations would be represented by (1),

TABLE 15.49 ANOVA Table for a Youden Square

Source of variation	Sum of squares	Degrees of freedom	Mean square	F	$\Pr > F$
Treatments (adjusted)	$SS_{A(adj)} = \frac{I-1}{J(\mathbf{b}-1)} \sum_{i=1}^{I} G_i^2$	$\nu_A = I - 1$	$MS_{A(adj)} = \frac{SS_{A(adj)}}{\nu_A}$	$F_{A(adj)} = \frac{MS_{A(adj)}}{MS_e}$	$\Pr[F(\nu_A, \nu_e) > F_{A(adj)}]$
Blocks	$SS_B = b \sum_{j=1}^{J} (X_{\bullet j} - X_{\bullet\bullet})^2$	$\nu_B = J - 1$	*		
Position within block	$SS_C = J \sum_{k=1}^{b} (X_{\bullet\bullet k} - X_{\bullet\bullet\bullet})^2$	$\nu_C = b - 1$	$MS_C = \frac{SS_C}{\nu_C}$	$F_C = \frac{MS_C}{MS_e}$	$\Pr[F(\nu_c, \nu_e) > F_c]$
Error	$SS_e = SS_{tot} - SS_{A(adj)} - SS_B - SS_C$	$\nu_e = N - I - J - b + 2$	$MS_e = \frac{SS_e}{\nu_e}$		
Total	$SS_{tot} = \sum_{(i,j,k) \in \mathcal{D}} (X_{ijk} - X_{\bullet\bullet})^2$	$N - 1$			

*SS_B cannot be used to test for block effects.

TABLE 15.50 ANOVA Table for Example 15.27

Source	SS	df	MS	F	Pr > F
Brands (adj)	139.2	4	34.8	6.16	0.015
Blocks	95.2	4	—	—	—
Positions	61.6	3	20.53	3.63	0.064
Error	45.2	8	5.65		
Total	341.2	19			

a, b, and ab. These designators are also used to represent the sum of the values of the observations obtained with the factors at the specified levels. Algebraically,

$$a^{i-1}\, b^{j-1} = \sum_{k=1}^{n} X_{ijk} \qquad \text{for } i, j = 1, 2.$$

Table 15.51 shows the correspondence (when $n = 1$) between this new notation and the X_{ij} notation used previously.

For a 2^2 design with factors A and B let

$A_{(1)}$ = average change in response due to A at the low level of B,

$$A_{(1)} = \frac{a - (1)}{n} \tag{15.141}$$

A_b = average change in response due to A at the high level of B,

$$A_b = \frac{ab - b}{n} \tag{15.142}$$

A = estimator of the effect of factor A,

$$A = \frac{A_{(1)} + A_b}{2} \tag{15.143}$$

$$= \frac{a - (1) + ab - b}{2n}. \tag{15.144}$$

TABLE 15.51 Correspondence Between the Two Notations for Observed Values in a 2^2 Factorial Experiment with $n = 1$ Observation per Cell

		Factor B 1	Factor B 2
Factor A	1	(1) X_{11}	b X_{12}
	2	a X_{21}	ab X_{22}

(Using A to represent both a factor and the estimator of the effect of that factor is somewhat regrettable, but standard, notation.)

Note that when a factor A has only two levels ($I = 2$), then $\hat{\alpha}_1^A = -\hat{\alpha}_2^A$ (the estimators of the two main effects are perfectly negatively correlated) and

$$A = 2\hat{\alpha}_2^A. \tag{15.145}$$

Similarly, for factor B

$$B = \frac{b - (1) + ab - a}{2n} \tag{15.146}$$

$$= 2\hat{\alpha}_2^B \tag{15.147}$$

where B is the estimator of the effect of factor B.

When factors A and B each have only two levels ($I = J = 2$), then

$$\hat{\alpha}_{11}^{AB} = -\hat{\alpha}_{12}^{AB} = -\hat{\alpha}_{21}^{AB} = \hat{\alpha}_{22}^{AB} \tag{15.148}$$

and

$$AB = \frac{(1) - a - b + ab}{2n} \tag{15.149}$$

$$= 2\hat{\alpha}_{22}^{AB} \tag{15.150}$$

where AB is the estimator of the AB interaction effect. Note that AB as defined in Eq. (15.149) is the average difference of A_b and $A_{(1)}$ [Eqs. (15.141) and (15.142)].

The estimators of the treatment effects [Eqs. (15.144), (15.146), and (15.149)] are all contrasts of the treatment combinations. That is, they are linear combinations of the treatment combinations where the coefficients of the treatment combinations sum to zero. Multiplying each treatment effect estimator by $2n$ (for convenience) makes all the coefficients either +1 or –1, and the appropriate signs can be determined as follows. For A any treatment combination with an a in it has a +1 coefficient, and all others have a –1 coefficient. Similarly for B. The coefficient for a particular treatment combination in the interaction effect estimator AB is obtained by multiplying that treatment combination's coefficient in A with its coefficient in B. Table 15.52 is an example of this for a 2^2 design. Everything is the same for a 2^k design except that one must multiply by $n2^{k-1}$ to make all the coefficients

TABLE 15.52 Coefficients of Treatment Combinations to Obtain the Effects in a 2^2 Factorial Experiment

Treatment combination	Effect		
	A	B	AB
(1)	–	–	+
a	+	–	–
b	–	+	–
ab	+	+	+

±1. For example, the treatment effect estimator A in a 2^3 design (with factors A, B, and C) is

$$A = \frac{-(1) + a - b + ab - c + ac - bc + abc}{n2^2}. \tag{15.151}$$

Alternatively, the estimators of the treatment effects can be obtained, say for a 2^3 design, by expanding

$$(a \pm 1)(b \pm 1)(c \pm 1)$$

where the sign in each term is –1 if that factor is present in the effect and +1 otherwise, and 1 is replaced by (1) in the expansion. The contrast obtained is then divided by $n2^{k-1}$. Thus, the interaction effect estimator AB in a 2^3 design is obtained from

$$(a - 1)(b - 1)(c + 1) = 1 - a - b + ab + c - ac - bc + abc$$

as

$$AB = \frac{(1) - a - b + ab + c - ac - bc + abc}{n2^2}. \tag{15.152}$$

Having the treatment effect estimators expressed as contrasts makes it easy to compute the associated sums of squares using

$$SS_{\text{contrast}} = \frac{(\text{contrast})^2}{n \sum c_i^2} \tag{15.153}$$

where the c_i are the coefficients on the terms of the contrast. Each contrast sum of squares has 1 degree of freedom. When one uses a contrast of the form $n2^{k-1}$ (effect) (that is, all coefficients are ±1), Eq. (15.153) reduces to

$$SS_{\text{effect}} = \frac{(\text{contrast})^2}{n2^k}. \tag{15.154}$$

Yates' Algorithm. Actual calculation of the estimates of the treatment effects is most easily done using *Yates' method*. First, all the treatment combinations are written in a column in standard order (sometimes referred to as Yates' order). Standard orders are obtained as follows. Starting with the ordered set {(1), a}, each element of the set is multiplied by the next factor designator (b in this case), and the new elements are added to the original set in the order they were obtained. This results in {(1), a, b, ab}. Repeating the process for a third factor gives {(1), a, b, ab, c, ac, bc, abc}, and so forth. With the treatment combinations in standard order one adds and subtracts their respective values in pairs, as indicated in Table 15.53, to produce column (1). The algorithm is then repeated on column (1) to produce column (2). For a 2^k design the process continues until column (k) is obtained. The first number in column (k) is the sum of all the observations, and the rest are $n2^{k-1}$ times the estimates of the treatment effects (see Table 15.53). A computer program for Yates' algorithm is given in L. S. Nelson (1982).

TABLE 15.53 Yates' Method Applied to a 2^2 Design with $n = 1$ Observation per Cell

Treatment combination	Response	(1)*	(2)
(1)	210	450	$830 = \Sigma X_{ij}$
a	240	380	$50 = 2A$
b	180	30	$-70 = 2B$
ab	200	20	$-10 = 2AB$

* → result of adding; -→ result of subtracting (bottom number minus top number).

Example 15.28. Using Yates' method on the thickening time of cement data (Example 15.23, Table 15.35) with the factors Temperature, Pressure, and Stirring Time represented by A, B, and C respectively, one obtains Table 15.54. The sums of squares are obtained using Eq. (15.154). For example,

$$SS_A = \frac{(-64)^2}{2^3} = 512.$$

Further analysis of these data would be the same as was done in Example 15.23.

15.6.2 3^k Factorial Designs

There are two possible reasons that one might want to consider designs with factors at three levels. First, there might be three qualitative (discrete) levels of a factor, such as three kinds of alcohol. When all the factors in an experiment are discrete, the experiment should be analyzed using the techniques discussed in Sec. 15.4. Second, for quantitative (continuous) factors, such as Relative Humidity and Temperature in Example 15.15, one might be interested in the quadratic effects of the factors as well as the linear effects. [With a (continuous) factor at just two levels only the linear effect can be estimated.] We will restrict our attention here to

TABLE 15.54 Yates' Method Applied to the 2^3 Design in Example 15.23 and the Resulting Sums of Squares (see Example 15.28)

Treatment combination	Response	(1)	(2)	(3)	SS
(1)	5	−6	10	0	—
a	−11	16	−10	−64	512
b	17	−18	−34	48	288
ab	−1	8	−30	−8	8
c	−3	−16	22	−20	50
ac	−15	−18	26	4	2
bc	13	−12	−2	4	2
abc	−5	−18	−6	−4	2

the situation with at least one continuous factor. The notation for 2^k designs, unfortunately, does not extend to 3^k designs, and therefore a different system is necessary for representing the treatment combinations. For 3^k designs the three levels of each factor (low to high) are represented by 0, 1, and 2. The various treatment combinations, say for a 3^3 design, are then represented by 001, 012, 102, and so forth. With a factor at three levels one can think of fitting a quadratic equation to the results and testing whether the linear and/or quadratic portions of the resulting curve are statistically significant. Separation of the linear and quadratic portions of the equation (the linear and quadratic effects of the factor) is done using orthogonal polynomials.

Orthogonal Polynomials. Since for factors at three levels one can fit at most a quadratic equation, we need be concerned with only polynomials of degree 2 (quadratics). For factors at more than three levels the results generalize to polynomials of higher order. If

$$P_2(x) = c_0 + c_1x + c_2x^2 \tag{15.155}$$

is an arbitrary polynomial of degree 2, then it is possible to express it as

$$P_2(x) = C_0\zeta_0(x) + C_1\zeta_1(x) + C_2\zeta_2(x) \tag{15.156}$$

where

$$\zeta_k(x) = \sum_{i=1}^{k} a_i x^i \tag{15.157}$$

and the $\zeta_k(x)$ are orthogonal. That is, $\zeta_k(x)$ is a polynomial of degree k, and

$$\sum_{\ell=0}^{2} \zeta_i(x_\ell)\zeta_j(x_\ell) = 0 \qquad \text{if } i \neq j. \tag{15.158}$$

We will consider the case where the three values of x (the levels of the factor) are equally spaced. One can also easily construct orthogonal polynomials for the situation where the values of x are not equally spaced using the Gram-Schmidt orthogonalization procedure [see, for example, Blum and Rosenblatt (1972)]. The following formulas, however, do not apply in that case. For three equally spaced values of x, chosen for convenience as $x_0 = -1$, $x_1 = 0$, and $x_2 = 1$,

$$\left.\begin{aligned} \zeta_0(x) &= 1 \\ \zeta_1(x) &= x \\ \zeta_2(x) &= 3x^2 - 2 \end{aligned}\right\} \tag{15.159}$$

and the values of these polynomials at the specified values of x are given in Table 15.55. Orthogonal polynomials of higher degree for equally spaced values of x can be found in Fisher and Yates (1963).

The values of the orthogonal polynomials given in Table 15.55 are used to separate the linear and quadratic effects of a factor (denoted with subscripts L and Q) by constructing tables of signs (similar to Table 15.52) for contrasts of the treatment combinations. For the linear effect of a factor A values of $\zeta_1(x_i)$ are used for the coefficients of the treatment combinations. Those treatment combinations with

TABLE 15.55 Orthogonal Polynomial Values for Use with Factors at Three Equally Spaced Levels

i	x_i	$\zeta_0(x_i)$	$\zeta_1(x_i)$	$\zeta_2(x_i)$
0	–1	1	–1	1
1	0	1	0	–2
2	1	1	1	1

A at level 0 are multiplied by –1, those with A at level 1 are multiplied by 0, and those with A at level 2 are multiplied by 1. Similarly, values of $\zeta_2(x_i)$ are used to obtain quadratic effects. Table 15.56 is an example of this for a 3^2 design. Note that the coefficients for the interaction effects are products of the coefficients for the two effects that make up the interaction. The contrasts, constructed in this manner, are themselves orthogonal (that is, the pairwise products of their coefficients sum to zero), and hence they are statistically independent. The sums of squares obtained from these contrasts using Eq. (15.153) are, therefore, also statistically independent.

Example 15.29. Reanalyzing the electronic component output data in Example 15.15 (Table 15.15) to test for linear and quadratic effects of Humidity (factor A) and Temperature (factor B), one would compute treatment contrasts using the coefficients in Table 15.56 and then compute the sums of squares using Eq. (15.153). The values of the treatment combinations (the cell totals) are given in Table 15.57. The contrast for the linear effect of A, for example, is

$$A_L = (-1)(338 + 347 + 370) + (1)(400 + 404 + 416) = 165$$

and the associated sum of squares is

$$SS_{A_L} = \frac{(165)^2}{2(6)} = 2268.75.$$

TABLE 15.56 Contrast Coefficients for a 3^2 Design with Continuous Factors together with Their Sums of Squares

	Treatment combinations									
Effect	00	01	02	10	11	12	20	21	22	Σc_i^2
A_L	–1	–1	–1	0	0	0	1	1	1	6
A_Q	1	1	1	–2	–2	–2	1	1	1	18
B_L	–1	0	1	–1	0	1	–1	0	1	6
B_Q	1	–2	1	1	–2	1	1	–2	1	18
A_LB_L	1	0	–1	0	0	0	–1	0	1	4
A_LB_Q	–1	2	–1	0	0	0	1	–2	1	12
A_QB_L	–1	0	1	2	0	–2	–1	0	1	12
A_QB_Q	1	–2	1	–2	4	–2	1	–2	1	36

TABLE 15.57 Treatment Combinations for Example 15.29

		Temperature, °C		
		10	20	30
	20	00 338	01 347	02 370
Relative humidity, %	30	10 378	11 402	12 407
	40	20 400	21 404	22 416

TABLE 15.58 ANOVA Table for Example 15.29

Source	SS	df	MS	F	Pr > F
A_L	2268.75	1	2268.75	85.42	0.000
A_Q	272.25	1	272.25	0.04	0.846
B_L	494.08	1	494.08	18.60	0.002
B_Q	0.25	1	0.25	0.01	0.923
A_LB_L	32	1	32	1.20	0.302
A_LB_Q	1.5	1	1.5	0.06	0.812
A_QB_L	4.17	1	4.17	0.16	0.698
A_QB_Q	50	1	50	1.88	0.204
Error	239	9	26.56		
Total	3362	17			

From the complete ANOVA table given in Table 15.58 one sees that only the linear effect of A (DLS = 0.000) and the linear effect of B (DLS = 0.002) are statistically significant, and therefore, the significance of Humidity and Temperature found in Example 15.15 is due solely to the linear effects of those factors. Note that (comparing Tables 15.17 and 15.58)

$$SS_A = SS_{A_L} + SS_{A_Q} \tag{15.160}$$

$$SS_{AB} = SS_{A_LB_L} + SS_{A_LB_Q} + SS_{A_QB_L} + SS_{A_QB_Q}. \tag{15.161}$$

Equation (15.160) also holds for factor B, and both it and Eq. (15.161) are true in general for continuous factors at three levels.

15.7 FRACTIONAL FACTORIAL DESIGNS

Even with factors at only two or three levels, complete factorial designs can require a large number of experimental trials. By assuming that some of the interactions do not exist (as was done for the Latin square), it is possible to construct designs requiring fewer experimental trials. Two important classes of designs of this kind are the 2^k and 3^k fractional factorial designs.

15.7.1 Fractions of 2^k Factorial Designs

By far the most useful designs in the physical sciences and engineering discussed thus far are the 2^k factorial designs. Fractions of 2^k factorial designs are possibly even more useful since they require fewer experimental trials. For example, consider the case of four factors, each at two levels. A complete 2^4 factorial design would require 16 runs, while using a one-half fraction of a 2^4 design (denoted by 2^{4-1}), the four factors could be studied in only eight runs. A 2^{4-1} fractional factorial design can be constructed as follows. Starting with a 2^3 design, one assumes the ABC interaction does not exist and uses it to measure the effect of a fourth factor, D. The effect of D is said to be confounded (confused) with the ABC interaction, and what is being measured is actually the effect of D plus the ABC interaction effect. The resulting design is shown in Table 15.59. The plus and minus signs represent the high and low levels of the factors, respectively; and the A, B, and C columns are the complete 2^3 factorial design. The $D = ABC$ column is obtained by multiplying the signs for the A, B, and C columns together [(+)(+) = + , (+)(–) = –, etc.]. The four columns of signs specify the design by indicating at what levels the factors should be set for each of the eight runs. For example, the third row (– + – +) indicates a run with A and C at the low level and B and D at the high level. The far right-hand column duplicates this information (for convenience) using the notation for treatment combinations.

Having confounded D with ABC, it is necessary to find out what other effects have been confounded. This is referred to as the *confounding pattern* or *alias structure* for the design, and is obtained as follows. Starting with $D = ABC$, which is called the *generator* for the design, both sides of the equation are multiplied by D to obtain $D^2 = ABCD$. For fractions of 2^k designs any factor raised to an even power is replaced by an identity operator I ($AI = IA = A$), and any factor raised to an odd power is replaced by that factor without an exponent. (One is actually considering the exponents modulo 2, which means each exponent is replaced with the remainder after dividing the exponent by 2.) Thus, $D^2 = ABCD$ becomes $I = ABCD$, and this is called the *defining relation* or *defining contrast* for the design. The confounding pattern is obtained by further multiplication using the same rule for exponents. For example, $AI = A^2BCD$ becomes $A = BCD$, indicating that A and BCD are confounded. The complete confounding pattern for the 2^{4-1} design in Table 15.59 is given in Table 15.60.

So far we have discussed only a half fraction of a 2^k design. To obtain smaller fractions ($\frac{1}{4}$, $\frac{1}{8}$, etc.) more than one generator is needed. For example, a one-quar-

TABLE 15.59 A2^{4-1} Fractional Factorial Design

	A	B	C	$D = ABC$	
(1)	–	–	–	–	(1)
a	+	–	–	+	*ad*
b	–	+	–	+	*bd*
ab	+	+	–	–	*ab*
c	–	–	+	+	*cd*
ac	+	–	+	–	*ac*
bc	–	+	+	–	*bc*
abc	+	+	+	+	*abcd*

TABLE 15.60 Confounding Pattern for the 2^{4-1} Design in Table 15.59

$I = ABCD$
$A = BCD$
$B = ACD$
$C = ABD$
$D = ABC$
$AB = CD$
$AC = BD$
$BC = AD$

ter fraction of a 2^5 design (a 2^{5-2} design) can be obtained by confounding D with AB ($D = AB$) and E with AC ($E = AC$), which leads to the defining relation

$$I = ABD = ACE = BCDE.$$

The last equality follows from

$$I = I^2 = (ABD)(ACE) = BCDE.$$

In general, if there are g generators, then there are $2^g - 1$ defining contrasts consisting of the g defining contrasts obtained from the generators and all possible multiplicative combinations of them. The effects in the defining relation (such as ABD, ACE, and $BCDE$) are referred to as *words*, and the length of the shortest word is called the *resolution* of the design. For example, the 2^{4-1} design in Table 15.59 is of resolution IV (denoted by 2_{IV}^{4-1}). The types of confounding patterns in fractions of 2^k factorial designs can be partially classified according to the resolution of the designs. Designs of resolution III confound main effects with two-factor interactions, designs of resolution IV confound two-factor interactions with other two-factor interactions, and designs of resolution V confound two-factor interactions with three-factor interactions. Tables of $1/2^p$ fractions of 2^k factorial designs (denoted by 2^{k-p}) can be found in Davies (1978), and a more extensive list was compiled by the National Bureau of Standards (1957).

Analysis of 2^{k-p} designs is easily accomplished using Yates' algorithm. Instead of writing the treatment combinations down in standard order, however, the order is obtained from the standard order for a complete 2^f factorial design, where $f = k - p$, by adding the designators for the p additional factors to obtain the set of treatment combinations that was actually run. An example of this is given in Table 15.59, where the leftmost column contains the treatment combinations for a 2^3 design and the rightmost column contains the actual runs in the correct order for Yates' algorithm. The analysis is then done as if one had a complete 2^f design. Each effect estimate and sum of squares resulting from Yates' algorithm measure the same effect they would have in the complete 2^f design plus whatever effects are confounded with it. For example, using the design in Table 15.59, the sum of squares in row $a(d)$ measures the effect of A plus the effect of the BCD interaction (see the confounding pattern in Table 15.60).

Example 15.30. Davies (1978, pp. 454-457) discusses an experiment run using the 2_{IV}^{4-1} design given in Table 15.59. Four factors in a filtration process were studied to assess their effect on the purity of the product. Purity was measured with a

TABLE 15.61 Yates' Method Applied to the 2^{4-1}_{IV} Design in Example 15.30 together with the Resulting Sums of Squares and Effects Being Measured

Treatment combination	Response	(1)	(2)	(3)	SS	Effect measured
(1)	107	221	473	952	—	$I + ABCD$
$a(d)$	114	252	479	42	220.5	$A + BCD$
$b(d)$	122	227	15	56	392	$B + ACD$
ab	130	252	27	–2	0.5	$AB + CD$
$c(d)$	106	7	31	6	4.5	$C + ABD$
ac	121	8	25	12	18	$AC + BD$
bc	120	15	1	–6	4.5	$BC + AD$
$abc(d)$	132	12	–3	–4	2	$ABC + D$

photoelectric instrument, and higher values correspond to higher purity. Table 15.61 contains the measured values, the analysis of the results using Yates' algorithm (including the sums of squares), and the effects being measured.

Testing for significant effects in a 2^{k-p} design (with $n = 1$) is not quite so straightforward as for a complete 2^k design. The highest-order (k-way) interaction may be confounded with an effect of interest, such as in Table 15.61, where ABC is confounded with D. Therefore, instead of using the k-way interaction as a measure of error, one uses the smallest sum of squares together with the other sums of squares that are of approximately the same order of magnitude. Daniel (1959) suggested a graphical procedure for determining which sums of squares should be used. He argued that if an effect did not really exist, and hence its estimator was actually measuring error, then that estimator should behave as a $N(0, \sigma^2)$ random variable. Thus when the (ordered) estimates of the effects are plotted on normal probability paper, the estimates of nonsignificant effects will tend to fall along a straight line, while the estimates of significant effects will not. The sums of squares associated with those estimates that tend to fall along a straight line should be used to form SS_e.

Example 15.31. The estimates of the effects in Example 15.30 [column (3) in Table 15.61 divided by 4] are plotted on normal probability paper in Fig. 15.20. The figure suggests that $A + BCD$ and $B + ACD$ are the only significant effects and that the sums of squares associated with all the other effects can be used to form SS_e. This results in the ANOVA table given in Table 15.62.

Alternative Fractions and Blocking. If a half fraction of a factorial design is obtained using the defining relation I = +Effect (this is referred to as the *principal fraction*), then the design obtained using the defining relation I = –Effect is called the *alternative half fraction* (also referred to as the *complementary half fraction*). The alternative half fraction to the 2^{4-1}_{IV} design given in Table 15.59 is obtained using the defining relation $I = -ABCD$ (that is, the generator $D = -ABC$) and is shown in Table 15.63. Its confounding pattern is the same as that shown in Table 15.60 except that there are minus signs in all the equations (that is, $I = -ABCD$, $A = -BCD$, etc.). The complementary half fraction together with the principal half fraction make up the entire 2^4 factorial design. Therefore, if after conducting an experiment using a half fraction of a factorial design one wished to deconfound the effects, it can be done by running additional trials using the complementary half fraction.

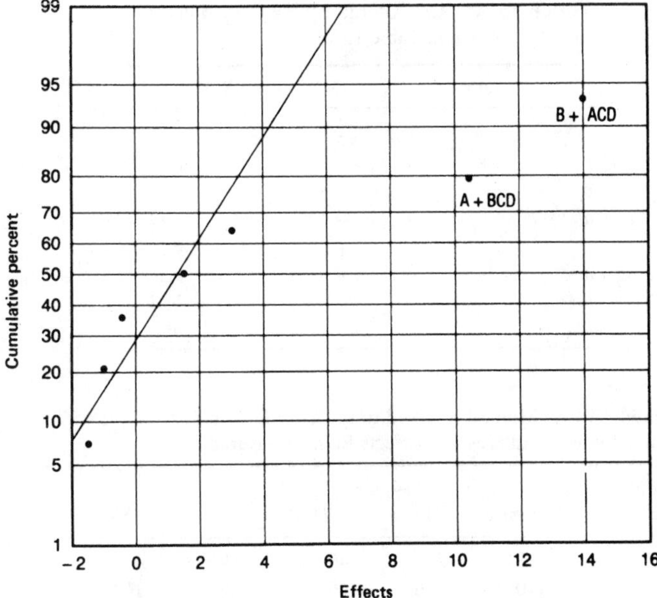

FIGURE 15.20 Normal probability plot of estimates of the effects in Example 15.30 (see Example 15.31).

TABLE 15.62 ANOVA Table for Example 15.31

Source	SS	df	MS	F	Pr > F
$A + BCD$	220.5	1	220.5	37.4	0.002
$B + ACD$	392	1	392	55.8	0.001
Error	29.5	5	5.9		
Total	642	7			

Example 15.32. Suppose that in Example 15.30 after running the 2^{4-1}_{IV} design the experimenter decided it was important to separate the AC effect from the BD effect and, therefore, to run the complementary half fraction (Table 15.63). The (fictitious) results are given in Table 15.64 along with the analysis of this half fraction. Adding and subtracting the corresponding values in column (3) of Table 15.61 and column (3) of Table 15.64 produces the values that would be obtained [column (4)] if the two sets of responses were combined and Yates' algorithm were applied to the resulting 2^4 design. For example,

$$(2^3)A = 42 + 36 = 78$$

$$(2^3)BCD = 42 - 36 = 6.$$

TABLE 15.63 The Alternative Half Fraction for the 2^{4-1}_{IV} Design in Table 15.59

	A	B	C	$D = ABC$	
(1)	–	–	–	+	d
a	+	–	–	–	a
b	–	+	–	–	b
ab	+	+	–	+	abd
c	–	–	+	–	c
ac	+	–	+	+	acd
bc	–	+	+	+	bcd
abc	+	+	+	–	abc

TABLE 15.64 Yates' Method Applied to the 2^{4-1}_{IV} Design in Example 15.32 together with the Resulting Sums of Squares and Effects Being Measured

Treatment combination	Response	(1)	(2)	(3)	SS	Effect measured
$(1)(d)$	113	223	469	934	—	$I - ABCD$
a	110	246	465	36	162	$A - BCD$
b	122	219	–1	50	312.5	$B - ACD$
$ab(d)$	124	246	37	4	2	$AB - CD$
c	100	–3	23	–4	2	$C - ABD$
$ac(d)$	119	2	27	38	180.5	$AC - BD$
$bc(d)$	114	19	5	4	2	$BC - AD$
abc	132	18	–1	–6	4.5	$ABC - D$

The only difference between the analysis of two half fractions and a complete factorial is that there is a half fraction effect (a block effect) confounded with the defining contrast. For this example $ABCD$ is confounded with blocks and

$$(2^3)ABCD = 952 - 934 = 18.$$

Table 15.65 shows the complete column (4) and the corresponding sums of squares for all the effects.

From Example 15.32 one sees that complementary half fractional factorial designs could be used to run a complete factorial in two blocks if, for example, there was only enough time to run half of the necessary trials on any one day. There is a general correspondence between running a complete 2^k factorial in 2^p blocks and a set of alternative 2^{k-p} fractional factorials. If, for example, one wished to run a 2^5 design in four blocks, then the four blocks could be obtained from the set of four alternative one-quarter fractions generated by $D = \pm AB$ and $E = \pm AC$. In general the block effect is confounded with the effects in the defining relation for the principal fraction. In this case the defining relation for the principal fraction is

$$I = ABD = ACE = BCDE$$

so that the block effect is confounded with $ABD + ACE + BCDE$.

TABLE 15.65 Effects and Corresponding Sums of Squares Obtained by Combining the Complementary Half Fractions in Examples 15.30 and 15.32

	(4)	SS
A	78	380.25
B	106	702.25
AB	2	0.25
C	2	0.25
AC	50	156.25
BC	–2	0.25
ABC	–10	6.25
D	2	0.25
AD	–10	6.25
BD	–26	42.25
ABD	10	6.25
CD	–6	2.25
ACD	6	2.25
BCD	6	2.25
ABCD + Blocks	18	20.25
Total		1327.75

Some Factors at Four Levels. It is sometimes convenient to represent a factor at four levels as two factors at only two levels. That way, for example, a 4×2^k factorial design can be looked at as a 2^{k+2} factorial design and fractional designs can be easily obtained. If one used *A* and *B* as the two factors in a 2^{k+2} design to represent a factor *Q* at four levels, then the four levels of *Q* would be represented by (1), *a*, *b*, and *ab*. Thus, *Q* at level two and a factor *C* at the high level would be *ac*, and so forth. The sum of squares for *Q* (with 3 degrees of freedom) is obtained using

$$SS_Q = SS_A + SS_B + SS_{AB}. \tag{15.162}$$

Note that *AB* does not measure an interaction effect but is part of the effect of *Q*. Similarly, *ABC* is only a two-way interaction, and so forth.

Example 15.33. Lukes and Nelson (1956) describe an experiment to study the yield of 1,2,5-pentanetriol using a half fraction of a 4×2^4 design. The five factors studied were

Amount of ion exchange resin (factors *A* and *B*)
Reaction temperature (factor *C*)
Ratio of isopropyl alcohol to water (factor *D*)
Amount of furfuryl alcohol (factor *E*)
Amount of Raney nickel (factor *F*)

The experimental design (a 2^{6-1} obtained using the generator $F = ABCDE$) is shown in Table 15.66 along with the results. Table 15.67 gives the treatment combinations in the correct order for Yates' algorithm together with the responses (from Table 15.66) and the resulting sums of squares computed using the program

TABLE 15.66 Experimental Design and Results (Percent Yield of 1,2,5-Pentanetriol) for Example 15.33

		Amount of Raney nickel							
		7.5 ml				15.0 ml			
		Amount of furfuryl alcohol							
		$\frac{3}{4}$ mole		1 mole		$\frac{3}{4}$ mole		1 mole	
		Ratio of isopropyl alcohol to water							
Amount of ion exchange resin	Reaction temperature, °C	0.5	1.0	0.5	1.0	0.5	1.0	0.5	1.0
10 g	75	20.6	—	—	16.1	—	19.7	19.9	—
10 g	100	—	23.8	27.5	—	22.3	—	—	23.4
20 g	75	—	22.6	18.9	—	20.9	—	—	18.5
20 g	100	24.5	—	—	30.8	—	26.4	25.7	—
30 g	75	—	21.1	21.8	—	17.8	—	—	19.5
30 g	100	18.0	—	—	25.7	—	25.9	26.7	—
40 g	75	18.4	—	—	16.5	—	16.2	19.4	—
40 g	100	—	17.4	16.7	—	22.2	—	—	25.0

in L. S. Nelson (1982). From Table 15.67 one can compute the sum of squares for Ion Exchange Resin

$$SS_Q = 2.94 + 34.65 + 49.25 = 86.84.$$

Since for chemical reasons it was felt that no three-way interactions should exist, $SS_{ADE + BCF} = 41.63$ and all those sums of squares less than 41.63 (not including those in SS_Q) were allocated to error, and the resulting ANOVA table is given in Table 15.68.

15.7.2 Fractions of 3^k Factorial Designs

The structure of fractions of 3^k designs (denoted by 3^{k-p}) is considerably more complicated than the structure of 2^{k-p} designs. In particular, to obtain 3^{k-p} designs the interactions are decomposed into orthogonal components that have no physical meaning, and hence the confounding pattern has no clear physical interpretation. Also, fairly large experiments are needed with even a small number of factors in order to be able to separate all the main effects and the two-factor interactions (for example, five factors would require 81 observations). For these reasons 3^{k-p} designs are not nearly as useful as 2^{k-p} designs, and they are usually best avoided. There are some situations, however, where 3^{k-p} designs might be useful, such as when studying three qualitative factors that are not expected to interact. Connor

TABLE 15.67 Treatment Combinations, Responses, and Sums of Squares for Example 15.33

Treatment combination	Response	SS	Treatment combination	Response	SS
(1)	20.6	—	*e(f)*	19.9	6.39
a(f)	20.9	2.94	*ae*	18.9	2.26
b(f)	17.8	34.65	*be*	21.8	6.39
ab	18.4	49.25	*abe(f)*	19.4	1.32
c(f)	22.3	171.59	*ce*	27.5	23.98
ac	24.5	0.01	*ace(f)*	25.7	0.23
bc	18.0	12.88	*bce(f)*	26.7	6.04
abc(f)	22.2	3.85	*abce*	16.7	1.95
d(f)	19.7	1.67	*de*	16.1	2.82
ad	22.6	1.16	*ade(f)*	18.5	41.63
bd	21.1	0.88	*bde(f)*	19.5	0.88
abd	16.2	19.69	*abde*	16.5	9.35
cd	23.8	15.54	*cde(f)*	23.4	5.36
acd(f)	26.4	1.24	*acde*	30.8	19.69
bcd(f)	25.9	1.40	*bcde*	25.7	1.95
abcd	17.4	0.03	*abcde(f)*	25.0	2.59

TABLE 15.68 ANOVA Table for Example 15.33

Source	SS	df	MS	F	Pr > F
Ion exchange resin (+ *BCDEF* + *ACDEF* + *CDEF*)	86.84	3	28.95	4.09	0.016
Temperature (+*ABDEF*)	171.59	1	171.59	24.23	0.000
Error	191.18	27	7.08		
Total	449.61	31			

and Zelen (1959) provide tables of 3^{k-p} designs, and a good description of the construction and analysis of such designs can be found in Davies (1978).

15.8 SCREENING DESIGNS

Screening designs are designs used to ascertain with as few experimental runs as possible which factors from a (generally) large number of factors have significant main effects. These designs are constructed to allow one to study $N-1$ factors with only N runs. In order to minimize the number of experimental runs needed, each factor is studied at only two levels and main effects are confounded with two-factor interactions. Therefore, in order to use these designs one must assume that no two-factor interactions exist, and since instances where this assumption is valid are limited (particularly with a large number of factors), these designs should be used with great caution. We will discuss two classes of screening designs: saturated 2^{k-p} designs and Plackett-Burman designs.

15.8.1 Saturated 2^{k-p} Designs

A 2^{k-p} design is said to be saturated if $k = 2^f - 1$, where $f = k - p$. Thus if $N = 2^f$, saturated 2^{k-p} designs can be used as screening designs. These are designs of resolution III and confound main effects with two-factor interactions. Their construction and analysis are the same as for any 2^{k-p} design.

15.8.2 Plackett-Burman Designs

Plackett and Burman (1946) introduced two-level fractional factorial designs for studying $N - 1$ factors in N runs, where N is a multiple of 4 for $N \leq 100$ (except $N = 92$). If $N = 2^f$, then the Plackett-Burman designs are saturated 2^{k-p} designs. The information necessary to construct designs for numbers of runs equal to 12, 20, and 24 is given in Table 15.69. The rows of plus and minus signs given are the first rows in each of the designs and represent the levels (high or low) at which each of the $N - 1$ factors should be set for one experimental trial. The remaining rows in each design are obtained as the $N - 2$ cyclic permutations of the first row, and a row of all minus signs. The Plackett-Burman design for $N = 12$ is given in Table 15.70. The actual order in which the runs are carried out should be randomized.

The analysis of Plackett-Burman designs is the same as for any two-level fractional factorial design. Namely, the effect of a factor is simply the average of the observations obtained at the high level of that factor minus the average of the observations obtained at the low level of that factor. Alternatively, $N/2$ times the effect of any factor is the contrast of the observations obtained using the column of plus and minus signs under that factor in the design. The sum of squares for any effect is obtained using Eq. (15.153), which in this case (all coefficients are ±1) reduces to

$$SS_{\text{effect}} = \frac{(\text{contrast})^2}{N}. \tag{15.163}$$

If the results of an experiment run using a Plackett-Burman design indicate that several factors have significant effects, then it may be necessary (depending on the confounding pattern) to deconfound the main effects from the two-factor interactions. One example would be if A, B, and F were found to be significant, and F was confounded with AB. This deconfounding can be accomplished by repeating the experiment using a design with all the signs reversed. The combined $2N$ runs will give estimates of main effects clear of any two-factor interactions.

Plackett-Burman designs for N runs can also be used to study fewer than $N - 1$ factors. If, for example, one was interested in studying only nine factors, the design in Table 15.70 ($N = 12$) could be used with two of the factors, say J and K, not being part of the actual experiment, but being used to measure error. That is, the

TABLE 15.69 Signs Used to Construct Plackett-Burman Designs for Numbers of Runs N = 12, 20, and 24

$N = 12$:	+ + − + + + − − − + −
$N = 20$:	+ + − − + + + + − + − + − − − − + + −
$N = 24$:	+ + + + + − + − + + − − + + − − + − + − − − −

TABLE 15.70 Plackett-Burman Design for $N = 12$ Runs

	Factor										
Run	*A*	*B*	*C*	*D*	*E*	*F*	*G*	*H*	*I*	*J*	*K*
1	+	+	−	+	+	+	−	−	−	+	−
2	−	+	+	−	+	+	+	−	−	−	+
3	+	−	+	+	−	+	+	+	−	−	−
4	−	+	−	+	+	−	+	+	+	−	−
5	−	−	+	−	+	+	−	+	+	+	−
6	−	−	−	+	−	+	+	−	+	+	+
7	+	−	−	−	+	−	+	+	−	+	+
8	+	+	−	−	−	+	−	+	+	−	+
9	+	+	+	−	−	−	+	−	+	+	−
10	−	+	+	+	−	−	−	+	−	+	+
11	+	−	+	+	+	−	−	−	+	−	+
12	−	−	−	−	−	−	−	−	−	−	−

experiment would be run using the levels indicated for the nine factors *A* through *I*, and the sums of squares obtained for *J* and *K* would be pooled to measure error. Note that an MS_e obtained in this fashion measures not only the error but also any interactions confounded with *J* or *K*. More information on screening designs can be found in Wheeler (1987).

15.9 RESPONSE SURFACE DESIGNS

Often when studying continuous factors, after finding their effects to be significant, one is interested in finding conditions (levels of the factors) that lead to a particular response, usually a maximum or a minimum. The responses of an experiment when considered as a function of the possible levels of the factors are called a *response surface*, and designs used to study a response surface are called *response surface designs*. Response surfaces are usually much too complex to be able to easily model the entire surface with a single function, so a simple model (one that is linear in the factor levels) is used to find the local slope (the slope in a specified area) of the surface and point the direction to the maximum (minimum) response. Then, when the desired area (say of the maximum) is reached, a more complex model with quadratic terms is used to provide a more accurate representation of the response surface in the area of its maximum value.

15.9.1 First-Order Designs

A first-order design is one that is constructed in order to fit a first-degree polynomial model (that is, a model that is linear in the factor levels). For k factors the first-degree polynomial model is

$$Y = \beta_0 + \beta_1 x_1 + \cdots + \beta_k x_k + \epsilon \qquad (15.164)$$

where Y represents the response, the β_i are unknown parameters (constants), x_i represents the level of factor i, and ϵ is a random error. For the purpose of conducting significance tests, it is usually assumed (and will be assumed here) that $\epsilon \sim N(0, \sigma^2)$. A model of this kind is used to estimate the local slope of a response surface. In addition to allowing for the fitting of model (15.164), a desirable first-order design should also provide an estimate of experimental error (an estimate of σ^2) and a means of assessing the adequacy of the model.

A good starting point for a first-order design is a 2^k factorial or a 2^{k-p} fractional factorial design. Model (15.164) is easily fitted in this case since when the levels of the x_i are coded as –1 and 1, estimators of the β_i (call them $\hat{\beta}_i$) are, for $i \geq 1$, the estimators of the corresponding factor effects divided by 2 (that is, $\hat{\beta}_1 = A/2, \hat{\beta}_2 = B/2$, and so on), and $\hat{\beta}_0$ is the grand mean. For a 2^k design the $\hat{\beta}_i$ can all be obtained by dividing the appropriate values in column (k) from Yates' algorithm by $n2^k$. A significant AB interaction effect would indicate that the cross-product term $\beta_{12}x_1x_2$ ($\hat{\beta}_{12} = AB/2$) is needed in the model, and the first-order model does not adequately describe that region of the response surface. The adequacy of the model can also be checked using residual analysis, and a convenient way to obtain the residuals is with the reverse Yates' algorithm.

Reverse Yates' Algorithm. To apply the reverse Yates' algorithm to a 2^k design, one takes column (k) from the (forward) Yates' algorithm and writes it down in reverse order. Any effects corresponding to terms that are not in the model of interest are replaced with zeros. In the case of model (15.164), all the interaction effects are replaced with zeros. These values are then added and subtracted in pairs as in the forward algorithm. The results in column (k) are 2^k times the predicted values $\hat{Y}$ at the 2^k treatment combination levels. Pairing the response column from the forward algorithm (in reverse order) with column (k) from the reverse algorithm divided by 2^k and subtracting produces the residuals. The computer program given in L. S. Nelson (1982) includes the reverse Yates' algorithm.

Example 15.34. Table 15.71 shows the results of applying the reverse Yates' algorithm to the data from the 2^3 design in Example 15.23 in order to fit the first-order model (15.164). Note that the value associated with the grand mean (the I

TABLE 15.71 Reverse Yates' Algorithm Applied to the Data in Example 15.23 (see Example 15.34)

Effect measured	(k)*	(1)	(2)	(3)	Predicted†	Response‡	Residual‖
ABC	0	0	– 20	– 36	– 4.5	– 5	– 0.5
BC	0	– 20	– 16	92	11.5	13	1.5
AC	0	48	– 20	– 132	– 16.5	– 15	1.5
C	– 20	– 64	112	– 4	– 0.5	– 3	– 2.5
AB	0	0	– 20	4	0.5	– 1	– 1.5
B	48	–20	– 112	132	16.5	17	0.5
A	– 64	48	– 20	– 92	– 11.5	– 11	0.5
I	0	64	16	36	4.5	5	0.5

*From forward Yates' algorithm (Table 15.54).
†Column (3) divided by 2^3.
‡From Table 15.54 (in reverse order).
‖ Response minus predicted.

effect) is zero because the responses in the experiment were coded so that they summed to zero, not because it was replaced with a zero.

Center Points. In order to obtain a pure estimate of the error (that is, an estimate that measures the error regardless of the correctness of the assumed model) and also another measure of the adequacy of the model, it is usually desirable to add to the 2^k (or 2^{k-p}) design several replicates (say c) of a *center point* (every factor at level zero in its coded metric). The sample variance of the center points (call it S_c^2) is a pure estimator of the error. Let

β_{ii} = coefficient of the model term x_i^2

$\overline{Y}_f$ = average response at points in the factorial design

$\overline{Y}_c$ = average response at the center point

and

$$\gamma = \sum_{i=1}^{k} \beta_{ii}.$$

If the true response function is second-order, then an unbiased estimator of γ is

$$\hat{\gamma} = \overline{Y}_f - \overline{Y}_c \tag{15.165}$$

and if $\hat{\gamma}$ is significantly different from zero, it indicates that the first-order model is not adequate.

An overall measure of lack of fit can be obtained by pooling the sums of squares for any interactions (call them SS_{int} and denote their degrees of freedom with ν_{int}) with

$$SS\gamma = \frac{(\alpha^2 - 1)^2(\hat{\gamma}_i)^2}{1/2^{k-p} + 1/c} \tag{15.166}$$

and computing an overall F value for lack of fit

$$F_{LF} = \frac{(SS_{int} + SS_{\gamma})/(\nu_{int} + 1)}{S_c^2}. \tag{15.167}$$

The first-order model is rejected as being adequate at level α if

$$\Pr[F(\nu_{int} + 1, c - 1) > F_{LF}] < \alpha. \tag{15.168}$$

If the first-order model appears to be adequate, the next step should be to confirm that all the factors have significant main effects. Otherwise either some of the terms can be dropped from the first-order model, or if none of the effects are significant, the model is of no use. The effect of a factor (say A) can be tested by computing the DLS

$$\Pr[F(1, c - 1) > F_A] \tag{15.169}$$

where $F_A = MS_A/S_c^2$.

TABLE 15.72 Initial 2^2 Design with Three Center Points for Studying the Percentage of Defective Welds as a Function of Current and Time of Contact (Example 15.35*a*)

Trial*	Current, amperes	Time, seconds	x_1	x_2	Percent defective
1	35	0.6	–1	–1	4.9
2	55	0.6	1	–1	4.2
3	35	1	–1	1	5.2
4	55	1	1	1	4.4
5	45	0.8	0	0	4.7
6	45	0.8	0	0	4.8
7	45	0.8	0	0	4.8

*Trials were run in random order.

The addition of center points does not change the estimators $\hat{\beta}_i$ for $i \geq 1$, but the center points must be included in the computation of $\hat{\beta}_0$ (the grand mean).

Example 15.35a. Fuses are welded into the bases of incandescent lamps using an electric welding machine, and previous experimentation had revealed that current and time of contact were the two factors that determined the quality of a weld. In order to decrease the percentage of defective welds, an experiment was performed to ascertain the optimum Current and Time settings. The initial ranges of the two variables were chosen as 35 to 55 amperes and 0.6 to 1 second. A 2^2 design with three replicates of the center point was run, and the results are given in Table 15.72. The values for the variables, coded so that the high and low values are 1 and –1, respectively, are obtained from

$$x_1 = \frac{\text{Current} - 45 \text{ amperes}}{10 \text{ amperes}}$$

$$x_2 = \frac{\text{Time} - 0.8 \text{ second}}{0.2 \text{ second}} \tag{15.170}$$

Analyzing the 2^2 part of the design results in Table 15.73, and dividing the values in column (2) by 4, one obtains $\hat{\beta}_0 = 4.675, \hat{\beta}_1 = -0.375, \hat{\beta}_2 = 0.125$, and $\hat{\beta}_{12} = -0.025$. Using the three center points, one computes $S_c^2 = 0.00333$; using Eq. (15.165), one computes

$$\hat{\gamma} = \frac{18.7}{4} - \frac{14.3}{3} = -0.092.$$

Applying Eqs. (15.166) and (15.167), one obtains

$$SS_\gamma = \frac{(-0.092)^2}{\frac{1}{4} + \frac{1}{3}} = 0.0145$$

$$F_{LF} = \frac{(0.0025 + 0.0145)/2}{0.00333} = 2.55$$

TABLE 15.73 Yates' Algorithm Applied to the Initial 2^2 Design in Example 15.35*a* and the Resulting Sums of Squares

Treatment combination	Response	(1)	(2)	SS
(1)	4.9	9.1	18.7	—
a	4.2	9.6	– 1.5	0.5625
b	5.2	– 0.7	0.5	0.0625
ab	4.4	– 0.8	– 0.1	0.0025

and the DLS for lack of fit [see inequality (15.168)] is 0.282. Thus, the first-order model appears to be adequate. Using Eq. (15.169), one finds that both Current (DLS = 0.006) and Time (DLS = 0.049) are needed in the model. The estimated first-order model (including the center points) is

$$\hat{Y} = 4.71 - 0.375x_1 + 0.125x_2.$$

Path of Steepest Ascent. The path of steepest ascent (the estimated direction to the maximum) is obtained using the $\hat{\beta}_i$. For k factors the response surface is $(k + 1)$-dimensional, and the estimated direction to the maximum is along the vector $\hat{\boldsymbol{\beta}} = (\hat{\beta}_1, \ldots, \hat{\beta}_k)$. Normalizing this vector so that it has unit length results in $\mathbf{u} = (u_i, \ldots, u_k)$, where

$$u_i = \hat{\beta}_i \left(\sum_{j=1}^{k} \hat{\beta}_j^2 \right)^{-\frac{1}{2}}. \tag{15.171}$$

Points along the path of steepest ascent are at $x_1 = \delta u_1, \ldots, x_k = \delta u_k$ for $\delta > 0$. (If one were interested in finding the minimum of the response surface, then the path of steepest descent would be used. It is obtained in a similar fashion using values of $\delta < 0$.) Several experimental trials should be run along the path of steepest ascent, say for $\delta = 2$, 4, 6, and 8, and continuing until the response starts to decrease. The experimental trial resulting in the maximum response is then used as the center point for a new 2^k design, and the entire process is repeated.

Example 15.35b. For the welding problem (Example 15.35*a*) the path of steepest descent is obtained as follows. Using the values in column (2) of Table 15.73 and Eq. (15.171), one computes

$$\hat{\beta}_1 = \frac{-1.5}{4} = -0.375$$

$$\hat{\beta}_2 = \frac{0.5}{4} = 0.125$$

$$u_1 = \frac{-0.375}{0.395} = -0.949$$

$$u_2 = \frac{0.125}{0.395} = 0.316.$$

TABLE 15.74 Experimental Trials Performed along the Path of Steepest Descent (Example 15.35*b*)

Trial	δ	x_1	x_2	Current, amperes	Time, seconds	Response
8	–2	1.898	–0.632	64	0.67	3.9
9	–4	3.796	–1.264	83	0.55	3.3
10	–6	5.694	–1.896	102	0.42	2.3
11	–8	7.592	–2.528	121	0.29	5.5

For $\delta = -2, -4, -6$, and -8, the resulting trials and responses are given in Table 15.74. The decoded Current and Time values were obtained using Eq. (15.170). Since at trial 11 the response had started to increase, a point near trial 10 with Current = 102 and Time = 0.4 was chosen as the center point for a second 2^2 design. It was decided to study a range of 20 amperes and 0.2 second around this center point, resulting in the coded values

$$x_1 = \frac{\text{Current} - 102 \text{ amperes}}{10 \text{ amperes}}$$
$$x_2 = \frac{\text{Time} - 0.4 \text{ second}}{0.1 \text{ second}}. \tag{15.172}$$

The second 2^2 design with three center points and the resulting responses are given in Table 15.75. Performing the same analysis as in Example 15.35*a*, one obtains $SS_{AB} = 0.0225$, $SS_\gamma = 1.075$, $F_{LF} = 164.8$, and DLS = 0.006 for lack of fit. Thus, it is clear that the first-order model is not adequate.

15.9.2 Central Composite Designs

When the first-order model is not adequate, a 2^k design can easily be augmented to allow for fitting the second-order model

$$Y = \beta_0 + \sum_{i=1}^{k} \beta_i x_i + \sum_{i=1}^{k} \sum_{j \geq i} \beta_{ij} x_i x_j + \epsilon. \tag{15.173}$$

Generally, the best way to do this is by adding $2k$ design points (with k factors the design points are k-dimensional) at

$$\begin{gathered}(\pm\alpha, 0, \ldots, 0)\\(0, \pm\alpha, 0, \ldots, 0)\\\vdots\\(0, \ldots, 0, \pm\alpha)\end{gathered} \tag{15.174}$$

TABLE 15.75 2^2 Design with Three Center Points Performed around the Minimum Response Found along the Path of Steepest Descent (Example 15.35*b*)

Trial*	x_1	x_2	Current, amperes	Time, seconds	Response
12	0	0	102	0.4	2.4
13	0	0	102	0.4	2.3
14	0	0	102	0.4	2.3
15	–1	–1	92	0.3	2.5
16	1	–1	112	0.3	3.4
17	–1	1	92	0.5	2.7
18	1	1	112	0.5	3.9

*Trials were run in random order.

and possibly some additional center points. The additional $2k$ points are called *star points* (or *axial points*), and the resulting design is called a *central composite design*. Central composite designs for $k = 2$ and $k = 3$ are shown in Fig. 15.21. If a 2^{k-p} design is being augmented and α is chosen such that

$$\alpha = (2^{k-p})^{1/4} \tag{15.175}$$

then the design has the desirable property of being rotatable. That is, the variance of the predicted response $\hat{Y}$ depends only on the distance from the center point and, hence, is unchanged if the design is rotated about the center point.

The adequacy of the second-order model (15.173) can be checked as follows. Let

$$\gamma_i = \beta_{iii} + \frac{1}{1-\alpha^2}\sum_{j\neq i}\beta_{ijj} \tag{15.176}$$

$$\overline{Y}_i(j) = \text{average response when } x_i = j.$$

An unbiased estimator of γ_i is

$$\hat{\gamma}_i = \frac{1}{\alpha^2 - 1}\left[\frac{\overline{Y}_i(\alpha) - \overline{Y}_i(-\alpha)}{2\alpha} - \frac{\overline{Y}_i(1) - \overline{Y}_i(-1)}{2}\right] \tag{15.177}$$

and its contribution to the sum of squares for lack of fit is

$$SS_{\gamma i} = \frac{(\alpha^2 - 1)^2(\hat{\gamma}_i)^2}{\dfrac{1}{2^{k-p}} + \dfrac{1}{2\alpha^2}}. \tag{15.178}$$

If additional center points were run with the star points, then two measures of overall curvature are available, and their difference should not be statistically significant if the second-order model is adequate. Specifically, the difference of the two measures of curvature is

$$d = \overline{Y}_f - \overline{Y}_{cf} - \frac{k}{\alpha^2}(\overline{Y}_s - \overline{Y}_{c_s}) \tag{15.179}$$

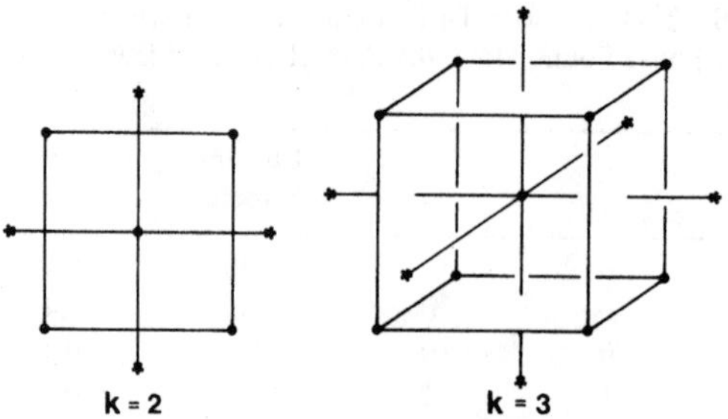

FIGURE 15.21 Central composite designs for $k = 2$ and $k = 3$ factors.

where $\overline{Y}_f$ = average response at the points in factorial design

$\overline{Y}_{c_f}$ = average response at the center points run with the factorial design

$\overline{Y}_s$ = average response at the star points

$\overline{Y}_{c_s}$ = average response at the center points run with the star points

and the associated sum of squares is

$$\mathrm{SS}_d = d^2 \Big/ \left(\frac{1}{2^{k-p}} + \frac{1}{c_f} + \frac{k}{2\alpha^4} + \frac{k^2}{\alpha^4 c_s} \right) \tag{15.180}$$

where c_f and c_s are the numbers of center points run with the factorial design and the star points, respectively. Therefore, the overall sum of squares for lack of fit (with $k + 1$ degrees of freedom) is

$$\mathrm{SS}_{LF} = \mathrm{SS}_d + \sum_{i=1}^{k} \mathrm{SS}_{\gamma_i}. \tag{15.181}$$

Also, a pure estimate of error can be obtained by pooling the sample variance of the center points run with the factorial design and the sample variance of the center points run with the star points. Namely,

$$S_p^2 = \frac{(c_f - 1)\, S_{c_f}^2 + (c_s - 1)\, S_{c_s}^2}{c_f + c_s - 2.} \tag{15.182}$$

Therefore, a test for overall lack of fit can be made using

$$F_{LF} = \frac{\mathrm{SS}_{LF}/(k+1)}{S_p^2} \tag{15.183}$$

and the DLS

$$\Pr[F(k+1, c_f + c_s - 2) > F_{LF}]. \tag{15.184}$$

Example 15.35c. Since in Example 15.35*b* the first-order model was found to be inadequate, the factorial design was augmented with star points and two addi-

TABLE 15.76 Star Points and Additional Center Points Added to the 2^2 Design in Table 15.75 to Form a Central Composite Design (Example 15.35*c*)

Trial*	x_1	x_2	Current, amperes	Time, seconds	Response
19	$-\sqrt{2}$	0	88	0.4	2.9
20	$\sqrt{2}$	0	116	0.4	4.3
21	0	$-\sqrt{2}$	102	0.26	2.8
22	0	$\sqrt{2}$	102	0.54	3.1
23	0	0	102	0.4	2.2
24	0	0	102	0.4	2.3

*Trials were run in random order.

tional center points to form a central composite design. In order to make the design rotatable, α was chosen [using Eq. (15.175)] as $\sqrt{2}$. The results of these trials are given in Table 15.76. The first step in the analysis is to check the adequacy of the second-order model. Using Eqs. (15.177) and (15.178), one computes

$$\hat{\gamma}_1 = \frac{4.3 - 2.9}{2\sqrt{2}} - \frac{(3.4 + 3.9)/2 - (2.5 + 2.7)/2}{2} = -0.0300$$

$$SS_{\gamma_1} = \frac{(-0.0300)^2}{\frac{1}{4} + \frac{1}{4}} = 0.0018$$

$$\hat{\gamma}_2 = \frac{3.1 - 2.8}{2} - \frac{(2.7 + 3.9)/2 - (2.5 + 3.4)/2}{2} = -0.0689$$

$$SS_{\gamma_2} = \frac{(-0.0689)^2}{\frac{1}{4} + \frac{1}{4}} = 0.0095$$

and using Eqs. (15.179) and (15.180), one computes

$$d = (3.125 - 2.35) - (3.275 - 2.25) = -0.25$$

$$SS_d = \frac{(-0.25)^2}{\frac{1}{4} + \frac{1}{2} + \frac{1}{4} + \frac{1}{2}} = 0.0417.$$

The pure estimate of error [Eq. (15.182)] is

$$S_p^2 = \frac{(2.4 - 2.35)^2 + \cdots + (2.3 - 2.25)^2}{2} = 0.005.$$

Combining these results using Eqs. (15.181) and (15.183) gives

$$SS_{LF} = 0.0018 + 0.0095 + 0.0417 = 0.053$$

$$F_{LF} = \frac{0.053/3}{0.005} = 3.53$$

and [formula (15.184)] DLS = 0.228. Thus, the second-order model is clearly adequate.

Fitting the second-order model (15.173) is best done using a computer and the general theory of least squares discussed in Chap. 14. Once the fitted model is obtained, a potential maximum (or minimum) point on the response surface in the region of the design points can be obtained by taking partial derivatives of the fitted model with respect to each x_i, setting the results equal to zero, and solving the system of equations. For $k = 2$ this results in

$$x_1 = \frac{2\hat{\beta}_1\hat{\beta}_{22} - \hat{\beta}_2\hat{\beta}_{12}}{(\hat{\beta}_{12})^2 - 4\hat{\beta}_{11}\hat{\beta}_{22}} \tag{15.185}$$

$$x_2 = \frac{2\hat{\beta}_2\hat{\beta}_{11} - \hat{\beta}_1\hat{\beta}_{12}}{(\hat{\beta}_{12})^2 - 4\hat{\beta}_{11}\hat{\beta}_{22}} \tag{15.186}$$

and for $k > 2$ the solution is most easily obtained numerically using a computer. Additional trials should be run at the potential maximum (minimum) point to establish the actual behavior there. Further information on response surfaces can be found in Box and Draper (1987), Cornell (1984), and Khuri and Cornell (1987).

Example 15.35d. Using least squares and a computer to fit the second-order model (15.173) to the central composite design (trials 12 to 24) in Tables 15.75 and 15.76, one obtains

$$\hat{Y} = 2.3 + 0.51x_1 + 0.1405x_2 + 0.075x_1x_2 + 0.6125x_1^2 + 0.2875x_2^2.$$

Using Eqs. (15.185) and (15.186), the minimum value of the estimated response surface is obtained when

$$x_1 = \frac{2(0.51)(0.2875) - 0.1405(0.075)}{(0.075)^2 - 4(0.6125)(0.2875)} = -0.405$$

$$x_2 = \frac{2(0.1405)(0.6125) - 0.51(0.075)}{(0.075)^2 - 4(0.6125)(0.2875)} = -0.192$$

and that minimum value is

$$\begin{aligned}\hat{Y} &= 2.3 + 0.51(-0.405) + 0.1405(-0.192) + 0.075(-0.405)(-0.192)\\ &\quad + 0.6125(-0.405)^2 + 0.2875(-0.192)^2\\ &= 2.18.\end{aligned}$$

The predicted minimum percent defective [using Eqs. (15.172)] occurs when

$$\text{Current} = 10(-0.405) + 102 = 97.95 \text{ amperes}$$

$$\text{Time} = 0.1(-0.192) + 0.4 = 0.3808 \text{ second}$$

Three additional trials were run at 98 amperes and 0.38 second and resulted in 2.2, 2.1, and 2.1 percent defective.

15.10 MIXTURE DESIGNS

When the response of an experiment depends only on the proportion of each factor and not on the total amount of all the factors, then one is said to be experimenting with a *mixture*. Some examples of mixtures are blends of gasoline, ingredients in concrete mix, and the components in glue. Studying response surfaces for mixture experiments requires either special kinds of designs or transformations, either of the factors or of the standard response surface designs.

Let $x_i \geq 0$ denote the proportion of component i, where proportions are usually measured by either weight, volume, or mole fraction. If there are k components (factors) in the mixture, then

$$\sum_{i=1}^{k} x_i = 1. \tag{15.187}$$

Thus, there are two major differences when it comes to designing mixture experiments. First, the levels of the different factors are not all independent; second, the experimental region is constrained. The k factors can be transformed to $k - 1$ factors whose levels are all independent, but the experimental region will (usually) still be constrained. Using the transformed factors, a standard response surface design can be chosen within the constrained experimental region, and then either the design points can be transformed back to the original factor space, or the experiment and its analysis can be carried out using the transformed factors. Transformation is useful for more complicated mixture experiments where the number of components cannot be reduced to at most three or four. For mixture experiments with fewer components, however, the special mixture designs are useful, and these designs are considered first. Subsequently a particularly simple type of transformation that is applicable in some situations will be discussed. More information on transformations can be found in Cornell (1981) and on mixture experiments in general in Cornell (1981, 1983).

15.10.1 Modeling over the Entire Simplex

When there are no constraints on the proportions x_i other than that they sum to 1 [Eq. (15.187)], then the constrained experimental region is a *simplex*. For two components (that is, in two dimensions) a simplex is a line, for three components a simplex is a triangle, and for four components a simplex is a tetrahedron. With three components the proportions x_i can be conveniently plotted on triangular graph paper, such as is shown in Fig. 15.22. The vertices of the triangle represent the three degenerate mixtures consisting of a single component, the sides of the triangle represent mixtures of only two components, and the interior points of the triangle represent mixtures in which all three components are present. Three useful designs for modeling over the entire simplex are simplex-lattice designs, simplex-centroid designs, and axial designs.

Simplex-Lattice Designs. Designs where the points are positioned uniformly over the simplex are called *simplex-lattice designs*. More specifically, a $\{k, m\}$ sim-

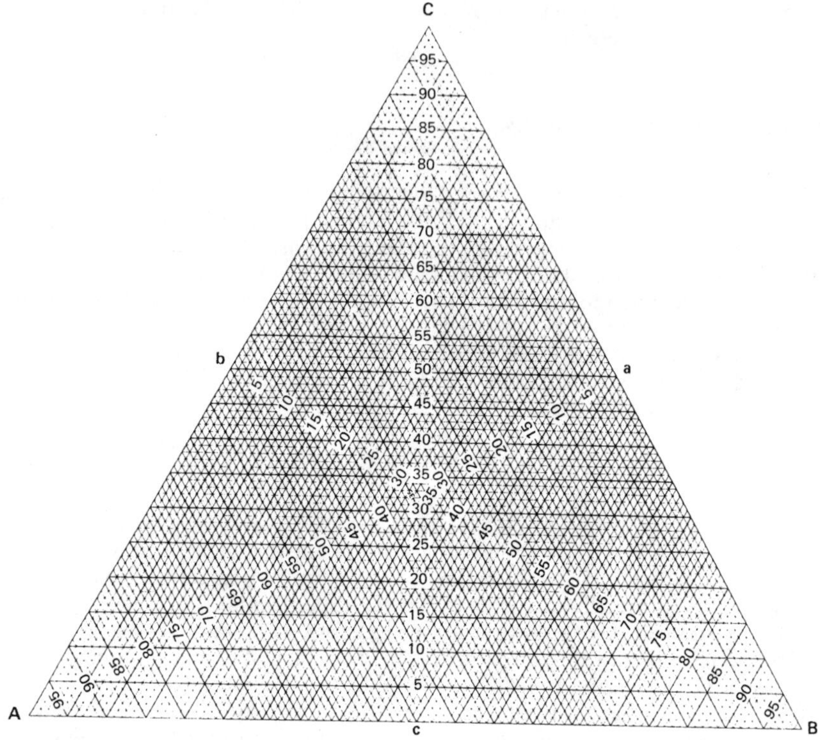

FIGURE 15.22 Triangular graph paper.

plex-lattice design is a design for studying mixtures with k potential components, where each component is studied at the $m + 1$ equally spaced values

$$x_i = 0, \frac{1}{m}, \frac{2}{m}, \cdots, 1.$$

Figure 15.23 shows a {3, 3} simplex-lattice design where, for example, $(\frac{2}{3}, \frac{1}{3}, 0)$ indicates $x_1 = \frac{2}{3}$, $x_2 = \frac{1}{3}$, and $x_3 = 0$. A {k, m] simples-lattice design consists of

$$\binom{k + m - 1}{m}$$

design points and can be used to fit an m-order polynomial model. Some advantages of simplex-lattice designs are that they give equal coverage to each component, they are simple to obtain, and it is easy to obtain least-squares estimates of the model coefficients. Two disadvantages are that the mixtures studied contain at most m of the k components and that a large number of design points are required when $k \geq 4$ and $m > 2$. Thus, since one would rarely want to use a design where none of the experimental trials involved all k components, and since the number of experimental trials it is possible to perform is usually limited, only {2, 2} and {3, 3} simplex-lattice designs can be recommended for general consideration.

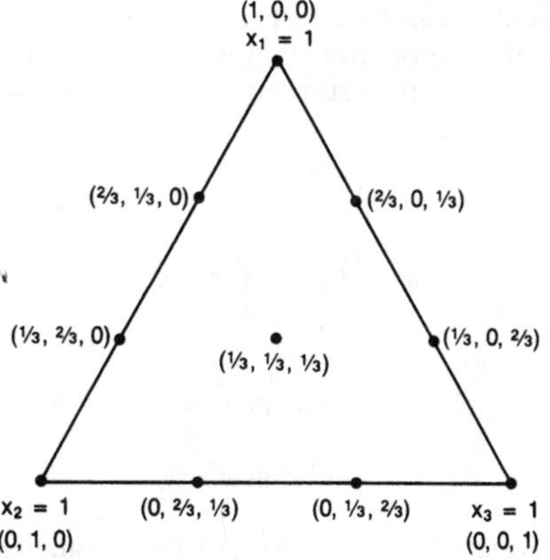

FIGURE 15.23 A{3, 3} Simplex-lattice design.

Fitting response surfaces to the results obtained using simplex-lattice (or other mixture) designs is most easily done using a canonical polynomial model. Scheffé (1958) introduced canonical polynomials specifically for this purpose. The canonical form of a polynomial model (the canonical polynomial model) is obtained by applying the constraint (15.187) to the standard polynomial model and simplifying. For the first-order model (15.164) this results in

$$\begin{aligned} Y &= \beta_0 + \sum_{i=1}^{k} \beta_i x_i + \epsilon \\ &= \beta_0 \sum_{i=1}^{k} x_i + \sum_{i=1}^{k} \beta_i x_i + \epsilon \\ &= \sum_{i=1}^{k} (\beta_0 + \beta_i) x_i + \epsilon \\ &= \sum_{i=1}^{k} \beta_i^* x_i + \epsilon. \end{aligned} \tag{15.188}$$

The second-order canonical polynomial model is

$$Y = \sum_{i=1}^{k} \beta_i^* x_i + \sum_{i=1}^{k} \sum_{j>i} \beta_{ij}^* x_i x_j + \epsilon \tag{15.189}$$

and the full third-order (cubic) canonical polynomial model is

$$\begin{aligned} Y = \sum_{i=1}^{k} \beta_i^* x_i + \sum_{i=1}^{k} \sum_{j>i} \beta_{ij}^* x_i x_j + \sum_{i=1}^{k} \sum_{j>i} \delta_{ij} x_i x_j (x_i - x_j) \\ + \sum_{i=1}^{k} \sum_{j>i} \sum_{\ell>j} \beta_{ij\ell}^* x_i x_j x_\ell + \epsilon. \end{aligned} \tag{15.190}$$

Note that the coefficient β_i^* can be interpreted as the expected response when the mixture consists only of component i (that is, when $x_i = 1$).

When using a {2, 2} design, the estimators of the parameters in the second-order model (15.189) are

$$b_i^* = Y_i \tag{15.191}$$

$$b_{ij}^* = 4Y_{ij} - 2(Y_i + Y_j) \tag{15.192}$$

where

$$Y_i = \text{response at } x_i = 1$$

$$Y_{ij} = \text{response at } x_i = x_j = \frac{1}{2}, \quad i < j.$$

When using a {3, 3} design, the estimators of the parameters in the cubic model (15.190) are given by Eq. (15.191) and

$$b_{ij}^* = \frac{9}{4}(Y_{iij} + Y_{ijj} - Y_i - Y_j) \tag{15.193}$$

$$d_{ij} = \frac{9}{4}(3Y_{iij} - 3Y_{ijj} - Y_i + Y_j) \tag{15.194}$$

$$b_{ij\ell}^* = 27Y_{ij\ell} - \frac{27}{4}(Y_{iij} + Y_{ijj} + Y_{ii\ell} + Y_{i\ell\ell} + Y_{jj\ell} + Y_{j\ell\ell}) + \frac{9}{2}(Y_i + Y_j + Y_\ell) \tag{15.195}$$

where d_{ij} is the estimator of δ_{ij}, Y_i and Y_{ij} are as previously defined, and

$$Y_{iij} = \text{response at } x_i = \frac{2}{3} \text{ and } x_j = \frac{1}{3}, \quad i < j$$

$$Y_{ijj} = \text{response at } x_i = \frac{1}{3} \text{ and } x_j = \frac{2}{3}, \quad i < j$$

$$Y_{ij\ell} = \text{response at } x_i = x_j = x_\ell = \frac{1}{3}, \quad i < j < \ell.$$

When using a $\{k, 1\}$ simplex-lattice design, estimators of the parameters in the first-order model (15.188) are given by Eq. (15.191). Model (15.188) is, however, of limited usefulness in that case since the estimated model is based entirely on mixtures consisting of a single component.

If replicate observations are available at any of the design points, then the estimators b^* are obtained by replacing the responses Y with the average responses $\overline{Y}$. For example, if multiple observations were taken at $x_i = 1$, then Eq. (15.191) becomes

$$b_i^* = \overline{Y}_i$$

where $\overline{Y}_i$ is the average response at $x_i = 1$.

Simplex-Centroid Designs. When the number of experimental trials is limited, an alternative to simplex-lattice designs is *simplex-centroid designs*. These designs were introduced by Scheffé (1963). A simplex-centroid design for k components requires $2^k - 1$ design points. These are the k mixtures consisting of only a single component, the $\binom{k}{2}$ mixtures where $x_i = x_j = \frac{1}{2}$, the $\binom{k}{3}$ mixtures where $x_i = x_j = x_\ell = \frac{1}{3}$, and so forth. A simplex-centroid design for three components (seven design points) is shown in Fig. 15.24. With only seven responses one can no longer fit the full cubic polynomial model (15.190), but one can fit the special cubic polynomial model

$$Y = \sum_{i=1}^{3} \beta_i^* x_i + \sum_{i=1}^{3} \sum_{j>i} \beta_{ij}^* x_i x_j + \beta_{123}^* x_1 x_2 x_3 + \epsilon. \qquad (15.196)$$

The estimators of the parameters in model (15.196) are given by Eqs. (15.191) and (15.192) and

$$b_{123}^* = 27Y_{123} - 12(Y_{12} + Y_{13} + Y_{23}) + 3(Y_1 + Y_2 + Y_3). \qquad (15.197)$$

Example 15.36. A simplex-centroid design was used to study the effect on mileage of different mixtures of three fuels. The results are given in Table 15.77. Using Eqs. (15.191), (15.192), and (15.197), one obtains

$$b_1^* = 25.4$$
$$b_2^* = 29.3$$
$$b_3^* = 35.5$$
$$b_{12}^* = 4(9.8) - 2(25.4 + 29.3) = -70.2$$
$$b_{13}^* = 4(40.7) - 2(25.4 + 35.5) = 41.0$$
$$b_{23}^* = 4(18.3) - 2(29.3 + 35.5) = -56.4$$
$$b_{123}^* = 27(15.2) - 12(9.8 + 40.7 + 18.3) + 3(25.4 + 29.3 + 35.5) = -144.6.$$

Axial Designs. One way of dealing with a large number of components is to screen them and then work only with the most important ones. Axial designs are useful for this purpose. They are designs with points positioned symmetrically on

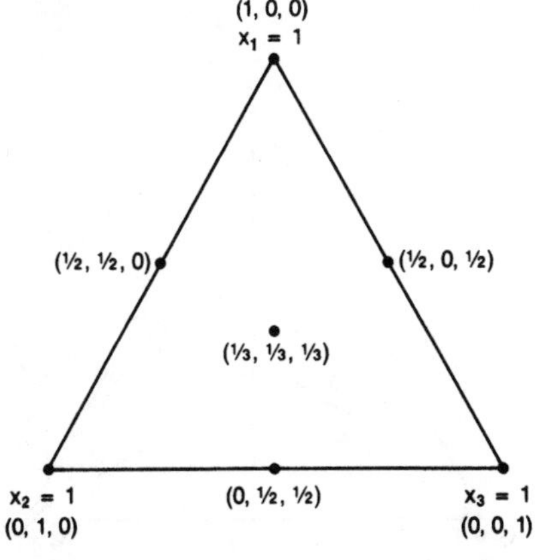

FIGURE 15.24 A simplex-centroid design for three components.

TABLE 15.77 Miles per Gallon Obtained with Different Mixtures of Three Fuels (Example 15.36)

Trial*	x_1	x_2	x_3	Miles per gallon
1	1	0	0	25.4
2	0	1	0	29.3
3	0	0	1	35.5
4	$\frac{1}{2}$	$\frac{1}{2}$	0	9.8
5	$\frac{1}{2}$	0	$\frac{1}{2}$	40.7
6	0	$\frac{1}{2}$	$\frac{1}{2}$	18.3
7	$\frac{1}{3}$	$\frac{1}{3}$	$\frac{1}{3}$	15.2

*Trials were run in random order.

the component axes. The axis of component i is the line connecting the point $x_i = 0$ and $x_j = 1/(k-1)$ for $j \neq i$ with the point $x_i = 1$ and $x_j = 0$ for $j \neq i$. Along this line the amount of component i varies from 0 to 1, and the remaining components are maintained in equal proportions. We consider only the simplest form of axial design where the points are positioned equidistant (a distance Δ) from the centroid $(1/k, \ldots, 1/k)$ and are located between the centroid and the vertices. A three-component axial design of this type is shown in Fig. 15.25.

The estimator of the effect of component i is

$$E_i = b_i^* - \sum_{j \neq i} \frac{b_j^*}{k-1} \tag{15.198}$$

which is the difference between the estimator of the response at $x_i = 1$ and the estimator of the response at $x_i = 0$ and $x_j = 1/(k-1)$ for $j \neq i$, when using the first-order model (15.188). Thus, E_i estimates the change in response when x_i changes from 1 to 0 along the axis of component i. The relative magnitudes of the E_i indicate the relative effects of the components.

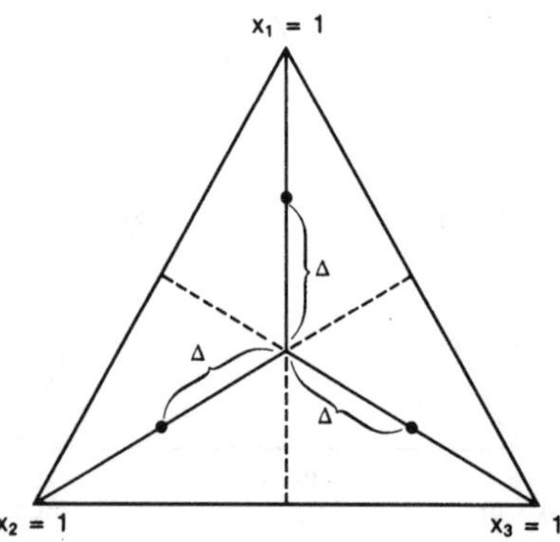

FIGURE 15.25 A three-component axial design.

If n replicate observations are taken at each of the points in the axial design, then the statistical significance of each component effect can be tested as follows. Let S_i^2 be the sample variance of the responses at distance Δ on axis i, and let

$$MS_e = \frac{1}{k}\sum_{i=1}^{k} S_i^2. \tag{15.199}$$

Then ANOM type decision lines for the component effects [similar to the decision lines (15.140)] are given by

$$0 \pm h(\alpha; k, k(n-1))\sqrt{MS_e}\sqrt{\frac{k-1}{n\Delta^2 k}} \tag{15.200}$$

and an effect E_i is statistically significant (at level α) if it falls outside the decision lines (15.200).

In order to compute the E_i, one must compute the estimates b_i^*, which can be easily obtained by numerically solving the system of equations

$$\overline{Y}_i = \Delta^* b_i^* + \frac{1-\Delta^*}{k-1}\sum_{j\neq i} b_j^*, \quad i = 1, \ldots, k \tag{15.201}$$

where $\overline{Y}_i$ is the average response at distance Δ on axis i and

$$\Delta^* = \Delta + \frac{1}{k}.$$

Example 15.37. A fertilizer for tomatoes was to be developed using five possible ingredients. In order to determine the effects of the ingredients, a mixture experiment was run using an axial design with $\Delta = 7/15$. Each mixture was used on two plants. The response recorded was pounds of tomatoes produced, and the coded results are given in Table 15.78. For $k = 5$ and Δ $7/15$ the system of Eqs. (15.201) is

$$\overline{Y}_i = \frac{2}{3}b_i^* + \sum_{j\neq i}\frac{b_j^*}{12}, \quad i = 1, \ldots, 5$$

which, when solved numerically, yields the results in Table 15.79. Also given in Table 15.79 are the component effects obtained using Eq. (15.198). The decision lines (15.200) with $\alpha = 0.1$ are

$$0 \pm h(0.1; 5, 5)\sqrt{MS_e}\sqrt{\frac{4}{10(7/10)^2}} = 0 \pm (3.05)\sqrt{4.168}\sqrt{1.837}$$
$$= 0 \pm 8.44$$

TABLE 15.78 Coded Amounts of Tomatoes Produced Using Fertilizers Consisting of Various Mixtures of Five Ingredients and an Axial Design (Example 15.37)

Axis	Plant 1	Plant 2	$\overline{Y}_i$	S_i^2
1	7.0	9.8	8.4	3.92
2	23.0	20.2	21.6	3.92
3	11.6	9.0	10.3	3.38
4	24.1	27.3	25.7	5.12
5	14.0	17.0	15.5	4.5

TABLE 15.79 Estimates of the Model Parameters and Component Effects for Example 14.37

i	b_i^*	E_i
1	2.757	–16.93
2	25.39	11.36
3	6.014	–12.86
4	32.41	20.14
5	14.93	–1.713

where $MS_e = 4.168$ was obtained by averaging the sample variances in Table 15.78 [Eq. (15.199)]. Comparing the component effects with these decision lines, one finds that ingredient 5 does not have a significant ($\alpha = 0.1$) effect.

15.10.2 Modeling over a Constrained Region of the Simplex

Very often one must have at least a certain proportion of some, or all, of the components in a mixture. For example, a cake (which is a mixture of ingredients) would not be a cake without flour. Similarly, too much of a component, such as too much shortening in a cake, might cause a problem. Thus, one needs to consider three types of constraints: component proportions with lower bounds, component proportions with upper bounds, and component proportions with both lower and upper bounds.

Lower Bounds on Component Proportions. Mixtures with lower bounds on the component proportions are most easily studied using pseudocomponents. Let $a_i \geq 0$ be the lower bound for component i, and let

$$L = \sum_{i=1}^{k} a_i. \tag{15.202}$$

Note that a_i could be zero for some components, and L must be less than 1, since otherwise there would be no region satisfying the constraints. The pseudocomponents

$$x_i' = \frac{x_i - a_i}{1 - L} \tag{15.203}$$

are linear transformations of the original components such that there are no constraints (other than $\Sigma_{i=1}^{k} x_i' = 1$) on the pseudocomponents. All the techniques for modeling over the entire simplex can be used with the pseudocomponents. Once a design has been chosen in the pseudocomponents, the settings of the original components can be obtained using the equation

$$x_i = a_i + (1 - L)x_i'. \tag{15.204}$$

Analysis of the experimental results is most easily done using the pseudocomponents.

If for some reason a model using the original components is desired, it can be obtained as follows. Suppose that using the pseudocomponents one has obtained the fitted model

$$\hat{Y} = \sum_{i=1}^{k} g_i x_i' + \sum_{i=1}^{k} \sum_{j>i} g_{ij} x_i' x_j'. \tag{15.205}$$

Then the corresponding model in the original components [using Eq. (15.203] would be

$$\hat{Y} = \sum_{i=1}^{k} g_i \left(\frac{x_i - a_i}{1-L}\right) + \sum_{i=1}^{k} \sum_{j>i} g_{ij} \left(\frac{x_i - a_i}{1-L}\right)\left(\frac{x_j - a_j}{1-L}\right)$$

$$= \sum_{i=1}^{k} b_i^* x_i + \sum_{i=1}^{k} \sum_{j>k} b_{ij}^* x_i x_j \tag{15.206}$$

and solving Eq. (15.206) for the b_i^* results in

$$b_i^* = \frac{1}{i-L} \sum_{j=1}^{k} g_j(\delta_{ij} - a_j) + \frac{1}{(1-L)^2} \sum_{j=1}^{k} \sum_{\ell>j} g_{j\ell}(a_j - \delta_{ij})(a_\ell - \delta_{i\ell}) \tag{15.207}$$

$$b_{ij}^* = \frac{g_{ij}}{(1-L)^2} \tag{15.208}$$

where

$$\delta_{ij} = \begin{cases} 1 & \text{if } i = j \\ 0 & \text{otherwise.} \end{cases}$$

Example 15.38. Suppose that in Example 15.36 the proportions of Fuels 1 and 3 had lower bounds of 0.1 and 0.3, respectively, and a mixture experiment was run using a simplex-centroid design in the pseudocomponents. Using Eq. (15.202), one obtains

$$L = 0.1 + 0 + 0.3 = 0.4$$

and the original component settings are obtained using Eq. (15.204). The pseudo-component settings, the original component settings, and the resulting miles per gallon are given in Table 15.80. Using Eqs. (15.191), (15.192), and (15.197) with the pseudocomponents, one obtains

$$g_1 = 36.9$$

$$g_2 = 15.0$$

$$g_3 = 38.1$$

$$g_{12} = 4(15.8) - 2(36.9 + 15.0) = -40.6$$

$$g_{13} = 4(41.3) - 2(36.9 + 38.1) = 15.2$$

$$g_{23} = 4(18.5) - 2(15.0 + 38.1) = -32.2$$

$$g_{123} = 27(23.1) - 12(15.8 + 41.3 + 18.5) + 3(36.9 + 15.0 + 38.1) = -13.5.$$

TABLE 15.80 Pseudocomponent Settings, Original Component Settings, and Miles per Gallon Obtained with These Settings (Example 15.38)

Pseudocomponents			Original components			Miles per gallon
x'_1	x'_2	x'_3	x_1	x_2	x_3	
1	0	0	0.7	0	0.3	36.9
0	1	0	0.1	0.6	0.3	15.0
0	0	1	0.1	0	0.9	38.1
$\frac{1}{2}$	$\frac{1}{2}$	0	0.4	0.3	0.3	15.8
$\frac{1}{2}$	0	$\frac{1}{2}$	0.4	0	0.6	41.3
0	$\frac{1}{2}$	$\frac{1}{2}$	0.1	0.3	0.6	18.5
$\frac{1}{3}$	$\frac{1}{3}$	$\frac{1}{3}$	0.3	0.2	0.5	23.1

Solving the fitted equation

$$\hat{Y} = 36.9x'_1 + 15.0x'_2 + 38.1x'_3 - 40.6x'_1x'_2 + 15.2x'_1x'_3 - 32.2x'_2x'_3 - 13.5x'_1x'_2x'_3$$

numerically for its maximum value, one finds that the maximum mileage per gallon of 41.32 occurs when $x'_1 = 0.46$, $x'_2 = 0$, and $x'_3 = 0.54$. In terms of the original components, this maximum occurs when [Eq. (15.204)]

$$x_1 = 0.1 + (1 - 0.4)(0.46) = 0.376$$

$$x_2 = 0 + (1 - 0.4)(0) = 0$$

$$x_3 = 0.3 + (1 - 0.4)(0.54) = 0.624.$$

Upper and Lower Bounds on the Component Proportions. While lower bounds on the component proportions can be easily handled, the situation is not quite so straightforward with upper bounds (or both upper and lower bounds). With both upper and lower bounds one approach is to start with a design using pseudocomponents that takes into account the lower bounds. (With only upper bounds, pseudocomponents would not be necessary.) Any design points having component proportions that exceed their upper bounds are modified by replacing those component proportions with their upper bounds and increasing the remaining component proportions by equal amounts. With several upper bounds the procedure may require some iteration since the component proportions that are raised could conceivably then exceed their upper bounds.

Example 15.39. Cornell (1983) discusses a fruit punch experiment where the three components (different fruit juices) have the restrictions

$$0.4 \le x_1 \le 0.8$$

$$0.1 \le x_2 \le 0.5$$

$$0.1 \le x_3 \le 0.3.$$

TABLE 15.81 Pseudocomponent Settings and Original Component Settings for Example 15.39

Pseudocomponents			Original components		
x_1'	x_2'	x_3'	x_1	x_2	x_3
1	0	0	0.8	0.1	0.1
0	1	0	0.4	0.5	0.1
0	0	1	0.4	0.1	0.5*
$\frac{1}{2}$	$\frac{1}{2}$	0	0.6	0.3	0.1
$\frac{1}{2}$	0	$\frac{1}{2}$	0.6	0.1	0.3
0	$\frac{1}{2}$	$\frac{1}{2}$	0.4	0.3	0.3
$\frac{1}{3}$	$\frac{1}{3}$	$\frac{1}{3}$	0.54	0.23	0.23

*This design point would be modified to (0.5, 0.2, 0.3).

Starting with a simplex-centroid design in the pseudocomponents, the corresponding original component settings are given in Table 15.81. Since the design point (0.4, 0.1, 0.5) is outside the region of interest, it is modified by replacing 0.5 with 0.3 (the upper bound for that component proportion) and increasing the proportions of the remaining two components by equal amounts. This results in the design point (0.5, 0.2, 0.3). The final design is shown in Fig. 15.26.

Actually, when there are no more than four components, the constrained region can be drawn fairly easily, and design points can be chosen graphically to cover this region. There is no disadvantage to this since regardless of the method used to choose the design points, the analysis of the results will have to be done using the general theory of least squares. Alternative algorithms for choosing the design points when there are both upper and lower bounds on the component proportions have been proposed by McLean and Anderson (1966), Snee and Marquardt (1974), and Saxena and Nigam (1977). Piepel (1988) presents a computer program

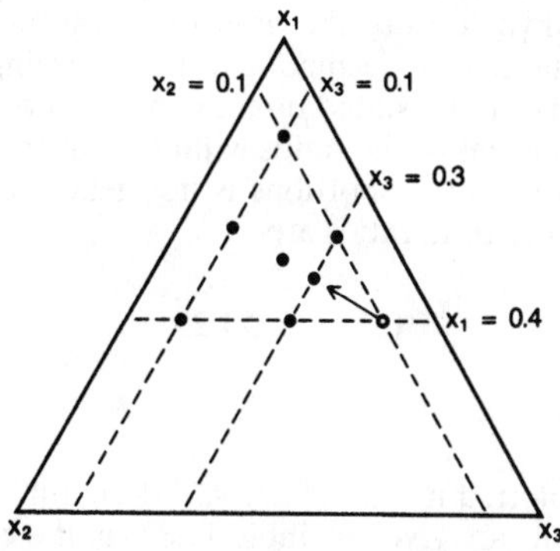

FIGURE 15.26 Design for Example 15.39.

that generates design points for a class of constrained experimental regions that includes upper and lower bounds on mixture component proportions.

15.10.3 Transformation Using Ratios

In practice, when there are more than four components, it seems preferable to transform the k mixture components to $k - 1$ mathematically independent factors. Kenworthy (1963) suggested a particularly simple type of transformation that uses the ratios of component proportions. Many different sets of ratios are possible. One need only have $k - 1$ ratios that involve all k component proportions. Four possible sets of ratios for three component mixtures are

$$r_1 = \frac{x_1}{x_2} \qquad r_2 = \frac{x_1}{x_3} \tag{15.209}$$

$$r_1 = \frac{x_1}{x_2} \qquad r_2 = \frac{x_2}{x_3} \tag{15.210}$$

$$r_1 = \frac{x_1}{x_1 + x_2} \qquad r_2 = \frac{x_2}{x_3} \tag{15.211}$$

$$r_1 = x_1 \qquad r_2 = \frac{x_2}{x_3}. \tag{15.212}$$

Which set of ratios should be used will depend on what makes sense physically for the particular mixture being studied. For example, when making glass using silica, lime, and soda, it makes sense to consider the ratio of silica to soda and the ratio of either silica to lime or soda to lime [Eq. (15.209) or Eq. (15.210)]. Using the $k - 1$ ratios, the response surface can be studied using a 2^{k-1} factorial design, a central composite design, or any other response surface design for $k - 1$ independent factors.

Example 15.40. Kenworthy (1963) discusses studying the resistance to weathering of a synthetic enamel. The enamel is a mixture of three resins: two alkyd resins and a urea resin. Denote the two alkyd resins as components 1 and 2 and the urea resin as component 3. Since in this problem the ratio of the two alkyd resins and the ratio of one alkyd resin to the urea resin are of physical interest, Eqs. (15.209) were used. Note that the technique of transforming to ratios is most useful when any restrictions can be stated directly in terms of the ratios themselves, since finding a region in terms of the ratios within a constrained region defined by restrictions on the component proportions is not generally a trivial task. In this case the restrictions on the resin ratios are

$$0.75 \le r_1 \le 1.25$$

$$0.6 \le r_2 \le 1.$$

These restrictions are plotted in Fig. 15.27. A 2^2 design in the ratio variables was run, and the coded results are given in Table 15.82 (in the correct order for Yates' algorithm). The component proportions (which were necessary for constructing the mixtures) corresponding to the desired ratios were obtained using

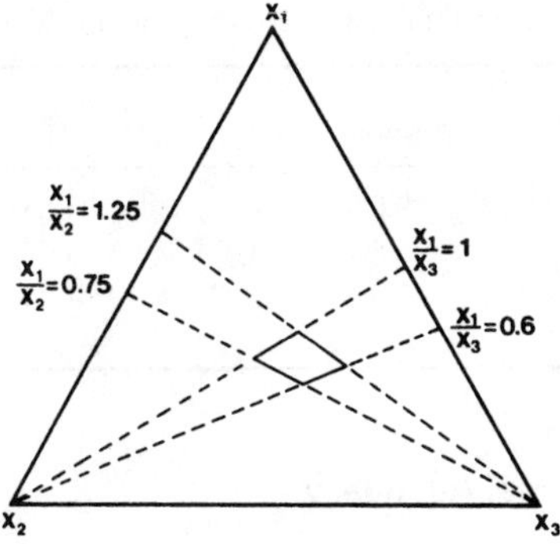

FIGURE 15.27 Constrained region in terms of ratios for Example 15.40.

TABLE 15.82 2^2 Design in the Ratio Variables and the Coded Responses for Example 15.40

Trial*	r_1	r_2	Response
1	0.75	0.6	9.0
2	1.25	0.6	3.5
3	0.75	1	7.0
4	1.25	1	4.5

*Trials were run in random order.

$$\left.\begin{aligned} x_1 &= \frac{r_1 r_2}{r_1 r_2 + r_1 + r_2} \\ x_2 &= \frac{r_2}{r_1 r_2 + r_1 + r_2} \\ x_3 &= \frac{r_1}{r_1 r_2 + r_1 + r_2}. \end{aligned}\right\} \tag{15.213}$$

Equations (15.213) were obtained from Eqs. (15.209) using the restriction that $x_1 + x_2 + x_3 = 1$.

Letting r_1' and r_2' denote the ratios coded so that their high and low values are 1 and –1, respectively, and applying Yates' algorithm (Table 15.83), one obtains the estimated response surface

$$\hat{Y} = \frac{24}{4} - \frac{8}{4} r_1' - \frac{1}{4} r_2' + \frac{3}{4} r_1' r_2'$$

$$= 6 - 1\left(\frac{r_1 - 1}{0.25}\right) - 0.25\left(\frac{r_2 - 0.8}{0.2}\right) + 0.75\left(\frac{r_1 - 1}{0.25}\right)\left(\frac{r_2 - 0.8}{0.2}\right)$$

$$= 27 - 20 r_1 - 16.25 r_2 + 15 r_1 r_2.$$

TABLE 15.83 Yates' Algorithm Applied to the Data in Table 15.82 (Example 15.40)

Treatment combination	Response	(1)	(2)
(1)	9.0	12.5	24
a	3.5	11.5	–8
b	7.0	–5.5	–1
ab	4.5	–2.5	3

15.10.4 Checking Model Adequacy

Having obtained a fitted model, before proceeding further it is a good idea to test whether this model is statistically significantly better at explaining the variability in the responses than a model that estimates every response as the average of all the responses obtained. In other words, one should test whether information on the mixture proportions is useful in predicting the response. In order to do this one must have more design points than there are estimated model parameters. Let

N = total number of experimental trials used to the fit model
Y_t = response at trial t
$\hat{Y}_t$ = estimator of response at trial t

and

$$Y_{\bullet} = \frac{1}{N}\sum_{t=1}^{N} Y_t$$

$$SS_e = \sum_{t=1}^{N} (Y_t - \hat{Y}_t)^2 \tag{15.214}$$

$$SS_{\text{model}} = \sum_{t=1}^{N} (\hat{Y}_t - Y_{\bullet})^2. \tag{15.215}$$

When using the second-order model (15.189), for example, the estimator of the response at trial t is

$$\hat{Y}_t = \sum_{i=1}^{k} b_i^* x_{ti} + \sum_{i=1}^{k}\sum_{j>i} b_{ij}^* x_{ti} x_{tj} \tag{15.216}$$

where x_{ti} is the proportion of component i for trial t. The significance of the fitted model [such as Eq. (15.216)] can be tested by computing

$$F_{\text{model}} = \frac{SS_{\text{model}}/(p-1)}{SS_e/(N-p)} \tag{15.217}$$

and the corresponding DLS

$$\Pr[F(p-1, N-p) > F_{\text{model}}] \tag{15.218}$$

where p is the number of terms (the number of estimated parameters) in the model. Small values of the DLS indicate a statistically significant (and potentially useful) model. The identity

$$\mathrm{SS}_{\mathrm{tot}} = \sum_{t=1}^{N} (Y_t - Y_{\bullet})^2 \tag{15.219}$$

$$= \mathrm{SS}_{\mathrm{model}} + \mathrm{SS}_e$$

is sometimes useful in these computations, which are conveniently summarized in the ANOVA table given in Table 15.84.

If one is fortunate enough to be able to use more design points than parameters that need to be estimated and to be able to obtain replicate observations, then the (statistical) adequacy of a model can be checked as follows. Let

U = number of mixtures for which replicate observations are obtained
n_u = number of replicates for mixture u ($u = 1, \ldots, U$)
$N_{r\bullet}$ = total number of trials associated with replicate observations,

$$N_r = \sum_{u=1}^{U} n_u \tag{15.220}$$

Y_{ui} = response for ith replicate of mixture u
$Y_{u\bullet}$ = average response for mixture u,

$$Y_{u\bullet} = \frac{1}{n_u} \sum_{i=1}^{n_u} Y_{ui} \tag{15.221}$$

S_u^2 = sample variance for the response of mixture u,

$$S_u^2 = \frac{1}{n_u - 1} \sum_{i=1}^{n_u} (Y_{ui} - Y_{u\bullet})^2 \tag{15.222}$$

$$\mathrm{SS}_{\text{pure error}} = \sum_{u=1}^{U} (n_u - 1) S_u^2 \tag{15.223}$$

$$\mathrm{SS}_{\mathrm{LF}} = \mathrm{SS}_e - \mathrm{SS}_{\text{pure error}} \cdot \tag{15.224}$$

One computes

$$F_{\mathrm{LF}} = \frac{\mathrm{SS}_{\mathrm{LF}}/(N - p - N_r + U)}{\mathrm{SS}_{\text{pure error}}/(N_r - U)} \tag{15.225}$$

and the model is accepted as being adequate at level α if

$$\Pr[F(N - p - N_r + U, N_r - U) > F_{\mathrm{LF}}] > \alpha. \tag{15.226}$$

Example 15.41. Suppose that for the fruit punch experiment described in Example 15.39 an experimental trial was run at each design point in Fig. 15.26. The result of an experimental trial is the average rating (based on three responses) for

TABLE 15.84 ANOVA Table for Testing the Significance of a Mixture Model

Source of variation	Sum of squares	Degrees of freedom	Mean square	F	$\Pr > F$
Model	$SS_{model} = \sum_{t=1}^{N}(\hat{Y}_t - Y_{\bullet})^2$	$p - 1$	$MS_{model} = \frac{SS_{model}}{p-1}$	$F_{model} = \frac{MS_{model}}{MS_e}$	$\Pr[F(p-1, N-p) > F_{model}]$
Error	$SS_e = \sum_{t=1}^{N}(Y_t - \hat{Y}_t)^2$	$N - p$	$MS_e = \frac{SS_e}{N-p}$		
Total	$SS_{tot} = \sum_{t=1}^{N}(Y_t - Y_{\bullet})^2$ $= SS_{model} + SS_e$	$N - 1$			

TABLE 15.85 Average Ratings for Fruit Punch Mixtures in Example 15.41 together with the Estimated Responses

Component settings				
x_1	x_2	x_3	Response	Estimated response
0.8	0.1	0.1	5.0	5.031
0.4	0.5	0.1	6.5	6.529
0.5	0.2	0.3	5.7	5.944
0.6	0.3	0.1	6.1	6.039
0.6	0.1	0.3	5.4	5.280
0.4	0.3	0.3	6.6	6.479
0.54	0.23	0.23	5.1	5.101

that fruit punch mixture, and these averages are shown in Table 15.85. Fitting the second-order model (15.189) to these values using the general theory of least squares and a computer, one obtains

$$\hat{Y} = 7.18x_1 + 7.86x_2 + 66.43x_3 + 6.47x_1x_2 - 97.47x_1x_3 - 86.22x_2x_3$$

and the estimated responses given in Table 15.85. Using the equations in Table 15.84, one obtains

$$SS_e = (5.0 - 5.031)^2 + \cdots + (5.1 - 5.101)^2 = 0.094$$

$$SS_{\text{tot}} = (5.0 - 5.771)^2 + \cdots + (5.1 - 5.771)^2 = 2.514$$

$$SS_{\text{model}} = 2.514 - 0.094 = 2.420$$

$$F_{\text{model}} = \frac{2.420/(6-1)}{0.094/(7-6)} = 5.15$$

and DLS = 0.322. Based on these results, the fitted second-order model does not appear to be useful, and one might try the special cubic polynomial model (15.196). Note that in this case fitting model (15.196) requires solving seven equations in seven unknowns, and the equations for the parameter estimates are not the same as those for a simplex-centroid design.

Example 15.42. An experiment was conducted to evaluate the effects of three components (a rubber latex and two tackifying resins) on the properties of an adhesive for pressure sensitive labels. All three components were necessary, and the lower bounds on the component proportions were

$$a_1 = 0.3, \qquad a_2 = 0.2, \qquad a_3 = 0.2.$$

A simplex-centroid design in the pseudocomponents was used, and for each mixture the amount of force needed to remove a label was measured. Replicate observations were taken at the vertices and the centroid, and the results are given in Table 15.86. The fitted second-order model (obtained using the general theory of least squares and a computer) is

$$\hat{Y} = 6.827x_1' + 6.727x_2' + 7.627x_3' - 0.36x_1'x_2' - 1.76x_1'x_3' - 0.76x_2'x_3'.$$

The fitted values, and for mixtures with replicates the average responses [Eq. (15.221)] and the sample variances of the responses [Eq. (15.222)], are given in Table 15.87. Using Eqs. (15.214) and (15.223) to (15.225), one obtains

$$SS_e = (6.7 - 6.827)^2 + \cdots + (6.5 - 6.740)^2 = 0.376$$

$$SS_{\text{pure error}} = 0.02 + 0.02 + 0.08 + 0 = 0.12$$

$$SS_{LF} = 0.376 - 0.12 = 0.256$$

$$F_{LF} = \frac{0.256/(11 - 6 - 8 + 4)}{0.12/(8 - 4)} = 8.53$$

and DLS = 0.043. Thus, there is significant lack of fit at $\alpha = 0.05$, and the second-order model does not describe the observed values adequately. One might next try the special cubic polynomial model (15.196), but since it has as many parameters as there are design points, one could not then test the statistical adequacy of the model.

TABLE 15.86 Pseudocomponent Settings, Original Component Settings, and the (Coded) Forces Needed with These Mixtures (Example 15.42)

Pseudocomponents			Original components			
x_1'	x_2'	x_3'	x_1	x_2	x_3	Force
1	0	0	0.6	0.2	0.3	6.7, 6.9
0	1	0	0.3	0.5	0.2	6.6, 6.8
0	0	1	0.3	0.2	0.5	7.4, 7.8
$\frac{1}{2}$	$\frac{1}{2}$	0	0.45	0.35	0.2	6.9
$\frac{1}{2}$	0	$\frac{1}{2}$	0.45	0.2	0.35	7.0
0	$\frac{1}{2}$	$\frac{1}{2}$	0.3	0.35	0.35	7.2
$\frac{1}{3}$	$\frac{1}{3}$	$\frac{1}{3}$	0.40	0.3	0.3	6.5, 6.5

TABLE 15.87 Fitted Values, and for Mixtures with Replicates the Average Responses and Sample Variances of Responses, in Example 15.42

Design point	Y	$\hat{Y}$	u	$Y_{u\bullet}$	S_u^2
1	6.7, 6.9	6.827	1	6.8	0.02
2	6.6, 6.8	6.727	2	6.7	0.02
3	7.4, 7.8	7.627	3	7.6	0.08
4	6.9	6.687			
5	7.0	6.787			
6	7.2	6.987			
7	6.5, 6.5	6.740	4	6.5	0

In addition to checking the statistical significance of the model and checking for any lack of fit, one should also check the practical value of the model. Therefore, it is a good idea to include some additional points in the design that are not used to fit the model. The responses at these points are compared with the predicted responses to see whether they are close enough for the fitted model to be of practical use.

15.10.5 Inclusion of Process Variables

In addition to mixture components, some experiments with mixtures involve other variables, which are referred to as process variables. For example, when baking a cake, one must consider not only the mixture of ingredients, but also the baking time and temperature. Process variables can also be qualitative, such as the kind of flour used in a cake. One can combine the mixture variables and the process variables into a single model [see Cornell (1981)], but this is advantageous only if one is interested in testing hypotheses on the model parameters. For the common situation in which finding a particular response (usually a maximum or a minimum) is of primary interest, a single model is not necessary. The different mixture experiments can be analyzed separately for each treatment combination of the process variables. For example, suppose that in the cake experiment two baking times (factor A) and two baking temperatures (factor B) are to be used in a 2^2 factorial design. A mixture experiment would be run for each of the four treatment combinations (1), a, b, ab; and the four experiments would initially be analyzed separately. If the height of a cake (higher is better) were the response of interest, then one would compare the estimates of the four local maxima (one from each mixture experiment) and choose from among these four the set of conditions corresponding to the highest value.

The only difference (and a very important one) between a mixture experiment with process variables and separate mixture experiments is the actual conduct of the experiment. If in the cake example there were three important mixture components, and a simplex-centroid design (seven points) was to be run at each of the four process variable treatment combinations, then the 28 trials should be run in random order. Randomization only within each mixture experiment would not be sufficient since that would amount to running the experiment in four blocks, and any block effects would be confounded with the effects of the process variables.

15.11 RANDOM EFFECTS

As was mentioned in Sec. 15.1, a factor is said to be random or to have a random effect if the levels of that factor constitute a random sample from the population of interest rather than the entire population itself. With a random factor the interest is in estimating the variability of the population rather than the effects of specific levels of the factor. For example, consider the process used to make $3\frac{1}{2}$-inch computer disks. Molten aluminum is poured into molds and cooled, and each mold supplies the substrate for a lot of 7500 disks. The sizes of the intermetallic particles in the substrate are important, and the sizes vary both within and among the lots. The variability between (among) lots is the random effect of Lots. Note that each lot is not of interest for its specific effect, but as a random sample from all lots produced.

We consider only one-way and two-way layouts for factors that have random effects, and only balanced designs. Information on higher-order layouts can be found, for example, in Neter, Wasserman, and Kutner (1985).

15.11.1 One-Way Layouts

The model for a one-way layout with a random factor is

$$X_{ij} = \mu + a_i + \epsilon_{ij} \tag{15.227}$$

which looks similar to model (15.9) for the fixed effects case (α_i has been replaced by a_i), but has different assumptions associated with it. Specifically, one assumes that

$$a_i \sim N(0, \sigma_A^2) \tag{15.228}$$

$$\epsilon_{ij} \sim N(0, \sigma_\epsilon^2) \tag{15.229}$$

and $$a_i \text{ and } \epsilon_{ij} \text{ are all independent.} \tag{15.230}$$

This implies that

$$E(X_{ij}) = \mu \tag{15.231}$$

$$\text{Var}(X_{ij}) = \sigma_A^2 + \sigma_\epsilon^2 \tag{15.232}$$

and amounts to assuming that all the observations have the same expected value and variance and that their common variance can be partitioned into two components. One component (σ_A^2) is the variability associated with the random factor, and the other component (σ_ϵ^2) is the variability due to replication. Because of Eq. (15.232), model (15.227) with assumptions (15.228) to (15.230) is sometimes referred to as a *variance components model.*

For a random factor the hypothesis corresponding to hypothesis (15.17) is

$$H_A\colon\ \sigma_A^2 = 0 \tag{15.233}$$

and can be tested using a one-way ANOVA. An ANOVA using a model with only random effects is sometimes referred to as a model II ANOVA. The sums of squares for an ANOVA table are computed in exactly the same way, regardless of whether the factors are fixed or random. However, except for SS_e, the sums of squares have different expected values depending on the assumptions [for example, (15.10) or (15.228) to (15.230)]. For the one-way layout even the different expected values for SS_A do not affect the analysis, and the equal sample size formulas for the one-way layout with fixed effects can be used to test hypothesis (15.233). These formulas are collected in Table 15.88 for convenience.

Actually, rather than testing hypothesis H_A, one is generally more interested in constructing a confidence interval for σ_A^2. Unfortunately, an exact procedure for this does not exist. Bulmer (1957) suggested the following approximate confidence limits. Since the procedure is also applicable with two-way and higher-order layouts, more general notation than is necessary for just the one-way layout will be used. Let $\nu_1 MS_1 \sim (c\phi + \theta)\,\chi^2(\nu_1)$ and $\nu_2 MS_2 \sim \theta\chi^2(\nu_2)$ with MS_1 and MS_2 independent. (For the one-way layout $MS_1 = MS_A$, $\nu_1 = I - 1$, $\phi = \sigma_A^2$, $c = n$, $\theta = \sigma_\epsilon^2$,

TABLE 15.88 ANOVA Table for a Balanced One-Way Layout with Random Effects

Source of variation	Sum of squares	Degrees of freedom	Mean square	Expected mean square	F	$\Pr > F$
Factor A	$SS_A = n\sum_{i=1}^{I}(X_{i\bullet} - X_{\bullet\bullet})^2$	$I - 1$	$MS_A = \frac{SS_A}{I-1}$	$\sigma_\epsilon^2 + n\sigma_A^2$	$F_A = \frac{MS_A}{MS_e}$	$\Pr[F(I-1, N-I) > F_A]$
Error	$SS_e = \sum_{i=1}^{I}\sum_{j=1}^{n}(X_{ij} - X_{i\bullet})^2$	$N - I$	$MS_e = \frac{SS_e}{N-I}$	σ_ϵ^2		
	$= (n-1)\sum_{i=1}^{i} S_i^2$		$= \frac{1}{n}\sum_{i=1}^{I} S_i^2$			
	$= SS_{tot} - SS_A$					
Total	$SS_{tot} = \sum_{i=1}^{I}\sum_{j=1}^{n}(X_{ij} - X_{\bullet\bullet})^2$	$N - 1$				

$MS_2 = MS_e$, and $\nu_2 = N - I$.) Approximate confidence limits for the parameter ϕ are obtained as follows. Let

$$L(\alpha_\ell; \nu_1, \nu_2, F_*) = \frac{F_*}{\ell_1} - 1 - \frac{\ell_2}{F_*}\left(\frac{\ell_2}{\ell_1} - 1\right) \tag{15.234}$$

$$U(\alpha_u; \nu_1, \nu_2, F_*) = u_1 F_* - 1 + \frac{1}{u_2 F_*}\left(1 - \frac{u_1}{u_2}\right) \tag{15.235}$$

where

$$\ell_1 = F(\alpha_\ell; \nu_1, x) \tag{15.236}$$

$$\ell_2 = F(\alpha_\ell; \nu_1, \nu_2) \tag{15.237}$$

$$u_1 = F(\alpha_u; x, \nu_1) \tag{15.238}$$

$$u_2 = F(\alpha_u; \nu_2, \nu_1) \tag{15.239}$$

$$F_* = \frac{MS_1}{MS_2}. \tag{15.240}$$

Then

$$\mathcal{L} = MS_2 \frac{L(\alpha_\ell; \nu_1, \nu_2, F_*)}{c} \tag{15.241}$$

is an approximate $(1 - \alpha_\ell)$ 100% lower confidence limit for ϕ,

$$\mathcal{U} = MS_2 \frac{U(\alpha_u; \nu_1, \nu_2, F_*)}{c} \tag{15.242}$$

is an approximate $(1 - \alpha_u)$100% upper confidence limit for ϕ, and $(\mathcal{L}, \mathcal{U})$ is an approximate $(1 - \alpha_\ell - \alpha_u)$100% confidence interval for ϕ. These confidence limits can be seriously invalidated if the normality assumptions [such as (15.228) and (15.229)] are violated. Otherwise, the approximation is fairly good if $\nu_2 \geq 6$, and it improves as ν_1/ν_2 decreases.

Thus, approximate confidence limits for σ_A^2 in the one-way layout are [Eqs. (15.241) and (15.242)]

$$\mathcal{L} = MS_e \frac{L(\alpha_\ell; I - 1, N - I, F_A)}{n} \tag{15.243}$$

$$\mathcal{U} = MS_e \frac{U(\alpha_u; I - 1, N - I, F_A)}{n}. \tag{15.244}$$

Example 15.43. In order to study the previously described process for making $3\frac{1}{2}$-inch disks, six lots were chosen at random, and from each lot three disks were chosen at random. For each disk the size of the largest intermetallic particle within a specified area was determined visually with the aid of a microscope, and the

TABLE 15.89 Sizes (Coded) of the Largest Intermetallic Particle on $3\frac{1}{2}$-inch Disks from Six Lots, together with the Sample Mean and Variance for Each Lot (Example 15.43)

Lot					
1	2	3	4	5	6
21	42	65	45	77	47
35	27	84	28	85	63
67	39	91	59	69	73
$X_{1\bullet} = 41$	$X_{2\bullet} = 36$	$X_{3\bullet} = 80$	$X_{4\bullet} = 44$	$X_{5\bullet} = 77$	$X_{6\bullet} = 61$
$S_1^2 = 556$	$S_2^2 = 63$	$S_3^2 = 181$	$S_4^2 = 241$	$S_5^2 = 64$	$S_6^2 = 172$

TABLE 15.90 ANOVA Table for Example 15.43

Source	SS	df	MS	F	Pr > F
A	5428.5	5	1085.7	5.10	0.0097
Error	2554	12	212.83		
Total	7982.5	17			

(coded) results are given in Table 15.89. Using the formulas in Table 15.88, one obtains the results given in Table 15.90. Using Eqs. (15.234) to (15.240), (15.243), and (15.244) to construct an approximate 95% confidence interval for σ_A^2, one obtains

$$\ell_1 = F(0.025; 5, x) = 2.57$$

$$\ell_2 = F(0.025; 5, 12) = 3.89$$

$$u_1 = F(0.025; x, 5) = 6.02$$

$$u_2 = F(0.025; 12, 5) = 6.52$$

$$L(0.025; 5, 12, 5.10) = \frac{5.10}{2.57} - 1 - \frac{3.89}{5.10}\left(\frac{3.89}{2.57} - 1\right) = 0.5927$$

$$U(0.025; 5, 12, 5.10) = 6.02(5.10) - 1 + \frac{1}{6.52(5.10)}\left(1 - \frac{6.02}{6.52}\right) = 29.70$$

$$\mathcal{L} = \frac{212.83(0.5927)}{3} = 42.0$$

$$\mathcal{U} = \frac{212.83(29.70)}{3} = 2107$$

and (42.0, 2107) is the required interval. Note that if only a 95% upper confidence limit was desired, one would obtain

$$u_1 = F(0.05; x, 5) = 4.36$$

$$u_2 = F(0.05; 12, 5) = 4.68$$

$$U(0.05; 5, 12, 5.10) = 4.36(5.10) - 1 + \frac{1}{4.68(5.10)}\left(1 - \frac{4.36}{4.68}\right) = 21.24$$

$$\mathcal{U} = \frac{212.83(21.24)}{3} = 15.07.$$

15.11.2 Two-Way Layouts

The model for a two-way layout where both factors are random is

$$X_{ijk} = u + a_i^A + a_j^B + a_{ij}^{AB} + \epsilon_{ijk}. \tag{15.245}$$

Again, this is similar to the fixed effects model with α replaced by a, but one assumes

$$\left.\begin{aligned} a_i^A &\sim N(0, \sigma_A^2) \\ a_j^B &\sim N(0, \sigma_B^2) \\ a_{ij}^{AB} &\sim N(0, \sigma_{AB}^2) \\ \epsilon_{ijk} &\sim N(0, \sigma_\epsilon^2) \end{aligned}\right\} \tag{15.246}$$

and that all these random variables are independent. In practice this independence is obtained by choosing the levels of both factors at random and randomizing with respect to any replication. The somewhat counterintuitive independence of the a_{ij}^{AB} is then a consequence of the assumption of normality. The hypotheses of potential interest in this situation are

$$H_A: \quad \sigma_A^2 = 0 \tag{15.247}$$

$$H_B: \quad \sigma_B^2 = 0 \tag{15.248}$$

$$H_{AB}: \quad \sigma_{AB}^2 = 0 \tag{15.249}$$

and can be tested using sums of squares that are computed in the same way as for the fixed effects case (see Table 15.13). However, since the expected values of the mean squares are different for the random effects model, different F ratios are used than in the fixed effects case. The expected values of the mean squares and the F ratios (used to test H_A, H_B, and H_{AB}) for the two-way layout with both factors random are given in Table 15.91. In contrast to the fixed effects situation, it is not necessary to start by testing H_{AB}. Similarly, when there is only one observation per cell ($n = 1$), even though it is not possible to test H_{AB} since $\nu_e = 0$, it is not necessary to assume that there are no interactions in order to test for main effects.

Approximate confidence limits for σ_A^2, σ_B^2, and σ_{AB}^2 in the two-way layout can be obtained using Eqs. (15.241) and (15.242) and the expected mean squares and F ratios in Table 15.91. For σ_A^2 one obtains

$$\mathcal{L} = \text{MS}_{AB} \frac{L(\alpha_\ell; I-1, (I-1)(J-1), F_A)}{nJ} \tag{15.250}$$

TABLE 15.91 Expected Mean Squares and F Ratios for a Two-Way Layout with Both Factors Random

Source of variation	Expected mean square	F	Pr > F
A	$\sigma_\epsilon^2 + n\sigma_{AB}^2 + nJ\sigma_A^2$	$F_A = \dfrac{\text{MS}_A}{\text{MS}_{AB}}$	$\Pr[F(\nu_A, \nu_{AB}) > F_A]$
B	$\sigma_\epsilon^2 + n\sigma_{AB}^2 + nI\sigma_B^2$	$F_B = \dfrac{\text{MS}_B}{\text{MS}_{AB}}$	$\Pr[F(\nu_B, \nu_{AB}) > F_B]$
AB	$\sigma_\epsilon^2 + n\sigma_{AB}^2$	$F_{AB} = \dfrac{\text{MS}_{AB}}{\text{MS}_e}$	$\Pr[F(\nu_{AB}, \nu_e) > F_{AB}]$
Error	σ_ϵ^2		

$$\mathcal{U} = \text{MS}_{AB} \frac{U(\alpha_u; I-1, (I-1)(J-1), F_A)}{nJ} \tag{15.251}$$

and for σ_B^2 one need only replace F_A with F_B and interchange I and J everywhere in Eqs. (15.250) and (15.251). Approximate confidence limits for σ_{AB}^2 are

$$\mathcal{L} = \text{MS}_e \frac{L(\alpha_\ell; (I-1)(J-1), IJ(n-1), F_{AB})}{n} \tag{15.252}$$

$$\mathcal{U} = \text{MS}_e \frac{U(\alpha_u; (I-1)(J-1), IJ(n-1), F_{AB})}{n}. \tag{15.253}$$

Example 15.44. In order to study the effect of operators (factor A) and machines (factor B) on a factory's output, three machines and four operators were chosen at random. A crossed design was used so that each operator worked with each machine, and the values in Table 15.92 are outputs for an 8-hour shift. Since interest centered on the variability associated with Operator and with Machine, but not on any interactions, only one shift per Operator/Machine combination was observed. Using the formulas for sums of squares given in Table 15.13 and the F ratios indicated in Table 15.91, one obtains the ANOVA table shown in Table 15.93. An upper 95% confidence limit on the variability associated with Operator is obtained using Eqs. (15.235) and (15.251). Namely,

$$u_1 = F(0.05; \infty, 3) = 8.53$$

$$u_2 = F(0.05; 6, 3) = 8.94$$

$$U(0.05; 3, 6, 6.59) = 8.53(6.59) - 1 + \frac{1}{8.94(6.59)}\left(1 - \frac{8.53}{8.94}\right) = 55.2$$

$$\mathcal{U} = \frac{84.67(55.2)}{3} = 1558.$$

TABLE 15.92 Outputs for an 8-Hour Shift with Various Operator/Machine Combinations

		Machine		
		1	2	3
Operator	1	83	90	88
	2	91	84	95
	3	100	103	121
	4	118	123	104

TABLE 15.93 ANOVA Table for Example 15.44

Source	SS	df	MS	F	$\Pr > F$
A (Operator)	1674	3	558	6.59	0.025
B (Machine)	32	2	16	0.19	0.832
AB	508	6	84.67		
Total	2214	11			

Mixed Models. When factor A in a two-way layout is random and factor B is fixed, then the model is

$$X_{ijk} = \mu + a_i^A + \alpha_j^B + a_{ij}^{AB} + \epsilon_{ijk} \tag{15.254}$$

where one assumes

$$\left.\begin{aligned} a_i^A &\sim N(0, \sigma_A^2) \\ a_{ij}^{AB} &\sim N(0, \sigma_{AB}^2) \\ \epsilon_{ijk} &\sim N(0, \sigma_\epsilon^2) \end{aligned}\right\} \tag{15.255}$$

and that all the random variables are independent. The α_j^B are unknown constants, exactly as they were in the fixed effects model (15.68). A model containing both fixed and random factors [such as Eq. (15.254)] is called a *mixed model.* The hypotheses of potential interest with model (15.254) are

$$H_A\colon\ \sigma_A^2 = 0 \tag{15.256}$$

$$H_B\colon\ \alpha_j^B = 0 \quad \text{for all } j \tag{15.257}$$

$$H_{AB}\colon\ \sigma_{AB}^2 = 0. \tag{15.258}$$

The sums of squares for the mixed model (15.254) are the same as for the two-way fixed effects and random effects models (Table 15.13), and the expected mean squares and F ratios are given in Table 15.94. With the mixed model (15.254) approximate confidence limits for σ_A^2 are

$$\mathcal{L} = \mathrm{MS}_e \frac{L(\alpha_\ell; I-1, IJ(n-1), F_A)}{nJ} \tag{15.259}$$

$$\mathcal{U} = \text{MS}_e \frac{U(\alpha_u; I-1, IJ(n-1), F_A)}{nJ} \tag{15.260}$$

and approximate confidence limits for σ^2_{AB} are given by Eqs. (15.252) and (15.253).

Example 15.45. If in Example 15.44 the three machines studied were not chosen at random, but were the only machines of interest, then Machines would be a fixed factor and the mixed model (15.254) would be the appropriate one. From Table 15.94 one finds that to test H_A (or do any inference about the variability associated with Operators) in this situation, more than one observation per cell would be needed in order to compute MS_e and hence F_A.

Nested Designs. In Sec. 15.4 a factor B was defined to be nested within a factor A if in the set of observations each level of B appears with only a single level of A. If neither factor is nested within the other, then the factors were defined to be crossed. With only two factors one refers to the design as being either *nested* or *crossed*, depending on whether the factors are nested or crossed. With more than two factors, designs involving both nested factors and crossed factors are possible [see, for example, Neter, Wasserman, and Kutner (1985)]. Generally, it is most realistic to consider nested factors as having a random effect. Also, it should be noted that replication is, in fact, a nested (random) factor, even in a fixed effects model.

Since in a nested design level j of factor B within level i of factor A is unrelated to level j of factor B within level i' of factor A. In order to avoid confusion one might denote the jth level of B within the ith level of A by $j(i)$. This notation is used in Table 15.95, which shows three levels of factor B nested within four levels of factor A as an incomplete layout. Cells containing observations are indicated with an $\times$. A nested design is said to be balanced if each level of factor A has the same number of levels (J) of factor B nested within it. Only balanced nested designs are considered here. Data from a balanced nested design are usually not displayed as in Table 15.95, but are displayed as an $I \times J$ table. Particular attention must, there-

TABLE 15.94 Expected Mean Squares and F Ratios for a Two-Way Layout with Factor A Random and Factor B Fixed

Source of variation	Expected mean square	F	Pr > F
A (random)	$\sigma^2_\epsilon + nJ\sigma^2_A$	$F_A = \dfrac{\text{MS}_A}{\text{MS}_e}$	$\Pr[F(\nu_A, \nu_e) > F_A]$
B (fixed)	$\sigma^2_\epsilon + n\sigma^2_{AB} + nI\phi_B$	$F_B = \dfrac{\text{MS}_B}{\text{MS}_{AB}}$	$\Pr[F(\nu_{B,}\ \nu_{AB}) > F_B]$
AB	$\sigma^2_\epsilon = n\sigma^2_{AB}$	$F_{AB} = \dfrac{\text{MS}_{AB}}{\text{MS}_e}$	$\Pr[F(\nu_{AB}, \nu_e) > F_{AB}]$
Error	σ^2_ϵ		

Note: $\phi_B = \dfrac{1}{J-1}\sum_{j=1}^{J} (\alpha_j^B)^2.$

TABLE 15.95 A Nested Design Shown as an Incomplete Layout

		Factor B											
		1(1)	2(1)	3(1)	1(2)	2(2)	3(2)	1(3)	2(3)	3(3)	1(4)	2(4)	3(4)
Factor A	1	×	×	×									
	2				×	×	×						
	3							×	×	×			
	4										×	×	×

fore, be paid as to how an experiment is conducted in order to ensure the correct analysis.

Since in a nested design each level of factor B appears with only a single level of factor A, it makes no sense to consider an AB interaction. Thus, the model for a nested design is

$$X_{ijk} = \mu + a_i^A + a_{j(i)}^B + \epsilon_{k(ij)} \tag{15.261}$$

where one assumes

$$\left.\begin{aligned} a_i^A &\sim N(0, \sigma_A^2) \\ a_{j(i)}^B &\sim N(0, \sigma_B^2) \\ \epsilon_{k(ij)} &\sim N(0, \sigma_\epsilon^2) \end{aligned}\right\} \tag{15.262}$$

and that all these random variables are independent. The notation $\epsilon_{k(ij)}$ has been used to indicate the nested feature of Replication. The ANOVA table for a balanced nested design is given in Table 15.96 and can be used to test H_A and H_B.

With a balanced nested design approximate confidence limits for σ_A^2 are

$$\mathcal{L} = \text{MS}_B \frac{L(\alpha_\ell; I-1, I(J-1), F_A)}{nJ} \tag{15.263}$$

$$\mathcal{U} = \text{MS}_B \frac{U(\alpha_u; I-1, I(J-1), F_A)}{nJ} \tag{15.264}$$

and for σ_B^2 they are

$$\mathcal{L} = \text{MS}_e \frac{L(\alpha_\ell; I(J-1), IJ(\text{n}-1), F_B)}{n} \tag{15.265}$$

$$\mathcal{U} = \text{MS}_e \frac{U(\alpha_u; I(J-1), IJ(n-1), F_B)}{n} \tag{15.266}$$

where the F ratios are obtained from Table 15.96.

TABLE 15.96 ANOVA Table for a Balanced Nested Design (B Nested within A)

Source of variation	Sum of squares	Degrees of freedom	Mean square	Expected mean square	F	$\Pr > F$
A	$SS_A = nJ \sum_{i=1}^{I} (X_{i\bullet\bullet} - X_{\bullet\bullet\bullet})^2$	$\nu_A = I - 1$	$MS_A = \frac{SS_A}{\nu_A}$	$\sigma_\epsilon^2 + n\sigma_B^2 + nJ\sigma_A^2$	$F_A = \frac{MS_A}{MS_B}$	$\Pr[F(\nu_A, \nu_B) > F_A]$
B	$SS_B = n \sum_{i=1}^{I} \sum_{j=1}^{J} (X_{ij\bullet} - X_{i\bullet\bullet})^2$	$\nu_B = I(J - 1)$	$MS_B = \frac{SS_B}{\nu_B}$	$\sigma_\epsilon^2 + n\sigma_B^2$	$F_B = \frac{MS_B}{MS_e}$	$\Pr[F(\nu_B, \nu_e) > F_A]$
Error	$SS_e = \sum_{i=1}^{I} \sum_{j=1}^{J} \sum_{k=1}^{n} (X_{ijk} - X_{ij\bullet})^2$	$\nu_e = IJ(n - 1)$	$MS_e = \frac{SS_e}{\nu_e} = \frac{1}{IJ} \sum_{i=1}^{I} \sum_{j=1}^{J} S_{ij}^2$	σ_ϵ^2		
Total	$SS_{tot} = \sum_{i=1}^{I} \sum_{j=1}^{J} \sum_{k=1}^{n} (X_{ijk} - X_{\bullet\bullet\bullet})^2$	$IJn - 1$				

Example 15.46. An experiment was conducted to study the variability in the strength of glass bulbs that are produced by a particular kind of machine with three heads. Four machines were chosen at random, and three randomly chosen bulbs from each head were tested for strength. The resulting coded strengths are given in Table 15.97. Since each machine had its own heads, and the heads were not moved from machine to machine, the experiment was a nested design. Using the formulas in Table 15.96 together with the sample means and variances in Tables 15.98, one obtains

$$SS_A = 3(3)[(8-9)^2 + \cdots + (10-9)^2] = 18$$

$$SS_B = 2[(2-8)^2 + \cdots + (15-10)^2] = 534$$

$$MS_e = \frac{4 + 9 + \cdots + 37}{12} = 13.5$$

TABLE 15.97 Strengths (Coded) of Glass Bulbs Produced by Machines with Three Heads (Example 15.46)

		Head		
		1	2	3
Machine	1	0 2 4	11 14 17	8 5 11
	2	12 12 3	8 16 12	10 5 3
	3	0 6 6	9 12 15	10 8 15
	4	8 5 2	8 10 12	11 22 12

TABLE 15.98*a* Cell Means, Column Means, Row Means, and the Grand Mean for the Data in Table 15.97 (Example 15.46)

		Head			
		1	2	3	
Machine	1	2	14	8	8
	2	9	12	6	9
	3	4	12	11	9
	4	5	10	15	10
		5	12	10	9

TABLE 15.98b Cell Variances for the Data in Table 15.97 (Example 15.46)

		Head		
		1	2	3
Machine	1	4	9	9
	2	27	16	13
	3	12	9	13
	4	9	4	37

TABLE 15.99 ANOVA Table for Example 15.46

Source	SS	df	MS	F	Pr > F
A	18	3	6	0.090	0.964
B (within A)	534	8	66.75	4.94	0.001
Error	324	24	13.5		
Total	876	35			

and the ANOVA table given in Table 15.99. Computing an upper 95% confidence limit on the variability associated with heads [Eq. (15.266)], one obtains

$$u_1 = F(0.05; \infty, 8) = 2.93$$

$$u_2 = F(0.05; 24, 8) = 3.12$$

$$U(0.05; 8, 24, 66.75) = 2.93(66.75) - 1 + \frac{1}{3.12(66.76)}\left(1 - \frac{2.93}{3.12}\right) = 194.6$$

$$\mathcal{U} = \frac{13.5(194.6)}{3} = 876.$$

ACKNOWLEDGMENTS

I would like to thank my father, Dr. Lloyd S. Nelson, for stimulating discussions and consultation on every aspect of this chapter and Dr. John A. Cornell for helpful comments on the section on mixture designs.

REFERENCES

Barnett, V.: "Probability Plotting Methods and Order Statistics," *Appl. Stat.*, vol. 24, pp. 95–108, 1975.

Blum, J., and J. I. Rosenblatt: *Probability and Statistics*, Saunders, Philadelphia, PA., 1972.

Box, G. E. P., and N. R. Draper: *Empirical Model-Building and Response Surfaces*, Wiley, New York, 1987

_____, W. G. Hunter, and J. S. Hunter: *Statistics for Experimenters*, Wiley, New York, 1978

Bulmer, M. G.: "Approximate Confidence Limits for Components of Variance," *Biometrika*, vol. 44, pp. 159–167, 1957.

Connor, W. S., and M. Zelen: *Fractional Factorial Designs for Factors at Three Levels*, Appl. Math. ser. 54, National Bureau of Standards, 1959.

Cornell, J. A.: *Experiments with Mixtures: Designs, Models, and the Analysis of Mixture Data*, Wiley, New York, 1981.

_____: *How to Run Mixture Experiments for Product Quality*, Am. Soc. Quality Control, Milwaukee, WI., 1983.

_____: *How to Apply Response Surface Methodology*, Am. Soc. Quality Control, Milwaukee, WI., 1984.

Craig, R. J.: "Normal Family Distribution Functions: FORTRAN and BASIC Programs," *J. Qual. Technol.*, vol. 16, pp. 232–236, 1984.

Daniel, C.: "Use of Half-Normal Plots in Interpreting Factorial Two Level Experiments," *Technometrics*, vol. 1, pp. 311–342, 1959.

_____ and F. S. Wood: *Fitting Equations to Data*, Wiley, New York, 1980.

Davies, O. L., Ed.: *The Design and Analysis of Industrial Experiments*, Longman, New York, 1978.

Dunn, O. J.: "Estimation of the Means of Dependent Variables," *Ann. Math. Stat.*, vol. 29, pp. 1095–1111, 1958.

Dunnett, C. W.: "Pairwise Multiple Comparisons in the Homogeneous Variance, Unequal Sample Size Case," *J. Am. Stat. Assoc.*, vol. 75, pp. 789–795, 1980.

Fisher, R. A.: *The Design of Experiments*, Hafner, New York, 1935.

_____ and F. Yates: *Statistical Tables for Biological, Agricultural, and Medical Research*, Oliver and Boyd, Edinburgh, U.K., 1963.

Harter, H. L.: "Expected Values of Normal Order Statistics," *Biometrika*, vol. 48, pp. 151–157, 1961.

Hayter, A. J.: "A Proof of the Conjecture that the Tukey-Kramer Multiple Comparison Procedure is Conservative," *Ann. Stat.*, vol. 12, pp. 61–75, 1984.

Hotelling, H.: "Some Problems in Weighing and Other Experimental Techniques," *Ann. Math. Stat.*, vol. 15, pp. 297–306, 1944.

Hunter, J. S.: "Statistical Design Applied to Product Design," *J. Qual. Technol.*, vol. 17, pp. 210–221, 1985.

Kenworthy, O. O.: "Factorial Experiments with Mixtures Using Ratios," *Ind. Qual. Control*, vol. 19, no. 12, pp. 24–26, 1963.

Khuri, A. I., and J. A. Cornell: *Response Surfaces*. Marcel Dekker, New York, 1987.

Kramer, C. Y.: "Extensions of Multiple Range Tests to Group Means with Unequal Numbers of Replications," *Biometrics*, vol. 12, pp. 307–310, 1956.

Little, R. J. A., and D. B. Rubin: *Statistical Analysis with Missing Data*, Wiley, New York, 1987.

Lukes, R. M., and L. S. Nelson: "The Concurrent Hydrogenation and Hydrolysis of Furfuryl Alcohol," *J. Organ. Chem.*, vol. 21, pp. 1096–1098, 1956.

McLean, R. A., and V. L. Anderson: "Extreme Vertices Design of Mixture Experiments," *Technometrics*, vol. 8, pp. 447–454, 1966.

Montgomery, D. C.: *Design and Analysis of Experiments*, Wiley, New York, 1984.

National Bureau of Standards: *Fractional Factorial Experiment Designs for Factors at Two Levels*, Appl. Math. ser. 48, National Bureau of Standards, 1957.

Nelson, L. S.: "A Nomograph of Student's *t*," *J. Qual. Technol.*, vol. 7, pp. 200–201, 1975.

_____: "Analysis of Two-Level Factorial Experiments," *J. Qual. Technol.*, vol. 14, pp. 95–98, 1982.

_____: "Transformations for Attribute Data," *J. Qual. Technol.*, vol. 15, pp. 55–56, 1983*a*.

_____: "Expected Normal Scores—A Useful Transformation," *J. Qual. Technol.*, vol. 15, pp. 144–146, 1983*b*.

_____: "Sample Size Tables for Analysis of Variance," *J. Qual. Technol.*, vol. 17. pp. 168–169, 1985.

_____: "Upper 10%, 5%, and 1% Points of the Maximum *F*-Ratio," *J. Qual. Technol.*, vol. 19, pp. 165–167, 1987.

Nelson, P. R.: "A Comparison of Sample Sizes for the Analysis of Means and the Analysis of Variance," *J. Qual. Technol.*, vol. 15, pp. 36–38, 1983.

_____: "Multiple Comparisons of Means Using Simultaneous Confidence Intervals," *J. Qual. Technol.*, vol. 21, pp. 232–241, 1989.

Neter, J., W. Wasserman, and M. H. Kutner: *Applied Linear Statistical Models*, Irwin, Homewood, Ill., 1985.

Pearson, E. S., and H. O. Hartley, Eds.: *Biometrika Tables for Statisticians*, vol. 1., Biometrika Trust, 1976.

Piepel, G. F.: "Programs for Generating Extreme Vertices and Centroids of Linearly Constrained Experimental Regions," *J. Qual. Technol*, vol. 20, pp. 125–139, 1988.

Plackett, R. L., and J. P. Burman: "The Design of Optimal Multifactor Experiments," *Biometrika*, vol. 33, pp. 305–325, 1946.

Saxena, S. K., and A. K. Nigam: "Restricted Exploration of Mixtures by Symmetric-Simplex Designs," *Technometrics*, vol. 19, pp. 47–52, 1977.

Scheffé, H.: "A Method of Judging All Contrasts in the Analysis of Variance," *Biometrika*, vol. 40, pp. 87–104, 1953.

_____: "Experiments with Mixtures," *J. R. Stat. Soc. B*, vol. 20, pp. 344–360, 1958.

_____: "The Simplex-Centroid Design for Experiments with Mixtures," *J. R. Stat. Soc. B*, vol. 25, pp. 235–263, 1963.

Shapiro, S. S., and R. S. Francia: "An Approximate Analysis of Variance Test for Normality," *J. Am. Stat. Assoc.*, vol. 67, pp. 215–216, 1972.

_____ and M. B. Wilk: "An Analysis of Variance Test for Normality (Complete Samples)," *Biometrika*, vol. 52, pp. 591–611, 1965.

Sidák, Z.: "Rectangular Confidence Regions for the Means of Multivariate Normal Distributions," *J. Am. Stat. Assoc.*, vol. 62, pp. 626–633, 1967.

Snee, R. D., and D. W. Marquardt: "Extreme Vertices Designs for Linear Mixture Models," *Technometrics*, vol. 16, pp. 399–408, 1974.

Tukey, J. W.: "The Problem of Multiple Comparisons," unpublished, 1953.

Wheeler, D. J.: *Tables of Screening Designs*, Statistical Process Controls, Knoxville, TN, 1987.

Wright, T., and H. Tsao: "Some Useful Notes on Simple Random Sampling," *J. Qual. Technol.*, vol. 17, pp. 67–73, 1985.

APPENDIX 15.1

ANOM Critical Values
$h(0.1; k, \text{df})$

	Number of contrasts k																		
df	2	3	4	5	6	7	8	9	10	11	12	13	14	15	16	17	18	19	20
1	6.31	9.52	11.0	12.0	12.8	13.4	14.0	14.4	14.8	15.2	15.5	15.7	16.0	16.2	16.5	16.7	16.8	17.0	17.2
2	2.92	4.05	4.56	4.92	5.19	5.41	5.60	5.76	5.91	6.03	6.14	6.25	6.34	6.43	6.51	6.58	6.65	6.72	6.78
3	2.35	3.16	3.50	3.75	3.94	4.09	4.23	4.34	4.44	4.53	4.61	4.68	4.74	4.81	4.86	4.91	4.96	5.01	5.05
4	2.13	2.81	3.09	3.29	3.45	3.58	3.69	3.78	3.86	3.93	4.00	4.06	4.11	4.17	4.21	4.26	4.30	4.34	4.37
5	2.02	2.63	2.88	3.05	3.19	3.30	3.40	3.48	3.55	3.62	3.68	3.73	3.78	3.83	3.87	3.91	3.94	3.98	4.01
6	1.94	2.52	2.74	2.91	3.03	3.13	3.22	3.30	3.36	3.42	3.48	3.53	3.57	3.61	3.65	3.69	3.72	3.75	3.78
7	1.89	2.44	2.65	2.81	2.92	3.02	3.10	3.17	3.24	3.29	3.34	3.39	3.43	3.47	3.51	3.54	3.57	3.60	3.63
8	1.86	2.39	2.59	2.73	2.85	2.94	3.02	3.08	3.14	3.19	3.24	3.29	3.33	3.36	3.40	3.43	3.46	3.49	3.52
9	1.83	2.34	2.54	2.68	2.79	2.87	2.95	3.01	3.07	3.12	3.17	3.21	3.25	3.28	3.32	3.35	3.38	3.40	3.43
10	1.81	2.31	2.50	2.64	2.74	2.83	2.90	2.96	3.02	3.06	3.11	3.15	3.19	3.22	3.25	3.28	3.31	3.34	3.36
11	1.80	2.29	2.47	2.60	2.70	2.79	2.86	2.92	2.97	3.02	3.06	3.10	3.14	3.17	3.20	3.23	3.26	3.28	3.31
12	1.78	2.27	2.45	2.57	2.67	2.75	2.82	2.88	2.93	2.98	3.02	3.06	3.10	3.13	3.16	3.19	3.21	3.24	3.26
13	1.77	2.25	2.43	2.55	2.65	2.73	2.79	2.85	2.90	2.95	2.99	3.03	3.06	3.09	3.12	3.15	3.18	3.20	3.23
14	1.76	2.23	2.41	2.53	2.63	2.70	2.77	2.83	2.88	2.92	2.96	3.00	3.03	3.06	3.09	3.12	3.15	3.17	3.19
15	1.75	2.22	2.39	2.51	2.61	2.68	2.75	2.80	2.85	2.90	2.94	2.97	3.01	3.04	3.07	3.09	3.12	3.14	3.17
16	1.75	2.21	2.38	2.50	2.59	2.67	2.73	2.79	2.83	2.88	2.92	2.95	2.99	3.02	3.04	3.07	3.10	3.12	3.14
17	1.74	2.20	2.37	2.49	2.58	2.65	2.71	2.77	2.82	2.86	2.90	2.93	2.97	3.00	3.02	3.05	3.08	3.10	3.12
18	1.73	2.19	2.36	2.47	2.56	2.64	2.70	2.75	2.80	2.84	2.88	2.92	2.95	2.98	3.01	3.03	3.06	3.08	3.10
19	1.73	2.18	2.35	2.46	2.55	2.63	2.69	2.74	2.79	2.83	2.87	2.90	2.93	2.96	2.99	3.02	3.04	3.06	3.08
20	1.72	2.18	2.34	2.45	2.54	2.62	2.68	2.73	2.78	2.82	2.86	2.89	2.92	2.95	2.98	3.00	3.03	3.05	3.07
24	1.71	2.15	2.31	2.43	2.51	2.58	2.64	2.69	2.74	2.78	2.81	2.85	2.88	2.91	2.93	2.96	2.98	3.00	3.02
30	1.70	2.13	2.29	2.40	2.48	2.55	2.61	2.66	2.70	2.74	2.77	2.81	2.84	2.86	2.89	2.91	2.93	2.96	2.97
40	1.68	2.11	2.26	2.37	2.45	2.52	2.57	2.62	2.66	2.70	2.73	2.77	2.79	2.82	2.84	2.87	2.89	2.91	2.93
60	1.67	2.09	2.24	2.34	2.42	2.49	2.54	2.59	2.63	2.66	2.70	2.73	2.75	2.78	2.80	2.82	2.84	2.86	2.88
120	1.66	2.07	2.22	2.32	2.39	2.45	2.51	2.55	2.59	2.62	2.66	2.69	2.71	2.74	2.76	2.78	2.80	2.82	2.84
∞	1.65	2.05	2.19	2.29	2.36	2.42	2.47	2.52	2.55	2.59	2.62	2.65	2.67	2.69	2.72	2.74	2.75	2.77	2.79

APPENDIX 15.1 (Continued)

ANOM Critical Values
$h(0.05; k, \text{df})$

	Number of contrasts k																		
df	2	3	4	5	6	7	8	9	10	11	12	13	14	15	16	17	18	19	20
1	12.7	19.1	22.0	24.1	25.7	26.9	28.0	28.9	29.7	30.4	31.0	31.6	32.1	32.5	33.0	33.4	33.8	34.1	34.4
2	4.30	5.89	6.60	7.10	7.49	7.81	8.07	8.30	8.50	8.68	8.84	8.99	9.12	9.24	9.36	9.46	9.56	9.65	9.74
3	3.18	4.18	4.60	4.91	5.15	5.34	5.50	5.65	5.77	5.88	5.98	6.08	6.16	6.24	6.31	6.38	6.44	6.50	6.55
4	2.78	3.56	3.89	4.12	4.30	4.45	4.58	4.69	4.79	4.87	4.95	5.02	5.09	5.15	5.21	5.26	5.31	5.35	5.40
5	2.57	3.25	3.53	3.72	3.88	4.00	4.11	4.21	4.29	4.36	4.43	4.49	4.55	4.60	4.65	4.69	4.73	4.77	4.81
6	2.45	3.07	3.31	3.49	3.62	3.73	3.83	3.91	3.99	4.05	4.11	4.17	4.22	4.26	4.31	4.35	4.39	4.42	4.45
7	2.36	2.95	3.17	3.33	3.45	3.56	3.64	3.72	3.79	3.85	3.90	3.95	4.00	4.04	4.08	4.12	4.15	4.19	4.22
8	2.31	2.86	3.07	3.21	3.33	3.43	3.51	3.58	3.64	3.70	3.75	3.80	3.84	3.88	3.92	3.95	3.99	4.02	4.05
9	2.26	2.79	2.99	3.13	3.24	3.33	3.41	3.48	3.54	3.59	3.64	3.68	3.72	3.76	3.80	3.83	3.86	3.89	3.92
10	2.23	2.74	2.93	3.07	3.17	3.26	3.33	3.40	3.45	3.51	3.55	3.59	3.63	3.67	3.70	3.73	3.76	3.79	3.82
11	2.20	2.70	2.88	3.01	3.12	3.20	3.27	3.33	3.39	3.44	3.48	3.52	3.56	3.60	3.63	3.66	3.69	3.71	3.74
12	2.18	2.67	2.85	2.97	3.07	3.15	3.22	3.28	3.33	3.38	3.42	3.46	3.50	3.53	3.57	3.59	3.62	3.65	3.67
13	2.16	2.64	2.81	2.94	3.03	3.11	3.18	3.24	3.29	3.33	3.38	3.42	3.45	3.48	3.51	3.54	3.57	3.59	3.62
14	2.14	2.62	2.79	2.91	3.00	3.08	3.14	3.20	3.25	3.30	3.34	3.37	3.41	3.44	3.47	3.50	3.52	3.55	3.57
15	2.13	2.60	2.76	2.88	2.97	3.05	3.11	3.17	3.22	3.26	3.30	3.34	3.37	3.40	3.43	3.46	3.49	3.51	3.53
16	2.12	2.58	2.74	2.86	2.95	3.02	3.09	3.14	3.19	3.23	3.27	3.31	3.34	3.37	3.40	3.43	3.45	3.48	3.50
17	2.11	2.57	2.73	2.84	2.93	3.00	3.06	3.12	3.16	3.21	3.25	3.28	3.31	3.34	3.37	3.40	3.42	3.45	3.47
18	2.10	2.55	2.71	2.82	2.91	2.98	3.04	3.10	3.14	3.18	3.22	3.26	3.29	3.32	3.35	3.37	3.40	3.42	3.44
19	2.09	2.54	2.70	2.81	2.89	2.96	3.02	3.08	3.12	3.16	3.20	3.24	3.27	3.30	3.32	3.35	3.37	3.40	3.42
20	2.09	2.53	2.68	2.79	2.88	2.95	3.01	3.06	3.11	3.15	3.18	3.22	3.25	3.28	3.30	3.33	3.35	3.38	3.40
24	2.06	2.50	2.65	2.75	2.83	2.90	2.96	3.01	3.05	3.09	3.13	3.16	3.19	3.22	3.24	3.27	3.29	3.31	3.33
30	2.04	2.47	2.61	2.71	2.79	2.85	2.91	2.96	3.00	3.04	3.07	3.10	3.13	3.16	3.18	3.20	3.23	3.25	3.26
40	2.02	2.43	2.57	2.67	2.75	2.81	2.86	2.91	2.95	2.98	3.01	3.04	3.07	3.10	3.12	3.14	3.16	3.18	3.20
60	2.00	2.40	2.54	2.63	2.70	2.76	2.81	2.86	2.90	2.93	2.96	2.99	3.01	3.04	3.06	3.08	3.10	3.12	3.14
120	1.98	2.37	2.50	2.59	2.66	2.72	2.77	2.81	2.85	2.88	2.91	2.93	2.96	2.98	3.00	3.02	3.04	3.06	3.08
∞	1.96	2.34	2.47	2.56	2.62	2.68	2.72	2.76	2.80	2.83	2.86	2.88	2.90	2.93	2.95	2.97	2.98	3.00	3.01

APPENDIX 15.1 (Continued)

ANOM Critical Values
$h(0.01; k, \text{df})$

	Number of contrasts k																		
df	2	3	4	5	6	7	8	9	10	11	12	13	14	15	16	17	18	19	20
1	63.7	95.7	110.	121.	129.	135.	140.	145.	149.	152.	155.	158.	160.	163.	165.	167.	169.	171.	172.
2	9.92	13.4	15.0	16.1	17.0	17.7	18.3	18.8	19.3	19.7	20.0	20.4	20.7	20.9	21.2	21.4	21.6	21.9	22.1
3	5.84	7.51	8.22	8.73	9.13	9.46	9.74	9.98	10.2	10.4	10.6	10.7	10.9	11.0	11.1	11.2	11.3	11.4	11.5
4	4.60	5.74	6.20	6.54	6.81	7.03	7.22	7.38	7.52	7.65	7.77	7.88	7.97	8.07	8.15	8.23	8.30	8.37	8.44
5	4.03	4.93	5.29	5.55	5.75	5.92	6.07	6.19	6.30	6.40	6.50	6.58	6.66	6.73	6.79	6.86	6.91	6.97	7.02
6	3.71	4.48	4.77	4.98	5.16	5.30	5.42	5.52	5.62	5.70	5.78	5.85	5.91	5.97	6.03	6.08	6.13	6.17	6.22
7	3.50	4.19	4.44	4.63	4.78	4.90	5.01	5.10	5.18	5.25	5.32	5.38	5.44	5.49	5.54	5.59	5.63	5.67	5.71
8	3.36	3.98	4.21	4.38	4.52	4.63	4.72	4.80	4.88	4.94	5.01	5.06	5.11	5.16	5.20	5.24	5.28	5.32	5.36
9	3.25	3.84	4.05	4.20	4.33	4.43	4.51	4.59	4.66	4.72	4.78	4.83	4.87	4.92	4.96	5.00	5.03	5.07	5.10
10	3.17	3.73	3.92	4.07	4.18	4.28	4.36	4.43	4.49	4.55	4.60	4.65	4.69	4.73	4.77	4.81	4.84	4.87	4.90
11	3.11	3.64	3.83	3.96	4.07	4.16	4.23	4.30	4.36	4.41	4.46	4.51	4.55	4.59	4.62	4.66	4.69	4.72	4.75
12	3.05	3.57	3.75	3.87	3.98	4.06	4.13	4.20	4.25	4.31	4.35	4.39	4.43	4.47	4.50	4.54	4.57	4.59	4.62
13	3.01	3.51	3.68	3.80	3.90	3.98	4.05	4.11	4.17	4.22	4.26	4.30	4.34	4.37	4.41	4.44	4.47	4.49	4.52
14	2.98	3.46	3.63	3.74	3.84	3.92	3.98	4.04	4.09	4.14	4.18	4.22	4.26	4.29	4.32	4.35	4.38	4.41	4.43
15	2.95	3.42	3.58	3.69	3.78	3.86	3.93	3.98	4.03	4.08	4.12	4.16	4.19	4.22	4.26	4.28	4.31	4.34	4.36
16	2.92	3.38	3.54	3.65	3.74	3.81	3.88	3.93	3.98	4.02	4.06	4.10	4.13	4.17	4.20	4.22	4.25	4.27	4.30
17	2.90	3.35	3.50	3.61	3.70	3.77	3.83	3.89	3.93	3.98	4.02	4.05	4.08	4.12	4.14	4.17	4.20	4.22	4.24
18	2.88	3.33	3.47	3.58	3.66	3.73	3.79	3.85	3.89	3.94	3.97	4.01	4.04	4.07	4.10	4.12	4.15	4.17	4.20
19	2.86	3.30	3.45	3.55	3.63	3.70	3.76	3.81	3.86	3.90	3.94	3.97	4.00	4.03	4.06	4.08	4.11	4.13	4.15
20	2.85	3.28	3.42	3.53	3.61	3.67	3.73	3.78	3.83	3.87	3.90	3.94	3.97	4.00	4.02	4.05	4.07	4.09	4.12
24	2.80	3.21	3.35	3.44	3.52	3.58	3.64	3.69	3.73	3.77	3.80	3.83	3.86	3.89	3.91	3.94	3.96	3.98	4.00
30	2.75	3.15	3.28	3.37	3.44	3.50	3.55	3.59	3.63	3.67	3.70	3.73	3.76	3.78	3.81	3.83	3.85	3.87	3.89
40	2.70	3.09	3.21	3.29	3.36	3.42	3.46	3.51	3.54	3.58	3.61	3.63	3.66	3.68	3.70	3.72	3.74	3.76	3.78
60	2.66	3.03	3.14	3.22	3.28	3.34	3.38	3.42	3.45	3.49	3.51	3.54	3.56	3.58	3.61	3.62	3.64	3.66	3.68
120	2.62	2.97	3.08	3.15	3.21	3.26	3.30	3.34	3.37	3.40	3.42	3.45	3.47	3.49	3.51	3.53	3.54	3.56	3.58
∞	2.58	2.91	3.01	3.08	3.14	3.19	3.22	3.26	3.29	3.32	3.34	3.36	3.38	3.40	3.42	3.44	3.45	3.47	3.48

APPENDIX 15.1 (Continued)

ANOM Critical Values
$h(0.001; k, \text{df})$

	Number of contrasts k																		
df	2	3	4	5	6	7	8	9	10	11	12	13	14	15	16	17	18	19	20
1	637	957	1103	1207	1285	1349	1401	1446	1485	1520	1551	1579	1605	1628	1650	1670	1689	1707	1724.
2	31.6	42.7	47.7	51.2	54.0	56.2	58.1	59.7	61.1	62.4	63.5	64.6	65.5	66.4	67.2	67.9	68.6	69.3	69.9
3	12.9	16.5	18.0	19.1	20.0	20.7	21.3	21.8	22.3	22.7	23.0	23.4	23.7	24.0	24.2	24.5	24.7	24.9	25.1
4	8.61	10.6	11.4	12.0	12.5	12.8	13.2	13.5	13.7	13.9	14.2	14.3	14.5	14.7	14.8	15.0	15.1	15.2	15.3
5	6.87	8.25	8.79	9.19	9.51	9.77	10.0	10.2	10.4	10.5	10.7	10.8	10.9	11.0	11.1	11.2	11.3	11.4	11.5
6	5.96	7.04	7.45	7.75	7.99	8.20	8.37	8.52	8.66	8.78	8.89	8.99	9.08	9.17	9.25	9.33	9.40	9.47	9.53
7	5.41	6.31	6.65	6.89	7.09	7.25	7.40	7.52	7.63	7.73	7.82	7.90	7.98	8.05	8.12	8.18	8.24	8.30	8.35
8	5.04	5.83	6.12	6.33	6.49	6.63	6.75	6.86	6.96	7.04	7.12	7.19	7.26	7.32	7.38	7.43	7.48	7.53	7.58
9	4.78	5.49	5.74	5.93	6.07	6.20	6.30	6.40	6.48	6.56	6.63	6.69	6.75	6.80	6.85	6.90	6.95	6.99	7.03
10	4.59	5.24	5.46	5.63	5.76	5.87	5.97	6.05	6.13	6.20	6.26	6.32	6.37	6.42	6.47	6.51	6.55	6.59	6.62
11	4.44	5.05	5.25	5.40	5.52	5.63	5.71	5.79	5.86	5.92	5.98	6.03	6.08	6.13	6.17	6.21	6.25	6.28	6.31
12	4.32	4.89	5.08	5.22	5.34	5.43	5.51	5.58	5.65	5.71	5.76	5.81	5.85	5.89	5.93	5.97	6.01	6.04	6.07
13	4.22	4.77	4.94	5.08	5.18	5.27	5.35	5.42	5.48	5.53	5.58	5.63	5.67	5.71	5.74	5.78	5.81	5.84	5.87
14	4.14	4.66	4.83	4.96	5.06	5.14	5.21	5.28	5.33	5.38	5.43	5.48	5.52	5.55	5.59	5.62	5.65	5.68	5.71
15	4.07	4.57	4.74	4.86	4.95	5.03	5.10	5.16	5.21	5.26	5.31	5.35	5.39	5.42	5.46	5.49	5.52	5.54	5.57
16	4.01	4.50	4.66	4.77	4.86	4.94	5.00	5.06	5.11	5.16	5.20	5.24	5.28	5.31	5.34	5.37	5.40	5.43	5.45
17	3.97	4.44	4.59	4.70	4.78	4.86	4.92	4.98	5.03	5.07	5.11	5.15	5.19	5.22	5.25	5.28	5.30	5.33	5.35
18	3.92	4.38	4.53	4.63	4.72	4.79	4.85	4.90	4.95	4.99	5.03	5.07	5.10	5.14	5.17	5.19	5.22	5.24	5.27
19	3.88	4.33	4.47	4.57	4.66	4.73	4.79	4.84	4.88	4.93	4.96	5.00	5.03	5.06	5.09	5.12	5.15	5.17	5.19
20	3.85	4.29	4.42	4.52	4.60	4.67	4.73	4.78	4.83	4.87	4.90	4.94	4.97	5.00	5.03	5.05	5.08	5.10	5.12
24	3.75	4.16	4.28	4.37	4.44	4.51	4.56	4.61	4.65	4.68	4.72	4.75	4.78	4.81	4.83	4.85	4.88	4.90	4.92
30	3.65	4.03	4.14	4.23	4.29	4.35	4.40	4.44	4.48	4.51	4.54	4.57	4.60	4.62	4.65	4.67	4.69	4.71	4.72
40	3.55	3.91	4.01	4.09	4.15	4.20	4.25	4.28	4.32	4.35	4.38	4.40	4.43	4.45	4.47	4.49	4.51	4.52	4.54
60	3.46	3.79	3.89	3.96	4.01	4.06	4.10	4.14	4.17	4.20	4.22	4.24	4.27	4.29	4.30	4.32	4.34	4.35	4.37
120	3.37	3.68	3.77	3.84	3.89	3.93	3.96	4.00	4.02	4.05	4.07	4.09	4.11	4.13	4.15	4.17	4.18	4.19	4.21
∞	3.29	3.58	3.66	3.72	3.76	3.80	3.84	3.86	3.89	3.91	3.93	3.95	3.97	3.99	4.00	4.02	4.03	4.04	4.06

APPENDIX 15.2

Studentized Maximum Modulus Upper Quantiles
$m(0.1; k, \text{df})$

	Number of contrasts k																		
df	2	3	4	5	6	7	8	9	10	11	12	13	14	15	16	17	18	19	20
1	8.96	10.5	11.6	12.5	13.2	13.7	14.2	14.6	15.0	15.3	15.6	15.8	16.1	16.3	16.5	16.7	16.9	17.1	17.2
2	3.83	4.38	4.77	5.06	5.30	5.50	5.67	5.82	5.96	6.08	6.18	6.28	6.37	6.45	6.53	6.60	6.67	6.74	6.80
3	2.99	3.37	3.64	3.84	4.01	4.15	4.27	4.38	4.47	4.55	4.63	4.70	4.76	4.82	4.88	4.93	4.98	5.02	5.07
4	2.66	2.98	3.20	3.37	3.51	3.62	3.72	3.81	3.89	3.96	4.02	4.08	4.13	4.18	4.23	4.27	4.31	4.35	4.38
5	2.49	2.77	2.96	3.12	3.24	3.34	3.43	3.51	3.58	3.64	3.69	3.75	3.79	3.84	3.88	3.92	3.95	3.99	4.02
6	2.38	2.64	2.82	2.96	3.07	3.17	3.25	3.32	3.38	3.44	3.49	3.54	3.58	3.62	3.66	3.70	3.73	3.76	3.79
7	2.31	2.56	2.73	2.86	2.96	3.05	3.13	3.19	3.25	3.31	3.35	3.40	3.44	3.48	3.51	3.55	3.58	3.61	3.63
8	2.26	2.49	2.66	2.78	2.88	2.96	3.04	3.10	3.16	3.21	3.26	3.30	3.34	3.37	3.41	3.44	3.47	3.50	3.52
9	2.22	2.45	2.60	2.72	2.82	2.90	2.97	3.03	3.09	3.13	3.18	3.22	3.26	3.29	3.32	3.35	3.38	3.41	3.44
10	2.19	2.41	2.56	2.68	2.77	2.85	2.92	2.98	3.03	3.08	3.12	3.16	3.20	3.23	3.26	3.29	3.32	3.34	3.37
11	2.17	2.38	2.53	2.64	2.73	2.81	2.88	2.93	2.98	3.03	3.07	3.11	3.15	3.18	3.21	3.24	3.26	3.29	3.31
12	2.15	2.36	2.50	2.61	2.70	2.78	2.84	2.90	2.95	2.99	3.03	3.07	3.10	3.14	3.17	3.19	3.22	3.24	3.27
13	2.13	2.34	2.48	2.59	2.67	2.75	2.81	2.87	2.91	2.96	3.00	3.04	3.07	3.10	3.13	3.16	3.18	3.21	3.23
14	2.12	2.32	2.46	2.57	2.65	2.72	2.79	2.84	2.89	2.93	2.97	3.01	3.04	3.07	3.10	3.13	3.15	3.17	3.20
15	2.11	2.31	2.44	2.55	2.63	2.70	2.76	2.82	2.87	2.91	2.95	2.98	3.01	3.04	3.07	3.10	3.12	3.15	3.17
16	2.10	2.29	2.43	2.53	2.62	2.69	2.75	2.80	2.85	2.89	2.93	2.96	2.99	3.02	3.05	3.08	3.10	3.12	3.15
17	2.09	2.28	2.42	2.52	2.60	2.67	2.73	2.78	2.83	2.87	2.91	2.94	2.97	3.00	3.03	3.06	3.08	3.10	3.12
18	2.08	2.27	2.41	2.51	2.59	2.66	2.72	2.77	2.81	2.85	2.89	2.92	2.96	2.99	3.01	3.04	3.06	3.08	3.10
19	2.07	2.26	2.40	2.50	2.58	2.64	2.70	2.75	2.80	2.84	2.88	2.91	2.94	2.97	3.00	3.02	3.04	3.07	3.09
20	2.07	2.26	2.39	2.49	2.57	2.63	2.69	2.74	2.79	2.83	2.86	2.90	2.93	2.96	2.98	3.01	3.03	3.05	3.07
24	2.05	2.23	2.36	2.46	2.53	2.60	2.66	2.70	2.75	2.79	2.82	2.85	2.88	2.91	2.94	2.96	2.98	3.01	3.03
30	2.03	2.21	2.33	2.43	2.50	2.57	2.62	2.67	2.71	2.75	2.78	2.81	2.84	2.87	2.89	2.92	2.94	2.96	2.98
40	2.01	2.18	2.30	2.40	2.47	2.53	2.58	2.63	2.67	2.71	2.74	2.77	2.80	2.82	2.85	2.87	2.89	2.91	2.93
60	1.99	2.16	2.28	2.37	2.44	2.50	2.55	2.59	2.63	2.67	2.70	2.73	2.76	2.78	2.80	2.83	2.85	2.87	2.88
120	1.97	2.14	2.25	2.34	2.41	2.47	2.52	2.56	2.60	2.63	2.66	2.69	2.72	2.74	2.76	2.78	2.80	2.82	2.84
∞	1.95	2.11	2.23	2.31	2.38	2.43	2.48	2.52	2.56	2.59	2.62	2.65	2.67	2.70	2.72	2.74	2.76	2.77	2.79

Source: From P. R. Nelson (1989).

APPENDIX 15.2 (Continued)

Studentized Maximum Modulus Upper Quantiles
$m(0.05; k, \text{df})$

	Number of contrasts k																		
df	2	3	4	5	6	7	8	9	10	11	12	13	14	15	16	17	18	19	20
1	18.0	21.1	23.4	25.0	26.4	27.5	28.5	29.3	30.0	30.6	31.3	31.8	32.3	32.7	33.1	33.5	33.9	34.2	34.5
2	5.57	6.34	6.89	7.31	7.65	7.93	8.17	8.38	8.57	8.74	8.89	9.03	9.16	9.28	9.39	9.49	9.59	9.68	9.77
3	3.96	4.43	4.76	5.02	5.23	5.41	5.56	5.69	5.81	5.92	6.01	6.10	6.18	6.26	6.33	6.39	6.45	6.51	6.57
4	3.38	3.74	4.00	4.20	4.37	4.50	4.62	4.72	4.82	4.90	4.97	5.04	5.11	5.17	5.22	5.27	5.32	5.37	5.41
5	3.09	3.40	3.62	3.79	3.93	4.04	4.14	4.23	4.31	4.38	4.45	4.51	4.56	4.61	4.66	4.70	4.74	4.78	4.82
6	2.92	3.19	3.39	3.54	3.66	3.77	3.86	3.94	4.01	4.07	4.13	4.18	4.23	4.28	4.32	4.36	4.39	4.43	4.46
7	2.80	3.06	3.24	3.38	3.49	3.59	3.67	3.74	3.80	3.86	3.92	3.96	4.01	4.05	4.09	4.13	4.16	4.19	4.22
8	2.72	2.96	3.13	3.26	3.36	3.45	3.53	3.60	3.66	3.71	3.76	3.81	3.85	3.89	3.93	3.96	3.99	4.02	4.05
9	2.66	2.89	3.05	3.17	3.27	3.36	3.43	3.49	3.55	3.60	3.65	3.69	3.73	3.77	3.80	3.84	3.87	3.90	3.92
10	2.61	2.83	2.98	3.10	3.20	3.28	3.35	3.41	3.47	3.52	3.56	3.60	3.64	3.68	3.71	3.74	3.77	3.80	3.82
11	2.57	2.78	2.93	3.05	3.14	3.22	3.29	3.35	3.40	3.45	3.49	3.53	3.57	3.60	3.63	3.66	3.69	3.72	3.74
12	2.54	2.75	2.89	3.00	3.09	3.17	3.24	3.29	3.34	3.39	3.43	3.47	3.51	3.54	3.57	3.60	3.63	3.65	3.68
13	2.51	2.72	2.86	2.97	3.06	3.13	3.19	3.25	3.30	3.34	3.39	3.42	3.46	3.49	3.52	3.55	3.57	3.60	3.62
14	2.49	2.69	2.83	2.94	3.02	3.09	3.16	3.21	3.26	3.30	3.34	3.38	3.41	3.45	3.48	3.50	3.53	3.55	3.58
15	2.47	2.67	2.81	2.91	2.99	3.06	3.13	3.18	3.23	3.27	3.31	3.35	3.38	3.41	3.44	3.46	3.49	3.51	3.54
16	2.46	2.65	2.78	2.89	2.97	3.04	3.10	3.15	3.20	3.24	3.28	3.31	3.35	3.38	3.40	3.43	3.46	3.48	3.50
17	2.44	2.63	2.77	2.87	2.95	3.02	3.08	3.13	3.17	3.21	3.25	3.29	3.32	3.35	3.38	3.40	4.43	3.45	3.47
18	2.43	2.62	2.75	2.85	2.93	3.00	3.05	3.11	3.15	3.19	3.23	3.26	3.29	3.32	3.35	3.38	3.40	3.42	3.44
19	2.42	2.61	2.73	2.83	2.91	2.98	3.04	3.09	3.13	3.17	3.21	3.24	3.27	3.30	3.33	3.35	3.38	3.40	3.42
20	2.41	2.59	2.72	2.82	2.90	2.96	3.02	3.07	3.11	3.15	3.19	3.22	3.25	3.28	3.31	3.33	3.36	3.38	3.40
24	2.38	2.56	2.68	2.77	2.85	2.91	2.97	3.02	3.06	3.10	3.13	3.16	3.19	3.22	3.25	3.27	3.29	3.31	3.33
30	2.35	2.52	2.64	2.73	2.80	2.87	2.92	2.96	3.00	3.04	3.07	3.11	3.13	3.16	3.18	3.21	3.23	3.25	3.27
40	2.32	2.49	2.60	2.69	2.76	2.82	2.87	2.91	2.95	2.99	3.02	3.05	3.08	3.10	3.12	3.14	3.17	3.18	3.20
60	2.29	2.45	2.56	2.65	2.72	2.77	2.82	2.86	2.90	2.93	2.96	2.99	3.02	3.04	3.06	3.08	3.10	3.12	3.14
120	2.26	2.42	2.53	2.61	2.67	2.73	2.77	2.81	2.85	2.88	2.91	2.94	2.96	2.98	3.01	3.02	3.04	3.06	3.08
∞	2.24	2.39	2.49	2.57	2.63	2.68	2.73	2.77	2.80	2.83	2.86	2.88	2.91	2.93	2.95	2.97	2.98	3.00	3.02

Source: From P. R. Nelson (1989).

APPENDIX 15.2 (Continued)

Studentized Maximum Modulus Upper Quantiles
$m(0.01; k, \text{df})$

	Number of contrasts k																		
df	2	3	4	5	6	7	8	9	10	11	12	13	14	15	16	17	18	19	20
1	90.0	106.	117.	125.	132.	138.	142.	146.	150.	153.	156.	159.	161.	164.	166.	168.	170.	171.	173.
2	12.7	14.4	15.7	16.6	17.4	18.0	18.5	19.0	19.4	19.8	20.1	20.5	20.8	21.0	21.3	21.5	21.7	21.9	22.1
3	7.13	7.91	8.48	8.92	9.28	9.58	9.84	10.1	10.3	10.4	10.6	10.8	10.9	11.0	11.2	11.3	11.4	11.5	11.6
4	5.46	5.99	6.36	6.66	6.90	7.10	7.27	7.43	7.57	7.69	7.80	7.91	8.00	8.09	8.17	8.25	8.32	8.39	8.45
5	4.70	5.11	5.40	5.63	5.81	5.97	6.11	6.23	6.33	6.43	6.52	6.60	6.67	67.4	6.81	6.87	6.93	6.98	7.03
6	4.27	4.61	4.86	5.05	5.20	5.33	5.45	5.55	5.64	5.72	5.80	5.86	5.93	5.99	6.04	6.09	6.14	6.18	6.23
7	4.00	4.30	4.51	4.68	4.81	4.93	5.03	5.12	5.20	5.27	5.34	5.39	5.45	5.50	5.55	5.60	5.64	5.68	5.72
8	3.81	4.08	4.27	4.42	4.55	4.65	4.74	4.82	4.89	4.96	5.02	5.07	5.12	5.17	5.21	5.25	5.29	5.33	5.36
9	3.67	3.92	4.10	4.24	4.35	4.45	4.53	4.61	4.67	4.73	4.79	4.84	4.88	4.92	4.96	5.00	5.04	5.07	5.10
10	3.57	3.80	3.97	4.10	4.20	4.29	4.37	4.44	4.50	4.56	4.61	4.66	4.70	4.74	4.78	4.81	4.84	4.88	4.91
11	3.48	3.71	3.87	3.99	4.09	4.17	4.25	4.31	4.37	4.42	4.47	4.51	4.55	4.59	4.63	4.66	4.69	4.72	4.75
12	3.42	3.63	3.78	3.90	4.00	4.08	4.15	4.21	4.26	4.31	4.36	4.40	4.44	4.48	4.51	4.54	4.57	4.60	4.63
13	3.36	3.57	3.71	3.83	3.92	4.00	4.06	4.12	4.18	4.22	4.27	4.31	4.34	4.38	4.41	4.44	4.47	4.50	4.52
14	3.32	3.52	3.66	3.77	3.85	3.93	3.99	4.05	4.10	4.15	4.19	4.23	4.26	4.30	4.33	4.36	4.39	4.41	4.44
15	3.28	3.47	3.61	3.71	3.80	3.87	3.93	3.99	4.04	4.08	4.12	4.16	4.20	4.23	4.26	4.29	4.31	4.34	4.36
16	3.25	3.43	3.57	3.67	3.75	3.82	3.88	3.94	3.99	4.03	4.07	4.11	4.14	4.17	4.20	4.23	4.25	4.28	4.30
17	3.22	3.40	3.53	3.63	3.71	3.78	3.84	3.89	3.94	3.98	4.02	4.06	4.09	4.12	4.15	4.17	4.20	4.22	4.25
18	3.19	3.37	3.50	3.60	3.68	3.74	3.80	3.85	3.90	3.94	3.98	4.01	4.04	4.07	4.10	4.13	4.15	4.18	4.20
19	3.17	3.35	3.47	3.57	3.65	3.71	3.77	3.82	3.86	3.90	3.94	3.97	4.01	4.03	4.06	4.09	4.11	4.13	4.16
20	3.15	3.32	3.45	3.54	3.62	3.68	3.74	3.79	3.83	3.87	3.91	3.94	3.97	4.00	4.03	4.05	4.07	4.10	4.12
24	3.09	3.25	3.37	3.46	3.53	3.59	3.64	3.69	3.73	3.77	3.80	3.83	3.86	3.89	3.91	3.94	3.96	3.98	4.00
30	3.03	3.18	3.29	3.38	3.45	3.50	3.55	3.60	3.64	3.67	3.70	3.73	3.76	3.78	3.81	3.83	3.85	3.87	3.89
40	2.97	3.12	3.22	3.30	3.37	3.42	3.47	3.51	3.54	3.58	3.61	3.63	3.66	3.68	3.71	3.73	3.74	3.76	3.78
60	2.91	3.05	3.15	3.23	3.29	3.34	3.38	3.42	3.46	3.49	3.51	3.54	3.56	3.59	3.61	3.63	3.64	3.66	3.68
120	2.86	2.99	3.09	3.16	3.21	3.26	3.30	3.34	3.37	3.40	3.43	3.45	3.47	3.49	3.51	3.53	3.55	3.56	3.58
∞	2.81	2.93	3.02	3.09	3.14	3.19	3.23	3.26	3.29	3.32	3.34	3.36	3.38	3.40	3.42	3.44	3.45	3.47	3.48

Source: From P. R. Nelson (1989).

APPENDIX 15.2 (Continued)

Studentized Maximum Modulus Upper Quantiles
$m(0.001; k, \text{df})$

	Number of contrasts k																		
df	2	3	4	5	6	7	8	9	10	11	12	13	14	15	16	17	18	19	20
1	900.	1058.	1169.	1253.	1320.	1375.	1423.	1464.	1501.	1533.	1562.	1589.	1614.	1636.	1657.	1677.	1695.	1712.	1728.
2	40.4	45.8	49.7	52.6	55.0	57.1	58.8	60.3	61.6	62.8	63.9	64.9	65.8	66.6	67.4	68.2	68.8	69.5	70.1
3	15.7	17.3	18.6	19.5	20.3	20.9	21.5	22.0	22.4	22.8	23.1	23.5	23.8	24.1	24.3	24.6	24.8	25.0	25.2
4	10.1	11.0	11.7	12.2	12.6	13.0	13.3	13.5	13.8	14.0	14.2	14.4	14.6	14.7	14.9	15.0	15.1	15.3	15.4
5	7.88	8.50	8.95	9.31	9.60	9.85	10.1	10.2	10.4	10.6	10.7	10.8	11.0	11.1	11.2	11.3	11.4	11.4	11.5
6	6.74	7.21	7.56	7.83	8.06	8.25	8.41	8.56	8.69	8.80	8.91	9.01	9.10	9.19	9.27	9.34	9.41	9.48	9.54
7	6.05	6.44	6.73	6.95	7.13	7.29	7.43	7.55	7.65	7.75	7.84	7.92	8.00	8.07	8.13	8.20	8.25	8.31	8.36
8	5.60	5.93	6.18	6.37	6.53	6.66	6.78	6.88	6.97	7.06	7.13	7.20	7.27	7.33	7.39	7.44	7.49	7.54	7.58
9	5.28	5.58	5.79	5.96	6.10	6.22	6.32	6.41	6.50	6.57	6.64	6.70	6.76	6.81	6.86	6.91	6.95	7.00	7.04
10	5.04	5.31	5.51	5.66	5.78	5.89	5.99	6.07	6.14	6.21	6.27	6.33	6.38	6.43	6.47	6.52	6.56	6.59	6.63
11	4.86	5.11	5.29	5.43	5.54	5.64	5.73	5.80	5.87	5.93	5.99	6.04	6.09	6.13	6.17	6.21	6.25	6.29	6.32
12	4.71	4.94	5.11	5.24	5.35	5.44	5.52	5.59	5.66	5.71	5.77	5.81	5.86	5.90	5.94	5.98	6.01	6.04	6.07
13	4.59	4.81	4.97	5.10	5.20	5.28	5.36	5.42	5.48	5.54	5.59	5.63	5.67	5.71	5.75	5.78	5.82	5.85	5.87
14	4.50	4.71	4.86	4.97	5.07	5.15	5.22	5.28	5.34	5.39	5.44	5.48	5.52	5.56	5.59	5.62	5.65	5.68	5.71
15	4.41	4.61	4.76	4.87	4.96	5.04	5.11	5.17	5.22	5.27	5.31	5.35	5.39	5.43	5.46	5.49	5.52	5.55	5.57
16	4.34	4.54	4.68	4.78	4.87	4.95	5.01	5.07	5.12	5.16	5.21	5.25	5.28	5.32	5.35	5.38	5.41	5.43	5.46
17	4.28	4.47	4.60	4.71	4.79	4.86	4.93	4.98	5.03	5.08	5.12	5.15	5.19	5.22	5.25	5.28	5.31	5.33	5.36
18	4.23	4.41	4.54	4.64	4.72	4.79	4.85	4.91	4.95	5.00	5.04	5.07	5.11	5.14	5.17	5.20	5.22	5.25	5.27
19	4.19	4.36	4.49	4.58	4.66	4.73	4.79	4.84	4.89	4.93	4.97	5.00	5.04	5.07	5.09	5.12	5.15	5.17	5.19
20	4.14	4.32	4.44	4.53	4.61	4.68	4.73	4.78	4.83	4.87	4.91	4.94	4.97	5.00	5.03	5.06	5.08	5.10	5.13
24	4.02	4.18	4.29	4.38	4.45	4.51	4.56	4.61	4.65	4.69	4.72	4.75	4.78	4.81	4.83	4.86	4.88	4.90	4.92
30	3.90	4.05	4.15	4.23	4.30	4.35	4.40	4.44	4.48	4.51	4.54	4.57	4.60	4.62	4.65	4.67	4.69	4.71	4.73
40	3.79	3.92	4.02	4.09	4.15	4.20	4.25	4.29	4.32	4.35	4.38	4.40	4.43	4.45	4.47	4.49	4.51	4.53	4.54
60	3.68	3.81	3.89	3.96	4.02	4.06	4.10	4.14	4.17	4.20	4.22	4.24	4.27	4.29	4.31	4.32	4.34	4.36	4.37
120	3.58	3.69	3.78	3.84	3.89	3.93	3.97	4.00	4.03	4.05	4.07	4.10	4.11	4.13	4.15	4.17	4.18	4.19	4.21
∞	3.48	3.59	3.66	3.72	3.76	3.80	3.84	3.87	3.89	3.91	3.93	3.95	3.97	3.99	4.00	4.02	4.03	4.04	4.06

***Source*:** From P. R. Nelson (1989).

APPENDIX 15.3

The tabled values are the sample sizes required for each treatment in order to detect differences among the treatments with specified power 1 – β when two of the treatments differ by at least Δ/σ and the test is performed at level of significance α.

ANOVA Sample Sizes

Number of Treatments = 2

	α = 0.10				α = 0.05				α = 0.01			
	β				β				β			
Δ/σ	0.50	0.20	0.10	0.05	0.50	0.20	0.10	0.05	0.50	0.20	0.10	0.05
0.4	35	78			49				85			
0.5	23	51	70	88	32	64	86		55	96		
0.6	16	36	49	61	23	45	60	74	39	67	85	
0.7	12	26	36	45	17	34	44	55	29	50	63	75
0.8	10	21	28	35	14	26	34	42	23	39	49	58
0.9	8	16	22	28	11	21	27	34	19	31	39	46
1.0	7	14	18	23	9	17	23	27	15	26	32	38
1.2	5	10	13	16	7	12	16	20	11	18	23	27
1.4	4	8	10	12	6	10	12	15	9	14	17	20
1.6	4	6	8	10	5	8	10	12	7	11	14	16
1.8	3	5	7	8	4	6	8	10	6	10	11	13
2.0	3	4	6	7	4	6	7	8	6	8	10	11
2.5	2	3	4	5	3	4	5	6	4	6	7	8
3.0	2	3	3	4	3	4	4	5	4	5	6	6

Number of Treatments = 3

	α = 0.10				α = 0.05				α = 0.01			
	β				β				β			
Δ/σ	0.50	0.20	0.10	0.05	0.50	0.20	0.10	0.05	0.50	0.20	0.10	0.05
0.4	46	98			63							
0.5	30	63	85		41	79			68			
0.6	21	44	59	74	29	55	72	87	48	79	99	
0.7	16	33	44	54	22	41	53	65	35	59	73	86
0.8	12	25	34	42	17	32	41	50	28	45	57	67
0.9	10	20	27	33	14	25	33	40	22	36	45	53
1.0	8	17	22	27	11	21	27	32	18	30	37	43
1.2	6	12	16	19	8	15	19	23	13	21	26	31
1.4	5	9	12	15	7	11	14	17	10	16	20	23
1.6	4	7	10	12	5	9	11	14	9	13	16	18
1.8	4	6	8	9	5	8	9	11	7	11	13	15
2.0	3	5	7	8	4	6	8	9	6	9	11	12
2.5	3	4	5	6	3	5	6	7	5	7	8	9
3.0	2	3	4	4	3	4	5	5	4	5	6	7

Number of Treatments = 4

	α = 0.10				α = 0.05				α = 0.01			
	β				β				β			
Δ/σ	0.50	0.20	0.10	0.05	0.50	0.20	0.10	0.05	0.50	0.20	0.10	0.05
0.4	54				73							
0.5	35	72	96		48	89			76			
0.6	25	50	67	82	33	62	80	97	54	88		
0.7	18	37	49	61	25	46	59	72	40	65	80	94
0.8	14	29	38	47	20	36	46	55	31	50	62	73
0.9	12	23	30	37	16	28	36	44	25	40	49	58
1.0	10	19	25	30	13	23	30	36	21	33	40	47
1.2	7	14	18	22	10	17	21	25	15	23	29	33
1.4	6	10	13	16	7	13	16	19	11	18	22	25
1.6	5	8	11	13	6	10	13	15	9	14	17	20
1.8	4	7	9	10	5	8	10	12	8	12	14	16
2.0	4	6	7	9	4	7	9	10	7	10	12	13
2.5	3	4	5	6	3	5	6	7	5	7	8	9
3.0	2	3	4	5	3	4	5	5	4	5	6	7

Number of Treatments = 5

	α = 0.10				α = 0.05				α = 0.01			
	β				β				β			
Δ/σ	0.50	0.20	0.10	0.05	0.50	0.20	0.10	0.05	0.50	0.20	0.10	0.05
0.4	60				82							
0.5	39	79			53	97			84			
0.6	27	55	73	89	37	68	87		59	95		
0.7	20	41	54	66	28	50	64	77	44	70	86	
0.8	16	32	42	51	22	39	50	59	34	54	67	78
0.9	13	25	33	40	17	31	39	47	27	43	53	62
1.0	11	21	27	33	14	25	32	39	22	35	43	50
1.2	8	15	19	23	10	18	23	27	16	25	31	36
1.4	6	11	14	17	8	14	17	20	12	19	23	27
1.6	5	9	11	14	7	11	14	16	10	15	18	21
1.8	4	7	9	11	5	9	11	13	8	12	15	17
2.0	4	6	8	9	5	7	9	11	7	10	12	14
2.5	3	4	5	6	4	5	6	7	5	7	9	10
3.0	3	4	4	5	3	4	5	6	4	6	7	7

Source: From L. S. Nelson (1985). Reproduced with permission from the American Society for Quality Control.

APPENDIX 15.3 (Continued)

ANOVA Sample Sizes

Number of Treatments = 6

	α = 0.10				α = 0.05				α = 0.01			
	β				β				β			
Δ/σ	0.50	0.20	0.10	0.05	0.50	0.20	0.10	0.05	0.50	0.20	0.10	0.05
0.4	65				89							
0.5	42	85			57				90			
0.6	30	59	78	95	40	73	93		63			
0.7	22	44	58	70	30	54	69	82	47	75	92	
0.8	17	34	44	54	23	42	53	63	36	58	71	82
0.9	14	27	35	43	19	33	42	50	29	46	56	65
1.0	12	22	29	35	15	27	34	41	24	38	46	53
1.2	8	16	20	25	11	19	24	29	17	27	32	38
1.4	7	12	15	19	9	15	18	22	13	20	24	28
1.6	5	9	12	15	7	11	14	17	10	16	19	22
1.8	5	8	10	12	6	9	12	14	9	13	15	18
2.0	4	7	8	10	5	8	10	11	7	11	13	15
2.5	3	5	6	7	4	6	7	8	5	8	9	10
3.0	3	4	4	5	3	4	5	6	4	6	7	8

Number of Treatments = 7

	α = 0.10				α = 0.05				α = 0.01			
	β				β				β			
Δ/σ	0.50	0.20	0.10	0.05	0.50	0.20	0.10	0.05	0.50	0.20	0.10	0.05
0.4	70				95							
0.5	46	90			61				96			
0.6	32	63	83		43	77	98		67			
0.7	24	47	61	74	32	57	73	87	50	79	96	
0.8	19	36	47	57	25	44	56	67	38	61	74	86
0.9	15	29	37	45	20	35	44	53	31	48	59	68
1.0	12	24	31	37	16	29	36	43	25	39	48	56
1.2	9	17	22	26	12	20	26	30	18	28	34	39
1.4	7	13	16	20	9	15	19	23	14	21	25	29
1.6	6	10	13	15	7	12	15	18	11	16	20	23
1.8	5	8	10	12	6	10	12	14	9	13	16	18
2.0	4	7	9	10	5	8	10	12	8	11	13	15
2.5	3	5	6	7	4	6	7	8	6	8	9	10
3.0	3	4	5	5	3	5	5	6	4	6	7	8

Number of Treatments = 8

	α = 0.10				α = 0.05				α = 0.01			
	β				β				β			
Δ/σ	0.50	0.20	0.10	0.05	0.50	0.20	0.10	0.05	0.50	0.20	0.10	0.05
0.4	75											
0.5	48	95			65							
0.6	34	67	87		46	81			71			
0.7	25	49	64	78	34	60	76	91	52	82		
0.8	20	38	49	60	26	46	59	70	40	63	77	90
0.9	16	30	39	48	21	37	47	55	32	51	62	71
1.0	13	25	32	39	17	30	38	45	26	41	50	58
1.2	9	18	23	27	12	21	27	32	19	29	35	41
1.4	7	13	17	20	10	16	20	24	14	22	26	30
1.6	6	10	13	16	8	13	16	18	11	17	21	24
1.8	5	9	11	13	6	10	13	15	9	14	17	19
2.0	4	7	9	11	5	9	11	12	8	12	14	16
2.5	3	5	6	7	4	6	7	8	6	8	9	11
3.0	3	4	5	5	3	5	6	6	4	6	7	8

Number of Treatments = 9

	α = 0.10				α = 0.05				α = 0.01			
	β				β				β			
Δ/σ	0.50	0.20	0.10	0.05	0.50	0.20	0.10	0.05	0.50	0.20	0.10	0.05
0.4	79											
0.5	51				69							
0.6	36	70	91		48	85			74			
0.7	27	52	67	81	36	63	79	94	55	86		
0.8	21	40	52	62	28	48	61	72	42	66	80	93
0.9	17	32	41	50	22	38	48	58	34	53	64	74
1.0	14	26	33	40	18	31	40	47	28	43	52	60
1.2	10	18	24	28	13	22	28	33	20	30	37	42
1.4	8	14	18	21	10	17	21	25	15	23	27	32
1.6	6	11	14	17	8	13	16	19	12	18	21	25
1.8	5	9	11	13	7	11	13	15	10	14	17	20
2.0	4	7	9	11	6	9	11	13	8	12	14	16
2.5	3	5	6	8	4	6	8	9	6	8	10	11
3.0	3	4	5	6	3	5	6	6	5	6	7	8

Source: From L. S. Nelson (1985). Reproduced with permission from the American Society for Quality Control.

APPENDIX 15.4

The tabled values are the sample sizes required for each treatment in order to detect with specified power among k treatments when two of the treatments differ by at least Δ (measured in units of the standard deviation) and the test is performed at level of significance α.

ANOM Sample Sizes
$\alpha = 0.1$

	Δ = 3.00 Power								Δ = 250 Power								Δ = 2.00 Power							
k	.50	.60	.70	.75	.80	.90	.95	.99	.50	.60	.70	.75	.80	.90	.95	.99	.50	.60	.70	.75	.80	.90	.95	.99
3	2	3	3	3	3	4	4	6	3	3	3	4	4	5	6	7	3	4	4	5	5	7	8	11
4	4	3	3	3	3	4	5	6	3	3	4	4	4	5	6	8	4	4	5	5	6	7	9	12
5	2	3	3	3	4	4	5	6	3	3	4	4	5	6	7	9	4	4	5	6	6	8	9	13
6	3	3	3	3	4	4	5	7	3	3	4	4	5	6	7	9	4	5	5	6	7	8	10	13
7	3	3	3	3	4	5	5	7	3	4	4	4	5	6	7	9	4	5	6	6	7	9	10	14
8	3	3	3	4	4	5	5	7	3	4	4	5	5	6	7	9	4	5	6	6	7	9	11	14
9	3	3	3	4	4	5	5	7	3	4	4	5	5	6	7	10	4	5	6	7	7	9	11	14
10	3	3	3	4	4	5	6	7	3	4	4	5	5	6	7	10	4	5	6	7	7	9	11	15
11	3	3	3	4	4	5	6	7	3	4	4	5	5	7	8	10	5	5	6	7	8	10	11	15
12	3	3	3	4	4	5	6	7	3	4	5	5	5	7	8	10	5	6	7	7	8	10	12	15
13	3	3	4	4	4	5	6	7	3	4	5	5	5	7	8	10	5	6	7	7	8	10	12	15
14	3	3	3	4	4	5	6	7	3	4	5	5	6	7	8	10	5	6	7	7	8	10	12	16
15	3	3	4	4	4	5	6	8	3	4	5	5	6	7	8	10	5	6	7	7	8	10	12	16
16	3	3	4	4	4	5	6	8	4	4	5	5	6	7	8	11	5	6	7	8	8	10	12	16
17	3	3	4	4	4	5	6	8	4	4	5	5	6	7	8	11	5	6	7	8	8	11	12	16
18	3	3	4	4	4	5	6	8	4	4	5	5	6	7	8	11	5	6	7	8	9	11	13	16
19	3	3	4	4	4	5	6	8	4	4	5	5	6	7	8	11	5	6	7	8	9	11	13	17
20	3	3	4	4	4	5	6	8	4	4	5	5	6	7	8	11	5	6	7	8	9	11	13	17
24	3	3	4	4	4	5	6	8	4	4	5	6	6	7	9	11	5	6	8	8	9	11	13	17
30	3	3	4	4	5	6	6	8	4	5	5	6	6	8	9	12	6	7	8	9	9	12	14	18
40	3	4	4	4	5	6	7	9	4	5	6	6	7	8	9	12	6	7	8	9	10	12	14	18
60	3	4	4	5	5	6	7	9	5	5	6	7	7	9	10	13	7	8	9	10	11	13	15	20

	Δ = 1.75 Power								Δ = 1.50 Power								Δ = 1.25 Power								Δ = 1.00 Power							
k	.50	.60	.70	.75	.80	.90	.95	.99	.50	.60	.70	.75	.80	.90	.95	.99	.50	.60	.70	.75	.80	.90	.95	.99	.50	.60	.70	.75	.80	.90	.95	.99
3	4	5	5	6	6	8	10	14	5	6	7	7	8	11	13	18	6	7	9	10	11	15	18	26	9	11	13	15	17	23	28	39
4	4	5	6	7	7	9	11	15	5	6	8	8	9	12	15	20	7	8	10	12	13	17	21	28	10	12	16	17	19	26	31	44
5	4	5	6	7	8	10	12	16	6	7	8	9	10	13	16	21	7	9	11	13	14	18	22	30	11	14	17	19	21	28	34	47
6	5	6	7	7	8	10	13	17	6	7	9	10	11	14	17	23	8	10	12	13	15	19	23	32	12	15	18	20	23	30	36	49
7	5	6	7	8	9	11	13	17	6	8	9	10	11	14	17	23	9	11	13	14	16	20	24	33	13	16	19	22	24	31	38	51
8	5	6	7	8	9	11	13	18	7	8	10	11	12	15	18	24	9	11	13	15	16	21	25	34	13	17	20	23	25	32	39	53
9	5	6	8	8	9	12	14	19	7	8	10	11	12	15	19	25	9	11	14	15	17	22	26	35	14	17	21	24	26	34	40	54
10	5	7	8	9	9	12	14	19	7	9	10	11	13	16	19	25	10	12	14	16	18	22	27	36	15	18	22	24	27	34	41	56
11	6	7	8	9	10	12	15	19	7	9	11	12	13	16	19	26	10	12	15	16	18	23	28	37	15	19	23	25	28	35	42	57
12	6	7	8	9	10	12	15	20	7	9	11	12	13	17	20	26	10	13	15	17	19	24	28	37	16	19	23	26	28	36	43	58
13	6	7	8	9	10	13	15	20	8	9	11	12	13	17	20	27	11	13	16	17	19	24	29	38	16	20	24	26	29	37	44	59
14	6	7	9	9	10	13	15	20	8	9	11	12	14	17	20	27	11	13	16	17	19	24	29	39	16	20	24	27	29	38	45	60
15	6	7	9	9	10	13	16	20	8	10	12	13	14	28	21	27	11	13	16	18	20	25	30	40	17	20	25	27	30	38	46	61
16	6	7	9	10	11	13	16	21	8	10	12	13	14	18	21	27	11	14	16	18	20	25	30	40	17	21	25	27	31	39	46	62
17	6	8	9	10	11	13	16	21	8	10	12	13	14	18	21	27	12	14	17	18	20	25	30	40	17	21	25	28	31	39	47	62
18	6	8	9	10	11	14	16	21	8	10	12	13	15	18	22	28	12	14	17	19	20	26	31	41	18	21	26	29	32	40	47	63
19	6	8	9	10	11	14	16	21	8	10	12	13	15	18	22	29	12	14	17	19	21	26	31	41	18	22	26	29	32	40	48	64
20	7	8	9	10	11	14	16	21	9	10	12	14	15	19	22	29	12	15	17	19	21	26	31	41	18	22	27	29	32	41	48	64
24	7	8	10	11	12	14	17	22	9	11	13	14	15	19	23	30	13	15	18	20	22	27	32	43	19	24	28	31	34	43	50	66
30	7	9	10	11	12	15	18	23	10	11	13	15	16	20	24	31	13	16	19	21	23	29	34	44	21	25	30	33	36	45	53	69
40	8	9	11	12	13	16	18	24	10	12	14	16	17	21	25	32	15	17	21	22	25	30	36	46	22	27	32	35	38	47	55	72
60	8	10	12	13	14	17	20	25	11	13	16	17	18	23	27	34	16	19	22	24	27	33	38	49	25	29	35	38	41	51	59	77

Source: From P. R. Nelson (1983).

APPENDIX 15.4 (Continued)

ANOM Sample Sizes
$\alpha = 0.05$

	Δ = 3.00								Δ = 2.50								Δ = 2.00							
	Power								Power								Power							
k	.50	.60	.70	.75	.80	.90	.95	.99	.50	.60	.70	.75	.80	.90	.95	.99	.50	.60	.70	.75	.80	.90	.95	.99
3	3	3	4	4	4	5	5	7	3	4	4	5	5	6	7	9	4	5	6	6	7	8	10	13
4	3	3	4	4	4	5	6	7	3	4	4	5	5	5	7	9	5	5	6	7	7	9	10	14
5	3	3	4	4	4	5	6	7	4	4	5	5	5	7	8	10	5	6	6	7	8	9	11	14
6	3	3	4	4	4	5	6	7	4	4	5	5	6	7	8	10	5	6	7	7	8	10	12	14
7	3	3	4	4	4	5	6	8	4	4	5	5	6	7	8	10	5	6	7	8	8	10	12	16
8	3	3	4	4	4	5	6	8	4	4	5	5	6	7	8	11	5	6	7	8	8	10	12	16
9	3	4	4	4	5	5	6	8	4	4	5	6	6	7	8	11	5	6	7	8	9	11	13	16
10	3	4	4	4	5	6	6	8	4	5	5	6	6	7	9	11	6	6	8	8	9	11	13	17
11	3	4	4	4	5	6	6	8	4	5	5	6	6	8	9	11	6	7	8	8	9	11	13	17
12	3	4	4	4	5	6	6	8	4	5	5	6	6	8	9	11	6	7	8	8	9	11	13	17
13	3	4	4	4	5	6	7	8	4	5	5	6	6	8	9	11	6	7	8	9	9	12	13	17
14	3	4	4	4	5	6	7	8	4	5	6	6	6	8	9	12	6	7	8	9	10	12	14	18
15	3	4	4	5	5	6	7	8	4	5	6	6	7	8	9	12	6	7	8	9	10	12	14	18
16	3	4	4	5	5	6	7	9	4	5	6	6	7	8	9	12	6	7	8	9	10	12	14	18
17	3	4	4	5	5	6	7	9	4	5	6	6	7	8	9	12	6	7	8	9	10	12	14	18
18	3	4	4	5	5	6	7	9	4	5	6	6	7	8	9	12	6	7	9	9	10	12	14	18
19	3	4	4	5	5	6	7	9	4	5	6	6	7	8	9	12	6	7	9	9	10	12	14	18
20	3	4	4	5	5	6	7	9	4	5	6	6	7	8	10	12	6	7	9	9	10	12	14	19
24	3	4	4	5	5	6	7	9	5	5	6	7	7	9	10	13	7	8	9	10	11	13	15	19
30	4	4	5	5	5	6	7	9	5	5	6	7	7	9	10	13	7	8	9	10	11	13	15	20
40	4	4	5	5	5	7	8	9	5	6	7	7	8	9	11	13	7	9	10	11	11	14	16	20
60	4	4	5	5	6	7	8	10	5	6	7	7	8	10	11	14	8	9	11	11	12	15	17	22

	Δ = 1.75								Δ = 1.50								Δ = 1.25								Δ = 1.00							
	Power								Power								Power								Power							
k	.50	.60	.70	.75	.80	.90	.95	.99	.50	.60	.70	.75	.80	.90	.95	.99	.50	.60	.70	.75	.80	.90	.95	.99	.50	.60	.70	.75	.80	.90	.95	.99
3	5	6	7	7	8	10	12	16	6	7	9	9	10	13	16	21	8	10	12	13	14	18	22	30	12	14	17	19	21	27	33	45
4	5	6	7	8	9	11	13	17	7	8	10	10	12	15	17	23	9	11	13	14	16	20	24	33	13	16	20	22	24	31	37	50
5	6	7	8	9	9	12	14	18	7	9	10	11	12	16	18	25	10	12	14	16	17	22	26	35	15	18	21	24	26	33	40	54
6	6	7	8	9	10	12	15	19	8	9	11	12	13	16	19	26	11	13	15	17	18	23	27	37	16	19	23	25	28	35	42	56
7	6	7	9	10	10	13	15	20	8	10	11	12	14	17	20	27	11	13	16	17	19	24	29	38	17	20	24	27	29	37	44	58
8	7	8	9	10	11	13	16	20	8	10	12	13	14	18	21	27	12	14	17	18	20	25	30	39	18	21	25	28	30	38	45	60
9	7	8	9	10	11	14	16	21	9	10	12	13	15	18	21	28	12	14	17	19	21	26	30	40	18	22	26	29	31	39	47	62
10	7	8	10	10	11	14	16	21	9	11	13	14	15	19	22	29	12	15	18	19	21	26	31	41	19	23	27	29	32	40	48	63
11	7	8	10	11	12	14	17	22	9	11	13	14	15	19	22	29	13	15	18	20	22	27	32	42	20	23	28	30	33	41	49	64
12	7	8	10	11	12	15	17	22	9	11	13	14	16	19	23	30	13	16	19	20	22	27	32	42	20	24	28	31	34	42	50	66
13	7	9	10	11	12	15	17	22	10	11	13	15	16	20	23	30	14	16	19	21	23	28	33	43	21	24	29	31	34	43	51	67
14	8	9	10	11	12	15	18	23	10	12	14	15	16	20	23	31	14	16	19	21	23	28	33	43	21	25	29	32	35	44	52	67
15	8	9	10	11	12	15	18	23	10	12	14	15	16	20	24	31	14	17	20	21	23	29	34	44	21	25	30	33	36	45	52	68
16	8	9	11	11	13	15	18	23	10	12	14	15	17	21	24	31	14	17	20	22	24	29	34	45	22	26	31	33	36	45	53	69
17	8	9	11	12	13	16	18	23	10	12	14	16	17	21	24	31	15	17	20	22	24	30	35	45	22	26	31	34	37	46	54	70
18	8	9	11	12	13	16	18	24	11	12	14	16	17	21	25	32	15	17	20	22	24	30	35	45	22	27	32	34	37	46	54	71
19	8	9	11	12	13	16	18	24	11	13	15	16	17	21	25	32	15	18	21	22	24	30	35	46	22	27	32	35	38	47	55	71
20	8	10	11	12	13	16	19	24	11	13	15	16	17	21	25	32	15	18	21	23	25	31	36	46	23	28	32	35	38	47	55	72
24	9	10	12	12	14	16	19	25	11	13	15	17	18	22	26	33	16	18	22	24	26	32	37	48	24	29	34	37	40	49	57	74
30	9	10	12	13	14	17	20	25	12	14	16	17	19	23	27	34	17	20	23	25	27	33	38	49	26	30	35	38	42	51	59	77
40	9	11	13	14	15	18	21	26	12	15	17	18	20	24	28	36	18	21	24	26	28	35	40	51	28	32	38	41	44	54	62	80
60	10	12	14	15	16	19	22	28	14	16	18	20	21	26	30	38	20	23	26	28	30	37	43	54	30	35	41	44	47	57	66	84

Source: From P. R. Nelson (1983).

APPENDIX 15.4 (Continued)

ANOM Sample Sizes
$\alpha = 0.01$

	$\Delta = 3.00$								$\Delta = 2.50$								$\Delta = 2.00$							
	Power								Power								Power							
k	.50	.60	.70	.75	.80	.90	.95	.99	.50	.60	.70	.75	.80	.90	.95	.99	.50	.60	.70	.75	.80	.90	.95	.99
3	4	4	5	5	5	6	7	9	5	5	6	6	7	8	9	11	6	7	8	9	9	11	13	16
4	4	5	5	5	6	7	7	9	5	6	6	7	7	8	10	12	7	8	9	9	10	12	14	18
5	4	5	5	5	6	7	8	9	5	6	7	7	7	9	10	13	7	8	9	10	11	13	15	18
6	4	5	5	5	6	7	8	9	5	6	7	7	8	9	10	13	7	8	10	10	11	13	15	19
7	4	5	5	6	6	7	8	10	5	6	7	7	8	9	11	13	8	9	10	10	11	13	15	20
8	4	5	5	6	6	7	8	10	5	6	7	7	8	9	11	13	8	9	10	11	12	14	16	20
9	4	5	5	6	6	7	8	10	6	6	7	8	8	10	11	14	8	9	10	11	12	14	16	20
10	4	5	5	6	6	7	8	10	6	6	7	8	8	10	11	14	8	9	10	11	12	14	16	21
11	4	5	5	6	6	7	8	10	6	6	7	8	8	10	11	14	8	9	11	11	12	14	17	21
12	4	5	5	6	6	7	8	10	6	7	7	8	8	10	11	14	8	9	11	11	12	15	17	21
13	4	5	6	6	6	7	8	10	6	7	7	8	8	10	11	14	8	10	11	12	12	15	17	22
14	4	5	6	6	6	7	8	10	6	7	7	8	8	10	11	14	8	10	11	12	13	15	17	22
15	4	5	6	6	6	7	8	10	6	7	7	8	9	10	12	14	9	10	11	12	13	15	17	22
16	4	5	6	6	6	7	8	10	6	7	8	8	9	10	12	15	9	10	11	12	13	15	18	22
17	5	5	6	6	6	7	8	10	6	7	8	8	9	10	12	15	9	10	11	12	13	16	18	22
18	5	5	6	6	6	7	8	11	6	7	8	8	9	10	12	15	9	10	11	12	13	16	18	22
19	5	5	6	6	6	8	9	11	6	7	8	8	9	10	12	15	9	10	12	12	13	16	18	23
20	5	5	6	6	6	8	9	11	6	7	8	8	9	11	12	15	9	10	12	12	13	16	18	23
24	5	5	6	6	7	8	9	11	6	7	8	8	9	11	12	15	9	11	12	13	14	16	19	23
30	5	5	6	6	7	8	9	11	6	7	8	9	9	11	13	16	10	11	12	13	14	17	19	24
40	5	5	6	7	7	8	9	11	7	8	9	9	10	11	13	16	10	11	13	14	15	17	20	25
60	5	6	6	7	7	8	10	12	7	8	9	10	10	12	14	17	11	12	14	15	16	18	21	26

	$\Delta = 1.75$								$\Delta = 1.50$								$\Delta = 1.25$								$\Delta = 1.00$							
	Power								Power								Power								Power							
k	.50	.60	.70	.75	.80	.90	.95	.99	.50	.60	.70	.75	.80	.90	.95	.99	.50	.60	.70	.75	.80	.90	.95	.99	.50	.60	.70	.75	.80	.90	.95	.99
3	8	9	10	11	12	14	16	21	10	11	13	14	15	18	21	27	13	15	17	19	21	25	29	38	19	22	26	28	31	38	45	59
4	8	9	11	12	13	15	18	22	11	12	14	15	16	20	23	30	14	17	19	21	23	28	32	42	21	25	29	31	34	42	49	65
5	9	10	11	12	13	16	18	23	11	13	15	16	17	21	24	31	15	18	21	22	24	29	34	44	23	27	31	34	37	45	53	69
6	9	10	12	13	14	17	19	24	12	14	16	17	18	22	25	33	16	19	22	23	25	31	36	46	24	28	33	36	39	48	55	71
7	9	11	12	13	14	17	20	25	12	14	16	17	19	23	26	34	17	19	23	24	26	32	37	48	26	30	35	37	41	49	57	74
8	10	11	13	13	15	17	20	26	13	14	17	18	19	23	27	35	17	20	23	25	27	33	38	49	27	31	36	39	42	51	59	76
9	10	11	13	14	15	18	21	26	13	15	17	18	20	24	28	35	18	21	24	26	28	34	39	50	28	32	37	40	43	52	61	78
10	10	11	13	14	15	18	21	27	13	15	17	19	20	24	28	36	19	21	25	27	29	35	40	51	28	33	38	41	44	54	62	79
11	10	12	13	14	15	19	21	27	14	16	18	19	21	25	29	36	19	22	25	27	29	35	41	52	29	33	38	41	45	55	63	80
12	10	12	14	15	16	19	22	27	14	16	18	20	21	25	29	37	19	22	26	28	30	36	41	53	30	34	40	43	46	56	64	82
13	11	12	14	15	16	19	22	28	14	16	18	20	21	26	29	37	20	23	26	28	30	36	42	53	30	35	40	43	47	56	65	83
14	11	12	14	15	16	19	22	28	14	16	19	20	22	26	30	38	20	23	26	28	31	37	43	54	30	36	41	44	47	57	66	84
15	11	12	14	15	16	20	22	28	15	17	19	20	22	26	30	38	20	23	27	29	31	37	43	54	32	36	42	45	48	58	67	85
16	11	13	14	15	17	20	23	29	15	17	19	21	22	27	30	38	21	24	27	29	31	38	43	55	32	37	42	45	49	59	68	86
17	11	13	15	16	17	20	23	29	15	17	19	21	22	27	31	39	21	24	27	30	32	38	44	56	32	37	43	46	49	59	68	86
18	11	13	15	16	17	20	23	29	15	17	20	21	23	27	31	39	21	24	28	30	32	39	44	56	33	38	43	46	50	60	69	87
19	11	13	15	16	27	20	23	29	15	17	20	21	23	27	31	39	21	24	28	30	33	39	45	56	33	38	44	47	50	61	69	88
20	12	13	15	16	17	20	23	29	15	17	20	21	23	28	32	40	21	25	28	30	33	39	45	57	34	38	44	47	51	61	70	88
24	12	13	15	16	18	21	24	30	16	18	21	22	24	28	32	41	23	26	29	31	34	40	46	58	35	40	46	49	52	63	72	91
30	12	14	16	17	18	22	25	31	17	19	21	23	25	29	33	42	24	27	31	33	35	42	48	60	36	42	47	51	54	67	74	93
40	13	15	17	18	19	23	26	32	17	20	22	24	26	30	35	43	25	28	32	34	37	44	50	62	38	44	50	53	57	68	77	97
60	14	16	18	19	20	24	27	33	19	21	24	25	27	32	36	45	27	30	34	36	39	46	52	65	41	47	53	56	60	72	81	101

***Source*:** From P. R. Nelson (1983).

APPENDIX 15.5

Maximum F Ratio Critical Values

$F_{max}(0.1; k, \nu)$

	k									
ν	3	4	5	6	7	8	9	10	11	12
2	42.48	69.13	98.18	129.1	161.7	195.6	230.7	266.8	303.9	341.9
3	16.77	23.95	30.92	37.73	44.40	50.94	57.38	63.72	69.97	76.14
4	10.38	13.88	17.08	20.06	22.88	25.57	28.14	30.62	33.01	35.33
5	7.68	9.86	11.79	13.54	15.15	16.66	18.08	19.43	20.71	21.95
6	6.23	7.78	9.11	10.30	11.38	12.38	13.31	14.18	15.01	15.79
7	5.32	6.52	7.52	8.41	9.20	9.93	10.60	11.23	11.82	12.37
8	4.71	5.68	6.48	7.18	7.80	8.36	8.88	9.36	9.81	10.23
9	4.26	5.07	5.74	6.31	6.82	7.28	7.70	8.09	8.45	8.78
10	3.93	4.63	5.19	5.68	6.11	6.49	6.84	7.16	7.46	7.74
12	3.45	4.00	4.44	4.81	5.13	5.42	5.68	5.92	6.14	6.35
15	3.00	3.41	3.74	4.02	4.25	4.46	4.65	4.82	4.98	5.13
20	2.57	2.87	3.10	3.29	3.46	3.60	3.73	3.85	3.96	4.06
30	2.14	2.34	2.50	2.62	2.73	2.82	2.90	2.97	3.04	3.10
60	1.71	1.82	1.90	1.96	2.02	2.07	2.11	2.14	2.18	2.21

$F_{max}(0.05; k, \nu)$

	k									
ν	3	4	5	6	7	8	9	10	11	12
2	87.49	142.5	202.4	266.2	333.2	403.1	475.4	549.8	626.2	704.4
3	27.76	39.51	50.88	61.98	72.83	83.48	93.94	104.2	114.4	124.4
4	15.46	20.56	25.21	29.54	33.63	37.52	41.24	44.81	48.27	51.61
5	10.75	13.72	16.34	18.70	20.88	22.91	24.83	26.65	28.38	30.03
6	8.36	10.38	12.11	13.64	15.04	16.32	17.51	18.64	19.70	20.70
7	6.94	8.44	9.70	10.80	11.80	12.70	13.54	14.31	15.05	15.74
8	6.00	7.19	8.17	9.02	9.77	10.46	11.08	11.67	12.21	12.72
9	5.34	6.31	7.11	7.79	8.40	8.94	9.44	9.90	10.33	10.73
10	4.85	5.67	6.34	6.91	7.41	7.86	8.27	8.64	8.99	9.32
12	4.16	4.79	5.30	5.72	6.09	6.42	6.72	6.99	7.24	7.48
15	3.53	4.00	4.37	4.67	4.94	5.17	5.38	5.57	5.75	5.91
20	2.95	3.28	3.53	3.74	3.92	4.08	4.22	4.35	4.46	4.57
30	2.40	2.61	2.77	2.90	3.01	3.11	3.19	3.27	3.34	3.40
60	1.84	1.96	2.04	2.11	2.16	2.21	2.25	2.29	2.32	2.35

Source: From L. S. Nelson (1987). Reproduced with permission from the American Society for Quality Control.

APPENDIX 15.5 (Continued)

Maximum F Ratio Critical Values
$F_{max}(0.01; k, \nu)$

	k									
ν	3	4	5	6	7	8	9	10	11	12
2	447.5	729.2	103.6	136.2	170.5	206.3	243.2	281.3	320.4	360.4
3	84.56	119.8	153.8	187.0	219.3	251.1	282.3	313.0	343.2	373.1
4	36.70	48.43	59.09	69.00	78.33	87.20	95.68	103.8	111.7	119.3
5	22.06	27.90	33.00	37.61	41.85	45.81	49.53	53.06	56.42	59.63
6	15.60	19.16	22.19	24.89	27.32	29.57	31.65	33.61	35.46	37.22
7	12.09	14.55	16.60	18.39	20.00	21.47	22.82	24.08	25.26	26.37
8	9.94	11.77	13.27	14.58	15.73	16.78	17.74	18.63	19.46	20.24
9	8.49	9.93	11.10	12.11	12.99	13.79	14.52	15.19	15.81	16.39
10	7.46	8.64	9.59	10.39	11.10	11.74	12.31	12.84	13.33	13.79
12	6.10	6.95	7.63	8.20	8.69	9.13	9.53	9.89	10.23	10.54
15	4.93	5.52	5.99	6.37	6.71	7.00	7.27	7.51	7.73	7.93
20	3.90	4.29	4.60	4.85	5.06	5.25	5.42	5.57	5.70	5.83
30	2.99	3.23	3.41	3.56	3.68	3.79	3.88	3.97	4.04	4.12
60	2.15	2.26	2.35	2.42	2.47	2.52	2.57	2.61	2.64	2.67

Source: From L. S. Nelson (1987). Reproduced with permission from the American Society for Quality Control.

APPENDIX 15.6

Studentized Range Upper Quantiles
$q(0.1; n, \nu)$

ν	n 2	3	4	5	6	7	8	9	10
1	8.93	13.44	16.36	18.49	20.15	21.51	22.64	23.62	24.48
2	4.13	5.73	6.77	7.54	8.14	8.63	9.05	9.41	9.72
3	3.33	4.47	5.20	5.74	6.16	6.51	6.81	7.06	7.29
4	3.01	3.98	4.59	5.03	5.39	5.68	5.93	6.14	6.33
5	2.85	3.72	4.26	4.66	4.98	5.24	5.46	5.65	5.82
6	2.75	3.56	4.07	4.44	4.73	4.97	5.17	5.34	5.50
7	2.68	3.45	3.93	4.28	4.55	4.78	4.97	5.14	5.28
8	2.63	3.37	3.83	4.17	4.43	4.65	4.83	4.99	5.13
9	2.59	3.32	3.76	4.08	4.34	4.54	4.72	4.87	5.01
10	2.56	3.27	3.70	4.02	4.26	4.47	4.64	4.78	4.91
11	2.54	3.23	3.66	3.96	4.20	4.40	4.57	4.71	4.84
12	2.52	3.20	3.62	3.92	4.16	4.35	4.51	4.65	4.78
13	2.50	3.18	3.59	3.88	4.12	4.30	4.46	4.60	4.72
14	2.49	3.16	3.56	3.85	4.08	4.27	4.42	4.56	4.68
15	2.48	3.14	3.54	3.83	4.05	4.23	4.39	4.52	4.64
16	2.47	3.12	3.52	3.80	4.03	4.21	4.36	4.49	4.61
17	2.46	3.11	3.50	3.78	4.00	4.18	4.33	4.46	4.58
18	2.45	3.10	3.49	3.77	3.98	4.16	4.31	4.44	4.55
19	2.45	3.09	3.47	3.75	3.97	4.14	4.29	4.42	4.53
20	2.44	3.08	3.46	3.74	3.95	4.12	4.27	4.40	4.51
24	2.42	3.05	3.42	3.69	3.90	4.07	4.21	4.34	4.44
30	2.40	3.02	3.39	3.65	3.85	4.02	4.16	4.28	4.38
40	2.38	2.99	3.35	3.60	3.80	3.96	4.10	4.21	4.32
60	2.36	2.96	3.31	3.56	3.75	3.91	4.04	4.16	4.25
120	2.34	2.93	3.28	3.52	3.71	3.86	3.99	4.10	4.19
∞	2.33	2.90	3.24	3.48	3.66	3.81	3.93	4.04	4.13

ν	n 11	12	13	14	15	16	17	18	19	20
1	25.24	25.92	26.54	27.10	27.62	28.10	28.54	28.96	29.35	29.71
2	10.01	10.26	10.49	10.70	10.89	11.07	11.24	11.39	11.54	11.68
3	7.49	7.67	7.83	7.98	8.12	8.25	8.37	8.48	8.58	8.68
4	6.49	6.65	6.78	6.91	7.02	7.13	7.23	7.33	7.41	7.50
5	5.97	6.10	6.22	6.34	6.44	6.54	6.63	6.71	6.79	6.86
6	5.64	5.76	5.87	5.98	6.07	6.16	6.25	6.32	6.40	6.47
7	5.41	5.53	5.64	5.74	5.83	5.91	5.99	6.06	6.13	6.19
8	5.25	5.36	5.46	5.56	5.64	5.72	5.80	5.87	5.93	6.00
9	5.13	5.23	5.33	5.42	5.51	5.58	5.66	5.72	5.79	5.85
10	5.03	5.13	5.23	5.32	5.40	5.47	5.54	5.61	5.67	5.73
11	4.95	5.05	5.15	5.23	5.31	5.38	5.45	5.51	5.57	5.63
12	4.89	4.99	5.08	5.16	5.24	5.31	5.37	5.44	5.49	5.55
13	4.83	4.93	5.02	5.10	5.18	5.25	5.31	5.37	5.43	5.48
14	4.79	4.88	4.97	5.05	5.12	5.19	5.26	5.32	5.37	5.43
15	4.75	4.84	4.93	5.01	5.08	5.15	5.21	5.27	5.32	5.38
16	4.71	4.81	4.89	4.97	5.04	5.11	5.17	5.23	5.28	5.33
17	4.68	4.77	4.86	4.93	5.01	5.07	5.13	5.19	5.24	5.30
18	4.65	4.75	4.83	4.90	4.98	5.04	5.10	5.16	5.21	5.26
19	4.63	4.72	4.80	4.88	4.95	5.01	5.07	5.13	5.18	5.23
20	4.61	4.70	4.78	4.85	4.92	4.99	5.05	5.10	5.16	5.20
24	4.54	4.63	4.71	4.78	4.85	4.91	4.97	5.02	5.07	5.12
30	4.47	4.56	4.64	4.71	4.77	4.83	4.89	4.94	4.99	5.03
40	4.41	4.49	4.56	4.63	4.69	4.75	4.81	4.86	4.90	4.95
60	4.34	4.42	4.49	4.56	4.62	4.67	4.73	4.78	4.82	4.86
120	4.28	4.35	4.42	4.48	4.54	4.60	4.65	4.69	4.74	4.78
∞	4.21	4.28	4.35	4.41	4.47	4.52	4.57	4.61	4.65	4.69

Source: From Pearson and Hartley (1976). Reproduced with permission from the Biometrika Trustees.

APPENDIX 15.6 (Continued)

Studentized Range Upper Quantiles
$q(0.05; n, \nu)$

ν	n = 2	3	4	5	6	7	8	9	10
1	17.97	26.98	32.82	37.08	40.41	43.12	45.40	47.36	49.07
2	6.08	8.33	9.80	10.88	11.74	12.44	13.03	13.54	13.99
3	4.50	5.91	6.82	7.50	8.04	8.48	8.85	9.18	9.46
4	3.93	5.04	5.76	6.29	6.71	7.05	7.35	7.60	7.83
5	3.64	4.60	5.22	5.67	6.03	6.33	6.58	6.80	6.99
6	3.46	4.34	4.90	5.30	5.63	5.90	6.12	6.32	6.49
7	3.34	4.16	4.68	5.06	5.36	5.61	5.82	6.00	6.16
8	3.26	4.04	4.53	4.89	5.17	5.40	5.60	5.77	5.92
9	3.20	3.95	4.41	4.76	5.02	5.24	5.43	5.59	5.74
10	3.15	3.88	4.33	4.65	4.91	5.12	5.30	5.46	5.60
11	3.11	3.82	4.26	4.57	4.82	5.03	5.20	5.35	5.49
12	3.08	3.77	4.20	4.51	4.75	4.95	5.12	5.27	5.39
13	3.06	3.73	4.15	4.45	4.69	4.88	5.05	5.19	5.32
14	3.03	3.70	4.11	4.41	4.64	4.83	4.99	5.13	5.25
15	3.01	3.67	4.08	4.37	4.59	4.78	4.94	5.08	5.20
16	3.00	3.65	4.05	4.33	4.56	4.74	4.90	5.03	5.15
17	2.98	3.63	4.02	4.30	4.52	4.70	4.86	4.99	5.11
18	2.97	3.61	4.00	4.28	4.49	4.67	4.82	4.96	5.07
19	2.96	3.59	3.98	4.25	4.47	4.65	4.79	4.92	5.04
20	2.95	3.58	3.96	4.23	4.45	4.62	4.77	4.90	5.01
24	2.92	3.53	3.90	4.17	4.37	4.54	4.68	4.81	4.92
30	2.89	3.49	3.85	4.10	4.30	4.46	4.60	4.72	4.82
40	2.86	3.44	3.79	4.04	4.23	4.39	4.52	4.63	4.73
60	2.83	3.40	3.74	3.98	4.16	4.31	4.44	4.55	4.65
120	2.80	3.36	3.68	3.92	4.10	4.24	4.36	4.47	4.56
∞	2.77	3.31	3.63	3.86	4.03	4.17	4.29	4.39	4.47

ν	n = 11	12	13	14	15	16	17	18	19	20
1	50.59	51.96	53.20	54.33	55.36	56.32	57.22	58.04	58.83	59.56
2	14.39	14.75	15.08	15.38	15.65	15.91	16.14	16.37	16.57	16.77
3	9.72	9.95	10.15	10.35	10.52	10.69	10.84	10.98	11.11	11.24
4	8.03	8.21	8.37	8.52	8.66	8.79	8.91	9.03	9.13	9.23
5	7.17	7.32	7.47	7.60	7.72	7.83	7.93	8.03	8.12	8.21
6	6.65	6.79	6.92	7.03	7.14	7.24	7.34	7.43	7.51	7.59
7	6.30	6.43	6.55	6.66	6.76	6.85	6.94	7.02	7.10	7.17
8	6.05	6.18	6.29	6.39	6.48	6.57	6.65	6.73	6.80	6.87
9	5.87	5.98	6.09	6.19	6.28	6.36	6.44	6.51	6.58	6.64
10	5.72	5.83	5.93	6.03	6.11	6.19	6.27	6.34	6.40	6.47
11	5.61	5.71	5.81	5.90	5.98	6.06	6.13	6.20	6.27	6.33
12	5.51	5.61	5.71	5.80	5.88	5.95	6.02	6.09	6.15	6.21
13	5.43	5.53	5.63	5.71	5.79	5.86	5.93	5.99	6.05	6.11
14	5.36	5.46	5.55	5.64	5.71	5.79	5.85	5.91	5.97	6.03
15	5.31	5.40	5.49	5.57	5.65	5.72	5.78	5.85	5.90	5.96
16	5.26	5.35	5.44	5.52	5.59	5.66	5.73	5.79	5.84	5.90
17	5.21	5.31	5.39	5.47	5.54	5.61	5.67	5.73	5.79	5.84
18	5.17	5.27	5.35	5.43	5.50	5.57	5.63	5.69	5.74	5.79
19	5.14	5.23	5.31	5.39	5.46	5.53	5.59	5.65	5.70	5.75
20	5.11	5.20	5.28	5.36	5.43	5.49	5.55	5.61	5.66	5.71
24	5.01	5.10	5.18	5.25	5.32	5.38	5.44	5.49	5.55	5.59
30	4.92	5.00	5.08	5.15	5.21	5.27	5.33	5.38	5.43	5.47
40	4.82	4.90	4.98	5.04	5.11	5.16	5.22	5.27	5.31	5.36
60	4.73	4.81	4.88	4.94	5.00	5.06	5.11	5.15	5.20	5.24
120	4.64	4.71	4.78	4.84	4.90	4.95	5.00	5.04	5.09	5.13
∞	4.55	4.62	4.68	4.74	4.80	4.85	4.89	4.93	4.97	5.01

Source: From Pearson and Hartley (1976). Reproduced with permission from the Biometrika Trustees.

APPENDIX 15.6 (Continued)

Studentized Range Upper Quantiles
$q(0.01; n, \nu)$

ν	n = 2	3	4	5	6	7	8	9	10
1	90.03	135.0	164.3	185.6	202.2	215.8	227.2	237.0	245.6
2	14.04	19.02	22.29	24.72	26.63	28.20	29.53	30.68	31.69
3	8.26	10.62	12.17	13.33	14.24	15.00	15.64	16.20	16.69
4	6.51	8.12	9.17	9.96	10.58	11.10	11.55	11.93	12.27
5	5.70	6.98	7.80	8.42	8.91	9.32	9.67	9.97	10.24
6	5.24	6.33	7.03	7.56	7.97	8.32	8.61	8.87	9.10
7	4.95	5.92	6.54	7.01	7.37	7.68	7.94	8.17	8.37
8	4.75	5.64	6.20	6.62	6.96	7.24	7.47	7.68	7.86
9	4.60	5.43	5.96	6.35	6.66	6.91	7.13	7.33	7.49
10	4.48	5.27	5.77	6.14	6.43	6.67	6.87	7.05	7.21
11	4.39	5.15	5.62	5.97	6.25	6.48	6.67	6.84	6.99
12	4.32	5.05	5.50	5.84	6.10	6.32	6.51	6.67	6.81
13	4.26	4.96	5.40	5.73	5.98	6.19	6.37	6.53	6.67
14	4.21	4.89	5.32	5.63	5.88	6.08	6.26	6.41	6.54
15	4.17	4.84	5.25	5.56	5.80	5.99	6.16	6.31	6.44
16	4.13	4.79	5.19	5.49	5.72	5.92	6.08	6.22	6.35
17	4.10	4.74	5.14	5.43	5.66	5.85	6.01	6.15	6.27
18	4.07	4.70	5.09	5.38	5.60	5.79	5.94	6.08	6.20
19	4.05	4.67	5.05	5.33	5.55	5.73	5.89	6.02	6.14
20	4.02	4.64	5.02	5.29	5.51	5.69	5.84	5.97	6.09
24	3.96	4.55	4.91	5.17	5.37	5.54	5.69	5.81	5.92
30	3.89	4.45	4.80	5.05	5.24	5.40	5.54	5.65	5.76
40	3.82	4.37	4.70	4.93	5.11	5.26	5.39	5.50	5.60
60	3.76	4.28	4.59	4.82	4.99	5.13	5.25	5.36	5.45
120	3.70	4.20	4.50	4.71	4.87	5.01	5.12	5.21	5.30
∞	3.64	4.12	4.40	4.60	4.76	4.88	4.99	5.08	5.16

ν	n = 11	12	13	14	15	16	17	18	19	20
1	253.2	260.0	266.2	271.8	277.0	281.8	286.3	290.4	294.3	298.0
2	32.59	33.40	34.13	34.81	35.43	36.00	36.53	37.03	37.50	37.95
3	17.13	17.53	17.89	18.22	18.52	18.81	19.07	19.32	19.55	19.77
4	12.57	12.84	13.09	13.32	13.53	13.73	13.91	14.08	14.24	14.40
5	10.48	10.70	10.89	11.08	11.24	11.40	11.55	11.68	11.81	11.93
6	9.30	9.48	9.65	9.81	9.95	10.08	10.21	10.32	10.43	10.54
7	8.55	8.71	8.86	9.00	9.12	9.24	9.35	9.46	9.55	9.65
8	8.03	8.18	8.31	8.44	8.55	8.66	8.76	8.85	8.94	9.03
9	7.65	7.78	7.91	8.03	8.13	8.23	8.33	8.41	8.49	8.57
10	7.36	7.49	7.60	7.71	7.81	7.91	7.99	8.08	8.15	8.23
11	7.13	7.25	7.36	7.46	7.56	7.65	7.73	7.81	7.88	7.95
12	6.94	7.06	7.17	7.26	7.36	7.44	7.52	7.59	7.66	7.73
13	6.79	6.90	7.01	7.10	7.19	7.27	7.35	7.42	7.48	7.55
14	6.66	6.77	6.87	6.96	7.05	7.13	7.20	7.27	7.33	7.39
15	6.55	6.66	6.76	6.84	6.93	7.00	7.07	7.14	7.20	7.26
16	6.46	6.56	6.66	6.74	6.82	6.90	6.97	7.03	7.09	7.15
17	6.38	6.48	6.57	6.66	6.73	6.81	6.87	6.94	7.00	7.05
18	6.31	6.41	6.50	6.58	6.65	6.73	6.79	6.85	6.91	6.97
19	6.25	6.34	6.43	6.51	6.58	6.65	6.72	6.78	6.84	6.89
20	6.19	6.28	6.37	6.45	6.52	6.59	6.65	6.71	6.77	6.82
24	6.02	6.11	6.19	6.26	6.33	6.39	6.45	6.51	6.56	6.61
30	5.85	5.93	6.01	6.08	6.14	6.20	6.26	6.31	6.36	6.41
40	5.69	5.76	5.83	5.90	5.96	6.02	6.07	6.12	6.16	6.21
60	5.53	5.60	5.67	5.73	5.78	5.84	5.89	5.93	5.97	6.01
120	5.37	5.44	5.50	5.56	5.61	5.66	5.71	5.75	5.79	5.83
∞	5.23	5.29	5.35	5.40	5.45	5.49	5.54	5.57	5.61	5.65

Source: From Pearson and Hartley (1976). Reproduced with permission from the Biometrika Trustees.

P • A • R • T • 4

ADVANCED STATISTICAL CONCEPTS

CHAPTER 16
MULTICOLLINEARITY AND BIASED ESTIMATION IN REGRESSION

Douglas C. Montgomery
Arizona State University, Tempe, AZ

Elizabeth A. Peck

The Coca-Cola Company, Atlanta, GA

James R. Simpson
United States Air Force Academy, Colorado Springs, CO

16.1 INTRODUCTION

Linear regression analysis is one of the most widely used statistical techniques in engineering, management, the sciences, and operations research. Surveys by Turban (1972), Ledbetter and Cox (1977), and Lane, Mansour, and Harpell (1993) place it either first or in the leading category of techniques among the analytical methods frequently employed by professional practitioners. Regression methods are used extensively in the areas of forecasting and economic modeling, planning and resource allocation, and cost or parameter estimation, among many others. Montgomery and Peck (1992) identify four generic uses of regression analysis:

1. Data description
2. Parameter estimation
3. Prediction and estimation, including forecasting
4. Control

All have important applications in management, engineering, and the sciences.

The method of least squares is generally used to estimate the coefficients in a linear regression model. These estimates have minimum variance in the class of unbiased estimators. While these properties are theoretically satisfying, in practice the results of a least-squares fit are often unacceptable, particularly when there are near-linear dependencies among the regressor variables. This is the problem of *multicollinearity* in regression analysis. (The terms "collinearity" and "ill-conditioning" are also widely used.) This chapter is a review of the multicollinearity

This work was supported in part by the Office of Naval Research under Contract N00014-78-C-0312.

problem oriented toward the engineering and management user of regression. We focus on the sources of multicollinearity, its harmful effects, available diagnostics, and suggested remedial measures, including a discussion of how these ideas are implemented in modern regression software packages. More theoretical reviews

TABLE 16.1 Residential Load Survey Data

Customer number	y, kW	x_1, ft^2/1000	x_2, \$/1000	x_3, tons	x_4, kW	x_5
1	7.518	3.164	34.990	7.0	7.789	4
2	3.579	1.929	21.446	1.5	5.251	5
3	5.910	2.613	28.731	6.5	6.325	3
4	4.790	2.337	25.058	4.0	5.733	4
5	4.997	2.757	30.358	4.0	6.216	1
6	2.242	1.398	15.464	1.0	3.113	6
7	7.427	3.366	37.267	5.0	9.415	1
8	4.533	2.378	25.939	3.0	6.142	2
9	5.990	2.881	32.362	3.5	7.700	5
10	4.101	2.098	22.395	2.5	5.222	1
11	1.685	1.178	12.531	0.0	2.575	1
12	4.560	2.360	25.784	4.0	5.536	2
13	4.657	2.236	25.152	2.5	6.208	2
14	3.151	1.771	19.106	1.0	5.213	3
15	2.976	1.852	20.677	1.0	4.659	1
16	2.867	1.823	20.037	1.5	4.453	3
17	2.662	1.578	18.154	0.5	3.978	4
18	4.363	2.117	23.951	2.5	6.236	1
19	2.991	2.052	22.069	1.5	4.892	4
20	2.766	1.715	18.324	1.5	3.960	5
21	5.323	2.333	25.942	5.0	5.038	1
22	6.553	2.887	32.236	5.5	7.815	3
23	3.736	1.972	22.123	2.0	4.432	5
24	6.796	2.886	32.161	6.0	7.039	4
25	4.496	1.874	21.070	3.0	5.254	1
26	2.831	1.408	15.957	1.5	2.967	6
27	5.495	2.526	27.687	4.5	6.481	2
28	6.656	2.821	31.145	5.5	7.284	4
29	2.349	1.328	14.160	0.5	3.652	4
30	6.824	2.856	31.812	6.0	7.186	1
31	5.354	2.223	24.788	5.0	5.965	3
32	5.802	2.489	26.661	6.0	6.862	4
33	4.790	2.455	27.203	3.0	6.007	5
34	6.283	2.720	29.524	5.5	6.715	4
35	3.400	2.201	23.424	1.0	5.625	1
36	6.083	2.694	29.096	5.0	6.949	2
37	5.740	2.456	27.076	3.5	6.143	7
38	3.599	1.772	19.177	2.0	4.864	1
39	5.010	2.253	24.535	5.0	4.975	2
40	4.625	2.398	25.949	2.5	6.947	3

directed toward the professional statistician are given by Gunst (1983) and Stewart (1987).

Before beginning our review, we illustrate with an example the types of problems that multicollinearity induces. The example is taken from Montgomery and Askin (1981). Table 16.1 contains data collected by an electric utility during an investigation of the factors that influence peak demand for electric energy by residential customers. Such information is used by utilities in developing a load-forecasting model for generation or capacity planning purposes. The response variable y is the electricity demand, or load, at the hour of system peak demand, in kilowatts (kW). The regressor variables are the size of the customer's house x_1, the annual family income x_2, tons of air-conditioning capacity x_3, the appliance index for the house x_4 (obtained by summing the kilowatt ratings for all major appliances), and the number of people x_5 typically in the house on a weekday.

Table 16.2 displays the results of a least-squares fit to these data, using the popular SAS PROC REG routine. The upper half of the table gives the usual overall

TABLE 16.2 Regression Results (from SAS PROC REG) for Residential Load Research Data

Least-squares regression on observations 1–40

Analysis of variance

Source	DF	Sum of squares	Mean square	F value	Prob > F
Model	5	90.27776	18.05555	313.918	0.0001
Error	34	1.95557	0.05752		
C total	39	92.23333			

Root MSE	0.23983	R^2	0.9788
Dep mean	4.63775	Adj R^2	0.9757
CV	5.17118		

Parameter estimates

Variable	DF	Parameter estimate	Standard error	T For H_0: Parameter = 0	Prob> |T|	Standardized estimate	Variance inflation
Intercep	1	–0.041600	0.25313688	–0.164	0.8704	0.00000000	0.00000000
X_1	1	–2.497561	1.02625271	–2.434	0.0204	–0.84056046	191.29677038
X_2	1	0.269473	0.08949196	3.011	0.0049	1.01070026	180.66514248
X_3	1	0.414298	0.03923619	10.559	0.0001	0.52462997	3.95865153
X_4	1	0.379315	0.08945856	4.240	0.0002	0.35889973	11.48903142
X_5	1	0.030351	0.02318262	1.309	0.1992	0.03359028	1.05563269

Collinearity diagnostics (intercept adjusted)

Number	Eigenvalue	Condition index	Var Prop X_1	Var Prop X_2	Var Prop X_3	Var Prop X_4	Var Prop X_5
1	3.69689	1.00000	0.0004	0.0004	0.0148	0.0058	0.0019
2	0.98379	1.93851	0.0000	0.0000	0.0017	0.0000	0.9517
3	0.26657	3.72400	0.0005	0.0005	0.6820	0.0742	0.0117
4	0.04999	8.59921	0.0203	0.0245	0.2909	0.9150	0.0207
5	0.00275	36.65435	0.9789	0.9745	0.0107	0.0050	0.0140

"summary" statistics from the regression model, the analysis of variance F test, R^2, and the coefficient of variation $CV = 100\sqrt{MS_E}/\bar{y}$. These statistics indicate a satisfactory fit. Residual plots do not reveal any disturbing problems. Therefore by many of the measures customarily used to assess the adequacy of a regression model, the least-squares fit is satisfactory.

The central portion of Table 16.2 presents the coefficient estimates. One is immediately disturbed by the negative estimate for x_1, house size. Intuition leads us to believe that as the house size increases, peak demand for electricity will increase, and thus the coefficient for x_1 should be positive. This unexpected "wrong sign" is not necessarily due to multicollinearity—it could be the result of small sample sizes, inadequate ranges for the regressors, and many other causes. (See Montgomery and Peck, 1992, chap. 9, for a discussion of the wrong-sign problem.) We shall see subsequently, however, that the problems identified in these data are due to strong multicollinearity.

16.2 LINEAR REGRESSION MODEL

The usual single-equation regression model

$$\mathbf{y} = \mathbf{X}\boldsymbol{\beta} + \boldsymbol{\epsilon} \tag{16.1}$$

is assumed, where $\mathbf{y}$ is an $n \times 1$ vector of observations, $\mathbf{X}$ is an $n \times p$ matrix of rank p, $\boldsymbol{\beta}$ is a $p \times 1$ vector of unknown parameters, and $\boldsymbol{\epsilon}$ is an $n \times 1$ vector of nonobservable errors. It is convenient to assume that the p regressor variables and the response y have been centered (sample means are zero) and scaled to unit length, so that in the least-squares normal equations

$$(\mathbf{X}'\mathbf{X})\hat{\beta} = \mathbf{X}'\mathbf{y} \tag{16.2}$$

the $\mathbf{X}'\mathbf{X}$ matrix and the $\mathbf{X}'\mathbf{y}$ vector consist of sample correlation coefficients.

It will occasionally be convenient to express the model in terms of orthogonal regressor variables. To accomplish this transformation, let $\boldsymbol{\Lambda} = \text{diag}(\lambda_1, \lambda_2, \ldots, \lambda_p)$ be the diagonal matrix of eigenvalues of $\mathbf{X}'\mathbf{X}$, and T the corresponding orthogonal matrix of eigenvectors so that $\mathbf{X}'\mathbf{X} = \mathbf{T}\boldsymbol{\Lambda}\mathbf{T}'$. Then letting $\mathbf{Z} = \mathbf{X}\mathbf{T}$ and $\boldsymbol{\beta} = \mathbf{T}\boldsymbol{\alpha}$, Eq. (16.1) becomes

$$\mathbf{y} = \mathbf{Z}\boldsymbol{\alpha} + \boldsymbol{\epsilon} \tag{16.3}$$

The normal equations for this model are

$$(\mathbf{Z}'\mathbf{Z})\hat{\alpha} = \mathbf{Z}'\mathbf{y}$$

or, since $\mathbf{Z}'\mathbf{Z} = \Lambda$,

$$\Lambda\hat{\alpha} = \mathbf{Z}'\mathbf{y} \tag{16.4}$$

with solution $\hat{\alpha} = \Lambda^{-1}\mathbf{Z}'\mathbf{y}$. The relationship between $\hat{\alpha}$ and $\hat{\beta}$ is

$$\hat{\boldsymbol{\beta}} = \mathbf{T}\hat{\alpha} \tag{16.5}$$

and the covariance matrix of $\hat{\alpha}$ is

$$V(\hat{\alpha}) = \sigma^2(\mathbf{Z}'\mathbf{Z})^{-1} = \sigma^2\Lambda^{-1} \tag{16.6}$$

The set of orthogonal regressors in the columns of Z are often called *principal components*.

16.3 SOURCES OF MULTICOLLINEARITY

There are several ways to define multicollinearity. A useful definition is given in Silvey (1969), Mason, Gunst, and Webster (1975), and Mandel (1989), in which multicollinearity is said to exist if there are one or more linear relationships of the form

$$\sum_{j=1}^{p} t_j \mathbf{X}_j = 0 \tag{16.7}$$

between the columns $\mathbf{X}_j$ $(j = 1, 2, \cdots, p)$ of the $\mathbf{X}$ matrix, where the t_j are a set of constants, not all of which are zero. If relationship (16.7) holds exactly, the rank of $\mathbf{X}'\mathbf{X}$ is less than p, so in practice multicollinearity is said to exist when relationship (16.7) holds only approximately. Thus, clearly, multicollinearity will exist to some degree in every set of regression data (except when the columns of $\mathbf{X}$ are orthogonal, which will not usually be the case unless the data come from a designed experiment).

The data analyst is faced with the task of diagnosing the extent of multicollinearity, assessing its potential harmful effects, and recommending or implementing appropriate remedial measures.

The multicollinearity problem occurs for at least one of the following reasons:

1. The data collection method
2. Constraints on the model or in the population
3. Model misspecification or choice of regressor variables

The first two sources of multicollinearity have been discussed extensively in Mason, Gunst, and Webster (1975) and Gunst (1983). The data collection method can lead to multicollinearity when the experimenter samples only a subspace of the region of the regressors defined approximately by relationship (16.7). In general, this implies that the data lie approximately along a hyperplane. For example, this has occurred in the acetylene data of Marquardt and Snee (1975) in which the authors investigate least-squares models relating conversion of *N*-neptane into acetylene to three regressor variables. The sample data for two of the three regressors, contact time x_1 and reactor temperature x_3, lie nearly along a straight line. This may have happened inadvertently, or it may have resulted from the second source of multicollinearity, physical constraints in the population being sampled; that is, longer contact times may be necessary at lower temperatures while shorter contact times may be required at higher temperatures. Other situations in which physical constraints are present in the model include experiments with mixtures, where a subset of the regressors (the mixture components) add to unity.

Model misspecification can either cause or intensify multicollinearity problems. For example, the addition of polynomial terms to a regression model causes ill-conditioning in $\mathbf{X}'\mathbf{X}$. In fact, if the range of x is small, even the addition of an x^2 term can result in significant multicollinearity. We often encounter situations where two or more regressors are nearly linearly dependent, and retaining all of these regressors in the model may contribute to multicollinearity. The acetylene data in

Marquardt and Snee (1975) serve to illustrate this source of multicollinearity. A full quadratic response surface is fitted using three regressors, two of which (x_1 and x_3) are highly correlated. The presence of both x_1 and x_3 in the model, along with the second-order terms involving these variables, intensifies the multicollinearity problem. In cases where multicollinearity is due to model specification, the usual recommendation is to redefine the model so as to minimize the problem. Often this is done by using only a subset of the original regressors. This solution is not always appropriate, however, and selection of the subset of regressors to use may not be straightforward.

16.4 DIAGNOSTIC TECHNIQUES

A number of techniques have been proposed for assessing the extent of multicollinearity in regression data. Desirable characteristics for a diagnostic are that it directly reflects the degree of the multicollinearity problem and that it provides information helpful in determining which regressors are involved. We now review briefly those diagnostics that are used most frequently.

16.4.1 Diagnostics Based on X′X or (X′X)$^{-1}$

A very simple measure of multicollinearity is inspection of the off-diagonal elements (say, r_{ij}) in $\mathbf{X}'\mathbf{X}$. Since $\mathbf{X}'\mathbf{X}$ is in correlation form, if regressors x_i and x_j are nearly linearly dependent, then $|r_{ij}|$ will be near unity. Unfortunately this diagnostic is only helpful in detecting a near-linear dependency between *pairs* of regressors. When more than two regressors are involved in a multicollinearity, there is no assurance that any of the pairwise correlations will be large. See Webster, Gunst, and Mason (1974) for an example as well as the discussions in Montgomery and Peck (1992) and Belsley, Kuh, and Welsch (1980).

The diagonal elements of the matrix $\mathbf{C} = (\mathbf{X}'\mathbf{X})^{-1}$ may be written as

$$C_{jj} = (1 - R_j^2)^{-1} \tag{16.8}$$

where R_j^2 is the coefficient of determination resulting from regressing x_j on the remaining $p - 1$ regressors. If x_j is nearly orthogonal to the remaining regressors, R^2 is small and C_{jj} is close to unity, while if x_j is nearly linearly dependent on a subset of the remaining regressors, R_j^2 is near unity and C_{jj} is large. Since $V(\hat{\beta}_j) = \sigma^2 C_{jj}$, we can view C_{jj} as the amount by which the variance of $\hat{\beta}_j$ is increased due to near-linear dependencies among the regressors. Consequently, Marquardt (1970) has called C_{jj} the *variance inflation factor* (VIF). The VIF for each term in the model measures the combined effect of the dependencies among the regressors on that term. The variance inflation factor may also be computed as the reciprocal of the tolerance statistic often reported in multiple-regression software packages. SAS PROC REG produces the VIFs as an optional output; refer to the middle portion of Table 16.2. Generally, if any of the VIFs exceed 10, this is an indication that the associated regression coefficients are poorly estimated because of multicollinearity. From Table 16.2, we note that VIF_1, VIF_2, and VIF_4 exceed 10, so we suspect multicollinearity involving these regressors.

The VIFs have another interesting interpretation. The length of the normal-theory confidence interval on the jth regression coefficient may be written as $L_j =$

$2(C_{jj}\sigma^2)^{1/2}t_{\alpha/2,n-p-1}$, and the length of the corresponding interval based on an *orthogonal* design with the same sample size and root mean square (rms) values, that is, $\Sigma_{i=1}^{b}[(x_{ij}-\bar{x}_j)^2/n]^{1/2}$, is $L^* = 2\sigma t_{\alpha/2,n-p-1}$. Thus the confidence interval ratio is $L_j/L^* = C_{jj}^{1/2}$ and the square root of VIF_j reflects the increase in length of the confidence interval for the *j*th regression coefficient due to multicollinearity. Stewart (1987) refers to the square roots of the VIFs as the collinearity indices and recommends them as simple collinearity diagnostics because they vary directly with the relative distance to exact collinearity.

We may also use $|\mathbf{X}'\mathbf{X}|$ as a measure of multicollinearity. Now $0 < |\mathbf{X}'\mathbf{X}| < 1$, with smaller values implying stronger near-linear dependencies. One problem with this diagnostic is that it is difficult to determine a cutoff or critical value that indicates when a serious multicollinearity problem exists. Farrar and Glauber (1967) observe that if $\mathbf{X}$ is viewed as a random sample from a *p*-variate normal distribution, and if the columns of $\mathbf{X}$ are orthogonal, then a function of $|\mathbf{X}'\mathbf{X}|$ is approximately distributed as χ^2, and a cutoff value may then be based on this result. However, the assumption that the elements of $\mathbf{X}$ are a random sample does not comfortably match the more frequent application of regression in which X is assumed to be fixed. A second difficulty is that while a small value of $|\mathbf{X}'\mathbf{X}|$ implies multicollinearity, it does not offer any insight into *which* regressors are involved.

Finally, Willan and Watts (1978) suggest another interesting interpretation of this diagnostic. The joint $100(1-\alpha)\%$ confidence region for $\boldsymbol{\beta}$ based on the observed data is

$$(\hat{\boldsymbol{\beta}}-\boldsymbol{\beta})'\mathbf{X}'\mathbf{X}(\hat{\boldsymbol{\beta}}-\boldsymbol{\beta}) \le p\sigma^2 F_{\alpha,p,n-p-1} \tag{16.9}$$

where $F_{\alpha,p,n-p-1}$ is the upper α percentage point of the F distribution with p and $n-p-1$ degrees of freedom, while the corresponding confidence region based on the orthogonal reference design described earlier is

$$(\hat{\boldsymbol{\beta}}-\hat{\boldsymbol{\beta}})'(\hat{\boldsymbol{\beta}}-\boldsymbol{\beta}) \le p\sigma^2 F_{\alpha;p,n-p-1} \tag{16.10}$$

Note that the orthogonal reference design produces the smallest joint confidence region for fixed n and rms values and a given α. The ratio of the volume of the confidence region [Eq. (16.10)] to the confidence region [Eq. (16.9)] is $|\mathbf{X}'\mathbf{X}|^{1/2}$, so that $|\mathbf{X}'\mathbf{X}|^{1/2}$ measures the loss of estimation power due to multicollinearity. Put another way, $100(|\mathbf{X}'\mathbf{X}|^{-1/2}-1)$ reflects the percentage increase in the volume of the joint confidence region because of the near-linear dependencies in $\mathbf{X}$.

16.4.2 Diagnostics Based on Eigensystem Analysis

The eigenvalues of $\mathbf{X}'\mathbf{X}$, $\lambda_1, \lambda_2, \ldots, \lambda_p$, can be used to measure the extent and nature of multicollinearity in the data. If there are one or more near-linear dependencies in the data, then one or more of the eigenvalues will be small. Numerical analysts have used the condition number

$$\kappa(\mathbf{X}'\mathbf{X}) = \frac{\lambda_{\max}}{\lambda_{\min}} \tag{16.11}$$

to describe the conditioning of $\mathbf{X}'\mathbf{X}$. If the columns of $\mathbf{X}$ are orthogonal, then $\kappa(\mathbf{X}'\mathbf{X}) = 1$. For each exact linear dependence in the columns of $\mathbf{X}$ there will be one zero eigenvalue of $\mathbf{X}'\mathbf{X}$. The presence of near-linear dependencies will result in small eigenvalues. The extent of the multicollinearity is measured by how small an eigenvalue is relative to λ_{max}. Thus it is useful to define the *p condition indices* of $\mathbf{X}'\mathbf{X}$ as

$$\kappa_j = \frac{\lambda_{max}}{\lambda_j}, \quad j = 1, 2, \ldots, p \tag{16.12}$$

and to observe that there are as many near-linear dependencies in the columns of X as there are large values of κ_j. Generally, if the condition index is less than 100, there is no serious multicollinearity problem. Condition indices between 100 and 1000 imply moderate to strong multicollinearity, while values of κ_j exceeding 1000 indicate severe multicollinearity.

The connection between this diagnostic and the VIFs is easily shown. Note that, using Eqs. (16.5) and (16.6), we may write

$$V(\hat{\boldsymbol{\beta}}) = V(\mathbf{T}\hat{\boldsymbol{\alpha}}) = \sigma^2\mathbf{T}\boldsymbol{\Lambda}^{-1}\mathbf{T}'$$

so that

$$V(\hat{\beta}_j) = \sigma^2 C_{jj} = \sigma^2 \left(\sum_{i=1}^{p} \frac{t_{ji}^2}{\lambda_i} \right) \tag{16.13}$$

Therefore, the *j*th VIF is a linear combination of the reciprocals of the eigenvalues of $\mathbf{X}'\mathbf{X}$. One or more small eigenvalues can result in a large VIF, unless the small λ_i is offset by a corresponding small value for t_{ji}^2 (the *j*th element of the *i*th eigenvector). The t_{ji} are equal to zero if the *j*th and *i*th columns of **X** are orthogonal, so t_{ji} will usually be small only when x_i and x_j are not involved in a multicollinearity. This implies that the VIF for a regressor not involved in a multicollinearity will be small, while those for regressors participating in one or more multicollinearities will be large.

The eigensystem structure of $\mathbf{X}'\mathbf{X}$ can be used in two ways to identify the regressors involved in the multicollinearity. The first of these is based directly on the definition $\mathbf{Z}'\mathbf{Z} = \boldsymbol{\Lambda}$. We can think of λ_i as the *variance* of the *i*th principal component $\mathbf{Z}_i$. Furthermore, since $\mathbf{Z} = \mathbf{XT}$, we have

$$\mathbf{Z}_i = \sum_{j=1}^{p} t_{ji}\mathbf{X}_j \tag{16.14}$$

where $\mathbf{X}_j$ is the *j*th column of **X** and t_{ji} are the elements of the *i*th column of **T**. Now suppose that the variance of $\mathbf{Z}_i(\lambda_i)$ is small, implying that the *i*th principal component $\mathbf{Z}_i$ is nearly a constant. Thus Eq. (16.14) indicates that there is a linear combination of the *original* regressors that is nearly constant. This is the definition of multicollinearity, that is, the t_{ji} are the constants in Eq. (16.7). Therefore, the elements of the eigenvectors corresponding to small eigenvalues (that is, large condition indices) provide information on which regressors are involved in the multicollinearity.

The second approach uses Eq. (16.13) directly. Following Belsley, Kuh, and Welsch (1980), define

$$\phi_{ji} = \frac{t_{ji}^2}{\lambda_i}, \quad j, i = 1, 2, \ldots, p \tag{16.15}$$

Then the proportion of the VIF of the *j*th regression coefficient associated with the *i*th component of its decomposition in Eq. (16.13) is

$$\pi_{ij} = \frac{\phi_{ji}}{C_{jj}}, \quad i, j = 1, 2, \ldots, p \tag{16.16}$$

If we array the π_{ij} in a $p \times p$ matrix $\boldsymbol{\pi}$, then the elements of each column are just the proportions of the variance of each β_j contributed by the ith eigenvalue λ_i. If a high proportion of the variance for two or more components (in excess of 0.5, say) is associated with a small eigenvalue (or large *condition index*), then multicollinearity is indicated.

SAS PROC REG computes several of these diagnostics; refer to the lower portion of Table 16.2. Notice that there is one small eigenvalue (in row 5). The column labled "condition index" in this table reports the square roots of the condition indices of $\mathbf{X}'\mathbf{X}$. We conclude that row 5 has a large condition index. In row 5, since π_{51} and π_{52} are large, we would suspect multicollinearity between the first two regressors.

In general, the analyst should also check for competing dependencies in which several condition indices of similar size mask individual dependencies and dominating dependencies in which a regressor is involved in dependencies of different strength. Belsley, Kuh, and Welsch (1980) suggest supplementing the $\boldsymbol{\pi}$ matrix with certain regression statistics for each x_j regressed on the other $p - 1$ regressors to assist in these analyses. For more details and examples illustrating the use of this diagnostic, see Belsley, Kuh, and Welsch (1980).

16.4.3 Other Diagnostics

There are many other techniques that are occasionally useful in diagnosing multicollinearity. These include "wrong" signs for regression coefficients or parameter estimates that seem suspiciously large, small t statistics for individual regressors accompanied by a large overall F statistic, and unusual sensitivity of the regression results to arbitrary removal of an observation or a regressor. Unfortunately these signals do not always accompany multicollinear data and, in fact, are neither necessary nor sufficient for the presence of near-linear dependencies in the data. They are best regarded as symptoms rather than diagnostics.

16.4.4 Discussion

Two aspects of these diagnostics deserve further discussion. The first is the choice of a diagnostic. From an overall viewpoint, the variance inflation factors are a reliable and informative diagnostic, and are easy to compute. Consequently, we recommend them as a general diagnostic. Variance inflation factors are becoming standard output in most regression software packages. However, the VIFs do not indicate the form of the multicollinearity, that is, they do not indicate which other regressors are involved. The eigensystem analysis is required to reveal this information. This may be very useful information to the analyst in cases where there are several dependencies, or where there are relatively complex interrelationships between the regressors.

A second point concerns the use of the spectral decomposition of $\mathbf{X}'\mathbf{X}$ in the diagnostic process. Belsley, Kuh, and Welsch (1980) suggest using the singular value decomposition of $\mathbf{X}$, say,

$$\mathbf{X} = \mathbf{UDT}'$$

where $\mathbf{U}$ is $n \times p$, $\mathbf{UU}' = \mathrm{I}$, and $\mathbf{D}$ is a $p \times p$ diagonal matrix with nonnegative diagonal elements μ_j, $j = 1, 2,..., p$, called the *singular values* of $\mathbf{X}$. Since $\mathbf{X}'\mathbf{X} = (\mathbf{UDT}')'\mathbf{UDT}' = \mathbf{TD}^2\mathbf{T}' = \mathbf{T\Lambda T}'$, say, the squares of the singular *values* of $\mathbf{X}$ are just the eigenvalues of $\mathbf{X}'\mathbf{X}$. The condition indices of $\mathbf{X}$ could be defined as $\eta_j = \mu_{\max}/\mu_j$, $j = 1, 2, ..., p$, and the square of the jth condition index of $\mathbf{X}$ is the jth condition index of $\mathbf{X}'\mathbf{X}$. Notice that it is the condition index of $\mathbf{X}$ that is reported in the SAS PROC REG output in Table 16.2. There are two potential advantages of this approach. First, we are now dealing directly with the data matrix $\mathbf{X}$, with which we are primarily concerned, not the matrix of sums of squares and cross products $\mathbf{X}'\mathbf{X}$. Second, algorithms for generating the singular-value decomposition are more numerically stable than those for eigensystem analysis. In practice, however, we have not found numerical stability to be a serious problem in the eigensystem analysis.

Belsley, Kuh, and Welsch (1980) suggest that the above analysis should be applied to the original $\mathbf{X}$ matrix with the columns scaled to unit length, but with the intercept included (thus the original regressors are *not* centered), so that the role of the intercept in near-linear dependencies can be diagnosed. Belsley (1984) gives a more detailed presentation of this approach and argues that centering the regressors actually demeans the interpretative value of multicollinearity diagnostics. Other authorities do not support this viewpoint. If $\delta_1, \delta_2,..., \delta_{p+1}$ are the eigenvalues of $\mathbf{X}'\mathbf{X}$ with the intercept included, then Hocking (1983) notes that the relationship between the δ's and the λ's is

$$\mathrm{diag}(\delta_1, \delta_2,..., \delta_{p+1}) = \mathbf{P}\ \mathrm{diag}(1, \lambda_1, \lambda_2,..., \lambda_p)\mathbf{P}'$$

where P is nonsingular. For the same degree of ill-conditioning the δ's will be larger than the λ's. While not centering the regressors may be desirable in some applications, the analyst should remember that any multicollinearity involving the intercept is nonessential, in that its effects can always be removed by centering the regressors. Furthermore, many analysts prefer to use the λ's, based on centered and scaled x's, because of the connection with principal components. Consequently we feel that unless specific information about multicollinearity involving the intercept is desired, the regressors should always be centered. For further details, see Belsley (1984), Gunst (1983), and Montgomery and Peck (1992, chap. 8).

16.5 HARMFUL EFFECTS OF MULTICOLLINEARITY

The presence of multicollinearity causes several potentially serious problems with the least-squares estimator. A major problem is stability of the regression coefficients, that is, since $V(\hat{\beta}_j) = \sigma^2 C_{jj} = \sigma^2(1 - R_j^2)^{-1}$, strong multicollinearity involving x_j will cause the variance of the least-squares regression coefficient $\hat{\beta}_j$ to be very large. Similarly the covariance between $\hat{\beta}_i$ and $\hat{\beta}_j$ may be large if either regressor is involved in a multicollinearity. This instability for some of the elements of $\hat{\beta}$ is potentially harmful, in that different samples (or modifications of the present sample, such as deletion of an observation) may lead to dramatically different estimates of β, with completely different associated inferences. On the other hand, if the analyst's interest centers on regressors *not* involved in the multicollinearity, then no problem exists. Furthermore, if σ^2 is small, the variance of $\hat{\beta}_j$ may be small enough for specific inference purposes (such as testing whether or not $\beta_j = 0$ or obtaining a confidence interval of reasonable width), regardless of the presence of multicollinearity.

Multicollinearity also tends to produce least-squares coefficients that are too large in absolute value. The expected squared distance from $\hat{\beta}$ to the vector of true parameters β is

$$E(L_1^2) = E(\hat{\boldsymbol{\beta}} - \boldsymbol{\beta})'(\hat{\boldsymbol{\beta}} - \boldsymbol{\beta}) \tag{16.17}$$

$$= \sigma^2 \sum_{j=1}^{p} \frac{1}{\lambda_j}$$

Thus if the X′X matrix is ill-conditioned because of multicollinearity, at least one of the λ_j will be small, and Eq. (16.17) implies that the distance from the least-squares estimate $\hat{\boldsymbol{\beta}}$ to $\boldsymbol{\beta}$ may be large. Equivalently, we can show that

$$E(\hat{\boldsymbol{\beta}}'\hat{\boldsymbol{\beta}}) = \boldsymbol{\beta}'\boldsymbol{\beta} + \sigma^2 \sum_{j=1}^{p} \frac{1}{\lambda_j} \tag{16.18}$$

that is, the squared length of $\hat{\boldsymbol{\beta}}$ is too long.

While least squares generally produces poor estimates of individual model parameters when multicollinearity is present, this does not always imply that the fitted model is a poor predictor. For example, consider prediction at the point $\mathbf{x}_i$, where x_i is the ith row of $\mathbf{X}$. The predicted value is $\hat{y}_{\mathrm{x}_i} = \mathrm{x}_i'\hat{\beta}$ and the variance of $\hat{y}_{\mathrm{x}_i}$ is

$$V(\hat{y}_{\mathrm{x}_i}) = \sigma^2 \mathrm{x}_i'(\mathbf{X}'\mathbf{X})^{-1}\mathbf{x}_i$$

$$= \sigma^2 \mathbf{x}_i' \left(\sum_{j=1}^{p} \frac{1}{\lambda_j} \mathbf{t}_j \mathbf{t}_j' \right) \mathbf{x}_i \tag{16.19}$$

Suppose that the eigenvalues $\lambda_1, \lambda_2, \ldots, \lambda_r$ $(r < p)$ describe the multicollinearity in the data. Thus λ_j, $j = 1, 2, \ldots, r$, are small and the corresponding $1/\lambda_j$ terms in Eq. (16.19) are large. However, for each of the eigenvectors $\mathbf{t}_1, \mathbf{t}_2, \ldots, \mathbf{t}_r$ associated with $\lambda_1, \lambda_2, \ldots, \lambda_r$ the product $x_i' t_j = 0$, which offsets the effect of the small eigenvalue. The net effect of this is that even though the individual elements of $\boldsymbol{\beta}$ are estimated poorly, the linear combination $x_i'\beta$ may be estimated quite well, provided that $\mathbf{x}_i$ is close to the original cloud of points used to fit the model. However, if $\mathbf{x}_i$ is far away from this cloud, the variance of $\hat{y}_{\mathrm{x}_i}$ may be large, and the predicted value may be a poor estimate of the true response.

For other discussions of the harmful effects of multicollinearity, see Farrar and Glauber (1967), Silvey (1969), Marquardt (1970), Marquardt and Snee (1975), Mason, Gunst, and Webster (1975), Belsley, Kuh, and Welsch (1980), Mandel (1989), and Montgomery and Peck (1992).

16.6 REMEDIAL MEASURES

There are three basic strategies for dealing with multicollinearity. The first is collecting additional data in a region of the x space designed to break up the near-linear dependencies in the original sample. Farrar and Glauber (1967) and Silvey (1969) suggest this as the best technique for treating the problem. Wang, Tse, and

Chow (1990) show that at least r additional data points need to be added in order to remove r multicollinearities. However, additional data are not always available, and in cases where the multicollinearity results from constraints on the model or in the population, the technique is not a viable alternative since the new data will closely resemble the old.

The second recommended approach is model respecification, that is, redefining the regressors or eliminating some of the regressors involved in the multicollinearity. Variable elimination is often a highly effective technique. However, it may not provide a satisfactory solution if the regressors removed have significant explanatory power relative to the response y. Thus removing regressors can in some situations damage the predictive performance of the model. Furthermore, care must be exercised in variable selection, because many selection procedures are seriously distorted by multicollinearity. In general, there is no assurance that the final model will exhibit any lesser degree of multicollinearity than was present in the original data. For example, see Montgomery and Peck (1992), who built several subset regression models for Marquardt and Snee's acetylene data. Generally, the all possible regressions procedure is the safest variable selection strategy.

Since multicollinearity can seriously distort the method of least squares, several alternative estimation methods have been suggested. The resulting estimators are biased, but may be preferable to least squares in a mean square error sense. Perhaps the most popular of these methods is ridge regression (Hoerl and Kennard, 1970*a*, *b*) in which one solves

$$\hat{\boldsymbol{\beta}}_R = (\mathbf{X}'\mathbf{X} + k\mathbf{I})^{-1}\mathbf{X}'\mathbf{y} \tag{16.20}$$

where $k > 0$ is a scalar constant specified by the analyst, often called the *biasing parameter*. (See Hoerl and Kennard, 1970*a*, *b*, 1976; Hoerl, Kennard, and Baldwin, 1975; McDonald and Galarneau, 1975; Dempster, Schatzoff, and Wermuth, 1977; Gibbons, 1979; He and Hayya, 1989; and Mason and Blaylock, 1991, for discussion of the choice of k.) An excellent review of the practical use of ridge regression is Marquardt and Snee (1975). Many analysts prefer to examine a ridge trace (a plot of $\hat{\boldsymbol{\beta}}_R$ versus k) and empirically select a value of k that stabilizes the regression coefficients. Generally, bias increases as k increases, so small values of k consistent with reasonable coefficient stability are preferred.

Other alternatives to least squares include generalized ridge regression in which kI in Eq. (16.20) is replaced by arbitrary **K**, principal components regression (see, for example, Hawkins, 1973; and Hill and Judge, 1990), fractional rank estimators (see Marquardt, 1970), latent root regression (see Webster, Gunst, and Mason, 1974), the shrunken least-squares estimator of Stein (1960), and the variant of Stein shrinkage discussed in Sclove (1968). Lee and Bilrch (1988) propose a fractional principal component estimator that uses different portions of the individual principal components. Hocking, Speed, and Lynn (1976) note that all of these estimators except latent root regression form a *class* of linear estimators of the form

$$\boldsymbol{\alpha}^* = \mathbf{B}\,\hat{\boldsymbol{\alpha}} \tag{16.21}$$

where B is a diagonal matrix. The choice of the diagonal elements of **B** determines the specific estimator. For example, if

$$b_{ii} = \frac{\lambda_i}{\lambda_i + k}$$

the ridge estimator results, while

$$b_{ii} = \frac{\lambda_i}{\lambda_i + k_i}$$

yields the generalized ridge estimator. Hocking, Speed, and Lynn (1976) show that closed-form estimates of the biasing parameters may be obtained in most cases, except for ridge regression.

A number of Monte Carlo simulation studies have been conducted to examine the effectiveness of biased estimators and to attempt to determine which procedures perform best. For example, see McDonald and Galarneau (1975), Hoerl and Kennard (1976), Hoerl, Kennard, and Baldwin (1975), Gunst and Mason (1977), Dempster, Schatzoff, and Wermuth (1977), and Gibbons (1979). While no single procedure emerges from these studies as best overall, there is considerable evidence indicating the superiority of biased estimation to least squares if multicollinearity is present. Our own preference in practice is for ordinary ridge regression, with the biasing parameter selected by inspection of the ridge trace. It is also useful to find the models associated with the biasing parameters suggested by Hoerl and Kennard (1976) and Hoerl, Kennard, and Baldwin (1975) and compare them with the model obtained via the ridge trace.

Biased estimation can also be viewed as constrained least-squares estimation, where the constraint is on some function of the length of the vector of estimates $\hat{\boldsymbol{\beta}}$. This is equivalent to incorporating prior information about $\boldsymbol{\beta}$ into the analysis. Consequently, biased estimation is at least computationally equivalent to bayesian estimation in the linear regression model (see Leamer, 1973, 1978; Lindley and Smith, 1972; Zellner, 1971; and Gruber, 1990) and the mixed estimation technique of Theil (1963) and Theil and Goldberger (1961). There is some difference in the viewpoint adopted. Bayesian estimation requires complete specification of a prior distribution for β, while mixed estimation is less formal, allowing introduction of prior information without complete specification of the prior. The mixed estimator is considered an unbiased estimator of $\boldsymbol{\beta}$.

The use of biased estimators in regression has sparked considerable controversy. Critical discussions of ridge regression are given in Conniffe and Stone (1973), Draper and Van Nostrand (1979), Smith and Campbell (1980), and Sundberg (1993). The objections voiced include the observation that ridge regression shrinks nonsignificant regression coefficients more severely than significant ones, and that ridge regression is really only appropriate in situations when external information can be added to a least-squares problem. Several technical details of the simulation studies of biased estimation are also criticized, suggesting that the simulations have been designed to favor biased estimators since the data are usually generated subject to a constraint on the length of $\boldsymbol{\beta}$, exactly the situation for which biased estimation is appropriate.

Despite the objections noted, we believe that biased estimation methods are useful techniques that the analyst should consider when dealing with multicollinearity. Biased estimation methods compare very favorably with other methods for handling multicollinearity, such as variable elimination. As Marquardt and Snee (1975) observe, it is often better to use some of the information in all of the regressors, as biased estimation does, than to use all of the information in some regressors and none of the information in others, as variable selection does. Furthermore, variable elimination often produces biased estimators if important regressors are deleted from the model. In effect, variable elimination procedures often shrink the vector of least-squares estimates, as does biased estimation. We do not recommend the mechanical use of biased estimation without thoughtful study of the data and careful analysis of the adequacy of the final model. Properly used, biased estimation methods are a useful tool in the data analyst's kit.

16.7 EXAMPLE

We now return to the electric utility data in Table 16.1 and apply the multicollinearity diagnostics discussed in Sec. 16.4. The **X′X** matrix in correlation form is

$$\mathbf{X'X} = \begin{array}{c} \\ \\ \\ \\ \\ \end{array}\begin{array}{ccccc} x_1 & x_2 & x_3 & x_4 & x_5 \end{array}$$

$$\mathbf{X'X} = \begin{bmatrix} 1.0000 & 0.9972 & 0.8488 & 0.9493 & -0.1161 \\ 0.9972 & 1.0000 & 0.8456 & 0.9479 & -0.1065 \\ 0.8488 & 0.8456 & 1.0000 & 0.7542 & -0.0797 \\ 0.9493 & 0.9479 & 0.7542 & 1.0000 & -0.1567 \\ -0.1161 & -0.1065 & -0.0797 & -0.1567 & 1.0000 \end{bmatrix} \begin{matrix} x_1 \\ x_2 \\ x_3 \\ x_4 \\ x_5 \end{matrix}$$

(columns x_1, x_2, x_3, x_4, x_5)

Note the high correlation between x_1 and x_2, x_1 (or x_2) and x_4, and x_1 (or x_2) and x_3. Pairwise correlation coefficients this large usually indicate strong multicollinearity. The variance inflation factors, eigenvalues, condition indices of **X′X**, and eigenvectors associated with the five-regressor least-squares model of these data are presented in Table 16.3. Notice that the variance inflation factors and eigenvalues agree with those computed by SAS PROC REG in Table 16.2, but the condition indices of **X′X** in Table 16.3 are the squares of the condition indices of **X** reported in Table 16.2. The eigenvalues were calculated assuming that the x's were centered and scaled so that the effect of the intercept on any multicollinearity has been removed. VIF_1, VIF_2, and VIF_4 exceed 10, indicating multicollinearity. As noted previously, both VIF_1 and VIF_2 exceed 150, so we suspect that the near-linear dependency between these regressors is moderately strong. The largest condition index is $\kappa_5 = 1343.54$, implying strong multicollinearity.

TABLE 16.3 Multicollinearity Diagnostics for the Residential Load Survey Data

	Variables				
	x_1	x_2	x_3	x_4	x_5
VIF_j	191.30	180.67	3.96	11.49	1.06
λ_j	3.697	0.9838	0.2666	0.04999	0.002752
$\kappa_j = \dfrac{\lambda_{max}}{\lambda_j}$	1.00	3.76	13.87	73.95	1343.54
	Eigenvectors				
	t_1	t_2	t_3	t_4	t_5
	– 0.51503	– 0.04531	0.15280	0.44056	0.71781
	– 0.51423	– 0.05474	0.16270	0.47081	– 0.69602
	– 0.46485	– 0.08093	– 0.84835	– 0.23992	– 0.01080
	– 0.49698	0.00907	0.47664	–0.72497	– 0.01251
	0.08509	– 0.99414	0.05749	– 0.03309	0.00638

If we use the elements of the 5th eigenvector t_5, which is associated with the smallest eigenvalue, and substitute into Eq. (16.7), we obtain

$$0.71781x_1 - 0.69602x_2 - 0.01080x_3 - 0.01251x_4 + 0.00638x_5 = 0$$

Assuming the last three coefficients to be approximately zero, this becomes

$$x_1 = 0.9696x_2 \approx x_2$$

Thus x_1 and x_2 are nearly linearly dependent. This was observable directly from the pairwise correlations, but in general, remember that the r_{ij} cannot be considered a reliable diagnostic. We also reached this conclusion by examination of the variance decomposition proportions (the π matrix) that were computed by SAS PROC REG and displayed in Table 16.2.

If we consider the second largest condition index (second smallest eigenvalue) $\kappa_4 = 73.95$ and the elements of the eigenvector t_4, we obtain

$$0.44056x_1 + 0.47081x_2 = 0.23992x_3 + 0.72497x_4$$

assuming that $t_{54} = -0.03309 \approx 0$. Thus, there may be a mild dependency between x_1, x_2, x_3, and x_4, which would account for VIF_4 exceeding 10. Considering the context of the problem, it would not be surprising to find that family income, house size, air-conditioning capacity, and total connected load of electric appliances are nearly linearly dependent.

Based on the above diagnostics, we conclude that the data exhibit moderately strong multicollinearity, and our expectations of developing an adequate least-squares model involving all five regressors should be modest. Table 16.4 summarizes several least-squares models of the residential load data. The usefulness of these models can best be ascertained by determining how well they perform in a forecasting or prediction environment. Table 16.5 presents residential load survey data collected for 20 additional customers. Row 1 of Table 16.4 summarizes estimation and prediction results for the least-squares fit to all five regressors using the first 40 observations (Table 16.1). The last two columns display the sum of prediction errors and the sum of squared prediction errors experienced when the full model is used to predict the 20 new observations in Table 16.5. The sum of squared prediction errors is relatively large. The mean-square prediction error is 49.4348/20 = 2.47, nearly two orders of magnitude larger than the residual mean square from the least-squares fit to the original data (0.0575). Multicollinearity has severely impacted the ability of this model to predict fresh data.

Table 16.4 also displays several subset regression models for these data. Rows 1 and 2 give the minimum residual mean square and minimum C_p models from all possible regressions. Rows 3, 4, and 5 present models obtained from the usual forward, backward, and stepwise algorithms. In the stepwise-type procedures, a nominal 10% level was used for inserting and removing variables. Notice that the all possible regressions model performs poorly in prediction because it contains x_1 and x_2. The predictive performance of the models obtained from the stepwise-type algorithms is considerably better, although all three procedures produce different final models. Two of the models still exhibit moderate multicollinearity, which would cause a prudent analyst to interpret the model coefficients with great care.

Table 16.6 presents estimation and prediction results for several biased estimators, including ordinary ridge, principal components, the fractional rank estimator, fully iterated generalized ridge, and a ridge procedure with the biasing parameter

TABLE 16.4 Fitted Least-Squares Models and Predictive Performance

		Coefficient estimates/VIF						Predictive performance	
Method	Model	$\hat{\beta}_0$	$\hat{\beta}_1$	$\hat{\beta}_2$	$\hat{\beta}_3$	$\hat{\beta}_4$	$\hat{\beta}_5$	$\sum_{i=41}^{60} e_i$	$\sum_{i=41}^{60} e_i^2$
Full model and MSE	x_1, x_2, x_3, x_4, x_5	– 0.0416	– 2.4976 (191.30)	0.2695 (180.67)	0.4143 (3.96)	0.3793 (11.49)	0.0304 (1.06)	14.4383	49.4348
Mnimum C_p	x_1, x_2, x_3, x_4	0.0729	– 2.6336 (189.32)	0.2851 (177.43)	0.4137 (3.96)	0.3614 (11.22)	—	14.8054	52.9357
Forward	x_2, x_3, x_4, x_5	– 0.2321	—	0.0618 (16.43)	0.3957 (3.81)	0.3361 (11.04)	0.0361 (1.04)	4.7570	2.9418
Backward	x_3, x_4	0.2627	—	—	0.4541 (2.32)	0.5029 (2.32)	—	4.3171	2.4734
Stepwise	x_2, x_3, x_4	– 0.1070	—	0.0670 (16.17)	0.3939 (2.54)	0.3117 (3.21)	—	4.5665	2.8983

TABLE 16.5 Additional Residential Load Survey Data

Customer number	y, kW	x_1, ft^2/1000	x_2, \$/1000	x_3, tons	x_4, kW	x_5
41	4.385	1.598	17.604	3.5	3.914	4
42	4.505	1.868	20.614	3.0	4.817	6
43	3.958	2.036	22.277	2.5	5.581	1
44	6.071	2.598	27.924	5.5	6.233	2
45	4.571	2.204	24.587	2.0	6.073	2
46	6.849	3.313	30.016	6.5	6.054	1
47	2.610	1.685	18.485	1.0	3.677	2
48	5.829	2.379	26.341	3.5	7.345	4
49	4.755	2.286	25.327	3.0	5.230	5
50	2.646	1.512	17.351	1.0	3.824	1
51	4.279	2.246	24.612	2.0	5.950	2
52	3.701	1.819	19.754	2.5	4.522	1
53	4.561	2.288	25.720	3.0	6.313	4
54	4.707	2.320	25.444	3.0	5.579	4
55	4.541	2.276	25.439	3.0	6.113	3
56	3.891	1.994	22.150	2.0	5.316	4
57	6.394	3.454	27.855	4.0	7.215	2
58	6.528	3.490	31.998	5.0	7.145	2
59	2.925	1.936	21.215	1.0	4.506	4
60	5.059	4.061	22.962	3.0	5.854	1

suggested by Hoerl, Kennard, and Baldwin (1975). Refer to Montgomery and Peck (1992) for detailed discussions of these estimators. The ridge trace for these data is shown in Fig. 16.1. The large changes in the regression coefficients for x_1 and x_2 are typical of models where strong multicollinearity is present. Notice that the coefficients stabilize rapidly, and for any biasing parameter in the range of $0.025 < k < 0.10$ the models should be very similar. Choosing $k = 0.032$ gives a positive coefficient for x_1, an intuitively appealing result. Inspection of Table 16.6 reveals that the predictive performance of this ordinary ridge model is dramatically better than least squares. The ridge model that uses the choice of k (row 4) of Hoerl, Kennard, and Baldwin (1975) also performs reasonably well. Neither the principal components, fractional rank, nor generalized ridge model improved predictive performance with respect to least squares. The fully iterated generalized ridge method often shrinks the regression coefficients too much, and this may be responsible for its poor performance. In any event, this example points out the need for a careful validation of regression models, particularly when the multicollinearity problem is present. Data splitting, such as used here, is an important validation technique.

It should be noted that careful examination of the "prediction" data (the last 20 customers) would reveal that several of these new customers are rather different than the original 40 customers that were used to build the regression model. Figure 16.2 presents a plot of all 60 customers in the x_1x_2 plane. Note that customers 46, 57, 58, and 60 appear "unusual" and, in fact, prediction at these points constitutes extrapolation with the regression model, something we are usually admonished to avoid. However, this application of regression is not unusual in the forecasting environment, where we are frequently required to generate predictions that we

TABLE 16.6 Predictive Performance for Biased Estimation Models

Model	Coefficient estimates						Predictive performance	
	$\hat{\beta}_0$	$\hat{\beta}_1$	$\hat{\beta}_2$	$\hat{\beta}_3$	$\hat{\beta}_4$	$\hat{\beta}_5$	$\sum_{i=41}^{60} e_i$	$\sum_{i=41}^{60} e_i^2$
Ordinary ridge, from ridge trace	– 0.3109	0.2369	0.0587	0.3644	0.2896	0.0332	3.8120	2.2114
Principal components	– 0.0416	– 2.4976	0.26947	0.41430	0.3793	0.030351	14.4421	49.4418
Fractional rank	– 0.09462	– 2.4255	0.27638	0.40387	0.33714	0.02871	14.2270	47.4952
Ridge ($k = p\sigma^2/\boldsymbol{\beta}'\boldsymbol{\beta}$)	– 0.17150	– 0.66305	0.11605	0.39918	0.35063	0.03433	7.3143	8.7582
Generalized ridge	– 0.11862	– 1.9881	0.23889	0.39457	0.33957	0.02351	12.4399	34.6022

know are extrapolations, yet the regression model is the only available forecasting technique. The example simply points out that extrapolation of a least-squares model in the presence of multicollinearity is often much more dangerous than if some reasonable alternative estimation technique has been used. To predict well in extrapolation we must have the correct model form and good coefficient estimates. This latter condition may not be met using least squares if multicollinearity is present.

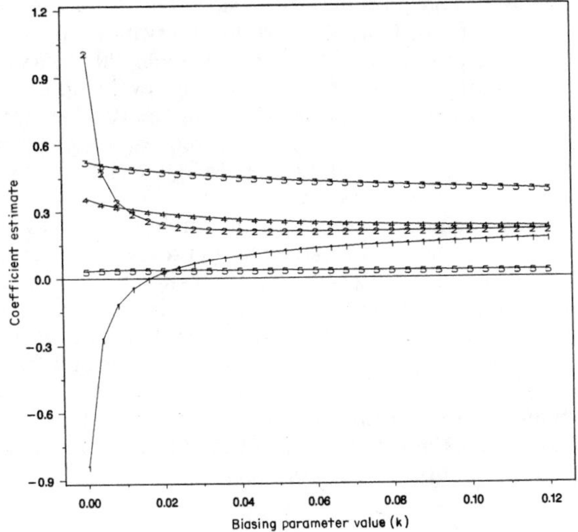

FIGURE 16.1 Ridge trace for residential load survey data.

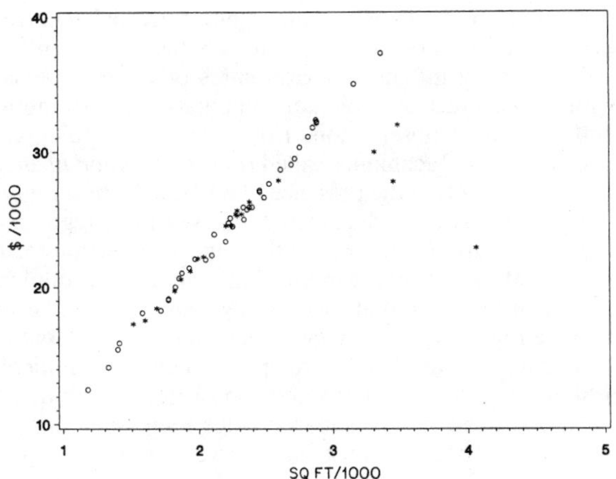

FIGURE 16.2 Scatter diagram of income versus square feet.
°=original 40 customers; *=20 new customers

16.8 COMPUTING

Although several software programs perform some portions of multicollinearity detection and biased estimation, SAS performs extensive multicollinearity diagnostics and several forms of biased estimation. SAS executes several variations of ridge regression using alternative biased estimation procedures, and also runs principal components regression. The SAS code shown in Fig. 16.3 performs several functions including data management, regression diagnostics, parameter estimation, biased estimation, and plotting. The residential load survey data is input including all 60 observations. Only the first 40 observations are used for model building and diagnostic purposes, so that the remaining 20 observations may be used for model validation and predictive performance evaluation.

Multicollinearity diagnostics are performed using the REG (regression) procedure. The procedure and associated options request output containing the ANOVA table and parameter estimates (Table 16.2), the $\mathbf{X}'\mathbf{X}$ matrix, VIFs, eigenvalues, condition numbers, and π matrix.

Data from the least-squares regression is then used to perform biased estimation. SAS first computes the biasing parameter (k) suggested by Hoerl, Kennard, and Baldwin (1975), and then performs ridge regression using their estimate of k. Another popular technique to determine k is analysis of the ridge trace. SAS is used to estimate regression models for various values of k. All the parameter estimates ($\hat{\boldsymbol{\beta}}_R$) are then plotted against k to determine the value of k that stabilizes $\hat{\boldsymbol{\beta}}_R$. A value for k is determined by inspection of the ridge trace and used in ridge regression estimation. SAS may also be used to determine the predictive performance of the least-squares model and each of the biased estimation models. The code for this analysis is not provided in this section.

16.9 ROBUST RIDGE REGRESSION

In addition to experiencing near linear dependencies among the regressors, practitioners often encounter observations that are significantly different from the majority of data. These distant observations, known as outliers, frequently occur in real data and can often heavily influence least-squares or ridge regression estimates. Outliers occur for reasons such as coding or computation errors, interchanging values, inadvertently including observations from different populations, equipment failures, or transient effects. Techniques called *robust regression methods* have been developed to estimate models in the presence of outliers. Robust regression estimation is designed to be less sensitive than least squares or ridge regression to outliers, resulting in improved fits to the majority of the data. This section focuses on methods designed to combat the combined multicollinearity-outlier problem.

Methods have been proposed that successfully combine biased estimation techniques, such as ridge regression, with robust techniques. A variety of these robust ridge regression techniques have been developed. When both multicollinearity and outliers are present, several of the proposed robust ridge techniques significantly improve on the alternatives of least squares, ridge regression, and robust regression. The suggested sequence of robust ridge regression estimation is outlined in the following steps.

1. Use robust estimation to obtain initial parameter estimates ($\hat{\boldsymbol{\beta}}$) and mean square error ($\hat{\sigma}^2$) for determining the ridge regression biasing parameter k.

```
/* data management section, input the data file */
FILENAME RESDAT DISK 'd:\docs\colnchap\resload.txt';

 /* read in the entire dataset */
DATA TEMP ;
 INFILE RESDAT;
 INPUT Y X1 X2 X3 X4 X5;

 /* restrict the data to the first 40 observations */
DATA TEMP2;
 SET TEMP (OBS=40);
 RUN;

 /* perform multicollinearity diagnostics */
PROC REG DATA=TEMP2 CORR OUTSTB OUTEST=REGOUT;
 LS: MODEL Y=X1-X5 / RIDGE=0.0 XPX STB VIF COLLINOINT;
 TITLE 'LEAST SQUARES REGRESSION ON OBSERVATIONS 1-40';
 RUN;

 /* find the Hoerl, Kennard, and Baldwin biasing constant */
DATA K_INIT;
 SET REGOUT;
 IF _TYPE_='RIDGESTB';
 NPARM=5;
 SUMBSQ=USS(X1,X2,X3,X4,X5);
 K_HKB=(NPARM*_RMSE_**2)/SUMBSQ;
 RUN;

 /* using the HKB constant in ridge regression */
PROC REG DATA=TEMP2 OUTSTB OUTEST=RIDGEHKB;
 RHKB: MODEL Y=X1-X5 / NOPRINT RIDGE=0.13481;
 RUN;

 /* form ridge regression models at various k values for the ridge trace */
PROC REG DATA=TEMP2 OUTSTB OUTEST=RIDGOUT;
 MODEL Y=X1-X5/ NOPRINT RIDGE=0 TO 0.12 BY 0.004;
 RUN;

/* set the options for the ridge trace plot */
GOPTIONS RESET=ALL FTEXT=SWISS DEVICE=HPLJ300 HSIZE=6.5 HORIGIN=0.75
     VSIZE=6.5 VORIGIN=4.0 HTEXT=1.5 HPOS=50 VPOS=80;
 SYMBOL1 V=1 C=BLACK I=JOIN L=1;
 SYMBOL2 V=2 C=BLACK I=JOIN L=1;
 SYMBOL3 V=3 C=BLACK I=JOIN L=1;
 SYMBOL4 V=4 C=BLACK I=JOIN L=1;
 SYMBOL5 V=5 C=BLACK I=JOIN L=1;

  /* plot the ridge trace, see figure 15.1 */
 PROC GPLOT DATA=RIDGOUT;
  TITLE1 H=3 'Ridge Trace Plot';
  TITLE2 'Residential Load Survey Data';
  FOOTNOTE J=L ' Note: only first 40 observations';
  PLOT (X1 X2 X3 X4 X5)*_RIDGE_
          / OVERLAY FRAME
           VREF=0 LVREF=1
           HMINOR=1 VMINOR=1
           VAXIS=AXIS1
           HAXIS=AXIS2;
  AXIS1 ORDER = (-0.9 TO 1.2 BY 0.30)
      LABEL = ('Coefficient' justify=right 'Estimate')
      WIDTH = 2;
  AXIS2 ORDER = (0 TO 0.12 BY 0.02)
      LABEL = ('Biasing Parameter Value (k)')
      WIDTH = 2
      OFFSET = (2);
  RUN;

  /* regression model for the k value selected using the ridge trace plot */
 PROC REG DATA=TEMP2 OUTEST=RTRACE;
  RTRACE: MODEL Y=X1-X5/ RIDGE=0.032 NOPRINT;
  TITLE 'RIDGE REGRESSION USING THE RIDGE TRACE BIASING PARAMETER';
  RUN;
```

FIGURE 16.3 SAS code to perform multicollinearity diagnostics and ridge regression.

2. Estimate the ridge regression biasing parameter. As is suggested in previous sections, the ridge trace, the one-step method of Hoerl, Kennard, and Baldwin (1975), and the iterative technique proposed by Hoerl and Kennard (1976) are reasonable approaches.
3. Augment the **X** matrix, using the estimated biasing parameter, and the vector of responses as shown in Montgomery and Peck (1992, pp. 334–336).
4. Perform augmented robust estimation using a computational approach (iteratively reweighted least squares) that can operate on an augmented **X** matrix.

Robust estimation is a compromise between fitting a model to the entire set of data and fitting a model without outliers. Recall that the objective function in least squares, summing the squared residuals, causes outliers to have excessive influence on the parameter estimates. Robust method objective functions are more gradual functions of the residuals, typically of the form

$$\min_{\boldsymbol{\beta}} \sum_{i=1}^{n} \rho(y_i - \mathbf{x}_i'\hat{\boldsymbol{\beta}}) \tag{16.22}$$

The function ρ is a function of the residuals that is often related to the likelihood function for an appropriate error distribution. The class of robust estimates using this approach is called *M*-estimation. A number of ρ functions have been proposed and several perform well in general. The objective function is normally minimized by using an iterative technique to solve the resulting nonlinear system of normal equations of the form

$$\sum_{i=1}^{n} \psi\,(e_i / s)\mathbf{x}_i = 0$$

where ψ is the derivate of ρ and x_i is the row vector of regressor variables of the ith observation. An estimate of scale s is needed to standardize the residuals because the solution to the normal equations is not equivariant with respect to a magnification of the y-axis. The most common approach for solving this system is iteratively reweighted least squares (IRLS), resulting in an estimator of the form

$$\hat{\boldsymbol{\beta}} = (\mathbf{X}'\mathbf{W}\mathbf{X})^{-1}\mathbf{X}'\mathbf{W}\mathbf{y}$$

where W $= diag(w_1, w_2, \ldots, w_i)$ and $w_i = \psi(e_i / s)(e_i / s)$. This approach is widely used because any ordinary least-squares package can be used to perform the estimation. The robust initial estimates from step 1 of the robust ridge sequence are found using a specified ρ-function and IRLS. Once the **X** matrix becomes augmented in step 3, the resulting robust ridge estimator is found by

$$\hat{\boldsymbol{\beta}} = (\mathbf{X}'\mathbf{W}\mathbf{X} + k\mathbf{I})^{-1}\mathbf{X}'\mathbf{W}\mathbf{y} \tag{16.23}$$

where $\mathbf{W} = diag(w_1, w_2, \ldots, w_n)$ and $w_i = \psi\,(e_i / s) / (e_i / s)$, as in *M*-estimation above. By using this approach of first finding the biasing parameter and then performing robust estimation, the weighting matrix is a function of the shrinkage parameter k. This distinction is important based on the findings of Walker and Birch (1988), who show that the influence of each observation depends largely on the value of k.

Although robust ridge techniques had been proposed previously, Askin and Montgomery (1980) were the first to propose the sequence of ridge followed by robust estimation. Subsequently, the proposals of Walker (1987) and Simpson (1995) have enhanced the robust aspects of the Askin-Montgomery combined estimator. Simulation studies by both authors demonstrate that their methods are nearly as efficient as least squares when neither problem exists, are comparable to ridge regression when only multicollinearity exists, and demonstrate substantial improvements over least squares and ridge regression when both multicollinearity and outliers are present.

16.10 CONCLUSION

We have discussed how the multicollinearity problem can impact the uses of regression. The effects of multicollinearity on the method of least squares can be quite serious, including poor precision of estimation and substantial losses in prediction accuracy. The user is faced with the problem of diagnosing the presence of multicollinearity, assessing its effect on the problem at hand, and applying appropriate remedial measures. Fortunately there are several techniques available to assist the data analyst in assessing the extent of multicollinearity in regression data and in model building. Through the careful use of these techniques, the harmful effects of multicollinearity can usually be minimized.

REFERENCES

Askin, R. G., and D. C. Montgomery: "Augmented Robust Estimators," *Technometrics*, vol. 22, pp. 333–341, 1980.

Belsley, D. A.: "Demeaning Regression Diagnostics through Centering (with Discussion)," *Am. Stat.*, vol. 38, pp. 73–93, 1984.

_____, E. Kuh, and R. E. Welsch: *Regression Diagnostics: Identifying Influential Data and Sources of Collinearity*, Wiley, New York, 1980.

Conniffe, D., and J. Stone: "A Critical View of Ridge Regression," *Appl. Stat.*, vol. 22, pp. 181–187, 1973.

Dempster, A. P., M. Schatzoff, and N. Wermuth: "A Simulation Study of Alternatives to Ordinary Least Squares," *J. Am. Stat. Assoc.*, vol. 72, pp. 77–90, 1977.

Draper, N. R., and R. C. Van Nostrand: "Ridge Regression and James-Stein Estimators: Review and Comments," *Technometrics*, vol. 21, pp. 451–466, 1979.

Farrar, D. E., and R. R. Glauber: "Multicollinearity in Regression Analysis: The Problem Revisited," *Rev. Econ. Stat.*, vol. 49, pp. 92–107, 1967.

Gibbons, D. G.: "A Simulation Study of Some Ridge Estimators," Rep. GMR-2659 (rev. ed.), General Motors Research Labs., Math. Dept., Warren, MI., 1979.

Gruber, M. H. J.: *Regression Estimators: A Comparative Study*, Academic Press, New York, 1990.

Gunst, R. F.: "Regression Analysis with Multicollinear Predictor Variables: Definition, Detection and Effects," *Commun. Stat.*, vol. A12, pp. 2217–2260, 1983.

_____ and R. L. Mason: "Biased Estimation in Regression: An Evaluation Using Mean Squared Error," *J. Am. Stat. Assoc.*, vol. 72, pp. 616–628, 1977.

Hawkins, D. M.: "On the Investigation of Alternative Regressions by Principal Components Analysis," *Appl. Stat.*, vol. 22, pp. 275–286, 1973.

He, X. X., and J. C. Hayya: "Selection of the Biasing Parameter in Ridge Regression," *Proceedings of the American Statistical Association, Business and Economic Statistics Section*, pp. 310–315, 1989.

Hill, R. C., and G. G. Judge: "Improved Estimation under Collinearity and Squared Error Loss," *J. Multivariate Analysis*, vol. 32, pp. 296–312, 1990.

Hocking, R. R.: "Developments in Linear Regression Methodology: 1959–1982," *Technometrics*, vol. 25, pp. 219–230, 1983.

_____, F. M. Speed, and M. J. Lynn: "A Class of Biased Estimators in Linear Regression," *Technometrics*, vol. 18, pp. 425–437, 1976.

Hoerl, A. E., and R. W. Kennard: "Ridge Regression: Biased Estimation for Nonorthogonal Problems," *Technometrics*, vol. 12, pp. 55–67, 1970*a*.

_____ and _____: "Ridge Regression: Applications to Nonorthogonal Problems," *Technometrics*, vol. 12, pp. 69–82, 1970*b*.

_____ and _____: "Ridge Regression: Iterative Estimation of the Biasing Parameter," *Commun. Stat.*, vol. A5, pp. 77–88, 1976.

_____, _____, and K. F. Baldwin: "Ridge Regression: Some Simulations," *Commun. Stat.*, vol. 4, pp. 105–123, 1975.

Lane, M. S., A. H. Mansour, and J. L. Harpell: "Operations Research Techniques: A Longitudinal Update 1973–1988," *Interfaces*, vol. 23, pp. 63–68, 1993.

Lawless, J. F.: "Ridge and Related Estimation Procedures: Theory and Practice," *Commun. Stat.*, vol. A7, pp. 139–164, 1978.

_____ and P. Wang: "A Simulation of Ridge and Other Regression Estimators," *Commun. Stat.*, vol. A5, pp. 307–323, 1976.

Leamer, E. E.: "Multicollinearity—A Bayesian Interpretation," *Rev. Econ. Stat.*, vol. 55, pp. 371–380, 1973.

_____: *Specification Searches: Ad Hoc Inference with Nonexperimental Data*, Wiley, New York, 1978.

Ledbetter, W. N., and J. F. Cox: "Are OR Techniques Being Used," *Ind. Eng.*, pp. 19–21, 1977.

Lee, W., and J. B. Birch: "Fractional Principal Components Regression: A General Approach to Biased Estimators," *Communications in Statistics–Simulation and Computation*, vol. 17, pp. 713–727, 1988.

Lindley, D. V., and A. F. M. Smith: "Bayes Estimates for the Linear Model (with Discussion)," *J. R. Stat. Soc.*, ser. B., vol. 34, pp. 1–41, 1972.

Mandel, J.: "The Nature of Collinearity," *J. Qual. Tech.*, vol. 21, pp. 268–276, 1989.

Marquardt, D. W.: "Generalized Inverses, Ridge Regression, Biased Linear Estimation, and Nonlinear Estimation," *Technometrics*, vol. 12, pp. 591–612, 1970.

_____ and R. D. Snee: "Ridge Regression in Practice," *Am. Stat.*, vol. 29, pp. 3–20, 1975.

Mason, R. L., and N. W. Blaylock: "Ridge Regression Estimator Comparisons Using Pitman's Measure of Closeness," *Communications in Statistics-Theory and Methods*, vol. 20, pp. 3629–3641, 1991.

_____, R. G. Gunst, and J. T. Webster: "Regression Analysis and Problems of Multicollinearity," *Commun. Stat.*, vol. 4, pp. 277–292, 1975.

McDonald, G. C., and D. I. Galarneau: "A Monte Carlo Evaluation of Some Ridge-Type Estimators," *J. Am. Stat. Assoc.*, vol. 71, pp. 596–604, 1975.

Montgomery, D. C., and R. G. Askin: "Problems of Nonnormality and Multicollinearity for Forecasting Methods Based on Least Squares," *AIIE Trans.*, vol. 13, pp. 102–115, 1981.

_____ and E. A. Peck: *Introduction to Linear Regression Analysis*, 2d ed., Wiley, New York, 1992.

Sclove, S. L.: "Improved Estimators for Coefficients in Linear Regression," *J. Am. Stat.* Assoc., vol. 63, pp. 596–606, 1968.

Silvey, S. D.: "Multicollinearity and Imprecise Estimation," *J. Res. Stat. Soc. Sci. B*, vol. 31, pp. 539–552, 1969.

Simpson, J. R.: "New Methods and Comparative Evaluations for Robust and Biased-Robust Regression Estimation," unpublished Ph.D. dissertation, Department of Industrial and Management Systems Engineering, Arizona State University, 1995.

Smith, G., and F. Campbell: "A Critique of Some Ridge Regression Methods (with Discussion)," *J. Am. Stat. Assoc.*, vol. 75, pp. 74–103, 1980.

Stein, C.: "Multiple Regression," in *Contributions to Probability and Statistics: Essays in Honor of Harold Hotelling*, I. Olkin, Ed., Stanford Univ. Press, Stanford, CA, 1960.

Stewart, G. W.: "Collinearity and Least Squares Regression," *Statistical Science*, vol. 2, pp. 68–100, 1987.

Sundberg, R.: "Continuum Regression and Ridge Regression," *J. R. Stat. Soc. S. B,* vol. 55, pp. 653–659, 1993.

Theil, H.: "On the Use of Incomplete Prior Information in Regression Analysis," *J. Am. Stat. Assoc.*, vol. 58, pp. 401–414, 1963.

_____ and A. S. Goldberger: "On Pure and Mixed Statistical Estimation in Economics," *Int. Econ. Rev.*, vol. 2, pp 65–78, 1961.

Turban, E.: "A Sample Survey of Operations Research Activities at the Corporate Level," *Oper. Res.*, vol. 20, pp. 708–721, 1972.

Walker, E.: "Augmented Bounded-Influence Estimators," *Statistical Computing Section Proceedings of the American Statistical Association*, pp. 158–164, 1987.

_____ and J. B. Birch: "Influence Measures in Ridge Regression," *Technometrics*, vol. 30, pp. 221–227, 1988.

Wang, S., S. Tse, and S. Chow: "On the Measures of Multicollinearity in Least Squares Regression," *Statistics and Probability Letters*, vol. 9, pp. 347–355, 1990.

Webster, J. T., R. F. Gunst, and R. L. Mason: "Latent Root Regression Analysis," *Technometrics*, vol. 16, pp. 513–522, 1974.

Willan, A. R., and D. G. Watts," Meaningful Multicollinearity Measures," *Technometrics*, vol. 20, pp. 407–412, 1978.

Zellner, A.: *An Introduction to Bayesian Inference in Economics*, Wiley, New York, 1971.

CHAPTER 17
ROBUST ESTIMATION AND IDENTIFYING OUTLIERS

Peter J. Rousseeuw
University of Antwerp (UIA)
Department of Mathematics
Antwerp, Belgium

17.1 INTRODUCTION

The least squares method is currently the most popular approach to estimation because of tradition and ease of computation. However, real data sets frequently contain outliers, which may be gross errors or exceptional observations. In this situation the least squares method becomes unreliable. Two things often happen: the estimates themselves become totally incorrect and the outliers are hidden, which means that one may not notice them at all. To remedy this problem, robust statistical techniques have been developed, which (1) still give a trustworthy answer when the data are contaminated and (2) allow us to easily identify the outliers at the same time.

The structure of this chapter is as follows. In Sec. 17.2 we introduce the notions of outliers and robustness in the special case of estimating a central value of a batch of numbers. The discussion focuses on the comparison of the sample mean with the sample median. Also, we consider the situation of very large data sets with application to robust "averaging" of curves and images, which is of special interest to engineers.

In Sec. 17.3 we arrive at the regression situation, starting with simple regression and continuing with multiple regression. We then consider some alternative regression estimators and finally mention other situations in which robust techniques have been made available.

Throughout this chapter we restrict our attention to robust methods which are both intuitively appealing and very powerful (in the sense that they can handle a sizable fraction of outliers if they have to). Our emphasis will be on the application of these methods rather than on their theoretical properties.

17.2 ROBUST ESTIMATION IN ONE DIMENSION

17.2.1 Outliers and Robustness

Suppose that we have five measurements of the weight of a stone,

$$5.59, \quad 5.66, \quad 5.63, \quad 5.57, \quad 5.60 \tag{17.1}$$

and we want to estimate its actual weight. To do this, one commonly computes the *sample mean*, which is merely the average of this batch of numbers,

$$\bar{x} = \frac{5.59 + 5.66 + 5.63 + 5.57 + 5.60}{5} = 5.61$$

A less well known estimator, especially in physics and engineering, is the *sample median*. We sort the observations from smallest to largest,

$$5.57 \le 5.59 \le 5.60 \le 5.63 \le 5.66$$

The sample median is then the middle observation, yielding 5.60. (If the number of observations is even, then we take the average of the two observations in the middle.) In this example the median and the mean are not equal, but they are close to each other. Both the sample mean and the sample median are said to be *location estimators*, because they measure the general position of the data.

Let us now suppose that someone has made an error in this data set (such as a recording error), yielding

$$5.59, \quad 5.66, \quad 5.63, \quad 55.7, \quad 5.60 \tag{17.2}$$

Outliers of this kind occur frequently. They may be due to transmission or copying errors, yielding a misplaced decimal point, or the permutation of two digits. It is even possible that the outlying observation is not an error at all but was made under exceptional circumstances (for example, a seismic quake) or belongs to another population (for example, it was a measurement of a different stone). Let us look at the effect of such an outlier on the estimate. For the arithmetic mean we find

$$\bar{x} = \frac{5.59 + 5.66 + 5.63 + 55.7 + 5.60}{5} = 15.64$$

which is utterly useless. For the median we sort the data again,

$$5.59 \le 5.60 \le 5.63 \le 5.66 \le 55.7$$

yielding the value of 5.63, which is still quite reasonable. The outlier has changed the median only slightly (and often it does not change it at all). We say that the sample median is a *robust* estimator, unlike the sample mean which is very sensitive to outliers.

In many classical statistics courses the sample median is not mentioned. The sample mean is typically preferred for several reasons, such as its ease of computation and the fact that it lends itself much more to elementary mathematical operations. Its usual justification is that it is optimal for normal distributions, but this is a circular reasoning because Gauss actually introduced the normal distribution as the best framework for the sample mean! In practice one notices that very few distributions (not even those of measurement errors) are perfectly normal. In the field of *robust statistics* we try to construct techniques that are not greatly affected by deviations from normality.

There are several ways to investigate how robust a procedure is. The *influence function* (Hampel, 1974) of an estimator describes the effect of one outlier; for a detailed description of this approach see Hampel et al. (1986). Another possibility is to simulate contaminated data sets and to try out how well the estimator does on them, as in Andrews et al. (1972). In this chapter we restrict attention to a more simple and yet far reaching tool, the *breakdown point*, which was developed by Hodges (1967), Hampel (1971), and Donoho and Huber (1983). The breakdown point of an estimator is the smallest fraction of the observations that have to be replaced to carry the estimator over all bounds. In this definition one can choose which observations to replace, as well as the magnitude of the outliers, in the least favorable way.

The breakdown point of the sample mean, applied to a sample $\{x_1, x_2, \ldots, x_n\}$ of n observations, is equal to $1/n$ because it is sufficient to replace a single observation by a large value. On the other hand, the sample median possesses the best possible breakdown point, namely, 50%. Indeed, we have to replace at least half of the observations by outlying values in order to be certain that the middle observation is among them. (Here the outliers are chosen in the least favorable way, that is, all on the same side of the original sample.) In fact, when fewer than half of the observations are replaced, the median even stays inside the *range* of the original data values, which is much stronger than saying that it remains bounded.

Of course, when considering breakdown points we do not forget about the usual criteria, such as the consistency of the estimator (meaning that the result becomes more and more precise as the number of observations increases). Also, any location estimator T should be *equivariant* when the data are multiplied by a constant, and when a constant is added to them,

$$T(\{cx_1 + d, \ldots, cx_n + d\}) = cT(\{x_1, \ldots, x_n\}) + d \tag{17.3}$$

There are quite a few other location estimators that satisfy these properties. Many classes exist, such as types *A*, *D*, *L*, *M*, *P*, *R*, *S*, and *W*. (For a survey see Hampel et al., 1986, pp. 100–116.) However, in this chapter we shall focus on the sample median, which belongs to several of these classes and is their "most robust" member in various ways.

Next to the sample's location, we often want to estimate its *scale* (or spread) as well. A scale estimator S should be equivariant in the sense that

$$S(\{cx_1 + d, \ldots, cx_n + d\}) = |c|S(\{x_1, \ldots, x_n\}) \tag{17.4}$$

where the absolute value is needed because a scale estimate is always positive. The classical scale estimator is the sample standard deviation,

$$S(\{x_1, \ldots, x_n\}) = \sqrt{\frac{1}{n-1}\sum_{i=1}^{n} (x_i - \bar{x})^2} \tag{17.5}$$

which is, however, notoriously nonrobust because it can become very big ("explosion") in the presence of even a single outlier. For our contaminated data set, Eq. (17.2), the standard deviation becomes 22.40, whereas it was only 0.0354 for the original data, Eq. (17.1). Therefore the breakdown point of the standard deviation is merely $1/n$. An extremely robust estimator of scale is the median of all absolute deviations (MAD) from the sample median,

$$S = 1.483 \operatorname*{median}_{j=1,\ldots,n} |x_j - \operatorname*{median}_{i=1,\ldots,n} (x_i)| \tag{17.6}$$

in which 1.483 is a correction factor to make the estimator consistent with the usual scale parameter of a normal distribution. Like the sample median, the MAD also has a breakdown point of 50%. The MAD for the contaminated sample, Eq. (17.2), is the same as that for the original sample, Eq. (17.1), namely, 0.0445. Instead of the MAD one can also use a recently developed scale estimator (Rousseeuw and Croux, 1993) which is more efficient and shares the MAD's advantages: a simple explicit formula, fast computation, and a 50% breakdown point.

When we have estimators of location and scale, we can build an outlier identifier. Indeed, what is an outlier? We have to specify with respect to what it is outlying. In a one dimensional sample, an outlier is a value that deviates from the majority of the points, that is, it lies far from the location T, compared to the scale S. We therefore have to compute the standardized observations,

$$z_i = \frac{x_i - T}{S} \tag{17.7}$$

These z_i, which are sometimes called z scores, then have to be compared to some cutoff value. If $|z_i|$ is larger than 2.5, we will identify the observation x_i. (Why do we take 2.5 for our cutoff value? If there are no outliers and the x_i come from a normal distribution, then the probability that $|z_i| > 2.5$ is very small. The exact choice of the cutoff value is to a certain extent arbitrary.) Which estimators T and S should we use? If we insert the sample mean for T and the usual standard deviation, Eq. (17.5), for S, then we obtain the classical studentized deviate, which is quite useless. For our contaminated data set, Eq. (17.2), we obtain the following values of z_i:

$$-0.45, \quad -0.45, \quad -0.45, \quad 1.79, \quad -0.45$$

none of which come even near the cutoff value. The classical z_i fail for two reasons: (1) because one subtracts a location estimate T which has moved toward the outlier, and (2) because one divides by a scale estimate S which has exploded. (Note that some people still use this method to guard against outliers and are reassured when they do not find any.) Both deficiencies are easily repaired by inserting robust estimators, such as the sample median for T and the MAD of Eq. (17.6) for S. This robust identifier correctly tells us that there are no outliers in the original sample (17.1). On the other hand, for the contaminated sample (17.2) we obtain the z_i values

$$-0.90, \quad 0.67, \quad 0.00, \quad 1125.17, \quad -0.67$$

one of which exceeds the cutoff 450 times. This example is but one illustration of a more general principle, which says that outliers can easily be identified by comparing data with a robust fit.

17.2.2 Large Data Sets and Averaging Curves and Images

In the preceding we assumed that we keep the data at our disposition and that we can go back to them several times. For instance, in order to compute the sample median we need to store all the values in the computer's memory so that we can sort them. This is a very natural assumption in statistics, but sometimes one encounters situations, especially in engineering, where there are so many data arriving in real time that they cannot be stored. In such situations one has nearly always restricted attention to the sample mean, which needs very little memory space. The mean can be computed with an updating mechanism, so only a single pass through the data is necessary. For instance, the following Fortran lines may be used:

```
   DO 10 I = 1,N
10 SUM = SUM + ENTER(I)
   XMEAN = SUM/N
```

where ENTER is a function that reads, records, generates, or otherwise accesses the ith observation. It is therefore not necessary to store the data in central memory. It is commonly thought that all robust estimators would need to store at least the data themselves, consuming n storage spaces. In many applications n storage spaces is too much, especially when a lot of estimations have to be carried out simultaneously.

To remedy this problem, Rousseeuw and Bassett (1990) introduced a new robust estimator, which can also be computed by means of a single pass updating mechanism without having to store all the observations. Let us assume that $n = b^k$, where b and k are integers. The *remedian* with base b proceeds by computing medians of groups of b observations, yielding b^{k-1} estimates on which this procedure is iterated, and so on, until only a single estimate remains. When implemented properly, this method merely needs k arrays of size b, which are continuously reused. Figure 17.1 illustrates the remedian with base 11 and exponent 4. The data enter at the top, and array 1 is filled with the first 11 observations. Then the median of these 11 observations is stored in the first element of array 2, and array 1 is used again for the second group of 11 observations, the median of which will be put in the second position of array 2. After some time array 2 is full too, and its median is stored in the first position of array 3, and so on. When 11^4 = 14,641 data values have passed by, array 4 is complete and its median becomes the final estimate. This method uses only 44 storage positions, and its speed is of the same order of magnitude as that of the sample mean or median. A physical analogy is the mileage recorder in a car: compare the first array with the rightmost wheel that counts the individual miles, the second array with the wheel indicating tens of miles, and so on. This also explains why b is called the *base*, as in the terminology of positional number systems. (We could take b = 10, but we prefer b odd because then the medians are easy to compute.)

The remedian with base b merely needs bk storage spaces for data sets with $n = b^k$ values. The total storage only increases as the logarithm of n, because $bk = b \log_b (n)$. (When n is not a power of b, Rousseeuw and Bassett compute a weighted median at the last step, which does not need more storage.) Due to its space efficiency, the remedian is very well adapted to robust estimation in real-time engineering applications in which the data are too numerous to be stored (such as control of electric power networks). The following lines are a Fortran implementation of the remedian of Fig. 17.1:

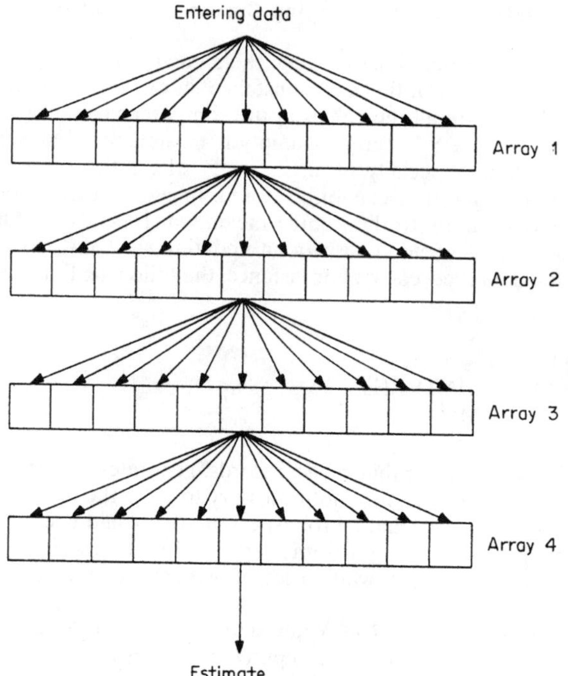

FIGURE 17.1 Mechanism of remedian with base 11 and exponent 4, using 44 storage spaces for data set of size $n = 11^4 = 14{,}641$.

```
CC      A PROGRAM FOR THE REMEDIAN
CC
        DIMENSION A1(11),A2(11),A3(11),A4(11)
        DO 40 M = 1,11
        DO 30 L = 1,11
        DO 20 K = 1,11
        DO 10 J = 1,11
        I = I + 1
   10   A1(J) = ENTER(I)
   20   A2(K) = FMED(A1)
   30   A3(L) = FMED(A2)
   40   A4(M) = FMED(A3)
        REMED = FMED(A4)
        WRITE(*,*)REMED
        STOP
        END
```

where FMED is a function that returns the median of an array of 11 numbers.

The remedian transforms properly when all observations x_i are replaced by $cx_i + d$. Like the sample median, it is even equivariant with respect to any *monotone* transformation of the x_i, such as a power function or an exponential. Rousseeuw and Bassett showed that the remedian is a consistent estimator of the underlying population median, investigated its sampling distribution, and computed its breakdown point. Alternative approaches are due to Martin and Masreliez (1975), Pearl (1981), Tierney (1983), and Tukey (1978).

An important application of the remedian is to curve averaging. Suppose we want to obtain a certain curve corresponding to a physical phenomenon. A curve can be registered by means of a list of its function values $x(t)$ at equally spaced arguments t (usually t stands for time). Unfortunately the observed values of $x(t)$ are subject to noise of various sources. Therefore one repeats the experiment several times, yielding n curves in all, so the data are of the form

$$\{x_i(t), t = 1, \ldots, M\} \quad \text{for } i = 1, \ldots, n \tag{17.8}$$

One wants to combine the n curves to estimate the true underlying shape. The classical approach is *averaging*, by which one obtains the curve

$$\bar{x}(t) = \frac{1}{n} \sum_{i=1}^{n} x_i(t), \qquad t = 1, \ldots, M \tag{17.9}$$

In the case of normal noise and no outliers, averaging makes good sense because then the noise goes down for large n. The averaging technique is very common in engineering and medicine. For instance, averaging is built into many special-purpose instruments used in hospitals (such as microprocessor-based average recorders).

Usually M and n are quite large, making it impractical to store all the observed curves in central memory. This precludes calculation of the "median curve,"

$$\operatorname*{median}_{i=1,\ldots,n} x_i(t), \qquad t = 1, \ldots, M \tag{17.10}$$

as well as many other robust summaries. We propose to compute the remedian curve

$$\operatorname*{remedian}_{i=1,\ldots,n} x_i(t), \qquad t = 1, \ldots, M \tag{17.11}$$

instead, because it is a robust single-pass method. The above Fortran program can be adapted quite easily to produce the remedian curve by replacing the arrays A_1, A_2, A_3, and A_4 of length 11 by matrices with 11 rows and M columns. For $b = 11$ and $k = 4$ the total storage becomes $44M$, whereas the plain median would have needed $14{,}641M$ positions.

Let us consider an example. The electroretinagram (ERG) is used in ophthalmology to examine disorders of the visual system. When the patient's eye is exposed to a white flash of light, it develops an electric potential that may be recorded by a contact lens electrode. The ERG curve shows the evolution of the evoked potential (expressed in microvolts) as a function of the time (in milliseconds) elapsed after the flash. The bottom curve in Fig. 17.2*b* is a standard ERG of a healthy patient (from Trau et al., 1983). The important features are the four peaks (denoted by a, b, OP_1, and OP_2) and in particular their t coordinates, which are used for medical diagnosis.

When the ERG curve is recorded only once, the noise typically dominates the signal so that no peak can be found. The current solution is to record many curves by repeating the stimulus flash of light, and then to average them. However, the average curve is often deformed and difficult to interpret, due to a high amount of contamination caused by electrical interference, involuntary eye movements, and other artifacts.

It is quite feasible to replace the averaging routine in the recording instrument by the remedian, because the latter is equally fast and does not need too much storage. To verify whether this replacement is worthwhile, computer simulations were performed in which both the average and the remedian were calculated for a bundle of curves, some of which were contaminated. The basic curve was the standard ERG of Fig. 17.2*b*, measured at $M = 320$ time units. Figure 17.2*a* contains $n = 81$ curves. (In opthalmology more curves are used, but this would make the display overcrowded.) The curves were generated as follows: with probability 0.7, curve i is the basic ERG plus some normal noise with modest scale. With probability 0.1, the $x(t)$ values are multiplied by a random factor greater than 1. With probability 0.2, the curve models a response at half the standard speed, again with magnified $x(t)$ values.

The upper curve in Fig. 17.2*b* is the average of the ERG curves in Fig. 17.2*a*. It was greatly affected by the contamination, which caused a substantial upward shift. What is worse, the average has one peak too many, rendering medical diagnosis dif-

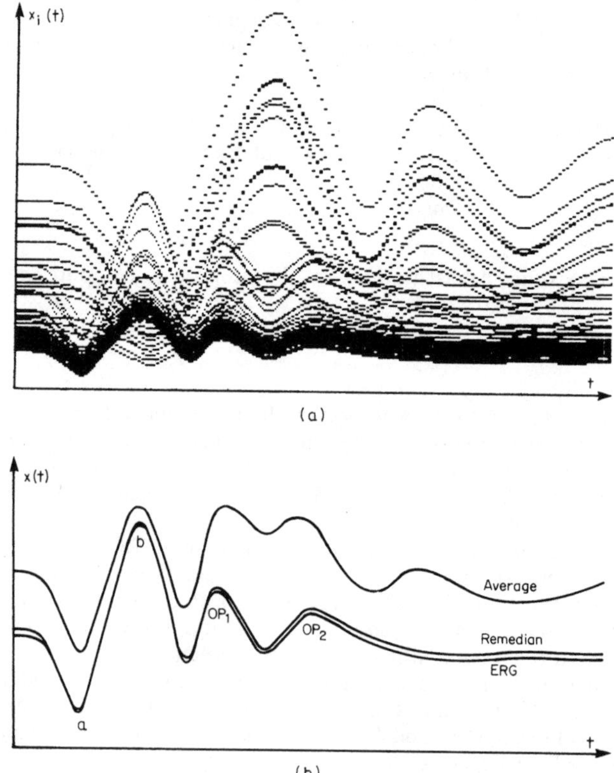

FIGURE 17.2 (*a*) Bundle of simulated electroretinagrams (ERG), some with normal noise and others with various kinds of contamination. (*b*) Plot with standard ERG (bottom), average of simulated ERGs (top), and their remedian (center).

ficult. Averaging often produces results like this in actual clinical practice. On the other hand, the 3^4 remedian lies very near the original ERG and is virtually undamaged by the contamination.

Many other applications of robust curve averaging could be envisaged, such as in spectroscopy. Median type procedures can also be used to estimate horizontal shifts between spectrograms (de Loore et al., 1987).

Averaging also occurs in image analysis. An image may be described as a rectangular grid of pixels, each with a corresponding number $x(r, c)$ indicating its grey intensity. When n images are recorded one after another, the data are

$$\{x_i(r, c), r = 1, \ldots, R \quad \text{and} \quad c = 1, \ldots, C\} \quad \text{for } i = 1, \ldots, n \qquad (17.12)$$

where R is the number of rows and C the number of columns. In one application a physicist recorded images of a crystallographic lattice by means of an electron microscope, with $R = 512$, $C = 512$, and $n \approx 10{,}000$. Usually such images are averaged to obtain a sharp result, but in this case averaging did not work well because in many images a part of the lattice was contaminated or even destroyed by the radiation of the microscope itself. Computing plain medians was not feasible because there were $nRC \approx 2{,}621{,}440{,}000$ data values in all, which could not be stored in central memory. One can, however, compute the remedian image given by

$$\underset{i = 1, \ldots, n}{\text{remedian}}\; x_i(r, c) \quad r = 1, \ldots, R \text{ and } c = 1, \ldots, C \qquad (17.13)$$

The computation of remedian curves (or images) may be speeded up if one has access to parallel computing facilities because one could let each processor work on a different portion of the curve.

17.3 ROBUST REGRESSION

17.3.1 Simple Regression

In simple regression one assumes a relation of the type

$$y_i = \beta_0 + \beta_1 x_i + \epsilon_i \qquad (17.14)$$

in which x_i is called the *explanatory variable* or *regressor*, and y_i is the *response variable*. The intercept β_0 and the slope β_1 have to be estimated from the data $\{(x_1, y_1), \ldots, (x_n, y_n)\}$. Classical theory assumes the deviations ϵ_i to be normally distributed.

In simple regression the observations (x_i, y_i) are two-dimensional, so they can be plotted. Regression users should always draw this plot first because it shows whether the data are roughly linear, and any unusual structures will be clearly visible. In this chapter we begin with the simple regression situation in which the phenomena are most easily visualized, after which we continue with multiple regression for which robust methods are much more necessary.

Applying a regression estimator to such a bivariate data set with n observations yields the regression coefficients $\hat{\beta}_0$ and $\hat{\beta}_1$. Although the true parameters β_0 and β_1 are unknown, one can insert these $\hat{\beta}_0$ and $\hat{\beta}_1$ in Eq. (17.14), yielding

$$\hat{y}_i = \hat{\beta}_0 + \hat{\beta}_1 x_i \qquad (17.15)$$

where $\hat{y}_i$ is the estimated value of y_i. The *residual* r_i of the *i*th case is the difference between the observed value and the estimated value,

$$r_i = y_i - \hat{y}_i \tag{17.16}$$

Note the distinction between the residual r_i and the deviation ϵ_i, because the latter equals $y_i - \beta_0 - \beta_1 x_i$, with the unknown true parameters β_0 and β_1. The residual can actually be computed, whereas the deviation cannot.

The most popular regression estimator (dating back to Gauss and Legendre, around 1800) is the *least squares method* given by

$$\underset{\hat{\beta}_0, \hat{\beta}_1}{\text{minimize}} \sum_{i=1}^{n} r_i^2 \tag{17.17}$$

The basic idea was to make all the residuals very small. Gauss preferred the least-squares criterion to other objective functions because in this way the regression coefficients could be computed *explicitly* from the data. (Note that there were no computers yet.) Even today most statistical packages still use least squares mainly because of tradition and speed of computation. Afterward, Gauss introduced the normal, or gaussian, distribution as the error distribution for which least squares is optimal. Since then the least-squares method has been theoretically justified in many other ways that are equally circular (such as the Gauss-Markov theorem).

More recently people began to realize that actual data often do not satisfy the gaussian assumptions, with dramatic effects on the least squares results. Let us look at some plots illustrating the effect of outliers. Figure 17.3*a* is the scatter plot of five points, $(x_1, y_1), \ldots, (x_5, y_5)$, which almost lie on a straight line. In such a situation the least-squares solution fits the data very well, as can be seen from the least-squares (LS) line $\hat{y} = \hat{\beta}_0 + \hat{\beta}_1 x$ in the plot. However, suppose that someone gets a wrong value of y_4 because of a recording or copying error. Then (x_4, y_4) may be rather far away from the "ideal" line. Figure 17.3*b* displays such a situation, where

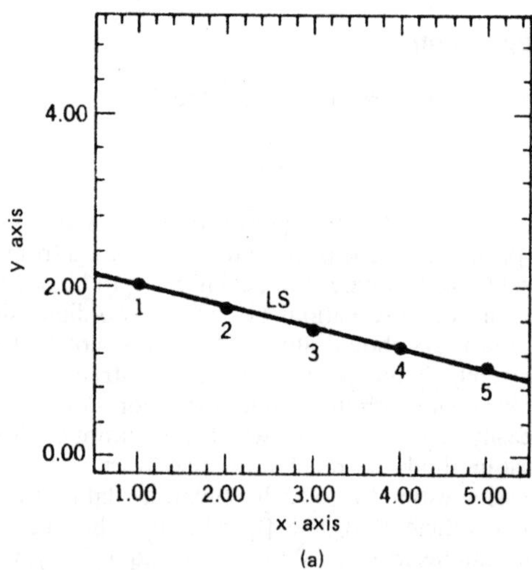

FIGURE 17.3*a* Five points and their least-squares regression line.

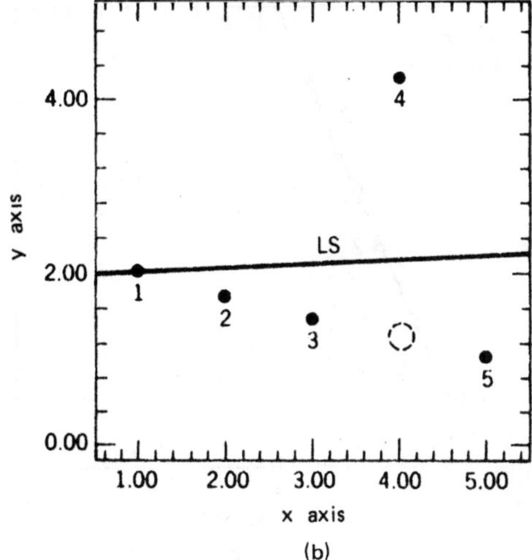

FIGURE 17.3*b* Same data with one outlier in *y* direction.

the fourth point has moved up from its original position (indicated by the dashed circle). This point is called an *outlier in the y direction*, and it has a rather large effect on the LS line in Fig. 17.3*b*. Such vertical outliers have received most attention in the literature because in designed experiments the values of x_i are preselected, and then one only expects outliers in the y_i.

In observational studies ("happenstance" data) the x_i are not fixed, but they are themselves subject to random variability. In that case outliers can also occur in the x_i. (Note that even in designed experiments one may have data entry errors in the x_i.) For the effect of such an outlier we turn to Fig. 17.4. In Fig. 17.4*a* we again see five points with a well-fitting LS line. If we now make an error in recording x_1, we obtain Fig. 17.4*b*. The resulting point is called an *outlier in the* x *direction*, and its effect on least squares is very large because it actually tilts the LS line. Therefore the point (x_1, y_1) is called a *leverage point*, in analogy to the notion of leverage in mechanics. This large "pull" on the least-squares estimator can be explained as follows. Because x_1 lies far away, the residual r_1 from the original line (the line in Fig. 17.4*a*) becomes a very large (negative) value, contributing an enormous amount to Σr_i^2 for that line. Therefore the original line cannot be selected from a least-squares perspective, and indeed the line in Fig. 17.4*b* possesses the smallest Σr_i^2 because it has tilted to reduce that larger r_1^2, even if the other terms, $r_2^2, \ldots, r_5^2$, have increased somewhat.

In general we call an observation (x_k, y_k) a leverage point whenever x_k lies far from the bulk of the observed x_i in the data. Note that this does not take y_k into account. In Fig. 17.5 the point (x_k, y_k) lies close to the linear pattern set by the majority of the data, so it can be considered a "good" leverage point. On the other hand, the point (x_4, y_4) in Fig. 17.4*b* is a bad leverage point. Therefore, to say that an observation (x_k, y_k) is a leverage point refers only to its *potential* for influencing the regression coefficients $\hat{\beta}_0$ and $\hat{\beta}_1$ (due to its outlying component x_k). Whether or not it will really exert this influence depends on whether it is in line with the trend set by the other data.

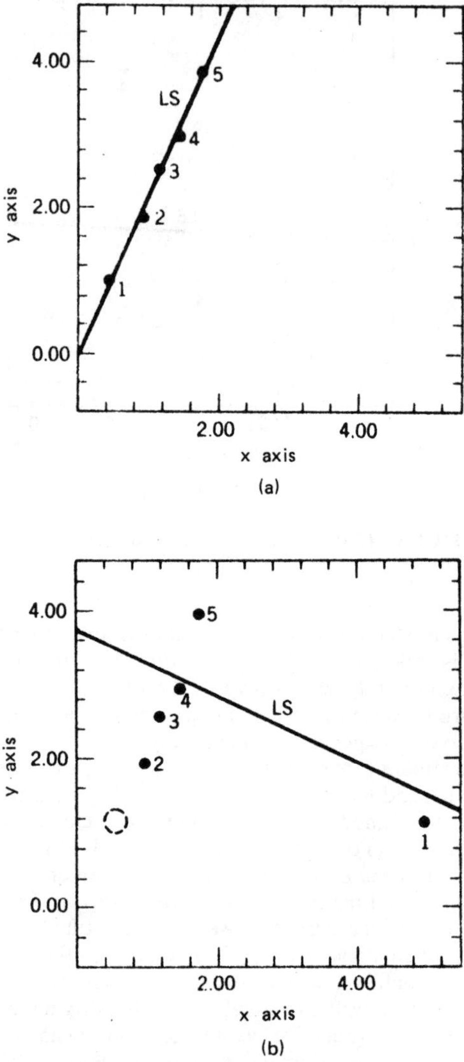

FIGURE 17.4 (*a*) Five points with their least-squares regression line. (*b*) Same data with one outlier in x direction.

When a point (x_i, y_i) deviates from the linear relation of the majority, we call it a *regression outlier*, taking into account both x_i and y_i simultaneously. In other words, a regression outlier is either a vertical outlier or a bad leverage point.

It is often believed that regression outliers can be identified by looking at the least squares residuals. Unfortunately things are not that simple. For example, consider again Fig. 17.4*b*. Case 1, being a bad leverage point, has tilted the LS line so much that it is now quite close to that line. Consequently the residual $r_1 = y_1 - \hat{y}_1$ is

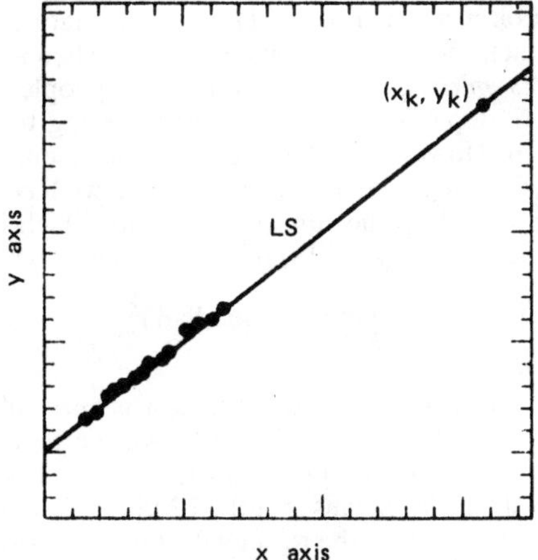

FIGURE 17.5 Point (x_k, y_k) is a leverage point because x_k is outlying. However, (x_k, y_k) is not a regression outlier because it matches the linear pattern set by the other data points.

a small (negative) number. The residuals r_2 and r_5 are much larger, although they correspond to good points. If one would apply a rule such as "delete the points with largest least squares residuals," then the good points would have to be deleted first. Of course, in simple regression there is really no problem at all because one can actually look at the data, but in Sec. 17.3.2 we shall see that in a multiple regression the outliers often remain invisible in spite of a careful inspection of least squares residuals.

From the examples in Figs. 17.3 and 17.4 we know that even a single regression outlier can totally offset the least squares estimator (provided it is far away). This implies that the breakdown point of the least squares method is merely $1/n$ (which is not so surprising, because the least squares estimator generalizes the sample mean we studied before). A first step toward a more robust regression estimator came from Edgeworth (1887), who argued that outliers have a very large effect on least squares because the residuals r_i are being squared in Eq. (17.17). Therefore he proposed the *least absolute values* method given by

$$\underset{\hat{\beta}_0, \hat{\beta}_1}{\text{minimize}} \sum_{i=1}^{n} |r_i| \tag{17.18}$$

This technique is often referred to as L_1 regression, whereas least squares is L_2. (For surveys on L_1, see Bloomfield and Steiger, 1983, and Dodge, 1987.) Unfortunately L_1 is only robust with respect to vertical outliers, but it does not protect against bad leverage points. Indeed, in the example of Fig. 17.4*b* the L_1 line is even more attracted by the fourth observation than the LS line. Therefore the breakdown point of the L_1 method is still no better than $1/n$.

There exist many other techniques with varying degrees of robustness. To avoid a digression at this stage we will concentrate on one method first, and leave descriptions of other approaches to Sec. 17.3.4. We want an estimator with a high breakdown point, so it can be used to analyze extremely messy data sets as well as clean ones. Let us look again at Eq. (17.17). A more natural name for the least squares method (in fact, the name you would expect when seeing its formula for the first time) is *least sum of squares*. Apparently few people have objected to the deletion of the word "sum," as if the only sensible thing to do with n numbers would be to add them. However, adding the r_i^2 is equivalent to using their mean (dividing by n does not change the minimization), and we have seen that the mean is not robust. Why not replace the sum by a median, which is very robust? This yields the *least median of squares* (LMS) method, given by

$$\underset{\hat{\beta}_0, \hat{\beta}_1}{\text{minimize}} \quad \underset{i=1, \ldots, n}{\text{median}} \; r_i^2 \tag{17.19}$$

(Rousseeuw, 1984). It turns out that this estimator is very robust with respect to outliers in y as well as outliers in x. In Fig. 17.6 we see that the LMS yields the desired fit for the examples of Figs. 17.3*b* and 17.4*b*. Its breakdown point is 50%, the highest possible value. This means that the LMS can cope with several outliers at the same time, in the sense that the result will still be trustworthy. Its basic principle is to fit the majority of the data, after which outliers may be identified as those points that lie far away from the robust fit, that is, the cases with large positive or large negative residuals. In Fig. 17.6*a* the fourth case possesses a considerable LMS residual, and that of case 1 in Fig. 17.6*b* is even more apparent.

The LMS line has an intuitive geometric interpretation, because it lies at the center of the narrowest strip covering half of the points. Note that this interpretation is simpler than that of least squares. It also provides some insight as to why the LMS is not much attracted by outliers. Using this geometric interpretation, it is possible to construct an algorithm for the LMS line, as described in Rousseeuw and Leroy (1987, chap. 5). This book is devoted to LMS and other high break down methods. It is also a user manual for the Fortran program PROGRESS (program for *robust regression*), which computes LMS regression on IBM PCs and other machines.

Let us look at an example from astronomy. The data in Table 17.1 form the Hertzsprung-Russell diagram of the star cluster CYG OB1, which contains 47 stars in the direction of Cygnus. Here x_i is the logarithm of the temperature at the surface of the star, and y_i is the logarithm of its light intensity. The data were given to us by C. Doom (personal communication).

The Hertzsprung-Russell diagram itself is shown in Fig. 17.7. It is the scatterplot of these points, where the log temperature is plotted from right to left. In the plot we see two groups of points: the majority, which seem to follow a steep band, and the four stars in the upper right corner. These parts in the diagram are well known in astronomy: the 43 stars are said to belong to the main sequence, whereas the four remaining stars are called giants. (The giants are the points with indices 11, 20, 30, and 34.)

Applying the LMS estimator to these data yields the solid line $\hat{y} = 3.898x - 12.298$, which fits the main sequence nicely. On the other hand, the least squares solution $\hat{y} = -0.409x + 6.78$ corresponds to the dashed line in Fig. 17.7 that has been pulled away by the four giant stars (which it does not fit well either). These are outliers but not errors. It would be more appropriate to say that the data came from two different populations. The two groups can easily be distinguished on the basis of their LMS residuals (the large residuals correspond to the giant stars), whereas the least squares residuals are rather homogeneous and do not allow us to separate the giants from the main sequence stars.

Several other examples from different fields can be found in Rousseeuw and Leroy (1987). Massart et al. (1986) computed LMS regression lines in analytical chemistry examples.

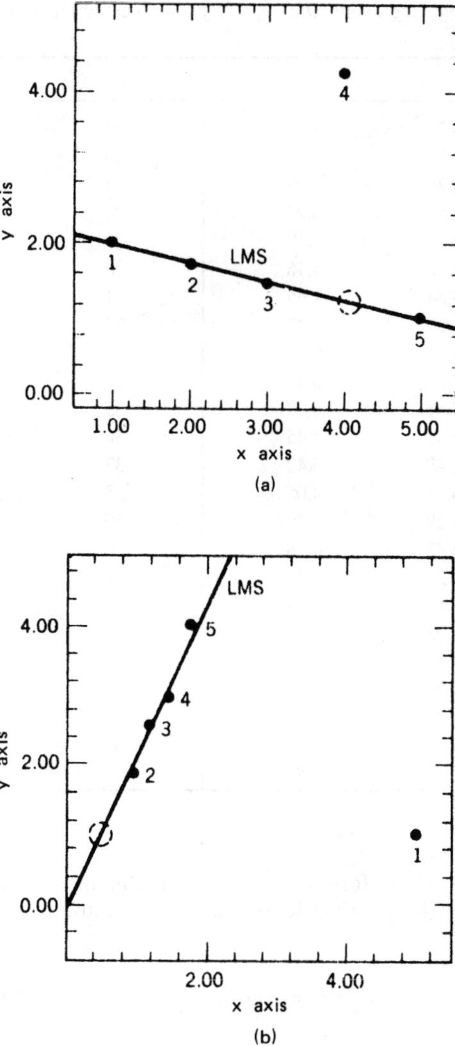

FIGURE 17.6 Robustness of LMS regression with respect to (*a*) an outlier in the *y* direction; (*b*) an outlier in the *x* direction.

17.3.2 Multiple Regression

By extending the simple regression framework, Eq. (17.14), to several explanatory variables we obtain the multiple regression model

$$y_i = \beta_0 + \beta_1 x_{i1} + \cdots + \beta_p x_{ip} + \epsilon_i \tag{17.20}$$

TABLE 17.1 Simple Regression Data: Hertzsprung-Russell Diagram of the Star Cluster CYG OB1

i	x_i	y_i	i	x_i	y_i
1	4.37	5.23	25	4.38	5.02
2	4.56	5.74	26	4.42	4.66
3	4.26	4.93	27	4.29	4.66
4	4.56	5.74	28	4.38	4.90
5	4.30	5.19	29	4.22	4.39
6	4.46	5.46	30	3.48	6.05
7	3.84	4.65	31	4.38	4.42
8	4.57	5.27	32	4.56	5.10
9	4.26	5.57	33	4.45	5.22
10	4.37	5.12	34	3.49	6.29
11	3.49	5.73	35	4.23	4.34
12	4.43	5.45	36	4.62	5.62
13	4.48	5.42	37	4.53	5.10
14	4.01	4.05	38	4.45	5.22
15	4.29	4.26	39	4.53	5.18
16	4.42	4.58	40	4.43	5.57
17	4.23	3.94	41	4.38	4.62
18	4.42	4.18	42	4.45	5.06
19	4.23	4.18	43	4.50	5.34
20	3.49	5.89	44	4.45	5.34
21	4.29	4.38	45	4.55	5.54
22	4.29	4.22	46	4.45	4.98
23	4.42	4.42	47	4.42	4.50
24	4.49	4.85			

Each data point is of the form $(x_{i1}, \ldots, x_{ip}, y_i)$, containing p regressors and one response variable. The classical least squares estimator of $\beta_0, \beta_1, \ldots, \beta_p$ corresponds to

$$\underset{\hat{\beta}_0, \ldots, \hat{\beta}_p}{\text{minimize}} \sum_{i=1}^{n} r_i^2 \tag{17.21}$$

in which the residuals r_i are given by

$$r_i = y_i - \hat{y}_i = y_i - \hat{\beta}_0 - \hat{\beta}_1 x_{i1} - \cdots - \hat{\beta}_p x_{ip} \tag{17.22}$$

The actual computation of $(\hat{\beta}_0, \ldots, \hat{\beta}_p)$ can be found in Chap. 14.

As before, the least squares estimator is very vulnerable to outliers in the response variable and to outliers in the regressors. Our main handicap is that the observations $(x_{i1}, \ldots, x_{ip}, y_i)$ have $p + 1$ dimensions so they cannot be plotted, and hence we cannot spot the outliers by eye. The worst problem is with observational studies, because they may contain leverage points, that is, points for which $(x_{k1}, \ldots, x_{kp})$ lies far from the bulk of the $(x_{i1}, \ldots, x_{ip})$ in the data. This is a major concern because

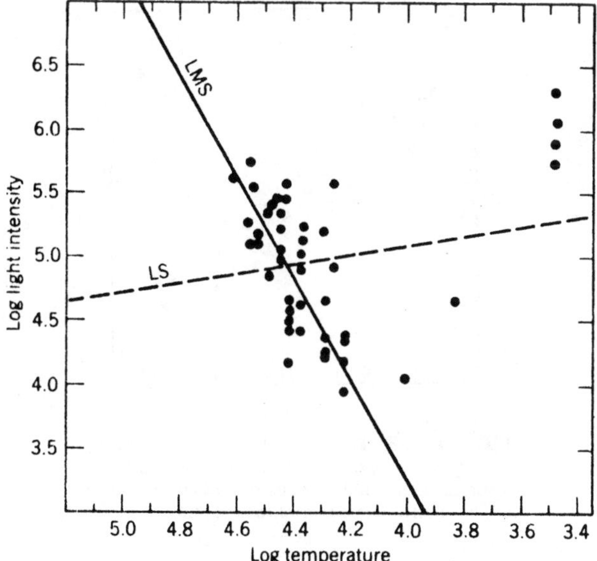

FIGURE 17.7 Hertzsprung-Russell diagram of star cluster CYG OB1 with least-squares (dashed line) and least-median-of-squares fit (solid line).

1. Least squares is affected more by leverage points than by vertical outliers, as was already seen in Fig. 17.4*b*.
2. There are p regressors, as opposed to only one response variable, so more outliers are likely to occur in them.
3. It becomes very difficult to *identify* leverage points due to the higher dimensionality.

An illustration of the third aspect is given in Fig. 17.8, which plots x_{i2} versus x_{i1} for some data set. In this plot we easily see two leverage points, which are, however, invisible when the variables x_{i1} and x_{i2} are considered separately. (Indeed, the one dimensional sample $\{x_{11}, x_{21}, \ldots, x_{n1}\}$ does not contain outliers, and neither does $\{x_{12}, x_{22}, \ldots, x_{n2}\}$. When there are more than two regressors, even more complicated configurations are possible. In general, it is not sufficient to look at each variable separately or at all plots of pairs of variables.

We are mostly concerned with regression outliers, that is, cases for which $(x_{i1}, \ldots, x_{ip}, y_i)$ deviates from the linear relation followed by the majority of the data, taking into account both the regressors and the response variable. Already in simple regression we saw that the outliers cannot always be identified by means of their least squares residuals. In order to identify the outliers we need to know what the bulk of the data are like. This calls for a high-breakdown estimator such as LMS, which is defined analogously by

$$\underset{\hat{\beta}_0, \ldots, \hat{\beta}_p}{\text{minimize}} \; \underset{i=1,\ldots,n}{\text{median}} \; r_i^2 \tag{17.23}$$

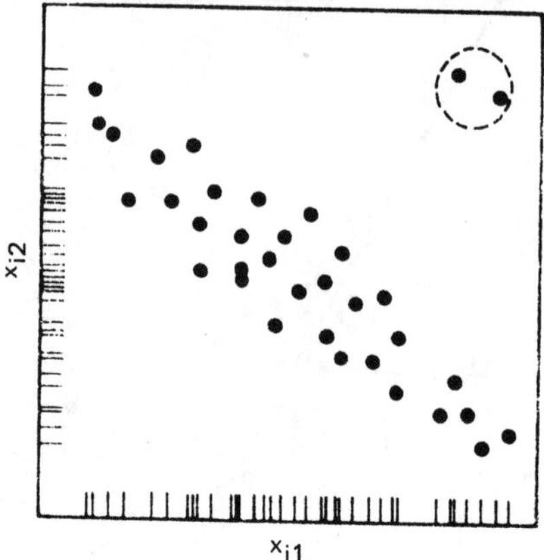

FIGURE 17.8 Plot of explanatory variables x_{i1} and x_{i2} of a regression data set. There are two leverage points (indicated by dashed circle), which are not outlying in either of the coordinates.

(Rousseeuw, 1984). The LMS has a breakdown point of 50% even in multiple regression, only its computation time goes up.

In order to find out whether a residual is unusually large, we have to compare it to something. For this purpose we need an estimate of the error scale, that is, the spread of the deviations ϵ_i. In classical theory the ϵ_i have a normal distribution with zero mean, and with an unknown standard deviation denoted by σ. When using least squares, σ is estimated by

$$\hat{\sigma} = \sqrt{\frac{1}{n-p-1}\sum_{i=1}^{n} r_i^2} \tag{17.24}$$

where the r_i are the least squares residuals. For least median of squares regression, the corresponding scale estimate becomes

$$\hat{\sigma} = 1.483 \sqrt{\underset{i=1,\ldots,n}{\text{median}}\ r_i^2} \tag{17.25}$$

where the r_i are the residuals from the LMS fit, and 1.483 is the appropriate constant to make $\hat{\sigma}$ a consistent estimator of σ. The LMS scale estimate has itself a 50% breakdown point, whereas the least squares scale has 0% breakdown because it explodes easily.

By analogy to the one dimensional z scores, Eq. (17.7), we consider the *standardized residuals*

$$\frac{r_i}{\hat{\sigma}} \tag{17.26}$$

If $|r_i/\hat{\sigma}|$ is larger than 2.5, we will consider the ith case to be a regression outlier. Inserting the least-squares residuals for r_i and the least squares scale for $\hat{\sigma}$ makes no sense, because leverage points often have small least squares residuals, and even when some least-squares residuals are large, they will blow up $\hat{\sigma}$ and hence be divided by a large denominator in Eq. (17.26). As in one dimension, outliers can be safely identified by inserting robust estimators, in this case by considering the residuals from the LMS fit and dividing them by the LMS scale.

Let us look at a real data example to illustrate the need for a robust regression method. The well known stackloss data of Brownlee (1965) describe the operation of a plant for the oxidation of ammonia to nitric acid. The data consist of 21 four dimensional observations, listed in Table 17.2. The stackloss y has to be explained by the rate of operation x_1, the cooling water inlet temperature x_2, and the acid concentration x_3. Applying least squares regression yields the equation

$$\hat{y} = 0.716x_1 + 1.295x_2 - 0.152x_3 - 39.9 \tag{17.27}$$

with corresponding scale estimate $\hat{\sigma} = 3.24$.

We cannot plot the data because they are four dimensional, but we can look at the standardized residuals, Eq. (17.26). The sixth column of Table 17.2 lists the standardized residuals obtained from least squares. From these results one would conclude that the data set does not contain any outliers, because all the standardized residuals are below 2.5. However, let us now consider the LMS fit

$$\hat{y} = 0.714x_1 + 0.357x_2 + 0.000x_3 - 34.5 \tag{17.28}$$

The equation itself is quite different, especially the coefficients of x_2 and x_3. The LMS scale estimate becomes 1.26, which is much smaller than the least squares scale. As to the search of outliers, we now look at the standardized residuals based

TABLE 17.2 Stackloss Data with Standardized Residuals

					Standardized residual	
i	x_{i1}	x_{i2}	x_{i3}	y_i	LS	LMS
1	80	27	89	42	1.00	7.70
2	80	27	88	37	– 0.59	3.74
3	75	25	90	37	1.40	7.14
4	62	24	87	28	1.76	7.64
5	62	22	87	18	– 0.53	0.28
6	62	23	87	18	– 0.93	0.00
7	62	24	93	19	– 0.74	0.51
8	62	24	93	20	– 0.43	1.30
9	58	23	87	15	– 0.97	– 0.11
10	58	18	80	14	0.39	0.51
11	58	18	89	14	0.81	0.51
12	58	17	88	13	0.86	0.00
13	58	18	82	11	– 0.44	– 1.87
14	58	19	93	12	– 0.02	– 1.36
15	50	18	89	8	0.73	0.28
16	50	18	86	7	0.28	– 0.51
17	50	19	72	8	– 0.47	0.00
18	50	19	79	8	– 0.14	0.00
19	50	20	80	9	– 0.18	0.51
20	56	20	82	15	0.44	1.87
21	70	20	91	15	– 2.23	– 6.06

on the LMS, which are given in the last column of the table. They indicate that observations 1, 3, 4, and 21 are outlying, because their $|r_i/\hat{\sigma}|$ is much larger than 2.5. Observation 2 is a borderline case. These conclusions correspond to the findings cited in the literature. Our robust regression technique has analyzed these data in a single blow, which should be contrasted to some of the earlier analyses of the same data set, which were long and laborious.

This example also illustrates the danger of merely looking at the least squares residuals. It is better to compare the standardized residuals of the least squares and the LMS in a given application. If the results of both procedures are in substantial agreement, least squares can be trusted. If they differ, the LMS technique can be used as a reliable tool for identifying the outliers, which may then be thoroughly investigated and perhaps corrected (if one has access to the original measurements) or deleted. Another possibility is to change the model, for example, by adding squared terms or transforming the response variable (see, for example, Carroll and Ruppert, 1988). In the stackloss example, the LMS analysis suggests selecting a model without the variable x_3.

The LMS regression equation is much more stable than the least squares one. Indeed, if we carry out least squares on the "cleaned" data set without observations 1, 2, 3, 4, and 21, then we find an equation and a scale estimate that are nearly the same as in Eq. (17.28). Such a "reweighted" analysis (where each point has a weight of 0 or 1, depending on whether or not LMS has identified it as an outlier) is useful because it is still robust and at the same time yields all the standard output, such as t values, confidence intervals, and a coefficient of determination. For this reason, the program PROGRESS described in Rousseeuw and Leroy (1987) computes the least squares fit, the LMS fit, and the reweighted analysis. In the same book many examples are treated.

It should be stressed that the LMS does *not* "throw away" 50% of the data. Rather, it finds a fit to the majority of the points, which can then be used to identify the actual outliers (of which there may be many, a few, or none at all). Also, note that we did not need any symmetry assumption anywhere.

In a multivariate point cloud, outliers can be identified by a robust estimator of its central location and its scatter matrix. For this we can use the *MVE (minimum volume ellipsoid) estimator* proposed by Rousseeuw (1985). The outliers are then detected by their *robust distances* (Rousseeuw and Leroy, 1987) relative to the MVE estimates. When a linear regression model is being fitted, it is recommended to apply the MVE to the data cloud formed by the n points $(x_{i1}, \ldots, x_{ip})$, since the outliers of that cloud are the leverage points of the regression. For examples of this approach, see Rousseeuw and van Zomeren (1990).

The LMS and MVE estimators are computed by similar algorithms. The standalone programs PROGRESS (for the LMS) and MINVOL (for the MVE) can be obtained free of charge from the author's website, http://win-www.uia.ae.be/w/statis. This includes the Fortran source code, which runs on most platforms. Both estimators are also available in the statistical package S-PLUS (Statistical Sciences, 1993) as the intrinsic functions *lmsreg* and *cov.mve*, and in SAS/IML as the calls LMS and MVE.

17.3.3 Applications

There have been quite a few substantive applications of the LMS and MVE methods. Here we will only list some applications in science and engineering, which may be of interest to readers of this book.

For instance, the LMS has been applied to measures of production efficiency (Seaver and Triantis, 1995), and is used in analytical chemistry following (Massart et al., 1986). Also, the LMS is an essential component of a new system for connecting optical fiber cables implemented at NIST (Wang et al., 1997). In large electric power systems, Mili et al. (1991) modified positive-breakdown methods to estimate the system's state variables. Faster algorithms needed to be constructed to allow real-time estimation (Mili et al., 1996), and are being implemented by several power utilities.

Positive-breakdown methods have opened new possibilities in the rapidly evolving field of *computer vision*. The LMS has been used for analyzing noisy images (Meer et al., 1991), for interpreting color images (Drew, 1994), for discontinuity-preserving surface reconstruction (Sinha and Schunck, 1992), for extracting geometric primitives (Roth and Levine, 1993; Stewart, 1994), for robot positioning (Kumar and Hanson, 1994), and for detecting moving objects in video from a mobile camera (Thompson et al., 1993; Abdel-Mottaleb et al., 1993). The MVE was applied to image segmentation (Jolion et al., 1991).

In geochemistry, Chork (1990) used the MVE to analyze data on surface rocks in New South Wales, for which concentrations of several chemical elements were measured. Outliers in this multivariate data set revealed mineralizations, yielding targets for mining prospecting. A larger study in Finland carried out an MVE based principal components analysis (Chork and Salminen, 1993). The same methods apply to environmetrics, since mineralizations in geochemistry are similar to contaminations of the environment.

17.3.4 Alternative Approaches

Many other robust regression estimators have been proposed. Huber (1973) introduced *M estimators* for regression, a survey of which can be found in Huber (1981). They are based on the idea of replacing the r_i^2 in Eq. (17.21) by another function of the residuals, yielding

$$\text{minimize} \sum_{i=1}^{n} \rho(r_i)$$

where ρ is an even function [that is, $\rho(-t) = \rho(t)$ for all t] with a unique minimum at zero. In practice, M estimators can be computed by means of iteratively reweighted least squares, or with Newton-Raphson-type algorithms. Unfortunately their breakdown point is again $1/n$ because of the effect of outlying $(x_{i1}, \ldots, x_{ip})$.

Because of this vulnerability to leverage points, *generalized M estimators* (GM estimators) were introduced, with the basic purpose of bounding the influence of outlying $(x_{i1}, \ldots, x_{ip})$ by means of some weight function. For this reason they are often called *bounded-influence estimators*. Particular types of GM estimators were studied by Mallows (1975) and Schweppe (see Hill, 1977); for a survey see Hampel et al. (1986, chap. 6). It turns out, however, that the breakdown point of all GM estimators decreases when the dimension increases (Maronna, Bustos, and Yohai, 1979). This is very unsatisfactory because it means that the breakdown point diminishes when there are more regressors, and hence more opportunities for outliers to occur.

Various other estimators have been proposed, such as the one in Brown and Mood (1951), the median of pairwise slopes (Theil, 1950), the resistant line (Tukey, 1970), R estimators (Jureckovà, 1971; Jaeckel, 1972), L estimators (Bickel, 1973;

Koenker and Bassett, 1978), and the technique of Andrews (1974). Unfortunately, even in simple regression none of these methods achieves a breakdown point of 30%.

The first robust regression method with a 50% breakdown point was the *repeated median* proposed by Siegel (1982). For any $p + 1$ observations with indices $\{i_1, \ldots, i_{p+1}\}$, it computes the coefficients $\beta_0 (i_1, \ldots, i_{p+1}), \ldots, \beta_p (i_1, \ldots, i_{p+1})$ such that the corresponding surface fits these $p + 1$ points exactly. The jth coefficient of the repeated median regression is then defined as

$$\hat{\beta}_j = \underset{i_1}{\text{median}} \, (\ldots(\underset{i_p}{\text{median}} \, (\underset{i_{p+1}}{\text{median}} \, \beta_j(i_1,\ldots, i_{p+1})))\ldots)$$

where the innermost median is over all choices of i_{p+1}, the next is over all choices of i_p, and so on. This estimator can be computed explicitly, but it requires consideration of all subsets of $p + 1$ points, which may cost a lot of time. It has been applied successfully to problems with small p. But unlike other regression estimators, the repeated median is not equivariant for linear transformations of the $(x_1, \ldots, x_p)$, which is due to its coordinatewise construction. More information can be found in Hössjer et al. (1994) and Rousseeuw et al. (1995).

The LMS method combines equivariance with a 50% breakdown point, but has a low asymptotic efficiency (which does not, however, detract from its ability to identify outliers). To obtain a better asymptotic efficiency, Rousseeuw (1984) also proposed the *least trimmed squares* (LTS) estimator, given by

$$\underset{\hat{\beta}_0,\ldots,\hat{\beta}_p}{\text{minimize}} \sum_{i=1}^{h} r^2_{(i)}$$

where $r^2_{(1)} \leq r^2_{(2)} \leq \ldots \leq r^2_{(n)}$ are the ordered squared residuals. (Note that the residuals are first squared and then ordered.) The LTS is also equivariant. The LTS formula is similar to least squares, the only difference being that the largest squared residuals are not used in the sum, therefore allowing the fit to stay away from the outliers. The best robustness properties are achieved when h is approximately $n/2$, in which case the breakdown point attains 50%. Another variant is the class of *S estimators* introduced by Rousseeuw and Yohai (1984), which are also equivariant and have 50% breakdown point, while sharing some of the nice mathematical properties of Huber's *M* estimators.

Another approach to the identification of aberrant points is the construction of *outlier diagnostics*. These are quantities computed from the data with the purpose of pinpointing influential points, after which these outliers are to be removed or corrected, followed by a least-squares analysis on the remaining cases. When there is only a single outlier, some of these methods work quite well by looking at the effect of deleting one point at a time (see, for example, Atkinson, 1985; Belsley, Kuh, and Welsch, 1980; Cook and Weisberg, 1982; and Hawkins, 1980). Unfortunately it is much more difficult to diagnose outliers when there are several of them, due to the so-called masking effect, which says that one outlier may mask another. The naive extensions of classical diagnostics to such multiple outliers often give rise to extensive computations (for example, the consideration of *all* subsets of points is an impossible task). Recent work by Atkinson (1986), Hawkins, Bradu, and Kass (1984), and Rousseeuw and van Zomeren (1990) indicates that one needs to use robust methods in one way or another to safely identify multiple outliers. This is not so surprising, because one needs to know with respect to which pattern the points are outlying.

Sometimes robust methods are confused with *nonparametric* methods. Basically there is no relation between them. The confusion stems from the coincidence that for one dimensional samples certain nonparametric methods (such as the Wilcoxon rank test) also happen to be relatively robust. In regression, however, they are quite different because nonparametric regression does *not* assume a linear model of the type of Eq. (17.20). Therefore it does not yield an explicit equation to describe the fit. Its main purpose is to compute interpolated $\hat{y}$ values corresponding to unobserved values of the regressors $x_1,\ldots,x_p$, which is achieved by local smoothing of the data. Until now, little has been done to construct nonparametric regression methods that are also insensitive to outliers.

17.3.5 Robust Methods for Other Situations

Robust methods are not only useful for estimation in one-dimensional samples or when fitting a linear regression model. Huber (1981), Hampel et al. (1986), and Rousseeuw and Leroy (1987) also cover robust tests, robust multivariate location and covariance matrices, the problem of unsuspected serial correlations in supposedly independent data, robustness in time series, and robust estimation for circular data.

Positive-breakdown regression such as LMS can be extended to regression models with several intercepts (Rousseeuw and Wagner, 1994) and to models including dummy regressors (Hubert and Rousseeuw, 1997). Rousseeuw and Leroy (1987, chap. 7) applied the LMS to the zero-intercept regression model, to autoregressive time series, to orthogonal regression, and to directional data. Other extensions were to nonparametric regression (van Hoorn, 1988), nonlinear regression (Stromberg and Ruppert, 1992; Stromberg, 1993), and logistic regression (Christmann, 1994).

In multivariate analysis one can replace the classical covariance matrix by a positive-breakdown scatter matrix such as the MVE, e.g., for discriminant analysis, correlation matrices, principal components, and factor analysis. More research needs to be done for other situations.

REFERENCES

Abdel-Mottaleb, M., R. Chellappa, and A. Rosenfeld: "Binocular Motion Stereo with Independently Moving Objects," *Report CAR-TR-679*, Center for Automation Research, University of Maryland, 1993.

Andrews, D. F.: "A Robust Method for Multiple Linear Regression," *Technometrics*, vol. 16, pp. 523–531, 1974.

_____, P. J. Bickel, F. R. Hampel, P. J. Huber, W. H. Rogers, and J. W. Tukey: *Robust Estimates of Location: Survey and Advances*, Princeton Univ. Press, Princeton, N J, 1972.

Atkinson, A. C.: *Plots, Transformations, and Regression*, Clarendon Press, Oxford, 1985.

_____, "Masking Unmasked," *Biometrika*, vol. 73, pp. 533–541, 1986.

Belsley, D. A., E. Kuh, and R. E. Welsch: *Regression Diagnostics*, Wiley, New York, 1980.

Bickel, P. J.: "On Some Analogues to Linear Combinations of Order Statistics in the Linear Model," *Ann. Stat.*, vol. 1, pp. 597–616, 1973.

Bloomfield, P., and W. L. Steiger: *Least Absolute Deviations: Theory, Applications, and Algorithms*, Birkhäuser Verlag, Boston, MA, 1983.

Brown, G. W., and A. M. Mood: "On Median Tests for Linear Hypotheses," in *Proc. 2d Berkeley Symp. on Mathematical Statistics and Probability*, Univ. of California Press, Berkeley and Los Angeles, 1951, pp. 159–166.

Brownlee, K. A.: *Statistical Theory and Methodology in Science and Engineering*, 2d ed, Wiley, New York, 1965.

Carroll, R. J., and D. Ruppert: *Transformation and Weighting in Regression*, Chapman and Hall, New York and London, 1988.

Chork, C. Y.: "Unmasking Multivariate Anomalous Observations in Exploration Geochemical Data from Sheeted-Vein Tin Mineralization Near Emmaville, N. S. W., Australia," *J. Geochemical Exploration*, vol. 37, pp. 205–223, 1990.

_____ and R. Salminen: "Interpreting Geochemical Data from Outokumpu, Finland: an MVE-robust Factor Analysis," *J. Geochemical Exploration*, vol. 48, pp. 1–20, 1993.

Christmann, A.: "Least Median of Weighted Squares in Logistic Regression with Large Strata," *Biometrika*, vol. 81, pp. 413–417, 1994.

Cook, R. D., and S. Weisberg: *Residuals and Influence in Regression*, Chapman and Hall, New York and London, 1982.

de Loore, C, P. Monderen, and P. J. Rousseeuw: "A New Statistical Method to Derive Radial Velocity Shifts from Stellar Spectra," *Astron. Astrophys*., vol. 178, pp. 307–309, 1987.

Dodge, Y.: *Statistical Data Analysis Based on the L_1-Norm and Related Methods*, North-Holland, Amsterdam, 1987.

Donoho, D. L., and P. J. Huber: "The Notion of Breakdown Point," in *Fetschrift for Erich Lehmann*, P. Bickel, K. Doksum, and J. L. Hodges, Jr., Eds., Wadsworth, Belmont, CA., 1983.

Drew, M. S.: "Robust Specularity Detection from a Single Multi-illuminant Color Image," *Computer Vision, Graphics and Image Processing: Image Understanding*, vol. 59, pp. 320–327, 1994.

Edgeworth, F. Y.: "On Observations Relating to Several Quantities," *Hermathena*, vol. 6, pp. 279–285, 1887.

Hampel, F. R.: "A General Qualitative Definition of Robustness," *Ann. Math. Stat*., vol. 42, pp. 1887–1896, 1971.

_____: "The Influence Curve and Its Role in Robust Estimation," *J. Am. Stat. Assoc*., vol. 69, pp. 383–393, 1974.

_____, E. M. Ronchetti, P. J. Rousseeuw, and W. A. Stahel: *Robust Statistics: The Approach Based on Influence Functions*, Wiley, New York, 1986.

Hawkins, D. M.: *Identification of Outliers*, Chapman and Hall, New York and London, 1980.

_____, D. Bradu, and G. V. Kass: "Location of Several Outliers in Multiple Regression Data Using Elemental Sets," *Technometrics*, vol. 26, pp. 197–208, 1984.

Hill, R. W.: "Robust Regression when There Are Outliers in the Carriers," Ph. D. dissertation, Harvard University, Cambridge, MA, 1977.

Hodges, J. L., Jr.: "Efficiency in Normal Samples and Tolerance of Extreme Values for Some Estimates of Location," in *Proc. 5th Berkeley Symp. on Mathematical Statistics and Probability*, vol. 1, Univ. of California Press, Berkeley and Los Angeles, 1967, pp. 163–168.

Hössjer, O., P. J. Rousseeuw, and C. Croux: "Asymptotics of the Repeated Median Slope Estimator," *Ann. Statist*., vol. 22, pp. 1478–1501, 1994.

Huber, P. J.: "*Robust Regression*: Asymptotics, Conjectures, and Monte Carlo," *Ann. Stat*., vol. 1, pp. 799–821, 1973.

_____: *Robust Statistics*, Wiley, New York, 1981.

Hubert, M., and P. J. Rousseeuw: "Robust Regression with Both Continuous and Binary Regressors," *J. Statist. Planning and Infer*., vol. 75, pp. 153–163, 1997.

Jaeckel, L. A.: "Estimating Regression Coefficients by Minimizing the Dispersion of Residuals," *Ann. Math. Stat*., vol. 5, pp. 1449–1458, 1972.

Jolion, J. -M., P. Meer, and S. Bataouche: "Robust Clustering with Applications in Computer Vision," *IEEE Trans. Pattern Anal. and Machine Intell.*, vol. 13, pp. 791–802, 1991.

Jureckovq, J.: "Nonparametric Estimate of Regression Coefficients," *Ann. Math. Stat.*, vol. 42, pp. 1328–1338, 1971.

Koenker, R., and G. W. Bassett: "Regression Quantiles," *Econometrica*, vol. 46, pp. 33–50, 1978.

Kumar, R., and A. R. Hanson: "Robust Methods for Pose Determination," *Proceedings of the NSF/ARPA Workshop on Performance versus Methodology in Computer Vision*, pp. 41–57, 1994.

Mallows, C. L.: "On Some Topics in Robustness," Tech. Memo., Bell Telephone Labs., Murray Hill, N J, 1975.

Maronna, R. A., O. Bustos, and V. Yohai: "Bias- and Efficiency-Robustness of General *M* Estimators for Regression with Random Carriers," in *Smoothing Techniques for Curve Estimation*, T. Gasser and M. Rosenblatt, Eds., Springer, New York, 1979, pp. 91–116.

Martin, R. D., and C. J. Masreliez: "Robust Estimation via Stochastic Approximation," *IEEE Trans. Inform. Theory*, vol. IT-21, pp. 263–271, 1975.

Massart, D., L. Kaufman, P. J. Rousseeuw, and A. M. Leroy: "Least Median of Squares: A Robust Method for Outlier and Model Error Detection in Regression and Calibration," *Anal. Chim. Acta*, vol. 187, pp. 171–179, 1986.

Meer, P., D. Mintz, A. Rosenfeld, and D. Y. Kim: "Robust Regression Methods in Computer Vision: A Review," *Int. J. Computer Vision*, vol. 6, pp. 59–70, 1991.

Mili, L., V. Phaniraj, and P. Rousseeuw: "Least Median of Squares Estimation in Power Systems (with discussion)," *IEEE Trans. Power Systems*, vol. 6, pp. 511–523, 1991.

_____, M. G. Cheniae, N. S. Vichare, and P. J. Rousseeuw: "Robust State Estimation Based on Projection Statistics," *IEEE Trans. Power Systems*, vol. 11, pp. 1118–1127, 1996.

Pearl, J.: "A Space-Efficient On-Line Method of Computing Quantile Estimates," *J. Algorithms*, vol. 2, pp. 164–177, 1981.

Roth, G., and M. D. Levine: "Extracting Geometric Primitives," *Computer Vision, Graphics and Image Processing: Image Understanding*, vol. 58, pp. 1–22, 1993.

Rousseeuw, P. J.: "Least Median of Squares Regression," *J. Am. Stat. Assoc.*, vol 79, pp. 871–880, 1984.

_____: "Multivariate Estimation with High Breakdown Point," *Mathematical Statistics and Applications*, vol. B, W. Grossmann, G. Pflug, I. Vincze, and W. Wertz, Eds., Reidel, Dordrecht, 1985, pp. 283–297.

_____ and G. W. Bassett, Jr.: "The Remedian: A Robust Averaging Method for Large Data Sets," *J. Am. Stat. Assoc.*, vol. 85, pp. 97–104, 1990.

_____ and C. Croux: "Alternatives to the Median Absolute Deviation," *J. Am. Stat. Assoc.*, vol. 88, pp. 1273–1283, 1993.

_____, C. Croux, and O. Hössjer: "Sensitivity Functions and Numerical Analysis of the Repeated Median Slope," *Computational Statistics*, vol. 10, pp. 71–90, 1995.

_____ and A. M. Leroy: *Robust Regression and Outlier Detection,* Wiley-Interscience, New York, 1987.

_____ and J. Wagner: "Robust Regression with a Distributed Intercept Using Least Median of Squares," *Comput. Statist. and Data Anal.*, vol. 17, pp. 65–76, 1994.

_____ and B. van Zomeren: "Unmasking Multivariate Outliers and Leverage Points," *J. Am. Stat. Assoc.*, vol. 85, pp. 633–639, 1990.

_____ and V. Yohai: "Robust Regression by Means of *S* Estimators," in *Robust and Nonlinear Time Series Analysis*, J. Franke, W. Härdle, and R. D. Martin, Eds., Lecture Notes in Statistics 26, Springer, New York, 1984, pp. 256–272.

Seaver, B. L., and K. P. Triantis: "The Impact of Outliers and Leverage Points for Technical Efficiency Measurement Using High Breakdown Procedures," *Management Science*, vol. 41, pp. 937–956, appear, 1995.

Siegel, A. F.: "Robust Regression Using Repeated Medians," *Biometrika*, vol. 69, pp. 242–244, 1982.

Sinha, S. S., and B. G. Schunck: "A Two-Stage Algorithm for Discontinuity-Preserving Surface Reconstruction," *IEEE Trans. Pattern Anal. and Machine Intell.*, vol. 14, pp. 36–55, 1992.

Statistical Sciences: *S-PLUS User's Manual, Version 3.2*, Seattle: Statistic, a Division of MathSoft, Inc., 1993.

Stewart, C. V.: "A New Robust Operator for Computer Vision: Theoretical Analysis," *Proceedings IEEE Conference on Computer Vision and Pattern Recognition*, 1994, pp. 1–8.

Stromberg, A. J.: "Computation of High Breakdown Nonlinear Regression," *J. Am. Stat. Assoc.*, vol. 88, pp. 237–244, 1993.

_____ and D. Ruppert: "Breakdown in Nonlinear Regression," *J. Am. Stat. Assoc.*, vol. 87, pp. 991–997, 1992.

Theil, H.: "A Rank-Invariant Method of Linear and Polynomial Regression Analysis, Pts. 1–3," *Ned. Akad. Wet. Proc., A*, vol. 53, pp. 386–392, 521–525, 1397–1412, 1950.

Thompson, W. B., P. Lechleider, and E. R. Stuck: "Detecting Moving Objects Using the Rigidity Constraint," *IEEE Trans. Pattern Anal. and Machine Intell.*, vol. 15, pp. 162–166, 1993.

Tierney, L.: "A Space-Efficient Recursive Procedure for Estimating a Quantile of an Unknown Distribution," *SIAM J. Sci. and Stat. Comput.*, vol. 4, pp. 706–711, 1983.

Trau, R., P. Salu, K. Wisnia, L. Kaufman, P. J. Rousseuw, and A. Pierreux: "Simultaneous ERG-VER Recording: Statistical Study," *Bull. Belg. Ophthalmol. Soc.*, vol. 206, pp. 61–67, 1983.

Tukey, J. W.: *Exploratory Data Analysis*, Addison-Wesley, Reading, MA, 1970.

_____: "The Ninther, a Technique for Low-Effort Robust (Resistant) Location in Large Samples," in *Contributions to Survey Sampling and Applied Statistics in Honor of H. O. Hartley*, H. A. David, Ed., Academic Press, New York, 1978, pp. 251–257.

van Hoorn, J.: "Robuste, nichtparametrische Schätzung von Regressionsfunktionen mit S-Schätzern," Ph. D. Thesis, Universität Essen, Germany, 1988.

Wang, C. M., D. F. Vecchia, M. Young, and N. A. Brilliant: "Robust Regression Applied to Optical-Fiber Dimensional Quality Control," *Technometrics*, vol. 39, pp. 25–33, 1997.

CHAPTER 18
NONLINEAR REGRESSION

James J. Swain
The University of Alabama in Huntsville, AL

18.1 INTRODUCTION

It is commonly observed in engineering practice that many process mechanisms can be usefully approximated using linear functional models, but that for a theoretical understanding and the prediction of extreme cases one must eventually deal with the inherent nonlinearity of the process. When investigating processes using statistical estimation, linear regression forms a useful basis for both empirical models and interpolation of function values over a limited range. Nonlinear regression is used where a precise determination of the mechanism or extrapolation beyond a limited region is needed.

Linear and nonlinear regression models can be viewed as two possible types of models whose relative usefulness depends on the knowledge, resources, and level of detail needed in a particular application. As summarized by Box and Draper (1987), response surface studies often begin with linear empiricial models, and as experience and experimental knowledge expand, the model may grow in detail and complexity until a theoretical, nonlinear model may be possible. The overall problem may be to determine a representation of the expected value $\eta = E[y]$ of a response y as a function of inputs x. The function η is called the *regression function*. Initially η is approximated locally using a linear, often first-order model, $g(\mathbf{x}; \boldsymbol{\theta})$. Later experience with the empirical model $g(\mathbf{x}; \boldsymbol{\theta})$ may lead to the suggestion of a model nonlinear in the parameters $\boldsymbol{\theta}$ $f(\mathbf{x}; \boldsymbol{\theta})$, or perhaps several models, $f_k(\mathbf{x}; \boldsymbol{\theta}_k)$. When there is one model to consider, the problem is essentially one of parameter estimation. When there are several candidates, the additional problem of model discrimination must also be addressed.

18.1.1 Linear and Nonlinear Regression Models

In both linear and nonlinear models, the data consist of observations $(X, \mathbf{y})$ of the independent variables $X = \{x_{ij}\}$, $i = 1, \ldots, n$ and $j = 1, \ldots, r$, together with n observations of the dependent variable $\mathbf{y} - \{y_i\}$. The two types of data are fit using a model of the responses y_i of the form $\boldsymbol{x}_i$, the ith row of x

$$y_i = \eta(\mathbf{x}_i; \boldsymbol{\theta}) + \epsilon_i \quad \text{for } i = 1, 2, \ldots, n. \tag{18.1}$$

The parameters $\boldsymbol{\theta}$ are chosen so that the predicted responses $\eta(\mathbf{x}_i; \boldsymbol{\theta})$ are as close as possible to the observed responses y_i. In the model of the observed responses y_i, the errors $\boldsymbol{\epsilon} = (\epsilon_1, \ldots, \epsilon_n)$ have mean zero and by assumption may be normally distributed with constant variances $\sigma^2 I$. When the response function is linear in the parameters $\boldsymbol{\theta}$ and the errors additive as in Eq. (18.1), the model is linear; otherwise the model is nonlinear. A linear model may be first-order,

$$g(\mathbf{x}; \boldsymbol{\theta}) = \theta_0 + \theta_1 x_1 + \theta_2 x_2$$

or of higher order,

$$g(\mathbf{x}; \boldsymbol{\theta}) = \theta_0 + \theta_1 x_1 + \theta_2 x_2 + \theta_{11} x_1^2 + \theta_{22} x_2^2 + \theta_{12} x_1 x_2.$$

The key idea is that the model is a linear combination of its parameters.

A model η may be transformable into a linear model, provided that the transformed model both is linear in the parameters and possesses additive errors. For instance, a model such as

$$y_i = \theta_0 \exp(-\theta_1 x_i + \epsilon_i)$$

can be transformed using the natural logarithm to obtain

$$\ln(y_i) = \ln \theta_0 - \theta_1 x_i + \epsilon_i$$

which is linear in $\boldsymbol{\theta}$ and additive in the errors $\boldsymbol{\epsilon}$. Models which can be transformed into linear form with additive errors are termed *intrinsically linear.* Another instance is the case $y = e^{\phi} x + \epsilon$, where substitution of the transformed parameter $\theta = e^{\phi}$ leads to a model linear in θ. By contrast,

$$y_i = \theta_1 \exp(-\theta_2 x_i) + \epsilon_i$$

cannot be so transformed while preserving the additive error term. Use of the logarithm to obtain a linear model would result in new errors ϵ', which depend on x, the parameter values, and the original errors ϵ. The usual assumption of constant variance and independence among errors is violated here, and application of linear least squares will lead to estimates with quite different properties than those expected. It is likely that estimates will be biased and have a larger variance than those obtained using least squares with the nonlinear model. The risk of poor estimates when using linear estimation of an intrinsically nonlinear model is illustrated in the example that closes this section.

Whether the model is linear or nonlinear, fitting is most often accomplished by minimizing some function of the residuals,

$$e_i(\boldsymbol{\theta}) = y_i - \eta(\mathbf{x}_i; \boldsymbol{\theta}) \tag{18.2}$$

which depend on the choice of parameters $\boldsymbol{\theta}$. The parameters values providing the best fit are denoted by $\hat{\boldsymbol{\theta}}$.

The differences in form between linear and nonlinear models lead to significant differences in estimation, difficulty of analysis, and the statistical properties of the estimators. When estimating linear models using least squares, for instance, solutions are closed form and the distribution theory of estimators is elegant and well developed. By contrast, even for least squares estimation, parameter estimation in the case of nonlinear models is iterative, not automatically successful, and the properties of estimators are accessible only for large samples or in cases where the model being estimated has properties close to those of a linear model.

18.1.2 Short Survey of Nonlinear Model Applications

Nonlinear models have been employed in a wide variety of problems. In this section we illustrate some of the many applications.

Nonlinear models are generally based on engineering and scientific principles and can take many forms. A commonly encountered form is variations on power laws, $\eta = \theta_0 + \theta_1 x + \theta_4 x^{\theta_3}$, sometimes used as a nonlinear empirical model and sometimes advanced from theoretical reasoning. For instance, Draper and Smith (1981) identify a model $\eta = \alpha + \beta x^{-\gamma}$ for predicted tensile strength as an example of this class, as is their chlorine shelf-life example, $\eta = \alpha + \beta \exp[\gamma(x - c)]$, Smith and Dubey (1964). Further models include one for steam pressure and temperature and for biological oxygen demand (BOD) in a stream. As a flexible model for data resembling a segmented line, Ratkowsky (1983) mentions the bent-hyperbola model (Griffiths and Miller, 1973),

$$\eta = \theta_1 + \theta_2(x - \theta_4) + \theta_3[(x - \theta_4)^2 + \theta_5]^{1/2}.$$

Spline models with variable knot points are also nonlinear models (Gallant and Fuller, 1973).

A source of models is the solution to the mth-order linear differential equation. Solution of the differential equations leads to responses of the type

$$\eta(\mathbf{x}; \boldsymbol{\theta}) = \sum_{j=1}^{m} \theta_{2j} \exp(-\theta_{2j-1} x). \tag{18.3}$$

An illustration of this model is given in Guttman and Meeter (1965), who consider linear reaction kinetics to model the reaction $A \rightarrow B \rightarrow C$ with initial conditions $[A] = 1$ and $[B] = [C] = 0$. The solution in terms of the three chemical species yields

$$\eta_A = \exp(-\theta_1 x)$$

$$\eta_B = \frac{\theta_1}{\theta_1 - \theta_2}[\exp(-\theta_2 x) - \exp(-\theta_1 x)]$$

$$\eta_C = 1 - \frac{1}{\theta_1 - \theta_2}[\theta_1 \exp(-\theta_2 x) - \theta_2 \exp(-\theta_1 x)].$$

Kinetics models are not limited to chemical engineering. Models of this form often arise in pharmacokinetic studies of drug transport through the body, where the body is modeled as a series of interconnected compartments through which the drug flows until it is absorbed or excreted. A particularly large pharmokinetic model of this type is described and fit for six individuals exposed to tetrachloroethylene by Gelman, Bois, and Jiang (1996); see also Seber and Wild (1989) for examples of compartment models. Ecological tracer studies use a similar model to model nutrient flows within a system, where a radioisotope is used to label a nutrient as it flows through the system.

When some of the characteristic roots associated with the differential equations are complex, periodic solutions arise. Equation (18.3) is then modified to include appropriate sine and cosine terms.

Chemical engineering has been a rich source of nonlinear models, particularly in the study of kinetic reactions. Blakemore and Hoerl (1963), for instance, list 18 possible models for a single reaction, depending on the form of the reaction rate. In some chemical reaction studies the reaction rate may be studied, as in the model given by

$$\eta(\mathbf{x}; \boldsymbol{\theta}) = \frac{\theta_1\theta_3 x_{i1}}{1 + \theta_1 x_{i1} + \theta_2 x_{i2}}. \tag{18.4}$$

More complicated systems of equations can be solved numerically even when the analytic solution is unknown or too complicated to be given explicitly. For instance, one may hypothesize the rate of reaction in chemical kinetics but actually observe the yield over a fixed period. Then the response function is defined implicitly through the integral equation

$$\int_{\eta_0}^{\eta_1} \frac{d\eta}{R(\eta; \mathbf{x}, \boldsymbol{\theta})} = t_1 - t_0$$

where η_0 and η_1 are the initial concentration ($t = t_0$) and the final concentration ($t = t_1$), respectively. Examples in which the implicit equations are solved explicitly are given in Hunter and Mezaki (1964) and Box and Hunter (1965). Two other examples, Sane, Eckert, and Woods (1974) and Ziegal and Gorman (1980), are expressed implicitly, though it is possible to represent the solution as an explicit function, $\eta(\mathbf{x}; \boldsymbol{\theta})$. Further examples of chemical models are given in Kittrell, Hunter, and Watson (1965), Peterson and Lapidus (1966), and Sutton and MacGregor (1977). Bates and Hunter (1985) provide an extensive bibliography of applications of nonlinear models, particularly kinetics models.

Beck and Arnold (1977) suggest a number of models associated with heat transfer in metal billets. In the usual textbook case the parameter $\alpha = hA/\rho cV$ is assumed to be constant, so that the solution to $\partial T/\partial t = \alpha(T_\infty - T)$ is $T(t) = \eta = T_\infty + (T_0 - T_\infty) e^{-\alpha t}$. More realistically, the parameter α can be modeled as varying with time, $\alpha(t) = \beta_1 + \beta_2 t + \beta_3 t^2$, or varying as a function of temperature, $\alpha(T) = \beta_1 + \beta_2(T - T_\infty)^n$, $n \neq 0$. The solutions to the differential equations are then

$$T(t) = T_\infty + (T_0 - T_\infty)e^{-\alpha(t)}$$

and

$$T(t) = T_\infty + (T_0 - T_\infty)e^{-\beta_1 t}\left[1 + \frac{\beta_2}{\beta_1}(T_0 - T_\infty)^n (1 - e^{-\beta_1 nt})\right]^{-1/n}.$$

Ratkowsky (1983) discusses a number of agricultural yield models, including the Bleasdale-Nelder model,

$$\eta = (\theta_1 + \theta_2 x)^{-1/\theta_3}$$

the Holliday (1960) model,

$$\eta = (\theta_1 + \theta_2 x + \theta_3 x^2)^{-1}$$

and the model of Farazdaghi and Harris (1968),

$$\eta = (\theta_1 + \theta_2 x^{\theta_3})^{-1}.$$

Variations on these models involving further transformations of the function or of choices in parameterizing the models are discussed further in Ratkowsky (1983). Ratkowsky also surveys sigmoidal growth models, such as the Gompertz,

$$\eta = \theta_1 \exp\left[-\exp\left(\theta_2 - \theta_3 x\right)\right]$$

the logistic,

$$\eta = \frac{\theta_1}{1 - \exp\left(\theta_2 - \theta_3 x\right)}$$

the Richards model (Richards, 1959),

$$\eta = \frac{\theta_1}{\left[1 - \exp\left(\theta_2 - \theta_3 x\right)\right]^{-1/\theta_4}}$$

and the Morgan-Mercer-Flodin model (Morgan, Mercer, and Flodin, 1975),

$$\eta = \frac{\theta_2 \theta_3 + \theta_1 x^{\theta_4}}{\theta_3 + x^{\theta_4}}.$$

Donaldson and Schnabel (1987) summarize a number of models, including an eight-parameter model associated with microwave adsorption lines. A case study of a model from the electronics industry uses nonlinear regression to calibrate a vector demodulator, used in signal processing (Kafadar, 1994). Draper and Smith (1981) have an extensive bibliography for nonlinear models plus problems taken from the literature. Nonlinear models are also employed in econometrics, and Gallant (1987) contains several examples.

Example 18.1: Using Linear Regression to Fit Nonlinear Models. To illustrate the effects of using a linearized model, consider the catalytic model given in Eq. (18.4). Ignoring the error term, one can obtain the linear model by taking the reciprocal of both sides and then dividing by x_1,

$$y_i' = \beta_0 + \beta_1 x_{i1} + \beta_2 x_{i2} + \epsilon_i'. \tag{18.5}$$

The model is given in the transformed parameters $\beta_0 = (\theta_1\theta_3)^{-1}$, $\beta_1 = 1/\theta_3$, $\beta_2 = \theta_2(\theta_1\theta_3)^{-1}$, and $y' = x_1/y$. We may form an estimate of the behavior of the two estimators based on nonlinear least squares using Eq. (18.4) and linear least squares using Eq. (18.5) by direct Monte Carlo sampling (see Sec. 18.3). Let the design be given by the five points (x_1, x_2): (1.5, 0), (3, 0), (2.25, 1.25), (1.5, 0.75), and (3, 1.5), and the error standard deviation by $\sigma = 0.005$. When the parameters are taken to be (2.9, 12.2, 0.69), 1000 simulated observations yield averages of (2.96, 12.4, 0.6900) for the nonlinear least squares estimator and (4.8, 19.3, 0.695) for the linear least squares estimator. Both solutions exhibit bias in the first two components, though the bias in the linear solution is much greater. Moreover, the linear solutions have a sampling distribution that is considerably more dispersed than that of the nonlinear least squares solution, in addition to being both asymmetric and heavy tailed, very different from a normal distribution.

The linear solution can be improved in this case by using weighted least squares, since the variance of the transformed errors is approximately $[x_1/\eta^2(\mathbf{x}; \boldsymbol{\theta})]^2\sigma^2$. However, derivation of the transformed error variance is not always easy to obtain. Since numerical software for the solution of the nonlinear problem is now readily available, resorting to a weighted linear formulation is not as advantageous as it once was.

The problem of obtaining estimates $\hat{\boldsymbol{\theta}}$ is considered in the next section, including a brief illustration from two commercially available programs. In Sec. 18.3 the distributional properties of estimators are considered so that the uncertainty about the precise value of the estimate can be quantified. In Sec. 18.4 the design of experiments for both model discrimination and precise estimation is briefly reviewed. The final section summarizes the material.

18.2 ESTIMATION OF PARAMETERS

In this section statistical criteria for choosing estimators $\hat{\boldsymbol{\theta}}$ of $\boldsymbol{\theta}$ are introduced, together with methods for computing the estimates. Most attention is focused upon *least-squares estimation*, but important alternatives such as the general classes of maximum-likelihood estimation, robust estimation, and bayesian estimation are discussed. Criteria for parameter estimation using multiple responses are discussed briefly.

When the response functions are nonlinear, the solutions of the estimation problems are not closed form, and the problems are solved by iterative algorithms. Numerical algorithms for least squares estimation are treated in detail. Important features of obtaining estimates, including termination criteria for the optimization procedure and methods of obtaining starting values, are included in the discussion. Finally, numerical examples obtained from some commercially available statistical packages are provided to illustrate their use.

18.2.1 Parameter Estimation Formulations

The presence of random errors $\boldsymbol{\epsilon}$ in Eq. (18.1) precludes the fitted model from exactly matching the observations $\mathbf{y}$. Thus one seeks parameter estimates which are best in some apparent or statistical sense, subject to any constraints on the feasible values of the parameters. These constraints usually arise from the physical theory. In most cases the estimate $\hat{\boldsymbol{\theta}}$ is chosen to minimize some function $\Psi(\mathbf{e}(\boldsymbol{\theta}))$ of the residuals. The function $\Psi(\cdot)$ is selected to quantify the discrepancy between $\mathbf{y}$ and $\boldsymbol{\eta}(\boldsymbol{\theta}) = (\eta(\mathbf{x}_1; \boldsymbol{\theta}), \ldots, \eta(\mathbf{x}_n; \boldsymbol{\theta}))$. The function selected usually weights the discrepancies to account for the functional form of the errors, the variances of the errors, or to include prior information in the computation.

The most common method of obtaining estimates is the least-squares (LSQ) procedure. This is accomplished by minimizing the sum of the squared residuals (18.2) in ordinary least squares (OLSQ),

$$\Psi(\boldsymbol{\theta}) = \mathbf{e}'(\boldsymbol{\theta})\mathbf{e}(\boldsymbol{\theta}) = \sum_{i=1}^{n} e_i^2(\boldsymbol{\theta}) \tag{18.6}$$

or by weighted least squares (WLSQ),

$$\Psi(\boldsymbol{\theta}) = \mathbf{e}'(\boldsymbol{\theta})W\mathbf{e}(\boldsymbol{\theta}) = \sum_{i=1}^{n}\sum_{j=1}^{n} w_{ij}\, e_i(\boldsymbol{\theta})\, e_j(\boldsymbol{\theta}) \tag{18.7}$$

for some positive definite matrix W of weighting constants.

A second method of estimation is given by maximum-likelihood estimation. In this method, estimates are chosen which are most likely to have generated the observed data under an assumed probability model for the errors $h(\boldsymbol{\epsilon})$. The likelihood is expressed by the density function of the errors using the residuals $\mathbf{e}(\boldsymbol{\theta})$, which depend on the parameters $\boldsymbol{\theta}$. When observations are independent, $L(\boldsymbol{\theta}) = h(\mathbf{e}(\boldsymbol{\theta}))$. Parameters obtained by maximizing the likelihood are termed maximum-likelihood estimates (MLEs). An equivalent formulation is that of the log likelihood, $\mathcal{L}(\boldsymbol{\theta}) = \log(L(\boldsymbol{\theta}))$, since log is a monotone function. An MLE is the solution to the maximization of either the likelihood or the log likelihood.

For the case that the error vector is normally distributed with mean $\mathbf{0}$ and variance matrix $V = \text{Var}\,[\boldsymbol{\epsilon}] = E[\boldsymbol{\epsilon}\boldsymbol{\epsilon}']$, the log likelihood can be written

$$\mathcal{L}(\boldsymbol{\theta}) = \frac{1}{2}\,\mathbf{n}\log(2\pi) - \frac{1}{2}\log|V| - \frac{1}{2}\,e'(\boldsymbol{\theta})V^{-1}e(\boldsymbol{\theta}).$$

Only the last term in this equation depends on $\boldsymbol{\theta}$ if V is known. When the errors are assumed to be independent and have common variance σ^2, then $V = \sigma^2 I$. The MLE is the solution to OLSQ, Eq. (18.6). When V is known, the function $\mathcal{L}(\boldsymbol{\theta})$ is maximized when the last term is minimized using WLSQ, Eq. (18.7), using $W = V^{-1}$.

When the variance matrix V is unknown, it must be estimated along with the parameter values. Bard (1974) gives a description of the MLEs for normal errors when V is either known partially or not known at all. The method of extended least squares (Peck, Sheiner, and Nichols, 1984) can also be used when the errors are normal, independent, and the variance is related to the mean by some power function. Least squares can be used for estimation even when the errors are not normally distributed. Provided that the errors do not depart heavily from the normal assumption, the least-squares solution will not differ greatly from the MLE. Care must be applied, as always, when least-squares estimates are applied with heavy tailed errors. As discussed in Sec. 18.3, however, when least squares are maximum likelihood, they possess the good properties of MLEs. In particular, as sample sizes increase indefinitely, MLEs are typically efficient, unbiased, and the distribution of the estimators $\hat{\boldsymbol{\theta}}$ is normal.

When errors differ greatly from the normal distributions, other estimators are preferred. For instance, the least absolute differences (LAD) estimator

$$\min_{\boldsymbol{\theta}} \sum_{i=1}^{n} |e_i(\boldsymbol{\theta})|$$

is the MLE for Laplace (double exponential) errors. Laplace errors have heavier tails than the normal, so extreme values are much more likely than with the normal. For errors of this type OLSQ usually performs poorly. Extreme observations or *outliers* are a common concern when using least squares, and LAD estimation is often used as an alternative since extreme observations are less influential in this case. Likewise, the minimax criterion

$$\min_{\boldsymbol{\theta}} \left[\min_{i} \left| \frac{e_i(\boldsymbol{\theta})}{b_i} \right| \right]$$

is the MLE for uniform errors on the intervals $(-b_i, +b_i)$. These two estimators are special cases of L_p-norm estimators, $\Sigma_{i=1}^{n} |e_i(\boldsymbol{\theta})|^p$, with LAD ($p = 1$), least squares ($p = 2$), and minimax ($p = \infty$), as discussed by Bard (1974) and Gonin and Money (1989). The choice of $p = 1.5$ is sometimes used as a robust compromise between LAD and least-squares estimation.

The sensitivity of least-squares estimators to certain types of nonnormal errors with heavy tails has led to the development of robust estimation criteria. Robust estimation procedures are chosen to be less sensitive to departures from normality while performing acceptably well against normal errors, for which least squares procedures are optimal. One alternative distribution is the contaminated normal, in which errors usually are drawn from a particular normal distribution with low variance, but infrequently are drawn from a second normal distribution with large variance. This distribution models the occurrence of outliers, observations with large residuals. One robust procedure is based on the Huber function (Huber, 1964), given by

$$\rho(e) = \begin{cases} \frac{1}{2}e^2, & \text{if } |e| \le c \\ c|e| - \frac{1}{2}c^2, & \text{if } |e| > c \end{cases}$$

and the estimators are computed by minimizing

$$\Psi(\boldsymbol{\theta}) = \sum_{i=1}^{n} \rho\left(\frac{e_i(\boldsymbol{\theta})}{s}\right).$$

The constant c is usually taken to be 2, while s is a scale estimate such as the median absolute deviation, the median of the values $| e_i(\boldsymbol{\theta}) - m |$, where m is the median from among the residuals $e_i(\boldsymbol{\theta})$. For smaller standardized residuals the method works like least squares, while for larger errors it performs like least absolute differences. Owing to the likelihood of outliers in electronic measurements, for instance, Kafadar (1994) used Tukey's biweight estimator in the formulation of the nonlinear regression problem. More detailed treatment of robust estimators is given in Huber (1981) and Rey (1983).

Bayes' estimators are based on the use of Bayes' theorem to combine direct information from the likelihood function with prior information in the form of a prior probability distribution $p^0(\boldsymbol{\theta})$ on the possible values of the unknown parameters. The resulting posterior distribution (distribution after the data have been taken), $p^*(\boldsymbol{\theta}) = k\, L(\boldsymbol{\theta})p^0(\boldsymbol{\theta})$, is used in a number of ways to form bayesian estimators. The constant k is taken to be 1 when an improper prior is used, and otherwise can be taken to be the inverse of $\int L(\boldsymbol{\theta})p^0(\boldsymbol{\theta})\, d\boldsymbol{\theta}$. For instance, the estimator can be defined to be the mode of the posterior density, so that $\hat{\boldsymbol{\theta}}$ is chosen to maximize

$p^*(\boldsymbol{\theta})$. The mean or median of the posterior distribution can also be taken as bayesian estimators; the mean is appropriate when a quadratic loss is assumed.

Bayes' estimators have a number of attractive features. The mode and the mean of the posterior distribution both converge to the MLE as the number of data points increases or when the "uninformative" prior distribution is used. Bayes' solutions can exist when MLEs are overdetermined and can be used to constrain solutions to certain regions. For instance, when parameters must be positive, the prior distribution can be set to 0 for all negative values. Informative prior distributions were employed by Gelman, Bois, and Jiang (1996), for instance, both to incorporate knowledge from the scientific literature and to constrain parameters to physcially reasonable values. On the other hand, Bayes' estimators introduce a degree of arbitrariness to the analysis through the introduction of the prior distribution. A locally uniform prior distribution is often given, so that many times the estimator is essentially an MLE.

Estimation of the parameters can also be performed when more than one response is being measured. For instance, in a series of chemical reactions, one may measure several components simultaneously. It is often advantageous to do so, since the information from the different responses can lead to estimators considerably better than those obtained by fitting each response separately. Box and Draper (1965) develop the determinant criterion

$$\min_{\boldsymbol{\theta}} \left| (\mathbf{y} - \boldsymbol{\eta}(\mathbf{x}; \boldsymbol{\theta}))' (\mathbf{y} - \boldsymbol{\eta}(\mathbf{x}; \boldsymbol{\theta})) \right|$$

for the $n \times s$ matrices of responses $\mathbf{y}$ and predicted values $\boldsymbol{\eta}(\mathbf{x}; \boldsymbol{\theta})$ using a Bayes argument for the case that the errors are normally distributed with unknown variance (within experiments) but independent across experiments. Both Bard (1974) and Seber and Wild (1989) detail least-squares estimators in the multivariate case under various assumptions about the error variances.

18.2.2 General Issues in Numerical Algorithms

For nonlinear response functions, optimization of the criterion $\Psi(\boldsymbol{\theta})$ is necessarily iterative. One can approach such problems by general algorithms for unconstrained optimization (Bard, 1974; Kennedy and Gentle, 1980; Thisted, 1988), but because of the special structure of least-squares problems, special algorithms are used for most applications. These algorithms are discussed in Sec. 18.2.3. Some preliminary issues are treated in this section, which are common to several of the algorithms to be introduced.

The iterative algorithms proceed by assuming a starting estimate $\hat{\boldsymbol{\theta}}^{(0)}$, from which an improved solution is obtained using $\hat{\boldsymbol{\theta}}^{(i+1)} = \hat{\boldsymbol{\theta}}^{(i)} + \mathbf{d}_i$, where the improving direction $\mathbf{d}_i$ depends on the residuals $\mathbf{e}(\hat{\boldsymbol{\theta}}^{(i)})$ at the current solution and derivatives of the response function with respect to $\boldsymbol{\theta}$ evaluated at $\hat{\boldsymbol{\theta}}^{(i)}$. When $\Psi(\hat{\boldsymbol{\theta}}^{(i+1)}) < \Psi(\hat{\boldsymbol{\theta}}^{(i)})$, the direction $\mathbf{d}_i$ is called *acceptable*. Sometimes the update in $\hat{\boldsymbol{\theta}}^{(i+1)}$ is made acceptable by searching along the direction $\mathbf{d}_i$ using a line search,

$$\min_{\alpha} \Psi(\hat{\boldsymbol{\theta}}^{(i)} + \alpha \mathbf{d}_i).$$

When the exact minimum α^* is obtained, the line search is termed exact, and otherwise it is inexact. Some algorithms work best with exact line searches, but most of the algorithms used with least-squares problems will work with inexact line searches.

Because of the nonlinearity of the response function, the solution of the estimation problem is complicated by the possibility of local minima. Each solution is termed a local minimum if all nearby values are strictly greater than the value at the solution. A global minimum is the least of the local minima. There is in general no method to determine whether local minima exist or whether a given minimum is the global one, except under constraints suchas convex sets or concave responses. These constraints rarely apply in nonlinear regression problems. To some extent these problems can be avoided when the sample size is large or when the distribution of the independent variables is consistent with good experimental design (Sec. 18.4). However, the existence of local minima can be checked by solving the problem from multiple starting points or by partitioning the feasible region T into subregions T_1, ..., and searching each one separately. The existence of local minima or unreasonable solutions can indicate problems in the model being fit, the design of the experiment, or the size of the errors. When the model is reasonable, the global minimum can be sought by using starting points far from starting points leading to the local minima.

A more serious problem due to nonlinearity is that no algorithm is guaranteed to converge to a solution in all cases. This is particularly true when the starting point is distant from the solution or the problem is ill-conditioned. This problem can sometimes be overcome by using several starting points until one is found that will lead to a solution. Sometimes scaling of variables or changing the parameterization of the model can help. Several of these points are related to statistical properties and are discussed further in Sec. 18.3.

Obtaining good starting values is important for successful estimation. Bard (1974) and Ratkowsky (1983) both discuss this problem in detail. Starting values can be obtained by the solution of a linearized version of the problem. Starting conditions, such as the value of the derivative of the response function at the initial conditions, can sometimes be used. Likewise, models with asymptotic behavior (increasing to a maximum value) can take advantage of this fact if the data suggest the approximate value of the asymptote. When the number of data points exceeds the number of parameters, a set of p expressions can be obtained by eliminating all but p of the data points. Solutions can then be obtained by solving the resultant simultaneous nonlinear equations. When these methods fail, points can also be obtained by grid searches and random sampling until a reasonable starting point can be found. Two of these methods are illustrated in Examples 2 and 3.

Two main strategies are employed for determining when the iterative process of optimization is to be terminated. The main approach is to terminate when the relative change in either the fitting criterion $\Psi(\boldsymbol{\theta})$ or the parameter estimates themselves $|\hat{\boldsymbol{\theta}}^{(i+1)} - \hat{\boldsymbol{\theta}}^{(i)}|$ becomes small. However, one can be misled using this, and Bates and Watts (1981) suggest an exact test based on the fact that at the optimum, the two vectors $\mathbf{e}(\hat{\boldsymbol{\theta}})$ and the tangent plane approximation to $\boldsymbol{\eta}(\boldsymbol{\theta})$ are orthogonal (for least squares). Their offset criteria halt iterations when the projection of the residual vector on the tangent plane is small relative to a statistical error estimate.

18.2.3 Algorithms for Least Squares

Special algorithms for least squares take advantage of the form of the objective function, a sum of squares, Eq. (18.6). In particular, it is easily shown that the gradient of $\Psi(\boldsymbol{\theta})$ is given by $\nabla_\theta \Psi = -F'(\boldsymbol{\theta})e(\boldsymbol{\theta})$, where $F(\boldsymbol{\theta})$ is the $n \times p$ Jacobian matrix of derivatives,

$$F(\boldsymbol{\theta}) = \begin{bmatrix} \frac{\partial \eta(\mathbf{x}_1; \boldsymbol{\theta})}{\partial \theta_1} & \frac{\partial \eta(\mathbf{x}_1; \boldsymbol{\theta})}{\partial \theta_2} & \cdots & \frac{\partial \eta(\mathbf{x}_1; \boldsymbol{\theta})}{\partial \theta_p} \\ \frac{\partial \eta(\mathbf{x}_2; \boldsymbol{\theta})}{\partial \theta_1} & \frac{\partial \eta(\mathbf{x}_2; \boldsymbol{\theta})}{\partial \theta_2} & \cdots & \frac{\partial \eta(\mathbf{x}_2; \boldsymbol{\theta})}{\partial \theta_p} \\ \vdots & \vdots & & \vdots \\ \frac{\partial \eta(\mathbf{x}_n; \boldsymbol{\theta})}{\partial \theta_1} & \frac{\partial \eta(\mathbf{x}_n; \boldsymbol{\theta})}{\partial \theta_2} & \cdots & \frac{\partial \eta(\mathbf{x}_n; \boldsymbol{\theta})}{\partial \theta_p} \end{bmatrix} \tag{18.8}$$

In addition, the Hessian matrix of second derivatives $\nabla^2_\theta \Psi$ of Ψ is given by

$$\nabla^2_\theta \Psi = F'(\boldsymbol{\theta})F(\boldsymbol{\theta}) + \sum_{i=1}^{n} e_i(\boldsymbol{\theta})\nabla^2_\theta \Psi_i \tag{18.9}$$

where $\nabla^2_\theta \Psi_i$ is the hessian of second derivatives of $\eta(\mathbf{x}_i; \boldsymbol{\theta})$ with respect to $\boldsymbol{\theta}$ given by $\{\partial^2\eta/\partial\theta_j\partial\theta_k\}$. Note that when the residuals are small, Eq. (18.9) reduces to the approximation $\nabla^2_\theta \Psi \approx F'(\boldsymbol{\theta})F(\boldsymbol{\theta})$.

The simplest algorithm for solution to OLSQ is *steepest descent*, using $\mathbf{d}_i = F'(\hat{\boldsymbol{\theta}}^{(i)})e(\hat{\boldsymbol{\theta}}^{(i)})$, since the negative of the gradient is the direction of greatest decrease in the value of the function $\Psi(\boldsymbol{\theta})$. However, since the gradient is a local property, a line search is most often employed to guarantee that the step taken is acceptable. Steepest descent often works well far from the optimum, but poorly near an optimum, since the direction of the gradient changes quickly in the vicinity of the optimum, where it vanishes altogether.

A class of algorithms used most often in practice for nonlinear least squares can be derived by approximating the response function about the current estimate of the solution with a first-order Taylor series,

$$\tilde{\boldsymbol{\eta}}(\boldsymbol{\theta}) = \boldsymbol{\eta}(\hat{\boldsymbol{\theta}}^{(i)}) + F(\hat{\boldsymbol{\theta}}^{(i)})(\boldsymbol{\theta} - \hat{\boldsymbol{\theta}}^{(i)}) \tag{18.10}$$

and substituting $\tilde{\boldsymbol{\eta}}(\boldsymbol{\theta})$ in place of $\boldsymbol{\eta}(\boldsymbol{\theta})$ in the least-squares formulation. Solution can be obtained in closed form, since the approximation is linear in the unknown parameters. This is the Gauss-Newton algorithm for solving the least-squares problem,

$$\hat{\boldsymbol{\theta}}^{(i+1)} = \hat{\boldsymbol{\theta}}^{(i)} + F'(\hat{\boldsymbol{\theta}}^{(i)})F(\hat{\boldsymbol{\theta}}^{(i)})]^{-1} F'(\hat{\boldsymbol{\theta}}^{(i)})e(\hat{\boldsymbol{\theta}}^{(i)}). \tag{18.11}$$

Observe that this is an approximation to the Newton algorithm,

$$\hat{\boldsymbol{\theta}}^{(i+1)} = \hat{\boldsymbol{\theta}}^{(i)} - [\nabla^2_\theta \Psi]^{-1} \nabla_\theta \Psi$$

where the Hessian term (in square brackets) is approximated with the leading terms in Eq. (18.9). The advantage to using the Gauss algorithm is that second derivatives need not be evaluated, while one gets some of the advantages of (partial) hessian information. The Gauss algorithm can fail, particularly when the Jacobian matrix is rank-deficient (similar to the collinearity problem of linear regression). The hessian approximation also breaks down for solutions far from the optimum where the residuals are large. The terms omitted from Eq. (18.9) are likely to be important when the residuals are large. Experience has also shown that a line search should be added to make each iteration acceptable. Thus if $\mathbf{d}_i$ [given by the right-hand side of Eq. (18.11)] does not result in a decrease in the sum of squares, a line search along the ray $\alpha\mathbf{d}_i$ is used to find an acceptable solution.

Example 18.2: Progress of a Gauss Algorithm. To illustrate the progress of finding an estimate using the Gauss algorithm, consider the problem of finding parameter estimates for the model

$$\eta(t; \boldsymbol{\theta}) = \theta_1(\theta_1 - \theta_2)^{-1}(e^{-\theta_2 t} - e^{-\theta_1 t})$$

which is a special case of Eq. (18.3). The observations given are

t	1	2	3	4	5	6
y	0.478	0.552	0.408	0.329	0.094	0.090

To begin the algorithm, starting values for the parameters are needed. Here they can be most readily approximated by using $t = 2$ as an estimate of the time that the response is greatest, and observing that one of the initial conditions is $\eta'(0; \boldsymbol{\theta}) = \boldsymbol{\theta}_1$. Approximating the initial slope of the response function by $0.50 \approx (0.478 - 0)/(1 - 0)$ leads to the choice $\theta_1 \approx 0.50$. From the condition that $\partial\eta/\partial t \approx 0$ when $t = 2$, $\log(\theta_2 \backslash \theta_1) \approx 2(\theta_2 - \theta_1)$ yields the solution $\theta_2 \approx 0.4$.

At the initial value of the parameters (0.5, 0.4), the residuals are given by

t	1	2	3	4	5	6
e	0.159	0.145	0.0178	– 0.004	– 0.172	– 0.115

from which the sum of squared errors $\Psi = 0.0894$ can be computed. Likewise, one obtains for the Jacobian matrix, Eq. (18.8),

$$F(\boldsymbol{\theta}^{(0)}) = \begin{bmatrix} 0.48108 & -0.16213 \\ 0.42081 & -0.42081 \\ 0.22439 & -0.61471 \\ 0.04426 & -0.70987 \\ -0.07789 & -0.72087 \\ -0.14362 & -0.67450 \end{bmatrix}$$

so that

$$-\nabla_\theta \Psi = F'(\boldsymbol{\theta}^{(0)})\mathbf{e} = \begin{bmatrix} 0.1711 \\ 0.1067 \end{bmatrix}$$

and

$$F'(\boldsymbol{\theta}^{(0)})F(\boldsymbol{\theta}^{(\theta)}) = \begin{bmatrix} 0.4875 & -0.2713 \\ -0.2713 & 2.0604 \end{bmatrix}$$

The initial direction leads to $\hat{\boldsymbol{\theta}}^{(1)} = (0.9098, 0.5058)$ and the improved sum of squares $\Psi(\hat{\boldsymbol{\theta}}^{(1)}) = 0.02612$. The progress of the algorithm is rapid and is summarized in Table 18.1.

At the end of the fourth iteration the derivative has a magnitude less than 0.0001 and the optimum has been reached. The solution is reached particularly quickly, due in part to the good starting estimates being in the neighborhood of the solution.

An interesting aspect of the Gauss algorithm is that it progresses essentially by solving a series of linear regressions based on the first-order approximation of $\boldsymbol{\eta}(\boldsymbol{\theta})$ at each successive solution $\hat{\boldsymbol{\theta}}^{(i)}$. A practical advantage of this is that the numerically stable algorithms used for linear regression can also be used in commercial codes for solving nonlinear regression problems.

A number of improvements have been made to the basic Gauss algorithm to improve its efficiency during line searches, to check for definiteness of the approximate Hessian, and to provide alternative courses of action when the Gauss step fails. See Bard (1974), Chambers (1973), and Kennedy and Gentle (1980). The most heavily used modification of the Gauss-Newton algorithm is the Marquardt (1963) algorithm, given by

$$\hat{\boldsymbol{\theta}}^{(i+1)} = \hat{\boldsymbol{\theta}}^{(i)} + [F'(\hat{\boldsymbol{\theta}}^{(i)})F(\hat{\boldsymbol{\theta}}^{(i)}) + \lambda I]^{-1}F'(\hat{\boldsymbol{\theta}}^{(i)})e(\hat{\boldsymbol{\theta}}^{(i)}).$$

This can be viewed as a further approximation of the Hessian matrix using the leading term from the Gauss approximation, plus a diagonal matrix λI. More importantly, it can be noted that as $\lambda > 0$ decreases, one obtains the basic Gauss algorithm, and as λ increases, steepest descent is obtained. The Marquardt algorithm includes provision for increasing and decreasing λ as the iterations progress.

A practical problem in all of these algorithms is that they require specification of at least the first derivatives of the response function. This specification is surprisingly prone to errors in derivation and coding. Some algorithms are available for "derivative-free" solution of the least-squares problem. One class consists of program-generated central or forward difference approximations to the first derivatives. Other solutions include DUD, "Doesn't use Derivatives" (Ralston and Jennrich, 1978), which uses information from previous iterations to build up derivativelike information. Other programs, such as the Nelder-Mead (1965) direct search, avoid derivatives by finding improving solutions using only evaluations of the function $\Psi(\boldsymbol{\theta})$. While such algorithms are quite robust and almost always converge, they tend to be slower than derivative-based algorithms when both converge.

One can use general optimization algorithms for finding the least-squares solution. In effect, one uses the general program and ignores the special structure of the

TABLE 18.1

Iteration i	$\hat{\theta}_1^{(i)}$	$\hat{\theta}_2^{(i)}$	$\Psi(\hat{\boldsymbol{\theta}}^{(i)})$	$\|\nabla_\theta \Psi\|$
0	0.5	0.4	0.0894	0.2016
1	0.9098	0.5058	0.0261	0.1149
2	1.0528	0.4263	0.0159	0.0264
3	1.0492	0.43709	0.01564847	0.00086
4	1.04838	0.43675	0.01564812	0.000063

problem. General algorithms sometimes work better; also there are a wide variety of routines to choosen from

The L_p-Norm estimators, which include LAD and minimax (L_∞) estimators, have a special structure of their own; optimization solutions are iterative even in the case of linear regression models. Special algorithms for these cases are given in Gonin and Money (1989). Many robust procedures can be solved as a series of iteratively reweighted least squares problems in which the weights are reevaluated at each iteration, using the loss function. One special algorithm of the Newton type is given in Welsch and Becker (1975). This is the approach taken in Kafadar (1994), for instance.

Example 18.3: Use of SAS NONLIN and BMDP AR Programs. In this example we consider fitting data using the model

$$\eta(x; \theta_1, \theta_2) = 1 - (\theta_1 e^{-\theta_2 x} - \theta_2 e^{-\theta_1 x})(\theta_1 - \theta_2)^{-1}$$

using two of the available commercial nonlinear least squares. The data to be used are given by

x	1	2	3	4	5	6
y	0.189	0.429	0.671	0.820	0.891	0.885

An initial problem is to obtain starting values. Because of the complicated form of this response function, a grid search of sum of squared errors is used to obtain starting estimates for θ. The SAS PROC NLIN in fact allows the specification of fixed grid points in lieu of specification of one set of initial conditions. Using a 4 × 4 grid of points over the intervals $1 \le \theta_1 \le 5$ and $0.10 \le \theta_2 \le 0.99$ yields a minimum sum of squares at (1, 0.6933) with a value of 0.01. A slightly finer grid yields the alternative estimate (1.25, 0.5) with sum of squares 0.0044. All of these pairs are in the same general neighborhood and should give rapid convergence to the solution.

An alternative approach is possible in this problem since the response is insensitive to the value of the first parameter. Thus the parameter θ_1 can be set arbitrarily and the remaining parameter θ_2 obtained by a line search.

Two codes are illustrated here, the BMDP AR, derivative-free nonlinear least squares, and the SAS PROC NLIN, also using the option for the program to compute numerical derivatives. The two program files are as follows:

1. *BMDP AR program file*

```
/PROBLEM          TITLE IS 'EXAMPLE 3 COMPUTATIONS'.
/INPUT            VARIABLES ARE 2.
                  FORMAT IS '(F3.0,F5.0)'.
                  FILE IS DATA3.
/VARIABLES        NAMES ARE X,Y.
/REGRESSION       DEPENDENT IS Y.
                  PARAMETERS ARE 2.
/PARAMETERS       INITIAL ARE 1.25,.5.
/FUN              F = 1.- (P1*EXP(- P2*X) - P2*EXP(-P1*X)/(P1 - P2).
/END
```

2. *SAS PROC NLIN program file*

```
TITLE 'EXAMPLE 3 COMPUTATIONS';
DATA A;
    INPUT X Y @@;
    CARDS;
1.   .189   2.   .429   3.   .671
4.   .820   5.   .891   6.   .885
;
PROC NLIN METHOD = GAUSS;
    PARMS B1 = 1.25 B2 = 0.5;
    MODEL = 1. - (B1*EXP(- B2*X) - B2*EXP(-B1*X)/(B1 - B2);
```

The results from the two programs are given in Figs. 18.1 and 18.2. In Fig. 18.1 the BMDP progress toward a solution is summarized by the decreasing value of the residual sum of squares and changes in the parameter values. Convergence is rapid, with the first three digits of the parameters established within the first five iterations of the algorithm. The final estimate $\hat{\boldsymbol{\theta}}' = (1.149501, 0.547389)$ yields a residual sum of squares of 0.003426160. The asymptotic standard deviations provided in the figure are described in the next section; they are obtained from the diagonal terms of $[F'(\hat{\boldsymbol{\theta}})F(\boldsymbol{\theta})]^{-1}\hat{\sigma}^2$, where $\hat{\sigma}^2$ is estimated from the mean-square error. The size of the standard errors indicates considerable uncertainty in the value of the estimated parameters, particularly the first one. This reflects the observation made earlier that the response is insentive to changes in the value of this parameter. The estimate of the correlation between the parameters is obtained from the asymptotic variance matrix using the off-diagonal terms. The correlations are high in this case, indicating the presence of dependence between the two estimates. At the bottom of the figure the residuals, predicted Y values, and other statistics are printed as a function of the independent variable X.

In Fig. 18.2 the progress of the SAS algorithm using DUD is shown. Note that although the GAUSS procedure was elected, since derivatives were not provided, the program switched to DUD. The decrease in the residual sum of squares is again rapid, and a similar solution to that of the BMDP program is obtained. The summary statistics are computed under the assumption that the parameters behave in a manner similar to those of a linear model, as with BMDP. The output gives the parameter values to 10 digits, though this apparent precision is misleading, given the size of the standard errors and the magnitude of the correlation between the terms.

18.3 STATISTICAL PROPERTIES OF ESTIMATORS

One of the complicating factors when using nonlinear models is the lack of precise information about the sampling distribution of the estimators. For instance, estimators in this case are typically biased by an unknown amount, and the sampling distribution may depart markedly from the normal distribution, particularly in the multivariate distribution of the estimators. This is in contrast to estimators arising from linear models, where the distributional properties are typically well known, particularly where the errors are normally distributed. However, in most cases, as the sample size increases, these differences become less pronounced. For *asymptotic* sample sizes, the properties of the nonlinear model estimators approach those of an equivalent linear model.

ITER. NO.	INCR. HALV.	RESIDUAL SUM OF SQUARES	PARAMETERS 1 P1	2 P2
0	0	.004417	1.250000	.500000
0	0	.004371	1.250000	.550000
0	0	.003685	1.375000	.500000
1	0	.003490	1.114983	.549787
2	0	.003427	1.142965	.549694
3	0	.003426	1.143165	.549297
4	0	.003426	1.152565	.546547
5	0	.003426	1.149688	.547362
6	3	.003426	1.149576	.547373
7	0	.003426	1.149406	.547414
8	0	.003426	1.149499	.547388
9	0	.003426	1.149498	.547389
10	0	.003426	1.149501	.547389

THE RESIDUAL SUM OF SQUARES (= 3.426160E-03) WAS SMALLEST WITH THE FOLLOWING PARAMETER VALUES

PARAMETER	ESTIMATE	ASYMPTOTIC STANDARD DEVIATION	COEFFICIENT OF VARIATION
P1	1.149501	.424847	.369593
P2	.547389	.111526	.203743

ESTIMATE OF ASYMPTOTIC CORRELATION MATRIX

		P1 1	P2 2
P1	1	1.0000	
P2	2	-.9727	1.0000

CASE NO.	NAME	RESIDUAL	OBSERVED 2 Y	PREDICTED 2 Y	COOK DISTANCE	STD. DEV. PREDICTED	1 X
1		.005341	.189000	.183659	.013532	.017235	1.0
2		-.023421	.429000	.452421	.559533	.020220	2.0
3		.011626	.671000	.659374	.043771	.015603	3.0
4		.024601	.820000	.795399	.208950	.015881	4.0
5		.011749	.891000	.879251	.057071	.016654	5.0
6		-.044393	.885000	.929393	.595649	.015302	6.0

FIGURE 18.1 Selected portion of output from BMDP AR program for Example 18.3.

```
EXAMPLE 3 COMPUTATIONS

NON-LINEAR LEAST SQUARES ITERATIVE PHASE

DEPENDENT VARIABLE: Y         METHOD: DUD

ITERATION               B1             B2           RESIDUAL SS

    -3          1.250000000    0.500000000    0.0044165234789
    -2          1.375000000    0.500000000    0.0036846867035
    -1          1.250000000    0.550000000    0.0043710549936
     0          1.375000000    0.500000000    0.0036846867035
     1          1.114983499    0.549787419    0.0034899759006
     2          1.142965194    0.549693543    0.0034268453998
     3          1.143165453    0.549297334    0.0034264547742
     4          1.152565459    0.546547460    0.0034262120176
     5          1.149688171    0.547361825    0.0034261603572
     6          1.149576447    0.547372816    0.0034261596393
     7          1.149406318    0.547414177    0.0034261596353

NOTE: CONVERGENCE CRITERION MET.

NON-LINEAR LEAST SQUARES SUMMARY STATISTICS      DEPENDENT VARIABLE Y

SOURCE                    DF  SUM OF SQUARES      MEAN SQUARE

REGRESSION                 2    2.9160828404     1.4580414202
RESIDUAL                   4    0.0034261596     0.0008565399
UNCORRECTED TOTAL          6    2.9195090000

(CORRECTED TOTAL)          5    0.4039715000

PARAMETER      ESTIMATE     ASYMPTOTIC              ASYMPTOTIC 95 %
                            STD. ERROR            CONFIDENCE INTERVAL
                                                  LOWER          UPPER
B1          1.149406318 0.42484659681 -.03014097559  2.3289536125
B2          0.547414177 0.11152157399 0.23778483845  0.8570435159

ASYMPTOTIC CORRELATION MATRIX OF THE PARAMETERS

CORR                  B1                  B2

B1                1.0000             -0.9726
B2               -0.9726              1.0000

NOTE: ALL ASYMPTOTIC STATISTICS ARE APPROXIMATE.
```

FIGURE 18.2 Output from SAS PROC NLIN for Example 18.3. (*From Ralston and Jennrich, 1978.*)

Since the properties of the estimators $\hat{\boldsymbol{\theta}}$ cannot be readily determined exactly, we resort to approximations. When the sample size is large, an approximation based on asymptotic properties is often used. Taylor series expansion of the response function can also be used as the basis for approximation of the properties of $\hat{\boldsymbol{\theta}}$. Such approximations are limited in the complexity that can be obtained, and in practice only linear and quadraticlike approximators are feasible. The series approximators are used to obtain moments and in the construction of confidence regions and likelihood ratio tests of hypotheses.

When approximators are used, the question of their appropriateness and accuracy is natural. Geometrical considerations have led to the derivation of *measures of nonlinearity.* As explained in Sec. 18.3.2, two sources can be used to explain how nonlinear estimators $\hat{\boldsymbol{\theta}}$ differ from the linear ones: there is an *intrinsic nonlinearity* corresponding to geometrical curvature of the response surface $\boldsymbol{\eta}(\boldsymbol{\theta})$, and there are effects due to *parameterization* that can in principle be reduced by transformations or changes in the parameterization of the nonlinear model.

18.3.1 Approximation Methods for Sampling Distribution of $\hat{\boldsymbol{\theta}}$

Under general conditions, an MLE possesses an asymptotic normal distribution. When these conditions are met, an MLE is consistent (asymptotically unbiased) and normally distributed with a variance matrix given by the inverse Fisher information matrix, $\mathcal{J}(\boldsymbol{\theta}) = -E_x[\nabla^2_\theta \mathcal{L}]$. For the case of least squares, these conditions require that the values of the independent variables be either from a repetition of a fixed design or taken from a distribution in such a way that the estimate is uniquely defined and that the Jacobian matrix is of full rank. Technical conditions include the continuity of first and second derivatives of η with respect to $\boldsymbol{\theta}$ and X over suitable regions. For instance, the regions should not include points where the preceding assumptions fail. For a full discussion see Gallant (1987), Wu (1981), and Jennrich (1969). For least squares the Fisher information is $\mathcal{J}(\boldsymbol{\theta}) = F'(\boldsymbol{\theta})V^{-1}F(\boldsymbol{\theta})$, so that $\hat{\boldsymbol{\theta}}$ is distributed asymptotically as $N(\boldsymbol{\theta}^*, {}^{-1}(\boldsymbol{\theta}^*))$; $\boldsymbol{\theta}^*$ denotes the underlying value of the parameters being estimated. The asymptotic approximation is often useful to give an indication of the joint distribution of the parameter estimator $\hat{\boldsymbol{\theta}}$.

For finite sample sizes, an alternative to the asymptotic approximation is to construct a series approximation to $\boldsymbol{\eta}(\boldsymbol{\theta})$, from which an approximator Δ to $\hat{\boldsymbol{\theta}}$ can be derived, in much the same way that in the Gauss update formula the series $\hat{\boldsymbol{\theta}}^{(i)}$ provides successive approximations to the value of $\hat{\boldsymbol{\theta}}$. The Gauss update is based on linearization of $\boldsymbol{\eta}(\boldsymbol{\theta})$. Using $\boldsymbol{\theta}^*$ for $\boldsymbol{\theta}$ and the knowledge that $\mathbf{e}(\boldsymbol{\theta}^*) = \boldsymbol{\epsilon}$, Eq. (18.11) yields a first-order approximator,

$$\Delta = \boldsymbol{\theta}^* + [F'(\boldsymbol{\theta}^*)F(\boldsymbol{\theta}^*)]^{-1}F'(\boldsymbol{\theta}^*)\boldsymbol{\epsilon} \tag{18.12}$$

whose distribution is $N(\boldsymbol{\theta}^*; \sigma^2[F'(\boldsymbol{\theta}^*)F(\boldsymbol{\theta}^*)]^{-1})$ whenever the errors are $N(0, \sigma^2\mathrm{I})$ in distribution. The linear approximator provides no information about the bias of $\hat{\boldsymbol{\theta}}$, but provides a variance approximation similar in form to that of the asymptotic approximation.

Higher-order approximators are possible but rarely completely feasible beyond the first order. Once the approximation of $\boldsymbol{\eta}(\boldsymbol{\theta})$ is constructed in terms of $(\boldsymbol{\theta} - \boldsymbol{\theta}^*)$, the series must be reexpressed (reverted) as a function of $\boldsymbol{\theta}$ in terms of $\boldsymbol{\epsilon}$. This can be done formally only in the univariate case (Shenton and Bowman, 1977). In Box (1971) the approximation is done through the terms of second order, and then terms of similar order are combined. The treatment in Clarke (1980) is more detailed and consists of essentially the first step of a successive substitutions algorithm. The Box bias approximation for univariate least squares is given by

$$\zeta = -\frac{\sigma^2}{2}[F'(\hat{\theta})F(\hat{\theta})]^{-1}\sum_{i=1}^{n}\nabla_\theta\eta(\mathbf{x}_i;\hat{\theta})\,\mathrm{tr}\{[F'(\hat{\theta})F(\hat{\theta})]^{-1}\nabla_\theta^2\eta(\mathbf{x}_i;\hat{\theta})\}$$

while the Clarke (1980) approximation for the variance matrix is given in summary by the expression

$$\mathrm{Var}[\hat{\theta}] = \sigma^2[F'(\hat{\theta})F(\hat{\theta})]^{-1} + \sigma^4K[V_a + V_b + V_c]K'$$

when estimation is univariate OLSQ. The expressions for the three matrices V_a, V_b, and V_c are complicated and are omitted here; K is a transformation matrix used in Clarke's development, where $K[F'(\hat{\theta})F(\hat{\theta})]^{-1}K' = I$. The two matrices V_b and V_c can be altered by transformation of the parameters, while the term V_a represents an effect due to the intrinsic curvature of the nonlinear response surface. Clarke includes formulas for the effects of transformations upon the Box bias formulation as well as the variance approximation.

The first-order approximation replaces the response surface with a tangent plane approximation and uses the properties of the parameters of the plane as the approximators. This versatile device is used in a number of papers to derive approximate confidence intervals and likelihood ratio tests for hypotheses concerning the parameters and subsets of the parameters (Gallant, 1987).

18.3.2 Measures of Nonlinearity

Measures of nonlinearity were introduced by Beale (1960) and extended by Bates and Watts (1980) for OLSQ estimation. These measures of nonlinearity attempt to quantify the degree of difference between the properties of the estimator distribution $\hat{\theta}$ and those of an estimator with the properties of a linear model, that is, these measures use the linear model case as a reference and measure the discrepancy between the two. The linear reference plane is given by Eq. (18.10), where $\hat{\theta}^{(i)} = \theta^*$.

The properties of least-squares estimators can be understood geometrically as the projection of the observed points $\mathbf{y}$ upon the response surface $\eta(\theta)$. This reflects the fact that the least-squares solution occurs when the residual vector is orthogonal to the response surface. We can think of the response surface being parameterized by the coordinates θ, and the least-squares solution is just the value of the coordinates at the point where $\mathbf{e}(\theta) = \mathbf{y} - \eta(\theta)$ is minimized. The properties of linear model estimators come from the fact that the response surface is a plane, that the coordinate "lines" on the plane are evenly spaced in each component of θ, and that the observations are normally distributed. The possible points $\mathbf{y}$ form a spherical region whose projection on the response surface plane leads to the familiar elliptical confidence regions for the parameters in linear least squares.

The geometrical situation is similar for the nonlinear response surface, except that the surface may be curved, and the parameter coordinates on the surface may no longer be uniformly spaced. These two departures from the linear model case are called, respectively, intrinsic and parameter effects nonlinearity, since the latter can be changed by a parameter transformation, while the shape of the surface is fixed by the model and the values of the independent variables and thus is intrinsic to any experiment being investigated. The response surface of the reference linear model given by $\tilde{\boldsymbol{\eta}}$ [Eq. (18.10)] is the tangent plane at the point $\boldsymbol{\eta}(\hat{\boldsymbol{\theta}})$.

To explore the differences between a linear and a nonlinear model, we can project the parameter lines of the nonlinear response surface onto the tangent plane for that surface. This is illustrated in Fig. 18.3, using the model from Example 18.2. In this particular case the response surface is fairly flat and the tangent plane is a reasonable approximation to it. However, as shown in the figure, the parameter lines depart from the parallel, equally spaced lines that would be characteristic of a linear model. The lines for varying θ_1 (θ_2 fixed) bend, and the spacing between both sets of lines changes. The circle superimposed on the figure represents the likely projection points for observations $\mathbf{y}$ when $\boldsymbol{\theta}^*$ is (1.048, 0.438). Since the spacing between the θ_2 coordinates (fixed θ_1) is reasonably constant throughout this region, we might expect the $\hat{\theta}_2$ to be distributed symmetrically and the $\hat{\boldsymbol{\theta}}_1$ estimates to be skewed to the right. The spacing is broad for the lower values of $\boldsymbol{\theta}_1$, but becomes increasingly close as θ_1 increases.

Beale (1960), and later Box (1971), worked out the algebraic discrepancy between the behavior of $\hat{\boldsymbol{\theta}}$ and $\boldsymbol{\Delta}$ as an average measure of nonlinearity. Beale's two terms were the current nonlinearity N_θ, which is parameter effects plus intrinsic nonlinearity, and N_ϕ, which is the measure of the average intrinsic nonlinearity. N_ϕ is the lower bound for any possible parameterization, so that when this measure is large, there is no way to reduce the sizable effects of nonlinearity. Bates and Watts (1980) used differential geometrical methods to redevelop and expand the analysis of nonlinearity. The revised measures can be shown to be equivalent to Beale's measures in terms of average nonlinearities. The new measures also shed light upon the form of the nonlinearity—the bending of parameter lines and the changes in distance between them, as illustrated in Fig. 18.3. This allowed Bates

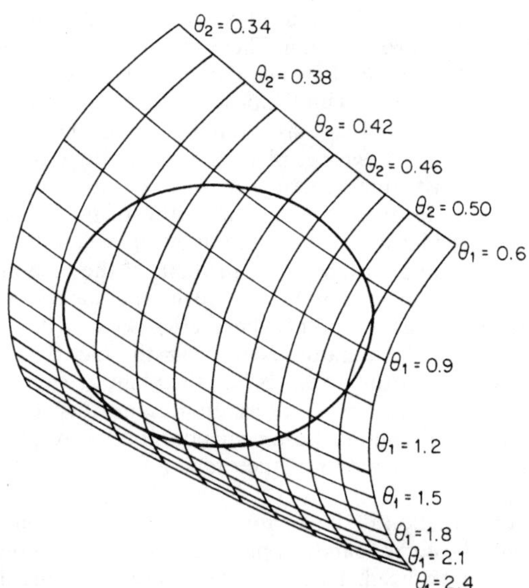

FIGURE 18.3 Projection of parameter lines from solution locus onto tangent plane using model from Example 18.2.

and Watts to quantify the worst-case behavior, that is, the biggest departures from the linear model as a function of direction. A particular advantage of the Bates and Watts approach is that it provides a good deal more information about the source and type of nonlinear behavior. Hamilton, Watts, and Bates (1982) have shown that in most published cases the parameter effects nonlinearity is the larger of the two nonlinearities, so that in principle many problems can be made more tractable by suitable reparameterization. However, they have also shown that in many cases such ideal transformations will be computationally impractical or impossible to obtain.

A number of practical observations can be made from the development of Bates and Watts. The nonlinearities depend on a number of factors, such as the form of the model and its parameterization, the choice of the independent variables, and the magnitude of the error variances. Nonlinearity can be reduced by replication, since replication lowers the effective variability of the errors. In Fig. 18.3, for example, replication would reduce the size of the error circle and the deviations from linear behavior would become less noticeable. Proper experimental design (Sec. 18.4) is particularly important for nonlinear models, and this can be used to ameliorate the effects of nonlinearity. Hamilton and Watts (1985) use geometrical considerations of this type to design experiments with low nonlinearity. Finally, for a given design, improvements can be made through changes of the model parameterization. Ross (1990) and Seber and Wild (1989) provide examples of improvements in parameter estimation and statistical inference due to parameter transformations. Cheng, Evans, and Iles (1992) identify embedded models which can help identify useful transformations. Finally, Chen and Jennrich (1995) and Tsai (1988) describe improvements that can be obtained through power transformations. Curvature methods are used by Cook and Tsai (1985) to examine the residuals obtained in nonlinear regression, and Cook and Witmer (1985) examine situations in which curvature measures are misleading.

18.3.3 Monte Carlo Approaches

Monte Carlo sampling can be used in many situations where the applicability of approximators is uncertain or where approximators have not been constructed. Most of the analytic approximators and measures of nonlinearity are derived for least squares estimators with normally distributed errors. Gillis and Ratkowsky (1978) indicate that when errors are skewed, Beale's measures can be misleading, while Box's bias provides a more reliable indication of nonlinearity. Thus many of the approximators have to be applied with caution when assumptions are varied. However, we can perform Monte Carlo sampling to achieve exact answers for cases with smaller sample sizes or in cases where the analytic approximations are likely to be poor. Monte Carlo sampling is also used to validate approximation techniques.

Direct Monte Carlo simulation of the distribution of $\hat{\boldsymbol{\theta}}$ is implemented using the general algorithm below. For fixed error distribution, choice of $\boldsymbol{\theta}^*$, design X, and sample size N, we obtain samples of $\boldsymbol{\epsilon}$ from which the $\hat{\boldsymbol{\theta}}$ can be computed:

1. For $u = 1, \ldots, N$,

a. Sample $\boldsymbol{\epsilon}$ from error distribution

b. Let $y_i = \eta(\boldsymbol{\theta}^*) + \epsilon_i$, $i = 1, 2, \ldots, n$

c. Solve estimation problem to obtain $\hat{\theta}_u$

2. Using the sample $\{\hat{\boldsymbol{\theta}}_u, u = 1, \ldots, N\}$ form sample statistics, histograms, and other descriptive measures

Standard simulation references can be consulted for methods of generating random variables for use in simulation. By picking the replication size N arbitrarily large, one can estimate any characteristic of the sampling distribution arbitrarily well. However, since standard errors decrease only as $\sqrt{N}$, extremely large sample sizes are needed for great accuracy using this method. Ratkowsky (1983) offers a Fortran program to perform Monte Carlo sampling to study the parameter estimator distribution and possible parameter transformations.

Simulation experiments can be improved by altering the sampling method or by changing the statistical estimator used to achieve the same end with greater precision. Methods of this type are called *variance reduction techniques*. It is possible to use the approximations given in Sec. 18.3.1 in conjunction with the direct sampling strategy to achieve such a variance reduction. This exploits the general correlation between $\hat{\boldsymbol{\theta}}$ and its approximator, such as Δ. For instance, add a step d to the algorithm above to compute an approximator Δ_u. Use Eq. (18.12) with $\boldsymbol{\epsilon}$ to compute Δ_u, so that $\hat{\boldsymbol{\theta}}_u$ and Δ_u have the same error. Then to compute an estimate of the mean $E[\hat{\boldsymbol{\theta}}]$ instead of the direct sample average

$$\hat{\mu}' = N^{-1} \sum_{u=1}^{N} \hat{\boldsymbol{\theta}}_u$$

use the control variate estimator which combines the direct estimator $\hat{\mu}'$ with the sample average of the Δ_u, $\overline{\Delta}$, used as a control variate:

$$\hat{\mu}'B = \hat{\mu}' - B(\overline{\Delta} - E[\overline{\Delta}])$$

where we can use the fact that the expectation of the approximator is known to be $\boldsymbol{\theta}^*$. The matrix B of control weights can be estimated in a number of ways, but experience has shown that $B = I$ is a good choice in practice. For further details and examples, see Swain and Schmeiser (1987).

18.4 DESIGN OF EXPERIMENTS

The design of experiments is concerned generally with the selection of sample size, the values of the independent variables to be used in experimentation, and the conduct of the actual experimentation to ensure the quality of observed responses. The choice of independent variables is considered here, depending on two purposes of experimentation, either the determination of the correct model, or the precise determination of the parameter values of a particular model. The general issues of selection of experimental units, randomization, independence of measurements, and so forth are common to all experiments and are not considered here.

For nonlinear modeling, the emphasis is often upon the selection of the appropriate theoretical or mechanistic model for a particular process. The discrimination problem arises since several plausible models often can be advanced to explain a particular process. This leads to designs that choose experimental runs to contrast the predicted responses as much as possible. Models whose responses are inconsistent with the observed responses can then be eliminated from further consideration. Once a model is selected, experimental design can then be used to estimate the unknown parameters as precisely as possible. Note that selection procedures such as stepwise regression and use of experimental designs which allow for statistical tests of lack-of-fit analysis play an analogous role in the discrimination for linear models.

The choice of the values of the independent variables is very important, since this choice often determines whether estimates can be obtained, the precision to which they can be estimated, or whether one can discriminate between two rival models. Box and Hunter (1965) observe that design is crucial, since a very poorly designed experiment cannot be undone by any amount of sophisticated analysis, while a good design can maximize the information from a fixed number of observations.

The theory of linear design is a convenient starting point, since under the local linearization of $\eta(\mathbf{x}; \boldsymbol{\theta})$, $F(\boldsymbol{\theta})$ assumes the central role of the design matrix of independent variable values X in the linear regression design problem. The design problem is complicated for the nonlinear case since the independent variables appear indirectly through the derivatives of the Jacobian matrix $F(\boldsymbol{\theta})$, $\{\partial\eta(\mathbf{x}_i; \boldsymbol{\theta})/\partial\theta_j\}$, and because these quantities depend on the value of the parameters $\boldsymbol{\theta}$. Thus in the nonlinear case one faces the apparent paradox that to estimate $\boldsymbol{\theta}^*$ precisely, one needs a good estimate of $\boldsymbol{\theta}^*$ beforehand. This feature only emphasizes what is generally true in all design problems, that design reflects what is known about a problem. However, because the best design depends on the value of $\boldsymbol{\theta}$ chosen, sequential designs are often employed to exploit refinements in our knowledge of $\boldsymbol{\theta}^*$, and as $\hat{\boldsymbol{\theta}}$ is updated, the design can be made with increasing precision. Since the actual value of $\boldsymbol{\theta}^*$ is unknown, the optimization implied by optimal design is dependent on the best guess available; Chernoff (1962) refers to this as local optimality.

18.4.1 Discrimination Design

In the initial stages of an investigation many plausible mechanisms can be hypothesized, depending on assumptions made and what factors are considered important. For instance, Blakemore and Hoerl (1963) suggest no less than 18 plausible models in a kinetics study, depending on the step of the reaction considered rate-controlling and the reaction order of the step. A central idea in this case is to determine values of the independent variables which will place one or more of the models in relative jeopardy, since in the face of conflicting predictions, one or the other must likely fail to mimic the observed response. Thus, for example, if one considered two models for a process given by $\eta_1 = 1 - \exp(-\alpha x)$ and $\eta_2 = \beta/(\beta - \gamma)\,[\exp(-\gamma x) - \exp(-\beta x)]$, one would select larger values of x because of their difference in predicted values, since asymptotically $\eta_1 = 1$, and $\eta_2 = 0$. Choosing a smaller value of x would not place either model in jeopardy, since both are roughly in agreement.

Where several models give good predictions, parsimony is used as a second criterion. Parsimony is the principle that the model with the least parameters be preferred whenever several give essentially similar predictions. The principle of placing models in jeopardy is used, of course, to determine whether the more parsimonious model is as accurate as the more complicated model.

Box and Hill (1967) have formulated a bayesian criterion for model discrimination using *information for discrimination* (Kullback and Leibler, 1951). Let $\eta^{(i)}(\mathbf{x}; \boldsymbol{\theta}^{(i)})$, $i = 1, \ldots m$, denote the possible models being considered. To each model we can assign a prior probability π_i of it being the correct one. This is often done using the noninformative assignment $\pi_i = m^{-1}$. Let $p^{(i)}(\mathbf{y}|\mathbf{x})$ be the likelihood of the observed responses **y**, given the experimental design X, assuming that the data arose from model i. The quantity

$$a_{ij}(\mathbf{x}) = \log \{p^{(i)}(\mathbf{y}|\mathbf{x})/p^{(i)}(\mathbf{y}|\mathbf{x})\}$$

is an expression for the relative likelihood of model i in favor of model j in view of the data **y**. Since the data **y** cannot be known prior to the actual experiment, one uses the expectation of $a_{ij}(\mathbf{x})$ over the possible values of **y** under the assumption that model i is correct. The Kullback-Leibler information for discrimination consists of the sum of the two expected values of $a_{ij}(\mathbf{x})$ and $a_{ji}(\mathbf{x})$,

$$J_{i,j}(\mathbf{x}) = \int \{p^{(i)}(\mathbf{y}|\mathbf{x}) - p^{(j)}(\mathbf{y}|\mathbf{x})\} \log \{p^{(i)}(\mathbf{y}|\mathbf{x})/p^{(j)}(\mathbf{y}|\mathbf{x})\}\, d\mathbf{y}$$

which is large whenever one is favored over the other. When there are only two models, one selects the next design point to maximize $J_{1,2}(\mathbf{x})$ as a function of the design point **x**. In the case of a single response, Box and Hill (1967) show that this reduces to

$$J_{1,2}(\mathbf{x}) = -1 + \frac{1}{2}\left(\frac{\sigma_2^2}{\sigma_1^2} + \frac{\sigma_1^2}{\sigma_2^2}\right) + \left(\frac{1}{\sigma_1^2} + \frac{1}{\sigma_2^2}\right)[\eta^{(2)}(\mathbf{x}; \hat{\boldsymbol{\theta}}^{(2)}) - \eta^{(1)}(\mathbf{x}; \hat{\boldsymbol{\theta}}^{(1)}]^2$$

where

$$\sigma_i^2 = \sigma^2 + \nabla_\theta \eta^{(i)} V^{-1} \nabla_\theta' \eta^{(i)}$$

is an estimate of the variance of a predicted observation at the current value of the parameters for model i, $\hat{\boldsymbol{\theta}}^{(i)}$. Similar equations can be derived in the multivariate case when the errors are assumed to be multivariate normal in distribution (Bard, 1974). When there are more than two models to choose from, a joint criterion is given by

$$J(\mathbf{x}) = \sum_{i \neq j}^{m} \pi_i \, \pi_j J_{ij}(\mathbf{x}).$$

After each observation is taken, the probabilities associated with each model can be updated using Bayes' rule,

$$\pi_i' = \frac{\pi_i p^{(i)}(\mathbf{y}|\mathbf{x})}{\sum_{j=1}^{m} \pi j p^{(j)}(\mathbf{y}|\mathbf{x})}.$$

The model discrimination criterion picks values of **x** to maximize the differences between models relative to their current likelihood of being correct. After one model has been selected, a different criterion can be used to select the design points. Hill, Hunter, and Wichern (1968) discuss a method that will adjust between the two types of designs to emphasize discrimination while there is uncertainty about the correct model, and gradually switch to precise estimation as one model assumes greater likelihood of being the correct model.

18.4.2 Precise Estimation

Once the model that generates the data can be specified, it is possible to design experiments to make parameter estimates as precise as possible. This will in general require information about the reasonable values of the parameters being estimated, but particularly if prior experimentation has been done to establish the form of the model, this information can be used to initialize the process. When starting *ab initio*, however, one must make some assumption about the value of the parameters, which can be refined as experimentation proceeds.

Whether proceeding sequentially or starting from scratch, one proceeds by predicting the uncertainty associated with the estimates as a function of the independent variables. For instance, under the assumptions that the response function can be locally approximated by a first-order Taylor series and with normally distributed errors, the predicted uncertainty in the estimates $\hat{\boldsymbol{\theta}}$ is $|[F'(\hat{\boldsymbol{\theta}})VF(\hat{\boldsymbol{\theta}})]^{-1}|$. One could select the n design points $\mathbf{x}_1, \ldots, \mathbf{x}_n$, by minimizing the determinant or, equivalently, by maximizing the determinant of the inverse. Under the assumption of normal errors and the linearity of the response, the confidence region associated with the estimates $\hat{\boldsymbol{\theta}}$ is minimized by this choice. This approach was originally developed by Box and Lucas (1959).

Maximization of the determinant $|F'(\tilde{\boldsymbol{\theta}})VF(\tilde{\boldsymbol{\theta}})|$ is a difficult problem even when the response function is linear in the unknown parameters. One approach is to make discrete the feasible values of $\boldsymbol{\theta}$ into a d^p grid of points, and perform the maximization over this restricted subset of the points. Empirical evidence (Box, 1968, 1970) suggests that many times the optimal choice for $n > p$ points is to replicate the points from an optimal set of p points. This will not always be the case, but it makes the problem more manageable. In this case Box and Lucas (1959) give an interesting geometrical interpretation of the criterion, together with the optimal p point designs for several models. Haines (1995) also uses a geometrical approach to determine optimal Bayes designs for models with a single parameter.

For exact maximization of the determinant, Bates (1983) has shown how to calculate the derivative of the determinant efficiently. Even so, the problem is still formidable. For instance, Bates uses the derivatives to maximize the determinant for an actual design problem, but Bohachevsky, Johnson, and Stein (1986) use simulated annealing to find a better design.

One drawback to this approach is that there are no degrees of freedom available for lack of fit when the design consists only of the repetition of p points. One can relax the minimum variance restriction, for instance, to include a few extra points for this purpose. A strategy of model building is advanced in Box and Hunter (1962), where the repetitions are actually made at different (factorial) levels of variables not currently included in the model for the purpose of examining whether the model must be changed. Here the model parameters are fit separately in each replication in an effort to determine whether the constants remain "constant" from replication to replication. The residuals are then used to suggest dependencies that must be added to the model.

When information about the values of parameters is available from previous experiments, one can treat the information as a prior distribution and design experiments sequentially to minimize the uncertainty in the predicted posterior distribution of the estimates from the next response, that is, one chooses the $(n + 1)$th design point using the first n points for the prior distribution and then using the posterior distribution to guide the decision. This method was developed in Box and Hunter (1963). Bard (1974) gives a multivariate derivation. For the case of a single response, the criterion becomes a maximization of the uncertainty in the value of $\hat{y}(\mathbf{x})$,

$$\text{Var}[\hat{y}(\mathbf{x})] = \nabla_\theta \eta(\mathbf{x}; \tilde{\boldsymbol{\theta}}) V_\theta \nabla'_\theta \eta(\mathbf{x}; \tilde{\boldsymbol{\theta}})$$

for the row vector $\nabla_\theta \eta(\mathbf{x}; \tilde{\boldsymbol{\theta}})$. This leads to the minimization of the uncertainty in the parameter estimates at the end of the next experiment.

Local linearization is used in the techniques cited to construct predictions of the uncertainty following the experiments to be conducted. However, this linearization ignores the effects of nonlinearity due to parameterization and curvature in the response surface, so the minimization of the confidence region can be considered an approximation only. Work by Hamilton and Watts (1985) has considered experimental design using higher-order approximations. The quadratic approximation leads to better designs for a fixed starting estimate $\tilde{\boldsymbol{\theta}}$ and is reported to be less sensitive to errors in the starting values than the first-order approximation. Ford, Titterington, and Kitsos (1989) incorporate these improvements for a number of well known examples for both static and sequential designs.

When parameters are fit using multiple responses, Box and Draper (1965) have constructed an estimator based on the determinant of the variance-covariance matrix of the responses. The method of Bates (1983) can be used to build an algorithm for multiresponse estimation. Bates and Watts (1984) provide an implementation.

Precise estimation is usually attempted after model discrimination has been used to eliminate competing models. When parameter estimation is necessary and there is uncertainty about the model to fit, a criterion suggested by Cook and Nachtsheim (1982) can be used to pick a design that is model robust, that is, a design is chosen in view of several alternative models (specified by the designer) that might be applicable, so that whatever the model, the parameters will be estimated adequately.

18.5 SUMMARY

Nonlinear models are a natural extension of linear models, and are both more flexible in functional form and more difficult to use than linear regression models. Linear regression techniques are well known and easy to apply, but not all models of engineering and scientific interest are linear in their parameters, and it is not always convenient to linearize models that are inherently nonlinear. The parameters of nonlinear models are more likely to have physical meaning, and the linear model is sometimes only suitable as a graduating function for a limited range of the input variables. In addition, the nonlinear model may require fewer parameters than an equivalent linearized model. Software for the estimation of parameters is now readily available, so that although estimation and interpretation of fitted parameters is more difficult in the nonlinear case, these problems are not as difficult as they once were.

Nonlinear models are encountered in a wide variety of applications. Their uses range from the purely empirical, where a nonlinear form (perhaps suggested by the transformation of a linear model) is used as a graduating function. Often extensive empirical work leads to a nonlinear model: response surface methods, especially as developed by Box and collaborators, are specifically designed to encourage the critical evaluation of models and their systematic development throughout the course of experimentation. Often empirical studies are designed with a view to likely theoretic models, or with the purpose of determining which model from a variety of possible models is the best representation.

Having obtained a model, one turns to the estimation of parameters. Two issues were addressed with respect to estimation, the choice of an estimator and the numerical methods of computing an estimate. The choice of an estimator for nonlinear models is similar to picking an estimator in any situation, for the choice reflects the assumed distribution for the random errors, or represents a hedge against the potential distributions of the errors. Least squares is often employed, under the assumption of normally distributed errors. Maximum-likelihood and Bayes estimators can be employed when other error distributions are assumed or prior information is to be used in determining an estimate. On the other hand, robust estimators are employed when the error distribution is uncertain, or when the errors may be contaminated by outliers that would adversely affect parametric estimators such as least squares. Whatever the estimator, the nonlinearity of the response function requires iterative numerical methods to effect a solution. Some elements of optimization have been introduced, though only least squares algorithms were discussed in any detail. As noted, it is no longer necessary to be an optimization expert to perform nonlinear estimation, since software for curve fitting is widely available. However, some familarity with numerical algorithms is useful.

Estimators for the parameters of linear and nonlinear models differ a great deal: the numerical estimates of nonlinear models require iterative algorithms and are more difficult to obtain, and because estimators are not linear or closed form, the statistical properties of the estimators are more varied and complicated than is the case for linear models. The properties of linear estimators provide a convenient point of reference for evaluating the degree and nature of "nonlinear" behavior in estimators. As noted earlier, the measures of Beale (1960) quantify the discrepancy between the actual solution locus and the linearized solution locus, given by the tangent plane.

Geometrical methods have been used to extend the approach of Beale to the properties of the nonlinear estimators by comparing their behavior to that of linear estimators. As Bates and Watts (1980) summarize, the characteristics of linear estimators can be summarized by a planar assumption about the solution locus $\boldsymbol{\eta}(\boldsymbol{\theta})$ and an equal spacing of parametric lines on the solution locus. These two characteristics can be quantified by measures of curvature on the tangent plane (at $\hat{\boldsymbol{\theta}}$) and normal to the plane. The normal component is associated with departures from the planar assumption and represents the intrinsic curvature of the problem. The tangential components represent parameter effects, the changes in spacing and curvature of lines projected onto the tangent plane, as illustrated in Fig. 18.3. Parameter effects nonlinearity can be altered by changes in parameterization, and parameter transformations are the primary method of changing the degree of nonlinearity. Intrinsic nonlinearity is fixed by the design and is unaffected by parameterization. It can, however, be altered by experimental design.

While generalizations about the statistical properties of estimators is difficult, some comments can be made. Box (1971) has shown that bias in estimators is proportional to error variance. Hamilton, Watts, and Bates (1982) have shown that nonlinear parameter effects are usually the major source of nonlinearity. Transformations that reduce nonlinearity are possible, though they are able to demonstrate that ideal transformations, those that eliminate parameter effects nonlinearity altogether, will often be impossible to obtain or difficult to use if available.

The degree of nonlinearity depends on the choice and parameterization of a model, the error distribution, and the experimental design for the independent variables. While model parameterizations can be changed, the experimenter's main control is through the design of the experiment. Due to the complexity of the problem, design is perhaps even more important here than in the linear case, though correspondingly more difficult.

Both discrimination design and precise estimation design were treated here. Discrimination design is used to choose experiments that will provide for the elimination of one or more candidate models from consideration. Ideal experiments have a large difference between predicted model responses relative to the predicted error of those responses. Designing experiments of this type can be computationally intensive, even when the choice of design points is reduced by discretization. Very often experiments are designed sequentially, picking each new experimental point on the basis of previous experiments. This approach reduces the computational effort to manageable levels. Once a model is chosen, the design effort can be shifted toward precise estimation. Here too, design can be computationally intensive, but experience can often be used to reduce the size of the problem. For instance, when p parameters are to be estimated, optimal n-point designs ($n \geq p$) often consist of repetitions of points from the optimal p-point design. Thus, only p points need to be found to make the optimal design. In addition, the solution locus for such a design is flat, so that these designs have no intrinsic nonlinearity. Finally, new optimization techniques such as simulated annealing can locate the designs more efficiently than was once possible.

REFERENCES

Bard, Y.: *Nonlinear Parameter Estimation*, Academic Press, New York, 1974.

Bates, D. M.: "The Derivative of $|X'X|$ and Its Uses," *Technometrics,* vol. 25, pp. 373–376, 1983.

______ and W. G. Hunter: "Nonlinear Models," in *Encyclopedia of Statistical Sciences,* S. Kotz and N. L. Johnson, Eds., Wiley, New York, 1985.

_____ and D. G. Watts: "Relative Curvature Measures of Nonlinearity," *J. R. Stat. Soc. B,* vol. 42, pp. 1–25, 1980.

_____ and _____: "A Relative Offset Orthogonality Criterion for Nonlinear Least Squares," *Technometrics,* vol. 23, pp. 179–183, 1981.

_____ and _____: "A Multi-Response Gauss-Newton Algorithm," *Commun. Stat.: Simulation and Computation,* vol. B13, pp. 705–715, 1984.

Beale, E. M. L.: "Confidence Regions in Nonlinear Estimation," *J. R. Stat. Soc. B,* vol. 22, pp. 41–88, 1960.

Beck, J. V., and K. J. Arnold: *Parameter Estimation in Engineering and Science,* Wiley, New York, 1977.

Blakemore, J. W., and A. E. Hoerl: "Fitting Nonlinear Reaction Rate Equations to Data," *Chem. Eng. Prog. Symp. Ser.,* vol. 59, no. 42, pp. 14–27, 1963.

Bohachevsky, I. O., M. E. Johnson, and M. L. Stein: "Generalized Simulated Annealing for Function Optimization," *Technometrics,* vol. 28, pp. 209–217, 1986.

Box, G. E. P., and N. R. Draper: "The Bayesian Estimation of Common Parameters from Several Responses," *Biometrika*, vol. 52, pp. 355–365, 1965.

_____ and _____: *Empirical Model-Building and Response Surfaces*, Wiley, New York, 1987.

______ and W. J. Hill: "Discrimination among Mechanistic Models," *Technometrics,* vol. 9, pp. 57–71, 1967.

_____ and W. G. Hunter: "A Useful Method for Model-Building," *Technometrics*, vol. 4, pp. 301–318, 1962.

____ and _____: "Sequential Design of Experiments for Nonlinear Models," in *Proc. IBM Scientific Computing Symp. in Statistics* (IBM, White Plains, NY, 1963).

_____ and _____: "The Experimental Study of Physical Mechanisms," *Technometrics*, vol. 7, pp. 23–42, 1965.

_____ and H. L. Lucas: "Design of Experiments in Nonlinear Situations," *Biometrika*, vol. 46, pp. 77–90, 1959.

Box, M. J.: "The Occurrence of Replications in Optimal Designs of Experiments to Estimate Parameters in Nonlinear Models," *J. R. Stat. Soc. B,* vol. 30, pp. 290–302, 1968.

_____: "Some Experiences with a Nonlinear Experimental Design Criterion," *Technometrics*, vol. 12, pp. 569–589, 1970.

_____: "Bias in Nonlinear Estimation," *J. R. Stat. Soc. B*, vol. 33, pp. 171–201, 1971.

Chambers, J. M.: "Fitting Nonlinear Models: Numerical Techniques," *Biometrika*, vol. 60, pp. 1–13, 1973.

Chen, J.-S., and R. I. Jennrich: "Transformations for Improving Linearization Confidence Intervals in Nonlinear Regression," *J. Am. Stat. Soc.*, vol. 90, pp. 1271–1276, 1995.

Cheng, R. C. H., B. E. Evans, and T. C. Iles: "Embedded Models in Non-linear Regression," *J. R. Stat. Soc. B*, vol. 54, pp. 877–888, 1992.

Chernoff, H.: "Optimal Accelerated Life Designs for Estimation," *Technometrics*, vol. 4, pp. 381–408, 1962.

Clarke, G. P. Y.: "Moments of the Least Squares Estimators in a Non-linear Regression Model," *J. R. Stat. Soc. B,* vol. 42, pp. 227–237, 1980.

Cook, R. D., and C. J. Nachtsheim: "Model Robust, Linear Optimal Designs," *Technometrics*, vol. 24, pp. 49–54, 1982.

_____ and C. L. Tsai: "Residuals in Nonlinear Regression," *Biometrika*, vol. 72, pp. 23–29, 1985.

_____ and J. A. Witmer: "A Note on Parameter-Effects Curvature," *J. Am. Stat. Assoc.,* vol. 80, pp. 872–878, 1985.

Donaldson, J. R., and R. B. Schnabel: "Computational Experience with Confidence Regions and Confidence Intervals for Nonlinear Least Squares," *Technometrics*, vol. 29, pp. 67–82, 1987.

Draper, N. R., and H. Smith: *Applied Regression Analysis,* 2d, ed., Wiley, New York, 1981.

Farazdaghi, H., and P. M. Harris: "Plant Competition and Crop Yield," *Nature,* vol. 217, pp. 289–290, 1968.

Ford, I., D. M. Titterington, and C. P. Kitsos: "Recent Advances in Nonlinear Experimental Design," *Technometrics*, vol. 31, pp. 49–60, 1989.

Gallant, A. R.: *Nonlinear Statistical Models*, Wiley, New York, 1987.

_____ and W. A. Fuller: "Fitting Segmented Polynomial Regression Models Whose Join Points Have to Be Estimated," *J. Am. Stat. Assoc.,* vol. 71, pp. 961–967, 1973.

Gelman, A., F. Bois, and J. Jiang: "Physiological Pharmacokinetic Analysis Using Population Modeling and Informative Prior Distributions," *J. Am. Stat. Soc.,* vol. 91, pp. 1400–1412, 1996.

Gillis, P. R., and D. A. Ratkowsky: "The Behaviour of Estimators of the Parameters of Various Yield-Density Relationships," *Biometrics*, vol. 34, pp. 191–198, 1978.

Gonin, R., and A. H. Money: *Nonlinear L_p-Norm Estimation*, Dekker, New York, 1989.

Griffiths, D. A., and A. J. Miller: "Hyperbolic Regression—A Model Based on Two-Phase Piecewise Linear Regression with a Smooth Transition between Regimes," *Commun. Stat.,* vol. 2, pp. 561–569, 1973.

Guttman, I., and D. A. Meeter: "On Beale's Measures of Non-Linearity," *Technometrics,* vol. 7, pp. 623–637, 1965.

Haines, M., "A Geometric Approach to Optimal Design for One-Parameter Non-Linear Models," *J. R. Stat. Soc. B.*, vol. 57, pp. 575–598, 1995.

Hamilton, D. C., and D. G. Watts: "A Quadratic Design Criterion for Precise Estimation in Nonlinear Regression Models," *Technometrics*, vol. 27, pp. 241–250, 1985.

____, _____, and D. M. Bates: "Accounting for Intrinsic Nonlinearity in Nonlinear Regression Parameter Inference Regions," *Ann. Stat.,* vol. 10, pp. 386–393, 1982.

Hill, W. J., W. G. Hunter, and D. W. Wichern: "A Joint Design Criterion for the Dual Problem of Model Discrimination and Parameter Estimation," *Technometrics,* vol. 10, pp. 145–160, 1968.

Holliday, R.: "Plant Population and Crop Yield," *Field Crop Abstr.*, vol. 13, pp. 159–167, 247–254, 1960.

Huber, P. J.: "Robust Estimation of a Location Parameter," *Ann. Math. Stat.,* vol. 43, pp. 1041–1067, 1964.

____: *Robust Statistics*, Wiley, New York, 1981.

Hunter, W. G., and R. Mezaki: "A Model Building Technique for Chemical Engineering Kinetics," *AIChE J.,* vol. 10, pp. 315–322, 1964.

____, J. R. Kittrell, and R. Mezaki: "Experimental Strategies for Mechanistic Modelling," *Trans. AIChE,* vol. 45, pp. T146-T152, 1967.

Jennrich, R. I.: "Asymptotic Properties of Non-Linear Least Squares Estimators," *Ann. Math. Stat.,* vol. 40, pp. 633–643, 1969.

Kafadar, K.: "An Application of Nonlinear Regression in Research and Development: A Case Study from the Electronics Industry," *Technometrics*, vol. 36, pp. 237–248, 1994.

Kittrell, J. R., W. G. Hunter, and C. C. Watson: "Nonlinear Least Squares Analysis of Catalytic Rate Models," *AIChE J.*, vol. 11, pp. 1051–1057, 1965.

Kullback, S., and R. A. Leibler: "On Information and Sufficiency," *Ann. Math. Stat.,* vol. 22, pp. 79–86, 1951.

Marquardt, D. W.: "An Algorithm for Least Squares Estimation of Nonlinear Parameters," *SIAM J.,* vol. 11, pp. 431–441, 1963.

Morgan, P. H., L. P. Mercer, and N. W. Flodin: "General Model for Nutritional Responses of Higher Organisms," *Proc. Nat. Acad. Sci.* (US), vol. 72, pp. 4327–4331, 1975.

Nelder, J. A., and R. Mead: "A Simplex Method for Function Minimization," *Comput. J.,* vol. 7, pp. 308–311, 1965.

Peck, C. C., L. B. Sheiner, and A. I. Nichols: "The Problem of Choosing Weights in Nonlinear Regression Analysis of Pharmokinetic Data," *Drug Metab. Rev.,* vol. 15, pp. 133–148, 1984.

Peterson, T. I., and L. Lapidus: "Nonlinear Estimation Analysis of the Kinetics of Catalytic Ethanol Dehydrogenation," *Chem. Eng. Sci.,* vol. 21, pp. 655–664, 1966.

Ralston, M. L., and R. I. Jennrich: "DUD, a Derivative-Free Algorithm for Nonlinear Least Squares," *Technometrics*, vol. 20, pp. 7–14, 1978.

Ratkowsky, D. A.: *Nonlinear Regression Modeling,* Dekker, New York, 1983.

Rey, W. J. J.: *Introduction to Robust and Quasi-Robust Statistical Methods,* Springer, New York, 1983.

Richards, F. J.: "A Flexible Growth Function for Empirical Use," *J. Exp. Biol.,* vol. 10, pp. 290–300, 1959.

Ross, J. S. G.: *Nonlinear Estimation*, Springer, New York, 1990.

Sane, P. P., R. E. Eckert, and J. M. Woods: "On Fitting Combined Integral and Differential Reaction Kinetic Data. Rate Modelling for Hydrogenation of Butadiene," *Ind. Eng. Chem. Fundam.*, vol. 13, pp. 52–56, 1974.

Seber, G. A. F., and C. J. Wild: *Nonlinear Regression*, Wiley, New York, 1989.

Shenton, L. R., and K. O. Bowman: *Maximum Likelihood Estimation in Small Samples,* Macmillan, New York, 1977.

Smith, H., and S. D. Dubey: "Some Reliability Problems in the Chemical Industry," *Ind. Qual. Contr.,* vol. 21, pp. 64–70, 1964.

Sutton, T. L., and J. F. MacGregor: "The Analysis and Design of Binary Vapour-Liquid Equilibrium Experiments. Part I: Parameter Estimation and Consistency Tests," *Can. J. Chem. Eng.*, vol. 55, pp. 602–608, 1977.

Swain, J. J., and B. W. Schmeiser: "Monte Carlo Estimation of the Sampling Distribution of Nonlinear Model Parameter Estimators," *Ann. Oper. Res.,* vol. 8, pp. 245–256, 1987.

Thisted, R. A.: *Elements of Statistical Computing*, Chapman and Hall, New York, 1988.

Tsai, C.-L.: "Power Transformations and Reparameterizations in Nonlinear Regression Models," *Technometrics,* vol. 30, pp. 441–448, 1988.

Welsch, R. E., and R. A. Becker: "Robust Nonlinear Regression Using DOGLEG Algorithm," in *Proc. Computer Science and Statistics: 8th Annual Symp. on the Interface,* J. W. Frane, Ed., Health Sci. Computing Facility, Univ. of California, Los Angeles, pp. 272–279, 1975.

Wu, C. F.: "Asymptotic Theory of Nonlinear Least Squares Estimation," *Ann. Stat.,* vol. 9, pp. 501–513, 1981.

Ziegal, E. R., and J. W. Gorman: "Kinetic Modelling with Multiresponse Data," *Technometrics*, vol. 22, pp. 139–151, 1980.

CHAPTER 19
TIME SERIES ANALYSIS

Frank B. Alt
University of Maryland, College Park, MD

Ken Hung
National Dong Hwa University
Hualien, Taiwan, R.O.C.

and

Lap-Ming Wun
National Institute of Health
Bethesda, MD

19.1 INTRODUCTION

An important part of statistics deals with the analysis of data that are collected sequentially over time. Two objectives of analyzing time series data are to model the historical data and to forecast or predict future values of the series. A feature of the techniques presented in this chapter is that only the values of one series are used to obtain the forecasts. Any relationship that this series has with other variables is ignored.

Sections 19.2 and 19.3 provide the necessary background for building probabilistic models for time series. Most of this chapter focuses on what is referred to as the Box-Jenkins approach to model building. This class of models, their properties, the model-building process, and examples illustrating their use are presented in Secs. 19.4 through 19.6. An introduction to simple exponential smoothing and its relationship to the Box-Jenkins models is found in Sec. 19.7. In describing the nature of a time series, the traditional approach has been to decompose it into four components (long-term trend, seasonal effect, cyclical effect, and irregular or residual effect) by assuming that they interact in either a multiplicative or an additive fashion. Section 19.8 introduces a special case of the additive model.

19.2 STATIONARITY

A requirement for using the Box-Jenkins approach is that the series be stationary. Let Z_t denote the observation at time t. A time series is *strictly stationary* if the joint probability distributions of $(Z_{t_1}, Z_{t_2}, \ldots, Z_{t_n})$ and $(Z_{t_1+k}, Z_{t_2+k}, \ldots, Z_{t_n+k})$ are the same for all $t_1, t_2, \ldots, t_n$ and k. Since this definition holds for any value of n, it certainly holds for $n = 1$. This implies that the univariate distribution of Z_{t_1} is the same as Z_{t_1+k}. It follows that $E(Z_t) = E(Z_{t+k})$ and $\text{Var}(Z_t) = \text{Var}(Z_{t+k})$, or that the mean μ and variance σ^2 are constant over time. Furthermore, for $n = 2$ the joint probability distributions of (Z_{t_1}, Z_{t_2}) and (Z_{t_1+k}, Z_{t_2+k}) are identical and must have the same covariance,

$$\text{Cov}(Z_{t_1}, Z_{t_2}) = \text{Cov}(Z_{t_1+k}, Z_{t_2+k})$$

which depends only on $t_2 - t_1$, the *lag*, and not on the actual values of t_1 and t_2. The covariance between Z_t and Z_{t+k} is called the *autocovariance* at lag k and is written as γ_k, where

$$\gamma_k = \text{Cov}(Z_t, Z_{t+k}) = E[(Z_t - \mu)(Z_{t+k} - \mu)]. \tag{19.1}$$

The prefix "auto" conveys the idea that γ_k is the covariance between different observations from the same series. Note that $\gamma_0 = \text{Var}(Z_t) = \sigma^2$. For interpretive purposes, it is useful to look at the autocorrelation function, denoted by ρ_k, where

$$\rho_k = \gamma_k/\gamma_0. \tag{19.2}$$

Note that $\rho_0 = 1$, $\rho_{-k} = \rho_k$, and $|\rho_k| \leq 1$. An informal assessment of γ_k can be obtained by looking at a scatterplot of Z_{t+k} and Z_t. For $k = 1$, the scatterplot would be obtained by plotting the following $n - 1$ pairs of points: $(Z_2, Z_1), (Z_3, Z_2), \ldots, (Z_n, Z_{n-1})$, where n denotes the number of observations.

A weaker, less restrictive form of stationarity is often useful in practice. A process is called *second-order stationary* if the mean is finite and constant over time and if the autocovariance is finite and depends only on the lag. This also implies that the variance is finite and constant over time.

Failure to account for autocorrelation can have a deleterious effect on the statistical analysis. As a specific example, consider the Shewhart control chart for the mean (refer to Chap. 7). When there is only one quality characteristic, which is normally distributed, with standard values specified for the process mean μ_0 and variance σ_0^2, Shewhart (1931) proposed that in order to monitor the process mean one should plot successive values of the sample mean $\overline{X}$ on a chart that has upper and lower control limits of the form

$$E(\overline{X}) \pm z_{\alpha/2}\sqrt{\text{Var}(\overline{X})} = \mu_0 \pm z_{\alpha/2}\frac{\sigma}{\sqrt{n}} \tag{19.3}$$

where $z_{\alpha/2}$ denotes the upper $\alpha/2$ percentage point of the standard normal random variable. Usually $z_{\alpha/2} = 3.0$. The control limits, presented in Eq. (19.3), were based on the assumption that the elements of the sample used to calculate $\overline{X}$ were uncorrelated. We will consider what the effects are by using the control limits given in Eq. (19.3) when, in fact, the observations have a first-order autocorrelation denoted by ρ.

Suppose that the n elements of the sample $X_1, X_2, \ldots, X_n$ are such that $E(X_t) = \mu_0$, $\mathrm{Var}(X_t) = \sigma^2$, and $\mathrm{Cov}(X_t, X_{t+k}) = \rho\sigma^2$ for $k = 1$, and 0 otherwise. Under these conditions,

$$\mathrm{Var}(\overline{X}) = \frac{\sigma^2}{n}\,[1 + 2\rho(1 - n^{-1})]. \tag{19.4}$$

This can be used to determine the true probability, denoted by α_0, of incorrectly concluding that the process mean is not μ_0 when using the control limits given in Eq. (19.3), assuming a nominal probability of $\alpha = 0.0027$. Specifically,

$$\alpha_0 = P\left[\,|Z| > \frac{3}{[1 + 2\rho(1 - n^{-1})]^{1/2}}\right]. \tag{19.5}$$

The use of Eq. (19.5) indicates that for $\rho < 0$, α_0 is less than the nominal value of 0.0027, while for $\rho > 0$, α_0 is greater than 0.0027. For example, with $n = 5$, $\alpha_0 = 0.0137$ for $\rho = 0.3$ while $\alpha_0 = 0.000032$ for $\rho = -0.3$. Both values of α_0 differ substantially from the nominal value of 0.0027.

19.3 DIFFERENCING

Recall that one of the requirements for using the Box-Jenkins approach is that the mean be constant over time. Obviously, this assumption is violated when a series exhibits trend, which can be defined as a long-term upward or downward movement. Fortunately, the effect of trend can usually be removed by *differencing*.

In differencing a time series, the first observation is subtracted from the second, the second from the third, and so on. To illustrate the concept, consider the integer sequence 1, 2, …, N. After differencing, we obtain

$$(2 - 1), (3 - 2), (4 - 3), \ldots, N - (N - 1)$$

or

$$1, 1, 1, \ldots, 1$$

where the differenced values are devoid of trend.

Before presenting the differencing operator ∇, we first introduce the *backshift operator,* denoted by B, where

$$BZ_t = Z_{t-1} \tag{19.6}$$

that is, B operates on Z_t to shift it back one time period. Two properties of the backshift operator are

$$B^n Z_t = Z_{t-n}$$

and

$$B^n B^m Z_t = B^{n+m} Z_t = Z_{t-n-m}$$

We also have $B^{-1}B = 1$. Note that the backshift operator has no effect on a constant. Thus for a constant k, $Bk = k$.

The backshift operator can be used to difference a time series since $(1 - B)Z_t = Z_t - Z_{t-1}$, for $t = 2, 3, \ldots, n$. It is customary to denote $(1 - B)$ by ∇, which is referred to as the *differencing operator*. While first differencing ∇ may be useful in removing linear trend, second differencing ∇^2 may be useful in removing quadratic trend. Note that

$$\nabla^2 Z_t = (1 - B)^2 Z_t = Z_t - 2Z_{t-1} + Z_{t-2}$$

for $t = 3, 4, \ldots, n$. In general, d is used to denote the degree of differencing necessary to induce stationarity. Occasionally it may be necessary to take a seasonal difference of degree D, denoted by ∇_S^D, for time series with seasonal variation. For quarterly data, $S = 4$; for monthly data, $S = 12$. Thus $\nabla_4 Z_t = (1 - B^4)Z_t = Z_t - Z_{t-4}$, and $\nabla_{12} Z_t = (1 - B^{12})Z_t = Z_t - Z_{t-12}$. The nonseasonal and seasonal differencing operators may be used together to achieve stationarity, in which case the transformed value W_t is given by

$$W_t = \nabla^d \nabla_S^D Z_t = (1 - B)^d (1 - B^S)^D Z_t.$$

Recall that $1/(1 - B) = 1 + B + B^2 + \cdots$. When $d = 1$ and $D = 0$, $W_t = (1 - B)Z_t$ and $Z_t = (1 + B + B^2 + \cdots)W_t = W_t + W_{t-1} + W_{t-2} + \cdots$. As Nelson (1973) states, "Z_t is the sum of all past changes… Z_t is referred to as the integration of the W_t series…." This is the origin of the term *integrated* in ARIMA.

A simple example of a time series is the *random walk* process, which has been used as a model for the movement of stock market prices (Fama, 1965). Let $a_1, a_2, \ldots$ be independent, identically distributed random variables, each with zero mean and variance σ_a^2. This is known as a *white noise process* for which $\rho_0 = 1$ and $\rho_k = 0$ for $k > 0$. The a_t notation will be used exclusively to denote white noise. The random walk process is determined by

$$Z_t = Z_{t-1} + a_t$$

or

$$Z_t - Z_{t-1} = a_t. \tag{19.7}$$

Although the random walk is nonstationary, the series that results from taking the first difference, obtained by letting $W_t = \nabla Z_t = Z_t - Z_{t-1}$, is stationary. Note that W_t is white noise since $W_t = a_t$.

A variation of Eq. (19.7), which allows for an upward or downward trend in the series, is the *random walk with drift* determined by

$$Z_t - Z_{t-1} = k + a_t$$

where k is a constant. By letting $W_t = Z_t - Z_{t-1}$, we see that W_t is a stationary process whose values would fluctuate around k.

Another requirement for using the Box-Jenkins approach is that the variance be constant over time. When this condition is not met, Cryer (1986) presents two scenarios where a transformation to natural logarithms and an analysis of the first differences in the natural logarithms, $\nabla[\ln Z_t]$, could be an appropriate remedy: (1) the level of the series is changing exponentially and the standard deviation of the series is directly proportional to the level of the series, or (2) the percentage changes in the observations from one time period to the next are approximately stable. The logarithmic transformation is a special case of a family of transformations called the Box-Cox (1964) or power transformation. For a given value of λ, the transformed value $Z_t^{(\lambda)}$ is given by

$$Z_t^{(\lambda)} = \begin{cases} \dfrac{Z_t^{(\lambda)} - 1}{\lambda}, & \lambda \neq 0 \\ \ln Z_t, & \lambda = 0 \end{cases}.$$

For all practical purposes, stationarity can be induced by a power transformation and/or differencing.

19.4 AUTOREGRESSIVE–INTEGRATED–MOVING–AVERAGE (ARIMA) MODELS

The ARIMA model can be described in terms of (p, d, q) notation, where p indicates the autoregressive order, q the moving-average order, and d the degree of differencing necessary to achieve stationarity.

19.4.1 Autoregressive Models

In *autoregressive processes* of order p, the current observation Z_t is expressed as the sum of three components: a linear combination of the p immediate past observations, a constant θ_0, and a random error component for the current period. The basic model is specified by

$$Z_t = \phi_1 Z_{t-1} + \phi_2 Z_{t-2} + \cdots + \phi_p Z_{t-p} + \theta_0 + a_t \tag{19.8}$$

where $\phi_1, \ldots, \phi_p$ are the autoregressive parameters. The term autoregressive is used since Z_t is regressed on observations from the same series. The model in Eq. (19.8) can be rewritten as

$$(1 - \phi_1 B - \phi_2 B^2 - \cdots - \phi_p B^p) Z_t = \theta_0 + a_t. \tag{19.9}$$

It is customary to let $(1 - \phi_1 B - \cdots - \phi_p B^p) = \phi(B)$. In order for an autoregressive process to be stationary, the roots of the equation, $\phi(B) = 0$, must lie outside the unit circle. This is discussed in Box and Jenkins (1976). If the process is stationary, then the mean μ is given by

$$E(Z_t) = \mu = \frac{\theta_0}{1 - \phi_1 - \phi_2 - \cdots - \phi_p} \tag{19.10}$$

and the variance γ_0 is given by

$$\mathrm{Var}(Z_t) = \gamma_0 = \frac{\sigma_a^2}{1 - \phi_1\rho_1 - \phi_2\rho_2 - \cdots - \phi_p\rho_p} \tag{19.11}$$

where σ_a^2 is the variance of the random error or white noise term a_t. The autocorrelations can be expressed in terms of the following recursive relation ship:

$$\rho_k = \phi_1\rho_{k-1} + \phi_2\rho_{k-2} + \cdots + \phi_p\rho_{k-p} \tag{19.12}$$

for $k \geq 1$. Cryer (1986) points out that a system of p linear equations, known as the *Yule-Walker equations*, are obtained from Eq. (19.12) by letting $k = 1, 2, \ldots,$ p. If values are specified for the autoregressive parameters, then the equations can be solved for $\rho_1, \rho_2, \ldots, \rho_p$, while values of ρ_k for $k > p$ can be obtained directly from Eq. (19.12).

An ARIMA (1, 0, 0) model, also known as AR(1), has only one autoregressive term (namely, $p = 1$) and can be written as

$$Z_t = \phi_1 Z_{t-1} + \theta_0 + a_t \tag{19.13}$$

or

$$(1 - \phi_1 B)Z_t = \theta_0 + a_t. \tag{19.14}$$

In order for the process to be stationary, the root of the equation $1 - \phi_1 B = 0$ must exceed 1 in absolute value. Equivalently, $|\phi_1| < 1$ since $B = 1/\phi_1$. The fact that $-1 < \phi_1 < +1$ is intuitively appealing since ϕ_1 is actually the correlation between adjacent observations. This follows from Eq. (19.12) by putting $k = 1$ and noting that $\rho_0 = 1$. Note that the autocorrelation function for an AR(1) process is characterized by exponential decay since $\rho_k = \phi^k$, for $k = 0, 1, 2, \ldots$.

An ARIMA (2, 0, 0) model, also known as AR(2), is specified by

$$Z_t = \phi_1 Z_{t-1} + \phi_2 Z_{t-2} + \theta_0 + a_t \tag{19.15}$$

or

$$(1 - \phi_1 B - \phi_2 B^2)Z_t = \theta_0 + a_t. \tag{19.16}$$

In this case, the stationarity constraints are

$$\begin{aligned} -1 < \phi_2 < +1 \\ \phi_1 + \phi_2 < +1 \\ \phi_2 - \phi_1 < +1. \end{aligned} \tag{19.17}$$

Although specific expressions for ρ_1, ρ_2, and ρ_k for higher lags can be obtained from Eq. (19.12), the pattern exhibited by the autocorrelation function is more important for modeling purposes than the actual values of ρ_k. The autocorrelation function of an AR(2) process is characterized by a mixture of damped exponentials or damped sine waves.

It could be difficult to determine the order of an autoregressive process from examination of the autocorrelation function since the autocorrelation functions of all autoregressive processes decay to zero. The *partial autocorrelation function* is useful in determining the order of a pure autoregressive process. As Cryer (1986) states, the partial autocorrelation at lag k, denoted by ϕ_{kk}, is "... the correlation between Z_t and Z_{t-k} after removing the effect of the intervening variables Z_{t-1}, Z_{t-2},

..., Z_{t-k+1}." Note that ϕ_{11} always equals ρ_1 since there are no intervening variables between Z_t and Z_{t-1}. An AR(p) process is characterized by the partial autocorrelation function cutting off after lag p, that is, $\phi_{kk} = 0$, $k > p$. For an AR(1) process, $\phi_{11} = \rho_1$, and $\phi_{kk} = 0$ for $k > 1$. For an AR(2) process, $\phi_{11} = \rho_1$, $\phi_{22} = (\rho_2 - \rho_1^2)/(1 - \rho_1^2)$, and $\phi_{kk} = 0$ for $k > 2$.

19.4.2 Moving-Average Models

In *moving-average processes* of order q, the current observation Z_t is the sum of a current and weighted, lagged random error terms for the last q periods, together with a constant θ_0. The ARIMA (0, 0, q) or MA(q) process is determined by

$$Z_t = \theta_0 + a_t - \theta_1 a_{t-1} - \theta_2 a_{t-2} - \cdots - \theta_q a_{t-q}$$

or (19.18)

$$Z_t = \theta_0 + (1 - \theta_1 B - \theta_2 B^2 - \cdots - \theta_q B^q) a_t$$

where $(1 - \theta_1 B - \theta_2 B^2 - \cdots - \theta_q B^q)$ is denoted by $\theta(B)$. A moving-average process is always stationary. Since the random error terms are white noise having mean zero and variance σ_a^2, it follows that

$$E(Z_t) = \mu = \theta_0 \tag{19.19}$$

$$\mathrm{Var}(Z_t) = \gamma_0 = \sigma_a^2(1 + \theta_1^2 + \theta_2^2 + \cdots + \theta_q^2) \tag{19.20}$$

and

$$p_k = \begin{cases} (-\theta_k + \theta_1\theta_{k+1} + \theta_2\theta_{k+2} + \cdots + \theta_{q-k}\theta_q)\,(\sigma_a^2/\gamma_0), & k = 1, 2, \ldots, q \\ 0, & k > q. \end{cases} \tag{19.21}$$

Thus the autocorrelation function for an MA(q) process cuts off after lag q. Furthermore, the partial autocorrelation function tails off. These are useful properties for identification purposes.

An ARIMA (0, 0, 1) or MA(1) process is specified by

$$Z_t = \theta_0 + a_t - \theta_1 a_{t-1} \tag{19.22}$$

or

$$Z_t = \theta_0 + (1 - \theta_1 B)a_t. \tag{19.23}$$

The autocorrelation function, obtained from Eq. (19.21) by putting $k = 1$, is given by

$$\rho_1 = \begin{cases} \dfrac{-\theta_1}{1 + \theta_1^2}, & k = 1 \\ 0, & k > 1. \end{cases} \tag{19.24}$$

Although there is no stationarity constraint on θ_1, there is an analagous invertibility constraint, namely, $-1 < \theta_1 < +1$. Without loss of generality, assume $\theta_0 = 0$. Since $Z_t = (1 - \theta_1 B)a_t$ and $(1 - \theta_1 B)^{-1} = 1 + \theta_1 B + \theta_2^1 B^2 + \cdots$, it follows that $a_t = (1 - \theta_1 B)^{-1} Z_t = Z_t + \theta_1 Z_{t-1} + \theta_1^2 Z_{t-2} + \cdots$ so that $Z_t = a_t - \theta_1 Z_{t-1} - \theta_1^2 Z_{t-2} - \cdots$. Unless $|\theta_1| < 1$, the weights given to the past observations would increase with age. Finally, note that the partial autocorrelation function tails off,

$$\phi_{kk} = -\phi_1^k \frac{1 - \theta_1^2}{1 - \theta_1^{2k+2}}$$

An ARIMA (0, 0, 2) or MA(2) process is determined by

$$Z_t = \theta_0 + a_t - \theta_1 a_{t-1} - \theta_2 a_{t-2} \tag{19.25}$$

or

$$Z_t = \theta_0 + (1 - \theta_1 B - \theta_2 B^2)a_t. \tag{19.26}$$

The invertibility constraints are

$$\begin{aligned} -1 < \theta_2 < +1 \\ \theta_1 + \theta_2 < +1 \\ \theta_2 - \theta_1 < +1. \end{aligned}$$

Values for ρ_1 and ρ_2 are easily obtained from Eq. (19.21). The important feature is that the autocorrelation function cuts off after lag 2 for an MA(2) process, while the partial autocorrelation function tails off.

19.4.3 Autoregressive–Moving-Average (ARMA) Models and Autoregressive–Integrated–Moving-Average (ARIMA) Models

A time series can have both autoregressive and moving-average terms. For example, consider the ARMA (1, 1) model given by

$$Z_t = \theta_0 + \phi_1 Z_{t-1} + a_t - \theta_1 a_{t-1} \tag{19.27}$$

or

$$(1 - \phi_1 B)Z_t = \theta_0 + (1 - \theta_1 B)a_t. \tag{19.28}$$

This model has certain features of the MA(1) and AR(1) models. For example, the mean is the same as that of an AR(1) process and can be obtained from Eq. (19.10). Furthermore, the autocorrelation function exhibits exponential decay but starting from ρ_1. Specifically, $\rho_k = \phi_1 \rho_{k-1}$ for $k > 1$. Similar to an MA(1) process, the partial autocorrelation function tails off to zero. Finally, the invertibility condition is that $|\theta_1| < 1$, while the stationarity condition is that $|\phi_1| < 1$.

The *mixed autoregressive–moving-average process* of order (p, q), denoted by ARMA (p, q), is represented by the equation

$$Z_t = \theta_0 + \phi_1 Z_{t-1} + \cdots + \phi_p Z_{t-p} + a_t - \theta_1 a_{t-1} - \cdots - \theta_q a_{t-q} \tag{19.29}$$

where the roots of $\theta(B) = 0$ and $\phi(B) = 0$ lie outside the unit circle. Also, $\theta(B)$ and $\phi(B)$ can have no common factors. The autocorrelation function and partial autocorrelation function both tail off as the lag increases. A more comprehensive discussion of the properties of the autocorrelation and partial autocorrelation functions is found in Abraham and Ledolter (1983). The *extended autocorrelation function* can be useful in identifying mixed processes, and is discussed in Sec. 19.5.1. A useful feature of mixed ARMA models is that time series data can be modeled by a mixed model with fewer parameters than a pure MA or AR model.

When it is necessary to first take a difference of degree d to achieve stationarity, this is accomplished by modifying the left-hand side of Eq. (19.29) to read $\nabla^d Z_t$. This change results in an ARIMA (p, d, q) model,

$$\phi(B)\nabla^d Z_t = \theta_0 + \theta(B)\, a_t \tag{19.30}$$

where

$$\phi(B) = 1 - \phi_1 B - \phi_2 B^2 - \cdots - \phi_p B^p \tag{19.31}$$

$$\theta(B) = 1 - \theta_1 B - \theta_2 B^2 - \cdots - \theta_q B^q. \tag{19.32}$$

19.4.4 Seasonal Autoregressive–Moving–Average (SARMA) Models and Seasonal Autoregressive–Integrated–Moving–Average (SARIMA) Models

For time series data that are observed on some regular calendar basis, a seasonal effect may be present. For example, ice cream sales are higher in the summer months of a particular year than at other times, and this periodic pattern would repeat in subsequent years. This type of behavior is an example of seasonal variation, and the periodicity is denoted by S. For monthly data, $S = 12$; for quarterly data, $S = 4$. If there is seasonal variation, observations S periods apart should be correlated. Suppose monthly observations are available for five years from January 1995 (J_{95}) through December 1999 (D_{99}). A method of detecting first-order seasonal autocorrelation would be the presence of a linear effect (either positive or negative) in the scatterplot of (J_{95}, J_{96}), (F_{95}, F_{96}), ..., (D_{98}, D_{99}). Similarly, second-order seasonal autocorrelation could be investigated by looking at a scatterplot of (J_{95}, J_{97}),..., (D_{97}, D_{99}). A more formal analysis could be conducted by looking at the sample autocorrelation function presented in Sec. 19.5.1.

In general, if a time series has seasonal variation, there will be autocorrelation at lag S and possibly at multiples of that lag, depending on the persistence and nature of that autocorrelation. This seasonal dependence can be described by seasonal ARMA (SARMA) models of order (P, Q) or seasonal ARIMA (SARIMA) models of order (P, D, Q) if it is necessary to take a seasonal difference. The model for a SARMA (P, Q) process is specified by

$$\Phi(B^S)Z_t = \Theta_0 + \Theta(B^S)a_t \tag{19.33}$$

where

$$\Phi(B^S) = 1 - \Phi_1 B^S - \Phi_2 B^{2S} - \cdots - \Phi_P B^{PS} \tag{19.34}$$

$$\Theta(B^S) = 1 - \Theta_1 B^S - \Theta_2 B^{2S} - \cdots - \Theta_Q B^{QS} \tag{19.35}$$

As Shumway (1988) states, "Series with these kinds of components tend to generate autocorrelation and partial autocorrelation functions that mimic the behavior observed for ordinary ARMA (p, q) processes, except that the peaks occur at multiples of the seasonal period S." When $P = 0$ or, equivalently, $\Phi(B^S) = 1$, the autocorrelation function is nonzero at lags $k = S, 2S, \ldots, QS$ and cuts off thereafter. When $Q = 0$ or $\Theta(B^S) = 1$, the partial autocorrelation function is nonzero at lags $k = S, 2S, \ldots, PS$ and cuts off thereafter. For stationarity, the roots of $\Phi(B^S) = 0$ must lie outside the unit circle; for invertibility, the roots of $\Theta(B^S) = 0$ must lie outside the unit circle. If seasonal differencing is required, Z_t in Eq. (19.33) is replaced by $\nabla_S^D Z_t$. This results in the SARIMA model of order $(P, D, Q)_S$,

$$\Phi(B^S)\nabla_S^D Z_t = \Theta_0 + \Theta(B^S)a_t. \tag{19.36}$$

19.4.5 General Multiplicative Model

The models (19.30) and (19.36) can be used to describe nonseasonal and seasonal dependencies, respectively, between the observations. However, there could be autocorrelation at both the seasonal and the nonseasonal lags. For example, there may be a relationship between ice cream sales in July 1998 and July 1999 as well as for observations in adjacent months, such as July and August of 1998. To account for both types of dependencies, Box and Jenkins (1976) presented the *multiplicative seasonal autoregressive–integrated–moving-average model* of order $(p, d, q) \times (P, D, Q)_S$. Specifically,

$$\phi(B)\Phi(B^S)W_t = \theta_0 + \theta(B)\Theta(B^S)a_t \tag{19.37}$$

where

$$W_t = \nabla^d \nabla_S^D Z_t = (1 - B)^d (1 - B^S)^D. \tag{19.38}$$

If differencing is not required, $d = D = 0$ and $W_t = Z_t$. $\phi(B)$ and $\theta(B)$ were defined in Eqs. (19.31) and (19.32); $\Phi(B^S)$ and $\Theta(B^S)$ were defined in Eqs. (19.34) and (19.35).

As a special case of the general multiplicative model, consider the ARIMA $(0, d, 1) \times (0, D, 1)_S$ model

$$W_t = \theta_0 + (1 - \theta_1 B)(1 - \Theta_1 B^S)a_t. \tag{19.39}$$

Here, there are nonzero autocorrelations at lags 1, $S - 1$, S, and $S + 1$. Specifically,

$$\rho_1 = \frac{-\theta_1}{1 + \theta_1^2}$$

$$\rho_S = \frac{-\Theta_1}{1 + \Theta_1^2}$$

$$\rho_{S-1} = \rho_{S+1} = \rho_1 \rho_S.$$

The autocovariances and special characteristics for some other multiplicative models can be found in Box and Jenkins (1976, appendix A9.1) and in Abraham and Ledolter (1983, Sec. 6.2).

In the remainder of the presentation it is assumed that the white noise term a_t is normally distributed.

19.5 MODEL BUILDING

In practice, an iterative three-stage approach (see Fig. 19.1), advocated by Box and Jenkins (1976), is employed in model building: identification, estimation, and diagnostic checking. In addition, there may be feedback from the forecasting phase for reference in future model specification.

19.5.1 Identification

At this stage it is necessary to identify the values of (p, d, q) and $(P, D, Q)_S$ if there is a seasonal component. This identification process is usually based solely on an examination of the data. However, if there is some underlying theory, this should not be discarded in the process. Hopefully, the underlying theory will be corroborated by the data-based identification. If not, new light may be shed on the existing theory and a revision may be in order.

A plot of the data is useful in determining the degree d of nonseasonal differencing. However, caution should be used in determining the presence or absence of a linear trend based solely on a plot of the data since scaling can affect visual perception. The sample autocorrelation function (SACF) is also useful in determining d. The SACF at lag k is denoted by r_k, where

$$r_k = \frac{\sum_{t=k+1}^{n} (Z_t - \overline{Z})(Z_{t-k} - \overline{Z})}{\sum_{t=1}^{n} (Z_t - \overline{Z})^2} \tag{19.40}$$

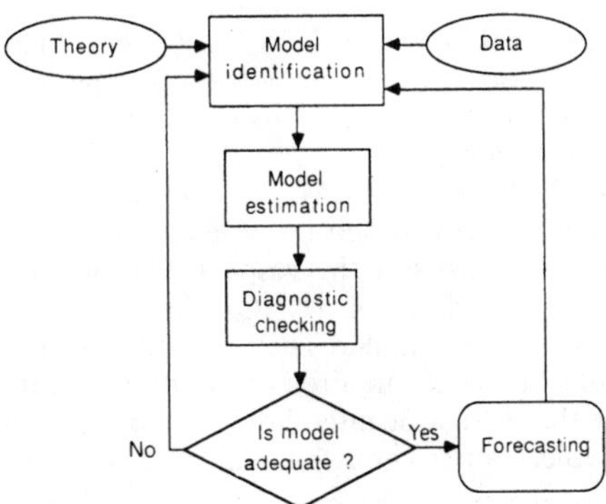

FIGURE 19.1 Iterative approach to model building. *(Adapted from Box and Jenkins, 1976.)*

$k = 0, 1, 2, \ldots$. The SACF at lag 0, r_0, equals 1. Note that if there is a trend, the sample mean $\bar{Z}$ essentially bisects the observations into two groups: those above the mean and those below it. Thus observations that are one lag apart tend to be on the same side of the mean. Furthermore, this tendency will persist for relatively large lags. Thus the SACF dies out very slowly. Stationarity can possibly be achieved by first ($d = 1$) or second ($d = 2$) differencing. Example 3 illustrates the use of first differencing. An example that demonstrates the danger of overdifferencing is presented in Nelson (1973). Where there is seasonal nonstationarity, the SACF decays very slowly at multiples of the seasonal lag: S, $2S$, $3S$, ... In this case, seasonal differencing would be used, either solely or in conjunction with nonseasonal differencing. Seasonal differencing is used in Example 2.

When stationarity has been achieved the SACF estimates the theoretical autocorrelation function defined in Eq. (19.2). It is also possible to estimate the partial autocorrelation function ϕ_{kk} by its sample counterpart, designated SPACF and denoted by $\hat{\phi}_{kk}$. A recursive formula for the SPACF can be found in Abraham and Ledolter (1983). The basic idea in choosing values for p and q is to match the behavior of the SACF and SPACF with that of the ACF and PACF for the class of ARMA models discussed previously. Thus a moving-average process of order q would be indicated if the SACF cuts off after lag q, while an autoregressive process of order p is in order if the SPACF cuts off after lag p. To determine whether or not the SACF and SPACF cut off, it is necessary to know their sampling distributions.

Barlett (1946) showed that if the process is MA(q), then the estimated standard deviation of r_k, denoted by $s(r_k)$, is given by

$$s(r_k) = n^{-1/2}\left(1 + 2\sum_{j=1}^{q} r_j^2\right)^{1/2} \tag{19.41}$$

for $k > q$. Therefore the hypothesis that $\rho_k = 0$ would be rejected at significance level α if $|r_k/s(r_k)| > z_{\alpha/2}$. The statistical significance of $\hat{\phi}_{kk}$ can be tested by using a result due to Quenouille (1949). For an AR(p) process, $\hat{\phi}_{kk}$ is approximately normally distributed with mean 0 and variance $1/n$ for $k > p$. Thus the hypothesis that $\phi_{kk} = 0$ would be rejected if $|\hat{\phi}_{kk}/\sqrt{n}| > z_{\alpha/2}$. Most computer programs plot the SACF and SPACF with horizontal bands at plus and minus 2 standard deviations. For example, if there are 56 observations, the horizontal bands for the SPACF would be at $\pm 2/\sqrt{56} = \pm 0.27$.

When the SACF is nonzero at lags S, $2S$, ..., QS and then cuts off, a seasonal moving-average model of order Q should be tried. Furthermore, when the SPACF is nonzero at lags S, $2S$, ..., PS and then cuts off, a seasonal autoregressive model of order P is a good candidate.

Since the ACF and PACF of an ARMA (p, q) process both tail off, the identification of such a mixed process based on an examination of the SACF and the SPACF could be difficult. To assist in this respect, Tsay and Tiao (1984) introduced the extended autocorrelation function (EACF), the values of which are used to specify the maximum orders of p and q. The procedure is simplified by looking for a "triangle of insignificant values" in a rectangular table of 0 and $\times$ entries, where an $\times$ indicates a value that is significantly different from zero by at least 2 standard deviations, and 0 depicts an insignificant value. This concept is illustrated in Examples 19.6.1 to 19.6.3.

In summary, statistics used for model identification include the sample autocorrelation function (SACF), the sample partial autocorrelation function (SPACF), and the extended sample autocorrelation function (ESACF). Another supplemental tool is a plot of the data.

19.5.2 Estimation

Once a model is tentatively identified, the parameters can be estimated by maximizing the corresponding likelihood function, assuming that the white noise term is normally distributed. There are no closed-form expressions for the *maximum-likelihood estimators* of the parameters in the general ARMA (p, q) model. Even for an AR(1) model, the maximum-likelihood estimate of ϕ_1 is the solution to a cubic equation. Because of the complexity of the maximum-likelihood estimation procedure for the general ARMA (p, q) model, further details are not given here, and the interested reader should consult Box and Jenkins (1976), Cryer (1986), or Abraham and Ledolter (1983).

Other estimation procedures include conditional least squares and unconditional least squares. The concept of conditional least squares (CLS) estimation can be explained by considering the ARMA (2, 1) model,

$$Z_t = \theta_0 + \phi_1 Z_{t-1} + \phi_2 Z_{t-2} + a_t - \theta_1 a_{t-1}$$

or

$$a_t = Z_t - \phi_1 Z_{t-1} - \phi_2 Z_{t-2} - \theta_0 + \theta_1 a_{t-1}.$$

The objective of CLS estimation is to find those values of θ_0, θ_1, ϕ_1, and ϕ_2 that minimize $\Sigma_{t=1}^{n} a_t^2$. Obviously, there are "start-up" problems since Z_0, Z_{-1}, and a_0 are needed when $t = 1$, while Z_0 is needed when $t = 2$. The remedy in this instance is to minimize $\Sigma_{t=3}^{n} a_t^2$ with $a_0 = 0$ (its expected value). In general, the lower index of summation is $t = p + 1$ with $a_p = a_{p-1} = \cdots = a_{1-q} = 0$.

In all cases, the parameter estimates must satisfy the stationarity and invertibility bounds required for the specified model.

19.5.3 Diagnostic Checking

Model adequacy is primarily determined by examining the residuals, denoted by $\hat{a}_t$. If an AR(1) model was fit, then

$$\hat{a}_t = Z_t - \hat{Z}_t = Z_t - (\hat{\phi}_1 Z_{t-1} + \hat{\theta}_0)$$

where $\hat{\phi}_1$ and $\hat{\theta}_0$ are the parameter estimates of ϕ_1 and θ_0, respectively.

One tool used at the diagnostic stage is the SACF of the residuals. If the model has been correctly specified, the residuals should be uncorrelated, and this property should be reflected in the absence of any significant values in the SACF of the residuals. Most computer packages give horizontal bands at ±2 standard deviations to facilitate the detection of significant autocorrelations for the residuals.

An overall measure of model adequacy is provided by the Ljung and Box (1978) statistic Q^*, also known as the modified Box-Pierce statistic, where

$$Q^* = n(n+2) \sum_{j=1}^{k} (n-j)^{-1} r_j^2. \tag{19.42}$$

Q^* is approximately distributed as a chi squared random variable with $k - p - q - P - Q$ degrees of freedom. If $Q^* > \chi^2_{\alpha, k-p-q-P-Q}$, then a further investigation regarding model adequacy is warranted with the understanding that Q^* is a global test.

If all criteria are satisfied, the once tentative model can now be adopted and used for forecasting.

19.5.4 Forecasting

Let $\hat{Z}_n(\xi)$ denote the forecast for observation $Z_{n+\xi}$ that is made at time origin n, where $\xi \geq 1$. The forecast will be optimally chosen in that it minimizes the mean-square forecast error,

$$E[Z_{n+\xi} - \hat{Z}_n(\xi)]^2. \tag{19.43}$$

Pankratz (1983) notes that "…some other (non-ARIMA) univariate model forecast could have a smaller forecast error than a properly constructed ARIMA model forecast in a particular instance, but not on average." Box and Jenkins (1976) show that if the specified model is the correct one, then the optimal forecast is given by the conditional expectation of $Z_{n+\xi}$. Specifically,

$$\hat{Z}_n(\xi) = E(Z_{n+\xi} | Z_1, Z_2, \ldots, Z_n).$$

The computation of the forecasts is illustrated for the AR(1) process [see Eq. (19.13)]. The one-period ahead forecast is given by

$$\begin{aligned} \hat{Z}_n(1) &= E(Z_{n+1} | Z_1, \ldots, Z_n) \\ &= E(\phi_1 Z_n + \theta_0 + a_{n+1}) = \phi_1 Z_n + \theta_0 \end{aligned}$$

since Z_n has been observed and a_{n+1} has expectation zero. Similarly, the two-period ahead forecast is given by

$$\begin{aligned} \hat{Z}_n(2) &= E(Z_{n+2} | Z_1, \ldots, Z_n) \\ &= E(\phi_1 Z_{n+1} + \theta_0 + a_{n+2}) \\ &= \phi_1(\phi_1 Z_n + \theta_0) + \theta_0 = \phi_1^2 Z_n + \theta_0(1 + \phi_1). \end{aligned}$$

In general, the ξ-period ahead forecast is

$$\hat{Z}_n(\xi) = \phi_1^{\xi} Z_n + \theta_0(1 + \phi_1 + \cdots + \phi_1^{\xi-1}).$$

Forecasts for other ARIMA models could be obtained in a similar fashion.

The standard deviation of the forecast error can be used to construct $100(1 - \alpha)$ percent confidence intervals around the point forecasts. Most computer packages have an option that plots the point forecasts and the corresponding confidence intervals.

Pankratz (1983) presents a collection of 32 practical rules for building Box-Jenkins models. These rules provide a convenient summary of the strategy necessary for successful modeling.

19.6 EXAMPLES

All computations were performed using the SCA (Scientific Computing Associates) statistical system.

19.6.1 Nuclear Plant Capacity

A capacity factor for an electric generating station is defined as the net electric energy generated ($\times$ 100) divided by the product of the hours in the period and the maximum dependable capacity of the plant. Capacity factors are usually computed on a monthly basis and range from 0 (the plant is shut down) to 100 (the plant operates at full capacity for the month). Previous studies have used capacity factors in regression analysis with independent parameters such as plant size, age, and type. The data set analyzed was the monthly capacity factors of a particular nuclear station from December 1969 through June 1979 (115 observations). The actual values are presented in Table 19.1.

A plot of the data indicates that differencing is unnecessary. This is reinforced by the behavior of the SACF in Fig. 19.2. Furthermore, since the t value of the mean is highly significant, a θ_0 term should be included in the model. The SPACF is presented in Fig. 19.3. Since the SACF cuts off at lag 1 and the SPACF tails off, an MA(1) model was tentatively specified. The estimation results showed that $\hat{\theta}_1 = -0.5005$ with a t value of -6.21 and $\hat{\theta}_0 = 66.7058$ with a t value of 16.13. Unfortunately the SACF of the residuals was significant at lag 3. As a result, a new model was specified with moving-average parameters at lags 1 and 3,

$$Z_t = \theta_0 + a_t - \theta_1 a_{t-1} - \theta_3 a_{t-3}$$

The estimation results and the SACF of the residuals are shown in Fig. 19.4. Although there is still a significant residual autocorrelation at lag 8 with a t value of $-0.19/0.10 = -1.9$, it is not too surprising to find one significant autocorrelation out of 36. Since this significant autocorrelation does not occur at a short (1, 2, 3) lag, a seasonal (12, 24) lag, or a near-seasonal (11, 13) lag, there is no reason to question the adequacy of the fitted model. This is supported by the EACF (see Fig. 19.4), which indicates that the maximum orders of p and q for the residuals are 0 and 0, respectively. Thus the fitted model has $\hat{\theta}_0 = 66.8179$, $\hat{\theta}_1 = -0.5314$, and $\hat{\theta}_3 = 0.1719$.

These data were also analyzed automatically using the PC version of AUTOBOX, licensed by Automatic Forecasting Systems, Inc. AUTOBOX specified a pure moving-average model with parameters at lags 1 and 3. The estimation results were $\hat{\theta}_1 = -0.5364$ and $\hat{\theta}_3 = 0.1779$.

19.6.2 Retail Sales of Jewelry Stores

The purpose of this example is to identify an appropriate model for monthly retail sales of jewelry stores in the United States for the period from January 1980 to June 1984. The data were collected from "Revised Monthly Retail Sales and Inventories," published by Bureau of the Census, U.S. Department of Commerce, in April 1985, and shown in Table 19.2. The AUTOBOX plot of the scaled data, obtained by dividing by 1000, is shown in Fig. 19.5. There is definitely seasonal variation, with December sales considerably higher than those in other months.

TABLE 19.1 Nuclear Power Plant Capacity Factors, December 1969 through June 1979

Z is a 115 × 1 variable	
10.00000	95.00000
8.00000	51.00000
51.00000	99.00000
80.00000	99.00000
12.00000	99.00000
51.00000	29.00000
19.00000	0.00000
83.00000	10.00000
94.00000	56.00000
86.00000	97.00000
49.00000	99.00000
83.00000	99.00000
87.00000	94.00000
82.00000	99.00000
95.00000	94.00000
2.00000	47.00000
0.00000	0.00000
33.00000	0.00000
87.00000	22.00000
90.00000	59.00000
97.00000	99.00000
91.00000	98.00000
98.00000	12.00000
95.00000	59.00000
97.00000	25.00000
96.00000	88.00000
95.00000	83.00000
79.00000	87.00000
65.00000	88.00000
0.00000	88.00000
0.00000	43.00000
84.00000	14.00000
89.00000	88.00000
91.00000	62.00000
33.00000	76.00000
37.00000	87.00000
90.00000	87.00000
88.00000	76.00000
94.00000	85.00000
94.00000	68.00000
95.00000	77.00000
95.00000	68.00000
94.00000	0.00000
62.00000	20.00000
87.00000	99.00000
90.00000	99.00000
75.00000	97.00000
93.00000	99.00000
85.00000	99.00000
0.00000	99.00000
0.00000	98.00000
0.00000	99.00000
6.00000	28.00000
68.00000	0.00000
54.00000	80.00000
63.00000	99.00000
80.00000	99.00000
94.00000	

```
TIME PERIOD ANALYZED . . . . . . . . . . .   1  TO    115
NAME OF THE SERIES . . . . . . . . . . . .              Z
EFFECTIVE NUMBER OF OBSERVATIONS . . .                115
STANDARD DEVIATION OF THE SERIES . . .            34.5728

MEAN OF THE (DIFFERENCED) SERIES . . .            66.7565
STANDARD DEVIATION OF THE MEAN . . . .             3.2239
T-VALUE OF MEAN (AGAINST ZERO) . . . .            20.7065
```

```
AUTOCORRELATIONS

 1- 12       .50  .07 -.18 -.17 -.11 -.10 -.20 -.21 -.05 -.00  .07  .04
 ST.E.       .09  .11  .11  .12  .12  .12  .12  .12  .13  .13  .13  .13
  Q         29.9 30.4 34.2 37.7 39.1 40.4 45.4 50.7 51.0 51.0 51.7 51.9

13- 24       .04  .11  .14  .02 -.06 -.11 -.05 -.04 -.03 -.01  .05  .12
 ST.E.       .13  .13  .13  .13  .13  .13  .13  .13  .13  .13  .13  .13
  Q         52.1 53.8 56.4 56.5 57.0 58.8 59.2 59.4 59.5 59.5 59.9 62.0

25- 36       .12  .01 -.14 -.09 -.02  .02 -.09 -.21 -.08  .10  .17  .16
 ST.E.       .13  .13  .13  .13  .13  .13  .13  .14  .14  .14  .14  .14
  Q         64.2 64.2 67.1 68.3 68.4 68.4 69.8 76.9 77.8 79.3 84.2 88.8
```

```
             -1.0 -0.8 -0.6 -0.4 -0.2  0.0  0.2  0.4  0.6  0.8  1.0
               +----+----+----+----+----+----+----+----+----+----+
                                        I
 1   0.50                          +    IXXXX+XXXXXXXX
 2   0.07                         +     IXX   +
 3  -0.18                         + XXXXI     +
 4  -0.17                         + XXXXI     +
 5  -0.11                         +  XXXI     +
 6  -0.10                         +  XXXI     +
 7  -0.20                         +XXXXXI     +
 8  -0.21                         +XXXXXI     +
 9  -0.05                         +    XI     +
10  -0.00                         +     I     +
11   0.07                         +     IXX   +
12   0.04                         +     IX    +
13   0.04                         +     IX    +
14   0.11                         +     IXXX  +
15   0.14                         +     IXXXX +
16   0.02                         +     IX    +
17  -0.06                         +   XXI     +
18  -0.11                         +  XXXI     +
19  -0.05                         +    XI     +
20  -0.04                         +    XI     +
21  -0.03                         +    XI     +
22  -0.01                         +     I     +
23   0.05                         +     IX    +
24   0.12                         +     IXXX  +
25   0.12                         +     IXXX  +
26   0.01                        +      I      +
27  -0.14                        +   XXXI      +
28  -0.09                        +    XXI      +
29  -0.02                        +     XI      +
30   0.02                        +      I      +
31  -0.09                        +    XXI      +
32  -0.21                        +  XXXXXI     +
33  -0.08                        +    XXI      +
34   0.10                        +      IXX    +
35   0.17                        +      IXXXX  +
```

FIGURE 19.2 SACF for nuclear power plant capacity.

```
PARTIAL AUTOCORRELATIONS

  1- 12     .50 -.25 -.13  .02 -.05 -.11 -.19 -.05  .08 -.15  .06 -.05
  ST.E.     .09  .09  .09  .09  .09  .09  .09  .09  .09  .09  .09  .09

 13- 24     .00  .10 -.01 -.09  .02 -.06  .08 -.12  .04  .07  .01  .07
  ST.E.     .09  .09  .09  .09  .09  .09  .09  .09  .09  .09  .09  .09

 25- 36     .02 -.11 -.05  .05 -.00 -.03 -.14 -.09  .15 -.03  .03  .07
  ST.E.     .09  .09  .09  .09  .09  .09  .09  .09  .09  .09  .09  .09

                -1.0 -0.8 -0.6 -0.4 -0.2  0.0  0.2  0.4  0.6  0.8  1.0
                  +----+----+----+----+----+----+----+----+----+----+
                                           I
 1   0.50                             +    IXXXX+XXXXXXXX
 2  -0.25                            X+XXXXI    +
 3  -0.13                             + XXXI    +
 4   0.02                             +    I    +
 5  -0.05                             +   XI    +
 6  -0.11                             + XXXI    +
 7  -0.19                             XXXXXI    +
 8  -0.05                             +   XI    +
 9   0.08                             +    IXX  +
10  -0.15                             +XXXXI    +
11   0.06                             +    IXX  +
12  -0.05                             +   XI    +
13   0.00                             +    I    +
14   0.10                             +    IXXX +
15  -0.01                             +    I    +
16  -0.09                             +  XXI    +
17   0.02                             +    I    +
18  -0.06                             +  XXI    +
19   0.08                             +    IXX  +
20  -0.12                             + XXXI    +
21   0.04                             +    IX   +
22   0.07                             +    IXX  +
23   0.01                             +    I    +
24   0.07                             +    IXX  +
25   0.02                             +    IX   +
26  -0.11                             + XXXI    +
27  -0.05                             +   XI    +
```

FIGURE 19.3 SPACF for nuclear power plant capacity.

The seasonal variation is further demonstrated in the SACF (Fig. 19.6) of sales with $r_{12} = 0.71$, $r_{24} = 0.50$, and $r_{36} = 0.28$. The SPACF (Fig. 19.7) shows that only $\hat{\phi}_{12,12} = 0.6650$ is significant. A determination needs to be made of whether the tailing off of the SACF and the cutting off of the SPACF at lag 12 are indicative of the need for a seasonal autoregressive factor or seasonal differencing. When a (0, 0, 0) $\times$ $(1, 0, 0)_{12}$ model was fit, $|\Phi_1| > 1$, which violates the stationarity constraint. As an alternative, seasonal differencing was employed. The SACF and the SPACF of the differenced series W_t are shown in Figs. 19.8 and 19.9, respectively. Since the SACF tails off and the SPACF cuts off after lag 2, an AR(2) model was fit to the nonseasonal component of the differenced data. Furthermore, a seasonal moving average at lag 12 was used even though it is the SPACF at lag 12 that has a t ratio (absolute value) of 2.47. Thus the tentative model is $(2, 0, 0) \times (0, 1, 1)_{12}$, or

$$(1 - \phi_1 B - \phi_2 B^2)(1 - B^{12})Z_t = \theta_0 + (1 - \Theta_{12} B^{12})a_t.$$

VARIABLE	TYPE OF VARIABLE	ORIGINAL OR CENTERED	DIFFERENCING
Z	RANDOM	ORIGINAL	NONE

PARAMETER LABEL		VARIABLE NAME	NUM./ DENOM.	FACTOR	ORDER	CONS- TRAINT	VALUE	STD ERROR	T VALUE
1	CNST		CNST	1	0	NONE	66.8179	3.7104	18.01
2	THETA1	Z	MA	1	1	NONE	-.5314	.0830	-6.41
3	THETA3	Z	MA	1	3	NONE	.1719	.0826	2.08

AUTOCORRELATIONS

1- 12	.08	.13	-.05	-.15	-.06	-.05	-.15	-.19	.06	-.09	.09	.04
ST.E.	.09	.09	.10	.10	.10	.10	.10	.10	.10	.10	.10	.10
Q	.7	2.6	2.9	5.5	5.9	6.2	8.9	13.4	13.9	15.0	16.0	16.2
13- 24	-.01	.10	.10	-.03	.01	-.12	.01	-.03	-.02	.00	.04	.05
ST.E.	.10	.10	.11	.11	.11	.11	.11	.11	.11	.11	.11	.11
Q	16.3	17.5	18.9	19.0	19.1	21.2	21.2	21.4	21.5	21.5	21.7	22.1
25- 36	.09	.03	-.15	.01	-.07	.01	-.01	-.20	-.02	.09	.06	.10
ST.E.	.11	.11	.11	.11	.11	.11	.11	.11	.11	.11	.12	.12
Q	23.3	23.5	27.1	27.1	27.9	27.9	27.9	34.6	34.6	35.8	36.5	38.2

SIMPLIFIED EXTENDED ACF TABLE (5% LEVEL)

(Q-->)	0	1	2	3	4	5	6	7	8	9	10	11	12
(P= 0)	O	O	O	O	O	O	O	O	O	O	O	O	O
(P= 1)	X	O	O	O	O	O	O	O	O	O	O	O	O
(P= 2)	X	X	O	O	O	O	O	O	O	O	O	O	O
(P= 3)	X	X	O	O	O	O	O	O	O	O	O	O	O
(P= 4)	O	O	X	X	O	O	O	O	O	O	O	O	O
(P= 5)	O	O	O	O	O	O	O	O	O	O	O	O	O
(P= 6)	O	O	O	X	O	O	O	O	O	O	O	O	O

FIGURE 19.4 Final estimation results, SACF of residuals, and EACF of residuals for nuclear power plant capacity.

TABLE 19.2 Monthly Retail Sales of Jewelry Stores, January 1980 through June 1984

Z is a 54×1 variable
479.00000
515.00000
521.00000
491.00000
583.00000
533.00000
492.00000
530.00000
526.00000
610.00000
769.00000
1992.00000

TABLE 19.2 (*Continued*)

Z is a 54×1 variable
514.00000
544.00000
539.00000
602.00000
717.00000
663.00000
613.00000
618.00000
583.00000
634.00000
766.00000
1946.00000
487.00000
528.00000
555.00000
618.00000
690.00000
626.00000
554.00000
558.00000
542.00000
597.00000
765.00000
1871.00000
526.00000
549.00000
583.00000
627.00000
731.00000
651.00000
591.00000
599.00000
624.00000
656.00000
855.00000
2163.00000
568.00000
628.00000
634.00000
663.00000
806.00000
743.00000

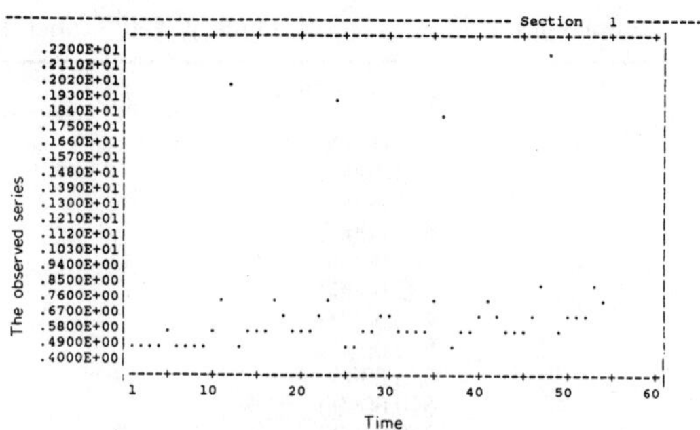

FIGURE 19.5 Plot of monthly retail sales of jewelry stores, January 1980 through June 1984.

```
TIME PERIOD ANALYZED . . . . . . . . . .   1  TO    54
NAME OF THE SERIES . . . . . . . . . . .             Z
EFFECTIVE NUMBER OF OBSERVATIONS . . .              54
STANDARD DEVIATION OF THE SERIES . . .        373.5747
MEAN OF THE (DIFFERENCED) SERIES . . .        710.5183
STANDARD DEVIATION OF THE MEAN . . . .         50.8371
T-VALUE OF MEAN (AGAINST ZERO) . . . .         13.9764


AUTOCORRELATIONS

  1- 12     .02  -.10  -.13  -.09  -.01   .00  -.01  -.10  -.14  -.12  -.04   .71
  ST.E.     .14   .14   .14   .14   .14   .14   .14   .14   .14   .14   .15   .15
   Q         .0    .6   1.6   2.1   2.2   2.2   2.2   2.8   4.0   5.1   5.2  41.0

 13- 24    -.01  -.09  -.11  -.08  -.02  -.01  -.02  -.08  -.10  -.09  -.03   .50
  ST.E.     .20   .20   .20   .20   .20   .20   .20   .20   .20   .20   .20   .20
   Q       41.0  41.6  42.6  43.1  43.1  43.1  43.2  43.8  44.8  45.6  45.7  70.6

 25- 36     .01  -.04  -.06  -.04   .00   .00  -.01  -.05  -.06  -.06  -.02   .28
  ST.E.     .23   .23   .23   .23   .23   .23   .23   .23   .23   .23   .23   .23
   Q       70.6  70.8  71.3  71.5  71.5  71.5  71.5  71.9  72.5  72.9  73.0  86.3

              -1.0 -0.8 -0.6 -0.4 -0.2  0.0  0.2  0.4  0.6  0.8  1.0
                +----+----+----+----+----+----+----+----+----+----+
                                         I
 1   0.02                         +      I      +
 2  -0.10                         +    XXI      +
 3  -0.13                         +   XXXI      +
 4  -0.09                         +    XXI      +
 5  -0.01                         +      I      +
 6   0.00                         +      I      +
 7  -0.01                         +      I      +
 8  -0.10                         +    XXI      +
 9  -0.14                         +   XXXI      +
10  -0.12                         +   XXXI      +
11  -0.04                         +     XI      +
12   0.71                         +      IXXXXXX+XXXXXXXXXXX
13  -0.01                      +         I         +
14  -0.09                      +       XXI         +
15  -0.11                      +      XXXI         +
16  -0.08                      +       XXI         +
17  -0.02                      +         I         +
18  -0.01                      +         I         +
19  -0.02                      +        XI         +
20  -0.08                      +       XXI         +
21  -0.10                      +      XXXI         +
22  -0.09                      +       XXI         +
23  -0.03                      +        XI         +
24   0.50                      +         IXXXXXXXXX+XX
25   0.01                     +          I          +
26  -0.04                     +         XI          +
27  -0.06                     +        XXI          +
28  -0.04                     +         XI          +
29   0.00                     +          I          +
30   0.00                     +          I          +
31  -0.01                     +          I          +
32  -0.05                     +         XI          +
33  -0.06                     +        XXI          +
34  -0.06                     +         XI          +
35  -0.02                     +         XI          +
36   0.28                     +          IXXXXXXX   +
```

FIGURE 19.6 SACF for monthly retail sales of jewelry stores.

```
PARTIAL AUTOCORRELATIONS

  1- 12    .02 -.10 -.13 -.10 -.03 -.04 -.05 -.12 -.16 -.18 -.15  .66
  ST.E.    .14  .14  .14  .14  .14  .14  .14  .14  .14  .14  .14  .14

 13- 24   -.13 -.04 -.02 -.02 -.06 -.06 -.06 -.02 -.01  .01  .02 -.05
  ST.E.    .14  .14  .14  .14  .14  .14  .14  .14  .14  .14  .14  .14

 25- 36    .02  .02  .02  .00  .02  .01  .02  .02  .03  .03  .01 -.12
  ST.E.    .14  .14  .14  .14  .14  .14  .14  .14  .14  .14  .14  .14

           -1.0 -0.8 -0.6 -0.4 -0.2  0.0  0.2  0.4  0.6  0.8  1.0
             +----+----+----+----+----+----+----+----+----+----+
                                      I
 1   0.02                     +      I      +
 2  -0.10                     +    XXI      +
 3  -0.13                     +   XXXI      +
 4  -0.10                     +   XXXI      +
 5  -0.03                     +     XI      +
 6  -0.04                     +     XI      +
 7  -0.05                     +     XI      +
 8  -0.12                     +   XXXI      +
 9  -0.16                     +  XXXXI      +
10  -0.18                     + XXXXXI      +
11  -0.15                     +  XXXXI      +
12   0.66                     +      IXXXXXX+XXXXXXXXXX
13  -0.13                     +   XXXI      +
14  -0.04                     +     XI      +
15  -0.02                     +      I      +
16  -0.02                     +     XI      +
17  -0.06                     +    XXI      +
18  -0.06                     +     XI      +
19  -0.06                     +     XI      +
20  -0.02                     +     XI      +
21  -0.01                     +      I      +
22   0.01                     +      I      +
23   0.02                     +      IX     +
24  -0.05                     +     XI      +
25   0.02                     +      IX     +
26   0.02                     +      IX     +
27   0.02                     +      I      +
28   0.00                     +      I      +
29   0.02                     +      I      +
30   0.01                     +      I      +
31   0.02                     +      IX     +
32   0.02                     +      I      +
33   0.03                     +      IX     +
34   0.03                     +      IX     +
35   0.01                     +      I      +
36  -0.12                     +   XXXI      +
```

FIGURE 19.7 SPACF for monthly retail sales of jewelry stores.

Since the t value for ϕ_2 is 0.73, this term was dropped from the model, and a $(1, 0, 0) \times (0, 1, 1)_{12}$ model with a constant term was specified. The estimation results are shown in Fig. 19.10 along with the autocorrelations and extended autocorrelations of the residuals. Thus the model fitted to the seasonally differenced data has $\hat{\theta}_0 = 12.55$, $\hat{\Theta}_{12} = 0.84$, and $\hat{\phi}_1 = 0.63$. Since none of the residual autocorrelations are significant and the triangle of insignificant values for the ESACF has a vertex at $p = 0$, $q = 0$, the $(1, 0, 0) \times (0, 1, 1)_{12}$ model sufficiently explains the stochastic behavior of the series and can be used for forecasting. The forecasts generated from this model, beginning with September 1983 (month 45), are shown in Fig. 19.11. For months 45

```
                DIFFERENCE ORDERS ARE  (1-B  )


TIME PERIOD ANALYZED . . . . . . . . .  13  TO    54
NAME OF THE SERIES . . . . . . . . . .           SIFZ
EFFECTIVE NUMBER OF OBSERVATIONS . . .             42
STANDARD DEVIATION OF THE SERIES . . .        65.8657
MEAN OF THE (DIFFERENCED) SERIES . . .        35.4524
STANDARD DEVIATION OF THE MEAN . . . .        10.1633
T-VALUE OF MEAN (AGAINST ZERO) . . . .         3.4883


AUTOCORRELATIONS

  1- 12      .57  .50  .39  .23  .18  .16  .08  .01  .01 -.03 -.09 -.29
 ST.E.       .15  .20  .23  .24  .25  .25  .25  .25  .25  .25  .25  .26
   Q        14.8 26.6 33.8 36.4 38.1 39.5 39.8 39.8 39.8 39.9 40.4 45.6

 13- 24     -.27 -.36 -.36 -.36 -.34 -.30 -.25 -.18 -.15 -.19 -.19 -.18
 ST.E.       .26  .27  .28  .29  .30  .31  .32  .32  .32  .33  .33  .33
   Q        50.2 58.9 67.7 76.7 85.4 92.3 97.2 99.9  102  105  109  112

 25- 36     -.08 -.02  .05  .10  .15  .15  .17  .13  .03  .05  .05  .05
 ST.E.       .33  .33  .33  .33  .34  .34  .34  .34  .34  .34  .34  .34
   Q         113  113  113  115  118  122  127  130  130  131  131  132

           -1.0 -0.8 -0.6 -0.4 -0.2  0.0  0.2  0.4  0.6  0.8  1.0
             +----+----+----+----+----+----+----+----+----+----+
                                      I
  1    0.57                  +       IXXXXXXX+XXXXXX
  2    0.50                +         IXXXXXXXXX+XXX
  3    0.39               +          IXXXXXXXXXX+
  4    0.23              +           IXXXXXX     +
  5    0.18              +           IXXXXX      +
  6    0.16              +           IXXXX       +
  7    0.08              +           IXX         +
  8    0.01              +           I           +
  9    0.01              +           I           +
 10   -0.03              +          XI           +
 11   -0.09              +         XXI           +
 12   -0.29              +    XXXXXXXI           +
 13   -0.27             +     XXXXXXXI            +
 14   -0.36             +   XXXXXXXXXI            +
 15   -0.36            +    XXXXXXXXXI             +
 16   -0.36            +    XXXXXXXXXI             +
 17   -0.34           +     XXXXXXXXXI              +
 18   -0.30           +       XXXXXXXI              +
 19   -0.25          +         XXXXXXI               +
 20   -0.18          +          XXXXXI               +
 21   -0.15          +           XXXXI               +
 22   -0.19          +          XXXXXI               +
 23   -0.19          +          XXXXXI               +
 24   -0.18          +          XXXXXI               +
 25   -0.08          +             XXI               +
 26   -0.02          +               I               +
 27    0.05          +               IX              +
 28    0.10          +               IXXX            +
 29    0.15          +               IXXXX           +
 30    0.15         +                IXXXX            +
 31    0.17         +                IXXXX            +
 32    0.13         +                IXXX             +
 33    0.03         +                IX               +
 34    0.05         +                IX               +
 35    0.05         +                IX               +
 36    0.05         +                IX               +
```

FIGURE 19.8 SACF of seasonal differences for monthly retail sales of jewelry stores.

```
PARTIAL AUTOCORRELATIONS

  1- 12    .57  .26  .04 -.12  .00  .07 -.06 -.10  .02  .00 -.10 -.37
  ST.E.    .15  .15  .15  .15  .15  .15  .15  .15  .15  .15  .15  .15

 13- 24    .00 -.06 -.04 -.13 -.02  .07  .01 -.03  .01 -.10 -.07 -.15
  ST.E.    .15  .15  .15  .15  .15  .15  .15  .15  .15  .15  .15  .15

 25- 36    .16  .03  .00 -.06  .04 -.04 -.05 -.07 -.13 -.02  .01 -.11
  ST.E.    .15  .15  .15  .15  .15  .15  .15  .15  .15  .15  .15  .15

              -1.0 -0.8 -0.6 -0.4 -0.2  0.0  0.2  0.4  0.6  0.8  1.0
                +----+----+----+----+----+----+----+----+----+----+
                                         I
  1   0.57                       +       IXXXXXXX+XXXXXX
  2   0.26                       +       IXXXXXXX+
  3   0.04                       +       IX      +
  4  -0.12                       +    XXXI       +
  5   0.00                       +       I       +
  6   0.07                       +       IXX     +
  7  -0.06                       +      XI       +
  8  -0.10                       +     XXI       +
  9   0.02                       +       IX      +
 10   0.00                       +       I       +
 11  -0.10                       +     XXI       +
 12  -0.37                      X+XXXXXXXI       +
 13   0.00                       +       I       +
 14  -0.06                       +      XI       +
 15  -0.04                       +      XI       +
 16  -0.13                       +    XXXI       +
 17  -0.02                       +      XI       +
 18   0.07                       +       IXX     +
 19   0.01                       +       I       +
 20  -0.03                       +      XI       +
 21   0.01                       +       I       +
 22  -0.10                       +    XXXI       +
 23  -0.07                       +     XXI       +
 24  -0.15                       +   XXXXI       +
 25   0.16                       +       IXXXX   +
 26   0.03                       +       IX      +
 27   0.00                       +       I       +
 28  -0.06                       +      XI       +
 29   0.04                       +       IX      +
 30  -0.04                       +      XI       +
 31  -0.05                       +      XI       +
 32  -0.07                       +     XXI       +
 33  -0.13                       +    XXXI       +
 34  -0.02                       +       I       +
 35   0.01                       +       I       +
 36  -0.11                       +    XXXI       +
```

FIGURE 19.9 SPACF of seasond differences for monthly retail sales of jewelry stores.

through 54, where the actual values of sales (denoted by D) are known, only the value for month 48 exceeds the 95 percent confidence limits (denoted by A and C).

19.6.3 Sales of Alcoholic Beverages

A basic task in marketing is to forecast the demand for a product. One level of forecasting is concerned with the total market demand or industry sales. In this

VARIABLE	TYPE OF VARIABLE	ORIGINAL OR CENTERED	DIFFERENCING
Z	RANDOM	ORIGINAL	$(1-B^{12})$

PARAMETER LABEL	VARIABLE NAME	NUM./ DENOM.	FACTOR	ORDER	CONS-TRAINT	VALUE	STD ERROR	T VALUE
1 CNST		CNST	1	0	NONE	12.5538	5.0488	2.49
2 THETA12	Z	MA	1	12	NONE	.8364	.1362	6.14
3 PHI1	Z	AR	1	1	NONE	.6307	.1272	4.96

AUTOCORRELATIONS

1- 12	-.06	.14	.08	-.19	-.04	-.10	-.04	-.05	-.02	.12	.17	-.03
ST.E.	.16	.16	.16	.16	.17	.17	.17	.17	.17	.17	.17	.17
Q	.2	1.1	1.4	3.1	3.2	3.7	3.8	3.9	4.0	4.7	6.4	6.4
13- 24	.17	-.09	-.06	-.15	-.17	-.15	-.10	-.01	-.03	-.02	.01	-.02
ST.E.	.17	.18	.18	.18	.18	.19	.19	.19	.19	.19	.19	.19
Q	8.3	8.9	9.2	10.7	12.8	14.5	15.3	15.3	15.4	15.4	15.4	15.4
25- 36	.06	-.03	-.03	-.03	.00	-.05	.05	.05	-.11	.05	.04	.07
ST.E.	.19	.19	.19	.19	.19	.19	.19	.19	.19	.19	.20	.20
Q	15.8	15.9	16.0	16.2	16.2	16.6	17.0	17.4	20.2	20.8	21.2	22.8

SIMPLIFIED EXTENDED ACF TABLE (5% LEVEL)

(Q-->)	0	1	2	3	4	5	6	7	8	9	10	11	12
(P= 0)	0	0	0	0	0	0	0	0	0	0	0	0	0
(P= 1)	X	0	0	0	0	0	0	0	0	0	0	0	0
(P= 2)	X	X	0	0	0	0	0	0	0	0	0	0	0
(P= 3)	X	X	0	0	0	0	0	0	0	0	0	0	0
(P= 4)	X	0	0	0	0	0	0	0	0	0	0	0	0
(P= 5)	X	0	0	0	0	0	0	0	0	0	0	0	0
(P= 6)	0	0	0	0	0	0	0	0	0	0	0	0	0

FIGURE 19.10 Estimation results, SACF of residuals, and EACF of residuals for monthly retail sales of jewelry stores.

example, the quarterly sales (in millions of dollars) of alcoholic beverages in the United States are analyzed for the time period from the first quarter of 1977 through the third quarter of 1988 for a total of 47 observations. The data were obtained from Standard and Poor's "Analysts Handbook." The observations are presented in Table 19.3. The corresponding plot (Fig. 19.12) shows the presence of trend and increasing variation. As a result, a logarithmic transformation was applied to the original data, and then a first difference was used on the transformed data. Since the SACF of $(1-B)(\ln Z_t)$ exhibits a tailing off behavior at the even lags (2, 4, 6, 8, ...), a seasonal difference was also used. Thus $W_t = (1-B)(1-B^4)(\ln Z_t)$. The SACF and the SPACF of the W_t are presented in Figs. 19.13 and 19.14, respec-

```
-----------------------------------
 15 FORECASTS, BEGINNING AT    44
-----------------------------------

 TIME      FORECAST    STD. ERROR    ACTUAL IF KNOWN

   45      594.1496       44.6298       624.0000
   46      665.6641       52.7654       656.0000
   47      824.3173       55.6723       855.0000
   48     1995.5224       56.7873      2163.0000
   49      572.5682       57.2248       568.0000
   50      609.3217       57.3980       628.0000
   51      628.6187       57.4667       634.0000
   52      666.6601       57.4940       663.0000
   53      763.5286       57.5049       806.0000
   54      701.4367       57.5092       743.0000
   55      645.6216       57.5109
   56      659.1102       57.5116
   57      644.6163       57.9959
   58      710.0485       58.1875
   59      864.8653       58.2635
--

UL=F+2*S.

--

LL=F-2*S.

--

SERIES     UL     IS REPRESENTED BY A
SERIES     F      IS REPRESENTED BY B
SERIES     LL     IS REPRESENTED BY C
SERIES     TZ     IS REPRESENTED BY D
```

FIGURE 19.11 Forecasted values of jewelry store sales, September 1983 through November 1984.

tively. A significant autocorrelation was found at lag 4, and a $(0, 1, 0) \times (0, 1, 1)_4$ model with a constant was tentatively identified. The constant was not significant, and thus this term was deleted and a revised estimate obtained for Θ_4. The estimation results for the final model are shown in Fig. 19.15. Thus the estimated value is $\Theta_4 = 0.6796$ with a t value of 6.02. The residual analysis shown in Fig. 19.15 clearly indicated that the $(0, 1, 0) \times (0, 1, 1)_4$ model applied to $\ln Z_t$ describes the behavior of the series.

19.7 EXPONENTIAL SMOOTHING

Suppose that the mean of the process is constant over time or, at the very worst, changing very slowly. Let $\hat{Z}_t(1)$ denote the forecast for Z_{t+1} that is made at time t.

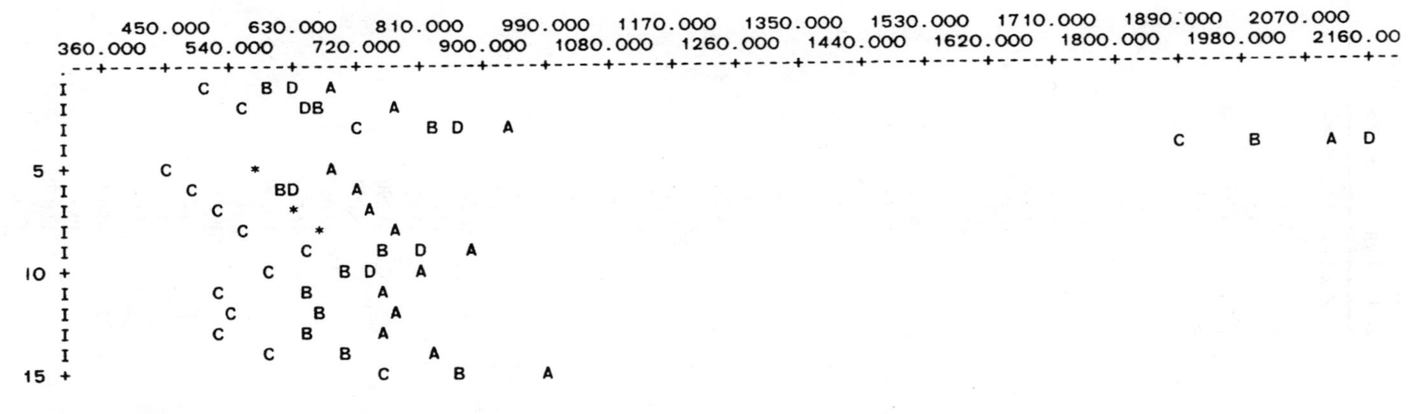

FIGURE 19.11 (*Continued*)

TABLE 19.3 Quarterly Sales (Millions of Dollars) of Alcoholic Beverages in the United States, 1st Quarter of 1977 through Third Quarter of 1988

Z is a 47 × 1 variable
16.110
20.540
20.290
15.930
16.530
21.680
23.650
18.680
20.240
24.860
25.670
19.510
23.460
28.240
29.850
24.100
25.760
31.360
31.690
27.090
28.080
29.030
34.860
30.490
36.980
45.070
46.470
40.000
39.440
46.580
49.640
42.020
43.290
48.090
49.050
42.550
43.920
51.810
49.880
45.490
47.630
56.750
58.680
46.470
42.970
52.620
55.540

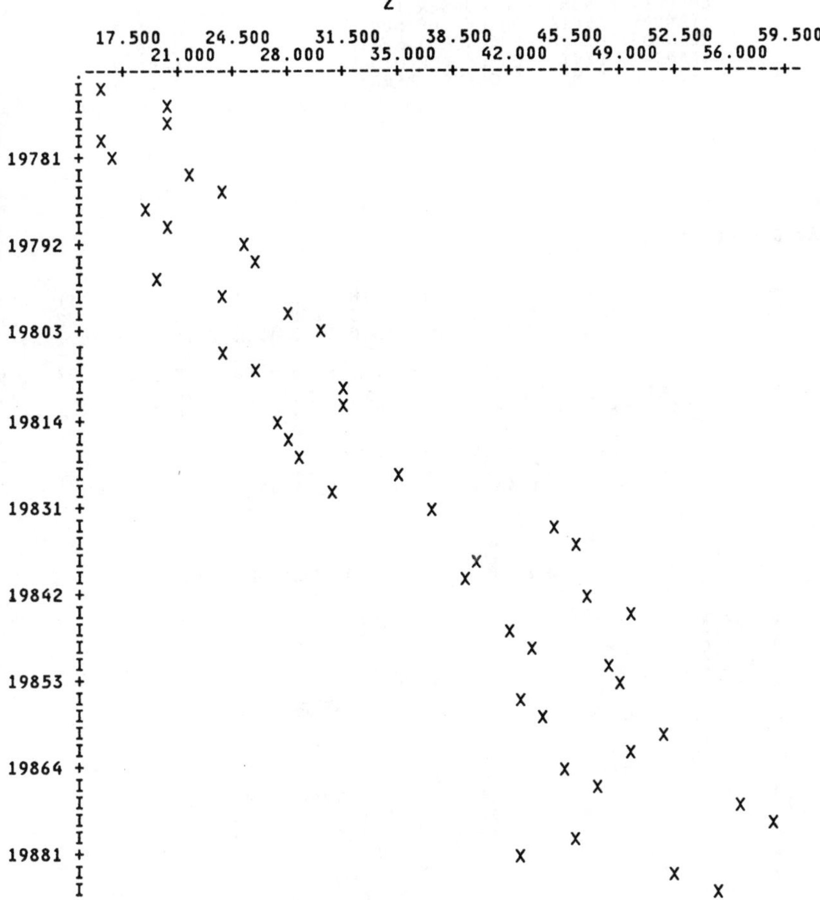

FIGURE 19.12 Plot of quarterly sales (millions of dollars) of alcoholic beverages in the United States, first quarter of 1977 through third quarter of 1988.

In *simple exponential smoothing,* $\hat{Z}_t(1)$ is a weighted sum of the current actual value Z_t and the forecast for time t, $\hat{Z}_{t-1}(1)$, that is made at time $t - 1$. Specifically,

$$\hat{Z}_t(1) = \alpha Z_t + (1 - \alpha)\hat{Z}_{t-1}(1) \tag{19.44}$$

Note that the weights in Eq. (19.44) add to 1. The fraction α is referred to as the *smoothing constant* where $0 \le \alpha \le 1$. Equivalently, Eq. (19.44) can be expressed as the forecast value of the previous period, $\hat{Z}_{t-1}(1)$, plus a fraction of the forecast error of the previous period,

$$\hat{Z}_t(1) = \hat{Z}_{t-1}(1) + \alpha(Z_t - \hat{Z}_{t-1}(1)) \tag{19.45}$$

```
DIFFERENCE ORDERS. . . . . . . . . . . . . (1-B  ) (1-B  )
TIME PERIOD ANALYZED . . . . . . . . . . .  1  TO     47
NAME OF THE SERIES . . . . . . . . . . . .             Z
EFFECTIVE NUMBER OF OBSERVATIONS . . .                42
STANDARD DEVIATION OF THE SERIES . . .            0.0837
MEAN OF THE (DIFFERENCED) SERIES . . .           -0.0019
STANDARD DEVIATION OF THE MEAN . . . .            0.0129
T-VALUE OF MEAN (AGAINST ZERO) . . . .           -0.1489
```

AUTOCORRELATIONS

```
 1- 12   -.06  .01 -.12 -.48  .22 -.18  .15 -.00 -.24  .28 -.11  .13
 ST.E.    .15  .15  .15  .16  .19  .19  .20  .20  .20  .21  .22  .22
  Q        .2   .2   .8 11.9 14.4 16.0 17.3 17.3 20.5 24.9 25.7 26.7

13- 24   -.03 -.21  .09 -.01  .29  .11 -.10 -.06 -.19 -.03  .07  .00
 ST.E.    .22  .22  .22  .23  .23  .23  .24  .24  .24  .24  .24  .24
  Q      26.8 29.6 30.2 30.2 36.5 37.4 38.2 38.5 41.8 41.9 42.4 42.4

25- 36    .01 -.04  .07 -.00 -.03  .00 -.02 -.05  .05  .08  .00 -.01
 ST.E.    .24  .24  .24  .24  .24  .24  .24  .24  .24  .24  .24  .24
  Q      42.4 42.6 43.1 43.1 43.3 43.3 43.4 43.9 44.3 45.8 45.8 45.8
```

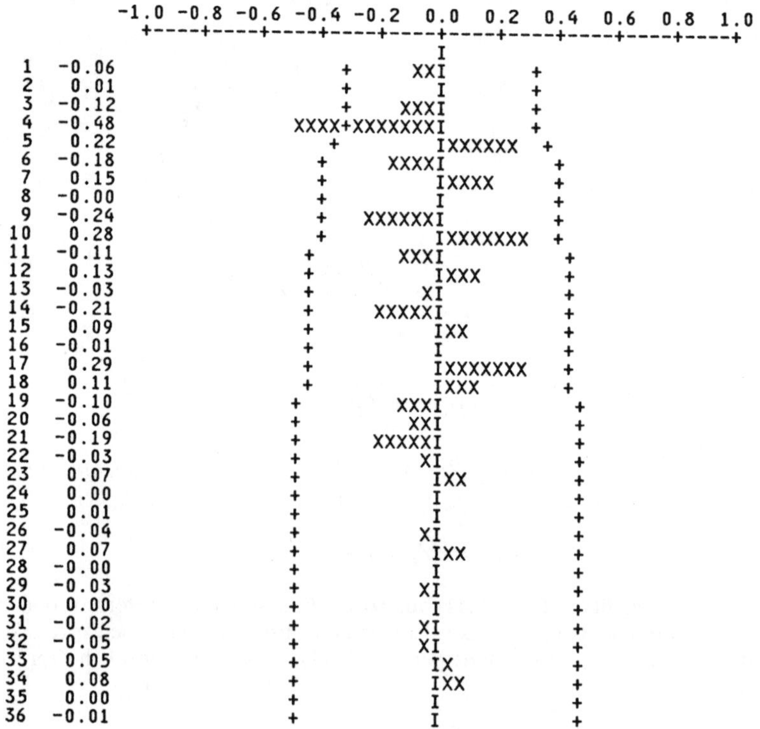

FIGURE 19.13 SACF of differenced data (seasonal and nonseasonal) after employing a logarithmic transformation.

PARTIAL AUTOCORRELATIONS

1- 12	-.06	.01	-.12	-.50	.19	-.22	.02	-.26	-.14	.09	-.01	-.10
ST.E.	.15	.15	.15	.15	.15	.15	.15	.15	.15	.15	.15	.15
13- 24	-.16	-.05	-.08	.13	.14	.11	.02	.03	.11	-.06	.11	-.03
ST.E.	.15	.15	.15	.15	.15	.15	.15	.15	.15	.15	.15	.15
25- 36	.02	-.02	.12	-.15	-.10	-.04	.07	-.17	-.12	-.04	-.07	-.09
ST.E.	.15	.15	.15	.15	.15	.15	.15	.15	.15	.15	.15	.15

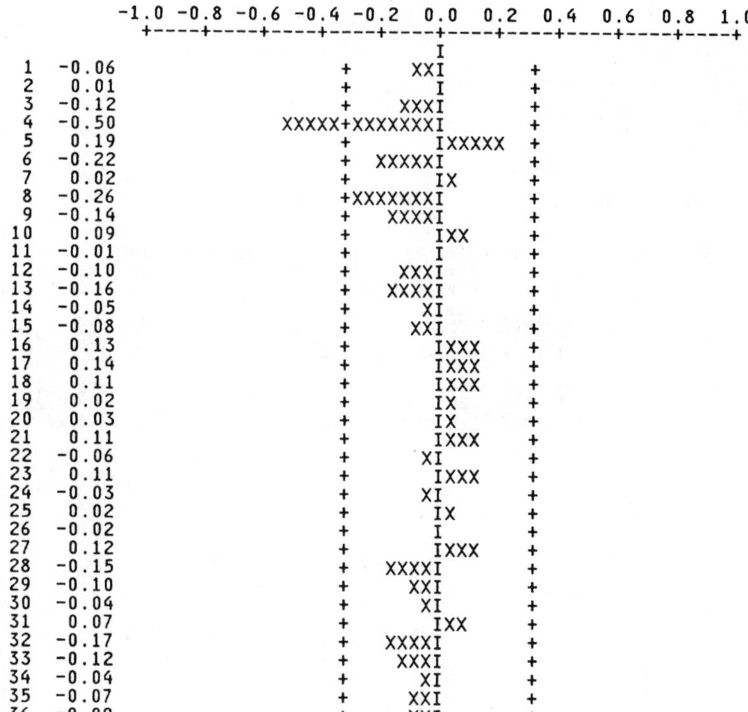

FIGURE 19.14 SPACF of differenced data (seasonal and nonseasonal) after employing a logarithmic transformation.

The formats of the forecasts presented in Eqs. (19.44) and (19.45) show why exponential smoothing techniques are often used in situations where forecasts are needed for a large number of series. Since only the smoothing constant, the current observation, and the most recent forecast need to be retained, the result is an economy of storage and minimal computational effort. However, the implementation does require specifying values for the smoothing constant and a forecast for the first period, $\hat{Z}_0(1)$.

Maximum smoothing is obtained when α is set equal to zero. In this case, $\hat{Z}_t(1) = \hat{Z}_{t-1}(1)$. If observations $Z_1, Z_2, \ldots, Z_n$ are available, then $\hat{Z}_n(1) = \hat{Z}_{n-1}(1) = \cdots = \hat{Z}_0(1)$, the initial value. Very little smoothing takes place when $\alpha = 1$ since the forecast is the most recent observation: $\hat{Z}_t(1) = Z_t$. In general, the exponential forecast profile is smoother for small values of α. Thus a small value of α is appropriate if

VARIABLE	TYPE OF VARIABLE	ORIGINAL OR CENTERED	DIFFERENCING
Z1	RANDOM	ORIGINAL	$(1-B^{1})(1-B^{4})$

PARAMETER LABEL	VARIABLE NAME	NUM./ DENOM.	FACTOR	ORDER	CONS- TRAINT	VALUE	STD ERROR
1 THETA4	Z1	MA	1	4	NONE	.6796	.1130

TOTAL SUM OF SQUARES	0.672972D+01
TOTAL NUMBER OF OBSERVATIONS	47
RESIDUAL SUM OF SQUARES.	0.189238D+00
R-SQUARE	0.969
EFFECTIVE NUMBER OF OBSERVATIONS . .	42
RESIDUAL VARIANCE ESTIMATE	0.450567D-02
RESIDUAL STANDARD ERROR.	0.671243D-01

AUTOCORRELATIONS

1- 12	-.07	-.07	-.12	-.13	.08	-.11	.11	.03	-.25	.19	-.03	.16
ST.E.	.15	.16	.16	.16	.16	.16	.16	.16	.16	.17	.18	.18
Q	.2	.5	1.2	2.0	2.3	2.9	3.5	3.5	7.1	9.1	9.2	10.7
13- 24	-.05	-.16	.06	-.02	.22	.03	-.06	-.11	-.14	-.02	.18	-.04
ST.E.	.18	.18	.19	.19	.19	.19	.19	.19	.19	.20	.20	.20
Q	10.8	12.5	12.7	12.7	16.3	16.4	16.7	17.8	19.4	19.5	22.7	22.9
25- 36	-.04	-.05	.08	-.06	-.02	-.01	-.01	-.10	.06	.05	.01	-.01
ST.E.	.20	.20	.20	.20	.20	.20	.20	.20	.20	.20	.20	.20
Q	23.0	23.4	24.2	24.6	24.6	24.6	24.7	26.5	27.2	27.8	27.8	27.9

SIMPLIFIED EXTENDED ACF TABLE (5% LEVEL)

(Q-->)	0	1	2	3	4	5	6	7	8	9	10	11	12
(P= 0)	0	0	0	0	0	0	0	0	0	0	0	0	0
(P= 1)	X	0	0	0	0	0	0	0	0	0	0	0	0
(P= 2)	X	X	0	0	0	0	0	0	0	0	0	0	0
(P= 3)	X	0	0	0	0	0	0	0	0	0	0	0	0
(P= 4)	0	X	0	0	0	0	0	0	0	0	0	0	0
(P= 5)	0	X	0	0	0	0	0	0	0	0	0	0	0
(P= 6)	0	0	0	0	0	0	0	0	0	0	0	0	0

FIGURE 19.15 Estimation results, SACF of residuals, and EACF of residuals for sales of alcoholic beverages data.

the data exhibit a pattern that changes just slightly. However, for large values of α, the forecast is more sensitive to the current value and will respond more quickly to shifts and short-term trends. In practice, small values of α (≤ 0.3) are used. If a set of data is available, an effective specification procedure is to choose the value of α that minimizes the average of the squared error deviations,

$$\sum_{t=1}^{n} \frac{[Z_t - \hat{Z}_{t-1}(1)]^2}{n}$$

or the average of the absolute deviations,

$$\frac{1}{n}\sum_{t=1}^{n} |Z_t - \hat{Z}_{t-1}(1)|$$

Bowerman and O'Connell (1979) suggest that the forecast for the first period be obtained in the following way. "In practice, the estimate [$\hat{Z}_0(1)$] would be obtained by calculating the average of an initial set of observations in the time series. If such an initial set of observations is not available, [$\hat{Z}_0(1)$] is commonly set equal to the first observed value of the time series."

Thus if $\bar{Z}$ denotes the average of an initial set of observations, then $\hat{Z}_1(1) = \alpha Z_1 + (1 - \alpha)\bar{Z}$. Otherwise $\hat{Z}_1(1) = Z_1$.

Since Eq. (19.44) holds for all t, it certainly holds for $t - 1$,

$$\hat{Z}_{t-1}(1) = \alpha Z_{t-1} + (1 - \alpha)\hat{Z}_{t-2}(1). \tag{19.46}$$

When this expression is substituted into Eq. (19.44), the result is

$$\hat{Z}_t(1) = \alpha Z_t + \alpha(1 - \alpha)Z_{t-1} + (1 - \alpha)^2\hat{Z}_{t-2}(1).$$

If this substitution procedure is continued for $\hat{Z}_{t-2}(1)$, $\hat{Z}_{t-3}(1)$, and so forth, this eventually results in the following expression for $\hat{Z}_t(1)$:

$$\hat{Z}_t(1) = \alpha Z_t + \alpha(1 - \alpha)Z_{t-1} + \alpha(1 - \alpha)^2 Z_{t-2} + \alpha(1 - \alpha)^3 Z_{t-3} + \cdots \tag{19.47}$$

The extended representation in Eq. (19.47) demonstrates that $\hat{Z}_t(1)$ depends on the values in all previous periods with greater weight given to the most recent values. Furthermore, this format can be used to establish a link between simple exponential smoothing and the ARIMA (0, 1, 1) model.

From Eq. (19.30) we see that the ARIMA (0, 1, 1) model is

$$(1 - B)Z_t = (1 - \theta B)a_t$$

or

$$a_t = (1 - B)(1 - \theta B)^{-1}Z_t. \tag{19.48}$$

Since $(1 - B)(1 - \theta B)^{-1} = (1 - B) + \theta B(1 - B) + \theta^2 B^2(1 - B) + \cdots$, Eq. (19.48) can be rewritten as

$$Z_t = (1 - \theta)Z_{t-1} + \theta(1 - \theta)Z_{t-2} + \theta^2(1 - \theta)Z_{t-3} + \cdots + a_t \tag{19.49}$$

Since Eq. (19.49) holds for all t, including $t + 1$, it follows that

$$\hat{Z}_{t+1} = \alpha Z_t + \alpha(1 - \alpha)Z_{t-1} + \alpha(1 - \alpha)^2 Z_{t-2} + \cdots + a_{t+1} \tag{19.50}$$

upon substituting α for $(1 - \theta)$. In Sec. 19.5.4, it was stated that the optimal forecast, $\hat{Z}_t(1)$, is given by the conditional expectation of Z_{t+1}. From Eq. (19.50) it is seen that

$$\hat{Z}_t(1) = \alpha Z_t + \alpha(1 - \alpha)Z_{t-1} + \alpha(1 - \alpha)^2 Z_{t-2} + \cdots \tag{19.51}$$

since $E(a_{t+1}) = 0$. Thus the forecast for an ARIMA (0, 1, 1) or, equivalently, an IMA (1, 1) process is identical with the extended version of the exponential smoothing forecast presented in Eq. (19.47). As Nelson (1973) states, "...exponential smoothing yields optimal forecasts if the series is an IMA (1, 1) process *and* the smoothing coefficient [α] is set equal to [$1 - \theta$]." Furthermore, for ARIMA (p, d, q) processes in general, exponential smoothing does not generate optimal forecasts.

At the beginning of this section it was stated that simple exponential smoothing is appropriate when there is no trend. In the presence of trend, other exponential smoothing procedures are required. Specifically, double exponential smoothing is suitable for linear trend, while triple exponential smoothing is called for when there is quadratic trend. A detailed discussion of these techniques, including examples, is found in Bowerman and O'Connell (1979).

19.8 THE ADDITIVE MODEL

Consider the additive model, briefly mentioned at the beginning of this chapter, in which there is only trend. Thus $Z_t = T_t + a_t$, where T_t denotes the value of the trend at time t. The functional forms usually employed to model trend are either linear, in which case $T_t = \beta_0 + \beta_1 t$, or quadratic, where $T_t = \beta_0 + \beta_1 t + \beta_2 t^2$. Here t is analogous to the x notation used in the linear regression models of Chap. 14. It is customary to represent the points in time by a sequence of increasing integers starting with $t = 1$.

Suppose there is also seasonal variation. If the seasonal effect remains constant over time and is not affected by the magnitude of T_t, this is indicative of additive seasonal variation, in which case $Z_t = T_t + S_t + a_t$. The seasonal pattern can be represented by $S = 1$ indicator variables, where S denotes the periodicity. For monthly data, $S = 12$; for quarterly data, $S = 4$. Suppose there is linear trend and additive seasonal variation for quarterly data. Then a tentative model is

$$Z_t = \beta_0 + \beta_1 t + \beta_{S1} X_{S1} + \beta_{S2} X_{S2} + \beta_{S3} X_{S3} + a_t \tag{19.52}$$

where $X_{sj} = 1$ for quarter j and 0 otherwise, $j = 1, 2, 3$. Taking the expected value of both sides of Eq. (19.52) shows that

$$E(Z_1) = \beta_0 + \beta_1 + \beta_{S1}$$

$$E(Z_2) = \beta_0 + 2\beta_1 + \beta_{S2}$$

$$E(Z_3) = \beta_0 + 3\beta_1 + \beta_{S3}$$

$$E(Z_4) = \beta_0 + 4\beta_1$$

and so forth. Thus for the fifth quarter, the magnitude of the seasonal effect is β_{S1}.

The method of least squares discussed in Chap. 14 is used to estimate the β. Recall that least-squares estimation requires that $E(a_t) = 0$, $V(a_t) = \sigma_a^2$, and the a_t be uncorrelated. In practice, the assumptions are usually not met, and least-squares estimation is primarily used for the practical reason that computer software is available. If the assumptions are not met, then any conclusions based on these assumptions are invalid, such as confidence intervals for the point forecasts. The validity of the assumptions would be checked by examination of the residuals a_t. If the SACF of the residuals has no significant values, it would be reasonable to conclude that the error terms are uncorrelated. If autocorrelation is present, then generalized least squares should be used (refer to Judge, Hill, Griffiths, Lükepohl, and Lee, 1982).

REFERENCES

Abraham, B., and J. Ledolter: *Statistical Methods for Forecasting*, Wiley, New York, 1983.

Bartlett, M. S.: "On the Theoretical Specification of Sampling Properties of an Autocorrelated Time Series," *J. R. Stat. Soc. B*, vol. 8, pp. 27–41, 1946

Bowerman, B. L., and R. T. O'Connell: *Time Series and Forecasting: An Applied Approach,* Duxbury, Belmont, CA., 1979.

Box, G. E. P., and D. R. Cox: "An Analysis of Transformations (with Discussion)," *J. R. Stat. Soc. B*, vol. 26, pp. 211–252, 1964.

_____ and G. M. Jenkins: *Time Series Analysis, Forecasting, and Control,* 2d ed., Holden Day, San Francisco, 1976.

Cryer, J. D.: *Time Series Analysis,* Duxbury, Boston, 1976.

Fama, E. F.: "Random Walks in Stock Market Prices," *Financ. Anal. J.,* vol. 21, pp. 55–59, 1965.

Judge, G. G., R. C. Hill, W. Griffiths, H. Lütkepohl, and T. C. Lee: *Introduction to the Theory and Practice of Econometrics,* Wiley, New York, 1983.

Ljung, G. M., and G. E. P. Box: "On a Measure of Lack of Fit in Time Series Models," *Biometrika*, vol. 65, pp. 297–303, 1978.

Nelson, C. R.: *Applied Time Series Analysis for Managerial Forecasting,* Holden Day, San Francisco, 1973.

Pankratz, A.: *Forecasting with Univariate Box-Jenkins Models,* Wiley, New York, 1983.

Quenouille, M. H.: "Approximate Tests of Correlation in Time-Series," *J. R. Stat. Soc. B*, vol. 11, pp. 68–84, 1949.

Shewhart, W. A.: *Economic Control of Quality of Manufactured Product,* Van Nostrand, New York, 1931.

Shumway, R. H.: *Applied Statistical Time Series Analysis,* Prentice-Hall, Englewood Cliffs, N, 1988.

Standard and Poor's Corp.: *The Analysts Handbook Monthly Suplements, Beverages, Alcoholic*, p. 8, Jan. 1981–Jan. 1989.

Tsay, R. S., and G. C. Tiao: "Consistent Estimates of Autoregressive Parameters and Extended Sample Autocorrelation Function for Stationary and Non-Stationary ARMA Models," *J. Am. Stat. Assoc.*, vol. 79, pp. 84–96, 1984.

CHAPTER 20
ROBUST DESIGN METHODS

Raghu N. Kacker
National Institute of Standards and Technology
Gaithersburg, MD

and

Subir Ghosh
University of California, Riverside, CA

20.1 INTRODUCTION

In a competitive economy, the demand for a product is greatly affected by its quality and selling price. The selling price is determined by the total production cost, the marketing cost (sales expenses), and the profit margin. Therefore the total production cost (hereafter referred to as cost) and the quality are very important characteristics of a product. Taguchi (1976, 1978, 1980, 1981, 1986) has provided great insights into the meaning of quality and cost. He interprets quality and cost from the viewpoint of their impact on the whole society, not just from the viewpoints of the producers and the customers.

Taguchi (1978) defines cost as the loss imparted to the society *before* the product is shipped to the customer. Accordingly, the purpose of cost control should be to reduce the total societal cost of producing the product. The components of total societal cost include (1) raw materials, energy, labor, and overhead expenses consumed in producing usable units of the product, (2) resources wasted in producing unusable units, and (3) harmful by-products produced during manufacturing. The cost to society is therefore reduced by reducing waste and production of harmful by-products. After the wastes are minimized, the major component of the remaining cost is raw materials, energy, labor, and overhead expenses consumed in producing usable units of the product. This component can be reduced only by developing and employing more cost-effective technologies and management practices.

Taguchi (1978) defines quality as the loss imparted to the society *after* the product has been shipped to the customer. This is a strange definition because in common language the word "quality" connotes desirability, whereas loss connotes undesirability. The essence of his definition is that the societal loss generated by a product, from the time the product is shipped to the customer, determines its desirability. The smaller the loss, the more desirable is the product. According to Taguchi, all societal losses that can be traced to the poor performance of a product should be attributed to the quality of the product. The examples of societal losses a product can generate include (1) loss due to harmful side effects of the product and (2) loss due to variations in the product's performance characteristics.

(Performance characteristics are also referred to as quality characteristics or functional characteristics).

1. *Loss due to harmful side effects of the product*: Many products produce harmful side effects. For example, most medicines have some therapeutic value, but many of them produce undesirable side effects. The side effects degrade the value of the medicine. Similarly an automobile is a very useful machine, but the air pollution and the noise produced by its use are a clear loss to society. Therefore efforts should be made to reduce the harmful side effects generated by a product.

2. *Loss due to variations in the product's performance characteristics*: Every product has numerous performance characteristics. Therefore quality should be defined in terms of the individual performance characteristics of the product. Since the actual value of a performance characteristic can be different for different units of the product, and since it can change over time for a given unit, each performance characteristic is a variable entity. Society can incur losses due to variations in the product's performance characteristics. These losses can range from inconvenience, at best, to monetary loss and physical harm. Therefore we should think of quality in terms of the societal loss due to variations in the product's performance characteristics during the product's life span and among different units of the product.

This chapter addresses the problem of evaluating and reducing the loss due to variations in the product's performance characteristics. Many of the ideas presented here are based on earlier works of this author (in particular, Kacker, 1985, with discussions and response, and Kacker, 1986) and his discussions with other researchers in the United States and Japan, including Dr. Taguchi himself.

Section 20.2 describes the concepts of loss function and expected quadratic loss due to variations in the product's performance characteristics. Section 20.3 illustrates the use of the loss function to determine manufacturer's tolerances from customer's tolerances. The variation in a performance characteristic can be reduced by identifying the causes of variation and implementing countermeasures against them. Sections 20.4 and 20.5 discuss Taguchi's framework to identify and implement countermeasures against the basic causes of variation. Section 20.5 includes a discussion of the concept of parameter design to reduce the sensitivity of engineering designs to the basic causes of variation. Section 20.6 describes Taguchi's approach to the use of statistically planned experiments for parameter design. Some of the details of statistical methods proposed by Taguchi are debatable. Section 20.6 also includes a discussion of competing viewpoints. Section 20.7 describes the dispersion models to determine the robust designs. A summary is given in Sec. 20.8.

20.2 LOSS DUE TO PERFORMANCE VARIATION

The variation in a performance characteristic cannot be defined satisfactorily unless we know its ideal value, which is often called the *target value*. Determination of the target value is not a trivial endeavor because it depends on many factors, including the intended use of the product. Once the target value is determined, however, we can define variation in relation to it. A high-quality product performs near the target value consistently throughout the product's life span and under all possible operating conditions. For example, a television set whose picture quality varies from the ideal state with weather conditions has poor quality. An automobile tire with an intended life span of 40,000 miles has poor quality if it wears out in 20,000 miles. Likewise parts of a machine that do not fit together have poor quality. The variation of a performance characteristic about its target value across different units of the product and for each unit during the product's life span under various operating conditions is referred to as *performance variation*. The smaller the performance variation about the target value, the better is the quality.

Performance variation can be evaluated most effectively when the performance characteristic is measured on a continuous scale, because continuous measurements can detect small changes in quality. Therefore it is important to identify performance characteristics of a product that can be measured on a continuous scale. For example, some of the performance characteristics of an automobile that can be measured on a continuous scale are the amount of carbon monoxide produced by the exhaust, the braking distance to make a complete stop for a given speed and brake pedal pressure, the time to accelerate from zero to 55 mi/h, and the amount of engine noise.

All target specifications of continuous performance characteristics should be stated in terms of the nominal levels and the tolerances about the nominal levels. It is a widespread practice in industry to state target values in terms of interval specifications only. This practice erroneously conveys the idea that the quality level remains the same for all values of the performance characteristic in the specification interval, and then quality suddenly deteriorates the moment the performance value slips out of the specification interval. The target value should be defined as the ideal state of the performance characteristic. Note that the target value need not be the midpoint of the tolerance interval. This would be the case, for example, when the performance characteristic is the amount of impurity and the target value is zero.

Some performance characteristics cannot be measured on a continuous scale, either because of their nature or because of the limitations of the measuring method. For example, performance characteristics that require subjective evaluation cannot be measured on a continuous scale. The next best thing to continuous measurements are measurements on an ordered categorical scale, such as poor, fair, good, and excellent. Ordered categorical measurements approximate continuous measurements in the same way as a histogram approximates a continuous distribution. When a measuring instrument has a finite range and some of the values are out of range, the measured data are a mixture of discrete and continuous values. Such data can be approximated by an ordered categorical distribution. When dealing with ordered categorical measurements, we can define the target value as the ideal category.

Although ordered categorical measurements are less effective in detecting small changes in quality than continuous measurements, they are more effective than binary measurements, such as good or bad. Actually binary measurements are the crudest form of discrimination. For example, a classification of solder bonds as good or bad provides very little information about the quality of the solder bonds. A more informative classification would be based on the amount of solder in the joint, such as no solder, insufficient solder, good solder, and excess solder.

Suppose we can evaluate the variation in a performance characteristic. We can then define the quality level with respect to this characteristic in terms of its variation. But the quality level so defined would depend on the particular scale of measurement, and we may not be able to compare the variations in different characteristics. Also the variations in different characteristics have different consequences. Therefore Taguchi suggests that a product's quality with respect to a characteristic should be defined in terms of the societal loss due to variations in the characteristic. But it is usually difficult to estimate the total societal loss. For example, it is difficult to estimate the loss to people other than the customer and the producer. As a practical alternative, the loss to customers is frequently substituted in place of the total societal loss. Thus quality is defined in terms of the loss to customers of the product due to variations in the product's performance characteristics.

Let Y be the value of a performance characteristic measured on a continuous scale. Suppose the target value of Y is τ. The value Y can be different for different units of the product, and also it can change over time for a given unit. In statistical terms Y is a random variable with some probability distribution. (The parameters of the probability distribution of Y are the design parameters of the product.) The variations of the random variable Y about the target value τ cause losses to the customers. Let $l(Y)$ represent losses in terms of dollars suffered by an arbitrary customer at an arbitrary time during the product's life span due to the deviation of Y from τ. Generally the larger the deviation of the performance characteristic Y from its target value τ, the larger is the customer's loss $l(Y)$. However, it is usually difficult to determine the actual form of $l(Y)$. Often a quadratic approximation to $l(Y)$ adequately represents economic losses due to the deviation of Y from τ. (The use of quadratic approximations is not new. Indeed, it is the basis of the statistical theory of least squares founded by K. F. Gauss in 1809.) The simplest quadratic loss function is

$$l(Y) = k(Y - \tau)^2 \tag{20.1}$$

where k is some constant (Fig. 20.1). Since this loss function involves only one constant, it can be completely specified when $l(Y)$ is known for a particular value of Y. For example, suppose $(\tau - \Delta, \tau + \Delta)$ is the customer's tolerance interval, and the product performs unsatisfactorily when Y slips out of this interval. If the average cost to the customer of repairing or discarding the product when $Y = \tau \pm \Delta$ is A dollars, then upon substituting $l(Y) = A$ and $Y = \tau \pm \Delta$ we get $A = k\Delta^2$ and $k = A/\Delta^2$. This version of the loss function is useful when a specific target value is the best and the loss increases symmetrically as the performance characteristic deviates from

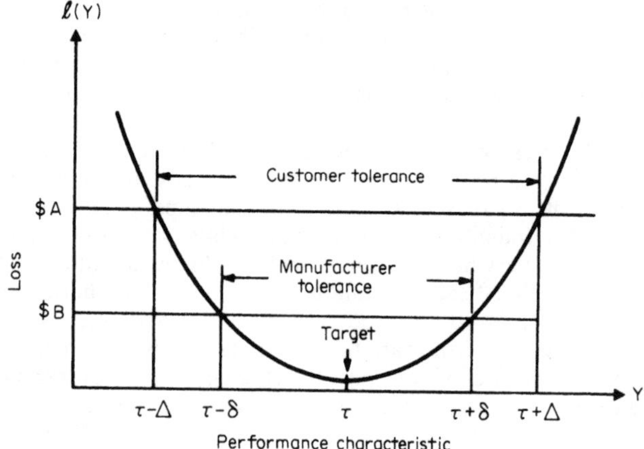

FIGURE 20.1 Loss to customer increases as the performance characteristic deviates from the target value.

the target. But the concept can be extended to many other situations. Two special cases to be referenced in Sec. 20.6.1 are (1) the smaller the better (this would be the case, for example, when the performance characteristic is the amount of impurity and the target value is zero; the smaller the impurity, the better it is), and (2) the larger the better (this would be the case, for example, when the performance characteristic is the strength; the larger the strength, the better it is).

Since the performance characteristic Y, and hence the quadratic loss $l(Y)$, is a random variable, the quality level should be defined by the expected value (weighted statistical average) of $l(Y)$. Since $l(Y) = k(Y - \tau)^2$ and the expected value of $(Y - \tau)^2$ is the mean-squared error $E[(Y - \tau)^2]$, the expected quadratic loss is

$$L = E[l(Y)] = kE[(Y - \tau)^2] \tag{20.2}$$

Taguchi suggests that a product's quality with respect to a characteristic Y should be defined by the expected quadratic loss L. Note that the expected quadratic loss is a function of the product design parameters.

The concept of expected quadratic loss implies that the fundamental measure of variability is the mean-squared error about the target value and not the variance. Since the mean-squared error is the sum of the variance and the square of the bias (that is, the difference between the mean and the target value), *both* variance and bias are important components of variability.

Furthermore, the concept of expected quadratic loss due to performance variation can be used to characterize process capability independently of the process specification limits. Process capability is often characterized by the "percentage defective," which depends on the process specification limits. But process specification limits are only tentative cut-off points used for standardizing a manufacturing process. As the process variability decreases, the specification limits are often tightened. The expected quadratic loss due to performance variation provides a measure of process variation that is independent of the tentative specification limits. The defect rates can be reduced to parts per million (ppm) levels only when the specification limits are interpreted as tentative cut-off points and the variation in the underlying performance characteristics is reduced continuously. The concept of expected quadratic loss emphasizes the importance of continuously reducing performance variation and not getting contented with meeting the "artificial" specification limits.

The importance of expected quadratic loss to define quality can be illustrated by the following example. On April 17, 1979, the prominent Japanese newspaper *Asahi Shimbun* compared the quality of color television sets produced by the Sony factory in Japan and by the Sony factory in San Diego, United States. One of the performance characteristics used for comparison was the color density. The distributions of color density for television sets made in Japan and the United States are shown in Fig. 20.2. Both distributions were symmetric about the target value m. The distribution for Sony USA was approximately uniform in the tolerance interval $m \pm 5$ with mean-squared error 100/12. (Here the mean is on target and therefore the variance is the mean-squared error.) On the other hand the distribution for Sony Japan was approximately normal with mean-squared error 100/36. The result was that the color density for about 0.27% of the television sets shipped by Sony Japan was out of tolerance. But Sony USA shipped no television sets whose color density was out of tolerance. Now whose color density was better—Sony Japan or Sony USA? Since a perfectly uniform distribution in the tolerance interval usually arises when the process is out of control and out-of-tolerance units are screened out, it appears that Sony USA had screened out out-of-tolerance television sets by

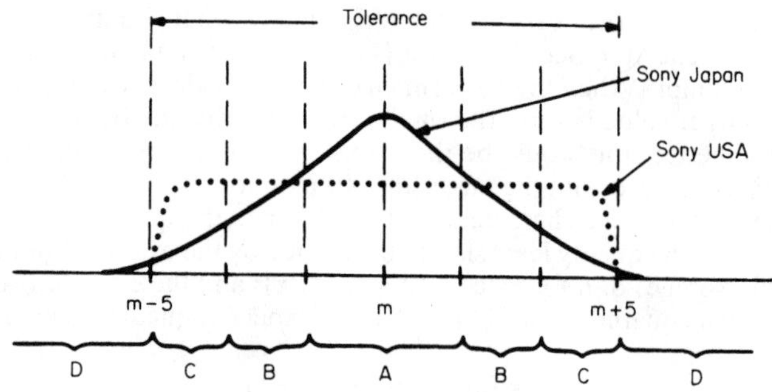

FIGURE 20.2 Distributions of color density in television sets produced by Sony Japan and Sony USA.

inspection. The distribution of color density for Sony Japan, on the other hand, was approximately normal and centered on target. So Sony Japan produced more better grade television sets than Sony USA. To see this, partition the tolerance interval into four grades *A*, *B*, *C*, and *D*. Grade *A* is better than grade *B* and grade *B* is better than grade *C*, based on their closeness to the target value; grade *D* is out of tolerance. It is clear that Sony Japan produced more grade *A* television sets than Sony USA. More specifically, the mean-squared error of the color density for Sony Japan was only one-third the mean-squared error for Sony USA. Therefore the expected quadratic loss to the customers (due to the variation of the color density) from the television sets made in Japan was only one-third the loss from the sets made in the United States. The moral of this story is that the tighter the distribution of a performance characteristic about its target value, the better is the quality.

In some situations, the loss function $l(Y)$ is not symmetric around τ. This would be the case, for example, if the performance characteristic Y had a nonnegative distribution and the target value of Y was zero. An asymmetric quadratic loss function has the form

$$l(Y) = \begin{cases} k_1(Y - \tau)^2 & \text{if } Y \leq \tau \\ k_2(Y - \tau)^2 & \text{if } Y > \tau \end{cases} \tag{20.3}$$

The unknown constants k_1 and k_2 can be determined if $l(Y)$ is known for a value of Y below τ and for a value of Y above τ. For example, suppose $(\tau - \Delta_1, \tau + \Delta_2)$ is the customer's tolerance interval, and the average cost of repairs to the customer is A_1 dollars when $Y = \tau - \Delta_1$ and A_2 dollars when $Y = \tau + \Delta_2$. Then $k_1 = A_1/\Delta_1^2$ and $k_2 = A_2/\Delta_2^2$. The expected value of the loss function given by Eq. (20.3) is

$$L = E[l(Y)] = k_1P_1E[(Y - \tau)^2 \,|Y \leq \tau] + k_2P_2E[(Y - \tau)^2 \mid Y > \tau] \tag{20.4}$$

where P_1 is the probability that $Y \leq \tau$ and $P_2 = 1 - P_1$. If Y is nonnegative, for example, and $\tau = 0$, then $k_1 = 0$, $k_2 = k$, say, and the loss function given by Eq. (20.3) reduces to

$$l(Y) = kY^2$$

The expected loss given by Eq. (20.4) reduces to

$$L = kE[Y^2] \tag{20.5}$$

20.3 DETERMINATION OF TOLERANCES BY USE OF LOSS FUNCTION

The customer's tolerance interval for a characteristic is based on the intended use of the product. Once this is known, the manufacturer's tolerance for the same characteristic and the manufacturer's tolerance for other related product characteristics can be determined by using the loss function. Here are three examples from Taguchi (1980, 1986): (1) manufacturer's tolerance, (2) tolerance for linearly related lower-level characteristic, and (3) tolerance for degradation coefficient.

20.3.1 Manufacturer's Tolerance

Suppose $(\tau - \Delta, \tau + \Delta)$ is the customer's tolerance interval for a characteristic Y, and the product performs unsatisfactorily when Y slips out of this interval. If the average cost to the customer of repairing or discarding the product when $Y = \tau \pm \Delta$ is A dollars, then the customer's loss function is defined by

$$l(Y) = \frac{A}{\Delta^2}(Y - \tau)^2$$

Now what should be the manufacturer's tolerance interval for inspecting the characteristic Y before the product is shipped to the customer? Suppose that $(\tau - \delta, \tau + \delta)$ is the manufacturer's tolerance interval and the average cost to the manufacturer of scrapping or reworking an item whose characteristic Y exceeds the customer's tolerance is B dollars. Taguchi suggests that δ should be determined by substituting $l(Y) = B$ and $Y = \tau \pm \delta$ in the customer's loss function. Thus

$$B = \frac{A}{\Delta^2}\delta^2$$

and

$$\delta = \sqrt{\frac{B}{A}} \times \Delta \tag{20.6}$$

Since B is usually much smaller than A, the manufacturer's tolerance interval will be narrower than the customer's tolerance interval (see Fig. 20.1).

Example. Suppose the customer's tolerance range for the output voltage Y of a power circuit is $(\tau - \Delta, \tau + \Delta)$, where $\tau = 115$ and $\Delta = 25$ volts. Suppose the average loss to the customer when the output voltage reaches 115 ± 25 volts is $A =$ \$150.00. Then the customer's loss function for the output voltage is

$$l(Y) = \frac{150}{(25)^2}(Y - 115)^2 = 0.24(Y - 115)^2$$

What should be the manufacturer's tolerance for the output voltage? Suppose that before the circuit is shipped to the customer, the manufacturer can adjust the output voltage when it exceeds the customer's tolerance by changing a resistor in the circuit at a cost of $B =$ \$0.50 per unit. Then from Eq. (20.6),

$$\delta = \sqrt{\frac{0.50}{150}} \times 25 = 1.4$$

Thus the manufacturer's tolerance for the output voltage should be $(115 - \delta, 115 + \delta)$, where $\delta = 1.4$ volts. Since it costs very little to the manufacturer to adjust the output voltage in the factory before the circuit is shipped to the customer, the manufacturer's tolerance interval is much narrower than the customer's tolerance interval.

20.3.2 Tolerance for Linearly Related Lower-Level Characteristic

Most characteristics of a product are functions of more basic, or lower-level, characteristics. For example, the "dimension" of a product that is stamped out from sheets of steel is a function of the hardness and the thickness of the sheet of steel. Thus in relation to the dimension of the product, the hardness and the thickness of the sheet of steel are lower-level characteristics. Further, many higher-level characteristics are approximately linear functions of the lower-level characteristics. In such cases the tolerance interval for the lower-level characteristic can be determined from the tolerance interval for the higher-level characteristic by using the loss function.

Suppose a characteristic Y is approximately a linear function of a lower-level characteristic X. More specifically,

$$Y = \tau + \beta(X - \mu) \tag{20.7}$$

where τ and μ are the target values of Y and X, respectively, and β is a physical constant. This model implies that when X is on target μ, Y is on target τ. When X deviates from μ by the amount $X - \mu$, Y deviates from τ by the amount $\beta(X - \mu)$. Now suppose that $(\tau - \Delta, \tau + \Delta)$ is the tolerance interval for Y, and the average loss when $Y = \tau \pm \Delta$ is A dollars. Then the loss function for Y is defined by

$$l(Y) = \frac{A}{\Delta^2}(Y - \tau)^2 \tag{20.8}$$

The loss function for the lower-level characteristic X can be obtained by substituting Eq. (20.7) in Eq. (20.8). Thus

$$l(X) = \frac{A}{\Delta^2}[\beta(X - \mu)]^2 \tag{20.9}$$

The tolerance interval for X can be obtained from this loss function. Suppose that $(\mu - \Delta', \mu + \Delta')$ is the desired tolerance interval for X, and the manufacturer incurs a loss of A' dollars per unit when the characteristic X reaches its tolerance limit. Then from the loss function given by Eq. (20.9),

$$A' = \frac{A}{\Delta^2}(\beta\Delta')^2$$

and

$$\Delta' = \sqrt{\frac{A'}{A}} \times \frac{\Delta}{\beta}$$

Example. Suppose Y is the dimension of a product that is stamped out from sheets of steel and suppose X is the hardness of the steel sheet. Now let the tolerance interval for dimension Y be $(\tau - \Delta, \tau + \Delta)$, where $\Delta = 300$ μm, and the cost of repairing or discarding a product whose dimension reaches $\tau \pm 300$ μm is $A = \$6.00$. Now assume that when the hardness X is on target μ, the dimension Y is on target τ, and when the hardness of the sheet steel deviates from the target μ by 1 unit, the dimension of the product deviates from its target τ by 180 μm. That is, $\beta = 180$ μm. If the cost of scrapping those sheets of steel whose hardness is out of tolerance is $A' = \$1.5$ per unit product, then the tolerance for the hardness of steel sheets is given by $(\mu - \Delta', \mu + \Delta')$, where

$$\Delta' = \sqrt{\frac{A'}{A}} \times \frac{\Delta}{\beta}$$

$$= \sqrt{\frac{1.5}{6.0}} \times \frac{300}{180} = 0.83 \text{ unit}$$

Thus, in this example the tolerance on the hardness of steel sheets should be $(\mu - \Delta', \mu + \Delta')$, where $\Delta' = 0.83$ unit. Steel sheets the hardness of which does not meet this specification should be screened out.

20.3.3 Tolerance for Degradation Coefficient

Some product characteristics degrade during the product's life span. The rate of degradation, hereafter referred to as the *degradation coefficient* and denoted by r, depends on many factors, including the properties of materials used in making the product, the engineering design of the product, and physical adjustments. Thus the product designer must set an upper limit on the degradation coefficient. Taguchi suggests that the upper limit r_0 should be determined by balancing the expected loss to the customers due to degradation against the manufacturer's cost of scrapping a unit whose degradation coefficient exceeds r_0.

Suppose the product is designed in such a way that the characteristic Y is on target τ initially (at time zero) and degrades at the rate of r per unit time uniformly throughout the product's intended life span. Thus the value of the characteristic Y at time t during the product's life span is

$$y_t = \tau + rt$$

If the intended life span of the product is T, then at the end of the product's intended life span the characteristic Y would deviate from τ by $(y_T - \tau) = rT$. The mean-squared error of Y about τ during the product's life span would be

$$E(Y - \tau)^2 = E(rt)^2$$

$$= \frac{r^2}{T}\int_0^r t^2\, dt$$

$$= \frac{r^2T^2}{3}$$

Now let the tolerance interval for Y be $(\tau - \Delta, \tau + \Delta)$ and let the average loss to the customer when $Y = \tau \pm \Delta$ be A dollars. Then the customer's expected loss associated with the degradation coefficient r is

$$L = \frac{A}{\Delta^2} E(Y - \tau)^2 = \frac{A}{\Delta^2}\left(\frac{r^2 T^2}{3}\right) \tag{20.10}$$

Now assume that the manufacturer's cost of scrapping a unit whose degradation coefficient exceeds the upper limit r_0 is C dollars. Taguchi suggests that r_0 should be obtained by substituting C for L and r_0 for r in the expected loss given in Eq. (20.10). Thus

$$C = \frac{A}{\Delta^2}\left(\frac{r_0^2 T^2}{3}\right)$$

and

$$r_0 = \sqrt{\frac{3C}{A}} \times \frac{\Delta}{T}$$

Example. Suppose the customer's tolerance interval for a characteristic Y is $(\tau - \Delta, \tau + \Delta)$, where $\Delta = 300$ μm, and the average cost to the customer for repairing or discarding the product when $Y = \tau \pm 300$ μm is $A = \$40$. Suppose further that the intended life span of the product is 10 years, the characteristic Y is on target at time zero, and it degrades at the rate of r micrometers per year uniformly throughout the product's life span. If the manufacturer's cost of scrapping a unit whose degradation coefficient exceeds the upper limit r_0 is $C = \$2.5$, then r_0 is given by

$$\begin{aligned} r_0 &= \sqrt{\frac{3C}{A}} \times \frac{\Delta}{T} \\ &= \sqrt{\frac{3(2.5)}{40}} \times \frac{300}{10} = 13 \ \mu\text{m per year} \end{aligned}$$

With this rate of degradation, the total degradation in 10 years will be 130 μm, which is smaller than the customer's limit of 300 μm.

20.4 IMPORTANCE OF PRODUCT AND PROCESS DESIGNS

The procedure to reduce the variation in a performance characteristic consists of identifying the causes of variation and implementing countermeasures against them. Taguchi (1976, 1986) has provided a framework to assist engineers and scientists in identifying the basic causes of variation. Most of the basic causes of performance variation can be classified into four categories: (1) environmental variables, (2) human variations in maintaining and operating the product, (3) product deterioration, and (4) manufacturing imperfections. We can clearly visualize these causes of performance variation when we think of the performance of an automobile. The performance of an automobile is affected by road, traffic, and weather conditions, maintenance of the automobile, driver skills, age of the vehicle, and manufacturing imperfections. Manufacturing imperfections are the deviations of the actual parameters of a manufactured product from their target values, and they are responsible for performance variations among different units of a product.

Manufacturing imperfections are caused by inevitable uncertainties in manufacturing processes. The causes of manufacturing imperfections include unstable process designs, variations in raw materials and components, variations in machines, lack of process controls, and other unidentified manufacturing variations. Note that the basic causes of a product's performance variation outlined here are chronic problems. They are inevitable, and they affect most products.

The countermeasures against most of the chronic causes of variation can be built into the product only at the early stages in a product's development cycle, which can be partitioned into three stages: product design, process design, and manufacturing. At the product design stage, engineers develop complete product design specifications, including the specification of materials, components, configurations, and features. Next, process engineers design a manufacturing process. It may involve the invention of a new process or modification of an existing process to produce the new product. The manufacturing department then uses the manufacturing process to mass produce the product. Since the performance variation caused by environmental variables, human variations in maintaining and operating the product, and product deterioration are primarily a function of the product design, the countermeasures against these causes of variation can be built into the product only at the product design stage. After the product has been designed, the only improvement that can be made is to reduce the manufacturing cost and the manufacturing imperfections in the product.

The manufacturing cost and the manufacturing imperfections in a product are largely determined by the design of the manufacturing process. With a given process design, increased process controls can reduce manufacturing imperfections (see Fig. 20.3). But process controls cost money. (Nevertheless, the cost of process controls is justified as long as the loss due to manufacturing imperfections is more than the cost of process controls.) Therefore it is necessary to reduce both manufacturing imperfections and the need for process controls. This can be accomplished only by improving the process design to move the cost curve down, as shown in Fig. 20.3.

Thus the degree of performance variation and the manufacturing cost are determined to a large extent by the engineering designs of the product and its manufacturing process. Indeed the dominance of a few corporations in high-technology products such as automobiles, industrial robots, microprocessors, optical devices, and machine tools can be traced to the strengths of the engineering designs of their products and manufacturing processes.

20.5 OFF-LINE QUALITY CONTROL

Because of the importance of product and process designs, quality control must begin with the first step in the product development cycle, and it must continue through all subsequent steps. The phrase "quality control," as used here, has a very broad meaning. It includes quality planning and quality improvement. *Off-line* quality control methods are technical aids for quality and cost control at the product and the process design stages in the product's development cycle. In contrast, *on-line* quality control methods are technical aids for quality and cost control in manufacturing. Off-line quality control methods are used to improve product quality and manufacturability, and to reduce product development, manufacturing, and lifetime costs.

Effective quality control requires off-line quality control methods that focus on quality improvement rather than quality evaluation. Some examples of off-line

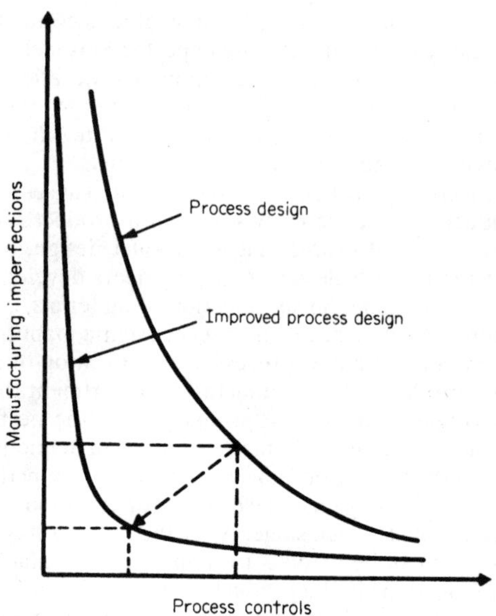

FIGURE 20.3 Improved process design reduces both manufacturing imperfections and the need for process controls.

quality control methods are sensitivity tests, prototype tests, accelerated life tests, and reliability tests. However, the primary function of these methods is quality evaluation. They are like a doctor's thermometer. A thermometer can indicate that a patient's temperature is too high, but it is not a medication. For example, most reliability tests are primarily concerned with reliability estimation. No doubt, reliability estimation is important. For example, good reliability estimates are necessary for quality assurance and for scheduling preventive maintenance. But often improvement of product reliability is the real issue. In such cases approximate reliability estimates often suffice, and greater emphasis on reliability improvement is needed.

As with performance characteristics, all specifications of product and process parameters should be stated in terms of their ideal values and tolerances around these ideal values. It is a widespread practice in industry to state these specifications in terms of tolerance intervals only. This practice can sometimes mislead a manufacturer into producing products whose parameters are barely inside the tolerance intervals. Such products are likely to be of poor quality. Even if all the parameters of a product are within their tolerance intervals, the product may not perform satisfactorily because of the interdependencies of the parameters. For example, if the size of a car door is near its lower tolerance limit and the size of the door frame is near its upper tolerance limit, the door may not close properly. A product performs best when all parameters of the product are at their ideal values. Further, the knowledge of the ideal values of product and process parameters encourages continuous quality improvement.

Taguchi (1978) has introduced a three-step approach to assign nominal values and tolerances to product and process parameters: (1) system design, (2) parameter design, and (3) tolerance design.

System design is the process of applying scientific and engineering knowledge to produce a basic functional prototype design. The prototype model defines the initial settings of product or process parameters. System design requires an understanding of both the customer's needs and the manufacturing environment. A product can satisfy the customer's needs only if it is designed to do so. Similarly, the ability to manufacture requires an understanding of the manufacturing environment.

Parameter design is the process of identifying the settings of product or process parameters that reduce the sensitivity of engineering designs to the sources of variation. For example, consider an electric circuit. Suppose the performance characteristic of interest is the output voltage of the circuit and its target value is y_0. Assume that the output voltage of the circuit is largely determined by the gain of a transistor X in the circuit, and the circuit designer is at liberty to choose the nominal value of this gain. Suppose also that the effect of transistor gain on the output voltage is nonlinear (Fig. 20.4). In order to obtain an output voltage of y_0, the circuit designer can select the nominal value of transistor gain to be x_0. If the actual transistor gain deviates from the nominal value x_0, the output voltage will deviate from y_0. The transistor gain can deviate from x_0 because of manufacturing imperfections in the transistor, deterioration during the circuit's life span, and environmental variables. If the distribution of transistor gain is as shown, the output voltage will have a large variation. One way of reducing the output variation is to use an expensive transistor whose gain has a very tight distribution around x_0. Another way of reducing output variation is to select a different value of transistor gain. For instance, if the nominal transistor gain is x_1, the output voltage will have a much smaller variance. But the mean value y_1 associated with the transistor gain x_1 is far from the target value y_0. Now suppose that there is another component in the circuit, such as a resistor, that has a linear effect on the output voltage and the circuit designer is at liberty to choose the nominal value of this component. The circuit designer can then adjust this component to move the mean value from y_1 to y_0.

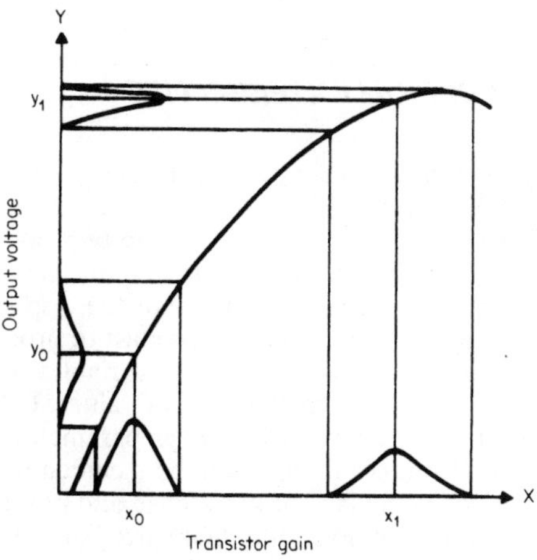

FIGURE 20.4 Effect of transistor gain on output voltage.

Adjustment of the mean value of a performance characteristic to its target value is usually a much easier engineering problem than the reduction of performance variation. When the circuit is designed in such a way that the nominal gain of transistor X is x_1, an inexpensive transistor having a wide distribution around x_1 can be used. Of course, this change would not necessarily improve the circuit design if it were accompanied by an increase in the variance of another performance characteristic of the circuit.

The utilization of nonlinear effects of product or process parameters on the performance characteristics, in order to reduce the sensitivity of engineering designs to the sources of variation, is the essence of parameter design. Because parameter design reduces performance variation by reducing the influence of the sources of variation rather than by controlling them, it is a cost-effective technique to improve engineering designs. Further, this technique can be applied to improve the designs of both the products and their manufacturing processes.

The concept of parameter design is not new to many circuit designers. In the words of Brayton, Hachtel, and Sangiovanni-Vincentelli (1981), "various influences, manufacturing, environmental and aging, will cause statistical variations among circuits which are nominally the same. Designers are driven by economical considerations to provide designs which are tolerant to statistical variations." The essence of parameter design is also familiar to many agronomists. As indicated by Eberhart and Russell (1966), many agronomists routinely conduct experiments to identify plant varieties that can tolerate a wide range of soil, moisture, and weather conditions. The significance of Taguchi's contribution, however, is that he has demonstrated that parameter design is applicable in most manufacturing situations.

Tolerance design is the process of determining tolerances around the nominal settings identified by parameter design. It is a common practice in industry to assign tolerances by convention rather than by scientific reasoning. Tolerances that are too narrow increase the manufacturing cost and tolerances that are too wide increase performance variation. Therefore tolerance design involves a trade-off between the customer's loss due to performance variation and the increase in manufacturing cost.

20.6 PARAMETER DESIGN EXPERIMENTS

Taguchi (1976) has proposed a novel approach to the use of statistically planned experiments for parameter design. He classifies the variables that affect the performance characteristics of a product or process into two categories, design parameters and sources of noise. The *design parameters* are those product or process parameters whose nominal settings can be chosen by the responsible engineer. The nominal settings of design parameters define a product or process design specification and vice versa. The *sources of noise* are those variables that cause the performance characteristics to deviate from their target values. (The sources of noise include the deviations of the actual values of design parameters from their nominal settings.) However, not all sources of noise can be included in a parameter design experiment, the limitations being lack of knowledge and physical constraints. The *noise factors* are those sources of noise and their surrogates that can be varied systematically in a parameter design experiment. The key noise factors, which represent the major sources of noise that affect a product's performance in the field and a process's performance in the manufacturing environment, should be identified and included in the experiment. The object of the experiment is then to identify the settings of design parameters at which the effect of noise factors on the performance characteristic is minimum. These settings are predicted by (1) systematically varying the settings of design parameters in the experiment and (2) comparing the effect of noise factors for each test run.

Taguchi-type parameter design experiments consist of two parts: a design parameter matrix and a noise factor matrix. The *design parameter matrix* specifies the test settings of design parameters. Its columns represent the design parameters and its rows represent different combinations of test settings. The *noise factor matrix* specifies the test levels of noise factors. Its columns represent the noise factors and its rows represent different combinations of noise levels. The complete experiment consists of a combination of the design parameter and noise factor matrices (see, for example, Fig. 20.5). Each test run of the design parameter matrix is crossed with every row of the noise factor matrix. The performance characteristic is evaluated for each noise level combination in the noise factor matrix. Thus the variation in multiple values of the performance characteristic mimics the product or process performance variation at the given design parameter settings.

In the case of continuous performance characteristics (see Fig. 20.5), multiple values from each test run of the design parameter matrix are used to compute a criterion called *performance statistic*. A performance statistic estimates the effect of noise factors. Some performance statistics are discussed later in this section. The computed values of a performance statistic are used to predict better settings of design parameters, and the prediction is subsequently verified by a confirmation experiment. The initial design parameter settings are not changed until the veracity of the prediction has been verified. Several iterations of such parameter design experiments may be required to identify design parameter settings at which the effect of noise factors is sufficiently small.

Parameter design experiments can be done in one of two ways: through physical experiments or through computer simulation trials. These experiments can be

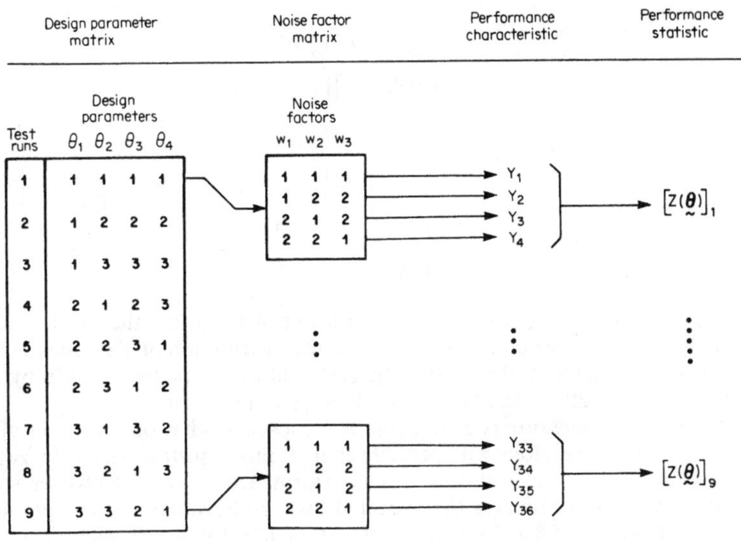

FIGURE 20.5 Plan of a Taguchi-type parameter design experiment.

done with a computer when the function $Y = f(\theta, w)$, relating performance characteristic Y to design parameters θ, and the noise factors w can be evaluated numerically. Frequently the function $Y = f(\theta, w)$ is not defined explicitly, but the implicit function can be evaluated numerically by solving a number of simultaneous differential equations. When computer simulation trials are possible, many alternative methods are available. For example, Brayton, Hachtel, and Sangiovanni-Vincentelli (1981) have surveyed a number of numerical algorithms for optimizing the design of integrated circuit devices. These methods and parameter design experiments have the same goal—reduce the sensitivity to chronic causes of performance variation. The list of algorithms surveyed by Brayton and his associates includes both the deterministic optimization techniques (such as the linear programming and the constrained quasi-Newton methods) and the statistical optimization techniques (such as the design centering and the Monte Carlo–based methods). Most of these optimization techniques are aimed at making trade-offs among multiple objectives while satisfying multiple constraints.

20.6.1 Performance Statistics

Taguchi recommends the use of criteria he calls signal-to-noise (s/n) ratios as performance statistics. A performance statistic estimates the effect of noise factors on the performance characteristic. An efficient performance statistic takes advantage of the prior engineering knowledge about the product, the loss function, and the distribution of the performance characteristic. Let $\theta = (\theta_1, \theta_2, \ldots, \theta_k)$ represent the design parameters and let $w = (w_1, w_2, \ldots, w_l)$ represent the noise factors included in the parameter design experiment. We assume that the performance characteristic Y is a function of θ and w, that is, $Y = f(\theta, w)$. The design parameters θ are the parameters of the distribution of Y, and for a given θ the noise factors generate the distribution of Y.

Let

$$\eta(\theta) = E[Y]$$

$$\sigma^2(\theta) = E[\{Y - \eta(\theta)\}^2]$$

$$\mathrm{CV}(\theta) = \frac{\sigma(\theta)}{\eta(\theta)}$$

$$\mathrm{MSE}(\theta) = E[(Y - \tau)^2]$$

$$\mathrm{B}(\theta) = \eta(\theta) - \tau$$

represent the mean, the variance, the coefficient of variation, the mean-squared error about the target value τ, and the bias of the distribution of Y given θ, respectively. It is clear that with this model, the expected quadratic losses given by Eqs. (20.2), (20.4), and (20.5) are functions of design parameters θ.

A *performance measure* is a function of θ chosen so that optimization of the performance measure, along with possible engineering adjustments, minimizes the expected loss. The expected loss itself is a performance measure. However, sometimes the problem of reducing the expected loss can be simplified by optimizing some other function of θ and making engineering adjustments. In such a case, this other function is the performance measure. The following examples illustrate that different engineering situations can lead to different performance measures.

While a performance measure is a function of θ, in general this function is not known. It must therefore be estimated. We use the term performance statistic for a statistical estimate of a performance measure. The performance statistic is then used as the actual criterion to be optimized. Taguchi has defined more than 60 different performance statistics, but only a fraction of them are available in the published literature.

When the performance characteristic Y is measured on a continuous scale, the loss function $l(Y)$ usually takes on one of three forms, depending on whether smaller is better, larger is better, or a specific target value is best. For the first two cases we discuss the performance statistics recommended by Taguchi. For the third case we show how two different engineering situations (one considered by Taguchi and one not) lead to different performance statistics.

The Smaller the Better. The performance characteristic Y has a nonnegative distribution, the target value is $\tau = 0$, and the loss function $l(Y)$ increases as Y increases from zero. In this case the expected loss given by Eq. (20.5) is proportional to

$$\text{MSE}(\theta) = E[(Y - 0)^2] = E[Y^2]$$

and Taguchi (1976) recommends the performance measure

$$\zeta(\theta) = -10 \log \text{MSE}(\theta) \tag{20.11}$$

The larger the performance measure, the smaller is the mean-squared error. Let y_1, y_2, …, y_n approximate a random sample from the distribution of Y for a given θ. The performance measure given by Eq. (20.11) can then be estimated by the performance statistic

$$Z(\theta) = -10 \log \left(\frac{1}{n} \sum y_i^2 \right) \tag{20.12}$$

This performance statistic is the *method of moments* estimator of the corresponding performance measure. Easterling (1985) points out that the performance statistic given by Eq. (20.12), evaluated from a parameter design experiment, does not really estimate the performance measure given by Eq. (20.11). If the performance statistic itself can be viewed as a legitimate estimate of the sensitivity to noise, however, we can analyze it meaningfully. Box (1988) suggests that the performance statistic given by Eq. (20.12) is undesirable because it lumps sample mean and sample variance together and thus hides useful information in the data.

The Larger the Better. The performance characteristic Y has a positive distribution, the target value is infinity, and the loss function $l(Y)$ decreases as Y increases from zero. Taguchi suggests that this case becomes a particular application of the smaller the better case when we treat the reciprocal $1/Y$ as the performance characteristic. The target value of $1/Y$ is zero, the performance measure given by Eq. (20.11) reduces to

$$\zeta(\theta) = -10 \log \text{MSE}(\theta)$$

where

$$\mathrm{MSE}(\theta) = E\left[\left(\frac{1}{Y}\right)^2\right]$$

and the performance statistic analogous to Eq. (20.12) becomes

$$Z(\theta) = -10 \log\left[\frac{1}{n}\sum\left(\frac{1}{y_i^2}\right)\right]$$

A Specific Target Value Is Best. The performance characteristic Y has a specific target value $\tau = \tau_0$, and the loss function $l(Y)$ increases symmetrically as Y deviates from τ_0 in either direction. In this case the expected loss given by Eq. (20.2) is proportional to $\mathrm{MSE}(\theta) = E[(Y - \tau_0)^2]$. Now

$$\mathrm{MSE}(\theta) = \sigma^2(\theta) + [B(\theta)]^2$$

and

$$\sigma(\theta) = \mathrm{CV}(\theta)\eta(\theta) \tag{20.13}$$

1. *Variance not linked to mean*: In many engineering situations $\sigma(\theta)$ and $\eta(\theta)$ are functionally independent of each other, and the bias $B(\theta) = \eta(\theta) - \tau_0$ can be reduced independently of the variance $\sigma^2(\theta)$. This would be the case, for example, if one of the design parameters, say θ_1, had a large effect on the mean value and hence the bias but almost no effect on the variance. The design parameter θ_1 could then be used to reduce the bias independently of the variance. Design parameters, such as θ_1, which have a large effect on the mean value but almost no effect on the variance are called *adjustment parameters*.

In general adjustment parameters are special design parameters that can be adjusted easily to reduce a component of the total variation about the target value without affecting the remaining nonadjustable component. This component is the adjustable component. When adjustment parameters are available, the total variation in the performance characteristic about the target value can be reduced in two stages: first the nonadjustable variation can be reduced by parameter design, and then the adjustable variation can be reduced by adjusting the adjustment parameters. Frequently adjustment parameters are known beforehand, and they are not even included in the parameter design experiment, while in other situations experiments are conducted to identify the adjustment parameters.

In the present case $B(\theta)$ is the adjustable component of the mean-squared error, $\sigma^2(\theta)$ is the nonadjustable component of the mean-squared error, and the adjustment parameters are those design parameters that have a large effect on $\eta(\theta)$ and hence $B(\theta)$, but almost no effect on $\sigma^2(\theta)$. When such adjustment parameters are available, $B(\theta)$ can be reduced independently of $\sigma^2(\theta)$. Therefore in this situation the primary purpose of parameter design is to reduce $\sigma^2(\theta)$. Thus an efficient performance measure should be a monotonic function of $\sigma^2(\theta)$.

Thus

$$\zeta(\theta) = -\log \sigma^2(\theta) \tag{20.14}$$

can be used as a performance measure. The larger the performance measure, the smaller is the variance $\sigma^2(\theta)$. Let $y_1, y_2, \ldots, y_n$ approximate a random sample from the distribution of Y for a given θ. Then

$$s^2 = \frac{1}{n-1} \sum (y_i - \bar{y})^2$$

where

$$\bar{y} = \frac{1}{n} \sum y_i$$

is an (unbiased) estimator of $\sigma^2(\theta)$ and

$$Z(\theta) = -\log(s^2) \tag{20.15}$$

is a reasonable performance statistic. (This is not a Taguchi signal-to-noise ratio.) Often the log transformation reduces curvatures and interactions in the s^2 response surface. In addition, the logarithm of s^2 is often less dependent on $\bar{y}$ than is s^2. The minus sign in the performance statistics given by Eq. (20.15) is used by convention so that we always maximize a performance statistic. Bartlett (1937) recommended the use of this performance statistic for testing the equality of several population variances.

2. *Variance linked to mean*: Sometimes $\sigma(\theta)$ is approximately proportional to $\eta(\theta)$, and thus the coefficient of variation $CV(\theta) = \sigma(\theta)/\eta(\theta)$ is a more meaningful measure of variability than $\sigma(\theta)$. In such cases, if the bias $B(\theta)$ can be reduced independently of the coefficient of variation $CV(\theta)$, the bias $B(\theta)$ is the adjustable variation, the coefficient of variation $CV(\theta)$ is the nonadjustable variation. The adjustment parameters are those design parameters that have a large effect on the mean $\eta(\theta)$ and hence $B(\theta)$ but almost no effect on $CV(\theta) = \sigma(\theta)/\eta(\theta)$. When such adjustment parameters are available, the bias $B(\theta)$ can be reduced independently of the coefficient of variation $CV(\theta)$, and the main purpose of parameter design is to reduce $CV(\theta)$. It can be seen from Eq. (20.13) that when the coefficient of variation $CV(\theta)$ remains small, any subsequent change in the mean value $\eta(\theta)$ does not bring about as large a change in $\sigma(\theta)$. So an efficient performance measure should be a monotonic function of $CV(\theta) = \sigma(\theta)/\eta(\theta)$. Taguchi recommends the performance measure

$$\zeta(\theta) = 10 \log \left[\frac{\eta(\theta)}{\sigma(\theta)} \right]^2 \tag{20.16}$$

The larger the performance measure, the smaller is the coefficient of variation $\sigma(\theta)/\eta(\theta)$. The quantity $n\bar{y}^2 - s^2$ is an (unbiased) estimator of $n[\eta(\theta)]^2$. Therefore Taguchi (1976) recommends the performance statistics

$$Z_1(\theta) = 10 \log \left(\frac{\bar{y}^2}{s^2} \right) \tag{20.17}$$

and

$$Z_2(\theta) = 10 \log \left(\frac{n\bar{y}^2 - s^2}{ns^2} \right) \tag{20.18}$$

Since we can write Eq. (20.18) as

$$Z_2(\theta) = 10 \log \left(\frac{\bar{y}^2}{s^2} - \frac{1}{n} \right)$$

both performance statistics are equivalent. The mean-squared error and the expected loss are minimized when the performance measures given by Eqs. (20.14) and (20.16) are maximized, and the bias $\eta(\theta) - \tau_0$ is minimized by the use of adjustment parameters.

Leon, Shoemaker, and Kacker (1987) have further explored the concept of performance measures and adjustment parameters. Box (1988) observed that the performance statistic given by Eq. (20.17) is almost equivalent to transforming the characteristic Y into log Y and then taking the log transformation of the sample variance of Y. The log transformation is the appropriate transformation for studying variability when $\sigma(\theta)$ is approximately proportional to $\eta(\theta)$. But in general the relationship between $\sigma(\theta)$ and $\eta(\theta)$ may not be known with certainty. For example, $\sigma(\theta)$ could be proportional to $\eta^2(\theta)$ or $\sigma^2(\theta)$ could be proportional to $\eta(\theta)$. Therefore the appropriate transformation for studying variability may not be the log transformation but some other transformation, such as the reciprocal or the square root. Box suggests that the experimental data should be used to identify the appropriate transformation. The data should then be analyzed by using the appropriate transformation.

20.6.2 Orthogonal Arrays

Taguchi (1976) recommends the use of orthogonal arrays, in particular three-level and mixed two- and three-level orthogonal arrays of strength 2, for constructing the design parameter and the noise factor matrices. Orthogonal arrays are generalized Graeco-Latin squares. The design parameter matrix in Fig. 20.5 is a Graeco-Latin square written in an orthogonal array form. This orthogonal array is a 9×4 matrix of three elements (levels) 1, 2, and 3, which has the balancing property that every pair of columns contains each of the nine pairs of elements (1, 1), (1, 2), (1, 3), (2, 1), (2, 2), (2, 3), (3, 1), (3, 2), and (3, 3) exactly once. Orthogonal arrays, such as this, in which the balancing property applies to *every pair* of columns, are said to have strength 2. A mixed-level orthogonal array of strength 2, denoted by $OA_N(s^m \times t^n)$, is a matrix of N rows and $m + n$ columns, in which the first m columns have s elements (levels) each, the next n columns have t elements (levels) each, and where in every pair of columns each possible ordered pair of elements occurs a constant number of times. The constant may, however, depend on the pair of columns selected. The general theory of orthogonal arrays was developed by Bose (1947), Rao (1947), and Bose and Bush (1952). Raghavarao (1971) gives several methods for constructing orthogonal arrays. Many researchers have developed catalogs of orthogonal arrays. Examples include Plackett and Burman (1946), National Bureau of Standards (1957, 1959, 1961), Addelman (1962, 1972), Hahn and Shapiro (1966), Taguchi (1976, 1987), Kacker (1982), Dey (1985), and others.

All common factorial and fractional factorial plans of experiments are orthogonal arrays (see, for example, Kacker and Tsui, 1990). But not all orthogonal arrays are common fractional factorial plans, the most popular exceptions being mixed-level orthogonal arrays of 18 and 36 runs. Tables 20.1 and 20.2 display these arrays. Kacker, Lagergren, and Filliben (1988) describe the construction of these arrays.

The orthogonal array experiments promoted by Taguchi are *orthogonal main-effect plans*, where the main effects are confounded with two-factor and higher-order interaction terms. Taguchi recommends the use of orthogonal main-effect plans to predict better settings of design parameters. He suggests that the prediction should be checked by follow-up confirmatory experiments. The initial settings of design parameters should not be changed until the veracity of the prediction has been verified.

Hunter (1985) points out that the use of orthogonal main-effect plans corresponds to fitting a first-order linear-effects-only model for the response statistic. And orthogonality is an important criterion when fitting first-order models. But the first-order models are often inadequate because of the presence of strong second-order effects, such as interactions and curvatures. Therefore we should fit second-order models involving interaction and curvature terms. But when we consider second-order models, the orthogonality property loses its importance, and the essential

TABLE 20.1 Orthogonal Arrays $OA_{18}(6^1 \times 3^6)$ and $OA_{18}(2^1 \times 3^7)$*

	Column								
Row	1	2	3	4	5	6	7	1′	2′
1	1	1	1	1	1	1	1	1	1
2	1	2	2	2	2	2	2	1	1
3	1	3	3	3	3	3	3	1	1
4	2	1	1	2	2	3	3	1	2
5	2	2	2	3	3	1	1	1	2
6	2	3	3	1	1	2	2	1	2
7	3	1	2	1	3	2	3	1	3
8	3	2	3	2	1	3	1	1	3
9	3	3	1	3	2	1	2	1	3
10	4	1	3	3	2	2	1	2	1
11	4	2	1	1	3	3	2	2	1
12	4	3	2	2	1	1	3	2	1
13	5	1	2	3	1	3	2	2	2
14	5	2	3	1	2	1	3	2	2
15	5	3	1	2	3	2	1	2	2
16	6	1	3	2	3	1	2	2	3
17	6	2	1	3	1	2	3	2	3
18	6	3	2	1	2	3	1	2	3

* Columns 1, 2, 3, 4, 5, 6, and 7 form OA_{18} $(6^1 \times 3^6)$; columns 1′, 2′, 2, 3, 4, 5, 6, and 7 form OA_{18} $(2^1 \times 3^7)$.

criteria for good prediction are the rotatability of the design and the minimization of bias from higher-order terms. Thus Hunter favors sequential 2^{k-p} fractional factorial plans, central composite plans, and Box-Behnken plans, rather than orthogonal main-effect plans.

The following discussions are our attempt to clarify Taguchi's reasoning for promoting the use of main-effects-only plans based on orthogonal arrays. First, Taguchi type parameter design experiments should be conducted in the laboratory environment. The main purpose of these experiments is to predict "optimal" settings of design parameters for the manufacturing environment (and the customer's use conditions), not the laboratory conditions. Taguchi points out that the optimal settings for the laboratory environment are likely to remain optimal for the manufacturing environment when the effects of design parameters are additive. An effi-

TABLE 20.2 Orthogonal Arrays $OA_{36}(2^{11} \times 3^{12})$, $OA_{36}(12^1 \times 3^{12})$, and $OA_{36}(2^3 \times 3^{13})$*

	Column																										
Row	1	2	3	4	5	6	7	8	9	10	11	12	13	14	15	16	17	18	19	20	21	22	23	1′	2′	3′	4′	1′
1	1	1	1	1	1	1	1	1	1	1	1	1	1	1	1	1	1	1	1	1	1	1	1	1	1	1	1	1
2	1	1	1	1	1	1	1	1	1	1	1	2	2	2	2	2	2	2	2	2	2	2	2	1	1	1	1	1
3	1	1	1	1	1	1	1	1	1	1	1	3	3	3	3	3	3	3	3	3	3	3	3	1	1	1	1	1
4	1	1	1	1	1	2	2	2	2	2	2	1	1	1	1	2	2	2	2	3	3	3	3	1	2	2	1	2
5	1	1	1	1	1	2	2	2	2	2	2	2	2	2	2	3	3	3	3	1	1	1	1	1	2	2	1	2
6	1	1	1	1	1	2	2	2	2	2	2	3	3	3	3	1	1	1	1	2	2	2	2	1	2	2	1	2
7	1	1	2	2	2	1	1	1	2	2	2	1	1	2	3	1	2	3	3	1	2	2	3	2	1	2	1	3
8	1	1	2	2	2	1	1	1	2	2	2	2	2	3	1	2	3	1	1	2	3	3	1	2	1	2	1	3
9	1	1	2	2	2	1	1	1	2	2	2	3	3	1	2	3	1	2	2	3	1	1	2	2	1	2	1	3
10	1	2	1	2	2	1	2	2	1	1	2	1	1	3	2	1	3	2	3	2	1	3	2	2	2	1	1	4
11	1	2	1	2	2	1	2	2	1	1	2	2	2	1	3	2	1	3	1	3	2	1	3	2	2	1	1	4
12	1	2	1	2	2	1	2	2	1	1	2	3	3	2	1	3	2	1	2	1	3	2	1	2	2	1	1	4
13	1	2	2	1	2	2	1	2	1	2	1	1	2	3	1	3	2	1	3	3	2	1	2	1	1	1	2	5
14	1	2	2	1	2	2	1	2	1	2	1	2	3	1	2	1	3	2	1	1	3	2	3	1	1	1	2	5
15	1	2	2	1	2	2	1	2	1	2	1	3	1	2	3	2	1	3	2	2	1	3	1	1	1	1	2	5
16	1	2	2	2	1	2	2	1	2	1	1	1	2	3	2	1	1	3	2	3	3	2	1	1	2	2	2	6
17	1	2	2	2	1	2	2	1	2	1	1	2	3	1	3	2	2	1	3	1	1	3	2	1	2	2	2	6
18	1	2	2	2	1	2	2	1	2	1	1	3	1	2	1	3	3	2	1	2	2	1	3	1	2	2	2	6
19	2	1	2	2	1	1	2	2	1	2	1	1	2	1	3	3	3	1	2	2	1	2	3	2	1	2	2	7
20	2	1	2	2	1	1	2	2	1	2	1	2	3	2	1	1	1	2	3	3	2	3	1	2	1	2	2	7
21	2	1	2	2	1	1	2	2	1	2	1	3	1	3	2	2	2	3	1	1	3	1	2	2	1	2	2	7
22	2	1	2	1	2	2	2	1	1	1	2	1	2	2	3	3	1	2	1	1	3	3	2	2	2	1	2	8
23	2	1	2	1	2	2	2	1	1	1	2	2	3	3	1	1	2	3	2	2	1	1	3	2	2	1	2	8
24	2	1	2	1	2	2	2	1	1	1	2	3	1	1	2	2	3	1	3	3	2	2	1	2	2	1	2	8

25	2	1	1	2	2	2	1	2	2	1	1	1	3	2	1	2	3	3	1	3	1	2	2	1	1	1	3	9
26	2	1	1	2	2	2	1	2	2	1	1	2	1	3	2	3	1	1	2	1	2	3	3	1	1	1	3	9
27	2	1	1	2	2	2	1	2	2	1	1	3	2	1	3	1	2	2	3	2	3	1	1	1	1	1	3	9
28	2	2	2	1	1	1	1	2	2	1	2	1	3	2	2	2	1	1	3	2	3	1	3	1	2	2	3	10
29	2	2	2	1	1	1	1	2	2	1	2	2	1	3	3	3	2	2	1	3	1	2	1	1	2	2	3	10
30	2	2	2	1	1	1	1	2	2	1	2	3	2	1	1	1	3	3	2	1	2	3	2	1	2	2	3	10
31	2	2	1	2	1	2	1	1	1	2	2	1	3	3	3	2	3	2	2	1	2	1	1	2	1	2	3	11
32	2	2	1	2	1	2	1	1	1	2	2	2	1	1	1	3	1	3	3	2	3	2	2	2	1	2	3	11
33	2	2	1	2	1	2	1	1	1	2	2	3	2	2	2	1	2	1	1	3	1	3	3	2	1	2	3	11
34	2	2	1	1	2	1	2	1	2	2	1	1	3	1	2	3	2	3	1	2	2	3	1	2	2	1	3	12
35	2	2	1	1	2	1	2	1	2	2	1	2	1	2	3	1	3	1	2	3	3	1	2	2	2	1	3	12
36	2	2	1	1	2	1	2	1	2	2	1	3	2	3	1	2	1	2	3	1	1	2	3	2	2	1	3	12

*Columns 1, 2, …, 22, and 23 form $OA_{36}(2^{11} \times 3^{12})$; columns 1′, 12, 13, …, 22, and 23 form $OA_{36}(12^1 \times 3^{12})$; columns 1′, 2′, 3′, 4′, 12, 13, …, 22, and 23 form $OA_{36}(2^3 \times 3^{13})$.

cient way of ascertaining that the effects of design parameters are additive is to check the prediction based on orthogonal main-effects-only plans by follow-up confirmatory experiments in the laboratory. When the results of confirmatory experiments approximately agree with the predicted results, it can be taken as an indication that the effects of design parameters are approximately additive. Thus optimal settings for the laboratory environment are likely to remain optimal for the manufacturing environment. On the other hand, when the predicted results disagree widely with the results of the confirmatory experiment, it is an indication that the effects of design parameters are not additive. In this situation it is difficult to determine optimal settings for the manufacturing environment (and the customer's use conditions) by laboratory experiments. So the engineer must return to the drawing board and redesign the product (or process) to determine design parameters whose effects are additive.

Second, the success of an experiment depends on whether it includes all of the important factors (design parameters and sources of noise). One way of increasing the probability that all of the important factors are included in the experiment is to maximize the number of factors included in the experiment. Now the required number of test runs increases with both the number of factors and the order of the experiment plan (a main-effects-only plan or a second-order plan involving interaction and curvature terms). Therefore for a given number of factors, the experimenter has to choose between the number of test runs and the order of the experiment plan. When the number of factors is very large, a main-effect plan may be the only practical choice.

20.6.3 Accumulation Analysis

Frequently performance characteristics cannot be measured on a continuous scale, but measurements on an ordered categorical scale are possible. Taguchi (1974) recommends accumulation analysis for analyzing ordered categorical data. Nair (1986) with discussions and response, and Hamada and Wu (1986) have criticized the accumulation analysis method, and they have discussed alternative data analysis techniques.

20.6.4 Binary Measurements

When the performance characteristic can be measured only on a binary scale such as good or bad, Taguchi (1976) recommends the following performance statistic:

$$Z(\theta) = 10 \log \left(\frac{p}{1-p} \right)$$

where p is proportion good. This performance statistic is the logistic transformation of the proportion p. Cox (1970) has discussed the use of logistic transformation for the data analysis of binary measurements.

20.6.5 Concluding Remarks

Taguchi has stimulated great interest in the application of statistically planned experiments to industrial product and process designs. Although the use of statistically planned experiments to improve industrial products and processes is not new [in fact, Tippett (1934) used statistically planned experiments in textile and wool industries more than 60 years ago], Taguchi has acquainted us with the breadth of the scope of statistically planned experiments for off-line quality control. From his writings it is clear that he uses statistically planned industrial experiments for at least four different purposes. (1) To identify the settings of design parameters at which the effect of the sources of noise on the performance characteristic is minimum. (2) To identify the settings of design parameters that reduce cost without hurting quality. (3) To identify those design parameters that have a large influence on the mean value of the performance characteristic, but have no effect on its variation. Such parameters can be used to adjust the mean value. (4) To identify those design parameters that have no detectable influence on the performance characteristics. The tolerances on such parameters can be relaxed. Taguchi's applications demonstrate that he has great insight into industrial quality and cost-control problems and the efficacy of statistically planned experiments. Since 1980 many U.S. companies have tried Taguchi's parameter design experiment approach and its modifications. The following papers describe some of the experiences: Phadke, Kacker, Speeney, and Grieco (1983), Pao, Phadke, and Sherrerd (1985), Pignatiello and Ramberg (1985), Lin and Kacker (1985), Kacker and Shoemaker (1986), Phadke (1986), Byrne and Taguchi (1986), Barker (1986), and American Supplier Institute (1984-1989).

20.7 DISPERSION MODELS OF ROBUST DESIGNS

Dispersion models of different kinds are considered in the literature for determining the robust design (see Box and Meyer, 1986; Nair and Pregibon, 1988; Ghosh and Lagergren, 1990, 1991; Shoemaker, Tsui, and Wu 1991). There are two setups,Taguchi and Standard.

20.7.1 Taguchi Setup

In Taguchi setup, the inner array and the outer array are considered. The inner array is in fact a matrix whose rows are level combinations of control factors and the outer array is also a matrix whose rows are level combinations of controllable noise factors. For each row of the inner array, the observations are taken at various rows of the noise array. These observations are assumed to have the same mean and variance. However, the means and variances are different for observations across the rows of the inner array. For details of the dispersion models, readers are referred to Box and Meyer (1986), Nair and Pregibon (1988), and Ghosh and Lagergren (1990).

20.7.2 Standard Setup

In Standard setup, the standard linear model with the homogeneity of variance assumption is considered. The vector of parameters consists of the general mean, main effects of both control and noise factors, and the two factor interactions between control and noise factors. The control-noise interactions play a very important role in this setup. For details, the readers are referred to Shoemaker,

Tsui, and Wu (1991), Ghosh and Derderian (1993), and Steinberg and Bursztyn (1994).

20.7.3 Robust Designs

The robust design is a set of level combinations of control factors so that the effect of the noise factors on response is minimum [see Kacker and Shoemaker (1986)]. The robust design is also called the parameter design (see Taguchi, 1987, Taguchi and Wu, 1980). In determining the robust design, fractional factorial experiments are performed using both control factors and controllable noise factors. The controllable noise factors have an important role in finding the robust design. Both Standard and Taguchi setups are considered in finding the robust designs. For details, readers are referred to Ghosh and Duh (1992) and Ghosh and Derderian (1995).

20.8 SUMMARY

This chapter provides an expository description of the philosophy and methodologies for quality control developed and promoted by Genichi Taguchi. He emphasizes that an important aspect of the quality of a product is the customer's loss due to variations in the product's performance characteristics during the product's life span and among different units of the product. Therefore the goal of quality control is to discover and implement effective and efficient countermeasures against the causes of variation. It turns out that effective and efficient countermeasures against most causes of variation can be built into the product only at the product and process design stages in the product's development cycle. Therefore quality control must start with the research and development of new product designs. Taguchi has introduced the terminology off-line quality control to refer to the technical aids for quality and cost control in product and process designs. One off-line quality control method, promoted by Taguchi, is called parameter design. The key idea underlying parameter design is that nonlinear effects of product or process parameters can be used to reduce the sensitivity of engineering designs to the causes of variation. Since parameter design reduces variation by reducing the sensitivity to the causes of variation rather than by controlling them, it is a very cost-effective technique to improve engineering designs. Frequently parameter design cannot be done analytically, and therefore an empirical approach based on statistically planned experiments is often used.

Taguchi has proposed some new and some not so new statistical methods for the design and analysis of parameter design experiments. Some of the details of statistical methods proposed by Taguchi are controversial. Over the next several years we can look forward to resolving these controversies. In the meantime we can combine Taguchi's quality philosophy and methodologies with the best of the engineering and statistical techniques we presently know.

REFERENCES

Addelman, S.: "Orthogonal Main Effect Plans for Asymmetrical Factorial Experiments," *Technometrics*, vol. 4, pp. 21–46, 1962.

_____: "Recent Developments in the Design of Factorial Experiments," *J. Am. Stat. Assoc.*, vol. 67, pp. 103–111, 1972.

American Supplier Institute: *Proc. ASI Symp. on Taguchi Methods*, vols. 1–6. Am. Supplier Inst., Dearborn, MI., 1984–1989.

Barker, T. B.: "Quality Engineering by Design: Taguchi's Philosophy," *Qual. Prog.*, pp. 32–42, Dec. 1986.

Bartlett, M. S.: "Properties of Sufficiency and Statistical Tests," *Proc. R. Soc. London A*, vol. 160, pp. 268–282, 1937.

Bose, R. C.: "Mathematical Theory of the Symmetrical Factorial Design," *Sankya*, vol. 8, pp. 107–166, 1947.

_____ and K. A. Bush: "Orthogonal Arrays of Strength Two through Three," *Ann. Math. Stat.*, vol. 23, pp. 508–524, 1952.

Box, G. E. P.: "Signal-to-Noise Ratios, Performance Criteria and Transformations," *Technometrics*, vol. 30, pp. 1–40, 1988.

_____ and R. D., Meyer: "Dispersion Effects from Fractional Designs," *Technometrics*, vol. 28, pp. 19–27, 1986.

Brayton, R. K., G. D. Hachtel, and A. L. Sangiovanni-Vincentelli: "A Survey of Optimization Techniques for Integrated-Circuit Design," *Proc. IEEE*, vol. 69, pp. 1334–1362, 1981.

Byrne, D. M., and S. Taguchi: "The Taguchi Approach to Parameter Design," in *ASQC Quality Congress Trans.* (Anaheim, CA, 1986).

Cox, D. R.: *The Analysis of Binary Data*, Chapman and Hall, London, 1970.

Dey, A.: *Orthogonal Fractional Factorial Designs*, Halstead Press, Wiley, New York, 1985.

Easterling, R. G.: "Discussion of Off-Line Quality Control, Parameter Design and Taguchi Method," *J. Qual. Technol.*, vol. 17, pp. 191–192, 1985.

Eberhart, S. A., and W. A. Russell: "Stability Parameters for Comparing Varieties," *Crop Sci.*, vol. 6, pp. 36–40, 1966.

Ghosh, S., Ed.: *Statistical Design and Analysis of Industrial Experiments*, Marcel Dekker, New York, 1990.

_____ and E. Derderian: "Robust Experimental Plan and Its Role in Determining Robust Design against Noise Factors," *Roy. Stat. Soc.*, Ser. D, *The Statistician*, vol. 42, pp. 19–28, 1993.

_____ and _____ "Determination of Robust Design Against Noise Factors and in Presence of Signal Factors," *Communications in Statistics, Simulation and Computation*, vol. 24, pp. 309–326, 1995.

_____ and Y. J. Duh: "Adjusting Residuals in Estimation of Dispersion Effects," *Australian Journal of Statistics*, vol. 33, no. 1, pp. 65–74, 1991.

_____ and _____ "Determination of Optimum Experimental Conditions Using Dispersion Main Effects and Interactions of Factors in Replicated Factorial Experiments", J. *Appl. Stat.*, vol. 19, no. 3, pp. 367–378, 1992.

_____ and E. Lagergren: "Measuring Dispersion Effects of Factors in Factorial Experiments," *Statistical Design and Analysis of Industrial Experiments*, chap. 16, pp. 459–478, Marcel Dekker, New York, 1990.

_____ and _____ "Dispersion Models and Estimation of Dispersion Effects in Replicated Factorial Experiments," *Journal of Statistical Planning and Inference*, vol. 26, no. 3, pp. 253–262, 1991.

Hahn, G. J., and S. S. Shapiro: "A Catalog and Computer Program for the Design and Analysis of Orthogonal Symmetric and Asymmetric Fractional Factorial Experiments," General Electric Co., CRD, Schenectady, N Y, 1966.

Hamada, M., and C. F. J. Wu: "A Critical Look at Accumulation Analysis and Related Methods," Tech. Rep., Dep. Statistics, University of Wisconsin, Madison, 1986.

Hunter, J. S.: "Statistical Design Applied to Product Design," *J. Qual. Technol.*, vol. 17, pp. 210–221, 1985.

Kacker, R. N.: "Some Orthogonal Arrays for Screening Designs," Tech. Memo., AT&T-Bell Labs., Holmdel, N J, 1982.

_____: "Off-Line Quality Control, Parameter Design, and the Taguchi Method," *J. Qual. Technol.*, vol. 17, pp. 176–209, 1985.

_____: "Taguchi's Quality Philosophy: Analysis and Commentary," *Qual. Prog.*, pp. 21–29, Dec. 1986.

_____ and A. C. Shoemaker: "Robust Design: A Cost-Effective Method for Improving Manufacturing Processes," *AT&T Tech. J.*, vol. 65, pp. 39–50, Mar./Apr. 1986.

_____ and K. L. Tsui: "Interaction Graphs: Graphic Aids for Planning Experiments," *J. Qual. Technol.*, vol. 22, pp. 1–14, 1990.

_____, E. S. Lagergren, and J. J. Filliben: "Taguchi's Fixed-Element Arrays Avre Fractional Factorials," *J. Qual. Technol.*, vol. 23, no. 2, pp. 107–116, 1991.

Leon, R. V., A. C. Shoemaker, and R. N. Kacker: "Performance Measures Independent of Adjustment: An Explanation and Extension of Taguchi's Signal-to-Noise Ratios," *Technometrics*, vol. 29, pp. 253–265, 1987.

Lin, K. M., and R. N. Kacker: "Optimizing the Wave Soldering Process," *Electron. Packag. Prod.*, pp. 108–115, Feb. 1986.

Nair, V. N.: "Testing in Industrial Experiments with Ordered Categorical Data," *Technometrics*, vol. 28, pp. 283–311, 1986.

_____ and D. Pregibon: "Analyzing Dispersion Effects from Replicated Factorial Experiments," *Technometrics*, vol. 30, pp. 247–257, 1988.

National Bureau of Standards: *Fractional Factorial Experiment Designs for Factors at Two Levels*, NBS Appl. Math. ser. 48, 1957.

_____: *Fractional Factorial Experiment Designs for Factors at Three Levels*, NBS Appl. Math. ser. 54, 1959.

_____: *Fractional Factorial Designs for Experiments with Factors at Two and Three Levels*, NBS Appl. Math. ser. 58, 1961.

Pao, T. W., M. S. Phadke, and C. S. Sherrerd: "Computer Response Time Optimization Using Orthogonal Array Experiments," *Proc. IEEE Int. Conf. on Communications*, pp. 890–895, June 1985.

Phadke, M. S.: "Design Optimization Case Studies," *AT&T Tech. J.*, vol. 65, pp. 51–58, Mar./Apr. 1986.

_____, R. N. Kacker, D. V. Speeney, and M. J. Grieco: "Off-Line Quality Control in Integrated Circuit Fabrication Using Experimental Design," *Bell Sys. Tech. J.*, vol. 62, pp. 1273–1309, May-June 1983.

Pignatiello, J. J., Jr., and J. S. Ramberg: "Discussion of Off-Line Quality Control, Parameter Design and the Taguchi Method," *J. Qual. Technol.*, vol. 17, pp. 198–206, 1985.

Plackett, R. L., and J. P. Burman: "The Design of Optimum Multifactorial Experiments," *Biometrika*, vol. 33, pp. 305–325, 1946.

Raghavarao, D.: *Constructions and Combinatorial Problems in Design of Experiments*, Wiley, New York, 1971.

Rao, C. R.: "Factorial Experiments Derivable from Combinatorial Arrangements of Arrays," *J. R. Stat. Soc.*, Suppl., vol. 9, pp. 128–139, 1947.

Shoemaker, A. C., K-L Tsui, and C. F. J. Wu: "Economical Experimentation Methods for Robust Design," *Technometrics*, vol. 33, pp. 415–427, 1991.

Steinberg, D. M., and D. Bursztyn: "Confounded Dispersion Effects in Robust Design Experiments with Noise Factors," *J. Qual. Technol.* vol. 26, pp. 12–20, 1994.

Taguchi, G.: "A New Statistical Analysis for Clinical Data, the Accumulating Analysis, in Contrast with the Chi-Square Test," *Saishin Igaku* (*The Newest Medicine*), vol. 29, pp. 806–813, 1974.

_____: *Experimental Designs*, 3d ed., vols. 1 and 2 (in Japanese). Maruzen Publ. Co., Tokyo, Japan, 1976. English transl., *The System of Experimental Design*, UNIPUB/Kraus Int. Publ. White Plains, N Y, 1987.

_____: "Off-Line and On-Line Quality Control Systems," in *Proc. Int. Conf. on Quality Control* Tokyo, Japan, 1978.

_____: *On-Line Quality Control during Production*, Japanese Standards Assoc., 1981; available from Am. Supplier Inst., Dearborn, MI.

_____: *Introduction to Quality Engineering*, Asian Productivity Organization, 1986; available from UNIPUB/Kraus Int. Publ., White Plains, N Y.

_____: *Orthogonal Arrays and Linear Graphs*, Am. Supplier Inst., Dearborn, MI, 1987.

_____ and Y. I. Wu: *Introduction to Off-Line Quality Control Systems*, Central Japan Quality Control Assoc., 1980; available from Am. Supplier Inst., Dearborn, MI.

Tippett, L. H. C.: "Applications of Statistical Methods to the Control of Quality in Industrial Production," *J. Manchester Stat. Soc.*, 1934.

CHAPTER 21
MULTIVARIATE QUALITY CONTROL

Frank B. Alt
University of Maryland, College Park, MD

Nancy D. Smith
Food and Drug Administration, Rockville, MD

and

Kamlesh Jain
Golden Gate University, San Francisco, CA

21.1 INTRODUCTION

In industrial situations control charts are often used to observe whether a process is in control. The procedures for doing this when there is one quality characteristic have been discussed in Chap. 7. Frequently it is desired to monitor several correlated quality characteristics simultaneously. In this case, multivariate quality control methods are necessary.

An example would be the situation described in Hatton (1983) concerning the manufacture of an automobile connecting rod. Of primary importance is the size of the large cylindrical hole at one end of the rod. Several measurements of the diameter of the hole are taken at specific angles, using a gauge that measures the deviation of the diameter of the rod from a standard. The minimum and maximum diameters over the entire top and bottom of the hole and the angles at which they occur are the key acceptance factors. Hatton gives data taken from 100 rods sampled from a day's production. We will only consider two of the eight variables, the maximum and minimum diameters measured at the top of the hole.

Assume the two characteristics are jointly distributed as a bivariate normal with a 2×1 mean vector $\boldsymbol{\mu}_0$ and a 2×2 covariance matrix $\boldsymbol{\Sigma}_0$, where the zero subscript indicates that these are standard values. Specifically,

$$\boldsymbol{\mu}_0 = \begin{bmatrix} \mu_{01} \\ \mu_{02} \end{bmatrix} = \begin{bmatrix} 0.365 \\ -1.080 \end{bmatrix}$$

$$\boldsymbol{\Sigma}_0 = \begin{bmatrix} \sigma_{01}^2 & \rho_0\sigma_{0_1}\sigma_{02} \\ \rho_0\sigma_{0_1}\sigma_{02} & \sigma_{02}^2 \end{bmatrix} = \begin{bmatrix} 0.969 & 0.519 \\ 0.519 & 1.019 \end{bmatrix} \tag{21.1}$$

In our situation, the correlation coefficient ρ_0 is 0.522.

Suppose we are interested in monitoring the mean of the process. A sample of, say, size 10 is drawn from the process at regular intervals and both characteristics are measured. If the two variables are monitored separately (ignoring the correlation), a univariate control chart could be constructed for each characteristic in the following way. For each sample of 10 items, the sample mean, denoted by $\bar{x}_i$, $i = 1, 2$, is calculated and plotted against time on an $\bar{x}$ chart with the upper and lower control limits UCL_i and LCL_i and the centerline CL_i given by

$$\begin{aligned} UCL_i &= \mu_{0i} + z_{\alpha/2}\left(\frac{\sigma_{0i}}{\sqrt{M}}\right) \\ CL_i &= \mu_{0i} \\ LCL_i &= \mu_{0i} - z_{\alpha/2}\left(\frac{\sigma_{0i}}{\sqrt{M}}\right) \end{aligned} \tag{21.2}$$

where M is the sample size and $z_{\alpha/2}$ is the standard normal percentile such that $P(Z > z_{\alpha/2}) = \alpha/2$. In our situation, using 3 Σ limits (which gives a type 1 error of 0.0027 for each chart) and setting $M = 10$, these limits would be

$$UCL_1 = 1.299 \qquad UCL_2 = -0.122$$
$$CL_1 = 0.365 \qquad CL_2 = -1.080$$
$$LCL_1 = -0.569 \qquad LCL_2 = -2.038$$

The process would be considered in control if both sample means plot within their respective control limits. This is equivalent to saying a process is in control if the sample means plot within a rectangle, as shown in Fig. 21.1. While the probability that either component will fall between its control limits is $1 - \alpha$, the overall significance level would not be α, but something smaller than α.

The true control region with significance level α is elliptical in shape, so use of the rectangular control region can lead to incorrect conclusions. The shape of the ellipse is determined by the correlation between the variables. Figure 21.2 shows the errors that can occur when simultaneous univariate control charts are used. However, individual $\bar{x}$ charts can be used in a supplemental fashion to help in determining which component caused an out-of-control signal. In this case, the significance level used for each chart should be α/p, where p is the number of variables. The use of this adjustment in Eqs. (21.2) yields what is commonly referred to as Bonferroni control limits.

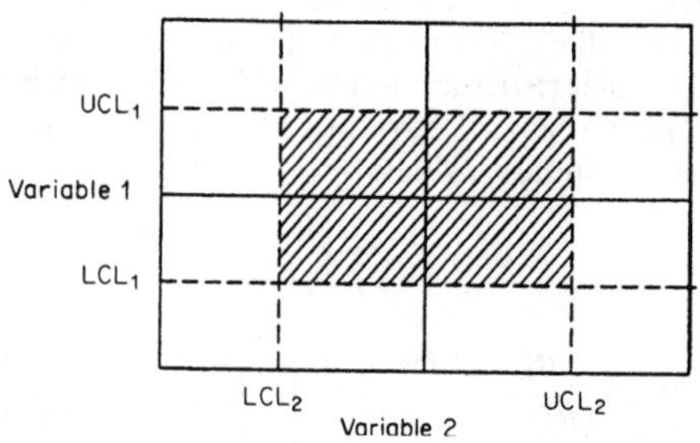

FIGURE 21.1 Rectangular control chart.

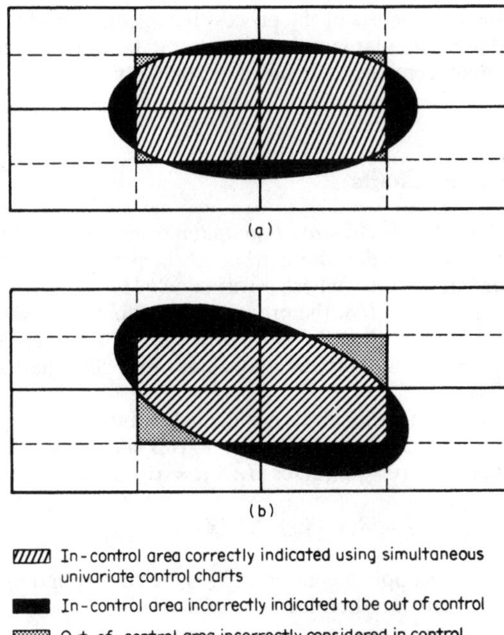

FIGURE 21.2 Elliptical control charts. (*a*) Independent variables with unequal variance. (*b*) Correlated variables with unequal variance

In the multivariate situation, the statistic that will be calculated will be the value of a quadratic form. Matrix multiplication of the form $\mathbf{x}'\mathbf{A}\mathbf{x}$, where $\mathbf{x}$ is a $p \times 1$ vector and $\mathbf{A}$ is a $p \times p$ matrix, will result in a scalar statistic which can be plotted on a chart or compared to predetermined limits. [Readers who are not familiar with matrix multiplication should consult a matrix algebra book such as Stewart (1973).] In the bivariate case, if

$$\mathbf{x} = \begin{bmatrix} x_1 \\ x_2 \end{bmatrix}, \qquad \mathbf{A} = \begin{bmatrix} a_{11} & a_{12} \\ a_{21} & a_{22} \end{bmatrix} \tag{21.3}$$

then $\mathbf{x}'\mathbf{A}\mathbf{x} = a_{11}x_1^2 + (a_{12} + a_{21})x_1x_2 + a_{22}\,x_2^2$. When $\mathbf{A}$ is a symmetric matrix, $\mathbf{x}'\mathbf{A}\mathbf{x} = a_{11}x_1^2 + 2a_{12}x_1x_2 + a_{22}x_2^2$ since $a_{12} = a_{21}$.

21.2 PHASE II CONTROL CHARTS FOR THE MEAN

When a process is first under observation, the values of the parameters in $\boldsymbol{\mu}_0$ and $\boldsymbol{\Sigma}_0$ are usually unknown. They must be estimated from samples, which are taken when the process is assumed to be in control. The control charts used at this stage are referred to as phase I charts. Phase II control charts are constructed under the assumption that the parameters of the process being considered are known. These can be estimates based on a large amount of past data, or they can be "target values" chosen by management to achieve certain objectives (for example, see Duncan, 1986).

21.2.1 Multivariate $\bar{x}$ Charts

Suppose that random samples of size M are taken from a process at specified intervals and an $\bar{x}$ chart is used to determine whether the process is in control. The control limits in the univariate case, where σ was known [Eq. (21.2)], can be viewed as repeated tests of significance that the process mean equals the standard value. The areas above the UCL and below the LCL correspond to the rejection regions for this test of hypothesis. A multivariate equivalent of the Shewhart $\bar{x}$ chart was presented in Alt (1973).

If the p-variate random variable $\mathbf{x}$ is normally distributed with known covariance matrix $\mathbf{\Sigma}_0$ and a random sample of M observations is taken, the likelihood ratio test (Anderson, 1984) of H_0: $\boldsymbol{\mu} = \boldsymbol{\mu}_0$ versus H_1: $\boldsymbol{\mu} \neq \boldsymbol{\mu}_0$ rejects the null hypothesis if

$$\chi_0^2 = M(\bar{\mathbf{x}} - \boldsymbol{\mu}_0)' \, \mathbf{\Sigma}_0^{-1} \, (\bar{\mathbf{x}} - \boldsymbol{\mu}_0) > \chi_{p,\alpha}^2 \tag{21.4}$$

where x is the vector of sample means and $\chi_{p,\alpha}^2$ is the corresponding chi square percentile. The control chart can therefore be defined by

$$\begin{aligned} \text{UCL} &= \chi_{p,\alpha}^2 \\ \text{LCL} &= 0 \end{aligned} \tag{21.5}$$

where the values plotted are $M(\bar{\mathbf{x}} - \boldsymbol{\mu}_0)' \, \mathbf{\Sigma}_0^{-1} \, (\bar{\mathbf{x}} - \boldsymbol{\mu}_0)$. This procedure will be referred to as the χ^2 chart.

In the bivariate case Eq. (21.4) is equivalent to

$$\begin{aligned} \chi_0^2 = M(1 - \rho_0^2)^{-1}[(\bar{x}_1 - \mu_{01})^2\sigma_{01}^{-2} &+ (\bar{x}_2 - \mu_{02})^2\sigma_{02}^{-2} \\ &- 2\rho_0\sigma_{01}^{-1}\sigma_{02}^{-1}(\bar{x}_1 - \mu_{01})(\bar{x}_2 - \mu_{02})] \end{aligned} \tag{21.6}$$

This is the equation of an ellipse with center at (μ_{01}, μ_{02}). Thus if a particular vector of sample means plots outside the ellipse of Eq. (21.6), the process would be considered out of control, and an attempt would be made to find the cause. For our example, we have

$$\chi_0^2 = \frac{10}{1 - (0.522)^2}\left[\frac{(\bar{x}_1 - 0.365)^2}{0.969} + \frac{(\bar{x}_2 + 1.080)^2}{1.019} - \frac{2(0.522)(\bar{x}_1 - 0.365)(\bar{x}_2 + 1.080)}{0.984(1.009)}\right] \tag{21.7}$$

The hypothesis that the process mean equals $\boldsymbol{\mu}_0$ would be rejected at the $\alpha = 0.0054$ level if χ_0^2 is greater than $\chi_{2,0.0054}^2 = 10.44$.

Alt (1973) discusses the power of this test for a single sample. Alt, Walker, and Goode (1980) investigated the effect on the power of the test in the bivariate case when one of the population standard deviations changes. They found that when the two characteristics are positively correlated and σ_2 is fixed, the power is not a monotonically decreasing function of σ_1 as it is in the univariate case.

The performance of a control chart is judged in terms of its average run length (ARL). The ARL of a procedure is the average number of samples taken before an out-of-control (action) signal is given. When a process is actually in control, an action signal is a false alarm. Therefore, we want to design quality control procedures with large in-control ARLs and relatively small out-of-control ARLs. With a significance level of 0.005, the in-control ARL for the χ^2 chart is 200.

To demonstrate the dependence of the ARL performance on ρ, assume that the two variables are standardized normal random variables with means 0, variances 1, and that the correlation coefficient between the two variables is ρ. Table 21.1 gives the theoretical ARLs for some shifts in the process mean vector and for various values of the correlation coefficient. The in-control ARL for this procedure is 200. For a given shift in the process mean vector, the out-of-control ARLs corresponding to $\rho = 0.9$ are much smaller than the ARLs for $\rho = 0$.

Example 21.1. The use of the χ^2 control chart is illustrated by assuming that the standard values are those specified by Eq. (21.1). The summary statistics for simulated data from such a process are presented in Table 21.2. The sample size M equals 10. Thus the values of χ_0^2 are obtained from Eq. (21.7) and are listed in Table 21.2. For samples 14 and 15, the process mean for the first variable (μ_{01}) was increased by 1 standard deviation to 1.35, and the process mean for the second variable was decreased by 1 standard deviation to –2.09. The χ^2 chart detected these shifts beginning with sample 14. However, this was not the case for the univariate $\bar{x}$ charts.

21.2.2 Multivariate CUSUM Charts

Cumulative sum (CUSUM) control charts were first introduced by Page in 1954. He proposed that a decision about changes in a process should not be based on a single observation or a statistic calculated from a few observations, but on all the observations that had been obtained up to the time of testing. The information from the new sample should be combined with past data to give an indication of a possible shift in the process level. Often successive observations cannot be considered independent, so combining past and present data should increase sensitivity and speed the detection of small shifts in a process.

When the observations are summed and these CUSUMs are plotted on a graph, changes in the mean are indicated by changes in the slope of the line connecting successive values. The value of the CUSUM after N observations is

$$S_N = \sum_{i=1}^{N} z_i \tag{21.8}$$

TABLE 21.1 Theoretical ARLs for the χ^2 Chart for Different Values of ρ

Mean vector	$\rho = 0$	$\rho = 0.5$	$\rho = 0.9$
(0.0,0.0)	200.0	200.0	200.0
(0.5,0.0)	114.6	98.5	30.6
(1.0,0.0)	41.1	30.2	4.8
(1.5,0.0)	15.9	10.8	1.6

TABLE 21.2 Statistics for Control Charts for the Mean and for Dispersion (Phase II)

Sample	$\bar{x}_1$	$\bar{x}_2$	χ_0^2	s_1	s_2	s_{12}	$\lvert S\rvert$	W^*
1	0.352	–1.05	0.02	0.829	1.238	0.5042	0.7991	1.86
2	0.562	–1.37	2.51	0.659	1.010	0.4786	0.2139	4.61
3	0.587	–1.68	7.48	1.018	1.259	0.9350	0.7684	1.71
4	0.172	–1.66	3.45	1.221	0.774	0.8175	0.2248	8.13
5	0.351	–1.90	8.91	0.889	1.101	0.6031	0.5943	0.67
6	0.142	–1.76	4.75	1.005	1.049	0.5238	0.7357	0.07
7	0.614	–1.73	8.92	1.207	0.696	0.6213	1.0708	5.69
8	0.885	–1.10	3.99	0.841	0.777	0.3975	0.5493	2.03
9	0.811	–0.83	2.05	0.817	0.612	0.2508	0.1871	3.91
10	0.152	–1.17	0.48	0.807	0.796	0.4893	0.1732	4.44
11	0.228	–1.25	0.32	1.156	1.320	0.9899	1.3485	1.68
12	0.786	–1.19	3.35	1.098	1.075	0.7577	0.8191	0.39
13	0.332	–1.24	0.28	0.767	1.331	0.8498	0.3200	7.25
14	1.308	–1.87	31.80	0.612	0.733	0.2833	0.1210	5.65
15	1.219	–1.76	24.98	1.179	0.970	0.5464	1.0093	1.00

Mean	Dispersion	
UCL = $\chi^2_{2,0.0054}$ = 10.44	$\lvert S\rvert^{1/2}$ chart (3σ limits)	UCL = 1.552
		CL = 0.753
		LCL = 0.00
	W^* chart (α = 0.01)	UCL = 12.38

where $z_i = (x_i - \mu)/\sigma$. The graph of the CUSUMs gives a clear visual indication of the process behavior, and even small shifts in the process level are quickly indicated.

A CUSUM chart can be regarded as the repeated application of a sequential sampling procedure "in reverse." Wald (1947) devised the sequential probability ratio test (SPRT), and it has been applied to many problems. Johnson and Leone (1962) showed that it was applicable to CUSUM charts.

If the random variable x is distributed normally with mean μ and known standard deviation σ, and if a sample of N mutually independent values are observed, the procedure would be as follows. Let

$$T_i = \frac{M^{1/2}}{\sigma}(\bar{x}_i - \mu_0)$$

$$\mathrm{SH}_i = \max(0, \mathrm{SH}_{i-1} + T_i - k) \tag{21.9}$$

$$\mathrm{SL}_i = \min(0, \mathrm{SH}_{i-1} + T_i - k)$$

where

$$\mathrm{SH}_0 = \mathrm{SL}_0 = 0$$

A signal is triggered if SH > h or if SL < $-h$. Here k is the reference value and h is the decision interval since it is the out-of-control signal.

Several other multivariate cumulative methods have been developed. Crosier (1986) developed a two-sided vector-valued procedure based on a single CUSUM. After each observation, he updates the CUSUM and then shrinks the updated CUSUM toward zero. He shrinks the CUSUM by multiplication rather than by addition or subtraction. After N observations, he defines C_N as

$$C_N = [(\mathbf{s}_{N-1} + \mathbf{x}_N - \boldsymbol{\mu}_0)'\boldsymbol{\Sigma}^{-1}(\mathbf{s}_{N-1} + \mathbf{x}_N - \boldsymbol{\mu}_0)]^{1/2} \tag{21.10}$$

Then

$$\mathbf{s}_N = \begin{cases} 0, & \text{if } C_N < k \\ (\mathbf{s}_{N-1} + \mathbf{x}_N - \boldsymbol{\mu}_0)\left(1 - \dfrac{k}{C_N}\right), & \text{if } C_N > k \end{cases} \tag{21.11}$$

where $s_0 = 0$ and k is the reference value of the scheme or the allowable slack in the process ($k > 0$). In this case, C_N is the generalized length of the CUSUM vector before shrinking. This method will indicate that the process is out of control at the first N such that $\mathbf{s}_N'\boldsymbol{\Sigma}^{-1}\,\mathbf{s}_N > h^2$, where h is the decision interval of the scheme ($h > 0$). Based on Monte Carlo methods, Crosier states that the average run length in the bivariate case is 200 using $k = 0.5$ and $h = 5.5$.

A procedure based on the likelihood ratio test has been developed by Smith (1987). This method works very well if the direction of a shift can be anticipated. If the deviation that the procedure desires to detect is given by **e**, the procedure is as follows:

1. Calculate the Cholesky decomposition of the known covariance matrix $\boldsymbol{\Sigma}$, $\boldsymbol{\Sigma} = \mathbf{RR}'$ (see Stewart, 1973).
2. Calculate $\boldsymbol{\delta}$ in the equation $\mathbf{R}\boldsymbol{\delta} = \mathbf{e}$ by back substitution.
3. For each successive sample of size M, beginning with $N = 1$, calculate the sum

$$X_N = M^{1/2} \sum_{i=1}^{N} \boldsymbol{\delta}' R^{-1}(\bar{x}_i - \boldsymbol{\mu}_0) \tag{21.12}$$

4. Let $S_N = \max(0, X_N - k)$, where $S_0 = 0$ and $k = \boldsymbol{\delta}'\boldsymbol{\delta}/2$. An out-of-control signal is given if $S_N \geq h$, where h is chosen to give the desired average run length when the process is in control.

This method is very sensitive to shifts in the specified direction. By varying the signs of the components of the vector **e**, the method can also quickly detect shifts in other directions. Healy (1987) has proposed a similar method.

21.2.3 Exponentially Weighted Moving Average Charts

Lowry, Woodall, Champ, and Rigdon (1992) developed a multivariate exponentially weighted moving average (MEWMA) procedure. Their procedure is as follows: Let

$$\mathbf{Z}_i = rx_i + (1 - r)\mathbf{Z}_{i-1} \tag{21.13}$$

where $\mathbf{Z}_0 = 0$ and $0 < r \leq 1$. The MEWMA procedure gives an out-of-control signal as soon as

$$\mathbf{T}^2 = \mathbf{Z}_i' \, \boldsymbol{\Sigma}_E^{-1} \, \mathbf{Z}_i > h_1 \tag{21.14}$$

where $\boldsymbol{\Sigma}_E = \{r[1-(1-r)^{2i}]/(2-r)\}\boldsymbol{\Sigma}$ and is the exact covariance matrix of $\mathbf{Z}_i$.

The asymptotic covariance matrix of $\mathbf{Z}_i$ is $\boldsymbol{\Sigma}_A = [r/(2-r)]\boldsymbol{\Sigma}$. If the asymptotic covariance matrix of $\mathbf{Z}_i$ is used in the computation of $\mathbf{T}^2$, the out-of-control signal is given when $\mathbf{T}^2$ is greater than h_2. The values of h_1 and h_2 are found by simulation to attain a given ARL. Based on simulations, it was found that smaller values of the smoothing constant r are more effective in detecting smaller shifts. Lowry et al. (1992) have shown that the performance of the MEWMA charts with $r = 0.1$ compares favorably with that of the multivariate CUSUM charts. If a fast initial response is desired, use of the exact covariance matrix of $\mathbf{Z}_i$ is recommended.

Using the exact covariance matrix of $\mathbf{Z}_i$, the MEWMA procedure appears to perform even better than the χ^2 procedure for detecting large shifts in the mean vector. The principal reason for this is that, even though for the first observation the MEWMA statistic $\mathbf{T}^2$ is the same as the χ^2 statistic, the decision interval for the MEWMA procedure is smaller than the decision interval for the Shewhart based procedure.

Several researchers have suggested that both univariate and multivariate CUSUM and EWMA procedures suffer from the problems of inertia. It has been suggested that the use of a Shewhart type rule along with the CUSUM or EWMA procedure could partially alleviate this problem. Lowry et al. (1992), however, point out that this solution involves a trade-off between protection from inertia and faster detection of small shifts in the mean vector.

21.2.4 Other Procedures

Hawkins (1991) indicated that an out-of-control signal for the Shewhart chart confounded mean shifts with variance shifts and required extensive analysis following a signal to determine the nature of the signal. He proposed Shewhart and CUSUM controls based on the vector $\mathbf{Z}$ of scaled residuals from the regression of each variable on the remaining variables. Each component of $\mathbf{Z}$ is the optimal single-test statistic for testing whether the mean of that variable has shifted. The Shewhart charts plot the components of $\mathbf{Z}$ and the CUSUM charts are based on the accumulation of components of $\mathbf{Z}$, leading one to anticipate good performance by the charts. The vector $\mathbf{Z}$ also has the valuable interpretive property that signals given are for shifts in the mean, or shifts in the variance, of particular variables rather than global signals indicating some unspecified departure from control.

Bayesian decision theory can be used to formally incorporate prior knowledge in monitoring a process. Jain (1993) proposed using a bayesian approach to monitor a multivariate process mean. She assumed that the multiple quality characteristics of interest have a multivariate normal distribution with a known covariance matrix $\boldsymbol{\Sigma}$ but unknown mean vector $\boldsymbol{\mu}$. Although the process mean vector is not known, a prior distribution for the process mean vector can be specified. Then, each time a sample is taken the posterior distribution is revised to incorporate the most recent sample data. The quadratic form used to determine if the process mean is in control is

(posterior mean vector)′ (posterior covariance matrix) (posterior mean vector)

If the value of the quadratic form for sample I, $I = 1,2,3, \ldots$, exceeds a simulation-determined decision interval (h), then it is deemed that the process is out of control and a special cause is sought. Based on simulation results, this multivariate bayesian procedure has much higher in-control ARLs than those for the existing multivariate procedures.

21.3 PHASE I CONTROL CHARTS FOR THE MEAN

21.3.1 Stage 1

Since the process parameters are unknown, they must be estimated from preliminary samples, which are referred to as rational subgroups. Hillier (1969) and Yang and Hillier (1970) developed a procedure consisting of two stages to determine first whether the data for the initial K subgroups came from an in-control process and then to determine whether future data from this process remained in control. This was extended to the multivariate case by Alt et al. (1976).

A random sample of K rational subgroups of observations are taken under identical conditions. In our example, K will be 8. We assume that statistical control existed within each subgroup, but not necessarily among the subgroups. Let $\mathbf{X}_j$ denote the $p \times N$ data matrix for group j and $\mathbf{x}_{ij}$ refer to the ith p-variable vector in group j. The sample mean vector $\bar{\mathbf{x}}_j$ and the sample covariance matrix $\mathbf{S}_j$ can be calculated for each group. If there is statistical control across the subgroups, then unbiased estimates for $\boldsymbol{\mu}$ and $\boldsymbol{\Sigma}$ are given by

$$\bar{\bar{\mathbf{x}}} = \frac{1}{K}\sum_{j=1}^{K} \bar{\mathbf{x}}_j, \qquad \mathbf{S}_p = \frac{1}{K}\sum_{j=1}^{K} \mathbf{S}_j \tag{21.15}$$

where

$$\bar{\mathbf{x}}_j = \frac{1}{N}\sum_{i=1}^{N} \mathbf{x}_{ij}, \qquad \mathbf{S}_j = \frac{1}{N-1}\mathbf{A}_j$$

$$\mathbf{A}_j = \sum_{i=1}^{N} (\mathbf{x}_{ij} - \bar{\mathbf{x}}_j)(\mathbf{x}_{ij} - \bar{\mathbf{x}}_j)'$$

$\mathbf{A}_j$ denoting the sums of squares and cross-products matrix for group j.

Alt (1982) showed that $N(\bar{\mathbf{x}}_j - \bar{\bar{\mathbf{x}}})'\mathbf{S}_p^{-1}(\bar{\mathbf{x}}_j - \bar{\bar{\mathbf{x}}})$, which is the phase I equivalent of χ_0^2 in Eq. (21.4), is distributed as

$$\frac{KNp - Kp - Np + p}{KN - K - p + 1} F_{p,\, KN - K - p + 1} \tag{21.16}$$

where $F_{\nu 1,\nu 2}$ is Snedecor's F with ν_1 and ν_2 degrees of freedom. In the bivariate case, the statistic calculated for each group, $j = 1, 2, \ldots, K$, would be

$$T_{0,1}^2 = N(\det S_p)^{-1}[(\bar{x}_{1,j} - \bar{\bar{x}}_1)^2 s_2^2 + (\bar{x}_{2,j} - \bar{\bar{x}}_2)^2 s_1^2 - 2(\bar{x}_{1,j} - \bar{\bar{x}}_1)(\bar{x}_{2,j} - \bar{\bar{x}}_2))s_{12}] \tag{21.17}$$

where

$$\mathbf{S}_p = \begin{bmatrix} s_1^2 & s_{12} \\ s_{12} & s_2^2 \end{bmatrix}$$

is defined as above and $\det \mathbf{S}_p = s_1^2 s_2^2 - s_{12}^2$. The control limits are given by

$$\text{UCL} = \frac{KNp - Kp - Np + p}{KN - K - p + 1} F_{p,\, KN-K-p+1,\, \alpha} \tag{21.18}$$

$$\text{LCL} = 0$$

If $T^2_{0,1}$ for any of the initial K rational subgroups plots out of control, that subgroup is discarded, the control limits are recalculated using the revised value of K, the $\bar{\bar{\mathbf{x}}}$ and $\mathbf{S}_p$ are recomputed, and the statistic

$$T^2_{0,1} = N(\bar{\mathbf{x}}_j - \bar{\bar{\mathbf{x}}})'\mathbf{S}_p^{-1}(\bar{\mathbf{x}}_j - \bar{\bar{\mathbf{x}}}) \tag{21.19}$$

is calculated for each of the K^* remaining subgroups, $K^* < K$. This procedure is continued until none of the values of $T^2_{0,1}$ exceeds the UCL in Eq. (21.18)

21.3.2 Stage 2

A similar procedure can be used to investigate future subgroups. Yang and Hillier (1970) refer to this as stage 2. Let $\mathbf{S}_p$ denote the sample variance-covariance matrix obtained from stage 1 and let $\bar{\mathbf{x}}_f$ denote the vector of sample means for each future subgroup obtained in stage 2. In this case, the test statistic is

$$T^2_{0,2} = N(\bar{\mathbf{x}}_f - \bar{\bar{\mathbf{x}}})'\mathbf{S}_p^{-1}(\bar{\mathbf{x}}_f - \bar{\bar{\mathbf{x}}}) \tag{21.20}$$

which is distributed as

$$\frac{pNK + pN - pK - p}{KN - K - p + 1} \mathrm{F}_{p,\, KN-K-p+1}$$

The control limits are given by

$$\text{UCL} = \frac{pNK + pN - pK - p}{KN - K - p + 1} F_{p,\, KN-K-p+1,\, \alpha} \tag{21.21}$$

$$\text{LCL} = 0$$

The data for those subgroups where $T^2_{0,2} < \text{UCL}$ in Eq. (21.21) can be used to update the pooled sample variance-covariance matrix $\mathbf{S}_p$ and the overall sample mean vector $\bar{\bar{\mathbf{x}}}$, both obtained from stage 1. When this updating occurs, the value of K in Eq. (21.21) needs to be revised to K^*, where K^* denotes the number of subgroups used to calculate $\mathbf{S}_p$.

Example. Again consider the deviations from a standard of the maximum and minimum diameters at one end of an automobile connecting rod. Altogether there are 8 rational subgroups ($K = 8$) in Table 21.3. For subgroups 4 and 8, each process mean was increased by 1 standard deviation. The summary statistics are presented in Table 21.3. Since $T^2_{0,1} > \text{UCL}$ for each of these two subgroups, $\bar{\bar{\mathbf{x}}}$ and $\mathbf{S}_p$ were recomputed using the remaining six subgroups, and the UCL in Eq. (21.18) was recalculated using $K = 6$. For each of the remaining six subgroups $T^2_{0,1}$ was less than the revised UCL.

Table 21.4 presents the summary statistics for future subgroups. In this case neither of the process means was intentionally altered. For the 10 subgroups consid-

TABLE 21.3 Statistics for Control Charts for the Mean (Phase I), Stage 1

Subgroup i	$\bar{x}_{1,i}$	$\bar{x}_{2,i}$	$s_{1,j}^2$	$s_{2,j}^2$	$s_{12,j}$	$T_{0,1}^2$	$T_{0,1}^2$
1	0.337	–0.936	0.953	1.08	0.389	1.03	0.31
2	0.370	–1.04	0.982	1.17	0.616	1.06	0.00
3	0.383	–0.885	1.02	0.962	0.352	0.77	0.13
4	1.645	0.194	1.64	0.735	0.881	10.83	—
5	0.310	–0.973	0.694	0.943	0.430	1.21	0.25
6	0.380	–1.29	1.24	0.840	0.740	2.98	0.95
7	0.376	–1.24	0.842	0.781	0.904	2.44	0.58
8	1.517	0.285	0.971	1.87	0.845	10.72	—
		UCL = 8.70*		Revised UCL = 8.66*			

*Using $\alpha = 0.01$.

ered, $T_{0,2}^2$ is less than the UCL in Eq. (21.21). At this time it may be appropriate to update $\bar{\bar{\mathbf{x}}}$, and $\mathbf{S}_p$, and $T_{0,2}^2$ for any other future subgroups would be computed using these updated values.

21.4 PHASE II CONTROL CHARTS FOR DISPERSION

Duncan (1986) gives an explanation of various univariate charts used to monitor dispersion. Also refer to Chap. 7. These include the Range chart, the s chart where s is the sample standard deviation, and the s^2 chart where s^2 is the sample variance. This latter chart is equivalent to repeated tests of significance that the process variance equals the standard value. This viewpoint carries over to the multivariate case and is discussed in Alt and Bedewi (1986).

TABLE 21.4 Statistics for Control Charts for the Mean (Phase I), Stage 2

Subgroup f	$\bar{x}_{1,f}$	$\bar{x}_{2,f}$	$T_{0,2}^2$
1	0.302	–1.09	0.03
2	0.380	–0.98	0.08
3	0.289	–1.08	0.06
4	0.403	–0.99	0.05
5	0.370	–1.05	0.00
6	0.396	–1.00	0.04
7	0.376	–0.95	0.21
8	0.421	–1.16	0.10
9	0.339	–1.24	0.45
10	0.358	–1.08	0.01
	UCL = 12.15		

Let W^* denote the likelihood ratio test statistic, modified to be unbiased, for testing H_0: $\Sigma = \Sigma_0$ versus H_1: $\Sigma \neq \Sigma_0$. Then, for a random sample of size M,

$$W^* = -p(M-1) - (M-1)\ln(|S|) + (M-1)\ln(|\Sigma_0|) + (M-1)\,\mathrm{tr}\,(\Sigma_0^{-1}S) \tag{21.22}$$

where tr $(\Sigma_0^{-1}S)$ is the trace of $\Sigma_0^{-1}S$. Although W^* is asymptotically distributed as χ^2 with $p(p+1)/2$ degrees of freedom, the upper 5% and 1% points of the exact distribution of W^* for $p = 2, \ldots, 10$ and various values of M have been tabulated (for example, see Anderson, 1984). For each sample of size M, the calculated value of W^* would be compared with either the 1% or the 5% tabled values to ascertain whether the process dispersion has gone awry.

The quantity $|S|$ is known as the *generalized sample variance* and has a geometrical interpretation. For a $p \times M$ data matrix, let $\mathbf{d}_i$, $i = 1, 2, \ldots, p$, denote the ith deviation vector defined by

$$\mathbf{d}_i = [x_{i1} - \bar{x}_1, x_{i2} - \bar{x}_2, \ldots, x_{iM} - \bar{x}_M]'$$

Then $|S|$ is proportional to the square of the volume of the parallelotope which has $\mathbf{d}_1, \mathbf{d}_2, \ldots, \mathbf{d}_p$ as principal edges (for example, see Anderson, 1984).

In the univariate case, the control limits for the s chart are motivated by the principle that most of the probability distribution of the sample standard deviation S is within 3 standard deviations of $E(S)$. Since $E(S) = c_4\sigma_0$, where

$$c_4 = \left(\frac{2}{M-1}\right)^{1/2} \frac{\Gamma(M/2)}{\Gamma((M-1)/2)}$$

and

$$\mathrm{Var}(S) = E(S^2) - [E(S)]^2 = \sigma_0^2 - c_4^2\sigma_0^2 = \sigma_0^2(1 - c_4^2)$$

it follows that the control limits for the s chart are given by

$$\begin{aligned} \mathrm{UCL} &= E(S) + 3\sqrt{\mathrm{Var}(S)} = c_4\sigma_0 + 3\sqrt{\sigma_0^2(1-c_4^2)} = B_6\sigma_0 \\ \mathrm{CL} &= c_4\sigma_0 \\ \mathrm{LCL} &= E(S) - 3\sqrt{\mathrm{Var}(S)} = c_4\sigma_0 - 3\sqrt{\sigma_0^2(1-c_4^2)} = B_5\sigma_0 \end{aligned} \tag{21.23}$$

The multivariate analog of the univariate s chart is obtained by looking at the square root of $|\mathbf{S}|$, since $|\mathbf{S}|^{1/2}$ equals the sample standard deviation when there is only one quality characteristic. Define b_1 and b_3 as follows:

$$b_1 = (M-1)^{-p} \prod_{i=1}^{p} (M - i) \tag{21.24}$$

$$b_3 = \left(\frac{2}{M-1}\right)^{p/2} \frac{\Gamma(M/2)}{\Gamma((M-p)/2)} \tag{21.25}$$

Since $E(|\mathbf{S}|^{1/2}) = b_3|\mathbf{\Sigma}_0|^{1/2}$ and $\mathrm{Var}(|\mathbf{S}|^{1/2}) = (b_1 - b_3^2)|\mathbf{\Sigma}_0|$, the 3σ control limits for an $|\mathbf{S}|^{1/2}$ chart are given by

$$\mathrm{UCL} = (b_3 + 3\sqrt{b_1 - b_3^2})\,|\boldsymbol{\Sigma}_0|^{1/2}$$
$$\mathrm{CL} = b_3\,|\boldsymbol{\Sigma}_0|^{1/2} \qquad (21.26)$$
$$\mathrm{LCL} = (b_3 - 3\sqrt{b_1 - b_3^2})\,|\boldsymbol{\Sigma}_0|^{1/2}$$

When there is only one quality characteristic, the control limits in Eqs. (21.26) reduce to those in Eqs. (21.23).

Other multivariate phase II control charts for dispersion are found in Alt (1985) and Alt and Smith (1988).

Example. To illustrate the use of the two phase II procedures for monitoring process variability, suppose the standard value of $\boldsymbol{\Sigma}_0$ is as stated in Eqs. (21.1). Recall that $M = 10$ (refer to Table 21.2). Neither the $|\boldsymbol{S}|^{1/2}$ chart nor the W^* chart indicates that the process dispersion has drifted from the standard as specified in Eqs. (21.1).

21.5 PHASE I CONTROL CHARTS FOR DISPERSION

The phase I approach used to monitor dispersion will differ from that used to monitor the mean where two different stages were considered. Instead, the unknown population parameters in Eqs. (21.22) and (21.26) will be replaced by unbiased estimates.

Examination of Eq. (21.22) reveals that unbiased estimates of $|\boldsymbol{\Sigma}_0|$ and $\boldsymbol{\Sigma}_0^{-1}$ are needed. Define $|\boldsymbol{V}| = (1/K)\Sigma_{i=1}^{K} |\boldsymbol{S}_i|$, which is the average of the generalized sample variances for the K subgroups. Since $E(|\boldsymbol{S}_i|) = b_1\,|\boldsymbol{\Sigma}_0|$, $i = 1, 2, \ldots, K$, it follows that $|\boldsymbol{V}|/b_1$ is an unbiased estimate of $|\boldsymbol{\Sigma}_0|$ in Eq. (21.22). Furthermore, Kshirsagar (1972) shows that $(M - p - 2)\boldsymbol{S}_i^{-1}/(M - 1)$ is an unbiased estimate of $\boldsymbol{\Sigma}_0^{-1}$. Since $\boldsymbol{S}_1, \boldsymbol{S}_2, \ldots, \boldsymbol{S}_K$ are available, it follows that $(M - p - 2)\,\boldsymbol{V}_*^{-1}/(M - 1)$ is an unbiased estimate of $\boldsymbol{\Sigma}_0^{-1}$ in Eq. (21.22), where $\mathbf{V}_*^{-1} = (1/K)\,\Sigma_{i=1}^{K}\,\boldsymbol{S}_i^{-1}$. Thus the phase I statistic is

$$W^* = -p(M-1) - (M-1)\ln(|\boldsymbol{S}_i|) + (M-1)\ln\frac{|\boldsymbol{V}|}{b_1} + (M-1)\,\mathrm{tr}\left[\frac{(M-p-2)\mathbf{V}_*^{-1}\,\mathbf{S}_i}{M-1}\right] \qquad (21.27)$$

for $i = 1, 2, \ldots, K$. Values of W^* are still compared with the 1% or 5% tabled values.

For the control limits presented in Eqs. (21.26), the only unknown parameter is $|\boldsymbol{\Sigma}_0|^{1/2}$. Since $E(|\boldsymbol{S}_i|^{1/2}) = b_3|\boldsymbol{\Sigma}_0|^{1/2}$, $i = 1, 2, \ldots, K$, it follows that $|\mathbf{U}|^{1/2}/b_3$ is an unbiased estimate of $|\boldsymbol{\Sigma}_0|^{1/2}$, where $|\mathbf{U}|^{1/2} = (1/K)\,\Sigma_{i=1}^{K}\,|\mathbf{S}_i|^{1/2}$. Thus $|\mathbf{U}|^{1/2}$ is merely the average of the square root of the generalized sample variance for each subgroup. Thus the phase I control limits are obtained by substituting $|\mathbf{U}|^{1/2}/b_3$ for $|\boldsymbol{\Sigma}_0|^{1/2}$ in Eqs. (21.26).

21.6 SUMMARY

When there is only one quality characteristic, it is appropriate to employ the univariate techniques presented in Chap. 7 to monitor the process mean and variance.

However, when product quality depends on two or more characteristics, which are correlated, it is necessary to employ the multivariate procedures presented in this chapter. The procedures that were presented include the multivariate equivalents of the univariate Shewhart charts for monitoring the process. Multivariate charts were presented for both phase I (rational subgroups) and phase II (standards given). When $p = 1$, the multivariate procedures yield the univariate procedures presented in Chap. 7.

For phase II a multivariate CUSUM chart for the mean vector was also presented. This CUSUM chart was based on the sequential probability ratio test. Crosier (1986, 1988) presented two heuristically derived CUSUM procedures for the mean vector and reviews these procedures as well as current research on this topic. The heuristically derived CUSUM procedures of Crosier (1986, 1988) were also reviewed. This multivariate CUSUM, together with the multivariate EWMA procedure of Lowry et al. (1992), consistently outperforms the χ^2 procedure in the sense of quickly detecting small shifts in the process mean vector. The multivariate bayesian procedure of Jain (1993) also has very desirable average run length properties. As statistical software becomes more readily available, it is conjectured that the multivariate CUSUM, EWMA, and bayesian procedures will become more frequently used.

REFERENCES

Alt, F. B.: "Aspects of Multivariate Quality Control Charts," M. S. thesis, Georgia Inst. of Technology, Atlanta, 1973.

______: "Multivariate Quality Control: State of the Art," in *ASQC Quality Congr. Trans.* (Detroit), pp. 886–893, 1982.

______: "Multivariate Control Charts," in *Encyclopedia of Statistical Sciences*, vol. 6, S. Kotz and N. L. Johnson, Eds., Wiley, New York, pp. 110–122, 1985.

______ and G. E. Bedewi: "SPC of Dispersion for Multivariate Data," in *ASQC Quality Congr. Trans.* (Anaheim), pp. 248–254, 1986.

______ and N. D. Smith: "Multivariate Process Control," in *Handbook of Statistics: Quality Control and Reliability*, vol. 7, P. R. Krishnaiah and C. R. Rao, Eds., North-Holland, Amsterdam, 1988, pp. 333–351.

______, J. J. Goode, and H. M. Wadsworth: "Small Sample Probability Limits for the Mean of a Multivariate Normal Process," in *ASQC Quality Congr. Trans.* (Toronto), pp. 170–176, 1976.

______, J. W. Walker, and J. J. Goode: "A Power Paradox for Testing Bivariate Normal Means," in *ASQC Quality Congr. Trans.* (Atlanta), pp. 754–759, 1980.

Anderson, T. W.: *An Introduction to Multivariate Statistical Analysis,* 2d ed., Wiley, New York, 1984.

Crosier, R. B.: "A New Two-Sided Cumulative Sum Quality Control Scheme," *Technometrics,* vol. 28, pp. 187–194, 1986.

______: "Multivariate Generalizations of Cumulative Sum Quality Control Schemes," *Technometrics,* vol. 30, pp. 291–303, 1988.

Duncan, A. J.: *Quality Control and Industrial Statistics,* 5th ed., Irwin, Homewood, IL, 1986.

Hatton, M. B.: "Effective Use of Multivariate Quality Control Charts," Res. Publ., General Motors Research Lab., GMR-4513, Warren, MI, 1983.

Hawkins, D. M.: "Multivariate Quality Control Based on Regression-Adjusted Variables," *Technometrics*, vol. 33, pp. 61–75, 1991.

Healy, J. D.: "A Note on Multivariate Cusum Procedures," *Technometrics*, vol. 29, pp. 409–412, 1987.

Hillier, F. S.: "$\overline{X}$ and *R*-Chart Control Limits Based on a Small Number of Subgroups," *J. Qual. Technol.,* vol. 1, pp. 17–26, 1969.

Jain, K.: "A Bayesian Approach to Multivariate Quality Control," Ph.D. Dissertation, University of Maryland, College Park, 1993.

Johnson, N. L., and F. C. Leone: "Cumulative Sum Control Charts–Mathematical Principles Applied to Their Construction and Use, Parts I–III," *Ind. Qual. Contr.,* vol. 18, pp. 15–21, 29–36, 1962; *ibid.*, vol. 19, pp. 22–28, 1963.

Kshirsagar, A. M.: *Multivariate Analysis,* Dekker, New York, 1972.

Lowry, C. A., W. H. Woodall, C. W. Champ and S. E. Rigdon: "A Multivariate Exponentially Weighted Moving Average Control Chart," *Technometrics*, vol. 34, pp. 46–53, 1992.

Page, E. S.: "Continuous Inspection Schemes," *Biometrika,* vol. 41, pp. 100–114, 1954.

Smith, N. D.: "Multivariate Cumulative Sum Control Charts," Ph. D. Dissertation, Univ. of Maryland, College Park, 1987.

Stewart, G. W.: *Introduction to Matrix Computations*, Academic Press, New York, 1973.

Wald, A.: *Sequential Analysis*, Wiley, New York, 1947.

Yang, C. H., and F. S. Hillier: "Mean and Variance Control Chart Limits Based on a Small Number of Subgroups," *J. Qual. Technol.,* vol. 2, pp. 9–16, 1970.

APPENDIX

TABLE A.1 Standard Normal Distribution*

z	0.00	0.01	0.02	0.03	0.04	0.05	0.06	0.07	0.08	0.09
0.0	0.5000	0.4960	0.4920	0.4880	0.4840	0.4801	0.4761	0.4721	0.4681	0.4641
0.1	0.4602	0.4562	0.4522	0.4483	0.4443	0.4404	0.4364	0.4325	0.4286	0.4247
0.2	0.4207	0.4168	0.4129	0.4090	0.4052	0.4013	0.3974	0.3936	0.3897	0.3859
0.3	0.3821	0.3783	0.3745	0.3707	0.3669	0.3632	0.3594	0.3557	0.3520	0.3483
0.4	0.3446	0.3409	0.3372	0.3336	0.3300	0.3264	0.3228	0.3192	0.3156	0.3121
0.5	0.3085	0.3050	0.3015	0.2981	0.2946	0.2912	0.2877	0.2843	0.2810	0.2776
0.6	0.2743	0.2709	0.2676	0.2643	0.2611	0.2578	0.2546	0.2514	0.2483	0.2451
0.7	0.2420	0.2389	0.2358	0.2327	0.2296	0.2266	0.2236	0.2206	0.2177	0.2148
0.8	0.2119	0.2090	0.2061	0.2033	0.2005	0.1977	0.1949	0.1922	0.1894	0.1867
0.9	0.1841	0.1814	0.1788	0.1762	0.1736	0.1711	0.1685	0.1660	0.1635	0.1611
1.0	0.1587	0.1562	0.1539	0.1515	0.1492	0.1469	0.1446	0.1423	0.1401	0.1379
1.1	0.1357	0.1335	0.1314	0.1292	0.1271	0.1251	0.1230	0.1210	0.1190	0.1170
1.2	0.1151	0.1131	0.1112	0.1093	0.1075	0.1056	0.1038	0.1020	0.1003	0.0985
1.3	0.0968	0.0951	0.0934	0.0918	0.0901	0.0885	0.0869	0.0853	0.0838	0.0823
1.4	0.0808	0.0793	0.0778	0.0764	0.0749	0.0735	0.0721	0.0708	0.0694	0.0681
1.5	0.0668	0.0655	0.0643	0.0630	0.0618	0.0606	0.0594	0.0582	0.0571	0.0559
1.6	0.0548	0.0537	0.0526	0.0516	0.0505	0.0495	0.0485	0.0475	0.0465	0.0455
1.7	0.0446	0.0436	0.0427	0.0418	0.0409	0.0401	0.0392	0.0384	0.0375	0.0367
1.8	0.0359	0.0351	0.0344	0.0336	0.0329	0.0322	0.0314	0.0307	0.0301	0.0294
1.9	0.0287	0.0281	0.0274	0.0268	0.0262	0.0256	0.0250	0.0244	0.0239	0.0233
2.0	0.0228	0.0222	0.0217	0.0212	0.0207	0.0202	0.0197	0.0192	0.0188	0.0183
2.1	0.0179	0.0174	0.0170	0.0166	0.0162	0.0158	0.0154	0.0150	0.0146	0.0143
2.2	0.0139	0.0136	0.0132	0.0129	0.0125	0.0122	0.0119	0.0116	0.0113	0.0110
2.3	0.0107	0.0104	0.0102	0.0099	0.0096	0.0094	0.0091	0.0089	0.0087	0.0084
2.4	0.0082	0.0080	0.0078	0.0075	0.0073	0.0071	0.0069	0.0068	0.0066	0.0064
2.5	0.0062	0.0060	0.0059	0.0057	0.0055	0.0054	0.0052	0.0051	0.0049	0.0048
2.6	0.0047	0.0045	0.0044	0.0043	0.0041	0.0040	0.0039	0.0038	0.0037	0.0036
2.7	0.0035	0.0034	0.0033	0.0032	0.0031	0.0030	0.0029	0.0028	0.0027	0.0026
2.8	0.0026	0.0025	0.0024	0.0023	0.0023	0.0022	0.0021	0.0021	0.0020	0.0019
2.9	0.0019	0.0018	0.0018	0.0017	0.0016	0.0016	0.0015	0.0015	0.0014	0.0014
3.0	0.0013	0.0013	0.0013	0.0012	0.0012	0.0011	0.0011	0.0011	0.0010	0.0010
3.1	0.0010	0.0009	0.0009	0.0009	0.0008	0.0008	0.0008	0.0008	0.0007	0.0007
3.2	0.0007	0.0007	0.0006	0.0006	0.0006	0.0006	0.0006	0.0005	0.0005	0.0005
3.3	0.0005	0.0005	0.0005	0.0004	0.0004	0.0004	0.0004	0.0004	0.0004	0.0003
3.4	0.0003	0.0003	0.0003	0.0003	0.0003	0.0003	0.0003	0.0003	0.0003	0.0002
3.5	0.0002	0.0002	0.0002	0.0002	0.0002	0.0002	0.0002	0.0002	0.0002	0.0002

*Table entries are the tail probabilities, right-tail from the value of z to $+\infty$ and left-tail from $-\infty$ to the value of $-z$ for all $P \leq 0.50$. z is the standardized normal variable, $z = (x - \mu)/\sigma$. Read down the first column to the correct first decimal value of z and over to the correct column for the second decimal value. The number at the intersection is the value of P.

Source: Adapted from Table 1 of Pearson, E. S., and H. O. Hartley, Eds.: *Biometrika Tables for Statisticians*, vol. 1, 3d ed. Cambridge Univ. Press, Cambridge, U.K., 1966. Reprinted by permission of the *Biometrika* Trustees.

TABLE A.2 Poisson Distribution*

	c									
λ	0	1	2	3	4	5	6	7	8	9
0.02	980	1000								
0.04	961	999	1000							
0.06	942	998	1000							
0.08	923	997	1000							
0.10	905	995	1000							
0.15	861	990	999	1000						
0.20	819	982	999	1000						
0.25	779	974	998	1000						
0.30	741	963	996	1000						
0.35	705	951	994	1000						
0.40	670	938	992	999	1000					
0.45	638	925	989	999	1000					
0.50	607	910	986	998	1000					
0.55	577	894	982	998	1000					
0.60	549	878	977	997	1000					
0.65	522	861	972	996	999	1000				
0.70	497	844	966	994	999	1000				
0.75	472	827	959	993	999	1000				
0.80	449	809	953	991	999	1000				
0.85	427	791	945	989	998	1000				
0.90	407	772	937	987	998	1000				
0.95	387	754	929	984	997	1000				
1.00	368	736	920	981	996	999	1000			
1.1	333	699	900	974	995	999	1000			
1.2	301	663	879	966	992	998	1000			
1.3	273	627	857	957	989	998	1000			
1.4	247	592	833	946	986	997	999	1000		
1.5	223	558	809	934	981	996	999	1000		
1.6	202	525	783	921	976	994	999	1000		
1.7	183	493	757	907	970	992	998	1000		
1.8	165	463	731	891	964	990	997	999	1000	
1.9	150	434	704	875	956	987	997	999	1000	
2.0	135	406	677	857	947	983	995	999	1000	
2.2	111	355	623	819	928	975	993	998	1000	
2.4	091	308	570	779	904	964	988	997	999	1000
2.6	074	267	518	736	877	951	983	995	999	1000

TABLE A.2 Poisson Distribution* (*Continued*)

	c									
λ	0	1	2	3	4	5	6	7	8	9
2.8	061	231	469	692	848	935	976	992	998	999
3.0	050	199	423	647	815	916	966	988	996	999
3.2	041	171	380	603	781	895	955	983	994	998
3.4	033	147	340	558	744	871	942	977	992	997
3.6	027	126	303	515	706	844	927	969	988	996
3.8	022	107	269	473	668	816	909	960	984	994
4.0	018	092	238	433	629	785	889	949	979	992
4.2	015	078	210	395	590	753	867	936	972	989
4.4	012	066	185	359	551	720	844	921	964	985
4.6	010	056	163	326	513	686	818	905	955	980
4.8	008	048	143	294	476	651	791	887	944	975
5.0	007	040	125	265	440	616	762	867	932	968
5.2	006	034	109	238	406	581	732	845	918	960
5.4	005	029	095	213	373	546	702	822	903	951
5.6	004	024	082	191	342	512	670	797	886	941
5.8	003	021	072	170	313	478	638	771	867	929
6.0	002	017	062	151	285	446	606	744	847	916
	10	11	12	13	14	15	16			
2.8	1000									
3.0	1000									
3.2	1000									
3.4	999	1000								
3.6	999	1000								
3.8	998	999	1000							
4.0	997	999	1000							
4.2	996	999	1000							
4.4	994	998	999	1000						
4.6	992	997	999	1000						
4.8	990	996	999	1000						
5.0	986	995	998	999	1000					
5.2	982	993	997	999	1000					
5.4	977	990	996	999	1000					
5.6	972	988	995	998	999	1000				
5.8	965	984	993	997	999	1000				
6.0	957	980	991	996	999	999				

TABLE A.2 Poisson Distribution* (*Continued*)

λ	c: 0	1	2	3	4	5	6	7	8	9
6.2	002	015	054	134	259	414	574	716	826	902
6.4	002	012	046	119	235	384	542	687	803	886
6.6	001	010	040	105	213	355	511	658	780	869
6.8	001	009	034	093	192	327	480	628	755	850
7.0	001	007	030	082	173	301	450	599	729	830
7.2	001	006	025	072	156	276	420	569	703	810
7.4	001	005	022	063	140	253	392	539	676	788
7.6	001	004	019	055	125	231	365	510	648	765
7.8	000	004	016	048	112	210	338	481	620	741
8.0	000	003	014	042	100	191	313	453	593	717
8.5	000	002	009	030	074	150	256	386	523	653
9.0	000	001	006	021	055	116	207	324	456	587
9.5	000	001	004	015	040	089	165	269	393	522
10.0	000	000	003	010	029	067	130	220	333	458
	10	11	12	13	14	15	16	17	18	19
6.2	949	975	989	995	998	999	1000			
6.4	939	969	986	994	997	999	1000			
6.6	927	963	982	992	997	999	999	1000		
6.8	915	955	978	990	996	998	999	1000		
7.0	901	947	973	987	994	998	999	1000		
7.2	887	937	967	984	993	997	999	999	1000	
7.4	871	926	961	980	991	996	998	999	1000	
7.6	854	915	954	976	989	995	998	999	1000	
7.8	835	902	945	971	986	993	997	999	1000	
8.0	816	888	936	966	983	992	996	998	999	1000
8.5	763	849	909	949	973	986	993	997	999	999
9.0	706	803	876	926	959	978	989	995	998	999
9.5	645	752	836	898	940	967	982	991	996	998
10.0	583	697	792	864	917	951	973	986	993	997
	20	21	22							
8.5	1000									
9.0	1000									
9.5	999	1000								
10.0	998	999	1000							

TABLE A.2 Poisson Distribution* (*Continued*)

λ	c: 0	1	2	3	4	5	6	7	8	9
10.5	000	000	002	007	021	050	102	179	279	397
11.0	000	000	001	005	015	038	079	143	232	341
11.5	000	000	001	003	011	028	060	114	191	289
12.0	000	000	001	002	008	020	046	090	155	242
12.5	000	000	000	002	005	015	035	070	125	201
13.0	000	000	000	001	004	011	026	054	100	166
13.5	000	000	000	001	003	008	019	041	079	135
14.0	000	000	000	000	002	006	014	032	062	109
14.5	000	000	000	000	001	004	010	024	048	088
15.0	000	000	000	000	001	003	008	018	037	070
	10	11	12	13	14	15	16	17	18	19
10.5	521	639	742	825	888	932	960	978	988	994
11.0	460	579	689	781	854	907	944	968	982	991
11.5	402	520	633	733	815	878	924	954	974	986
12.0	347	462	576	682	772	844	899	937	963	979
12.5	297	406	519	628	725	806	869	916	948	969
13.0	252	353	463	573	675	764	835	890	930	957
13.5	211	304	409	518	623	718	798	861	908	942
14.0	176	260	358	464	570	669	756	827	883	923
14.5	145	220	311	413	518	619	711	790	853	901
15.0	118	185	268	363	466	568	664	749	819	875
	20	21	22	23	24	25	26	27	28	29
10.5	997	999	999	1000						
11.0	995	998	999	1000						
11.5	992	996	998	999	1000					
12.0	988	994	997	999	999	1000				
12.5	983	991	995	998	999	999	1000			
13.0	975	986	992	996	998	999	1000			
13.5	965	980	989	994	997	998	999	1000		
14.0	952	971	983	991	995	997	999	999	1000	
14.5	936	960	976	986	992	996	998	999	999	1000
15.0	917	947	967	981	989	994	997	998	999	1000

TABLE A.2 Poisson Distribution* (*Continued*)

λ	c = 4	5	6	7	8	9	10	11	12	13
16	000	001	004	010	022	043	077	127	193	275
17	000	001	002	005	013	026	049	085	135	201
18	000	000	001	003	007	015	030	055	092	143
19	000	000	001	002	004	009	018	035	061	098
20	000	000	000	001	002	005	011	021	039	066
21	000	000	000	000	001	003	006	013	025	043
22	000	000	000	000	001	002	004	008	015	028
23	000	000	000	000	000	001	002	004	009	017
24	000	000	000	000	000	000	001	003	005	011
25	000	000	000	000	000	000	001	001	003	006

λ	14	15	16	17	18	19	20	21	22	23
16	368	467	566	659	742	812	868	911	942	963
17	281	371	468	564	655	736	805	861	905	937
18	208	287	375	469	562	651	731	799	855	899
19	150	215	292	378	469	561	647	725	793	849
20	105	157	221	297	381	470	559	644	721	787
21	072	111	163	227	302	384	471	558	640	716
22	048	077	117	169	232	306	387	472	556	637
23	031	052	082	123	175	238	310	389	472	555
24	020	034	056	087	128	180	243	413	392	473
25	012	022	038	060	092	134	185	247	318	394

λ	24	25	26	27	28	29	30	31	32	33
16	978	987	993	996	998	999	999	1000		
17	959	975	985	991	995	997	999	999	1000	
18	932	955	972	983	990	994	997	998	999	1000
19	893	927	951	969	980	988	993	996	998	999
20	843	888	922	948	966	978	987	992	995	997
21	782	838	883	917	944	963	976	985	991	994
22	712	777	832	877	913	940	959	973	983	989
23	635	708	772	827	873	908	936	956	971	981
24	554	632	704	768	823	868	904	932	953	969
25	473	553	629	700	763	818	863	900	929	950

λ	34	35	36	37	38	39	40	41	42	43
19	999	1000								
20	999	999	1000							
21	997	998	999	999	1000					
22	994	996	998	999	999	1000				
23	988	993	996	997	999	999	1000			
24	979	987	992	995	997	998	999	999	1000	
25	966	978	985	991	994	997	998	999	999	1000

*1000 × probability of *c* or less occurrences of event that has average number of occurrences equal to λ.

Source: From Grant, E. L., and R. S. Leavenworth: *Statistical Quality Control*, 5th ed., McGrawHill, New York, 1980. Reprinted by permission of the publisher.

TABLE A.3 Chi Square Distribution*

	α										
ν	0.995	0.990	0.975	0.950	0.900	0.500	0.100	0.050	0.025	0.010	0.005
1	0.00+	0.00+	0.00+	0.00+	0.02	0.45	2.71	3.84	5.02	6.63	7.88
2	0.01	0.02	0.05	0.10	0.21	1.39	4.61	5.99	7.38	9.21	10.60
3	0.07	0.11	0.22	0.35	0.58	2.37	6.25	7.81	9.35	11.34	12.84
4	0.21	0.30	0.48	0.71	1.06	3.36	7.78	9.49	11.14	13.28	14.86
5	0.41	0.55	0.83	1.15	1.61	4.35	9.24	11.07	12.83	15.09	16.75
6	0.68	0.87	1.24	1.64	2.20	5.35	10.65	12.59	14.45	16.81	18.55
7	0.99	1.24	1.69	2.17	2.83	6.35	12.02	14.07	16.01	18.48	20.28
8	1.34	1.65	2.18	2.73	3.49	7.34	13.36	15.51	17.53	20.09	21.96
9	1.73	2.09	2.70	3.33	4.17	8.34	14.68	16.92	19.02	21.67	23.59
10	2.16	2.56	3.25	3.94	4.87	9.34	15.99	18.31	20.48	23.21	25.19
11	2.60	3.05	3.82	4.57	5.58	10.34	17.28	19.68	21.92	24.72	26.76
12	3.07	3.57	4.40	5.23	6.30	11.34	18.55	21.03	23.34	26.22	28.30
13	3.57	4.11	5.01	5.89	7.04	12.34	19.81	22.36	24.74	27.69	29.82
14	4.07	4.66	5.63	6.57	7.79	13.34	21.06	23.68	26.12	29.14	31.32
15	4.60	5.23	6.27	7.26	8.55	14.34	22.31	25.00	27.49	30.58	32.80
16	5.14	5.81	6.91	7.96	9.31	15.34	23.54	26.30	28.85	32.00	34.27
17	5.70	6.41	7.56	8.67	10.09	16.34	24.77	27.59	30.19	33.41	35.72
18	6.26	7.01	8.23	9.39	10.87	17.34	25.99	28.87	31.53	34.81	37.16
19	6.84	7.63	8.91	10.12	11.65	18.34	27.20	30.14	32.85	36.19	38.58
20	7.43	8.26	9.59	10.85	12.44	19.34	28.41	31.41	34.17	37.57	40.00
21	8.03	8.90	10.28	11.59	13.24	20.34	29.62	32.67	35.48	38.93	41.40
22	8.64	9.54	10.98	12.34	14.04	21.34	30.81	33.92	36.78	40.29	42.80
23	9.26	10.20	11.69	13.09	14.85	22.34	32.01	35.17	38.08	41.64	44.18
24	9.89	10.86	12.40	13.85	15.66	23.34	33.20	36.42	39.36	42.98	45.56
25	10.52	11.52	13.12	14.61	16.47	24.34	34.28	37.65	40.65	44.31	46.93
26	11.16	12.20	13.84	15.38	17.29	25.34	35.56	38.89	41.92	45.64	48.29
27	11.81	12.88	14.57	16.15	18.11	26.34	36.74	40.11	43.19	46.96	49.65
28	12.46	13.57	15.31	16.93	18.94	27.34	37.92	41.34	44.46	48.28	50.99
29	13.12	14.26	16.05	17.71	19.77	28.34	39.09	42.56	45.72	49.59	52.34
30	13.79	14.95	16.79	18.49	20.60	29.34	40.26	43.77	46.98	50.89	53.67
40	20.71	22.16	24.43	26.51	29.05	39.34	51.81	55.76	59.34	63.69	66.77
50	27.99	29.71	32.36	34.76	37.69	49.33	63.17	67.50	71.42	76.15	79.49
60	35.53	37.48	40.48	43.19	46.46	59.33	74.40	79.08	83.30	88.38	91.95
70	43.28	45.44	48.76	51.74	55.33	69.33	85.53	90.53	95.02	100.42	104.22
80	51.17	53.54	57.15	60.39	64.28	79.33	96.58	101.88	106.63	112.33	116.32
90	59.20	61.75	65.65	69.13	73.29	89.33	107.57	113.14	118.14	124.12	128.30
100	67.33	70.06	74.22	77.93	82.36	99.33	118.50	124.34	129.56	135.81	140.17

*ν = degrees of freedom.

Source: Adapted from Hines, W. W., and D. C. Montgomery: *Probability and Statistics in Engineering and Management Science*, 2d ed., Wiley, New York, 1980. Reprinted by permission of the publisher.

TABLE A.4 Student's t Distribution

	α									
ν	0.40	0.25	0.10	0.05	0.025	0.01	0.005	0.0025	0.001	0.0005
1	0.325	1.000	3.078	6.314	12.706	31.821	63.657	127.32	318.31	636.62
2	0.289	0.816	1.886	2.920	4.303	6.965	9.925	14.089	23.326	31.598
3	0.277	0.765	1.638	2.353	3.182	4.541	5.841	7.453	10.213	12.924
4	0.271	0.741	1.533	2.132	2.776	3.747	4.604	5.598	7.173	8.610
5	0.267	0.727	1.476	2.015	2.571	3.365	4.032	4.773	5.893	6.869
6	0.265	0.718	1.440	1.943	2.447	3.143	3.707	4.317	5.208	5.959
7	0.263	0.711	1.415	1.895	2.365	2.998	3.499	4.029	4.785	5.408
8	0.262	0.706	1.397	1.860	2.306	2.896	3.355	3.833	4.501	5.041
9	0.261	0.703	1.383	1.833	2.262	2.821	3.250	3.690	4.297	4.781
10	0.260	0.700	1.372	1.812	2.228	2.764	3.169	3.581	4.144	4.587
11	0.260	0.697	1.363	1.796	2.201	2.718	3.106	3.497	4.025	4.437
12	0.259	0.695	1.356	1.782	2.179	2.681	3.055	3.428	3.930	4.318
13	0.259	0.694	1.350	1.771	2.160	2.650	3.012	3.372	3.852	4.221
14	0.258	0.692	1.345	1.761	2.145	2.624	2.977	3.326	3.787	4.140
15	0.258	0.691	1.341	1.753	2.131	2.602	2.947	3.286	3.733	4.073
16	0.258	0.690	1.337	1.746	2.120	2.583	2.921	3.252	3.686	4.015
17	0.257	0.689	1.333	1.740	2.110	2.567	2.898	3.222	3.646	3.965
18	0.257	0.688	1.330	1.734	2.101	2.552	2.878	3.197	3.610	3.922
19	0.257	0.688	1.328	1.729	2.093	2.539	2.861	3.174	3.579	3.883
20	0.257	0.687	1.325	1.725	2.086	2.528	2.845	3.153	3.552	3.850
21	0.257	0.686	1.323	1.721	2.080	2.518	2.831	3.135	3.527	3.819
22	0.256	0.686	1.321	1.717	2.074	2.508	2.819	3.119	3.505	3.792
23	0.256	0.685	1.319	1.714	2.069	2.500	2.807	3.104	3.485	3.767
24	0.256	0.685	1.318	1.711	2.064	2.492	2.797	3.091	3.467	3.745
25	0.256	0.684	1.316	1.708	2.060	2.485	2.787	3.078	3.450	3.725
26	0.256	0.684	1.315	1.706	2.056	2.479	2.779	3.067	3.435	3.707
27	0.256	0.684	1.314	1.703	2.052	2.473	2.771	3.057	3.421	3.690
28	0.256	0.683	1.313	1.701	2.048	2.467	2.763	3.047	3.408	3.674
29	0.256	0.683	1.311	1.699	2.045	2.462	2.756	3.038	3.396	3.659
30	0.256	0.683	1.310	1.697	2.042	2.457	2.750	3.030	3.385	3.646
40	0.255	0.681	1.303	1.684	2.021	2.423	2.704	2.971	3.307	3.551
60	0.254	0.679	1.296	1.671	2.000	2.390	2.660	2.915	3.232	3.460
120	0.254	0.677	1.289	1.658	1.980	2.358	2.617	2.860	3.160	3.373
∞	0.253	0.674	1.282	1.645	1.960	2.326	2.576	2.807	3.090	3.291

Source: Adapted from Pearson, E. S., and H. O. Hartley, Eds.: *Biometrika Tables for Statisticians,* vol. 1, 3d ed. Cambridge Univ. Press, Cambridge, U. K., 1966. Reprinted by permission of the *Biometrika* Trustees.

TABLE A.5 *F* Distribution. $F_{0.25, \nu 1, \nu 2}$

Degrees of freedom for denominator ν_2	Degrees of freedom for numerator ν_1																		
	1	2	3	4	5	6	7	8	9	10	12	15	20	24	30	40	60	120	∞
1	5.83	7.50	8.20	8.58	8.82	8.98	9.10	9.19	9.26	9.32	9.41	9.49	9.58	9.63	9.67	9.71	9.76	9.80	9.85
2	2.57	3.00	3.15	3.23	3.28	3.31	3.34	3.35	3.37	3.38	3.39	3.41	3.43	3.43	3.44	3.45	3.46	3.47	3.48
3	2.02	2.28	2.36	2.39	2.41	2.42	2.43	2.44	2.44	2.44	2.45	2.46	2.46	2.46	2.47	2.47	2.47	2.47	2.47
4	1.81	2.00	2.05	2.06	2.07	2.08	2.08	2.08	2.08	2.08	2.08	2.08	2.08	2.08	2.08	2.08	2.08	2.08	2.08
5	1.69	1.85	1.88	1.89	1.89	1.89	1.89	1.89	1.89	1.89	1.89	1.89	1.88	1.88	1.88	1.88	1.87	1.87	1.87
6	1.62	1.76	1.78	1.79	1.79	1.78	1.78	1.78	1.77	1.77	1.77	1.76	1.76	1.75	1.75	1.75	1.74	1.74	1.74
7	1.57	1.70	1.72	1.72	1.71	1.71	1.70	1.70	1.70	1.69	1.68	1.68	1.67	1.67	1.66	1.66	1.65	1.65	1.65
8	1.54	1.66	1.67	1.66	1.66	1.65	1.64	1.64	1.63	1.63	1.62	1.62	1.61	1.60	1.60	1.59	1.59	1.58	1.58
9	1.51	1.62	1.63	1.63	1.62	1.61	1.60	1.60	1.59	1.59	1.58	1.57	1.56	1.56	1.55	1.54	1.54	1.53	1.53
10	1.49	1.60	1.60	1.59	1.59	1.58	1.57	1.56	1.56	1.55	1.54	1.53	1.52	1.52	1.51	1.51	1.50	1.49	1.48
11	1.47	1.58	1.58	1.57	1.56	1.55	1.54	1.53	1.53	1.52	1.51	1.50	1.49	1.49	1.48	1.47	1.47	1.46	1.45
12	1.46	1.56	1.56	1.55	1.54	1.53	1.52	1.51	1.51	1.50	1.49	1.48	1.47	1.46	1.45	1.45	1.44	1.43	1.42
13	1.45	1.55	1.55	1.53	1.52	1.51	1.50	1.49	1.49	1.48	1.47	1.46	1.45	1.44	1.43	1.42	1.42	1.41	1.40
14	1.44	1.53	1.53	1.52	1.51	1.50	1.49	1.48	1.47	1.46	1.45	1.44	1.43	1.42	1.41	1.41	1.40	1.39	1.38
15	1.43	1.52	1.52	1.51	1.49	1.48	1.47	1.46	1.46	1.45	1.44	1.43	1.41	1.41	1.40	1.39	1.38	1.37	1.36
16	1.42	1.51	1.51	1.50	1.48	1.47	1.46	1.45	1.44	1.44	1.43	1.41	1.40	1.39	1.38	1.37	1.36	1.35	1.34
17	1.42	1.51	1.50	1.49	1.47	1.46	1.45	1.44	1.43	1.43	1.41	1.40	1.39	1.38	1.37	1.36	1.35	1.34	1.33
18	1.41	1.50	1.49	1.48	1.46	1.45	1.44	1.43	1.42	1.42	1.40	1.39	1.38	1.37	1.36	1.35	1.34	1.33	1.32
19	1.41	1.49	1.49	1.47	1.46	1.44	1.43	1.42	1.41	1.41	1.40	1.38	1.37	1.36	1.35	1.34	1.33	1.32	1.30
20	1.40	1.49	1.48	1.47	1.45	1.44	1.43	1.42	1.41	1.40	1.39	1.37	1.36	1.35	1.34	1.33	1.32	1.31	1.29
21	1.40	1.48	1.48	1.46	1.44	1.43	1.42	1.41	1.40	1.39	1.38	1.37	1.35	1.34	1.33	1.32	1.31	1.30	1.28
22	1.40	1.48	1.47	1.45	1.44	1.42	1.41	1.40	1.39	1.39	1.37	1.36	1.34	1.33	1.32	1.31	1.30	1.29	1.28
23	1.39	1.47	1.47	1.45	1.43	1.42	1.41	1.40	1.39	1.38	1.37	1.35	1.34	1.33	1.32	1.31	1.30	1.28	1.27
24	1.39	1.47	1.46	1.44	1.43	1.41	1.40	1.39	1.38	1.38	1.36	1.35	1.33	1.32	1.31	1.30	1.29	1.28	1.26
25	1.39	1.47	1.46	1.44	1.42	1.41	1.40	1.39	1.38	1.37	1.36	1.34	1.33	1.32	1.31	1.29	1.28	1.27	1.25
26	1.38	1.46	1.45	1.44	1.42	1.41	1.39	1.38	1.37	1.37	1.35	1.34	1.32	1.31	1.30	1.29	1.28	1.26	1.25
27	1.38	1.46	1.45	1.43	1.42	1.40	1.39	1.38	1.37	1.36	1.35	1.33	1.32	1.31	1.30	1.28	1.27	1.26	1.24
28	1.38	1.46	1.45	1.43	1.41	1.40	1.39	1.38	1.37	1.36	1.34	1.33	1.31	1.30	1.29	1.28	1.27	1.25	1.24
29	1.38	1.45	1.45	1.43	1.41	1.40	1.38	1.37	1.36	1.35	1.34	1.32	1.31	1.30	1.29	1.27	1.26	1.25	1.23
30	1.38	1.45	1.44	1.42	1.41	1.39	1.38	1.37	1.36	1.35	1.34	1.32	1.30	1.29	1.28	1.27	1.26	1.24	1.23
40	1.36	1.44	1.42	1.40	1.39	1.37	1.36	1.35	1.34	1.33	1.31	1.30	1.28	1.26	1.25	1.24	1.22	1.21	1.19
60	1.35	1.42	1.41	1.38	1.37	1.35	1.33	1.32	1.31	1.30	1.29	1.27	1.25	1.24	1.22	1.21	1.19	1.17	1.15
120	1.34	1.40	1.39	1.37	1.35	1.33	1.31	1.30	1.29	1.28	1.26	1.24	1.22	1.21	1.19	1.18	1.16	1.13	1.10
∞	1.32	1.39	1.37	1.35	1.33	1.31	1.29	1.28	1.27	1.25	1.24	1.22	1.19	1.18	1.16	1.14	1.12	1.08	1.00

TABLE A.5 *F* Distribution (*Continued*). $F_{0.10,\, \nu 1,\, \nu 2}$

Degrees of freedom for denominator ν_2	Degrees of freedom for numerator ν_1: 1	2	3	4	5	6	7	8	9	10	12	15	20	24	30	40	60	120	∞
1	39.86	49.50	53.59	55.83	57.24	58.20	58.91	59.44	59.86	60.19	60.71	61.22	61.74	62.00	62.26	62.53	62.79	63.06	63.33
2	8.53	9.00	9.16	9.24	9.29	9.33	9.35	9.37	9.38	9.39	9.41	9.42	9.44	9.45	9.46	9.47	9.47	9.48	9.49
3	5.54	5.46	5.39	5.34	5.31	5.28	5.27	5.25	5.24	5.23	5.22	5.20	5.18	5.18	5.17	5.16	5.15	5.14	5.13
4	4.54	4.32	4.19	4.11	4.05	4.01	3.98	3.95	3.94	3.92	3.90	3.87	3.84	3.83	3.82	3.80	3.79	3.78	3.76
5	4.06	3.78	3.62	3.52	3.45	3.40	3.37	3.34	3.32	3.30	3.27	3.24	3.21	3.19	3.17	3.16	3.14	3.12	3.10
6	3.78	3.46	3.29	3.18	3.11	3.05	3.01	2.98	2.96	2.94	2.90	2.87	2.84	2.82	2.80	2.78	2.76	2.74	2.72
7	3.59	3.26	3.07	2.96	2.88	2.83	2.78	2.75	2.72	2.70	2.67	2.63	2.59	2.58	2.56	2.54	2.51	2.49	2.47
8	3.46	3.11	2.92	2.81	2.73	2.67	2.62	2.59	2.56	2.54	2.50	2.46	2.42	2.40	2.38	2.36	2.34	2.32	2.29
9	3.36	3.01	2.81	2.69	2.61	2.55	2.51	2.47	2.44	2.42	2.38	2.34	2.30	2.28	2.25	2.23	2.21	2.18	2.16
10	3.29	2.92	2.73	2.61	2.52	2.46	2.41	2.38	2.35	2.32	2.28	2.24	2.20	2.18	2.16	2.13	2.11	2.08	2.06
11	3.23	2.86	2.66	2.54	2.45	2.39	2.34	2.30	2.27	2.25	2.21	2.17	2.12	2.10	2.08	2.05	2.03	2.00	1.97
12	3.18	2.81	2.61	2.48	2.39	2.33	2.28	2.24	2.21	2.19	2.15	2.10	2.06	2.04	2.01	1.99	1.96	1.93	1.90
13	3.14	2.76	2.56	2.43	2.35	2.28	2.23	2.20	2.16	2.14	2.10	2.05	2.01	1.98	1.96	1.93	1.90	1.88	1.85
14	3.10	2.73	2.52	2.39	2.31	2.24	2.19	2.15	2.12	2.10	2.05	2.01	1.96	1.94	1.91	1.89	1.86	1.83	1.80
15	3.07	2.70	2.49	2.36	2.27	2.21	2.16	2.12	2.09	2.06	2.02	1.97	1.92	1.90	1.87	1.85	1.82	1.79	1.76
16	3.05	2.67	2.46	2.33	2.24	2.18	2.13	2.09	2.06	2.03	1.99	1.94	1.89	1.87	1.84	1.81	1.78	1.75	1.72
17	3.03	2.64	2.44	2.31	2.22	2.15	2.10	2.06	2.03	2.00	1.96	1.91	1.86	1.84	1.81	1.78	1.75	1.72	1.69
18	3.01	2.62	2.42	2.29	2.20	2.13	2.08	2.04	2.00	1.98	1.93	1.89	1.84	1.81	1.78	1.75	1.72	1.69	1.66
19	2.99	2.61	2.40	2.27	2.18	2.11	2.06	2.02	1.98	1.96	1.91	1.86	1.81	1.79	1.76	1.73	1.70	1.67	1.63
20	2.97	2.59	2.38	2.25	2.16	2.09	2.04	2.00	1.96	1.94	1.89	1.84	1.79	1.77	1.74	1.71	1.68	1.64	1.61
21	2.96	2.57	2.36	2.23	2.14	2.08	2.02	1.98	1.95	1.92	1.87	1.83	1.78	1.75	1.72	1.69	1.66	1.62	1.59
22	2.95	2.56	2.35	2.22	2.13	2.06	2.01	1.97	1.93	1.90	1.86	1.81	1.76	1.73	1.70	1.67	1.64	1.60	1.57
23	2.94	2.55	2.34	2.21	2.11	2.05	1.99	1.95	1.92	1.89	1.84	1.80	1.74	1.72	1.69	1.66	1.62	1.59	1.55
24	2.93	2.54	2.33	2.19	2.10	2.04	1.98	1.94	1.91	1.88	1.83	1.78	1.73	1.70	1.67	1.64	1.61	1.57	1.53
25	2.92	2.53	2.32	2.18	2.09	2.02	1.97	1.93	1.89	1.87	1.82	1.77	1.72	1.69	1.66	1.63	1.59	1.56	1.52
26	2.91	2.52	2.31	2.17	2.08	2.01	1.96	1.92	1.88	1.86	1.81	1.76	1.71	1.68	1.65	1.61	1.58	1.54	1.50
27	2.90	2.51	2.30	2.17	2.07	2.00	1.95	1.91	1.87	1.85	1.80	1.75	1.70	1.67	1.64	1.60	1.57	1.53	1.49
28	2.89	2.50	2.29	2.16	2.06	2.00	1.94	1.90	1.87	1.84	1.79	1.74	1.69	1.66	1.63	1.59	1.56	1.52	1.48
29	2.89	2.50	2.28	2.15	2.06	1.99	1.93	1.89	1.86	1.83	1.78	1.73	1.68	1.65	1.62	1.58	1.55	1.51	1.47
30	2.88	2.49	2.28	2.14	2.03	1.98	1.93	1.88	1.85	1.82	1.77	1.72	1.67	1.64	1.61	1.57	1.54	1.50	1.46
40	2.84	2.44	2.23	2.09	2.00	1.93	1.87	1.83	1.79	1.76	1.71	1.66	1.61	1.57	1.54	1.51	1.47	1.42	1.38
60	2.79	2.39	2.18	2.04	1.95	1.87	1.82	1.77	1.74	1.71	1.66	1.60	1.54	1.51	1.48	1.44	1.40	1.35	1.29
120	2.75	2.35	2.12	1.99	1.90	1.82	1.77	1.72	1.68	1.65	1.60	1.55	1.48	1.45	1.41	1.37	1.32	1.26	1.19
∞	2.71	2.30	2.08	1.94	1.85	1.77	1.72	1.67	1.63	1.60	1.55	1.49	1.42	1.38	1.34	1.30	1.24	1.17	1.00

TABLE A.5 *F* Distribution (*Continued*). $F_{0.05,\ \nu 1,\ \nu 2}$

Degrees of freedom for denominator ν_2	Degrees of freedom for numerator ν_1																		
	1	2	3	4	5	6	7	8	9	10	12	15	20	24	30	40	60	120	∞
1	161.4	199.5	215.7	224.6	230.2	234.0	236.8	238.9	240.5	241.9	243.9	245.9	248.0	249.1	250.1	251.1	252.2	253.3	254.3
2	18.51	19.00	19.16	19.25	19.30	19.33	19.35	19.37	19.38	19.40	19.41	19.43	19.45	19.45	19.46	19.47	19.48	19.49	19.50
3	10.13	9.55	9.28	9.12	9.01	8.94	8.89	8.85	8.81	8.79	8.74	8.70	8.66	8.64	8.62	8.59	8.57	8.55	8.53
4	7.71	6.94	6.59	6.39	6.26	6.16	6.09	6.04	6.00	5.96	5.91	5.86	5.80	5.77	5.75	5.72	5.69	5.66	5.63
5	6.61	5.79	5.41	5.19	5.05	4.95	4.88	4.82	4.77	4.74	4.68	4.62	4.56	4.53	4.50	4.46	4.43	4.40	4.36
6	5.99	5.14	4.76	4.53	4.39	4.28	4.21	4.15	4.10	4.06	4.00	3.94	3.87	3.84	3.81	3.77	3.74	3.70	3.67
7	5.59	4.74	4.35	4.12	3.97	3.87	3.79	3.73	3.68	3.64	3.57	3.51	3.44	3.41	3.38	3.34	3.30	3.27	3.23
8	5.32	4.46	4.07	3.84	3.69	3.58	3.50	3.44	3.39	3.35	3.28	3.22	3.15	3.12	3.08	3.04	3.01	2.97	2.93
9	5.12	4.26	3.86	3.63	3.48	3.37	3.29	3.23	3.18	3.14	3.07	3.01	2.94	2.90	2.86	2.83	2.79	2.75	2.71
10	4.96	4.10	3.71	3.48	3.33	3.22	3.14	3.07	3.02	2.98	2.91	2.85	2.77	2.74	2.70	2.66	2.62	2.58	2.54
11	4.84	3.98	3.59	3.36	3.20	3.09	3.01	2.95	2.90	2.85	2.79	2.72	2.65	2.61	2.57	2.53	2.49	2.45	2.40
12	4.75	3.89	3.49	3.26	3.11	3.00	2.91	2.85	2.80	2.75	2.69	2.62	2.54	2.51	2.47	2.43	2.38	2.34	2.30
13	4.67	3.81	3.41	3.18	3.03	2.92	2.83	2.77	2.71	2.67	2.60	2.53	2.46	2.42	2.38	2.34	2.30	2.25	2.21
14	4.60	3.74	3.34	3.11	2.96	2.85	2.76	2.70	2.65	2.60	2.53	2.46	2.39	2.35	2.31	2.27	2.22	2.18	2.13
15	4.54	3.68	3.29	3.06	2.90	2.79	2.71	2.64	2.59	2.54	2.48	2.40	2.33	2.29	2.25	2.20	2.16	2.11	2.07
16	4.49	3.63	3.24	3.01	2.85	2.74	2.66	2.59	2.54	2.49	2.42	2.35	2.28	2.24	2.19	2.15	2.11	2.06	2.01
17	4.45	3.59	3.20	2.96	2.81	2.70	2.61	2.55	2.49	2.45	2.38	2.31	2.23	2.19	2.15	2.10	2.06	2.01	1.96
18	4.41	3.55	3.16	2.93	2.77	2.66	2.58	2.51	2.46	2.41	2.34	2.27	2.19	2.15	2.11	2.06	2.02	1.97	1.92
19	4.38	3.52	3.13	2.90	2.74	2.63	2.54	2.48	2.42	2.38	2.31	2.23	2.16	2.11	2.07	2.03	1.98	1.93	1.88
20	4.35	3.49	3.10	2.87	2.71	2.60	2.51	2.45	2.39	2.35	2.28	2.20	2.12	2.08	2.04	1.99	1.95	1.90	1.84
21	4.32	3.47	3.07	2.84	2.68	2.57	2.49	2.42	2.37	2.32	2.25	2.18	2.10	2.05	2.01	1.96	1.92	1.87	1.81
22	4.30	3.44	3.05	2.82	2.66	2.55	2.46	2.40	2.34	2.30	2.23	2.15	2.07	2.03	1.98	1.94	1.89	1.84	1.78
23	4.28	3.42	3.03	2.80	2.64	2.53	2.44	2.37	2.32	2.27	2.20	2.13	2.05	2.01	1.96	1.91	1.86	1.81	1.76
24	4.26	3.40	3.01	2.78	2.62	2.51	2.42	2.36	2.30	2.25	2.18	2.11	2.03	1.98	1.94	1.89	1.84	1.79	1.73
25	4.24	3.39	2.99	2.76	2.60	2.49	2.40	2.34	2.28	2.24	2.16	2.09	2.01	1.96	1.92	1.87	1.82	1.77	1.71
26	4.23	3.37	2.98	2.74	2.59	2.47	2.39	2.32	2.27	2.22	2.15	2.07	1.99	1.95	1.90	1.85	1.80	1.75	1.69
27	4.21	3.35	2.96	2.73	2.57	2.46	2.37	2.31	2.25	2.20	2.13	2.06	1.97	1.93	1.88	1.84	1.79	1.73	1.67
28	4.20	3.34	2.95	2.71	2.56	2.45	2.36	2.29	2.24	2.19	2.12	2.04	1.96	1.91	1.87	1.82	1.77	1.71	1.65
29	4.18	3.33	2.93	2.70	2.55	2.43	2.35	2.28	2.22	2.18	2.10	2.03	1.94	1.90	1.85	1.81	1.75	1.70	1.64
30	4.17	3.32	2.92	2.69	2.53	2.42	2.33	2.27	2.21	2.16	2.09	2.01	1.93	1.89	1.84	1.79	1.74	1.68	1.62
40	4.08	3.23	2.84	2.61	2.45	2.34	2.25	2.18	2.12	2.08	2.00	1.92	1.84	1.79	1.74	1.69	1.64	1.58	1.51
60	4.00	3.15	2.76	2.53	2.37	2.25	2.17	2.10	2.04	1.99	1.92	1.84	1.75	1.70	1.65	1.59	1.53	1.47	1.39
120	3.92	3.07	2.68	2.45	2.29	2.17	2.09	2.02	1.96	1.91	1.83	1.75	1.66	1.61	1.55	1.55	1.43	1.35	1.25
∞	3.48	3.00	2.60	2.37	2.21	2.10	2.01	1.94	1.88	1.83	1.75	1.67	1.57	1.52	1.46	1.39	1.32	1.22	1.00

TABLE A.5 *F* Distribution (*Continued*). $F_{0.025,\ \nu1,\ \nu2}$

Degrees of freedom denominator ν_2	Degrees of freedom for numerator ν_1																		
	1	2	3	4	5	6	7	8	9	10	12	15	20	24	30	40	60	120	∞
1	647.8	799.5	864.2	899.6	921.8	937.1	948.2	956.7	963.3	968.6	976.7	948.9	993.1	997.2	1001	1006	1010	1014	1018
2	38.51	39.00	39.17	39.25	39.30	39.33	39.36	39.37	39.39	39.40	39.41	39.43	39.45	39.46	39.46	39.47	39.48	39.49	39.50
3	17.44	16.04	15.44	15.10	14.88	14.73	14.62	14.54	14.47	14.42	14.34	14.25	14.17	14.12	14.08	14.04	13.99	13.95	13.90
4	12.22	10.65	9.98	9.60	9.36	9.20	9.07	8.98	8.90	8.84	8.75	8.66	8.56	8.51	8.46	8.41	8.36	8.31	8.26
5	10.01	8.43	7.76	7.39	7.15	6.98	6.85	6.76	6.68	6.62	6.52	6.43	6.33	6.28	6.23	6.18	6.12	6.07	6.02
6	8.81	7.26	6.60	6.23	5.99	5.82	5.70	5.60	5.52	5.46	5.37	5.27	5.17	5.12	5.07	5.01	4.96	4.90	4.85
7	8.07	6.54	5.89	5.52	5.29	5.12	4.99	4.90	4.82	4.76	4.67	4.57	4.47	4.42	4.36	4.31	4.25	4.20	4.14
8	7.57	6.06	5.42	5.05	4.82	4.65	4.53	4.43	4.36	4.30	4.20	4.10	4.00	3.95	3.89	3.84	3.78	3.73	3.67
9	7.21	5.71	5.08	4.72	4.48	4.32	4.20	4.10	4.03	3.96	3.87	3.77	3.67	3.61	3.56	3.51	3.45	3.39	3.33
10	6.94	5.46	4.83	4.47	4.24	4.07	3.95	3.85	3.78	3.72	3.62	3.52	3.42	3.37	3.31	3.26	3.20	3.14	3.08
11	6.72	5.26	4.63	4.28	4.04	3.88	3.76	3.66	3.59	3.53	3.43	3.33	3.23	3.17	3.12	3.06	3.00	2.94	2.88
12	6.55	5.10	4.47	4.12	3.89	3.73	3.61	3.51	3.44	3.37	3.28	3.18	3.07	3.02	2.96	2.91	2.85	2.79	2.72
13	6.41	4.97	4.35	4.00	3.77	3.60	3.48	3.39	3.31	3.25	3.15	3.05	2.95	2.89	2.84	2.78	2.72	2.66	2.60
14	6.30	4.86	4.24	3.89	3.66	3.50	3.38	3.29	3.21	3.15	3.05	2.95	2.84	2.79	2.73	2.67	2.61	2.55	2.49
15	6.20	4.77	4.15	3.80	3.58	3.41	3.29	3.20	3.12	3.06	2.96	2.86	2.76	2.70	2.64	2.59	2.52	2.46	2.40
16	6.12	4.69	4.08	3.73	3.50	3.34	3.22	3.12	3.05	2.99	2.89	2.79	2.68	2.63	2.57	2.51	2.45	2.38	2.32
17	6.04	4.62	4.01	3.66	3.44	3.28	3.16	3.06	2.98	2.92	2.82	2.72	2.62	2.56	2.50	2.44	2.38	2.32	2.25
18	5.98	4.56	3.95	3.61	3.38	3.22	3.10	3.01	2.93	2.87	2.77	2.67	2.56	2.50	2.44	2.38	2.32	2.26	2.19
19	5.92	4.51	3.90	3.56	3.33	3.17	3.05	2.96	2.88	2.82	2.72	2.62	2.51	2.45	2.39	2.33	2.27	2.20	2.13
20	5.87	4.46	3.86	3.51	3.29	3.13	3.01	2.91	2.84	2.77	2.68	2.57	2.46	2.41	2.35	2.29	2.22	2.16	2.09
21	5.83	4.42	3.82	3.48	3.25	3.09	2.97	2.87	2.80	2.73	2.64	2.53	2.42	2.37	2.31	2.25	2.18	2.11	2.04
22	5.79	4.38	3.78	3.44	3.22	3.05	2.93	2.84	2.76	2.70	2.60	2.50	2.39	2.33	2.27	2.21	2.14	2.08	2.00
23	5.75	4.35	3.75	3.41	3.18	3.02	2.90	2.81	2.73	2.67	2.57	2.47	2.36	2.30	2.24	2.18	2.11	2.04	1.97
24	5.72	4.32	3.72	3.38	3.15	2.99	2.87	2.78	2.70	2.64	2.54	2.44	2.33	2.27	2.21	2.15	2.08	2.01	1.94
25	5.69	4.29	3.69	3.35	3.13	2.97	2.85	2.75	2.68	2.61	2.51	2.41	2.30	2.24	2.18	2.12	2.05	1.98	1.91
26	5.66	4.27	3.67	3.33	3.10	2.94	2.82	2.73	2.65	2.59	2.49	2.39	2.28	2.22	2.16	2.09	2.03	1.95	1.88
27	5.63	4.24	3.65	3.31	3.08	2.92	2.80	2.71	2.63	2.57	2.47	2.36	2.25	2.19	2.13	2.07	2.00	1.93	1.85
28	5.61	4.22	3.63	3.29	3.06	2.90	2.78	2.69	2.61	2.55	2.45	2.34	2.23	2.17	2.11	2.05	1.98	1.91	1.83
29	5.59	4.20	3.61	3.27	3.04	2.88	2.76	2.67	2.59	2.53	2.43	2.32	2.21	2.15	2.09	2.03	1.96	1.89	1.81
30	5.57	4.18	3.59	3.25	3.03	2.87	2.75	2.65	2.57	2.51	2.41	2.31	2.20	2.14	2.07	2.01	1.94	1.87	1.79
40	5.42	4.05	3.46	3.13	2.90	2.74	2.62	2.53	2.45	2.39	2.29	2.18	2.07	2.01	1.94	1.88	1.80	1.72	1.64
60	5.29	3.93	3.34	3.01	2.79	2.63	2.51	2.41	2.33	2.27	2.17	2.06	1.94	1.88	1.82	1.74	1.67	1.58	1.48
120	5.15	3.80	3.23	2.89	2.67	2.52	2.39	2.30	2.22	2.16	2.05	1.94	1.82	1.76	1.69	1.61	1.53	1.43	1.31
∞	5.02	3.69	3.12	2.79	2.57	2.41	2.29	2.19	2.11	2.05	1.94	1.83	1.71	1.64	1.57	1.48	1.39	1.27	1.00

TABLE A.5 *F* Distribution (*Continued*). $F_{0.01, \nu 1, \nu 2}$

Degrees of freedom for numerator ν_1 (columns); degrees of freedom for denominator ν_2 (rows)

ν_2	1	2	3	4	5	6	7	8	9	10	12	15	20	24	30	40	60	120	∞
1	4052	499.5	5403	5625	5764	5859	5928	5982	6022	6056	6106	6157	6209	6235	6261	6287	6313	6339	6366
2	98.50	999.00	99.17	99.25	99.30	99.33	99.36	99.37	99.39	99.40	99.42	99.43	99.45	99.46	99.47	99.47	99.48	99.49	99.50
3	34.12	30.82	29.46	28.71	28.24	27.91	27.67	27.49	27.35	27.23	27.05	26.87	26.69	26.00	26.50	26.41	26.32	26.22	26.13
4	21.20	18.00	16.69	15.98	15.52	15.21	14.98	14.80	14.66	14.55	14.37	14.20	14.02	13.93	13.84	13.75	13.65	13.56	13.46
5	16.26	13.27	12.06	11.39	10.97	10.67	10.46	10.29	10.16	10.05	9.89	9.72	9.55	9.47	9.38	9.29	9.20	9.11	9.02
6	13.75	10.92	9.78	9.15	8.75	8.47	8.26	8.10	7.98	7.87	7.72	7.56	7.40	7.31	7.23	7.14	7.06	6.97	6.88
7	12.15	9.55	8.45	7.85	7.46	7.19	6.99	6.84	6.72	6.62	6.47	6.31	6.16	6.07	5.99	5.91	5.82	5.74	5.65
8	11.26	8.65	7.59	7.01	6.63	6.37	6.18	6.03	5.91	5.81	5.67	5.52	5.36	5.28	5.20	5.12	5.03	4.95	4.86
9	10.56	8.02	6.99	6.42	6.06	5.80	5.61	5.47	5.35	5.26	5.11	4.96	4.81	4.73	4.65	4.57	4.48	4.40	4.31
10	10.04	7.56	6.55	5.99	5.64	5.39	5.20	5.06	4.94	4.85	4.71	4.56	4.41	4.33	4.25	4.17	4.08	4.00	3.91
11	9.65	7.21	6.22	5.67	5.32	5.07	4.89	4.74	4.63	4.54	4.40	4.25	4.10	4.02	3.94	3.86	3.78	3.69	3.60
12	9.33	6.93	5.95	5.41	5.06	4.82	4.64	4.50	4.39	4.30	4.16	4.01	3.86	3.78	3.70	3.62	3.54	3.45	3.36
13	9.07	6.70	5.74	5.21	4.86	4.62	4.44	4.30	4.19	4.10	3.96	3.82	3.66	3.59	3.51	3.43	3.34	3.25	3.17
14	8.86	6.51	5.56	5.04	4.69	4.46	4.28	4.14	4.03	3.94	3.80	3.66	3.51	3.43	3.35	3.27	3.18	3.09	3.00
15	8.68	6.36	5.42	4.89	4.36	4.32	4.14	4.00	3.89	3.80	3.67	3.52	3.37	3.29	3.21	3.13	3.05	2.96	2.87
16	8.53	6.23	5.29	4.77	4.44	4.20	4.03	3.89	3.78	3.69	3.55	3.41	3.26	3.18	3.10	3.02	2.93	2.84	2.75
17	8.40	6.11	5.18	4.67	4.34	4.10	3.93	3.79	3.68	3.59	3.46	3.31	3.16	3.08	3.00	2.92	2.83	2.75	2.65
18	8.29	6.01	5.09	4.58	4.25	4.01	3.84	3.71	3.60	3.51	3.37	3.23	3.08	3.00	2.92	2.84	2.75	2.66	2.57
19	8.18	5.93	5.01	4.50	4.17	3.94	3.77	3.63	3.52	3.43	3.30	3.15	3.00	2.92	2.84	2.76	2.67	2.58	2.49
20	8.10	5.85	4.94	4.43	4.10	3.87	3.70	3.56	3.46	3.37	3.23	3.09	2.94	2.86	2.78	2.69	2.61	2.52	2.42
21	8.02	5.78	4.87	4.37	4.04	3.81	3.64	3.51	3.40	3.31	3.17	3.03	2.88	2.80	2.72	2.64	2.55	2.46	2.36
22	7.95	5.72	4.82	4.31	3.99	3.76	3.59	3.45	3.35	3.26	3.12	2.98	2.83	2.75	2.67	2.58	2.50	2.40	2.31
23	7.88	5.66	4.76	4.26	3.94	3.71	3.54	3.41	3.30	3.21	3.07	2.93	2.78	2.70	2.62	2.54	2.45	2.35	2.26
24	7.82	5.61	4.72	4.22	3.90	3.67	3.50	3.36	3.26	3.17	3.03	2.89	2.74	2.66	2.58	2.49	2.40	2.31	2.21
25	7.77	5.57	4.68	4.18	3.85	3.63	3.46	3.32	3.22	3.13	2.99	2.85	2.70	2.62	2.54	2.45	2.36	2.27	2.17
26	7.72	5.53	4.64	4.14	3.82	3.59	3.42	3.29	3.18	3.09	2.96	2.81	2.66	2.58	2.50	2.42	2.33	2.23	2.13
27	7.68	5.49	4.60	4.11	3.78	3.56	3.39	3.26	3.15	3.06	2.93	2.78	2.63	2.55	2.47	2.38	2.29	2.20	2.10
28	7.64	5.45	4.57	4.07	3.75	3.53	3.36	3.23	3.12	3.03	2.90	2.75	2.60	2.52	2.44	2.35	2.26	2.17	2.06
29	7.60	5.42	4.54	4.04	3.73	3.50	3.33	3.20	3.09	3.00	2.87	2.73	2.57	2.49	2.41	2.33	2.23	2.14	2.03
30	7.56	5.39	4.51	4.02	3.70	3.47	3.30	3.17	3.07	2.98	2.84	2.70	2.55	2.47	2.39	2.30	2.21	2.11	2.01
40	7.31	5.18	4.31	3.83	3.51	3.29	3.12	2.99	2.89	2.80	2.66	2.52	2.37	2.29	2.20	2.11	2.02	1.92	1.80
60	7.08	4.98	4.13	3.65	3.34	3.12	2.95	2.82	2.72	2.63	2.50	2.35	2.20	2.12	2.03	1.94	1.84	1.73	1.60
120	6.85	4.79	3.95	3.48	3.17	2.96	2.79	2.66	2.56	2.47	2.34	2.19	2.03	1.95	1.86	1.76	1.66	1.53	1.38
∞	6.63	4.61	3.78	3.32	3.02	2.80	2.64	2.51	2.41	2.32	2.18	2.04	1.88	1.79	1.70	1.59	1.47	1.32	1.00

Source: Adapted from Pearson, E. S., and H. O. Hartley, Eds.: *Biometrika Tables for Statisticians,* vol. 1, 3d ed., Cambridge Univ. Press, Cambridge, U. K., 1966. Reprinted by permission of the *Biometrika* Trustees.

TABLE A.6 Binomial Curves*
Confidence Coefficient $1 - 2\alpha = 0.95$

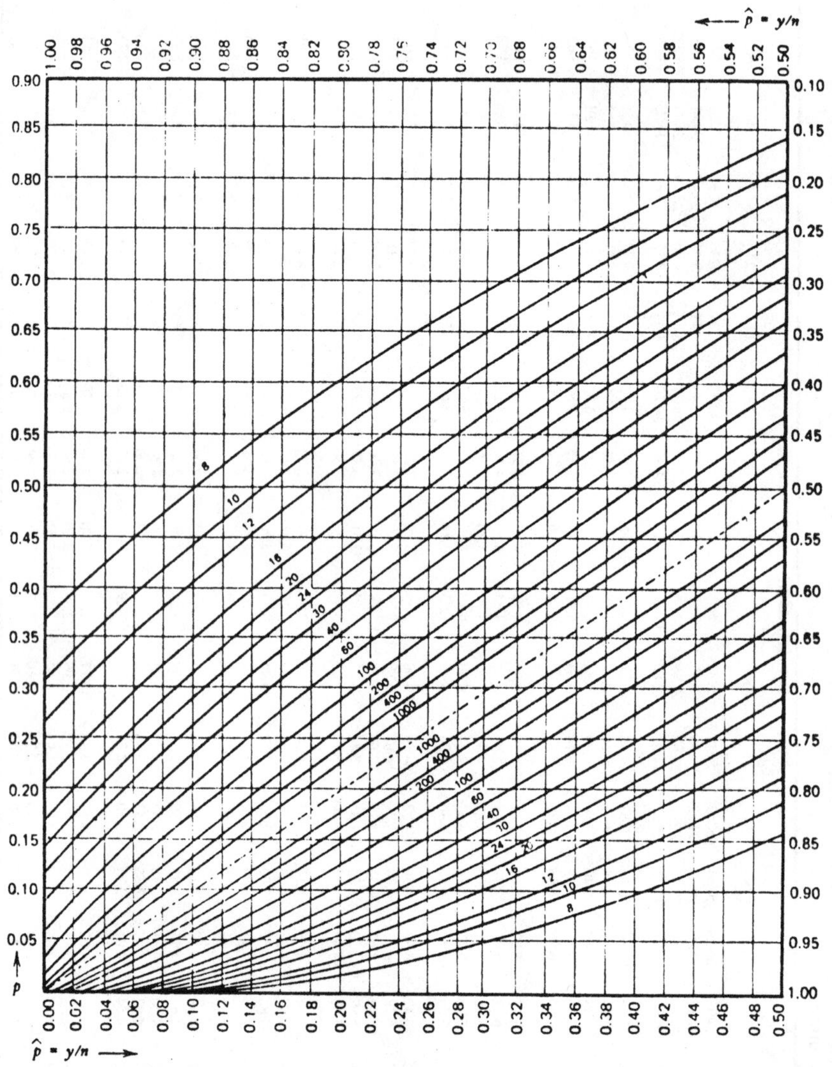

*The numbers printed along the curves indicate the sample size n. If for a given value of the abscissa y/n, L and U are the ordinates, read from (or interpolate between) the appropriate lower and upper curves. Then $P(L \leq p \leq U) \geq 1 - 2\alpha$. The process of reading from the curves can be simplified with the help of the right-angled corner of a loose sheet of paper or thin card, along the edges of which are marked off the scales shown in the top left-hand corner of each chart.

Source: Adapted from Pearson, E. S., and H. O. Hartley, Eds.: *Biometrika Tables for Statisticians,* vol. 1, 3d ed., Cambridge Univ. Press, Cambridge, U. K., 1966. Reprinted by permission of the *Biometrika* Trustees.

TABLE A.6 Binomial Curves* (*Continued*)
Confidence Coefficient $1 - 2\alpha = 0.99$

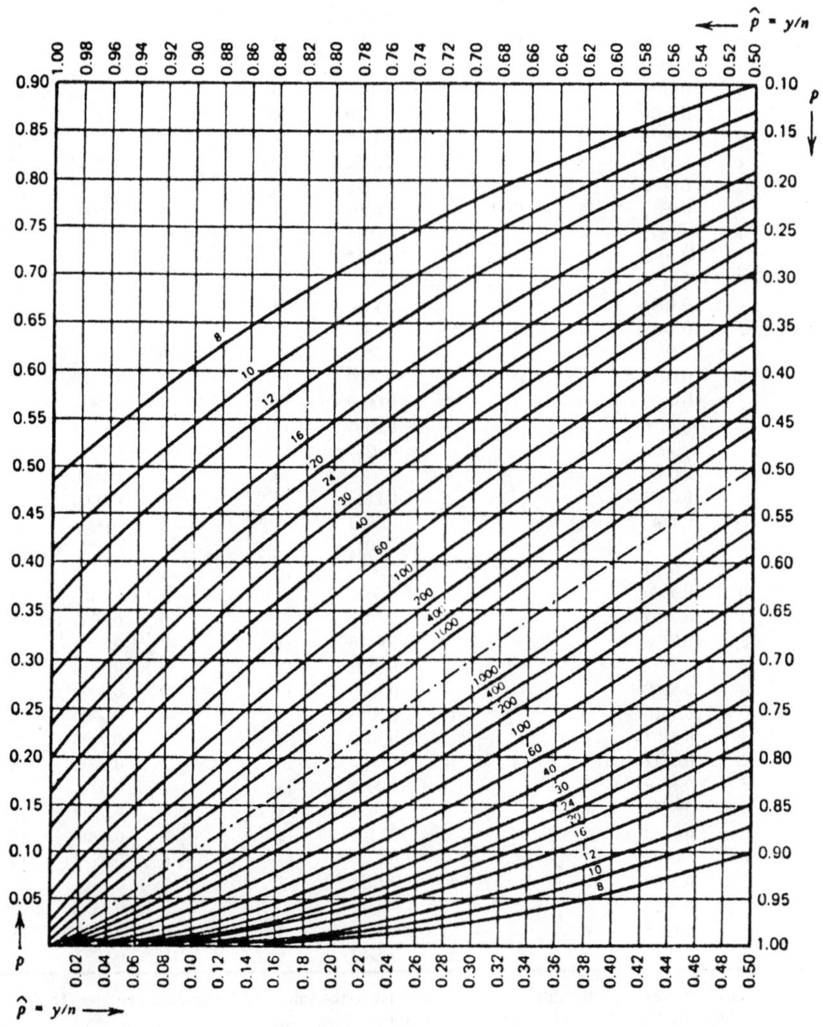

*The numbers printed along the curves indicate the sample size n. If for a given value of the abscissa y/n, L and U are the ordinates, read from (or interpolate between) the appropriate lower and upper curves. Then $P(L \leq p \leq U) \geq 1 - 2\alpha$. The process of reading from the curves can be simplified with the help of the right-angled corner of a loose sheet of paper or thin card, along the edges of which are marked off the scales shown in the top left-hand corner of each chart.

Source: Adapted from Pearson, E. S., and H. O. Hartley, Eds.: *Biometrika Tables for Statisticians,* vol. 1, 3d ed., Cambridge Univ. Press, Cambridge, U. K., 1966. Reprinted by permission of the *Biometrika* Trustees.

TABLE A.7 Binomial Distribution for $\theta = 0.5$*

n	Left S	P	Right S	n	Left S	P	Right S	n	Left S	P	Right S
1	0	0.5000	1	11	5	0.5000	6	17	0	0.0000	17
2	0	0.2500	2	12	0	0.0002	12		1	0.0001	16
	1	0.7500	1		1	0.0032	11		2	0.0012	15
3	0	0.1250	3		2	0.0193	10		3	0.0064	14
	1	0.5000	2		3	0.0730	9		4	0.0245	13
4	0	0.0625	4		4	0.1938	8		5	0.0717	12
	1	0.3125	3		5	0.3872	7		6	0.1662	11
	2	0.6875	2		6	0.6128	6		7	0.3145	10
5	0	0.0312	5	13	0	0.0001	13		8	0.5000	9
	1	0.1875	4		1	0.0017	12	18	0	0.0000	18
	2	0.5000	3		2	0.0112	11		1	0.0001	17
6	0	0.0156	6		3	0.0461	10		2	0.0007	16
	1	0.1094	5		4	0.1334	9		3	0.0038	15
	2	0.3438	4		5	0.2905	8		4	0.0154	14
	3	0.6562	3		6	0.5000	7		5	0.0481	13
7	0	0.0078	7	14	0	0.0000	14		6	0.1189	12
	1	0.0625	6		1	0.0009	13		7	0.2403	11
	2	0.2266	5		2	0.0065	12		8	0.4073	10
	3	0.5000	4		3	0.0287	11		9	0.5927	9
8	0	0.0039	8		4	0.0898	10	19	0	0.0000	19
	1	0.0352	7		5	0.2120	9		1	0.0000	18
	2	0.1445	6		6	0.3953	8		2	0.0004	17
	3	0.3633	5		7	0.6047	7		3	0.0022	16
	4	0.6367	4	15	0	0.0000	15		4	0.0096	15
9	0	0.0020	9		1	0.0005	14		5	0.0318	14
	1	0.0195	8		2	0.0037	13		6	0.0835	13
	2	0.0898	7		3	0.0176	12		7	0.1796	12
	3	0.2539	6		4	0.0592	11		8	0.3238	11
	4	0.5000	5		5	0.1509	10		9	0.5000	10
10	0	0.0010	10		6	0.3036	9	20	0	0.0000	20
	1	0.0107	9		7	0.5000	8		1	0.0000	19
	2	0.0547	8	16	0	0.0000	16		2	0.0002	18
	3	0.1719	7		1	0.0003	15		3	0.0013	17
	4	0.3770	6		2	0.0021	14		4	0.0059	16
	5	0.6230	5		3	0.0106	13		5	0.0207	15
11	0	0.0005	11		4	0.0384	12		6	0.0577	14
	1	0.0059	10		5	0.1051	11		7	0.1316	13
	2	0.0327	9		6	0.2272	10		8	0.2517	12
	3	0.1133	8		7	0.4018	9		9	0.4119	11
	4	0.2744	7		8	0.5982	8		10	0.5881	10

*Entries P are the tail probabilities from each extreme to the value of S for given n when $\theta = 0.5$. Left-tail probabilities are given for $S \leq 0.5n$, and right-tail for $S \geq 0.5n$.

For $n > 20$, the probabilities are found from Table A.1 as follows:

$$z_R = \frac{S - 0.5 - 0.5n}{0.5\sqrt{n}} \qquad z_L = \frac{S + 0.5 - 0.5n}{0.5\sqrt{n}}$$

Desired:	Approximated by:
Right-tail probability for S	Right-tail probability for Z_R
Left-tail probability for S	Left-tail probability for Z_L

Source: Adapted from Table 2 of *Tables of the Binomial Distribution*, National Bureau of Standards, U.S. Government Printing Office, Washington, D C, Jan. 1950, with *Corrigenda* 1952 and 1958. Reprinted with permission.

TABLE A.8 Probabilities for Wilcoxon Signed Rank Test*

	Left		Right		Left		Right		Left		Right
n	*T*	*P*	*T*	*n*	*T*	*P*	*T*	*n*	*T*	*P*	*T*
2	0	0.250	3	8	0	0.004	36	10	4	0.007	51
	1	0.500	2		1	0.008	35		5	0.010	50
3	0	0.125	6		2	0.012	34		6	0.014	49
	1	0.250	5		3	0.020	33		7	0.019	48
	2	0.375	4		4	0.027	32		8	0.024	47
	3	0.625	3		5	0.039	31		9	0.032	46
4	0	0.062	10		6	0.055	30		10	0.042	45
	1	0.125	9		7	0.074	29		11	0.053	44
	2	0.188	8		8	0.098	28		12	0.065	43
	3	0.312	7		9	0.125	27		13	0.080	42
	4	0.438	6		10	0.156	26		14	0.097	41
	5	0.562	5		11	0.191	25		15	0.116	40
5	0	0.031	15		12	0.230	24		16	0.138	39
	1	0.062	14		13	0.273	23		17	0.161	38
	2	0.094	13		14	0.320	22		18	0.188	37
	3	0.156	12		15	0.371	21		19	0.216	36
	4	0.219	11		16	0.422	20		20	0.246	35
	5	0.312	10		17	0.473	19		21	0.278	34
	6	0.406	9		18	0.527	18		22	0.312	33
	7	0.500	8	9	0	0.002	45		23	0.348	32
6	0	0.016	21		1	0.004	44		24	0.385	31
	1	0.031	20		2	0.006	43		25	0.423	30
	2	0.047	19		3	0.010	42		26	0.461	29
	3	0.078	18		4	0.014	41		27	0.500	28
	4	0.109	17		5	0.020	40	11	0	0.000	66
	5	0.156	16		6	0.027	39		1	0.001	65
	6	0.219	15		7	0.037	38		2	0.001	64
	7	0.281	14		8	0.049	37		3	0.002	63
	8	0.344	13		9	0.064	36		4	0.003	62
	9	0.422	12		10	0.082	35		5	0.005	61
	10	0.500	11		11	0.102	34		6	0.007	60
7	0	0.008	28		12	0.125	33		7	0.009	59
	1	0.016	27		13	0.150	32		8	0.012	58
	2	0.023	26		14	0.180	31		9	0.016	57
	3	0.039	25		15	0.213	30		10	0.021	56
	4	0.055	24		16	0.248	29		11	0.027	55
	5	0.078	23		17	0.285	28		12	0.034	54
	6	0.109	22		18	0.326	27		13	0.042	53
	7	0.148	21		19	0.367	26		14	0.051	52
	8	0.188	20		20	0.410	25		15	0.062	51
	9	0.234	19		21	0.455	24		16	0.074	50
	10	0.289	18		22	0.500	23		17	0.087	49
	11	0.344	17	10	0	0.001	55		18	0.103	48
	12	0.406	16		1	0.002	54		19	0.120	47
	13	0.469	15		2	0.003	53		20	0.139	46
	14	0.531	14		3	0.005	52		21	0.160	45

TABLE A.8 Probabilities for Wilcoxon Signed Rank Test* *(Continued)*

n	Left T	P	Right T
11	22	0.183	44
	23	0.207	43
	24	0.232	42
	25	0.260	41
	26	0.289	40
	27	0.319	39
	28	0.350	38
	29	0.382	37
	30	0.416	36
	31	0.449	35
	32	0.483	34
	33	0.517	33
12	0	0.000	78
	1	0.000	77
	2	0.001	76
	3	0.001	75
	4	0.002	74
	5	0.002	73
	6	0.003	72
	7	0.005	71
	8	0.006	70
	9	0.008	69
	10	0.010	68
	11	0.013	67
	12	0.017	66
	13	0.021	65
	14	0.026	64
	15	0.032	63
	16	0.039	62
	17	0.046	61
	18	0.055	60
	19	0.065	59
	20	0.076	58
	21	0.088	57
	22	0.102	56
	23	0.117	55
	24	0.133	54
	25	0.151	53
	26	0.170	52
	27	0.190	51
	28	0.212	50
	29	0.235	49
	30	0.259	48
	31	0.285	47
	32	0.311	46
	33	0.339	45

n	Left T	P	Right T
12	34	0.367	44
	35	0.396	43
	36	0.425	42
	37	0.455	41
	38	0.485	40
	39	0.515	39
13	0	0.000	91
	1	0.000	90
	2	0.000	89
	3	0.001	88
	4	0.001	87
	5	0.001	86
	6	0.002	85
	7	0.002	84
	8	0.003	83
	9	0.004	82
	10	0.005	81
	11	0.007	80
	12	0.009	79
	13	0.011	78
	14	0.013	77
	15	0.016	76
	16	0.020	75
	17	0.024	74
	18	0.029	73
	19	0.034	72
	20	0.040	71
	21	0.047	70
	22	0.055	69
	23	0.064	68
	24	0.073	67
	25	0.084	66
	26	0.095	65
	27	0.108	64
	28	0.122	63
	29	0.137	62
	30	0.153	61
	31	0.170	60
	32	0.188	59
	33	0.207	58
	34	0.227	57
	35	0.249	56
	36	0.271	55
	37	0.294	54
	38	0.318	53
	39	0.342	52

n	Left T	P	Right T
13	40	0.368	51
	41	0.393	50
	42	0.420	49
	43	0.446	48
	44	0.473	47
	45	0.500	46
14	0	0.000	105
	1	0.000	104
	2	0.000	103
	3	0.000	102
	4	0.000	101
	5	0.001	100
	6	0.001	99
	7	0.001	98
	8	0.002	97
	9	0.002	96
	10	0.003	95
	11	0.003	94
	12	0.004	93
	13	0.005	92
	14	0.007	91
	15	0.008	90
	16	0.010	89
	17	0.012	88
	18	0.015	87
	19	0.018	86
	20	0.021	85
	21	0.025	84
	22	0.029	83
	23	0.034	82
	24	0.039	81
	25	0.045	80
	26	0.052	79
	27	0.059	78
	28	0.068	77
	29	0.077	76
	30	0.086	75
	31	0.097	74
	32	0.108	73
	33	0.121	72
	34	0.134	71
	35	0.148	70
	36	0.163	69
	37	0.179	68
	38	0.196	67
	39	0.213	66

TABLE A.8 Probabilities for Wilcoxon Signed Rank Test* *(Continued)*

	Left		Right		Left		Right		Left		Right
n	T	P	T	n	T	P	T	n	T	P	T
14	40	0.232	65	15	12	0.002	108	15	37	0.104	83
	41	0.251	64		13	0.003	107		38	0.115	82
	42	0.271	63		14	0.003	106		39	0.126	81
	43	0.292	62		15	0.004	105		40	0.138	80
	44	0.313	61		16	0.005	104		41	0.151	79
	45	0.335	60		17	0.006	103		42	0.165	78
	46	0.357	59		18	0.008	102		43	0.180	77
	47	0.380	58		19	0.009	101		44	0.195	76
	48	0.404	57		20	0.011	100		45	0.211	75
	49	0.428	56		21	0.013	99		46	0.227	74
	50	0.452	55		22	0.015	98		47	0.244	73
	51	0.476	54		23	0.018	97		48	0.262	72
	52	0.500	53		24	0.021	96		49	0.281	71
15	0	0.000	120		25	0.024	95		50	0.300	70
	1	0.000	119		26	0.028	94		51	0.319	69
	2	0.000	118		27	0.032	93		52	0.339	68
	3	0.000	117		28	0.036	92		53	0.360	67
	4	0.000	116		29	0.042	91		54	0.381	66
	5	0.000	115		30	0.047	90		55	0.402	65
	6	0.000	114		31	0.053	89		56	0.423	64
	7	0.001	113		32	0.060	88		57	0.445	63
	8	0.001	112		33	0.068	87		58	0.467	62
	9	0.001	111		34	0.076	86		59	0.489	61
	10	0.001	110		35	0.084	85		60	0.511	60
	11	0.002	109		36	0.094	84				

*Entries P are the tail probabilities from each extreme to the value of the signed rank statistic for given n. Left-tail probabilities are given for $T \le n(n + 1)/4$, and right-tail for $T \ge n(n + 1)/4$. T is interpreted as either T_+ or T_-.

For $n > 15$, the probabilities are found from Table .1 as follows:

$$z_R = \frac{T_+ - 0.5 - n(n+1)/4}{\sqrt{n(n+1)(2n+1)/24}} \qquad z_R = \frac{T_- - 0.5 - n(n+1)/4}{\sqrt{n(n+1)(2n+1)/24}}$$

Desired:	Approximated by:
Right-tail probability for T_+	Right-tail probability for z_R
Right-tail probability for T_-	Right-tail probability for z_R'

Source: Adapted from Table II of Harter, H. L., and D. B. Owen, Eds.: *Selected Tables in Mathematical Statistics*, vol. 1., Markham Publishing, Chicago, 1972. Reprinted by permission of the Institute of Mathematical Statistics.

TABLE A.9 Mann-Whitney-Wilcoxon Distribution*

n	Left T_x	P	Right T_x
	$m = 1$		
1	1	0.500	2
2	1	0.333	3
	2	0.667	2
3	1	0.250	4
	2	0.500	3
4	1	0.200	5
	2	0.400	4
	3	0.600	3
	3	0.600	3
	3	0.600	3
	3	0.500	4
6	1	0.143	7
	2	0.286	6
	3	0.429	5
	4	0.571	4
7	1	0.125	8
	2	0.250	7
	3	0.375	6
	4	0.500	5
8	1	0.111	9
	2	0.222	8
	3	0.333	7
	4	0.444	6
	5	0.556	5
9	1	0.100	10
	2	0.200	9
	3	0.300	8
	4	0.400	7
	5	0.500	6
10	1	0.091	11
	2	0.182	10
	3	0.273	9
	4	0.364	8
	5	0.455	7
	6	0.545	6
	$m = 2$		
2	3	0.167	7
	4	0.333	6
	5	0.667	5
3	3	0.100	9
	4	0.200	8
	5	0.400	7
	6	0.600	6
4	3	0.067	11
	4	0.133	10
	5	0.267	9
	6	0.400	8
	7	0.600	7
5	3	0.048	13

n	Left T_x	P	Right T_x
5	4	0.095	12
	5	0.190	11
	6	0.286	10
	7	0.429	9
	8	0.571	8
6	3	0.036	15
	4	0.071	14
	5	0.143	13
	6	0.214	12
	7	0.321	11
	8	0.429	10
	9	0.571	9
7	3	0.028	17
	4	0.056	16
	5	0.111	15
	6	0.167	14
	7	0.250	13
	8	0.333	12
	9	0.444	11
	10	0.556	10
8	3	0.022	19
	4	0.044	18
	5	0.089	17
	6	0.133	16
	7	0.200	15
	8	0.267	14
	9	0.356	13
	10	0.444	12
	11	0.556	11
9	3	0.018	21
	4	0.036	20
	5	0.073	19
	6	0.109	18
	7	0.164	17
	8	0.218	16
	9	0.291	15
	10	0.364	14
	11	0.455	13
	12	0.545	12
10	3	0.015	23
	4	0.030	22
	5	0.061	21
	6	0.091	20
	7	0.136	19
	8	0.182	18
	9	0.242	17
	10	0.303	16
	11	0.379	15
	12	0.455	14
	13	0.545	13

n	Left T_x	P	Right T_x
	$m = 3$		
3	6	0.050	15
	7	0.100	14
	8	0.200	13
	9	0.350	12
	10	0.500	11
4	6	0.029	18
	7	0.057	17
	8	0.114	16
	9	0.200	15
	10	0.314	14
	11	0.429	13
	12	0.571	12
5	6	0.018	21
	7	0.036	20
	8	0.071	19
	9	0.125	18
	10	0.196	17
	11	0.286	16
	12	0.393	15
	13	0.500	14
6	6	0.012	24
	7	0.024	23
	8	0.048	22
	9	0.083	21
	10	0.131	20
	11	0.190	19
	12	0.274	18
	13	0.357	17
	14	0.452	16
	15	0.548	15
7	6	0.008	27
	7	0.017	26
	8	0.033	25
	9	0.058	24
	10	0.092	23
	11	0.133	22
	12	0.192	21
	13	0.258	20
	14	0.333	19
	15	0.417	18
	16	0.500	17
8	6	0.006	30
	7	0.012	29
	8	0.024	28
	9	0.042	27
	10	0.067	26
	11	0.097	25
	12	0.139	24
	13	0.188	23

TABLE A.9 Mann-Whitney-Wilcoxon Distribution* (*Continued*)

n	Left T_x	P	Right T_x
	$m = 3$		
8	14	0.248	22
	15	0.315	21
	15	0.388	20
	17	0.461	19
	18	0.539	18
9	6	0.005	33
	7	0.009	32
	8	0.018	31
	9	0.032	30
	10	0.050	29
	11	0.073	28
	12	0.105	27
	13	0.141	26
	14	0.186	25
	15	0.241	24
	16	0.300	23
	17	0.364	22
	18	0.432	21
	19	0.500	20
10	6	0.003	36
	7	0.007	35
	8	0.014	34
	9	0.024	33
	10	0.038	32
	11	0.056	31
	12	0.080	30
	13	0.108	29
	14	0.143	28
	15	0.185	27
	16	0.234	26
	17	0.287	25
	18	0.346	24
	19	0.406	23
	20	0.469	22
	21	0.531	21
	$m = 4$		
4	10	0.014	26
	11	0.029	25
	12	0.057	24
	13	0.100	23
	14	0.171	22
	15	0.243	21
	16	0.343	20
	17	0.443	19
	18	0.557	18
5	10	0.008	30
	11	0.016	29
	12	0.032	28
5	13	0.056	27
	14	0.095	26
	15	0.143	25
	16	0.206	24
	17	0.278	23
	18	0.365	22
	19	0.452	21
	20	0.548	20
6	10	0.005	34
	11	0.010	33
	12	0.019	32
	13	0.033	31
	14	0.057	30
	15	0.086	29
	16	0.129	28
	17	0.176	27
	18	0.238	26
	19	0.305	25
	20	0.381	24
	21	0.457	23
	22	0.543	22
7	10	0.003	38
	11	0.006	37
	12	0.012	36
	13	0.021	35
	14	0.036	34
	15	0.055	33
	16	0.082	32
	17	0.115	31
	18	0.158	30
	19	0.206	29
	20	0.264	28
	21	0.324	27
	22	0.394	26
	23	0.464	25
	24	0.536	24
8	10	0.002	42
	11	0.004	41
	12	0.008	40
	13	0.014	39
	14	0.024	38
	15	0.036	37
	16	0.055	36
	17	0.077	35
	18	0.107	34
	19	0.141	33
	20	0.184	32
	21	0.230	31
	22	0.285	30
	23	0.341	29
	24	0.404	28
8	25	0.467	27
	26	0.533	26
9	10	0.001	46
	11	0.003	45
	12	0.006	44
	13	0.010	43
	14	0.017	42
	15	0.025	41
	16	0.038	40
	17	0.053	39
	18	0.074	38
	19	0.099	37
	20	0.130	36
	21	0.165	35
	22	0.207	34
	23	0.252	33
	24	0.302	32
	25	0.355	31
	26	0.413	30
	27	0.470	29
	28	0.530	28
10	10	0.001	50
	11	0.002	49
	12	0.004	48
	13	0.007	47
	14	0.012	46
	15	0.018	45
	16	0.027	44
	17	0.038	43
	18	0.053	42
	19	0.071	41
	20	0.094	40
	21	0.120	39
	22	0.152	38
	23	0.187	37
	24	0.227	36
	25	0.270	35
	26	0.318	34
	27	0.367	33
	28	0.420	32
	29	0.473	31
	30	0.527	30
	$m = 5$		
5	15	0.004	40
	16	0.008	39
	17	0.016	38
	18	0.028	37
	19	0.048	36
	20	0.075	35
	21	0.111	34

TABLE A.9 Mann-Whitney-Wilcoxon Distribution* (*Continued*)

n	Left T_x	P	Right T_x	n	Left T_x	P	Right T_x	n	Left T_x	P	Right T_x
$m = 5$				8	24	0.064	46	10	31	0.155	49
5	22	0.155	33		25	0.085	45		32	0.185	48
	23	0.210	32		26	0.111	44		33	0.220	47
	24	0.274	31		27	0.142	43		34	0.257	46
	25	0.345	30		28	0.117	42		35	0.297	45
	26	0.421	29		29	0.218	41		36	0.339	44
	27	0.500	28		30	0.262	40		37	0.384	43
6	15	0.002	45		31	0.311	39		38	0.430	42
	16	0.004	44		32	0.362	38		39	0.477	41
	17	0.009	43		33	0.416	37		40	0.523	40
	18	0.015	42		34	0.472	36	$m = 6$			
	19	0.026	41		35	0.528	35	6	21	0.001	57
	20	0.041	40	9	15	0.000	60		22	0.002	56
	21	0.063	39		16	0.001	59		23	0.004	55
	22	0.089	38		17	0.002	58		24	0.008	54
	23	0.123	37		18	0.003	57		25	0.013	53
	24	0.165	36		19	0.006	56		26	0.021	52
	25	0.214	35		20	0.009	55		27	0.032	51
	26	0.268	34		21	0.014	54		28	0.047	50
	27	0.331	33		22	0.021	53		29	0.066	49
	28	0.396	32		23	0.030	52		30	0.090	48
	29	0.465	31		24	0.041	51		31	0.120	47
	30	0.535	30		25	0.056	50		32	0.155	46
7	15	0.001	50		26	0.073	49		33	0.197	45
	16	0.003	49		27	0.095	48		34	0.242	44
	17	0.005	48		28	0.120	47		35	0.294	43
	18	0.009	47		29	0.149	46		36	0.350	42
	19	0.015	46		30	0.182	45		37	0.409	41
	20	0.024	45		31	0.219	44		38	0.469	40
	21	0.037	44		32	0.259	43		39	0.531	39
	22	0.053	43		33	0.303	42	7	21	0.001	63
	23	0.074	42		34	0.350	41		22	0.001	62
	24	0.101	41		35	0.399	40		23	0.002	61
	25	0.134	40		36	0.449	39		24	0.004	60
	26	0.172	39		37	0.500	38		25	0.007	59
	27	0.216	38	10	15	0.000	65		26	0.011	58
	28	0.265	37		16	0.001	64		27	0.017	57
	29	0.319	36		17	0.001	63		28	0.026	56
	30	0.378	35		18	0.002	62		29	0.037	55
	31	0.438	34		19	0.004	61		30	0.051	54
	32	0.500	33		20	0.006	60		31	0.069	53
8	15	0.001	55		21	0.010	59		32	0.090	52
	16	0.002	54		22	0.014	58		33	0.117	51
	17	0.003	53		23	0.020	57		34	0.147	50
	18	0.005	52		24	0.028	56		35	0.183	49
	19	0.009	51		25	0.038	55		36	0.223	48
	20	0.015	50		26	0.050	54		37	0.267	47
	21	0.023	49		27	0.065	53		38	0.314	46
	22	0.033	48		28	0.082	52		39	0.365	45
	23	0.047	47		29	0.103	51		40	0.418	44
					30	0.127	50				

TABLE A.9 Mann-Whitney-Wilcoxon Distribution* (*Continued*)

n	Left T_x	P	Right T_x	n	Left T_x	P	Right T_x	n	Left T_x	P	Right T_x
$m = 6$				9	43	0.303	53	7	40	0.064	65
7	41	0.473	43		44	0.344	52		41	0.082	64
	42	0.527	42		45	0.388	51		42	0.104	63
8	21	0.000	69		46	0.432	50		43	0.130	62
	22	0.001	68		47	0.477	49		44	0.159	61
	23	0.001	67		48	0.523	48		45	0.191	60
	24	0.002	66	10	21	0.000	81		46	0.228	59
	25	0.004	65		22	0.000	80		47	0.267	58
	26	0.006	64		23	0.000	79		48	0.310	57
	27	0.010	63		24	0.001	78		49	0.355	56
	28	0.015	62		25	0.001	77		50	0.402	55
	29	0.021	61		26	0.002	76		51	0.451	54
	30	0.030	60		27	0.004	75		52	0.500	53
	31	0.041	59		28	0.005	74	8	28	0.000	84
	32	0.054	58		29	0.008	73		29	0.000	83
	33	0.071	57		30	0.011	72		30	0.001	82
	34	0.091	56		31	0.016	71		31	0.001	81
	35	0.114	55		32	0.021	70		32	0.002	80
	36	0.141	54		33	0.028	69		33	0.003	79
	37	0.172	53		34	0.036	68		34	0.005	78
	38	0.207	52		35	0.047	67		35	0.007	77
	39	0.245	51		36	0.059	66		36	0.010	76
	40	0.286	50		37	0.074	65		37	0.014	75
	41	0.331	49		38	0.090	64		38	0.020	74
	42	0.377	48		39	0.110	63		39	0.027	73
	43	0.426	47		40	0.132	62		40	0.036	72
	44	0.475	46		41	0.157	61		41	0.047	71
	45	0.525	45		42	0.184	60		42	0.060	70
9	21	0.000	75		43	0.214	59		43	0.076	69
	22	0.000	74		44	0.246	58		44	0.095	68
	23	0.001	73		45	0.281	57		45	0.116	67
	24	0.001	72		46	0.318	56		46	0.140	66
	25	0.002	71		47	0.356	55		47	0.168	65
	26	0.004	70		48	0.396	54		48	0.198	64
	27	0.006	69		49	0.437	53		49	0.232	63
	28	0.009	68		50	0.479	52		50	0.268	62
	29	0.013	67		51	0.521	51		51	0.306	61
	30	0.018	66	$m = 7$					52	0.347	60
	31	0.025	65	7	28	0.000	77		53	0.389	59
	32	0.033	64		29	0.001	76		54	0.433	58
	33	0.044	63		30	0.001	75		55	0.478	57
	34	0.057	62		31	0.002	74		56	0.522	56
	35	0.072	61		32	0.003	73	9	28	0.000	91
	36	0.091	60		33	0.006	72		29	0.000	90
	37	0.112	59		34	0.009	71		30	0.000	89
	38	0.136	58		35	0.013	70		31	0.001	88
	39	0.164	57		36	0.019	69		32	0.001	87
	40	0.194	56		37	0.027	68		33	0.002	86
	41	0.228	55		38	0.036	67		34	0.003	85
	42	0.264	54		39	0.049	66		35	0.004	84
									36	0.006	83

TABLE A.9 Mann-Whitney-Wilcoxon Distribution* (*Continued*)

n	Left T_x	P	Right T_x
	$m = 7$		
9	37	0.008	82
	38	0.011	81
	39	0.016	80
	40	0.021	79
	41	0.027	78
	42	0.036	77
	43	0.045	76
	44	0.057	75
	45	0.071	74
	46	0.087	73
	47	0.105	72
	48	0.126	71
	49	0.150	70
	50	0.176	69
	51	0.204	68
	52	0.235	67
	53	0.268	66
	54	0.303	65
	55	0.340	64
	56	0.379	63
	57	0.419	62
	58	0.459	61
	59	0.500	60
10	28	0.000	98
	29	0.000	97
	30	0.000	96
	31	0.000	95
	32	0.001	94
	33	0.001	93
	34	0.002	92
	35	0.002	91
	36	0.003	90
	37	0.005	89
	38	0.007	88
	39	0.009	87
	40	0.012	86
	41	0.017	85
	42	0.022	84
	43	0.028	83
	44	0.035	82
	45	0.044	81
	46	0.054	80
	47	0.067	79
	48	0.081	78
	49	0.097	77
	50	0.115	76
	51	0.135	75
	52	0.157	74
	53	0.182	73
10	54	0.209	72
	55	0.237	71
	56	0.268	70
	57	0.300	69
	58	0.335	68
	59	0.370	67
	60	0.406	66
	61	0.443	65
	62	0.481	64
	63	0.519	63
	$m = 8$		
8	36	0.000	100
	37	0.000	99
	38	0.000	98
	39	0.001	97
	40	0.001	96
	41	0.001	95
	42	0.002	94
	43	0.003	93
	44	0.005	92
	45	0.007	91
	46	0.010	90
	47	0.014	89
	48	0.019	88
	49	0.025	87
	50	0.032	86
	51	0.041	85
	52	0.052	84
	53	0.065	83
	54	0.080	82
	55	0.097	81
	56	0.117	80
	57	0.139	79
	58	0.164	78
	59	0.191	77
	60	0.221	76
	61	0.253	75
	62	0.287	74
	63	0.323	73
	64	0.360	72
	65	0.399	71
	66	0.439	70
	67	0.480	69
	68	0.520	68
9	36	0.000	108
	37	0.000	107
	38	0.000	106
	39	0.000	105
	40	0.000	104
	41	0.001	103
9	42	0.001	102
	43	0.002	101
	44	0.003	100
	45	0.004	99
	46	0.006	98
	47	0.008	97
	48	0.010	96
	49	0.014	95
	50	0.018	94
	51	0.023	93
	52	0.030	92
	53	0.037	91
	54	0.046	90
	55	0.057	89
	56	0.069	88
	57	0.084	87
	58	0.100	86
	59	0.118	85
	60	0.138	84
	61	0.161	83
	62	0.185	82
	63	0.212	81
	64	0.240	80
	65	0.271	79
	66	0.303	78
	67	0.336	77
	68	0.371	76
	69	0.407	75
	70	0.444	74
	71	0.481	73
	72	0.519	72
10	36	0.000	116
	37	0.000	115
	38	0.000	114
	39	0.000	113
	40	0.000	112
	41	0.000	111
	42	0.001	110
	43	0.001	109
	44	0.002	108
	45	0.002	107
	46	0.003	106
	47	0.004	105
	48	0.006	104
	49	0.008	103
	50	0.010	102
	51	0.013	101
	52	0.017	100
	53	0.022	99
	54	0.027	98
	55	0.034	97

TABLE A.9 Mann-Whitney-Wilcoxon Distribution* (*Continued*)

n	Left T_x	P	Right T_x
	$m = 8$		
10	56	0.042	96
	57	0.051	95
	58	0.061	94
	59	0.073	93
	60	0.086	92
	61	0.102	91
	62	0.118	90
	63	0.137	89
	64	0.158	88
	65	0.180	87
	66	0.204	86
	67	0.230	85
	68	0.257	84
	69	0.286	83
	70	0.317	82
	71	0.348	81
	72	0.381	80
	73	0.414	79
	74	0.448	78
	75	0.483	77
	76	0.517	76
	$m = 9$		
9	45	0.000	126
	46	0.000	125
	47	0.000	124
	48	0.000	123
	49	0.000	122
	50	0.000	121
	51	0.001	120
	52	0.001	119
	53	0.001	118
	54	0.002	117
	55	0.003	116
	56	0.004	115
	57	0.005	114
	58	0.007	113
	59	0.009	112
	60	0.012	111
	61	0.016	110
	62	0.020	109
	63	0.025	108
	64	0.031	107
	65	0.039	106
	66	0.047	105
	67	0.057	104
	68	0.068	103
	69	0.081	102
	70	0.095	101
9	71	0.111	100
	72	0.129	99
	73	0.149	98
	74	0.170	97
	75	0.193	96
	76	0.218	95
	77	0.245	94
	78	0.273	93
	79	0.302	92
	80	0.333	91
	81	0.365	90
	82	0.398	89
	83	0.432	88
	84	0.466	87
	85	0.500	86
10	45	0.000	135
	46	0.000	134
	47	0.000	133
	48	0.000	132
	49	0.000	131
	50	0.000	130
	51	0.000	129
	52	0.000	128
	53	0.001	127
	54	0.001	126
	55	0.001	125
	56	0.002	124
	57	0.003	123
	58	0.004	122
	59	0.005	121
	60	0.007	120
	61	0.009	119
	62	0.011	118
	63	0.014	117
	64	0.017	116
	65	0.022	115
	66	0.027	114
	67	0.033	113
	68	0.039	112
	69	0.047	111
	70	0.056	110
	71	0.067	109
	72	0.078	108
	73	0.091	107
	74	0.106	106
	75	0.121	105
	76	0.139	104
	77	0.158	103
	78	0.178	102
	79	0.200	101
	80	0.223	100
10	81	0.248	99
	82	0.274	98
	83	0.302	97
	84	0.330	96
	85	0.360	95
	86	0.390	94
	87	0.421	93
	88	0.452	92
	89	0.484	91
	90	0.516	90
	$m = 10$		
10	55	0.000	155
	56	0.000	154
	57	0.000	153
	58	0.000	152
	59	0.000	151
	60	0.000	150
	61	0.000	149
	62	0.000	148
	63	0.000	147
	64	0.001	146
	65	0.001	145
	66	0.001	144
	67	0.001	143
	68	0.002	142
	69	0.003	141
	70	0.003	140
	71	0.004	139
	72	0.006	138
	73	0.007	137
	74	0.009	136
	75	0.012	135
	76	0.014	134
	77	0.018	133
	78	0.022	132
	79	0.026	131
	80	0.032	130
	81	0.038	129
	82	0.045	128
	83	0.053	127
	84	0.062	126
	85	0.072	125
	86	0.083	124
	87	0.095	123
	88	0.109	122
	89	0.124	121
	90	0.140	120
	91	0.157	119
	92	0.176	118
	93	0.197	117

TABLE A.9 Mann-Whitney-Wilcoxon Distribution* (*Continued*)

n	Left T_x	P	Right T_x	n	Left T_x	P	Right T_x	n	Left T_x	P	Right T_x
	$m = 10$			10	97	0.289	113	10	102	0.427	108
10	94	0.218	116		98	0.315	112		103	0.456	107
	95	0.241	115		99	0.342	111		104	0.485	106
	96	0.264	114		100	0.370	110		105	0.515	105
					101	0.398	109				

*Entries P are the tail probabilities from each extreme to the value of T_x for the given sample sizes $m \le n$ (m is the size of the x sample). Left-tail probabilities are given for $T_x \le m(N+1)/2$ and right-tail for $T_x \ge m(N+1)/2$, where $N = m + n$. For m or $n > 10$, the probabilities are found from Table A.1 as follows:

$$z_L = \frac{T_x + 0.5 - m(N+1)/2}{\sqrt{mn(N+1)/12}} \qquad z_R = \frac{T_x - 0.5 - m(N+1)/2}{\sqrt{mn(N+1)/12}}$$

Desired:	Approximated by:
Left-tail probability for T_x	Left-tail probability for z_L
Right-tail probability for T_x	Right-tail probability for z_R

Source: Adapted from Table B of Kraft, C. H., and C. Van Eeden: *A Nonparametric Introduction to Statistics*, Macmillan, New York, 1969. Reprinted by permission of the publisher.

TABLE A.10 Kruskal-Wallis Test Table of Critical Values*

	Right-tail probability for Q				
n_1, n_2, n_3	0.100	0.050	0.020	0.010	0.001
2, 2, 2	4.571	—	—	—	—
3, 2, 1	4.286	—	—	—	—
3, 2, 2	4.500	4.714	—	—	—
3, 3, 1	4.571	5.143	—	—	—
3, 3, 2	4.556	5.361	6.250	—	—
3, 3, 3	4.622	5.600	6.489	7.200	—
4, 2, 1	4.500	—	—	—	—
4, 2, 2	4.458	5.333	6.000	—	—
4, 3, 1	4.056	5.208	—	—	—
4, 3, 2	4.511	5.444	6.144	6.444	—
4, 3, 3	4.709	5.791	6.564	6.745	—
4, 4, 1	4.167	4.967	6.667	6.667	—
4, 4, 2	4.555	5.455	6.600	7.036	—
4, 4, 3	4.545	5.598	6.712	7.144	8.909
4, 4, 4	4.654	5.692	6.962	7.654	9.269
5, 2, 1	4.200	5.000	—	—	—
5, 2, 2	4.373	5.160	6.000	6.533	—
5, 3, 1	4.018	4.960	6.044	—	—
5, 3, 2	4.651	5.251	6.124	6.909	—
5, 3, 3	4.533	5.648	6.533	7.079	8.727
5, 4, 1	3.987	4.985	6.431	6.955	—
5, 4, 2	4.541	5.273	6.505	7.205	8.591
5, 4, 3	4.549	5.656	6.676	7.445	8.795
5, 4, 4	4.668	5.657	6.953	7.760	9.168
5, 5, 1	4.109	5.127	6.145	7.309	—
5, 5, 2	4.623	5.338	6.446	7.338	8.938
5, 5, 3	4.545	5.705	6.866	7.578	9.284
5, 5, 4	4.523	5.666	7.000	7.823	9.606
5, 5, 5	4.560	5.780	7.220	8.000	9.920

*Each table entry is the smallest value of the Kruskal-Wallis Q such that its right-tail probability is less than or equal to the given value on the top row for $k = 3$; each sample size is less than or equal to 5. For $k > 3$, right-tail probabilities on Q are found from Table A.3 with $k - 1$ degrees of freedom.

Source: Adapted from Iman, R. L., D. Quade, and D. A. Alexander: "Exact Probability Levels for the Kruskal-Wallis Test," in *Selected Tables in Mathematical Statistics*, vol. 3., Institute of Mathematical Statistics, American Mathematical Society, Providence, RI, 1975, pp. 329–384. Reprinted by permission of the publisher.

TABLE A.11 Friedman Analysis of Variance Statistic*

k	S	P	k	S	P	k	S	P	k	S	P
$n = 3$			6	24	0.184	8	72	0.010	4	80	0.000
2	8	0.167		18	0.252		62	0.018		78	0.001
	6	0.500		14	0.430		56	0.030		76	0.001
3	18	0.028	7	98	0.000		54	0.038		74	0.001
	14	0.194		96	0.000		50	0.047		72	0.002
	8	0.361		86	0.000		42	0.079		70	0.003
4	32	0.005		78	0.001		38	0.120		68	0.003
	26	0.042		74	0.003		32	0.149		66	0.006
	24	0.069		72	0.004		26	0.236		64	0.007
	18	0.125		62	0.008		24	0.285		62	0.012
	14	0.273		56	0.016		18	0.355		58	0.014
	8	0.431		54	0.021	$n = 4$				56	0.019
5	50	0.001		50	0.027	2	20	0.042		54	0.033
	42	0.008		42	0.051		18	0.167		52	0.036
	38	0.024		38	0.085		16	0.208		50	0.052
	32	0.039		32	0.112		14	0.375		48	0.054
	26	0.093		26	0.192		12	0.458		46	0.068
	24	0.124		24	0.237	3	45	0.002		44	0.077
	18	0.182		18	0.305		43	0.002		42	0.094
	14	0.367		14	0.486		41	0.017		40	0.105
6	72	0.000	8	128	0.000		37	0.033		38	0.141
	62	0.002		126	0.000		35	0.054		36	0.158
	56	0.006		122	0.000		33	0.075		34	0.190
	54	0.008		114	0.000		29	0.148		32	0.200
	50	0.012		104	0.000		27	0.175		30	0.242
	42	0.029		98	0.001		25	0.207		26	0.324
	38	0.052		96	0.001		21	0.300		24	0.355
	32	0.072		86	0.002		19	0.342		22	0.389
	26	0.142		78	0.005		17	0.446		20	0.432
				74	0.008						

*Entries P are the tail probabilities, right-tail from the value of S to its maximum value, for all $P \le 0.50$, $k \le 8$ for $n = 3$, $k \le 4$ for $n = 4$. For n and k outside the range of this table, right-tail probabilities are found from Table A.3 as follows:

$$Q = \frac{12S}{kn(n+1)}$$

Desired:	Approximated by:
Right-tail probability for S or W	Right-tail probability for Q with $n - 1$ degrees of freedom

Source: Adapted from Table 46 of Pearson, E. S., and H. O. Hartley, Eds.: *Biometrika Tables for Statisticians*, vol. 1, 3d ed., Cambridge Univ. Press, Cambridge, U.K., 1966. Reprinted by permission of the *Biometrika* Trustees. Also from Table 5 of Kendall, M. G.: *Rank Correlation Methods*, Charles Griffin, London and High Wycombe, U.K., 1948; 4th ed. 1970. Reprinted by permission of the author and publisher.

TABLE A.12 Spearman Rank Correlation Statistic*

n	R	P	n	R	P	n	R	P	n	R	P
3	1.000	0.167	7	1.000	0.000	8	0.810	0.011	9	0.967	0.000
	0.500	0.500		0.964	0.001		0.786	0.014		0.950	0.000
4	1.000	0.042		0.929	0.003		0.762	0.018		0.933	0.000
	0.800	0.167		0.893	0.006		0.738	0.023		0.917	0.001
	0.600	0.208		0.857	0.012		0.714	0.029		0.900	0.001
	0.400	0.375		0.821	0.017		0.690	0.035		0.883	0.002
	0.200	0.458		0.786	0.024		0.667	0.042		0.867	0.002
	0.000	0.542		0.750	0.033		0.643	0.048		0.850	0.003
5	1.000	0.008		0.714	0.044		0.619	0.057		0.833	0.004
	0.900	0.042		0.679	0.055		0.595	0.066		0.817	0.005
	0.800	0.067		0.643	0.069		0.571	0.076		0.800	0.007
	0.700	0.117		0.607	0.083		0.548	0.085		0.783	0.009
	0.600	0.175		0.571	0.100		0.524	0.098		0.767	0.011
	0.500	0.225		0.536	0.118		0.500	0.108		0.750	0.013
	0.400	0.258		0.500	0.133		0.476	0.122		0.733	0.016
	0.300	0.342		0.464	0.151		0.452	0.134		0.717	0.018
	0.200	0.392		0.429	0.177		0.429	0.150		0.700	0.022
	0.100	0.475		0.393	0.198		0.405	0.163		0.683	0.025
	0.000	0.525		0.357	0.222		0.381	0.180		0.667	0.029
6	1.000	0.001		0.321	0.249		0.357	0.195		0.650	0.033
	0.943	0.008		0.286	0.278		0.333	0.214		0.633	0.038
	0.886	0.017		0.250	0.297		0.310	0.231		0.617	0.043
	0.829	0.029		0.214	0.331		0.286	0.250		0.600	0.048
	0.771	0.051		0.179	0.357		0.262	0.268		0.583	0.054
	0.714	0.068		0.143	0.391		0.238	0.291		0.567	0.060
	0.657	0.088		0.107	0.420		0.214	0.310		0.550	0.066
	0.600	0.121		0.071	0.453		0.190	0.332		0.533	0.074
	0.543	0.149		0.036	0.482		0.167	0.352		0.517	0.081
	0.486	0.178		0.000	0.518		0.143	0.376		0.500	0.089
	0.429	0.210	8	1.000	0.000		0.119	0.397		0.483	0.097
	0.371	0.249		0.976	0.000		0.095	0.420		0.467	0.106
	0.314	0.282		0.952	0.001		0.071	0.441		0.450	0.115
	0.257	0.329		0.929	0.001		0.048	0.467		0.433	0.125
	0.200	0.357		0.905	0.002		0.024	0.488		0.417	0.135
	0.143	0.401		0.881	0.004		0.000	0.512		0.400	0.146
	0.086	0.460		0.857	0.005	9	1.000	0.000		0.383	0.156
	0.029	0.500		0.833	0.008		0.983	0.000		0.367	0.168

TABLE A.12 Spearman Rank Correlation Statistic* (*Continued*)

n	*R*	*P*	*n*	*R*	*P*	*n*	*R*	*P*	*n*	*R*	*P*
9	0.350	0.179	10	0.939	0.000	10	0.624	0.030	10	0.309	0.193
	0.333	0.193		0.927	0.000		0.612	0.033		0.297	0.203
	0.317	0.205		0.915	0.000		0.600	0.037		0.285	0.214
	0.300	0.218		0.903	0.000		0.588	0.040		0.273	0.224
	0.283	0.231		0.891	0.001		0.576	0.044		0.261	0.235
	0.267	0.247		0.879	0.001		0.564	0.048		0.248	0.246
	0.250	0.260		0.867	0.001		0.552	0.052		0.236	0.257
	0.233	0.276		0.855	0.001		0.539	0.057		0.224	0.268
	0.217	0.290		0.842	0.002		0.527	0.062		0.212	0.280
	0.200	0.307		0.830	0.002		0.515	0.067		0.200	0.292
	0.183	0.322		0.818	0.003		0.503	0.072		0.188	0.304
	0.167	0.339		0.806	0.004		0.491	0.077		0.176	0.316
	0.150	0.354		0.794	0.004		0.479	0.083		0.164	0.328
	0.133	0.372		0.782	0.005		0.467	0.089		0.152	0.341
	0.117	0.388		0.770	0.007		0.455	0.096		0.139	0.354
	0.100	0.405		0.758	0.008		0.442	0.102		0.127	0.367
	0.083	0.422		0.745	0.009		0.430	0.109		0.115	0.379
	0.067	0.440		0.733	0.010		0.418	0.116		0.103	0.393
	0.050	0.456		0.721	0.012		0.406	0.124		0.091	0.406
	0.033	0.474		0.709	0.013		0.394	0.132		0.079	0.419
	0.017	0.491		0.697	0.015		0.382	0.139		0.067	0.433
	0.000	0.509		0.685	0.017		0.370	0.148		0.055	0.446
10	1.000	0.000		0.673	0.019		0.358	0.156		0.042	0.459
	0.988	0.000		0.661	0.022		0.345	0.165		0.030	0.473
	0.976	0.000		0.648	0.025		0.333	0.174		0.018	0.486
	0.964	0.000		0.636	0.027		0.321	0.184		0.006	0.500
	0.952	0.000									

*Entries P are the tail probabilities, right-tail from the value of R to its maximum value 1 for all $R \geq 0$, $n \leq 10$. The same probability is a cumulative left-tail probability, from the minimum value –1 to the value $-R$.

TABLE A.12 Spearman Rank Correlation Statistic (*Continued*)

For $10 < n \leq 30$, the following table gives the smallest value of R (largest value of $-R$) for which the right-tail (left-tail) probability for a one-sided test is less than or equal to selected values 0.100, 0.050, 0.025, 0.010, 0.005, 0.001 (top row). These same values apply to $|R|$ for a two-sided test with tail probabilities 0.200, 0.100, 0.050, 0.020, 0.010, 0.002 (bottom row).

	Right-tail (left-tail) probability on R ($-R$) for one-sided test					
n	0.100	0.050	0.025	0.010	0.005	0.001
11	0.427	0.536	0.618	0.709	0.764	0.855
12	0.406	0.503	0.587	0.678	0.734	0.825
13	0.385	0.484	0.560	0.648	0.703	0.797
14	0.367	0.464	0.538	0.626	0.679	0.771
15	0.354	0.446	0.521	0.604	0.657	0.750
16	0.341	0.429	0.503	0.585	0.635	0.729
17	0.329	0.414	0.488	0.566	0.618	0.711
18	0.317	0.401	0.474	0.550	0.600	0.692
19	0.309	0.391	0.460	0.535	0.584	0.675
20	0.299	0.380	0.447	0.522	0.570	0.660
21	0.292	0.370	0.436	0.509	0.556	0.647
22	0.284	0.361	0.425	0.497	0.544	0.633
23	0.278	0.353	0.416	0.486	0.532	0.620
24	0.275	0.344	0.407	0.476	0.521	0.608
25	0.265	0.337	0.398	0.466	0.511	0.597
26	0.260	0.331	0.390	0.457	0.501	0.586
27	0.255	0.324	0.383	0.449	0.492	0.576
28	0.250	0.318	0.376	0.441	0.483	0.567
29	0.245	0.312	0.369	0.433	0.475	0.557
30	0.241	0.307	0.363	0.426	0.467	0.548
	0.200	0.100	0.050	0.020	0.010	0.002
	Tail probability on $\|R\|$ for two-sided test					

For $n > 30$, the probabilities are found from Table A.1 by calculating $z = R\sqrt{n-1}$. The left or right-tail probability for R can be approximated by the left or right-tail probability for z.

Source: Part I ($n \leq 10$) is adapted from Table 44 of Pearson, E. S., and H. O. Hartley, Eds.: *Biometrika Tables for Statisticians*, vol. 1, 3d ed, Cambridge Univ. Press, Cambridge, U.K., 1966. Reprinted by permission of the *Biometrika* Trustees. Also from Table 2 of Kendall, M. G.: *Rank Correlation Methods*, Charles Griffin, London and High Wycombe, U. K., 1948; 4th ed. 1970. Reprinted by permission of the author and publisher. Part II ($10 < n \leq 30$) is adapted from Glasser, G. J., and R. F. Winter: "Critical Values of the Rank Correlation Coefficient for Testing the Hypothesis of Independence," *Biometrika*, vol. 48, pp. 444–448, 1961. Reprinted by permission of the Editor of *Biometrika* Auxiliary Publications and the authors.

TABLE A.13 Kendall's τ Statistic*

n	T	P	n	T	P	n	T	P
3	1.000	0.167	7	0.048	0.500	9	0.278	0.179
	0.333	0.500	8	1.000	0.000		0.222	0.238
4	1.000	0.042		0.929	0.000		0.167	0.306
	0.667	0.167		0.857	0.001		0.111	0.381
	0.333	0.375		0.786	0.003		0.056	0.460
	0.000	0.625		0.714	0.007		0.000	0.540
5	1.000	0.008		0.643	0.016	10	1.000	0.000
	0.800	0.042		0.571	0.031		0.956	0.000
	0.600	0.117		0.500	0.054		0.911	0.000
	0.400	0.242		0.429	0.089		0.867	0.000
	0.200	0.408		0.357	0.138		0.822	0.000
	0.000	0.592		0.286	0.199		0.778	0.000
6	1.000	0.001		0.214	0.274		0.733	0.001
	0.867	0.008		0.143	0.360		0.689	0.002
	0.733	0.028		0.071	0.452		0.644	0.005
	0.600	0.068		0.000	0.548		0.600	0.008
	0.467	0.136	9	1.000	0.000		0.556	0.014
	0.333	0.235		0.944	0.000		0.511	0.023
	0.200	0.360		0.889	0.000		0.467	0.036
	0.067	0.500		0.833	0.000		0.422	0.054
7	1.000	0.000		0.778	0.001		0.378	0.078
	0.905	0.001		0.722	0.003		0.333	0.108
	0.810	0.005		0.667	0.006		0.289	0.146
	0.714	0.015		0.611	0.012		0.244	0.190
	0.619	0.035		0.556	0.022		0.200	0.242
	0.524	0.068		0.500	0.038		0.156	0.300
	0.429	0.119		0.444	0.060		0.111	0.364
	0.333	0.191		0.389	0.090		0.067	0.431
	0.238	0.281		0.333	0.130		0.022	0.500
	0.143	0.386						

*Entries P are the tail probabilities, right-tail from the value of T to its maximum value 1 for all $T \geq 0$, $n \leq 10$. The same probability is a cumulative left-tail probability, from the minimum value -1 to the value $-T$.

TABLE A.13 Kendall's τ Statistic (*Continued*)

For $10 < n \leq 30$, the following table gives the smallest value of T (largest value of $-T$) for which the right-tail (left-tail) probability for a one-sided test is less than or equal to selected values 0.100, 0.050, 0.025, 0.010, 0.005 (top row). These same values apply to $|T|$ for a two-sided test with tail probabilities 0.200, 0.100, 0.050, 0.020, 0.010 (bottom row).

	Right-tail (left-tail) probability on T ($-T$) for one-sided test				
n	0.100	0.050	0.025	0.010	0.005
11	0.345	0.418	0.491	0.564	0.600
12	0.303	0.394	0.455	0.545	0.576
13	0.308	0.359	0.436	0.513	0.564
14	0.275	0.363	0.407	0.473	0.516
15	0.276	0.333	0.390	0.467	0.505
16	0.250	0.317	0.383	0.433	0.483
17	0.250	0.309	0.368	0.426	0.471
18	0.242	0.294	0.346	0.412	0.451
19	0.228	0.287	0.333	0.392	0.439
20	0.221	0.274	0.326	0.379	0.421
21	0.210	0.267	0.314	0.371	0.410
22	0.203	0.264	0.307	0.359	0.394
23	0.202	0.257	0.296	0.352	0.391
24	0.196	0.246	0.290	0.341	0.377
25	0.193	0.240	0.287	0.333	0.367
26	0.188	0.237	0.280	0.329	0.360
27	0.179	0.231	0.271	0.322	0.356
28	0.180	0.228	0.265	0.312	0.344
29	0.172	0.222	0.261	0.310	0.340
30	0.172	0.218	0.255	0.301	0.333
	0.200	0.100	0.050	0.020	0.010
	Tail probability on $|T|$ for two-sided test				

For $n > 30$, the probabilities are found from Table A.1 by calculating $z = 3T\sqrt{n(n-1)}/\sqrt{2(2n+5)}$. The left or right-tail probability for T can be approximated by the left or right-tail probability for z.

Source: Part I ($n \leq 10$) is adapted from Table 45 of Pearson, E. S., and H. O. Hartley, Eds.: *Biometrika Tables for Statisticians*, vol. 1, 3d ed., Cambridge Univ. Press, Cambridge, U.K., 1966. Reprinted by permission of the *Biometrika* Trustees. Also from Table 1 of Kendall, M. G.: *Rank Correlation Methods*, Charles Griffin, London and High Wycombe, U.K., 1948; 4th ed. 1970. Reprinted by permission of the author and publisher. Part II ($10 < n \leq 30$) is adapted from Kaarsemaker, L., and A. van Wijngaarden: "Tables for Use in Rank Correlation," *Statistica Neerlandica*, vol. 7, pp. 41-54, 1953. Reprinted by permission.

TABLE A.14 Number of Runs Distribution*

Left-tail probabilities

m	n	U	P	m	n	U	P	m	n	U	P	m	n	U	P
2	2	2	0.333	2	18	2	0.011	3	14	3	0.025	4	10	3	0.014
	3	2	0.200			3	0.105			4	0.101			4	0.068
		3	0.500			4	0.284			5	0.350			5	0.203
	4	2	0.133	3	3	2	0.100		15	2	0.002			6	0.419
		3	0.400			3	0.300			3	0.022		11	2	0.001
	5	2	0.095		4	2	0.057			4	0.091			3	0.011
		3	0.333			3	0.200			5	0.331			4	0.055
	6	2	0.071		5	2	0.036		16	2	0.002			5	0.176
		3	0.286			3	0.143			3	0.020			6	0.374
	7	2	0.056			4	0.429			4	0.082		12	2	0.001
		3	0.250		6	2	0.024			5	0.314			3	0.009
	8	2	0.044			3	0.107		17	2	0.002			4	0.045
		3	0.222			4	0.345			3	0.018			5	0.154
	9	2	0.036		7	2	0.017			4	0.074			6	0.335
		3	0.200			3	0.083			5	0.298		13	2	0.001
		4	0.491			4	0.283	4	4	2	0.029			3	0.007
	10	2	0.030		8	2	0.012			3	0.114			4	0.037
		3	0.182			3	0.067			4	0.371			5	0.136
		4	0.455			4	0.236		5	2	0.016			6	0.302
	11	2	0.026		9	2	0.009			3	0.071		14	2	0.001
		3	0.167			3	0.055			4	0.262			3	0.006
		4	0.423			4	0.200			5	0.500			4	0.031
	12	2	0.022			5	0.491		6	2	0.010			5	0.121
		3	0.154		10	2	0.007			3	0.048			6	0.274
		4	0.396			3	0.045			4	0.190		15	2	0.001
	13	2	0.019			4	0.171			5	0.405			3	0.005
		3	0.143			5	0.455		7	2	0.006			4	0.027
		4	0.371		11	2	0.005			3	0.033			5	0.108
	14	2	0.017			3	0.038			4	0.142			6	0.249
		3	0.133			4	0.148			5	0.333		16	2	0.000
		4	0.350			5	0.423		8	2	0.004			3	0.004
	15	2	0.015		12	2	0.004			3	0.024			4	0.023
		3	0.125			3	0.033			4	0.109			5	0.097
		4	0.331			4	0.130			5	0.279			6	0.227
	16	2	0.013			5	0.396		9	2	0.003	5	5	2	0.008
		3	0.118		13	2	0.004			3	0.018			3	0.040
		4	0.314			3	0.029			4	0.085			4	0.167
	17	2	0.012			4	0.114			5	0.236			5	0.357
		3	0.111			5	0.371			6	0.471		6	2	0.004
		4	0.298		14	2	0.003		10	2	0.002			3	0.024

TABLE A.14 Number of Runs Distribution* (*Continued*)

Left-tail probabilities

m	n	U	P	m	n	U	P	m	n	U	P	m	n	U	P
5	6	4	0.110	5	13	7	0.330	6	10	4	0.013	7	8	3	0.002
		5	0.262		14	2	0.000			5	0.047			4	0.015
	7	2	0.003			3	0.002			6	0.137			5	0.051
		3	0.015			4	0.011			7	0.287			6	0.149
		4	0.076			5	0.044			8	0.497			7	0.296
		5	0.197			6	0.125		11	2	0.000		9	2	0.000
		6	0.424			7	0.299			3	0.001			3	0.001
	8	2	0.002			8	0.496			4	0.009			4	0.010
		3	0.010		15	2	0.000			5	0.036			5	0.035
		4	0.054			3	0.001			6	0.108			6	0.108
		5	0.152			4	0.009			7	0.242			7	0.231
		6	0.347			5	0.037			8	0.436			8	0.427
	9	2	0.001			6	0.108		12	2	0.000		10	2	0.000
		3	0.007			7	0.272			3	0.001			3	0.001
		4	0.039			8	0.460			4	0.007			4	0.006
		5	0.119	6	6	2	0.002			5	0.028			5	0.024
		6	0.287			3	0.013			6	0.087			6	0.080
	10	2	0.001			4	0.067			7	0.205			7	0.182
		3	0.005			5	0.175			8	0.383			8	0.355
		4	0.029			6	0.392		13	2	0.000		11	2	0.000
		5	0.095		7	2	0.001			3	0.001			3	0.001
		6	0.239			3	0.008			4	0.005			4	0.004
		7	0.455			4	0.043			5	0.022			5	0.018
	11	2	0.000			5	0.121			6	0.070			6	0.060
		3	0.004			6	0.296			7	0.176			7	0.145
		4	0.022			7	0.500			8	0.338			8	0.296
		5	0.077		8	2	0.001		14	2	0.000			9	0.484
		6	0.201			3	0.005			3	0.001		12	2	0.000
		7	0.407			4	0.028			4	0.004			3	0.000
	12	2	0.000			5	0.086			5	0.017			4	0.003
		3	0.003			6	0.226			6	0.058			5	0.013
		4	0.017			7	0.413			7	0.151			6	0.046
		5	0.063		9	2	0.000			8	0.299			7	0.117
		6	0.170			3	0.003	7	7	2	0.001			8	0.247
		7	0.365			4	0.019			3	0.004			9	0.428
	13	2	0.000			5	0.063			4	0.025		13	2	0.000
		3	0.002			6	0.175			5	0.078			3	0.000
		4	0.013			7	0.343			6	0.209			4	0.002
		5	0.053		10	2	0.000			7	0.383			5	0.010
		6	0.145			3	0.002		8	2	0.000			6	0.035

TABLE A.14 Number of Runs Distribution* (*Continued*)

Left-tail probabilities

m	*n*	*U*	*P*	*m*	*n*	*U*	*P*	*m*	*n*	*U*	*P*	*m*	*n*	*U*	*P*
7	13	7	0.095	8	12	3	0.000	9	12	3	0.000	10	12	9	0.142
		8	0.208			4	0.001			4	0.001			10	0.271
		9	0.378			5	0.006			5	0.003			11	0.425
8	8	2	0.000			6	0.025			6	0.014	11	11	2	0.000
		3	0.001			7	0.067			7	0.040			3	0.000
		4	0.009			8	0.159			8	0.103			4	0.000
		5	0.032			9	0.297			9	0.205			5	0.002
		6	0.100			10	0.480			10	0.362			6	0.007
		7	0.214	9	9	2	0.000	10	10	2	0.000			7	0.023
		8	0.405			3	0.000			3	0.000			8	0.063
	9	2	0.000			4	0.003			4	0.001			9	0.135
		3	0.001			5	0.012			5	0.004			10	0.260
		4	0.005			6	0.044			6	0.019			11	0.410
		5	0.020			7	0.109			7	0.051		12	2	0.000
		6	0.069			8	0.238			8	0.128			3	0.000
		7	0.157			9	0.399			9	0.242			4	0.000
		8	0.319		10	2	0.000			10	0.414			5	0.001
		9	0.500			3	0.000		11	2	0.000			6	0.005
	10	2	0.000			4	0.002			3	0.000			7	0.015
		3	0.000			5	0.008			4	0.001			8	0.044
		4	0.003			6	0.029			5	0.003			9	0.099
		5	0.013			7	0.077			6	0.012			10	0.202
		6	0.048			8	0.179			7	0.035			11	0.335
		7	0.117			9	0.319			8	0.092	12	12	2	0.000
		8	0.251		11	2	0.000			9	0.185			3	0.000
		9	0.419			3	0.000			10	0.335			4	0.000
	11	2	0.000			4	0.001			11	0.500			5	0.001
		3	0.000			5	0.005		12	2	0.000			6	0.003
		4	0.002			6	0.020			3	0.000			7	0.009
		5	0.009			7	0.055			4	0.000			8	0.030
		6	0.034			8	0.135			5	0.002			9	0.070
		7	0.088			9	0.255			6	0.008			10	0.150
		8	0.199			10	0.430			7	0.024			11	0.263
		9	0.352		12	2	0.000			8	0.067			12	0.421
	12	2	0.000												

TABLE A.14 Number of Runs Distribution* (*Continued*)

Right-tail probabilities															
m	n	U	P	m	n	U	P	m	n	U	P	m	n	U	P
2	2	4	0.333	4	10	9	0.126	5	14	10	0.234	7	7	10	0.209
	3	5	0.100			8	0.294		15	11	0.129			9	0.383
		4	0.500		11	9	0.154			10	0.258		8	15	0.000
	4	5	0.200			8	0.330	6	6	12	0.002			14	0.002
	5	5	0.286		12	9	0.181			11	0.013			13	0.012
	6	5	0.357			8	0.363			10	0.067			12	0.051
	7	5	0.417		13	9	0.208			9	0.175			11	0.133
	8	5	0.467			8	0.393			8	0.392			10	0.296
3	3	6	0.100		14	9	0.234		7	13	0.001			9	0.486
		5	0.300			8	0.421			12	0.008		9	15	0.001
	4	7	0.029		15	9	0.258			11	0.034			14	0.006
		6	0.200			8	0.446			10	0.121			13	0.025
		5	0.457		16	9	0.282			9	0.267			12	0.084
	5	7	0.071			8	0.470			8	0.500			11	0.194
		6	0.286	5	5	10	0.008		8	13	0.002			10	0.378
	6	7	0.119			9	0.040			12	0.016		10	15	0.002
		6	0.357			8	0.167			11	0.063			14	0.010
	7	7	0.167			7	0.357			10	0.179			13	0.043
		6	0.417		6	11	0.002			9	0.354			12	0.121
	8	7	0.212			10	0.024		9	13	0.006			11	0.257
		6	0.467			9	0.089			12	0.028			10	0.451
	9	7	0.255			8	0.262			11	0.098		11	15	0.004
	10	7	0.294			7	0.478			10	0.238			14	0.017
	11	7	0.330		7	11	0.008			9	0.434			13	0.064
	12	7	0.363			10	0.045		10	13	0.010			12	0.160
	13	7	0.393			9	0.146			12	0.042			11	0.318
	14	7	0.421			8	0.348			11	0.136		12	15	0.007
	15	7	0.446		8	11	0.016			10	0.294			14	0.025
	16	7	0.470			10	0.071		11	13	0.017			13	0.089
	17	7	0.491			9	0.207			12	0.058			12	0.199
4	4	8	0.029			8	0.424			11	0.176			11	0.376
		7	0.114		9	11	0.028			10	0.346		13	15	0.010
		6	0.371			10	0.098		12	13	0.025			14	0.034
	5	9	0.008			9	0.266			12	0.075			13	0.116
		8	0.071			8	0.490			11	0.217			12	0.238
		7	0.214		10	11	0.042			10	0.395			11	0.430
		6	0.500			10	0.126		13	13	0.034	8	8	16	0.000
	6	9	0.024			9	0.322			12	0.092			15	0.001
		8	0.119		11	11	0.058			11	0.257			14	0.009
		7	0.310			10	0.154			10	0.439			13	0.032
	7	9	0.045			9	0.374		14	13	0.044			12	0.100
		8	0.167		12	11	0.075			12	0.111			11	0.214
		7	0.394			10	0.181			11	0.295			10	0.405
	8	9	0.071			9	0.421			10	0.480		9	17	0.000
		8	0.212		13	11	0.092	7	7	14	0.001			16	0.001
		7	0.467			10	0.208			13	0.004			15	0.004
	9	9	0.098			9	0.465			12	0.025			14	0.020
		8	0.255		14	11	0.111			11	0.078			13	0.061

TABLE A.14 Number of Runs Distribution* (*Continued*)

Right-tail probabilities															
m	n	U	P	m	n	U	P	m	n	U	P	m	n	U	P
8	9	12	0.157	9	9	13	0.109	10	10	13	0.128	11	11	17	0.023
		11	0.298			12	0.238			12	0.242			16	0.063
		10	0.500			11	0.399			11	0.414			15	0.135
	10	17	0.000		10	19	0.000		11	21	0.000			14	0.260
		16	0.002			18	0.000			20	0.000			13	0.410
		15	0.010			17	0.001			19	0.000		12	23	0.000
		14	0.036			16	0.008			18	0.003			22	0.000
		13	0.097			15	0.026			17	0.010			21	0.000
		12	0.218			14	0.077			16	0.035			20	0.001
		11	0.379			13	0.166			15	0.085			19	0.004
	11	17	0.001			12	0.319			14	0.185			18	0.015
		16	0.004			11	0.490			13	0.320			17	0.041
		15	0.018		11	19	0.000			12	0.500			16	0.099
		14	0.057			18	0.001		12	21	0.000			15	0.191
		13	0.138			17	0.003			20	0.000			14	0.335
		12	0.278			16	0.015			19	0.001			13	0.493
		11	0.453			15	0.045			18	0.006	12	12	24	0.000
	12	17	0.001			14	0.115			17	0.020			23	0.000
		16	0.007			13	0.227			16	0.056			22	0.000
		15	0.029			12	0.395			15	0.125			21	0.001
		14	0.080	10	10	20	0.000			14	0.245			20	0.003
		13	0.183			19	0.000			13	0.395			19	0.009
		12	0.337			18	0.000	11	11	22	0.000			18	0.030
9	9	18	0.000			17	0.001			21	0.000			17	0.070
		17	0.000			16	0.004			20	0.000			16	0.150
		16	0.003			15	0.019			19	0.002			15	0.263
		15	0.012			14	0.051			18	0.007			14	0.421
		14	0.044												

*Entries P are the tail probabilities from each extreme to the value of U for given $m \le n$. Tail probabilities are given only for $P \le 0.50$.

For $m + n = N > 20$ and $m > 12$, $n > 12$, the probabilities are found from Table A.1 as follows:

$$z_L = \frac{U + 0.5 - 1 - 2mn/N}{\sqrt{\dfrac{2m\,n(2m\,n - N)}{N^2(N-1)}}} \qquad z_R = \frac{U - 0.5 - 1 - 2mn/N}{\sqrt{\dfrac{2m\,n(2m\,n - N)}{N^2(N-1)}}}$$

Desired:	Approximated by:
Left-tail probability for U	Left-tail probability for z_L
Right-tail probability for U	Right-tail probability for z_R

Source: Adapted from Swed, F. S., and C. Eisenhart: "Tables for Testing Randomness of Grouping in a Sequence of Alternatives," *Annals of Mathematical Statistics*, vol. 14, pp. 66–87, 1943. Reprinted by permission.

TABLE A.15 Number of Runs Up and Down Distribution*

N	V	Left-tail P	V	Right-tail P	N	V	Left-tail P	V	Right-tail P
3	1	0.3333	2	0.6667	13	3	0.0001	12	0.0072
4			3	0.4167		4	0.0026	11	0.0568
	1	0.0833	2	0.9167		5	0.0213	10	0.2058
5	1	0.0167	4	0.2667		6	0.0964	9	0.4587
	2	0.2500	3	0.7500		7	0.2749	8	0.7251
6	1	0.0028			14	1	0.0000		
	2	0.0861	5	0.1694		2	0.0000		
	3	0.4139	4	0.5861		3	0.0000		
7	1	0.0004	6	0.1079		4	0.0007	13	0.0046
	2	0.0250	5	0.4417		5	0.0079	12	0.0391
	3	0.1909	4	0.8091		6	0.0441	11	0.1536
8	1	0.0000				7	0.1534	10	0.3722
	2	0.0063	7	0.0687		8	0.3633	9	0.6367
	3	0.0749	6	0.3250	15	1	0.0000		
	4	0.3124	5	0.6876		2	0.0000		
9	1	0.0000				3	0.0000		
	2	0.0014				4	0.0002		
	3	0.0257	8	0.0437		5	0.0027	14	0.0029
	4	0.1500	7	0.2347		6	0.0186	13	0.0267
	5	0.4347	6	0.5653		7	0.0782	12	0.1134
10	1	0.0000				8	0.2216	11	0.2970
	2	0.0003	9	0.0278		9	0.4520	10	0.5480
	3	0.0079	8	0.1671	16	1	0.0000		
	4	0.0633	7	0.4524		2	0.0000		
	5	0.2427	6	0.7573		3	0.0000		
11	1	0.0000				4	0.0001	15	0.0019
	2	0.0001				5	0.0009	14	0.0182
	3	0.0022	10	0.0177		6	0.0072	13	0.0828
	4	0.0239	9	0.1177		7	0.0367	12	0.2335
	5	0.1196	8	0.3540		8	0.1238	11	0.4631
	6	0.3438	7	0.6562		9	0.2975	10	0.7025
12	1	0.0000			17	1	0.0000		
	2	0.0000				2	0.0000		
	3	0.0005				3	0.0000		
	4	0.0082	11	0.0113		4	0.0000		
	5	0.0529	10	0.0821		5	0.0003	16	0.0012
	6	0.1918	9	0.2720		6	0.0026	15	0.0123
	7	0.4453	8	0.5547		7	0.0160	14	0.0600
13	1	0.0000				8	0.0638	13	0.1812
	2	0.0000				9	0.1799	12	0.3850
						10	0.3770	11	0.6230

TABLE A.15 Number of Runs Up and Down Distribution* (*Continued*)

N	*V*	Left-tail *P*	*V*	Right-tail *P*	*N*	*V*	Left-tail *P*	*V*	Right-tail *P*
18	1	0.0000			21	4	0.0000		
	2	0.0000				5	0.0000		
	3	0.0000				6	0.0000		
	4	0.0000				7	0.0003	20	0.0002
	5	0.0001				8	0.0023	19	0.0025
	6	0.0009	17	0.0008		9	0.0117	18	0.0154
	7	0.0065	16	0.0083		10	0.0431	17	0.0591
	8	0.0306	15	0.0431		11	0.1202	16	0.1602
	9	0.1006	14	0.1389		12	0.2622	15	0.3293
	10	0.2443	13	0.3152		13	0.4603	14	0.5397
	11	0.4568	12	0.5432	22	1	0.0000		
19	1	0.0000				2	0.0000		
	2	0.0000				3	0.0000		
	3	0.0000				4	0.0000		
	4	0.0000				5	0.0000		
	5	0.0000	18	0.0005		6	0.0000	21	0.0001
	6	0.0003	17	0.0056		7	0.0001	20	0.0017
	7	0.0025	16	0.0308		8	0.0009	19	0.0108
	8	0.0137	15	0.1055		9	0.0050	18	0.0437
	9	0.0523	14	0.2546		10	0.0213	17	0.1251
	10	0.1467	13	0.4663		11	0.0674	16	0.2714
	11	0.3144	12	0.6856		12	0.1661	15	0.4688
20	1	0.0000				13	0.3276	14	0.6724
	2	0.0000			23	1	0.0000		
	3	0.0000				2	0.0000		
	4	0.0000				3	0.0000		
	5	0.0000				4	0.0000		
	6	0.0001	19	0.0003		5	0.0000		
	7	0.0009	18	0.0038		6	0.0000		
	8	0.0058	17	0.0218		7	0.0000	22	0.0001
	9	0.0255	16	0.0793		8	0.0003	21	0.0011
	10	0.0821	15	0.2031		9	0.0021	20	0.0076
	11	0.2012	14	0.3945		10	0.0099	19	0.0321
	12	0.3873	13	0.6127		11	0.0356	18	0.0968
21	1	0.0000				12	0.0988	17	0.2211
	2	0.0000				13	0.2188	16	0.4020
	3	0.0000				14	0.3953	15	0.6047

TABLE A.15 Number of Runs Up and Down Distribution* (*Continued*)

N	V	Left-tail P	V	Right-tail P	N	V	Left-tail P	V	Right-tail P
24	1	0.0000			25	1	0.0000		
	2	0.0000				2	0.0000		
	3	0.0000				3	0.0000		
	4	0.0000				4	0.0000		
	5	0.0000				5	0.0000		
	6	0.0000				6	0.0000		
	7	0.0000				7	0.0000	24	0.0000
	8	0.0001	23	0.0000		8	0.0000	23	0.0005
	9	0.0008	22	0.0007		9	0.0003	22	0.0037
	10	0.0044	21	0.0053		10	0.0018	21	0.0170
	11	0.0177	20	0.0235		11	0.0084	20	0.0564
	12	0.0554	19	0.0742		12	0.0294	19	0.1423
	13	0.1374	18	0.1783		13	0.0815	18	0.2852
	14	0.2768	17	0.3405		14	0.1827	17	0.4708
	15	0.4631	16	0.5369		15	0.3384	16	0.6616

*Entries P are the tail probabilities from each extreme to the value of V. Left-tail probabilities are given only for V less than or equal to its median; right-tail probabilities are given for all other values of V. N is the number of observations for which successive differences are nonzero, and hence one more than the number of symbols of signs of differences.

For $N > 25$, the probabilities are found from Table A.1 as follows:

$$z_L = \frac{V + 0.5 - (2N - 1)/3}{\sqrt{(16N - 29)/90}} \qquad z_R = \frac{V - 0.5 - (2N - 1)/3}{\sqrt{(16N - 29)/90}}$$

Desired:	Approximated by:
Left-tail probability for V	Left-tail probability for z_L
Right-tail probability for V	Right-tail probability for z_R

Source: Adapted from II Edgington, E. S.: "Probability Table for Number of Runs of Signs of First Differences," *Journal of the American Statistical Association*, vol. 56, pp. 156–159, 1961. Reprinted by permission.

TABLE A.16 Kolmogorov-Smirnov One-Sample Statistic*

N	0.200	0.100	0.050	0.020	0.010	N	0.200	0.100	0.050	0.020	0.010
1	0.900	0.950	0.975	0.990	0.995	21	0.226	0.259	0.287	0.321	0.344
2	0.684	0.776	0.842	0.900	0.929	22	0.221	0.253	0.281	0.314	0.337
3	0.565	0.636	0.708	0.785	0.829	23	0.216	0.247	0.275	0.307	0.330
4	0.493	0.565	0.624	0.689	0.734	24	0.212	0.242	0.269	0.301	0.323
5	0.447	0.509	0.563	0.627	0.669	25	0.208	0.238	0.264	0.295	0.317
6	0.410	0.468	0.519	0.577	0.617	26	0.204	0.233	0.259	0.290	0.311
7	0.381	0.436	0.483	0.538	0.576	27	0.200	0.229	0.254	0.284	0.305
8	0.358	0.410	0.454	0.507	0.542	28	0.197	0.225	0.250	0.279	0.300
9	0.339	0.387	0.430	0.480	0.513	29	0.193	0.221	0.246	0.275	0.295
10	0.323	0.369	0.409	0.457	0.489	30	0.190	0.218	0.242	0.270	0.290
11	0.308	0.352	0.391	0.437	0.468	31	0.187	0.214	0.238	0.266	0.285
12	0.296	0.338	0.375	0.419	0.449	32	0.184	0.211	0.234	0.262	0.281
13	0.285	0.325	0.361	0.404	0.432	33	0.182	0.208	0.231	0.258	0.277
14	0.275	0.314	0.349	0.390	0.418	34	0.179	0.205	0.227	0.254	0.273
15	0.266	0.304	0.338	0.377	0.404	35	0.177	0.202	0.224	0.251	0.269
16	0.258	0.295	0.327	0.366	0.392	36	0.174	0.199	0.221	0.247	0.265
17	0.250	0.286	0.318	0.355	0.381	37	0.172	0.196	0.218	0.244	0.262
18	0.244	0.279	0.309	0.346	0.371	38	0.170	0.194	0.215	0.241	0.258
19	0.237	0.271	0.301	0.337	0.361	39	0.168	0.191	0.213	0.238	0.255
20	0.232	0.265	0.294	0.329	0.352	40	0.165	0.189	0.210	0.235	0.252

*Table entries for any sample size N are the values of a Kolmogorov-Smirov one-sample random variable for which the right-tail probability for a two-sided test is as given in the top row.

For $N > 40$, the table entries based on the asymptotic distribution are approximated by calculating the following for the appropriate value of N:

0.200	0.100	0.050	0.020	0.010
$1.07/\sqrt{N}$	$1.22/\sqrt{N}$	$1.36/\sqrt{N}$	$1.52/\sqrt{N}$	$1.63/\sqrt{N}$

Source: Adapted from Miller, L. H.: "Table of Percentage Points of Kolmogorov Statistics," *Journal of the American Statistical Association*, vol. 51, pp. 111–121, 1956. Reprinted by permission.

TABLE A. 17 Kolmogorov-Smirnov Two-Sample Statistic*

n_1	n_2	n_1n_2D	P	n_1	n_2	n_1n_2D	P	n_1	n_2	n_1n_2D	P
2	2	4	0.333	3	9	18	0.236	4	10	30	0.046
	3	6	0.200		10	30	0.007			28	0.084
	4	8	0.133			27	0.028			26	0.126
	5	10	0.095			24	0.070		11	44	0.001
		8	0.286			21	0.140			40	0.007
	6	12	0.071		11	33	0.005			36	0.022
		10	0.214			30	0.022			33	0.035
	7	14	0.056			27	0.055			32	0.063
		12	0.167			24	0.110			29	0.098
	8	16	0.044		12	36	0.004			28	0.144
		14	0.133			33	0.018		12	48	0.001
	9	18	0.036			30	0.044			44	0.005
		16	0.109			27	0.088			40	0.016
	10	20	0.030			24	0.189			36	0.048
		18	0.091	4	4	16	0.029			32	0.112
		16	0.182			12	0.229	5	5	25	0.008
	11	22	0.026		5	20	0.016			20	0.079
		20	0.077			16	0.079			15	0.357
		18	0.154			15	0.143		6	30	0.004
	12	24	0.022		6	24	0.010			25	0.026
		22	0.066			20	0.048			24	0.048
		20	0.132			18	0.095			20	0.108
3	3	9	0.100			16	0.181		7	35	0.003
	4	12	0.057		7	28	0.006			30	0.015
		9	0.229			24	0.030			28	0.030
	5	15	0.036			21	0.067			25	0.066
		12	0.143			20	0.121			23	0.116
	6	18	0.024		8	32	0.004		8	40	0.002
		15	0.095			28	0.020			35	0.009
		12	0.333			24	0.085			32	0.020
	7	21	0.017			20	0.222			30	0.042
		18	0.067		9	36	0.003			27	0.079
		15	0.167			32	0.014			25	0.126
	8	24	0.012			28	0.042		9	45	0.001
		21	0.048			27	0.062			40	0.006
		18	0.121			24	0.115			36	0.014
	9	27	0.009		10	40	0.002			35	0.028
		24	0.036			36	0.010			31	0.056
		21	0.091			32	0.030			30	0.086

TABLE A. 17 Kolmogorov-Smirnov Two-Sample Statistic* (*Continued*)

n_1	n_2	n_1n_2D	P	n_1	n_2	n_1n_2D	P	n_1	n_2	n_1n_2D	P
5	9	27	0.119	6	8	36	0.023	7	7	28	0.212
	10	50	0.001			34	0.043		8	56	0.000
		45	0.004			32	0.061			49	0.002
		40	0.019			30	0.093			48	0.005
		35	0.061			28	0.139			42	0.013
		30	0.166		9	54	0.000			41	0.024
	11	55	0.000			48	0.003			40	0.033
		50	0.003			45	0.006			35	0.056
		45	0.010			42	0.014			34	0.087
		44	0.014			39	0.028			33	0.118
		40	0.029			36	0.061		9	63	0.000
		39	0.044			33	0.095			56	0.001
		35	0.074			30	0.176			54	0.003
		34	0.106		10	60	0.000			49	0.008
6	6	36	0.002			54	0.002			47	0.015
		30	0.026			50	0.004			45	0.021
		24	0.143			48	0.009			42	0.034
	7	42	0.001			44	0.019			40	0.055
		36	0.008			42	0.031			38	0.079
		35	0.015			40	0.042			36	0.098
		30	0.038			38	0.066			35	0.127
		29	0.068			36	0.092	8	8	64	0.000
		28	0.091			34	0.125			56	0.002
		24	0.147	7	7	49	0.001			48	0.019
	8	48	0.001			42	0.008			40	0.087
		42	0.005			35	0.053			32	0.283
		40	0.009								

*Entries P are the tail probabilities, right tail from the value of n_1n_2D to $+\infty$ for small P, all $2 \le n_1 \le n_2 \le 12$ or $n_1 + n_2 \le 16$, whichever occurs first.

TABLE A.17 Kolmogorov-Smirnov Two-Sample Statistic (*Continued*)

For $9 \le (n_1 = n_2) \le 20$, the table entries are the smallest values of $n_1 n_2 D$ for which the right-tail probability is less than or equal to selected values 0.010, 0.020, 0.050, 0.100, 0.200.

$n_1 = n_2$	0.200	0.100	0.050	0.020	0.010
9	45	54	54	63	63
10	50	60	70	70	80
11	66	66	77	88	88
12	72	72	84	96	96
13	78	91	91	104	117
14	84	98	112	112	126
15	90	105	120	135	135
16	112	112	128	144	160
17	119	136	136	153	170
18	126	144	162	180	180
19	133	152	171	190	190
20	140	160	180	200	220

For sample sizes outside the range of this table, the quantile points based on the asymptotic distribution are approximated by calculating the following for the appropriate values of n_1, n_2, and $N = n_1 + n_2$.

0.200	0.100	0.050	0.020	0.010
$1.07\sqrt{N/n_1 n_2}$	$1.22\sqrt{N/n_1 n_2}$	$1.36\sqrt{N/n_1 n_2}$	$1.52\sqrt{N/n_1 n_2}$	$1.63\sqrt{N/n_1 n_2}$

Source: Adapted from Table I of Harter, H. L., and D. B. Owen, Eds.: *Selected Tables in Mathematical Statistics,* vol. 2., Markham Publishing, Chicago, L, 1970. Reprinted by permission of the Institute of Mathematical Statistics.

TABLE A.18 One-Tail Percentage Points of *W* Test for Normality

n	100α				
	1%	2%	5%	10%	50%
3	0.753	0.756	0.767	0.789	0.959
4	0.687	0.707	0.748	0.792	0.935
5	0.686	0.715	0.762	0.806	0.927
6	0.713	0.743	0.788	0.826	0.927
7	0.730	0.760	0.803	0.838	0.928
8	0.749	0.788	0.818	0.851	0.932
9	0.764	0.791	0.829	0.859	0.935
10	0.781	0.806	0.842	0.869	0.938
11	0.792	0.817	0.850	0.876	0.940
12	0.805	0.828	0.859	0.883	0.943
13	0.814	0.837	0.866	0.889	0.945
14	0.825	0.846	0.874	0.895	0.947
15	0.835	0.855	0.881	0.901	0.950
16	0.844	0.863	0.887	0.906	0.952
17	0.851	0.869	0.892	0.910	0.954
18	0.858	0.874	0.897	0.914	0.956
19	0.863	0.879	0.901	0.917	0.957
20	0.868	0.884	0.905	0.920	0.959
21	0.873	0.888	0.908	0.923	0.960
22	0.878	0.892	0.911	0.926	0.961
23	0.881	0.895	0.914	0.928	0.962
24	0.884	0.898	0.916	0.930	0.963
25	0.888	0.901	0.918	0.931	0.964
26	0.891	0.904	0.920	0.933	0.965
27	0.894	0.906	0.923	0.935	0.965
28	0.896	0.908	0.924	0.936	0.966
29	0.898	0.910	0.926	0.937	0.966
30	0.900	0.912	0.927	0.939	0.967
31	0.902	0.914	0.929	0.940	0.967
32	0.904	0.915	0.930	0.941	0.968
33	0.906	0.917	0.931	0.942	0.968
34	0.908	0.919	0.933	0.943	0.969
35	0.910	0.920	0.934	0.944	0.969
36	0.912	0.922	0.935	0.945	0.970
37	0.914	0.924	0.936	0.946	0.970
38	0.916	0.925	0.938	0.947	0.971
39	0.917	0.927	0.939	0.948	0.971
40	0.919	0.928	0.940	0.949	0.972
41	0.920	0.929	0.941	0.950	0.972
42	0.922	0.930	0.942	0.951	0.972
43	0.923	0.932	0.943	0.951	0.973
44	0.924	0.933	0.944	0.952	0.973
45	0.926	0.934	0.945	0.953	0.973
46	0.927	0.935	0.945	0.953	0.974
47	0.928	0.936	0.946	0.954	0.974
48	0.929	0.937	0.947	0.954	0.974
49	0.929	0.937	0.947	0.955	0.974
50	0.930	0.938	0.947	0.955	0.974

Source: Shapiro, S. S.: *How to Test Normality and Other Distributional Assumptions,* American Society for Quality Control, Milwaukee, WI., 1980. Reprinted by permission of the publisher.

TABLE A.19 Constants b_{n-i+1} for the W' Test for Normality

	n						
i	50	51	52	53	54	55	56
1	0.32660	0.32433	0.32212	0.31997	0.31786	0.31580	0.31379
2	0.26935	0.26784	0.26636	0.26491	0.26348	0.26209	0.26071
3	0.23650	0.23545	0.23441	0.23338	0.23237	0.23137	0.23038
4	0.21256	0.21185	0.21114	0.21044	0.20974	0.20904	0.20834
5	0.19329	0.19288	0.19245	0.19201	0.19157	0.19112	0.19067
6	0.17694	0.17677	0.17659	0.17639	0.17618	0.17595	0.17570
7	0.16257	0.16263	0.16267	0.16268	0.16267	0.16264	0.16259
8	0.14963	0.14991	0.15016	0.15037	0.15054	0.15070	0.15082
9	0.13779	0.13827	0.13871	0.13911	0.13946	0.13978	0.14007
10	0.12680	0.12748	0.12810	0.12867	0.12920	0.12968	0.13013
11	0.11650	0.11736	0.11816	0.11890	0.11959	0.12023	0.12083
12	0.10675	0.10779	0.10877	0.10967	0.11052	0.11132	0.11206
13	0.09746	0.09868	0.09982	0.10089	0.10190	0.10284	0.10373
14	0.08856	0.08995	0.09126	0.09249	0.09365	0.09474	0.09576
15	0.07998	0.08155	0.08302	0.08441	0.08572	0.08695	0.08811
16	0.07167	0.07341	0.07505	0.07659	0.07805	0.07943	0.08073
17	0.06359	0.06550	0.06731	0.06901	0.07062	0.07214	0.07357
18	0.05570	0.05779	0.05976	0.06162	0.06338	0.06504	0.06661
19	0.04797	0.05024	0.05238	0.05440	0.05631	0.05811	0.05982
20	0.04038	0.04283	0.04513	0.04731	0.04938	0.05133	0.05318
21	0.03290	0.03552	0.03801	0.04035	0.04257	0.04466	0.04665
22	0.02550	0.02831	0.03097	0.03348	0.03586	0.03811	0.04023
23	0.01817	0.02118	0.02401	0.02670	0.02923	0.03163	0.03390
24	0.01085	0.01409	0.01711	0.01997	0.02267	0.02522	0.02765
25	0.00362	0.00704	0.01025	0.01329	0.01616	0.01887	0.02144
26			0.00341	0.00664	0.00968	0.01256	0.01529
27					0.00322	0.00627	0.00916
28							0.00305

TABLE A.19 Constants b_{n-i+1} for the W' Test for Normality *(Continued)*

	n						
i	57	58	59	60	61	62	63
1	0.31183	0.30991	0.30803	0.30619	0.30483	0.30262	0.30089
2	0.25937	0.25805	0.25675	0.25548	0.25422	0.25299	0.25178
3	0.22941	0.22845	0.22750	0.22656	0.22564	0.22473	0.22383
4	0.20765	0.20696	0.20628	0.20560	0.20492	0.20426	0.20359
5	0.19021	0.18975	0.18928	0.18882	0.18835	0.18788	0.18740
6	0.17545	0.17518	0.17491	0.17463	0.17433	0.17404	0.17373
7	0.16252	0.16243	0.16233	0.16221	0.16208	0.16194	0.16178
8	0.15092	0.15099	0.15105	0.15108	0.15110	0.15110	0.15109
9	0.14033	0.14056	0.14076	0.14094	0.14110	0.14123	0.14135
10	0.13054	0.13091	0.13126	0.13157	0.13186	0.13212	0.13235
11	0.12138	0.12190	0.12238	0.12282	0.12323	0.12362	0.12397
12	0.11275	0.11340	0.11401	0.11458	0.11511	0.11561	0.11608
13	0.10456	0.10534	0.10608	0.10677	0.10742	0.10803	0.10860
14	0.09673	0.09764	0.09850	0.09931	0.01007	0.10080	0.10148
15	0.08921	0.09025	0.09123	0.09216	0.09304	0.09386	0.09465
16	0.08196	0.08312	0.08422	0.08527	0.08626	0.08719	0.08808
17	0.07493	0.07622	0.07744	0.07860	0.07970	0.08074	0.08174
18	0.06810	0.06952	0.07086	0.07213	0.07334	0.07449	0.07558
19	0.06144	0.06298	0.06444	0.06583	0.06715	0.06840	0.06960
20	0.05493	0.05659	0.05817	0.05967	0.06110	0.06246	0.06376
21	0.04854	0.05033	0.05203	0.05365	0.05519	0.05665	0.05805
22	0.04226	0.04417	0.04600	0.04773	0.04938	0.05095	0.05245
23	0.03606	0.03811	0.04006	0.04191	0.04367	0.04535	0.04695
24	0.02994	0.03212	0.03420	0.03617	0.03804	0.03983	0.04154
25	0.02388	0.02620	0.02840	0.03049	0.03249	0.03438	0.03620
26	0.01787	0.02033	0.02266	0.02487	0.02699	0.02900	0.03091
27	0.01190	0.01449	0.01696	0.01930	0.02153	0.02366	0.02569
28	0.00594	0.00868	0.01129	0.01376	0.01612	0.01836	0.02050
29		0.00289	0.00564	0.00825	0.01073	0.01309	0.01535
30				0.00275	0.00536	0.00785	0.01022
31						0.00261	0.00510

TABLE A.19 Constants b_{n-i+1} for the W' Test for Normality (*Continued*)

	n						
i	64	65	66	67	68	69	70
1	0.29920	0.29754	0.29591	0.29432	0.29275	0.29121	0.28971
2	0.25060	0.24943	0.24828	0.24715	0.24604	0.24494	0.24387
3	0.22295	0.22207	0.22121	0.22036	0.21952	0.21868	0.21787
4	0.20293	0.20228	0.20163	0.20099	0.20035	0.19971	0.19909
5	0.18693	0.18646	0.18599	0.18551	0.18504	0.18457	0.18410
6	0.17342	0.17311	0.17279	0.17247	0.17214	0.17181	0.17148
7	0.16162	0.16145	0.16127	0.16108	0.16088	0.16068	0.16047
8	0.15106	0.15102	0.15096	0.15089	0.15081	0.15073	0.15063
9	0.14144	0.14152	0.14158	0.14163	0.14166	0.14168	0.14169
10	0.13257	0.13276	0.13294	0.13309	0.13323	0.13335	0.13345
11	0.12430	0.12460	0.12488	0.12513	0.12537	0.12559	0.12578
12	0.11651	0.11692	0.11730	0.11766	0.11799	0.11830	0.11859
13	0.10914	0.10965	0.11013	0.11058	0.11101	0.11140	0.11178
14	0.10212	0.10273	0.10331	0.10385	0.10436	0.10485	0.10531
15	0.09540	0.09610	0.09677	0.09741	0.09801	0.09858	0.09912
16	0.08893	0.08973	0.09049	0.09121	0.09190	0.09255	0.09318
17	0.08268	0.08357	0.08443	0.08524	0.08601	0.08675	0.08745
18	0.07662	0.07761	0.07855	0.07945	0.08031	0.08113	0.08191
19	0.07073	0.07182	0.07285	0.07384	0.07478	0.07567	0.07653
20	0.06499	0.06617	0.06729	0.06836	0.06939	0.07037	0.07130
21	0.05938	0.06065	0.06186	0.06302	0.06413	0.06519	0.06620
22	0.05388	0.05524	0.05655	0.05779	0.05898	0.06012	0.06121
23	0.04848	0.04994	0.05133	0.05266	0.05394	0.05516	0.05632
24	0.04316	0.04472	0.04620	0.04762	0.04898	0.05028	0.05152
25	0.03792	0.03957	0.04115	0.04266	0.04410	0.04548	0.04680
26	0.03274	0.03449	0.03616	0.03776	0.03929	0.04075	0.04215
27	0.02762	0.02947	0.03123	0.03292	0.03453	0.03608	0.03756
28	0.02254	0.02449	0.02635	0.02813	0.02983	0.03146	0.03303
29	0.01749	0.01955	0.02151	0.02338	0.02517	0.02689	0.02853
30	0.01248	0.01463	0.01669	0.01866	0.02055	0.02235	0.02408
31	0.00748	0.00974	0.01191	0.01398	0.01595	0.01785	0.01966
32	0.00249	0.00487	0.00714	0.00931	0.01138	0.01336	0.01527
33			0.00238	0.00465	0.00682	0.00890	0.01089
34					0.00227	0.00445	0.00653
35							0.00218

TABLE A.19 Constants b_{n-i+1} for the W' Test for Normality (*Continued*)

	n						
i	71	72	73	74	75	76	77
1	0.28823	0.28677	0.28534	0.28394	0.28256	0.28120	0.27987
2	0.24281	0.24177	0.24074	0.23973	0.23876	0.23775	0.23678
3	0.21706	0.21626	0.21547	0.21469	0.21392	0.21316	0.21241
4	0.19846	0.19785	0.19724	0.19663	0.19603	0.19544	0.19485
5	0.18364	0.18317	0.18271	0.18224	0.18178	0.18132	0.18087
6	0.17114	0.17081	0.17047	0.17013	0.16979	0.16945	0.16910
7	0.16025	0.16003	0.15980	0.15958	0.15934	0.15910	0.15886
8	0.15052	0.15041	0.15029	0.15016	0.15002	0.14988	0.14973
9	0.14168	0.14167	0.14164	0.14160	0.14156	0.14151	0.14144
10	0.13354	0.13362	0.13368	0.13373	0.13377	0.13380	0.13382
11	0.12596	0.12613	0.12628	0.12641	0.12653	0.12664	0.12674
12	0.11885	0.11910	0.11933	0.11955	0.11975	0.11993	0.12010
13	0.11213	0.11246	0.11277	0.11307	0.11334	0.11360	0.11384
14	0.10574	0.10615	0.10654	0.10690	0.10725	0.10758	0.10788
15	0.09963	0.10012	0.10058	0.10102	0.10144	0.10183	0.10221
16	0.09377	0.09433	0.09487	0.09538	0.09586	0.09632	0.09676
17	0.08812	0.08875	0.08936	0.08994	0.09049	0.09102	0.09153
18	0.08265	0.08336	0.08404	0.08469	0.08531	0.08590	0.08647
19	0.07735	0.07813	0.07888	0.07960	0.08029	0.08094	0.08157
20	0.07220	0.07305	0.07387	0.07465	0.07541	0.07613	0.07682
21	0.06717	0.06810	0.06898	0.06984	0.07065	0.07143	0.07219
22	0.06225	0.06325	0.06421	0.06513	0.06601	0.06686	0.06767
23	0.05744	0.05851	0.05954	0.06053	0.06147	0.06238	0.06325
24	0.05272	0.05386	0.05496	0.05601	0.05702	0.05799	0.05892
25	0.04807	0.04929	0.05045	0.05158	0.05265	0.05369	0.05468
26	0.04350	0.04479	0.04602	0.04721	0.04835	0.04945	0.05051
27	0.03898	0.04035	0.04166	0.04291	0.04412	0.04528	0.04640
28	0.03452	0.03596	0.03734	0.03867	0.03994	0.04117	0.04234
29	0.03011	0.03162	0.03308	0.03447	0.03581	0.03710	0.03834
30	0.02574	0.02733	0.02886	0.03032	0.03173	0.03308	0.03439
31	0.02140	0.02307	0.02467	0.02621	0.02768	0.02911	0.03047
32	0.01709	0.01883	0.02051	0.02212	0.02367	0.02516	0.02659
33	0.01280	0.01463	0.01638	0.01807	0.01968	0.02124	0.02274
34	0.00852	0.01043	0.01227	0.01403	0.01572	0.01735	0.01891
35	0.00426	0.00626	0.00817	0.01001	0.01178	0.01347	0.01510
36		0.00208	0.00408	0.00600	0.00784	0.00961	0.01131
37				0.00200	0.00392	0.00576	0.00754
38						0.00192	0.00377

TABLE A.19 Constants b_{n-i+1} for the W' Test for Normality (*Continued*)

	n						
i	78	79	80	81	82	83	84
1	0.27856	0.27727	0.27600	0.27475	0.27353	0.27232	0.27113
2	0.23583	0.23489	0.23396	0.23305	0.23215	0.23126	0.23039
3	0.21167	0.21094	0.21021	0.20950	0.20879	0.20809	0.20740
4	0.19427	0.19369	0.19312	0.19255	0.19199	0.19143	0.19088
5	0.18041	0.17996	0.17951	0.17906	0.17862	0.17818	0.17774
6	0.16876	0.16842	0.16807	0.16773	0.16739	0.16705	0.16670
7	0.15862	0.15837	0.15812	0.15787	0.15762	0.15737	0.15711
8	0.14958	0.14942	0.14926	0.14909	0.14892	0.14875	0.14857
9	0.14137	0.14130	0.14122	0.14113	0.14103	0.14093	0.14083
10	0.13383	0.13383	0.13382	0.13381	0.13378	0.13375	0.13372
11	0.12682	0.12690	0.12696	0.12701	0.12706	0.12709	0.12712
12	0.12026	0.12040	0.12053	0.12065	0.12076	0.12086	0.12094
13	0.11406	0.11427	0.11446	0.11465	0.11482	0.11498	0.11512
14	0.10818	0.10845	0.10871	0.10895	0.10918	0.10940	0.10960
15	0.10256	0.10290	0.10322	0.10352	0.10381	0.10408	0.10434
16	0.09718	0.09758	0.09796	0.09832	0.09867	0.09900	0.09931
17	0.09201	0.09247	0.09290	0.09332	0.09372	0.09411	0.09447
18	0.08701	0.08753	0.08803	0.08850	0.08896	0.08939	0.08981
19	0.08217	0.08275	0.08330	0.08383	0.08434	0.08483	0.08530
20	0.07748	0.07811	0.07872	0.07931	0.07987	0.08041	0.08092
21	0.07291	0.07360	0.07426	0.07490	0.07552	0.07611	0.07667
22	0.06845	0.06920	0.06992	0.07061	0.07128	0.07192	0.07253
23	0.06409	0.06490	0.06567	0.06642	0.06713	0.06783	0.06849
24	0.05982	0.06069	0.06152	0.06231	0.06308	0.06382	0.06454
25	0.05564	0.05655	0.05744	0.05829	0.05911	0.05990	0.06067
26	0.05152	0.05250	0.05344	0.05434	0.05521	0.05606	0.05687
27	0.04747	0.04850	0.04950	0.05046	0.05138	0.05227	0.05313
28	0.04348	0.04456	0.04562	0.04663	0.04761	0.04855	0.04946
29	0.03954	0.04069	0.04179	0.04286	0.04389	0.04488	0.04584
30	0.03564	0.03685	0.03801	0.03913	0.04022	0.04126	0.04226
31	0.03179	0.03305	0.03427	0.03545	0.03659	0.03768	0.03874
32	0.02797	0.02929	0.03057	0.03180	0.03299	0.03414	0.03525
33	0.02418	0.02557	0.02690	0.02819	0.02943	0.03063	0.03179
34	0.02042	0.02187	0.02326	0.02461	0.02590	0.02716	0.02837
35	0.01667	0.01819	0.01964	0.02105	0.02240	0.02371	0.02497
36	0.01295	0.01453	0.01605	0.01751	0.01892	0.02028	0.02159
37	0.00924	0.01088	0.01246	0.01399	0.01545	0.01687	0.01824
38	0.00554	0.00725	0.00889	0.01048	0.01201	0.01348	0.01490
39	0.00185	0.00362	0.00533	0.00698	0.00857	0.01010	0.01157
40			0.00178	0.00349	0.00514	0.00673	0.00826
41					0.00171	0.00336	0.00495
42							0.00165

TABLE A.19 Constants b_{n-i+1} for the W' Test for Normality (*Continued*)

	n						
i	85	86	87	88	89	90	91
1	0.26995	0.26880	0.26766	0.26654	0.26544	0.26435	0.26328
2	0.22952	0.22867	0.22783	0.22700	0.22618	0.22537	0.22457
3	0.20672	0.20605	0.20538	0.20472	0.20407	0.20343	0.20279
4	0.19033	0.18979	0.18925	0.18872	0.18819	0.18767	0.18715
5	0.17730	0.17687	0.17643	0.17601	0.17558	0.17516	0.17474
6	0.16636	0.16602	0.16568	0.16534	0.16500	0.16466	0.16432
7	0.15685	0.15659	0.15633	0.15607	0.15581	0.15555	0.15529
8	0.14839	0.14820	0.14802	0.14783	0.14763	0.14744	0.14724
9	0.14072	0.14060	0.14048	0.14036	0.14023	0.14010	0.13997
10	0.13367	0.13362	0.13357	0.13351	0.13344	0.13337	0.13329
11	0.12714	0.12715	0.12715	0.12715	0.12714	0.12713	0.12711
12	0.12102	0.12109	0.12115	0.12121	0.12125	0.12129	0.12132
13	0.11526	0.11538	0.11550	0.11561	0.11570	0.11579	0.11587
14	0.10980	0.10998	0.11014	0.11030	0.11045	0.11059	0.11072
15	0.10459	0.10482	0.10504	0.10525	0.10545	0.10563	0.10581
16	0.09961	0.09989	0.10016	0.10042	0.10066	0.10089	0.10111
17	0.09482	0.09515	0.09547	0.09577	0.09606	0.09634	0.09661
18	0.09021	0.09059	0.09095	0.09130	0.09164	0.09196	0.09227
19	0.08574	0.08617	0.08659	0.08698	0.08736	0.08773	0.08808
20	0.08142	0.08190	0.08236	0.08280	0.08322	0.08363	0.08402
21	0.07722	0.07774	0.07825	0.07873	0.07920	0.07965	0.08008
22	0.07313	0.07370	0.07425	0.07478	0.07528	0.07578	0.07625
23	0.06913	0.06975	0.07034	0.07092	0.07147	0.07200	0.07252
24	0.06523	0.06589	0.06653	0.06714	0.06774	0.06831	0.06887
25	0.06140	0.06211	0.06279	0.06345	0.06409	0.06471	0.06530
26	0.05765	0.05840	0.05913	0.05983	0.06051	0.06117	0.06180
27	0.05396	0.05476	0.05553	0.05628	0.05700	0.05770	0.05837
28	0.05033	0.05118	0.05200	0.05279	0.05355	0.05429	0.05500
29	0.04676	0.04765	0.04852	0.04935	0.05015	0.05093	0.05168
30	0.04324	0.04417	0.04508	0.04596	0.04680	0.04762	0.04842
31	0.03976	0.04074	0.04169	0.04261	0.04350	0.04436	0.04519
32	0.03631	0.03735	0.03834	0.03931	0.04024	0.04114	0.04201
33	0.03291	0.03399	0.03503	0.03604	0.03701	0.03796	0.03887
34	0.02953	0.03066	0.03175	0.03280	0.03382	0.03481	0.03576
35	0.02618	0.02736	0.02850	0.02959	0.03066	0.03168	0.03268
36	0.02286	0.02408	0.02527	0.02641	0.02752	0.02859	0.02963
37	0.01956	0.02083	0.02206	0.02325	0.02440	0.02552	0.02660
38	0.01627	0.01760	0.01888	0.02011	0.02131	0.02247	0.02359
39	0.01300	0.01438	0.01571	0.01699	0.01823	0.01944	0.02060
40	0.00974	0.01117	0.01255	0.01388	0.01517	0.01642	0.01763
41	0.00649	0.00797	0.00940	0.01079	0.01212	0.01342	0.01467
42	0.00324	0.00478	0.00626	0.00770	0.00908	0.01043	0.01172
43		0.00159	0.00313	0.00462	0.00605	0.00744	0.00878
44				0.00154	0.00303	0.00446	0.00585
45						0.00149	0.00292

TABLE A.19 Constants b_{n-i+1} for the W' Test for Normality (*Continued*)

	n						
i	92	93	94	95	96	97	98
1	0.26222	0.26118	0.26015	0.25914	0.25814	0.25715	0.25618
2	0.22378	0.22301	0.22224	0.22148	0.22073	0.21999	0.21926
3	0.20216	0.20154	0.20092	0.20031	0.19971	0.19911	0.19852
4	0.18664	0.18613	0.18563	0.18513	0.18463	0.18414	0.18366
5	0.17432	0.17390	0.17349	0.17308	0.17268	0.17227	0.17187
6	0.16399	0.16365	0.16332	0.16299	0.16266	0.16233	0.16200
7	0.15502	0.15476	0.15449	0.15423	0.15397	0.15370	0.15344
8	0.14705	0.14685	0.14665	0.14645	0.14624	0.14604	0.14583
9	0.13983	0.13969	0.13955	0.13940	0.13926	0.13911	0.13896
10	0.13321	0.13313	0.13304	0.13295	0.13285	0.13276	0.13265
11	0.12708	0.12705	0.12701	0.12697	0.12692	0.12687	0.12682
12	0.12135	0.12136	0.12137	0.12138	0.12138	0.12138	0.12137
13	0.11595	0.11601	0.11607	0.11612	0.11617	0.11621	0.11624
14	0.11084	0.11095	0.11105	0.11115	0.11123	0.11132	0.11139
15	0.10597	0.10613	0.10627	0.10641	0.10654	0.10666	0.10678
16	0.10132	0.10152	0.10171	0.10189	0.10206	0.10222	0.10237
17	0.09686	0.09710	0.09733	0.09755	0.09776	0.09796	0.09815
18	0.09256	0.09284	0.09311	0.09337	0.09362	0.09386	0.09409
19	0.08841	0.08874	0.08905	0.08934	0.08963	0.08990	0.09017
20	0.08440	0.08476	0.08511	0.08544	0.08577	0.08608	0.08637
21	0.08050	0.08090	0.08129	0.08166	0.08202	0.08236	0.08270
22	0.07671	0.07715	0.07757	0.07798	0.07838	0.07876	0.07912
23	0.07301	0.07349	0.07395	0.07440	0.07483	0.07524	0.07564
24	0.06940	0.06992	0.07042	0.07090	0.07137	0.07182	0.07225
25	0.06587	0.06643	0.06696	0.06748	0.06798	0.06847	0.06893
26	0.06241	0.06301	0.06358	0.06413	0.06467	0.06519	0.06569
27	0.05902	0.05965	0.06026	0.06085	0.06142	0.06197	0.06251
28	0.05569	0.05636	0.05700	0.05763	0.05823	0.05882	0.05939
29	0.05241	0.05312	0.05380	0.05446	0.05510	0.05572	0.05632
30	0.04918	0.04993	0.05064	0.05134	0.05201	0.05267	0.05330
31	0.04600	0.04678	0.04754	0.04827	0.04898	0.04966	0.05033
32	0.04286	0.04368	0.04447	0.04524	0.04598	0.04670	0.04740
33	0.03975	0.04061	0.04144	0.04224	0.04302	0.04377	0.04451
34	0.03668	0.03758	0.03844	0.03928	0.04009	0.04088	0.04165
35	0.03364	0.03457	0.03548	0.03635	0.03720	0.03803	0.03883
36	0.03063	0.03160	0.03254	0.03345	0.03434	0.03520	0.03603
37	0.02764	0.02865	0.02963	0.03058	0.03150	0.03239	0.03326
38	0.02467	0.02573	0.02674	0.02773	0.02869	0.02962	0.03052
39	0.02173	0.02282	0.02388	0.02490	0.02589	0.02686	0.02779
40	0.01880	0.01993	0.02103	0.02209	0.02312	0.02412	0.02509
41	0.01588	0.01706	0.01819	0.01929	0.02036	0.02140	0.02240
42	0.01298	0.01419	0.01537	0.01651	0.01762	0.01869	0.01973
43	0.01008	0.01134	0.01256	0.01374	0.01489	0.01600	0.01708
44	0.00720	0.00850	0.00976	0.01098	0.01217	0.01332	0.01443
45	0.00432	0.00566	0.00697	0.00823	0.00945	0.01064	0.01179
46	0.00144	0.00283	0.00418	0.00548	0.00675	0.00797	0.00916
47			0.00139	0.00274	0.00405	0.00531	0.00654
48					0.00135	0.00266	0.00392
49							0.00131

Source: Shapiro, S. S.: *How to Test Normality and Other Distributional Assumptions*, American Society for Quality Control, Milwaukee, WI., 1980. Reprinted by permission of the publisher.

TABLE A.20 Coefficients a_{n-i+1} Used in W Test for Normality for $n = 3(1)50$

i \ n	3	4	5	6	7	8	9	10	11	12	13	14	15	16	17	18
1	0.7071	0.6872	0.6646	0.6431	0.6233	0.6052	0.5888	0.5739	0.5601	0.5475	0.5359	0.5251	0.5150	0.5056	0.4968	0.4886
2		0.1677	0.2413	0.2806	0.3031	0.3164	0.3244	0.3291	0.3315	0.3325	0.3325	0.3318	0.3306	0.3290	0.3273	0.3253
3				0.0875	0.1401	0.1743	0.1976	0.2141	0.2260	0.2347	0.2412	0.2460	0.2495	0.2521	0.2540	0.2553
4						0.0561	0.0947	0.1224	0.1429	0.1586	0.1707	0.1802	0.1878	0.1939	0.1988	0.2027
5								0.0399	0.0695	0.0922	0.1099	0.1240	0.1353	0.1447	0.1524	0.1587
6										0.0303	0.0539	0.0727	0.0880	0.1005	0.1109	0.1197
7												0.0240	0.0433	0.0593	0.0725	0.0837
8														0.0196	0.0359	0.0496
9																0.0163

i \ n	19	20	21	22	23	24	25	26	27	28	29	30	31	32	33	34
1	0.4808	0.4734	0.4643	0.4590	0.4542	0.4493	0.4450	0.4407	0.4366	0.4328	0.4291	0.4254	0.4220	0.4188	0.4156	0.4127
2	0.3232	0.3211	0.3185	0.3156	0.3126	0.3098	0.3069	0.3043	0.3018	0.2992	0.2968	0.2944	0.2921	0.2898	0.2876	0.2854
3	0.2561	0.2565	0.2578	0.2571	0.2563	0.2554	0.2543	0.2533	0.2522	0.2510	0.2499	0.2487	0.2475	0.2463	0.2451	0.2439
4	0.2059	0.2085	0.2119	0.2131	0.2139	0.2145	0.2148	0.2151	0.2152	0.2151	0.2150	0.2148	0.2145	0.2141	0.2137	0.2132
5	0.1641	0.1686	0.1736	0.1764	0.1787	0.1807	0.1822	0.1836	0.1848	0.1857	0.1864	0.1870	0.1874	0.1878	0.1880	0.1882
6	0.1271	0.1334	0.1399	0.1443	0.1480	0.1512	0.1539	0.1563	0.1584	0.1601	0.1616	0.1630	0.1641	0.1651	0.1660	0.1667
7	0.0932	0.1013	0.1092	0.1150	0.1201	0.1245	0.1283	0.1316	0.1346	0.1372	0.1395	0.1415	0.1433	0.1449	0.1463	0.1475
8	0.0612	0.0711	0.0804	0.0878	0.0941	0.0997	0.1046	0.1089	0.1128	0.1162	0.1192	0.1219	0.1243	0.1265	0.1284	0.1301
9	0.0303	0.0422	0.0530	0.0618	0.0696	0.0764	0.0823	0.0876	0.0923	0.0965	0.1002	0.1036	0.1066	0.1093	0.1118	0.1140
10		0.0140	0.0263	0.0368	0.0459	0.0539	0.0610	0.0672	0.0728	0.0778	0.0822	0.0862	0.0899	0.0931	0.0961	0.0988
11				0.0122	0.0228	0.0321	0.0403	0.0476	0.0540	0.0598	0.0650	0.0697	0.0739	0.0777	0.0812	0.0844
12						0.0107	0.0200	0.0284	0.0358	0.0424	0.0483	0.0537	0.0585	0.0629	0.0669	0.0706
13								0.0094	0.0178	0.0253	0.0320	0.0381	0.0435	0.0485	0.0530	0.0572
14										0.0084	0.0159	0.0227	0.0289	0.0344	0.0395	0.0441
15												0.0076	0.0144	0.0206	0.0262	0.0314
16														0.0068	0.0131	0.0187
17																0.0062

i \ n	35	36	37	38	39	40	41	42	43	44	45	46	47	48	49	50
1	0.4096	0.4068	0.4040	0.4015	0.3989	0.3964	0.3940	0.3917	0.3894	0.3872	0.3850	0.3830	0.3808	0.3789	0.3770	0.3751
2	0.2834	0.2813	0.2794	0.2774	0.2755	0.2737	0.2719	0.2701	0.2684	0.2667	0.2651	0.2635	0.2620	0.2604	0.2589	0.2574
3	0.2427	0.2415	0.2403	0.2391	0.2380	0.2368	0.2357	0.2345	0.2334	0.2323	0.2313	0.2302	0.2291	0.2281	0.2271	0.2260
4	0.2127	0.2121	0.2116	0.2110	0.2104	0.2098	0.2091	0.2085	0.2078	0.2072	0.2065	0.2058	0.2052	0.2045	0.2038	0.2032
5	0.1883	0.1883	0.1883	0.1881	0.1880	0.1878	0.1876	0.1874	0.1871	0.1868	0.1865	0.1862	0.1859	0.1855	0.1851	0.1847
6	0.1673	0.1678	0.1683	0.1686	0.1689	0.1691	0.1693	0.1694	0.1695	0.1695	0.1695	0.1695	0.1695	0.1693	0.1692	0.1691
7	0.1487	0.1496	0.1505	0.1513	0.1520	0.1526	0.1531	0.1535	0.1539	0.1542	0.1545	0.1548	0.1550	0.1551	0.1553	0.1554
8	0.1317	0.1331	0.1344	0.1356	0.1366	0.1376	0.1384	0.1392	0.1398	0.1405	0.1410	0.1415	0.1420	0.1423	0.1427	0.1430
9	0.1160	0.1179	0.1196	0.1211	0.1225	0.1237	0.1249	0.1259	0.1269	0.1278	0.1286	0.1293	0.1300	0.1306	0.1312	0.1317
10	0.1013	0.1036	0.1056	0.1075	0.1092	0.1108	0.1123	0.1136	0.1149	0.1160	0.1170	0.1180	0.1189	0.1197	0.1205	0.1212
11	0.0873	0.0900	0.0924	0.0947	0.0967	0.0986	0.1004	0.1020	0.1035	0.1049	0.1062	0.1073	0.1085	0.1095	0.1105	0.1113
12	0.0739	0.0770	0.0798	0.0824	0.0848	0.0870	0.0891	0.0909	0.0927	0.0943	0.0959	0.0972	0.0986	0.0998	0.1010	0.1020
13	0.0610	0.0645	0.0677	0.0706	0.0733	0.0759	0.0782	0.0804	0.0824	0.0842	0.0860	0.0876	0.0892	0.0906	0.0919	0.0932
14	0.0484	0.0523	0.0559	0.0592	0.0622	0.0651	0.0677	0.0701	0.0724	0.0745	0.0765	0.0783	0.0801	0.0817	0.0832	0.0846
15	0.0361	0.0404	0.0444	0.0481	0.0515	0.0546	0.0575	0.0602	0.0628	0.0651	0.0673	0.0694	0.0713	0.0731	0.0748	0.0764
16	0.0239	0.0287	0.0331	0.0372	0.0409	0.0444	0.0476	0.0506	0.0534	0.0560	0.0584	0.0607	0.0628	0.0648	0.0667	0.0685
17	0.0119	0.0172	0.0220	0.0264	0.0305	0.0343	0.0379	0.0411	0.0442	0.0471	0.0497	0.0522	0.0546	0.0568	0.0588	0.0608
18		0.0057	0.0110	0.0158	0.0203	0.0244	0.0283	0.0318	0.0352	0.0383	0.0412	0.0439	0.0465	0.0489	0.0511	0.0532
19				0.0053	0.0101	0.0146	0.0188	0.0227	0.0263	0.0296	0.0328	0.0357	0.0385	0.0411	0.0436	0.0459
20						0.0049	0.0094	0.0136	0.0175	0.0211	0.0245	0.0277	0.0307	0.0335	0.0361	0.0386
21								0.0045	0.0087	0.0126	0.0163	0.0197	0.0229	0.0259	0.0288	0.0314
22										0.0042	0.0081	0.0118	0.0153	0.0185	0.0215	0.0244
23												0.0039	0.0076	0.0111	0.0143	0.0174
24														0.0037	0.0071	0.0104
25																0.0035

Source: Shapiro, S. S.: *How to Test Normality and Other Distributional Assumptions*, American Society for Quality Control, Milwaukee, WI, 1980. Reprinted by permission of the publisher.

TABLE A.21 One-Tail Percentage Points of W' Test for Normality*

	100α			
n	1%	2.5%	5%	10%
51	0.943	0.951	0.958	0.965
53	0.944	0.953	0.959	0.966
55	0.945	0.954	0.960	0.967
57	0.947	0.955	0.961	0.968
59	0.948	0.957	0.962	0.969
61	0.949	0.958	0.964	0.970
63	0.951	0.959	0.965	0.970
65	0.952	0.960	0.966	0.971
67	0.954	0.961	0.967	0.972
69	0.955	0.962	0.968	0.973
71	0.956	0.963	0.968	0.973
73	0.958	0.964	0.969	0.974
75	0.959	0.965	0.970	0.975
77	0.960	0.966	0.971	0.975
79	0.961	0.967	0.972	0.976
81	0.963	0.968	0.972	0.976
83	0.964	0.969	0.973	0.977
85	0.965	0.970	0.974	0.977
87	0.965	0.970	0.974	0.978
89	0.966	0.971	0.975	0.978
91	0.967	0.972	0.975	0.979
93	0.968	0.972	0.976	0.979
95	0.968	0.973	0.976	0.980
97	0.968	0.973	0.977	0.980
99	0.969	0.974	0.977	0.980

*Percentiles based on 9000 simulations and smoothed by regression using polynomial fits.

Source: Shapiro, S. S.: *How to Test Normality and Other Distributional Assumptions*, American Society for Quality Control, Milwaukee, WI, 1980. Reprinted by permission of the publisher.

TABLE A.22 Percentage Points for *WE* Test Statistic for Exponentiality

	α										
n	0.005	0.01	0.025	0.05	0.10	0.50	0.90	0.95	0.975	0.99	0.995
3	0.2519	0.2538	0.2596	0.2697	0.2915	0.5714	0.9709	0.9926	0.9981	0.9997	0.99993
4	0.1241	0.1302	0.1434	0.1604	0.1891	0.3768	0.7514	0.8581	0.9236	0.9680	0.9837
5	0.0845	0.0905	0.1048	0.1187	0.1442	0.2875	0.5547	0.6682	0.7590	0.8600	0.9192
6	0.0610	0.0665	0.0802	0.0956	0.1173	0.2276	0.4292	0.5089	0.5842	0.6775	0.7501
7	0.0514	0.0591	0.0700	0.0810	0.0986	0.1874	0.3474	0.4162	0.4852	0.5706	0.6426
8	0.0454	0.0512	0.0614	0.0710	0.0852	0.1625	0.2934	0.3497	0.4033	0.4848	0.5428
9	0.0404	0.0442	0.0537	0.0633	0.0751	0.1415	0.2553	0.3005	0.3454	0.4015	0.4433
10	0.0369	0.0404	0.0487	0.0568	0.0678	0.1225	0.2178	0.2525	0.2879	0.3391	0.3701
11	0.0339	0.0380	0.0447	0.0528	0.0616	0.1112	0.1934	0.2265	0.2619	0.3039	0.3314
12	0.0311	0.0358	0.0410	0.0494	0.0567	0.1009	0.1723	0.2019	0.2364	0.2716	0.2978
13	0.0287	0.0337	0.0382	0.0460	0.0528	0.0925	0.1563	0.1829	0.2113	0.2422	0.2642
14	0.0265	0.0317	0.0362	0.0428	0.0496	0.0847	0.1417	0.1647	0.1862	0.2131	0.2315
15	0.0247	0.0298	0.0344	0.0398	0.0466	0.0778	0.1285	0.1485	0.1669	0.1926	0.2123
16	0.0233	0.0280	0.0326	0.0374	0.0438	0.0728	0.1187	0.1355	0.1542	0.1770	0.1931
17	0.0222	0.0264	0.0310	0.0352	0.0412	0.0684	0.1099	0.1257	0.1423	0.1614	0.1794
18	0.0212	0.0250	0.0294	0.0332	0.0388	0.0640	0.1015	0.1164	0.1311	0.1483	0.1668
19	0.0203	0.0238	0.0278	0.0314	0.0368	0.0600	0.0935	0.1071	0.1199	0.1374	0.1452
20	0.0196	0.0227	0.0264	0.0302	0.0352	0.0570	0.0884	0.1002	0.1121	0.1286	0.1369
21	0.0190	0.0217	0.0250	0.0290	0.0337	0.0540	0.0839	0.0948	0.1054	0.1198	0.1288
22	0.0185	0.0208	0.0238	0.0278	0.0323	0.0516	0.0794	0.0894	0.0988	0.1118	0.1213
23	0.0181	0.0201	0.0230	0.0266	0.0310	0.0492	0.0749	0.0836	0.0933	0.1043	0.1142
24	0.0177	0.0194	0.0224	0.0256	0.0298	0.0468	0.0704	0.0788	0.0882	0.0984	0.1071
25	0.0173	0.0188	0.0218	0.0248	0.0286	0.0447	0.0668	0.0749	0.0836	0.0927	0.1000
26	0.0169	0.0182	0.0213	0.0240	0.0274	0.0426	0.0636	0.0712	0.0791	0.0885	0.0948
27	0.0165	0.0177	0.0208	0.0232	0.0264	0.0407	0.0606	0.0678	0.0747	0.0843	0.0896
28	0.0161	0.0172	0.0203	0.0225	0.0256	0.0391	0.0576	0.0649	0.0706	0.0801	0.0859
29	0.0157	0.0168	0.0198	0.0219	0.0249	0.0377	0.0555	0.0621	0.0671	0.0759	0.0822
30	0.0153	0.0164	0.0193	0.0213	0.0242	0.0364	0.0536	0.0593	0.0643	0.0719	0.0786
31	0.0149	0.0160	0.0188	0.0207	0.0235	0.0352	0.0518	0.0569	0.0615	0.0686	0.0753
32	0.0145	0.0156	0.0183	0.0201	0.0229	0.0340	0.0491	0.0547	0.0591	0.0661	0.0722
33	0.0141	0.0152	0.0178	0.0195	0.0223	0.0329	0.0475	0.0527	0.0573	0.0636	0.0691
34	0.0137	0.0148	0.0173	0.0190	0.0217	0.0319	0.0459	0.0507	0.0555	0.0611	0.0660
35	0.0133	0.0144	0.0168	0.0185	0.0211	0.0309	0.0444	0.0488	0.0537	0.0588	0.0639
36	0.0129	0.0141	0.0164	0.0180	0.0205	0.0300	0.0429	0.0470	0.0519	0.0567	0.0608
37	0.0125	0.0138	0.0160	0.0176	0.0200	0.0291	0.0414	0.0454	0.0501	0.0546	0.0578
38	0.0122	0.0135	0.0156	0.0172	0.0195	0.0283	0.0400	0.0440	0.0483	0.0525	0.0553
39	0.0120	0.0133	0.0152	0.0168	0.0190	0.0275	0.0386	0.0426	0.0465	0.0512	0.0531
40	0.0118	0.0131	0.0148	0.0164	0.0186	0.0267	0.0375	0.0414	0.0447	0.0499	0.0510
41	0.0116	0.0129	0.0144	0.0161	0.0182	0.0260	0.0364	0.0402	0.0430	0.0476	0.0493
42	0.0114	0.0127	0.0140	0.0158	0.0178	0.0253	0.0355	0.0389	0.0417	0.0464	0.0482
43	0.0112	0.0125	0.0137	0.0155	0.0174	0.0248	0.0346	0.0379	0.0405	0.0452	0.0471
44	0.0110	0.0123	0.0134	0.0152	0.0170	0.0243	0.0338	0.0369	0.0394	0.0440	0.0460
45	0.0108	0.0121	0.0131	0.0149	0.0166	0.0238	0.0329	0.0359	0.0385	0.0423	0.0449
46	0.0106	0.0119	0.0129	0.0146	0.0162	0.0233	0.0320	0.0349	0.0376	0.0416	0.0438
47	0.0104	0.0117	0.0127	0.0143	0.0158	0.0228	0.0311	0.0340	0.0367	0.0394	0.0427
48	0.0103	0.0115	0.0125	0.0141	0.0155	0.0223	0.0303	0.0332	0.0358	0.0382	0.0416
49	0.0102	0.0113	0.0123	0.0139	0.0152	0.0218	0.0295	0.0324	0.0349	0.0371	0.0405
50	0.0101	0.0111	0.0122	0.0137	0.0149	0.0213	0.0288	0.0317	0.0340	0.0360	0.0394
51	0.0100	0.0109	0.0120	0.0135	0.0147	0.0209	0.0282	0.0310	0.0331	0.0349	0.0383
52	0.0099	0.0107	0.0119	0.0133	0.0145	0.0205	0.0276	0.0303	0.0323	0.0341	0.0373
53	0.0097	0.0106	0.0118	0.0131	0.0143	0.0201	0.0270	0.0296	0.0315	0.0332	0.0363

TABLE A.22 Percentage Points for *WE* Test Statistic for Exponentiality (*Continued*)

	α										
n	0.005	0.01	0.025	0.05	0.10	0.50	0.90	0.95	0.975	0.99	0.995
54	0.0095	0.0104	0.0116	0.0129	0.0141	0.0197	0.0264	0.0289	0.0307	0.0329	0.0353
55	0.0094	0.0103	0.0115	0.0127	0.0139	0.0193	0.0258	0.0282	0.0299	0.0321	0.0343
56	0.0093	0.0102	0.0113	0.0125	0.0137	0.0189	0.0252	0.0275	0.0292	0.0313	0.0333
57	0.0092	0.0101	0.0112	0.0123	0.0135	0.0185	0.0247	0.0268	0.0285	0.0306	0.0324
58	0.0091	0.0100	0.0110	0.0121	0.0133	0.0182	0.0242	0.0262	0.0279	0.0301	0.0318
59	0.0090	0.0098	0.0109	0.0119	0.0131	0.0179	0.0238	0.0257	0.0274	0.0296	0.0312
60	0.0089	0.0095	0.0108	0.0117	0.0129	0.0176	0.0234	0.0252	0.0270	0.0291	0.0306
61	0.0088	0.0093	0.0107	0.0115	0.0127	0.0173	0.0230	0.0247	0.0266	0.0286	0.0301
62	0.0087	0.0092	0.0105	0.0113	0.0125	0.0170	0.0226	0.0242	0.0262	0.0281	0.0296
63	0.0086	0.0091	0.0104	0.0112	0.0123	0.0167	0.0222	0.0238	0.0257	0.0276	0.0291
64	0.0085	0.0090	0.0102	0.0111	0.0121	0.0164	0.0218	0.0234	0.0252	0.0271	0.0286
65	0.0084	0.0089	0.0101	0.0109	0.0119	0.0161	0.0215	0.0230	0.0247	0.0266	0.0281
66	0.0082	0.0088	0.0099	0.0108	0.0117	0.0159	0.0211	0.0225	0.0242	0.0261	0.0276
67	0.0081	0.0087	0.0098	0.0107	0.0115	0.0157	0.0207	0.0221	0.0237	0.0256	0.0271
68	0.0080	0.0086	0.0096	0.0105	0.0114	0.0155	0.0204	0.0217	0.0232	0.0251	0.0266
69	0.0079	0.0085	0.0095	0.0104	0.0113	0.0152	0.0198	0.0213	0.0227	0.0246	0.0261
70	0.0078	0.0084	0.0094	0.0103	0.0111	0.0150	0.0194	0.0209	0.0222	0.0241	0.0256
71	0.0077	0.0083	0.0093	0.0102	0.0109	0.0147	0.0191	0.0205	0.0218	0.0237	0.0251
72	0.0076	0.0082	0.0092	0.0101	0.0108	0.0145	0.0188	0.0201	0.0214	0.0232	0.0246
73	0.0075	0.0081	0.0091	0.0100	0.0107	0.0143	0.0185	0.0198	0.0211	0.0228	0.0241
74	0.0074	0.0080	0.0090	0.0098	0.0106	0.0141	0.0182	0.0195	0.0208	0.0224	0.0236
75	0.0073	0.0079	0.0089	0.0097	0.0105	0.0139	0.0179	0.0192	0.0205	0.0220	0.0231
76	0.0073	0.0078	0.0088	0.0096	0.0104	0.0137	0.0176	0.0189	0.0202	0.0217	0.0227
77	0.0072	0.0077	0.0087	0.0095	0.0103	0.0135	0.0173	0.0186	0.0199	0.0214	0.0223
78	0.0071	0.0077	0.0086	0.0093	0.0101	0.0134	0.0170	0.0183	0.0196	0.0211	0.0219
79	0.0070	0.0076	0.0085	0.0092	0.0100	0.0132	0.0168	0.0180	0.0193	0.0208	0.0215
80	0.0070	0.0075	0.0084	0.0091	0.0099	0.0131	0.0166	0.0177	0.0190	0.0205	0.0211
81	0.0069	0.0074	0.0083	0.0090	0.0098	0.0129	0.0164	0.0175	0.0187	0.0202	0.0207
82	0.0068	0.0074	0.0082	0.0088	0.0097	0.0128	0.0162	0.0173	0.0184	0.0199	0.0203
83	0.0067	0.0073	0.0081	0.0087	0.0096	0.0126	0.0160	0.0170	0.0181	0.0196	0.0199
84	0.0067	0.0073	0.0080	0.0086	0.0095	0.0125	0.0158	0.0168	0.0178	0.0193	0.0196
85	0.0066	0.0072	0.0079	0.0085	0.0094	0.0123	0.0156	0.0166	0.0174	0.0190	0.0193
86	0.0066	0.0071	0.0078	0.0085	0.0093	0.0122	0.0154	0.0164	0.0172	0.0187	0.0190
87	0.0065	0.0071	0.0077	0.0084	0.0092	0.0120	0.0152	0.0162	0.0170	0.0184	0.0187
88	0.0065	0.0070	0.0077	0.0084	0.0091	0.0119	0.0150	0.0160	0.0168	0.0181	0.0185
89	0.0064	0.0070	0.0076	0.0083	0.0090	0.0117	0.0148	0.0158	0.0166	0.0179	0.0183
90	0.0064	0.0069	0.0075	0.0082	0.0089	0.0016	0.0147	0.0156	0.0164	0.0176	0.0181
91	0.0063	0.0068	0.0075	0.0082	0.0088	0.0114	0.0145	0.0154	0.0162	0.0173	0.0179
92	0.0063	0.0068	0.0074	0.0081	0.0087	0.0113	0.0143	0.0153	0.0160	0.0171	0.0177
93	0.0062	0.0067	0.0073	0.0081	0.0086	0.0112	0.0141	0.0151	0.0158	0.0168	0.0175
94	0.0062	0.0067	0.0073	0.0080	0.0085	0.0110	0.0139	0.0149	0.0156	0.0165	0.0173
95	0.0061	0.0066	0.0072	0.0079	0.0084	0.0109	0.0138	0.0147	0.0154	0.0163	0.0171
96	0.0061	0.0065	0.0072	0.0078	0.0083	0.0108	0.0136	0.0145	0.0153	0.0161	0.0169
97	0.0060	0.0065	0.0071	0.0077	0.0082	0.0107	0.0134	0.0143	0.0152	0.0159	0.0167
98	0.0060	0.0064	0.0070	0.0076	0.0081	0.0105	0.0133	0.0142	0.0151	0.0157	0.0165
99	0.0059	0.0064	0.0070	0.0075	0.0080	0.0104	0.0132	0.0140	0.0150	0.0155	0.0163
00	0.0059	0.0063	0.0069	0.0074	0.0079	0.0103	0.0131	0.0139	0.0149	0.0153	0.0161

Source: Shapiro, S. S.: *How to Test Normality and Other Distributional Assumptions*, American Society for Quality Control, Milwaukee, WI, 1980. Reprinted by permission of the publisher.

TABLE A.23 Random Numbers

10480	15011	01536	02011	81647	91646	69179	14194	62590
22368	46573	25595	85393	30995	89198	27982	53402	93965
24130	48360	22527	97265	76393	64809	15179	24830	49340
42167	93093	06243	61680	07856	16376	39440	53537	71341
37570	39975	81837	16656	06121	91782	60468	81305	49684
77921	06907	11008	42751	27756	53498	18602	70659	90655
99562	72905	56420	69994	98872	31016	71194	18738	44013
96301	91977	05463	07972	18876	20922	94595	56869	69014
89579	14342	63661	10281	17453	18103	57740	84378	25331
85475	36857	53342	53988	53060	59533	38867	62300	08158
28918	69578	88231	33276	70997	79936	56865	05859	90106
63553	40961	48235	03427	49626	69445	18663	72695	52180
09429	93969	52636	92737	88974	33488	36320	17617	30015
10365	61129	87529	85689	48237	52267	67689	93394	01511
07119	97336	71048	08178	77233	13916	47564	81056	97735
51085	12765	51821	51259	77452	16308	60756	92144	49442
02368	21382	52404	60268	89368	19885	55322	44819	01188
01011	54092	33362	94904	31273	04146	18594	29852	71585
52162	53916	46369	58586	23216	14513	83149	98736	23495
07056	97628	33787	09998	42698	06691	76988	13602	51851
48663	91245	85828	14346	09172	30168	90229	04734	59193
54164	58492	22421	74103	47070	25306	76468	26384	58151
32639	32363	05597	24200	13363	38005	94342	28728	35806
29334	27001	87637	87308	58731	00256	45834	15398	46557
02488	33062	28834	07351	19731	92420	60952	61280	50001
81525	72295	04839	96423	24878	82651	66566	14778	76797
29676	20591	68086	26432	46901	20849	89768	81536	86645
00742	57392	39064	66432	84673	40027	32832	61362	98947
05366	04213	25669	26422	44407	44048	37937	63904	45766
91921	26418	64117	94305	26766	25940	39972	22209	71500
00582	04711	87917	77341	42206	35126	74087	99547	81817
00725	69884	62797	56170	86324	88072	76222	36086	84637
69011	65795	95876	55293	18988	27354	26575	08625	40801
25976	57976	29888	88604	67917	48708	18912	82271	65424
09763	83473	73577	12908	30883	18317	28290	35797	05998
91567	42595	27958	30134	04024	86385	29880	99730	55536
17955	56349	90999	49127	20044	59931	06115	20542	18059
46503	18584	18845	49618	02304	51038	20655	58727	28168
92157	89634	94824	78171	84610	82834	09922	25417	44137
14577	62765	35605	81263	39667	47358	56873	56307	61607
98247	07523	33362	64270	01638	92477	66969	98420	04880
34914	63976	88720	82765	34476	17032	87589	40836	32427
70060	28277	39475	46473	23219	53416	94970	25832	69975
53976	54914	06990	67245	68350	82948	11398	42878	80287
76072	29515	40980	07391	58745	25774	22987	80059	39911
90725	52210	83974	29992	65831	38857	50490	83765	55657
64364	67412	33339	31926	14883	24413	59744	92351	97473
08962	00358	31662	25388	61642	34072	81249	35648	56891
95012	68379	93526	70765	10592	04542	76463	54328	02349
15664	10493	20492	38391	91132	21999	59516	81652	27195

Source: Table XV of Hines, W. W., and D. C. Montgomery: *Probability and Statistics in Engineering and Management Science*, 2d ed., Wiley, New York, 1980. Reprinted by permission of the publisher.

NAME INDEX

Numbers in **bold** refer to chapter numbers.

SUBJECT INDEX

Numbers in **bold** refer to chapter numbers.

ABOUT THE EDITOR

Harrison M. Wadsworth is a Profesor Emeritus in the School of Industrial and Systems Engineering at the Georgia Institute of Technology. He is the author of numerous publications and technical reports, as well as co-author of the book, Modern Methods for Quality Control and Improvement (JohnWiley). A former editor of the Journal of Quality Technology, Dr. Wadsworth is a Fellow of the American Society for Quality, a member of the American Statistical Association, and a Life Member of the Institute of Industrial Engineers. He chairs subcommittee 1 of ISO Technical Committee 69 on Statistical Methods. He also chairs the statistical Committee of the American Society for Quality and is a U.S. delegate to ISO Technical Committee 176 on Quality Assurance. He has received many awards, including the Shewhart Medal of The American Society for Quality.